AF572545

STUDENT'S SOLUTIONS MANUAL

SAL SCIANDRA

Niagara County Community College

MATHEMATICS WITH APPLICATIONS IN THE MANAGEMENT, NATURAL, AND SOCIAL SCIENCES

TWELFTH EDITION

Margaret L. Lial

American River College

Thomas Hungerford

Saint Louis University

John Holcomb

Cleveland State University

Bernadette Mullins

Birmingham-Southern College

Pearson

The author and publisher of this book have used their best efforts in preparing this book. These efforts include the development, research, and testing of the theories and programs to determine their effectiveness. The author and publisher make no warranty of any kind, expressed or implied, with regard to these programs or the documentation contained in this book. The author and publisher shall not be liable in any event for incidental or consequential damages in connection with, or arising out of, the furnishing, performance, or use of these programs.

Reproduced by Pearson from electronic files supplied by the author.

Copyright © 2019, 2015, 2011 Pearson Education, Inc.
Publishing as Pearson, 330 Hudson Street, NY NY 10013

All rights reserved. No part of this publication may be reproduced, stored in a retrieval system, or transmitted, in any form or by any means, electronic, mechanical, photocopying, recording, or otherwise, without the prior written permission of the publisher. Printed in the United States of America.

1 18

ISBN-13: 978-0-13-477267-7
ISBN-10: 0-13-477267-9

Contents

Chapter 1: Algebra and Equations

Section 1.1 The Real Numbers

1. True. This statement is true, since every integer can be written as the ratio of the integer and 1. For example, $5 = \frac{5}{1}$.

3. Answers vary with the calculator, but $\frac{2,508,429,787}{798,458,000}$ is the best.

5. $6(t+4) = 6t + 6 \cdot 4$
This illustrates the distributive property.

7. $(-5) + 0 = -5$
This illustrates the identity property of addition.

9. $8 + (12 + 6) = (8 + 12) + 6$
This illustrates the associative property of addition.

11. Answers vary. One possible answer: The sum of a number and its additive inverse is the additive identity. The product of a number and its multiplicative inverse is the multiplicative identity.

For Exercises 13–16, let $p = -2$, $q = 3$ and $r = -5$.

13. $-3(p+5q) = -3[-2+5(3)] = -3[-2+15]$
$= -3(13) = -39$

15. $\frac{q+r}{q+p} = \frac{3+(-5)}{3+(-2)} = \frac{-2}{1} = -2$

17. Let $r = 3.8$.
$APR = 12r = 12(3.8) = 45.6\%$

19. Let $APR = 11$.
$APR = 12r$
$11 = 12r$
$\frac{11}{12} = r$
$r \approx .9167\%$

21. $3 - 4 \cdot 5 + 5 = 3 - 20 + 5 = -17 + 5 = -12$

23. $(4-5) \cdot 6 + 6 = -1 \cdot 6 + 6 = -6 + 6 = 0$

25. $8 - 4^2 - (-12)$
Take powers first.
$8 - 16 - (-12)$
Then add and subtract in order from left to right.
$8 - 16 + 12 = -8 + 12 = 4$

27. $\frac{2(-3) + \frac{3}{(-2)} - \frac{2}{(-\sqrt{16})}}{\sqrt{64} - 1}$
Work above and below fraction bar. Take roots.
$\frac{2(-3) + \frac{3}{(-2)} - \frac{2}{(-4)}}{8 - 1}$
Do multiplications and divisions.
$\frac{-6 - \frac{3}{2} + \frac{1}{2}}{8 - 1}$
Add and subtract.
$\frac{-\frac{12}{2} - \frac{3}{2} + \frac{1}{2}}{7} = \frac{-\frac{14}{2}}{7} = \frac{-7}{7} = -1$

29. $\frac{2040}{523}, \frac{189}{37}, \sqrt{27}, \frac{4587}{691}, 6.735, \sqrt{47}$

31. 12 is less than 18.5.
$12 < 18.5$

33. x is greater than or equal to 5.7.
$x \geq 5.7$

35. z is at most 7.5.
$z \leq 7.5$

37. $-6 < -2$

39. $3.14 < \pi$

41. a lies to the right of b or is equal to b.

43. $c < a < b$

45. $(-8, -1)$
This represents all real numbers between –8 and –1, not including –8 and –1. Draw parentheses at –8 and –1 and a heavy line between them. The parentheses at –8 and –1 show that neither of these points belongs to the graph.

–8 –1

Copyright © 2019 Pearson Education, Inc.

47. $(-2,3]$
All real numbers x such that $-2 < x \le 3$
Draw a heavy line from –2 to 3. Use a parenthesis at –2 since it is not part of the graph. Use a bracket at 3 since it is part of the graph.

–4 –2 0 2 4

49. $(-2,\infty)$

All real numbers x such that $x > -2$
Start at –2 and draw a heavy line to the right. Use a parenthesis at –2 since it is not part of the graph.

–2

51. $|-9|-|-12| = 9-(12) = -3$

53. $-|-4|-|-1-14| = -(4)-|-15|$
$= -(4)-15 = -19$

55. $|5|\_\_|-5|$
$5\_\_5$
$5 = 5$

57. $|10-3|\_\_|3-10|$
$|7|\_\_|-7|$
$7\_\_7$
$7 = 7$

59. $|-2+8|\_\_|2-8|$
$|6|\_\_|-6|$
$6\_\_6$
$6 = 6$

61. $|3-5|\_\_|3|-|5|$
$|-2|\_\_3-5$
$2\_\_-2$
$2 > -2$

63. When $a < 7$, $a-7$ is negative.
So $|a-7| = -(a-7) = 7-a$.

Answers will vary for exercises 65–67. Sample answers are given.

65. No, it is not always true that $|a+b| = |a|+|b|$. For example, let $a = 1$ and $b = -1$. Then,
$|a+b| = |1+(-1)| = |0| = 0$, but
$|a|+|b| = |1|+|(-1)| = 1+1 = 2$.

67. $|2-b| = |2+b|$ only when $b = 0$. Then each side of the equation is equal to 2. If b is any other value, subtracting it from 2 and adding it to 2 will produce two different values.

69. 1; 30062007

71. 9; 2006, 2008, 2009, 2010, 2011, 2012, 2013, 2014, 2015

73. 4; 2006, 2007, 2009, 2011

75. $|10.6-14.9| = |-4.3| = 4.3$

77. $|-1.0-63.1| = |-64.1| = 64.1$

79. $|-5.7-(-8.0)| = |2.3| = 2.3$

81. 3; 30062010, 2015, 2016

83. 6; 2010, 2011, 2012, 2013, 2014, 2015

Section 1.2 Polynomials

1. $11.2^6 \approx 1{,}973{,}822.685$

3. $\left(-\frac{18}{7}\right)^6 \approx 289.0991339$

5. -3^2 is negative, whereas $(-3)^2$ is positive. Both -3^3 and $(-3)^3$ are negative.

7. $4^2 \cdot 4^3 = 4^{2+3} = 4^5$

9. $(-6)^2 \cdot (-6)^5 = (-6)^{2+5} = (-6)^7$

11. $\left[(5u)^4\right]^7 = (5u)^{4\cdot 7} = (5u)^{28}$

13. degree 4; coefficients: 6.2, –5, 4, –3, 3.7; constant term 3.7.

15. Since the highest power of x is 3, the degree is 3.

Copyright © 2019 Pearson Education, Inc.

17. $(3x^3+2x^2-5x)+(-4x^3-x^2-8x)$
$=(3x^3-4x^3)+(2x^2-x^2)+(-5x-8x)$
$=-x^3+x^2-13x$

19. $(-4y^2-3y+8)-(2y^2-6y+2)$
$=(-4y^2-3y+8)+(-2y^2+6y-2)$
$=-4y^2-3y+8-2y^2+6y-2$
$=(-4y^2-2y^2)+(-3y+6y)+(8-2)$
$=-6y^2+3y+6$

21. $(2x^3+2x^2+4x-3)-(2x^3+8x^2+1)$
$=(2x^3+2x^2+4x-3)+(-2x^3-8x^2-1)$
$=2x^3+2x^2+4x-3-2x^3-8x^2-1$
$=(2x^3-2x^3)+(2x^2-8x^2)+(4x)+(-3-1)$
$=-6x^2+4x-4$

23. $-9m(2m^2+6m-1)$
$=(-9m)(2m^2)+(-9m)(6m)+(-9m)(-1)$
$=-18m^3-54m^2+9m$

25. $(3z+5)(4z^2-2z+1)$
$=(3z)(4z^2-2z+1)+(5)(4z^2-2z+1)$
$=12z^3-6z^2+3z+20z^2-10z+5$
$=12z^3+14z^2-7z+5$
$=12z^3+20z^2-6z^2-10z+3z+5$
$=12z^3+14z^2-7z+5$

27. $(6k-1)(2k+3)$
$=(6k)(2k+3)+(-1)(2k+3)$
$=12k^2+18k-2k-3$
$=12k^2+16k-3$

29. $(3y+5)(2y+1)$
Use FOIL.
$=6y^2+3y+10y+5$
$=6y^2+13y+5$

31. $(9k+q)(2k-q)$
$=18k^2-9kq+2kq-q^2$
$=18k^2-7kq-q^2$

33. $(6.2m-3.4)(.7m+1.3)$
$=4.34m^2+8.06m-2.38m-4.42$
$=4.34m^2+5.68m-4.42$

35. $5k-[k+(-3+5k)]$
$=5k-[6k-3]$
$=5k-6k+3$
$=-k+3$

37. $R=5(1000x)=5000x$
$C=200{,}000+1800x$
$P=(5000x)-(200{,}000+1800x)$
$=3200x-200{,}000$

39. $R=9.75(1000x)=9750x$
$C=260{,}000+(-3x^2+3480x-325)$
$=-3x^2+3480x+259{,}675$
$P=(9750x)-(-3x^2+3480x+259{,}675)$
$=3x^2+6270x-259{,}675$

41. **a.** According to the bar graph, the net earnings in 2007 were \$673 million.

b. Let $x=7$.
$4.79x^3-122.5x^2+1104x-2863$
$=4.79(7)^3-122.5(7)^2+1104(7)-2863$
$=505.47$
According to the polynomial, the net earnings in 2007 were approximately \$505 million.

43. **a.** According to the bar graph, the net earnings in 2012 were \$1384 million.

b. Let $x=12$.
$4.79x^3-122.5x^2+1104x-2863$
$=4.79(12)^3-122.5(12)^2+1104(12)-2863$
$=1022.12$

According to the polynomial, the net earnings in 2012 were approximately \$1022 million.

Copyright © 2019 Pearson Education, Inc.

45. Let $x = 17$.

$4.79x^3 - 122.5x^2 + 1104x - 2863$

$= 4.79(17)^3 - 122.5(17)^2 + 1104(17) - 2863$

$= 4035.77$

According to the polynomial, the net earnings in 2017 will be approximately \$4036 million.

47. Let $x = 19$.

$4.79x^3 - 122.5x^2 + 1104x - 2863$

$= 4.79(19)^3 - 122.5(19)^2 + 1104(19) - 2863$

$= 6745.11$

According to the polynomial, the net earnings in 2019 will be approximately \$6745 million.

For exercises 49–52, we use the polynomial $9.5x^3 - 401.6x^2 + 6122x - 25{,}598$.

49. Let $x = 10$.

$9.5(10)^3 - 401.6(10)^2 + 6122(10) - 25{,}598$

$= 4962$

Thus, the costs were approximately \$4962 million in 2010. The statement is false.

51. Let $x = 12$.

$9.5(12)^3 - 401.6(12)^2 + 6122(12) - 25{,}598$

$= 6451.6$

Let $x = 15$.

$9.5(15)^3 - 401.6(15)^2 + 6122(15) - 25{,}598$

$= 7934.5$

Thus, the costs were \$6451.6 million in 2012 and \$7934.5 million in 2015. The statement is false.

For exercises 53–58, we use the polynomial $-72.85x^3 + 2082x^2 - 16{,}532x + 59{,}357$.

53. Let $x = 7$.

$-72.85x^3 + 2082x^2 - 16{,}532x + 59{,}357$

$= -72.85(7)^3 + 2082(7)^2 - 16{,}532(7) + 59{,}357$

$= 20{,}663.45$

Thus, the profit for PepsiCo Inc in 2007 was approximately \$20,663 million.

55. Let $x = 12$.

$-72.85x^3 + 2082x^2 - 16{,}532x + 59{,}357$

$= -72.85(12)^3 + 2082(12)^2 - 16{,}532(12) + 59{,}357$

$= 34{,}896.2$

Thus, the profit for PepsiCo Inc in 2012 was approximately \$34,896 million.

57. Let $x = 13$.

$-72.85x^3 + 2082x^2 - 16{,}532x + 59{,}357$

$= -72.85(13)^3 + 2082(13)^2 - 16{,}532(13) + 59{,}357$

$= 36{,}247.55$

Thus, the profit for PepsiCo Inc in 2013 was approximately \$36,248 million.

Let $x = 9$.

$-72.85x^3 + 2082x^2 - 16{,}532x + 59{,}357$

$= -72.85(9)^3 + 2082(9)^2 - 16{,}532(9) + 59{,}357$

$= 26{,}103.35$

Thus, the profit for PepsiCo Inc in 2009 was approximately \$26,104 million. Therefore, the profit was higher in 2013.

59. $P = 7.2x^2 + 5005x - 230{,}000$. Here is part of the screen capture.

X	Y1
0	-2.3E5
5	-2E5
10	-1.8E5
15	-1.5E5
20	-1.3E5
25	-1E5
30	-73370

X=0

For 25,000, the loss will be \$100,375;

X	Y1
10	-1.8E5
15	-1.5E5
20	-1.3E5
25	[illegible]
30	-73370
35	-46005
40	-18280

Y1=-100375

For 60,000, there profit will be \$96,220.

X	Y1
45	9805
50	38250
55	67055
60	[illegible]
65	125745
70	155630
75	185875

Y1=96220

There is a loss at the beginning because of large fixed costs. When more items are made, these costs become a smaller portion of the total costs.

Copyright © 2019 Pearson Education, Inc.

61. Let $x = 100$ (in thousands)
$7.2(100)^2 + 5005(100) - 230{,}000 = 342{,}500$

d. The profit for selling 100,000 calculators is \$342,500.

Section 1.3 Factoring

1. $12x^2 - 24x = 12x \cdot x - 12x \cdot 2 = 12x(x-2)$

3. $r^3 - 5r^2 + r = r\left(r^2\right) - r(5r) + r(1)$
$= r\left(r^2 - 5r + 1\right)$

5. $6z^3 - 12z^2 + 18z$
$= 6z\left(z^2\right) - 6z(2z) + 6z(3)$
$= 6z\left(z^2 - 2z + 3\right)$

7. $3(2y-1)^2 + 7(2y-1)^3$
$= (2y-1)^2(3) + (2y-1)^2 \cdot 7(2y-1)$
$= (2y-1)^2[3 + 7(2y-1)]$
$= (2y-1)^2(3 + 14y - 7)$
$= (2y-1)^2(14y - 4)$
$= 2(2y-1)^2(7y-2)$

9. $3(x+5)^4 + (x+5)^6$
$= (x+5)^4 \cdot 3 + (x+5)^4(x+5)^2$
$= (x+5)^4\left[3 + (x+5)^2\right]$
$= (x+5)^4\left(3 + x^2 + 10x + 25\right)$
$= (x+5)^4\left(x^2 + 10x + 28\right)$

11. $x^2 + 5x + 4 = (x+1)(x+4)$

13. $x^2 + 7x + 12 = (x+3)(x+4)$

15. $x^2 + x - 6 = (x+3)(x-2)$

17. $x^2 + 2x - 3 = (x+3)(x-1)$

19. $x^2 - 3x - 4 = (x+1)(x-4)$

21. $z^2 - 9z + 14 = (z-2)(z-7)$

23. $z^2 + 10z + 24 = (z+4)(z+6)$

25. $2x^2 - 9x + 4 = (2x-1)(x-4)$

27. $15p^2 - 23p + 4 = (3p-4)(5p-1)$

29. $4z^2 - 16z + 15 = (2z-5)(2z-3)$

31. $6x^2 - 5x - 4 = (2x+1)(3x-4)$

33. $10y^2 + 21y - 10 = (5y-2)(2y+5)$

35. $6x^2 + 5x - 4 = (2x-1)(3x+4)$

37. $3a^2 + 2a - 5 = (3a+5)(a-1)$

39. $x^2 - 81 = x^2 - (9)^2 = (x+9)(x-9)$

41. $9p^2 - 12p + 4 = (3p)^2 - 2(3p)(2) + 2^2$
$= (3p-2)^2$

43. $r^2 + 3rt - 10t^2 = (r-2t)(r+5t)$.

45. $m^2 - 8mn + 16n^2 = (m)^2 - 2(m)(4n) + (4n)^2$
$= (m-4n)^2$

47. $4u^2 + 12u + 9 = (2u+3)^2$

49. $25p^2 - 10p + 4$
This polynomial cannot be factored

51. $4r^2 - 9v^2 = (2r+3v)(2r-3v)$

53. $x^2 + 4xy + 4y^2 = (x+2y)^2$

55. $3a^2 - 13a - 30 = (3a+5)(a-6)$.

57. $21m^2 + 13mn + 2n^2 = (7m+2n)(3m+n)$

59. $y^2 - 4yz - 21z^2 = (y-7z)(y+3z)$

61. $121x^2 - 64 = (11x+8)(11x-8)$

63. $a^3 - 64 = a^3 - (4)^3 = (a-4)\left(a^2 + 4a + 16\right)$

Copyright © 2019 Pearson Education, Inc.

65. $8r^3 - 27s^3$
$= (2r)^3 - (3s)^3$
$= (2r - 3s)\left[(2r)^2 + (2r)(3s) + (3s)^2\right]$
$= (2r - 3s)\left(4r^2 + 6rs + 9s^2\right)$

67. $64m^3 + 125$
$= (4m)^3 + (5)^3$
$= (4m + 5)\left[(4m)^2 - (4m)(5) + (5)^2\right]$
$= (4m + 5)\left(16m^2 - 20m + 25\right)$

69. $1000y^3 - z^3$
$= (10y)^3 - (z)^3$
$= (10y - z)\left[(10y)^2 + (10y)(z) + (z)^2\right]$
$= (10y - z)\left(100y^2 + 10yz + z^2\right)$

71. $x^4 + 5x^2 + 6 = \left(x^2 + 2\right)\left(x^2 + 3\right)$

73. $b^4 - b^2 = b^2\left(b^2 - 1\right) = b^2(b + 1)(b - 1)$

75. $x^4 - x^2 - 12 = \left(x^2 - 4\right)\left(x^2 + 3\right)$
$= (x + 2)(x - 2)\left(x^2 + 3\right)$

77. $16a^4 - 81b^4 = \left(4a^2 - 9b^2\right)\left(4a^2 + 9b^2\right)$
$= (2a + 3b)(2a - 3b)\left(4a^2 + 9b^2\right)$

79. $x^8 + 8x^2 = x^2\left(x^6 + 8\right) = x^2\left(\left(x^2\right)^3 + 2^3\right)$
$= x^2\left(x^2 + 2\right)\left(x^4 - 2x^2 + 4\right)$

81. $6x^4 - 3x^2 - 3 = \left(2x^2 + 1\right)\left(3x^2 - 3\right)$ is not the correct complete factorization because $3x^2 - 3$ contains a common factor of 3. This common factor should be factored out as the first step. This will reveal a difference of two squares, which requires further factorization. The correct factorization is

$6x^4 - 3x^2 - 3 = 3\left(2x^4 - x^2 - 1\right)$
$= 3\left(2x^2 + 1\right)\left(x^2 - 1\right)$
$= 3\left(2x^2 + 1\right)(x + 1)(x - 1)$

83. $(x + 2)^3 = (x + 2)(x + 2)^2$
$= (x + 2)(x^2 + 4x + 4)$
$= x^3 + 4x^2 + 2x^2 + 8x + 4x + 8$
$= x^3 + 6x^2 + 12x + 8,$

which is not equal to $x^3 + 8$. The correct factorization is $x^3 + 8 = (x + 2)(x^2 - 2x + 4)$.

Section 1.4 Rational Expressions

1. $\dfrac{8x^2}{56x} = \dfrac{x \cdot 8x}{7 \cdot 8x} = \dfrac{x}{7}$

3. $\dfrac{25p^2}{35p^3} = \dfrac{5 \cdot 5p^2}{7p \cdot 5p^2} = \dfrac{5}{7p}$

5. $\dfrac{5m + 15}{4m + 12} = \dfrac{5(m + 3)}{4(m + 3)} = \dfrac{5}{4}$

7. $\dfrac{4(w - 3)}{(w - 3)(w + 6)} = \dfrac{4}{w + 6}$

9. $\dfrac{3y^2 - 12y}{9y^3} = \dfrac{3y(y - 4)}{3y\left(3y^2\right)}$
$= \dfrac{y - 4}{3y^2}$

11. $\dfrac{m^2 - 4m + 4}{m^2 + m - 6} = \dfrac{(m - 2)(m - 2)}{(m + 3)(m - 2)} = \dfrac{m - 2}{m + 3}$

13. $\dfrac{x^2 + 2x - 3}{x^2 - 1} = \dfrac{(x + 3)(x - 1)}{(x + 1)(x - 1)} = \dfrac{x + 3}{x + 1}$

Copyright © 2019 Pearson Education, Inc.

15. $\frac{3a^2}{64}\cdot\frac{8}{2a^3}=\frac{3a^2\cdot 8}{64\cdot 2a^3}=\frac{3}{16a}$

17. $\frac{7x}{11}\div\frac{14x^3}{66y}=\frac{7x}{11}\cdot\frac{66y}{14x^3}=\frac{7x\cdot 66y}{11\cdot 14x^3}=\frac{3y}{x^2}$

19. $\frac{2a+b}{3c}\cdot\frac{15}{4(2a+b)}=\frac{(2a+b)\cdot 15}{(2a+b)\cdot 12c}=\frac{15}{12c}=\frac{5}{4c}$

21. $\frac{15p-3}{6}\div\frac{10p-2}{3}=\frac{15p-3}{6}\cdot\frac{3}{10p-2}$
$=\frac{3(5p-1)\cdot 3}{3\cdot 2\cdot 2\cdot(5p-1)}$
$=\frac{3(5p-1)\cdot 3}{3(5p-1)\cdot 2\cdot 2}=\frac{3}{4}$

23. $\frac{9y-18}{6y+12}\cdot\frac{3y+6}{15y-30}=\frac{9(y-2)}{6(y+2)}\cdot\frac{3(y+2)}{15(y-2)}$
$=\frac{27(y-2)(y+2)}{90(y+2)(y-2)}=\frac{27}{90}=\frac{3}{10}$

25. $\frac{4a+12}{2a-10}\div\frac{a^2-9}{a^2-a-20}=\frac{4a+12}{2a-10}\cdot\frac{a^2-a-20}{a^2-9}$
$=\frac{4(a+3)}{2(a-5)}\cdot\frac{(a-5)(a+4)}{(a+3)(a-3)}$
$=\frac{4(a+3)(a-5)(a+4)}{2(a-5)(a+3)(a-3)}$
$=\frac{2(a+4)}{a-3}$

27. $\frac{k^2-k-6}{k^2+k-12}\cdot\frac{k^2+3k-4}{k^2+2k-3}$
$=\frac{(k-3)(k+2)}{(k+4)(k-3)}\cdot\frac{(k+4)(k-1)}{(k+3)(k-1)}$
$=\frac{(k-3)(k+2)(k+4)(k-1)}{(k+4)(k-3)(k+3)(k-1)}=\frac{k+2}{k+3}$

Answers will vary for exercises 29 and 30. Sample answers are given.

29. To find the least common denominator for two fractions, factor each denominator into prime factors, multiply all unique prime factors raising each factor to the highest frequency it occurred.

31. The common denominator is $35z$.
$\frac{2}{7z}-\frac{1}{5z}=\frac{2\cdot 5}{7z\cdot 5}-\frac{1\cdot 7}{5z\cdot 7}=\frac{10}{35z}-\frac{7}{35z}=\frac{3}{35z}$

33. $\frac{r+2}{3}-\frac{r-2}{3}=\frac{(r+2)-(r-2)}{3}$
$=\frac{r+2-r+2}{3}=\frac{4}{3}$

35. The common denominator is $5x$.
$\frac{4}{x}+\frac{1}{5}=\frac{4\cdot 5}{x\cdot 5}+\frac{1\cdot x}{5\cdot x}=\frac{20}{5x}+\frac{x}{5x}=\frac{20+x}{5x}$

37. The common denominator is $m(m-1)$.
$\frac{1}{m-1}+\frac{2}{m}=\frac{m\cdot 1}{m\cdot(m-1)}+\frac{(m-1)\cdot 2}{(m-1)\cdot m}$
$=\frac{m}{m(m-1)}+\frac{2(m-1)}{m(m-1)}$
$=\frac{m+2(m-1)}{m(m-1)}=\frac{m+2m-2}{m(m-1)}$
$=\frac{3m-2}{m(m-1)}$

39. The common denominator is $5(b+2)$.
$\frac{7}{b+2}+\frac{2}{5(b+2)}=\frac{7\cdot 5}{(b+2)\cdot 5}+\frac{2}{5(b+2)}$
$=\frac{35+2}{5(b+2)}=\frac{37}{5(b+2)}$

41. The common denominator is $20(k-2)$.
$\frac{2}{5(k-2)}+\frac{5}{4(k-2)}=\frac{8}{20(k-2)}+\frac{25}{20(k-2)}$
$=\frac{8+25}{20(k-2)}=\frac{33}{20(k-2)}$

43. First factor the denominators in order to find the common denominator.
$x^2-4x+3=(x-3)(x-1)$
$x^2-x-6=(x-3)(x+2)$
The common denominator is $(x-3)(x-1)(x+2)$.
$\frac{2}{x^2-4x+3}+\frac{5}{x^2-x-6}$
$=\frac{2}{(x-3)(x-1)}+\frac{5}{(x-3)(x+2)}$
$=\frac{2(x+2)}{(x-3)(x-1)(x+2)}+\frac{5(x-1)}{(x-3)(x+2)(x-1)}$
$=\frac{2(x+2)+5(x-1)}{(x-3)(x+2)(x-1)}=\frac{2x+4+5x-5}{(x-3)(x-1)(x+2)}$
$=\frac{7x-1}{(x-3)(x-1)(x+2)}$

Copyright © 2019 Pearson Education, Inc.

45. First factor the denominators in order to find the common denominator.

$$y^2+7y+12=(y+3)(y+4)$$
$$y^2+5y+6=(y+3)(y+2)$$

The common denominator is $(y+4)(y+3)(y+2)$.

$$\frac{2y}{y^2+7y+12}-\frac{y}{y^2+5y+6}$$
$$=\frac{2y}{(y+4)(y+3)}-\frac{y}{(y+3)(y+2)}$$
$$=\frac{2y(y+2)}{(y+4)(y+3)(y+2)}-\frac{y(y+4)}{(y+4)(y+3)(y+2)}$$
$$=\frac{2y(y+2)-y(y+4)}{(y+4)(y+3)(y+2)}=\frac{2y^2+4y-y^2-4y}{(y+4)(y+3)(y+2)}$$
$$=\frac{y^2}{(y+4)(y+3)(y+2)}$$

47. $\dfrac{1+\frac{1}{x}}{1-\frac{1}{x}}$

Multiply both numerator and denominator of this complex fraction by the common denominator, x.

$$\frac{1+\frac{1}{x}}{1-\frac{1}{x}}=\frac{x\left(1+\frac{1}{x}\right)}{x\left(1-\frac{1}{x}\right)}=\frac{x\cdot 1+x\left(\frac{1}{x}\right)}{x\cdot 1-x\left(\frac{1}{x}\right)}=\frac{x+1}{x-1}$$

49. $\dfrac{\frac{1}{x+h}-\frac{1}{x}}{h}$

The common denominator in the numerator is $x(x+h)$

$$\frac{\frac{1}{x+h}-\frac{1}{x}}{h}=\frac{\frac{x-(x+h)}{x(x+h)}}{h}=\frac{\frac{x-x-h}{x(x+h)}}{h}=\frac{\frac{-h}{x(x+h)}}{h}$$
$$=\frac{-h}{x(x+h)}\div h=\frac{-h}{x(x+h)}\cdot\frac{1}{h}$$
$$=\frac{-1}{x(x+h)}\text{ or }-\frac{1}{x(x+h)}$$

51. The length of each side of the dartboard is $2x$, so the area of the dartboard is $4x^2$. The area of the shaded region is πx^2.

a. The probability that a dart will land in the shaded region is $\dfrac{\pi x^2}{4x^2}$.

b. $\dfrac{\pi x^2}{4x^2}=\dfrac{\pi}{4}$

53. The length of each side of the dartboard is $5x$, so the area of the dartboard is $25x^2$. The area of the shaded region is x^2.

a. The probability that a dart will land in the shaded region is $\dfrac{x^2}{25x^2}$.

b. $\dfrac{x^2}{25x^2}=\dfrac{1}{25}$

55. Average cost = total cost C divided by the number of calculators produced.

$$\frac{-7.2x^2+6995x+230{,}000}{1000x}$$

57. Let $x = 13$. Then

$$\frac{.505(13)^2-4.587(13)+27.6}{13+1}\approx 3.8$$

The ad cost approximately \$3.8 million in 2013

59. Let $x = 45$. Then

$$\frac{.072(45)^2+.744(45)+1.2}{45+2}=3.84$$

Let $x = 20$. Then

$$\frac{.505(20)^2-4.587(20)+27.6}{20+1}\approx 6.6$$

The cost of an ad will not reach \$7 million in 2020.

61. Let $x = 11$. Then

$$\frac{.049(11)^2+2.40(11)+.83}{11+2}\approx 2.55$$

The hourly insurance cost in 2011 was \$2.55.

63. Let $x = 20$. Then

$$\frac{.049(20)^2+2.40(20)+.83}{20+2}\approx 3.11$$

The hourly insurance cost in 2020 will be \$3.11. The annual cost will be 3.11(2100) = \$6531.

Section 1.5 Exponents and Radicals

1. $\dfrac{7^5}{7^3}=7^{5-3}=7^2=49$

3. $(4c)^2=4^2c^2=16c^2$

Copyright © 2019 Pearson Education, Inc.

5. $\left(\frac{2}{x}\right)^5 = \frac{2^5}{x^5} = \frac{32}{x^5}$

7. $(3u^2)^3(2u^3)^2 = (27u^6)(4u^6) = 108u^{12}$

9. $7^{-1} = \frac{1}{7^1} = \frac{1}{7}$

11. $-6^{-5} = -\frac{1}{6^5} = -\frac{1}{7776}$

13. $(-y)^{-3} = \frac{1}{(-y)^3} = -\frac{1}{y^3}$

15. $\left(\frac{4}{3}\right)^{-2} = \left(\frac{3}{4}\right)^2 = \frac{9}{16}$

17. $\left(\frac{a}{b^3}\right)^{-1} = \left(\frac{b^3}{a}\right)^1 = \frac{b^3}{a}$

19. $49^{1/2} = 7$ because $7^2 = 49$.

21. $(5.71)^{1/4} = (5.71)^{.25} \approx 1.55$ Use a calculator.

23. $-64^{2/3} = -\left(64^{1/3}\right)^2 = -(4)^2 = -16$

25. $\left(\frac{8}{27}\right)^{-4/3} = \left(\frac{27^{1/3}}{8^{1/3}}\right)^4 = \left(\frac{3}{2}\right)^4 = \frac{3^4}{2^4} = \frac{81}{16}$

27. $\frac{5^{-3}}{4^{-2}} = \frac{4^2}{5^3}$

29. $4^{-3} \cdot 4^6 = 4^3$

31. $\frac{4^{10} \cdot 4^{-6}}{4^{-4}} = 4^{10} \cdot 4^{-6} \cdot 4^4 = 4^8$

33. $\frac{z^6 \cdot z^2}{z^5} = \frac{z^8}{z^5} = z^{8-5} = z^3$

35. $\frac{3^{-1}(p^{-2})^3}{3p^{-7}} = \frac{3^{-1}p^{-6}}{3^1p^{-7}} = 3^{-1-1}p^{-6-(-7)}$

$= 3^{-2}p^1 = \frac{1}{3^2} \cdot p = \frac{p}{9}$

37. $\left(q^{-5}r^3\right)^{-1} = q^5r^{-3} = q^5 \cdot \frac{1}{r^3} = \frac{q^5}{r^3}$

39. $\left(2p^{-1}\right)^3 \cdot \left(5p^2\right)^{-2} = 2^3\left(p^{-1}\right)^3(5)^{-2}\left(p^2\right)^{-2}$

$= 2^3\left(p^{-3}\right)\left(\frac{1}{5^2}\right)\left(p^{-4}\right)$

$= 2^3\left(\frac{1}{p^3}\right)\left(\frac{1}{5^2}\right)\left(\frac{1}{p^4}\right)$

$= \frac{8}{25p^7}$

41. $(2p)^{1/2} \cdot \left(2p^3\right)^{1/3} = 2^{1/2}p^{1/2} \cdot 2^{1/3} \cdot \left(p^3\right)^{1/3}$

$= 2^{1/2}p^{1/2} \cdot 2^{1/3} \cdot p^1$

$= 2^{5/6}p^{3/2}$

43. $p^{2/3}\left(2p^{1/3} + 5p\right) = p^{2/3}\left(2p^{1/3}\right) + p^{2/3}(5p)$

$= 2p + 5p^{5/3}$

45. $\frac{\left(x^2\right)^{1/3}\left(y^2\right)^{2/3}}{3x^{2/3}y^2} = \frac{(x)^{2/3}(y)^{4/3}}{3x^{2/3}y^2}$

$= \frac{1}{3y^{2-4/3}} = \frac{1}{3y^{2/3}}$

47. $\frac{(7a)^2(5b)^{3/2}}{(5a)^{3/2}(7b)^4} = \frac{7^2a^25^{3/2}b^{3/2}}{5^{3/2}a^{3/2}7^4b^4} = \frac{a^{2-\frac{3}{2}}}{7^2b^{4-\frac{3}{2}}}$

$= \frac{a^{1/2}}{49b^{5/2}}$

49. $x^{1/2}\left(x^{2/3} - x^{4/3}\right) = x^{1/2}x^{2/3} - x^{1/2}x^{4/3}$

$= x^{7/6} - x^{11/6}$

51. This is a difference of two squares.

$\left(x^{1/2} + y^{1/2}\right)\left(x^{1/2} - y^{1/2}\right) = \left(x^{1/2}\right)^2 - \left(y^{1/2}\right)^2$

$= x - y$

53. $(-3x)^{1/3} = \sqrt[3]{-3x}$, (f)

55. $(-3x)^{-1/3} = \frac{1}{(-3x)^{1/3}} = \frac{1}{\sqrt[3]{-3x}}$, (h)

57. $(3x)^{1/3} = \sqrt[3]{3x}$, (g)

Copyright © 2019 Pearson Education, Inc.

59. $(3x)^{-1/3} = \dfrac{1}{(3x)^{1/3}} = \dfrac{1}{\sqrt[3]{3x}}$, (c)

61. $\sqrt[3]{125} = 125^{1/3} = 5$

63. $\sqrt[4]{625} = 625^{1/4} = 5$

65. $\sqrt{63}\cdot\sqrt{7} = 3\sqrt{7}\cdot\sqrt{7} = 3\cdot 7 = 21$

67. $\sqrt{81-4} = \sqrt{77}$

69. $\sqrt{5}\sqrt{15} = \sqrt{75} = \sqrt{25\cdot 3} = \sqrt{25}\sqrt{3} = 5\sqrt{3}$

71. $\sqrt{50} - \sqrt{72} = 5\sqrt{2} - 6\sqrt{2} = -\sqrt{2}$

73. $5\sqrt{20} - \sqrt{45} + 2\sqrt{80}$
$= 5\cdot 2\sqrt{5} - 3\sqrt{5} + 2\cdot 4\sqrt{5}$
$= 10\sqrt{5} - 3\sqrt{5} + 8\sqrt{5} = 15\sqrt{5}$

75. $\left(\sqrt{5}+\sqrt{2}\right)\left(\sqrt{5}-\sqrt{2}\right) = \left(\sqrt{5}\right)^2 - \left(\sqrt{2}\right)^2$
$= 5 - 2 = 3$

77. $\dfrac{3}{1-\sqrt{2}} = \dfrac{3}{1-\sqrt{2}}\cdot\dfrac{1+\sqrt{2}}{1+\sqrt{2}} = \dfrac{3\left(1+\sqrt{2}\right)}{(1)^2 - \left(\sqrt{2}\right)^2}$
$= \dfrac{3\left(1+\sqrt{2}\right)}{1-2} = \dfrac{3\left(1+\sqrt{2}\right)}{-1}$
$= -3\left(1+\sqrt{2}\right) = -3 - 3\sqrt{2}$

79. $\dfrac{9-\sqrt{3}}{3-\sqrt{3}} = \dfrac{9-\sqrt{3}}{3-\sqrt{3}}\cdot\dfrac{3+\sqrt{3}}{3+\sqrt{3}} = \dfrac{27+9\sqrt{3}-3\sqrt{3}-3}{3^2 - \left(\sqrt{3}\right)^2}$
$= \dfrac{24+6\sqrt{3}}{9-3} = \dfrac{24+6\sqrt{3}}{6} = 4 + \sqrt{3}$

81. $\dfrac{3-\sqrt{2}}{3+\sqrt{2}} = \dfrac{3-\sqrt{2}}{3+\sqrt{2}}\cdot\dfrac{3+\sqrt{2}}{3+\sqrt{2}}$
$= \dfrac{9-2}{9+6\sqrt{2}+2} = \dfrac{7}{11+6\sqrt{2}}$

83. $x = \sqrt{\dfrac{kM}{f}}$

Note that because x represents the number of units to order, the value of x should be rounded to the nearest integer.

a. $k = \$1, f = \$500, M = 100{,}000$
$x = \sqrt{\dfrac{1\cdot 100{,}000}{500}} = \sqrt{200} \approx 14.1$
The number of units to order is 14.

b. $k = \$3, f = \$7, M = 16{,}700$
$x = \sqrt{\dfrac{3\cdot 16{,}700}{7}} \approx 84.6$
The number of units to order is 85.

c. $k = \$1, f = \$5, M = 16{,}800$
$x = \sqrt{\dfrac{1\cdot 16{,}800}{5}} = \sqrt{3360} \approx 58.0$
The number of units to order is 58.

For exercises 85–88, we use the model revenue $= 6.67x^{.188}$, $x \ge 5$, $x = 5$ corresponds to 2005.

85. Let $x = 15$. Then $6.67(15)^{.188} \approx 11.1$
The domestic revenue for 2015 were about \$11.1 billion.

87. Let $x = 20$. Then $6.67(20)^{.188} \approx 11.7$
The domestic revenue for 2020 will be about \$11.7 billion.

For exercises 89–92, we use the model revenue $= 1.08x^{.527}$, $x \ge 1$, $x = 1$ corresponds to the first quarter of the year 2013.

89. Let $x = 13$. Then $1.08(13)^{.527} \approx 4.2$
The advertising revenue in the first quarter of 2016 was approximately \$4.2 billion.

91. Let $x = 24$. Then $1.08(24)^{.527} \approx 5.8$
The advertising revenue in the fourth quarter of 2018 will be approximately \$5.8 billion.

Copyright © 2019 Pearson Education, Inc.

For exercises 93–96, we use the model Part-time Students $= 2.72x^{.238}; x \geq 10, x = 10$ corresponds to 1990.

93. Let $x = 29$. Then $2.72(29)^{.238} \approx 6.1$
According to the model, there were approximately 6.1 million part-time students attending college or university in 2009.

95. Let $x = 38$. Then $2.72(38)^{.238} \approx 6.5$
According to the model, there will be approximately 6.5 million part-time students attending college or university in 2018.

Section 1.6 First-Degree Equations

1.
$$\begin{aligned} 3x+8&=20\\ 3x+8-8&=20-8\\ 3x&=12\\ \frac{1}{3}(3x)&=\frac{1}{3}(12)\\ x&=4 \end{aligned}$$

3.
$$\begin{aligned} .6k-.3&=.5k+.4\\ .6k-.5k-.3&=.5k-.5k+.4\\ .1k-.3&=.4\\ .1k-.3+.3&=.4+.3\\ .1k&=.7\\ \frac{.1k}{.1}&=\frac{.7}{.1}\Rightarrow k=7 \end{aligned}$$

5.
$$\begin{aligned} 2a-1&=4(a+1)+7a+5\\ 2a-1&=4a+4+7a+5\\ 2a-1&=11a+9\\ 2a-2a-1&=11a-2a+9\\ -1&=9a+9\\ -1-9&=9a+9-9\\ -10&=9a\\ \frac{-10}{9}&=\frac{9a}{9}\Rightarrow -\frac{10}{9}=a \end{aligned}$$

7.
$$\begin{aligned} 2[x-(3+2x)+9]&=3x-8\\ 2(x-3-2x+9)&=3x-8\\ 2(-x+6)&=3x-8\\ -2x+12&=3x-8\\ 12&=5x-8\\ 20&=5x\Rightarrow 4=x \end{aligned}$$

9. $\frac{3x}{5}-\frac{4}{5}(x+1)=2-\frac{3}{10}(3x-4)$
Multiply both sides by the common denominator, 10.
$$\begin{aligned} 10\left(\frac{3x}{5}\right)-10\left(\frac{4}{5}\right)(x+1)\\ =(10)(2)-(10)\left(\frac{3}{10}\right)(3x-4)\\ 2(3x)-8(x+1)&=20-3(3x-4)\\ 6x-8x-8&=20-9x+12\\ -2x-8&=32-9x\\ -2x+9x&=32+8\\ 7x&=40\\ \frac{1}{7}(7x)&=\frac{1}{7}(40)\Rightarrow x=\frac{40}{7} \end{aligned}$$

11. $\frac{5y}{6}-8=5-\frac{2y}{3}$
$$\begin{aligned} 6\left(\frac{5y}{6}-8\right)&=6\left(5-\frac{2y}{3}\right)\\ 6\left(\frac{5y}{6}\right)-6(8)&=6(5)-6\left(\frac{2y}{3}\right)\\ 5y-48&=30-4y\\ 9y-48&=30\\ 9y&=78\\ y&=\frac{78}{9}=\frac{26}{3} \end{aligned}$$

13.
$$\begin{aligned} \frac{m}{2}-\frac{1}{m}&=\frac{6m+5}{12}\\ 12m\left(\frac{m}{2}-\frac{1}{m}\right)&=12m\left(\frac{6m+5}{12}\right)\\ (12m)\left(\frac{m}{2}\right)-(12m)\left(\frac{1}{m}\right)&=m(6m)+m(5)\\ 6m^2-12&=6m^2+5m\\ -12&=5m\\ \frac{1}{5}(-12)&=\frac{1}{5}(5m)\Rightarrow -\frac{12}{5}=m \end{aligned}$$

Copyright © 2019 Pearson Education, Inc.

15. $\frac{4}{x-3}-\frac{8}{2x+5}+\frac{3}{x-3}=0$

$\frac{4}{x-3}+\frac{3}{x-3}-\frac{8}{2x+5}=0$

$\frac{7}{x-3}-\frac{8}{2x+5}=0$

Multiply each side by the common denominator, $(x-3)(2x+5)$.

$$(x-3)(2x+5)\left(\frac{7}{x-3}\right)-(x-3)(2x+5)\left(\frac{8}{2x+5}\right)=(x-3)(2x+5)(0)$$

$$7(2x+5)-8(x-3)=0$$
$$14x+35-8x+24=0$$
$$6x+59=0$$
$$6x=-59\Rightarrow x=-\frac{59}{6}$$

17. $\frac{3}{2m+4}=\frac{1}{m+2}-2$

$\frac{3}{2(m+2)}=\frac{1}{m+2}-2$

$$2(m+2)\left(\frac{3}{2(m+2)}\right)=2(m+2)\left(\frac{1}{m+2}\right)-2(m+2)(2)$$

$$=2(m+2)\left(\frac{1}{m+2}\right)-2(m+2)(2)$$

$$3=2-4(m+2)$$
$$3=2-4m-8$$
$$3=-6-4m\Rightarrow 9=-4m\Rightarrow m=-\frac{9}{4}$$

19.
$$9.06x+3.59(8x-5)=12.07x+.5612$$
$$9.06x+28.72x-17.95=12.07x+.5612$$
$$9.06x+28.72x-12.07x=17.95+.5612$$
$$25.71x=18.5112$$
$$x=\frac{18.5112}{25.71}=.72$$

21. $\frac{2.63r-8.99}{1.25}-\frac{3.90r-1.77}{2.45}=r$

Multiply by the common denominator (1.25)(2.45) to eliminate the fractions.

$$(2.45)(2.63r-8.99)-(1.25)(3.90r-1.77)=(2.45)(1.25)r$$
$$6.4435r-22.0255-4.875r+2.2125=3.0625r$$
$$1.5685r-19.813=3.0625r$$
$$-19.813=1.494r$$
$$\frac{19.813}{1.494}=\frac{1.494r}{1.494}$$
$$r\approx-13.26$$

23. $4(a+x)=b-a+2x$

$$4a+4x=b-a+2x$$
$$4a=b-a-2x$$
$$5a-b=-2x$$
$$\frac{5a-b}{-2}=\frac{-2x}{-2}$$
$$-\frac{5a-b}{2}=x \text{ or } x=\frac{b-5a}{2}$$

25. $5(b-x)=2b+ax$

First, use the distributive property.

$$5b-5x=2b+ax$$
$$5b=2b+ax+5x$$
$$3b=ax+5x$$
$$3b=(a+5)x$$
$$\frac{3b}{a+5}=\frac{(a+5)x}{a+5}\Rightarrow\frac{3b}{a+5}=x$$

Now use the distributive property on the right.

$$3b=(a+5)x$$
$$\frac{3b}{a+5}=\frac{(a+5)x}{a+5}$$
$$\frac{3b}{a+5}=x$$

27. $PV=k$ for V

$$\frac{1}{P}(PV)=\frac{1}{P}(k)$$
$$V=\frac{k}{P}$$

Copyright © 2019 Pearson Education, Inc.

29. $V = V_0 + gt$ for g

$$V - V_0 = gt$$
$$\frac{V - V_0}{t} = \frac{gt}{t}$$
$$\frac{V - V_0}{t} = g$$

31. $A = \frac{1}{2}(B+b)h$ for B

$$A = \frac{1}{2}Bh + \frac{1}{2}bh$$
$$2A = Bh + bh \quad \text{Multiply by 2.}$$
$$2A - bh = Bh$$
$$\frac{2A - bh}{h} = \frac{Bh}{h} \quad \text{Multiply by } \frac{1}{h}.$$
$$\frac{2A - bh}{h} = \frac{2A}{h} - b = B$$

33. $|2h - 1| = 5$

$2h - 1 = 5$ or $2h - 1 = -5$

$2h = 6$ or $2h = -4$

$h = 3$ or $h = -2$

35. $|6 + 2p| = 10$

$6 + 2p = 10$ or $6 + 2p = -10$

$2p = 4$ or $2p = -16$

$p = 2$ or $p = -8$

37. $\left|\frac{5}{r-3}\right| = 10$

$\frac{5}{r-3} = 10$ or $\frac{5}{r-3} = -10$

$5 = 10(r-3)$ or $5 = -10(r-3)$

$5 = 10r - 30$ or $5 = -10r + 30$

$35 = 10r$ or $-25 = -10r$

$\frac{35}{10} = \frac{7}{2} = r$ or $\frac{-25}{-10} = \frac{5}{2} = r$

39. $-5 = \frac{5}{9}(F - 32)$

$$-5\left(\frac{9}{5}\right) = \left(\frac{9}{5}\right)\left(\frac{5}{9}\right)(F - 32)$$
$$-9 = F - 32 \Rightarrow 23 = F$$

The temperature –5°C = 23°F.

41. $22 = \frac{5}{9}(F - 32)$

$$22\left(\frac{9}{5}\right) = \left(\frac{9}{5}\right)\left(\frac{5}{9}\right)(F - 32)$$
$$39.6 = F - 32 \Rightarrow 71.6 = F$$

The temperature 22°C = 71.6°F.

43. Let $x = 10$.

$$y = 1.15x + 1.62$$
$$y = 1.15(10) + 1.62$$
$$y = 13.12$$

Therefore, the gross federal debt in 2010 was $13.12 trillion.

45. $y = 1.15x + 1.62$

Substitute 22.32 for y.

$$22.32 = 1.15x + 1.62$$
$$20.7 = 1.15x \Rightarrow 18 = x$$

Therefore, the federal deficit will be $22.32 trillion in 2018.

47. $y = 1.15x + 1.62$

Substitute 25.77 for y.

$$25.77 = 1.15x + 1.62$$
$$24.15 = 1.15x \Rightarrow 21 = x$$

Therefore, the federal deficit will be $825.77 trillion in 2021.

49. $E = .108x + 1.517$

Substitute $2.7052422 in for E.

$$2.705 = .108x + 1.517$$
$$1.188 = .108x \Rightarrow 11 = x$$

The health care expenditures were at $2.7052250 trillion in 2011.

51. $E = .108x + 1.517$

Substitute $3.461 in for E.

$$3.461 = .108x + 1.517$$
$$1.944 = .108x \Rightarrow 18 = x$$

The health care expenditures will be $3.4612422 trillion in 2018.

53. $114.8(x - 2010) = 5y - 3390.5$

Substitute 746.98 for y and solve for x.

$$114.8(x - 2010) = 5(746.98) - 3390.5$$
$$114.8x - 230{,}748 = 344.4$$
$$114.8x = 231{,}092.4$$
$$x = 2013$$

The amount of income was $746.98 billion in 2013.

Copyright © 2019 Pearson Education, Inc.

55. $114.8(x-2010)=5y-3390.5$

Substitute 907.7 for y and solve for x.

$$114.8(x-2010)=5(907.7)-3390.5$$
$$114.8x-230,748=1148$$
$$114.8x=231,896$$
$$x=2020$$

The amount of income will be \$907.7 billion in 2020.

57. $f=800$, $n=18$, $q=36$

$$u=f\cdot\frac{n(n+1)}{q(q+1)}=800\cdot\frac{18(19)}{36(37)}$$
$$=800\cdot\frac{342}{1332}\approx 205.41$$

The amount of unearned interest is \$205.41.

59. Let x = the number invested at 5%.

$$.05x+.04(52,000-x)=2290$$
$$.05x+2080-.04x=2290$$
$$.01x+2080=2290$$
$$.01x+2080-2080=2290-2080$$
$$.01x=210$$
$$\frac{.01x}{.01x}=\frac{210}{.01}$$
$$x=21,000$$

Joe invested \$21,000 at 5%.

61. Let x = price of first plot.
Then $120,000-x$ = price of second plot.
$.15x$ = profit from first plot
$-.10(120,000-x)$ = loss from second plot.

$$.15x-.10(120,000-x)=5500$$
$$.15x-12,000+.10x=5500$$
$$.25x=17,500$$
$$x=70,000$$

She paid \$70,000 for the first plot and 120,000 – 70,000, or \$50,000 for the second plot.

63. Let x = average rate of growth of Twitter.
Then $5.7+x$ = average rate of growth of Instagram.
$35x$ = visitors to Twitter
$21(5.7+x)$ = visitors to Instagram.

$$35x=21(5.7+x)$$
$$35x=119.7+21x$$
$$14x=119.7$$
$$x\approx 8.6$$

Since x represents the average rate of growth of Twitter, there was an average growth of approximately 8.6 million visitorsShe .

65. The number of visitors to Twitter was $35x=35(8.6)=301$ million.

67. Let x = the number of liters of 94 octane gas;
200 = the number of liters of 99 octane gas;
$200+x$ = the number of liters of 97 octane gas.

$$94x+99(200)=97(200+x)$$
$$94x+19,800=19,400+97x$$
$$400=3x$$
$$\frac{400}{3}=x$$

Thus, $\frac{400}{3}$ liters of 94 octane gas are needed.

Section 1.7 Quadratic Equations

1. $(x+4)(x-14)=0$

$$x+4=0 \quad\text{or}\quad x-14=0$$
$$x=-4 \quad\text{or}\quad x=14$$

The solutions are –4 and 14.

3. $x(x+6)=0$

$$x=0 \quad\text{or}\quad x+6=0$$
$$x=-6$$

The solutions are 0 and –6.

5. $2z^2=4z$

$$2z^2-4z=0$$
$$2z(z-2)=0$$
$$2z=0 \quad\text{or}\quad z-2=0$$
$$z=0 \quad\text{or}\quad z=2$$

The solutions are 0 and 2.

Copyright © 2019 Pearson Education, Inc.

7. $y^2 + 15y + 56 = 0$

$(y+7)(y+8) = 0$

$y + 7 = 0$ or $y + 8 = 0$

$y = -7$ or $y = -8$

The solutions are –7 and –8.

9. $2x^2 = 7x - 3$

$2x^2 - 7x + 3 = 0$

$(2x-1)(x-3) = 0$

$2x - 1 = 0$ or $x - 3 = 0$

$x = \frac{1}{2}$ or $x = 3$

The solutions are $\frac{1}{2}$ and 3.

11. $6r^2 + r = 1$

$6r^2 + r - 1 = 0$

$(3r-1)(2r+1) = 0$

$3r - 1 = 0$ or $2r + 1 = 0$

$r = \frac{1}{3}$ or $r = -\frac{1}{2}$

The solutions are $\frac{1}{3}$ and $-\frac{1}{2}$.

13. $2m^2 + 20 = 13m$

$2m^2 - 13m + 20 = 0$

$(2m-5)(m-4) = 0$

$2m - 5 = 0$ or $m - 4 = 0$

$m = \frac{5}{2}$ or $m = 4$

The solutions are $\frac{5}{2}$ and 4.

15. $m(m+7) = -10$

$m^2 + 7m + 10 = 0$

$(m+5)(m+2) = 0$

$m + 5 = 0$ or $m + 2 = 0$

$m = -5$ or $m = -2$

The solutions are –5 and –2.

17. $9x^2 - 16 = 0$

$(3x+4)(3x-4) = 0$

$3x + 4 = 0$ or $3x - 4 = 0$

$3x = -4$ $\quad$ $3x = 4$

$x - \frac{4}{3}$ or $x = \frac{4}{3}$

The solutions are $-\frac{4}{3}$ and $\frac{4}{3}$.

19. $16x^2 - 16x = 0$

$16x(x-1) = 0$

$16x = 0$ or $x - 1 = 0$

$x = 0$ or $x = 1$

The solutions are 0 and 1.

21. $(r-2)^2 = 7$

$r - 2 = \sqrt{7}$ or $r - 2 = -\sqrt{7}$

$r = 2 + \sqrt{7}$ or $r = 2 - \sqrt{7}$

We abbreviate the solutions as $2 \pm \sqrt{7}$.

23. $(4x-1)^2 = 20$

Use the square root property.

$4x - 1 = \sqrt{20}$ or $4x - 1 = -\sqrt{20}$

$4x - 1 = 2\sqrt{5}$ or $4x - 1 = -2\sqrt{5}$

$4x = 1 + 2\sqrt{5}$ or $4x = 1 - 2\sqrt{5}$

The solutions are $\frac{1 \pm 2\sqrt{5}}{4}$.

25. $2x^2 + 7x + 1 = 0$

Use the quadratic formula with $a = 2$, $b = 7$, and $c = 1$.

$$x = \frac{-b \pm \sqrt{b^2 - 4ac}}{2a}$$

$$= \frac{-7 \pm \sqrt{7^2 - 4(2)(1)}}{2(2)}$$

$$= \frac{-7 \pm \sqrt{49 - 8}}{4} = \frac{-7 \pm \sqrt{41}}{4}$$

The solutions are $\frac{-7 + \sqrt{41}}{4}$ and $\frac{-7 - \sqrt{41}}{4}$, which are approximately –.1492 and –3.3508.

Copyright © 2019 Pearson Education, Inc.

27. $4k^2+2k=1$

Rewrite the equation in standard form.

$4k^2+2k-1=0$

Use the quadratic formula with $a=4$, $b=2$, and $c=-1$.

$$k=\frac{-2\pm\sqrt{2^2-4(4)(-1)}}{2(4)}=\frac{-2\pm\sqrt{4+16}}{8}$$
$$=\frac{-2\pm\sqrt{20}}{8}=\frac{-2\pm2\sqrt{5}}{8}=\frac{2\left(-1\pm\sqrt{5}\right)}{2\cdot4}$$
$$k=\frac{-1\pm\sqrt{5}}{4}$$

The solutions are $\frac{-1+\sqrt{5}}{4}$ and $\frac{-1-\sqrt{5}}{4}$, which are approximately .3090 and –.8090.

29. $5y^2+5y=2$

$5y^2+5y-2=0$

$a=5, b=5, c=-2$

$$y=\frac{-5\pm\sqrt{5^2-4(5)(-2)}}{2(5)}=\frac{-5\pm\sqrt{25+40}}{10}$$
$$=\frac{-5\pm\sqrt{65}}{10}=\frac{-5\pm\sqrt{65}}{10}$$

The solutions are $\frac{-5+\sqrt{65}}{10}$ and $\frac{-5-\sqrt{65}}{10}$, which are approximately .3062 and –1.3062.

31. $6x^2+6x+4=0$

$a=6, b=6, c=4$

$$x=\frac{-6\pm\sqrt{6^2-4(6)(4)}}{2(6)}=\frac{-6\pm\sqrt{36-96}}{12}$$
$$=\frac{-6\pm\sqrt{-60}}{12}$$

Because $\sqrt{-60}$ is not a real number, the given equation has no real number solutions.

33. $2r^2+3r-5=0$

$a=2, b=3, c=-5$

$$r=\frac{-(3)\pm\sqrt{9-4(2)(-5)}}{2(2)}=\frac{-3\pm\sqrt{9+40}}{4}$$
$$=\frac{-3\pm\sqrt{49}}{4}=\frac{-3\pm7}{4}$$
$$r=\frac{-3+7}{4}=\frac{4}{4}=1 \text{ or } r=\frac{-3-7}{4}=\frac{-10}{4}=\frac{-5}{2}$$

The solutions are $-\frac{5}{2}$ and 1.

35. $2x^2-7x+30=0$

$a=2, b=-7, c=30$

$$x=\frac{-(-7)\pm\sqrt{49-4(2)(30)}}{2(2)}=\frac{7\pm\sqrt{-191}}{4}$$

Since $\sqrt{-191}$ is not a real number, there are no real solutions.

37. $1+\frac{7}{2a}=\frac{15}{2a^2}$

To eliminate fractions, multiply both sides by the common denominator, $2a^2$.

$2a^2+7a=15 \Rightarrow 2a^2+7a-15=0$

$a=2, b=7, c=-15$

$$a=\frac{-7\pm\sqrt{7^2-4(2)(-15)}}{2(2)}=\frac{-7\pm\sqrt{49+120}}{4}$$
$$=\frac{-7\pm\sqrt{169}}{4}\Rightarrow a=\frac{-7+13}{4}=\frac{6}{4}=\frac{3}{2} \text{ or}$$
$$a=\frac{-7-13}{4}=\frac{-20}{4}=-5$$

The solutions are $\frac{3}{2}$ and –5.

39. $25t^2+49=70t$

$25t^2-70t+49=0$

$$b^2-4ac=(-70)^2-4(25)(49)$$
$$=4900-4900$$
$$=0$$

The discriminant is 0.

There is one real solution to the equation.

41. $13x^2+24x-5=0$

$$b^2-4ac=(24)^2-4(13)(-5)$$
$$=576+260=836$$

The discriminant is positive.

There are two real solutions to the equation.

For Exercises 43–46 use the quadratic formula: $x=\frac{-b\pm\sqrt{b^2-4ac}}{2a}$.

43. $4.42x^2-10.14x+3.79=0$

$$x=\frac{-(-10.14)\pm\sqrt{(-10.14)^2-4(4.42)(3.79)}}{2(4.42)}$$
$$\approx\frac{10.14\pm5.9843}{8.84}\approx .4701 \text{ or } 1.8240$$

Copyright © 2019 Pearson Education, Inc.

45. $7.63x^2 + 2.79x = 5.32$

$7.63x^2 + 2.79x - 5.32 = 0$

$$x = \frac{-2.79 \pm \sqrt{(2.79)^2 - 4(7.63)(-5.32)}}{2(7.63)}$$

$$\approx \frac{-2.79 \pm 13.0442}{15.26} \approx -1.0376 \text{ or } .6720$$

47. Let $x = 10$.

$P = .396x^2 - 7.09x + 74.9$

$P = .396(10)^2 - 7.09(10) + 74.9$

$P = 43.6$

The average annual pay for truckers in 2010 was \$43.6 thousand

49. Let $P = 48$.

$P = .396x^2 - 7.09x + 74.9$

$48 = .396x^2 - 7.09x + 74.9$

$0 = .396x^2 - 7.09x + 26.9$

$$x = \frac{7.09 \pm \sqrt{(-7.09)^2 - 4(.396)(26.9)}}{2(.396)}$$

≈ 5.46 or 12.45

The first full year the average annual pay for truckers was below \$48,000 was 2006.

51. Let $P = 55$.

$P = .396x^2 - 7.09x + 74.9$

$55 = .396x^2 - 7.09x + 74.9$

$0 = .396x^2 - 7.09x + 19.9$

$$x = \frac{7.09 \pm \sqrt{(-7.09)^2 - 4(.396)(19.9)}}{2(.396)}$$

≈ 3.49 or 14.42

The first full year the average annual pay for truckers recovered to be above \$55,000 was 2015.

53. Let $x = 3$.

$O = .867x^2 - 9.31x + 63.32$

$O = .867(3)^2 - 9.31(3) + 63.32$

$O = 43.19$

The price of crude oil for the third quarter of 2015 was \$43.19.

55. Let $O = 45$.

$O = .867x^2 - 9.31x + 63.32$

$45 = .867x^2 - 9.31x + 63.32$

$0 = .867x^2 - 9.31x + 18.32$

$$x = \frac{9.31 \pm \sqrt{(-9.31)^2 - 4(.867)(18.32)}}{2(.867)}$$

≈ 2.59 or 8.14

The most recent quarter where the price of crude oil was at least \$45 was the first quarter of 2017.

57. Triangle *ABC* represents the original position of the ladder, while triangle *DEC* represents the position of the ladder after it was moved.

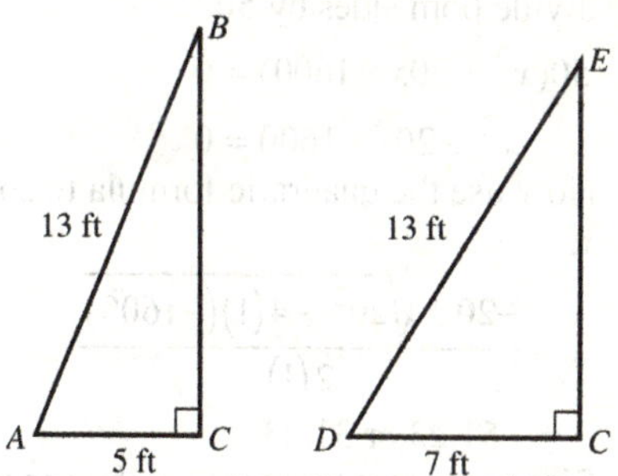

Use the Pythagorean theorem to find the distance from the top of the ladder to the ground.
In triangle *ABC*,

$13^2 = 5^2 + BC^2 \Rightarrow 169 - 25 = BC^2 \Rightarrow$

$144 = BC^2 \Rightarrow 12 = BC$

Thus, the top of the ladder was originally 12 feet from the ground.
In triangle *DEC*,

$13^2 = 7^2 + EC^2 \Rightarrow 169 - 49 = EC^2 \Rightarrow$

$120 = EC^2 \Rightarrow EC = \sqrt{120}$

The top of the ladder was $\sqrt{120}$ feet from the ground after the ladder was moved. Therefore, the ladder moved down $12 - \sqrt{120} \approx 1.046$ feet.

59. **a.** The eastbound train travels at a speed of $x + 20$.

b. The northbound train travels a distance of $5x$ in 5 hours.
The eastbound train travels a distance of $5(x + 20) = 5x + 100$ in 5 hours.

Copyright © 2019 Pearson Education, Inc.

c. B

300 mi

$5x$

C A

$5(x+20)$

By the Pythagorean theorem,

$$(5x)^2+(5x+100)^2=300^2$$

d. Expand and combine like terms.

$$25x^2+25x^2+1000x+10{,}000=90{,}000$$
$$50x^2+1000x-80{,}000=0$$

Factor out the common factor, 50, and divide both sides by 50.

$$50(x^2+20x-1600)=0$$
$$x^2+20x-1600=0$$

Now use the quadratic formula to solve for x.

$$x=\frac{-20\pm\sqrt{20^2-4(1)(-1600)}}{2(1)}$$
$$\approx -51.23 \text{ or } 31.23$$

Since x cannot be negative, the speed of the northbound train is $x\approx 31.23$ mph, and the speed of the eastbound train is $x+20\approx$ 51.23 mph.

61. a. Let x represent the length. Then, $\frac{300-2x}{2}$ or $150-x$ represents the width.

b. Use the formula for the area of a rectangle.
$LW=A \Rightarrow x(150-x)=5000$

c. $150x-x^2=5000$
Write this quadratic equation in standard form and solve by factoring.

$$0=x^2-150x+5000$$

$$x^2-150x+5000=0$$
$$(x-50)(x-100)=0$$
$$x-50=0 \quad\text{or}\quad x-100=0$$
$$x=50 \quad\text{or}\quad x=100$$

Choose $x=100$ because the length is the larger dimension. The length is 100 m and the width is $150-100=50$ m.

63. Let x = the width of the uniform strip around the rug.

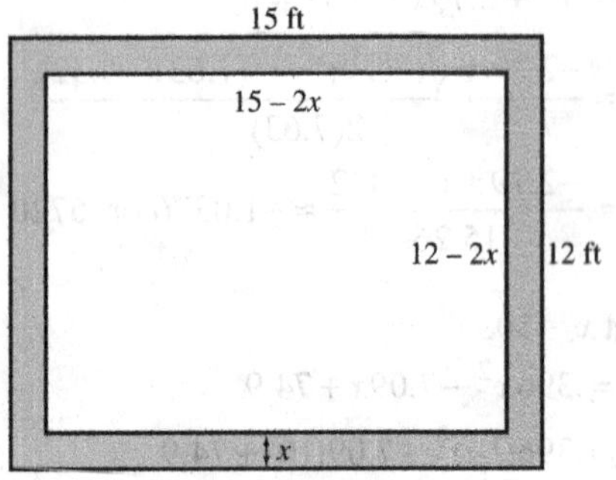

The dimensions of the rug are $15-2x$ and $12-2x$. The area, 108, is the length times the width.

Solve the equation.

$$(15-2x)(12-2x)=108$$
$$180-54x+4x^2=108$$
$$4x^2-54x+72=0$$
$$2x^2-27x+36=0$$
$$(x-12)(2x-3)=0$$
$$x-12=0 \quad\text{or}\quad 2x-3=0$$
$$x=12 \quad\text{or}\quad x=\frac{3}{2}$$

Discard $x=12$ since both $12-2x$ and $15-2x$ would be negative.

If $x=\frac{3}{2}$, then

$$15-2x=15-2\left(\frac{3}{2}\right)=12$$

and $12-2x=12-2\left(\frac{3}{2}\right)=9$.

The dimensions of the rug should be 9 ft by 12 ft.

Copyright © 2019 Pearson Education, Inc.

For exercises 65–70, use the formula $h = -16t^2 + v_0 t + h_0$, where h_0 is the height of the object when $t = 0$, and v_0 is the initial velocity at time $t = 0$.

65. $v_0 = 0,\ h_0 = 625,\ h = 0$

$0 = -16t^2 + (0)t + 625 \Rightarrow 16t^2 - 625 = 0 \Rightarrow$

$(4t-25)(4t+25) = 0 \Rightarrow t = \frac{25}{4} = 6.25$ or

$t = -\frac{25}{4} = -6.25$

The negative solution is not applicable. It takes 6.25 seconds for the baseball to reach the ground.

67. a. $v_0 = 0,\ h_0 = 200,\ h = 0$

$0 = -16t^2 - (0)t + 200 \Rightarrow 16t^2 = 200 \Rightarrow$

$t^2 = \frac{200}{16} \Rightarrow t = \pm\frac{\sqrt{200}}{4} \approx \pm 3.54$

The negative solution is not applicable. It will take about 3.54 seconds for the rock to reach the ground if it is dropped.

b. $v_0 = -40,\ h_0 = 200,\ h = 0$

$0 = -16t^2 - 40t + 200$

Using the quadratic formula, we have

$t = \frac{-(-40) \pm \sqrt{(-40)^2 - 4(-16)(200)}}{2(-16)}$

$= -5$ or 2.5

The negative solution is not applicable. It will take about 2.5 seconds for the rock to reach the ground if it is thrown with an initial velocity of 40 ft/sec.

c. $v_0 = -40,\ h_0 = 200,\ t = 2$

$h = -16(2)^2 - 40(2) + 200 = 56$

After 2 seconds, the rock is 56 feet above the ground. This means it has fallen $200 - 56 = 144$ feet.

69. a. $v_0 = 64,\ h_0 = 0,\ h = 64$

$64 = -16t^2 + 64t + 0 \Rightarrow$

$16t^2 - 64t + 64 = 0 \Rightarrow$

$t^2 - 4t + 4 = 0 \Rightarrow (t-2)^2 = 0 \Rightarrow t = 2$

The ball will reach 64 feet after 2 seconds.

b. $v_0 = 64,\ h_0 = 0,\ h = 39$

$39 = -16t^2 + 64t + 0 \Rightarrow$

$16t^2 - 64t + 39 = 0 \Rightarrow (4t-13)(4t-3) \Rightarrow$

$t = \frac{13}{4} = 3.25$ or $t = \frac{3}{4} = .75$

The ball will reach 39 feet after .75 seconds and after 3.25 seconds.

c. Two answers are possible because the ball reaches the given height twice, once on the way up and once on the way down.

In exercises 71–76, we discard negative roots since all variables represent positive real numbers.

71. $S = \frac{1}{2}gt^2$ for t

$2S = gt^2$

$\frac{2S}{g} = t^2$

$\sqrt{\frac{2S}{g}} \cdot \frac{\sqrt{g}}{\sqrt{g}} = t$

$\frac{\sqrt{2Sg}}{g} = t$

73. $L = \frac{d^4k}{h^2}$ for h

$Lh^2 = d^4k$

$h^2 = \frac{d^4k}{L}$

$h = \sqrt{\frac{d^4k}{L}} \cdot \frac{\sqrt{L}}{\sqrt{L}} = \frac{\sqrt{d^4kL}}{L}$

$h = \frac{d^2\sqrt{kL}}{L}$

Copyright © 2019 Pearson Education, Inc.

75.

$$P = \frac{E^2R}{(r+R)^2} \text{ for } R$$

$$P(r+R)^2 = E^2R$$

$$P\left(r^2+2rR+R^2\right) = E^2R$$

$$Pr^2+2PrR+PR^2 = E^2R$$

$$PR^2+\left(2Pr-E^2\right)R+Pr^2 = 0$$

Solve for R by using the quadratic formula with $a = P$, $b = 2\text{Pr} - E^2$, and $c = \text{Pr}^2$.

$$R = \frac{-\left(2\text{Pr}-E^2\right)\pm\sqrt{\left(2\text{Pr}-E^2\right)^2-4P\cdot\text{Pr}^2}}{2P}$$

$$= \frac{-2\text{Pr}+E^2\pm\sqrt{4P^2r^2-4\text{Pr}\,E^2+E^4-4P^2r^2}}{2P}$$

$$= \frac{-2\text{Pr}+E^2\pm\sqrt{E^4-4\text{Pr}\,E^2}}{2P}$$

$$= \frac{-2\text{Pr}+E^2\pm\sqrt{E^2\left(E^2-4\text{Pr}\right)}}{2P}$$

$$R = \frac{-2\text{Pr}+E^2\pm E\sqrt{E^2-4\text{Pr}}}{2P}$$

Chapter 1 Review Exercises

1. 0 and 6 are whole numbers.

2. $-12, -6, -\sqrt{4}$, 0, and 6 are integers.

3. $-12, -6, -\frac{9}{10}, -\sqrt{4}, 0, \frac{1}{8}$, and 6 are rational numbers.

4. $-\sqrt{7}, \frac{\pi}{4}, \sqrt{11}$ are irrational numbers.

5. $9[(-3)4] = 9[4(-3)]$
Commutative property of multiplication

6. $7(4+5) = (4+5)7$
Commutative property of multiplication

7. $6(x+y-3) = 6x+6y+6(-3)$
Distributive property

8. $11+(5+3) = (11+5)+3$
Associative property of addition

9. x is at least 9.
$x \geq 9$

10. x is negative.
$x < 0$

11. $|6-4| = 2$, $-|-2| = -2$, $|8+1| = |9| = 9$,
$-|3-(-2)| = -|3+2| = -|5| = -5$
Since $-5, -2, 2, 9$ are in order, then
$-|3-(-2)|, -|-2|, |6-4|, |8+1|$ are in order.

12. $-\left|\sqrt{16}\right| = -4, -\sqrt{8}, \sqrt{7}, \left|-\sqrt{12}\right| = \sqrt{12}$

13. $7-|-8| = 7-8 = -1$

14. $|-3|-|-9+6| = 3-|-3| = 3-3 = 0$

15. $x \geq -3$
Start at -3 and draw a ray to the right. Use a bracket at -3 to show that -3 is a part of the graph.

−3

16. $-4 < x \leq 6$
Put a parenthesis at -4 and a bracket at 6. Draw a line segment between these two endpoints.

−4 6

17. $$\frac{-9+(-6)(-3)\div 9}{6-(-3)} = \frac{-9+18\div 9}{6+3} = \frac{-9+2}{9} = -\frac{7}{9}$$

18. $$\frac{20\div 4\cdot 2\div 5-1}{-9-(-3)-12\div 3} = \frac{5\cdot 2\div 5-1}{-9-(-3)-4} = \frac{10\div 5-1}{-9+3-4} = \frac{2-1}{-6-4} = -\frac{1}{10}$$

19. $$\left(3x^4-x^2+5x\right)-\left(-x^4+3x^2-6x\right)$$
$$= 3x^4-x^2+5x+x^4-3x^2+6x$$
$$= \left(3x^4+x^4\right)+\left(-x^2-3x^2\right)+(5x+6x)$$
$$= 4x^4-4x^2+11x$$

20. $$\left(-8y^3+8y^2-5y\right)-\left(2y^3+4y^2-10\right)$$
$$= -8y^3+8y^2-5y-2y^3-4y^2+10$$
$$= -8y^3-2y^3+8y^2-4y^2-5y+10$$
$$= -10y^3+4y^2-5y+10$$

Copyright © 2019 Pearson Education, Inc.

21. $(5k-2h)(5k+2h)=(5k)^2-(2h)^2=25k^2-4h^2$

22. $(2r-5y)(2r+5y)=(2r)^2-(5y)^2$
$=4r^2-25y^2$

23. $(3x+4y)^2=(3x)^2+2(3x)(4y)+(4y)^2$
$=9x^2+24xy+16y^2$

24. $(2a-5b)^2=(2a)^2-2(2a)(5b)+(5b)^2$
$=4a^2-20ab+25b^2$

25. $2kh^2-4kh+5k=k\left(2h^2-4h+5\right)$

26. $2m^2n^2+6mn^2+16n^2=2n^2\left(m^2+3m+8\right)$

27. $5a^4+12a^3+4a^2=a^2\left(5a^2+12a+4\right)$
$=a^2(5a+2)(a+2)$

28. $24x^3+4x^2-4x=4x\left(6x^2+x-1\right)$
$=4x(3x-1)(2x+1)$

29. $144p^2-169q^2=(12p)^2-(13q)^2$
$=(12p-13q)(12p+13q)$

30. $81z^2-25x^2=(9z)^2-(5x)^2$
$=(9z+5x)(9z-5x)$

31. $27y^3-1=(3y)^3-1^3$
$=(3y-1)\left[(3y)^2+(3y)(1)+1^2\right]$
$=(3y-1)\left(9y^2+3y+1\right)$

32. $125a^3+216=(5a)^3+(6)^3$
$=(5a+6)\left[(5a)^2-5a(6)+6^2\right]$
$=(5a+6)\left(25a^2-30a+36\right)$

33. $\frac{3x}{5}\cdot\frac{45x}{12}=\frac{3x\cdot 45x}{5\cdot 12}=\frac{3\cdot 5\cdot 3\cdot 3x^2}{4\cdot 5\cdot 3}=\frac{9x^2}{4}$

34. $\frac{5k^2}{24}-\frac{70k}{36}=\frac{5k^2\cdot 3}{24\cdot 3}-\frac{70k\cdot 2}{36\cdot 2}=\frac{15k^2}{72}-\frac{140k}{72}$
$=\frac{15k^2-140k}{72}=\frac{5k(3k-28)}{72}$

35. $\frac{c^2-3c+2}{2c(c-1)}\div\frac{c-2}{8c}=\frac{(c-1)(c-2)}{2c(c-1)}\cdot\frac{8c}{(c-2)}$
$=\frac{8c(c-1)(c-2)}{2c(c-1)(c-2)}=\frac{8}{2}=4$

36. $\frac{p^3-2p^2-8p}{3p\left(p^2-16\right)}\div\frac{p^2+4p+4}{9p^2}$
$=\frac{p\left(p^2-2p-8\right)}{3p(p+4)(p-4)}\cdot\frac{9p^2}{(p+2)(p+2)}$
$=\frac{p(p-4)(p+2)\cdot 9p^2}{3p(p+4)(p-4)(p+2)(p+2)}$
$=\frac{3p(p-4)(p+2)\cdot 3p^2}{3p(p-4)(p+2)\cdot(p+4)(p+2)}$
$=\frac{3p^2}{(p+4)(p+2)}$

37. $\frac{2m^2-4m+2}{m^2-1}\div\frac{6m+18}{m^2+2m-3}$
$=\frac{2\left(m^2-2m+1\right)}{(m+1)(m-1)}\cdot\frac{m^2+2m-3}{6m+18}$
$=\frac{2(m-1)^2}{(m+1)(m-1)}\cdot\frac{(m+3)(m-1)}{6(m+3)}$
$=\frac{2(m-1)(m+3)\cdot(m-1)^2}{2(m-1)(m+3)\cdot 3(m+1)}=\frac{(m-1)^2}{3(m+1)}$

38. $\frac{x^2+6x+5}{4\left(x^2+1\right)}\cdot\frac{2x(x+1)}{x^2-25}$
$=\frac{(x+5)(x+1)\cdot 2x(x+1)}{4\left(x^2+1\right)\cdot(x+5)(x-5)}$
$=\frac{2(x+5)\cdot x(x+1)^2}{2(x+5)\cdot 2\left(x^2+1\right)(x-5)}$
$=\frac{x(x+1)^2}{2\left(x^2+1\right)(x-5)}$

39. $5^{-3}=\frac{1}{5^3}$ or $\frac{1}{125}$

40. $10^{-2}=\frac{1}{10^2}$ or $\frac{1}{100}$

41. $-8^0=-\left(8^0\right)=-1$

Copyright © 2019 Pearson Education, Inc.

42. $\left(-\frac{5}{6}\right)^{-2} = \left(-\frac{6}{5}\right)^{2} = \frac{36}{25}$

43. $4^6 \cdot 4^{-3} = 4^{6+(-3)} = 4^3$

44. $7^{-5} \cdot 7^{-2} = 7^{-5+(-2)} = 7^{-7} = \frac{1}{7^7}$

45. $\frac{8^{-5}}{8^{-4}} = 8^{-5-(-4)} = 8^{-5+4} = 8^{-1} = \frac{1}{8}$

46. $\frac{6^{-3}}{6^4} = 6^{-3-4} = 6^{-7} = \frac{1}{6^7}$

47. $5^{-1} + 2^{-1} = \frac{1}{5} + \frac{1}{2} = \frac{7}{10}$

48. $5^{-2} + 5^{-1} = \frac{1}{5^2} + \frac{1}{5} = \frac{1}{25} + \frac{1}{5} = \frac{6}{25}$

49. $\frac{5^{1/3}5^{1/2}}{5^{3/2}} = 5^{1/3+1/2-3/2} = 5^{-2/3} = \frac{1}{5^{2/3}}$

50. $\frac{2^{3/4} \cdot 2^{-1/2}}{2^{1/4}} = \frac{2^{1/4}}{2^{1/4}} = 1$

51. $\left(3a^2\right)^{1/2} \cdot \left(3^2 a\right)^{3/2} = 3^{1/2} a \cdot 3^3 a^{3/2} = 3^{7/2} a^{5/2}$

52. $(4p)^{2/3} \cdot \left(2p^3\right)^{3/2} = 4^{2/3} p^{2/3} \cdot 2^{3/2} \cdot p^{9/2}$

$= \left(2^2\right)^{2/3} p^{2/3} \cdot 2^{3/2} p^{9/2}$

$= 2^{4/3} \cdot 2^{3/2} p^{2/3} p^{9/2}$

$= 2^{17/6} p^{31/6}$

53. $\sqrt[3]{27} = 3$

54. $\sqrt[6]{-64}$ is not a real number.

55. $\sqrt[3]{54p^3q^5} = \sqrt[3]{27 \cdot 2p^3q^3q^2}$

$= \sqrt[3]{27p^3q^3} \cdot \sqrt[3]{2q^2} = 3pq\sqrt[3]{2q^2}$

56. $\sqrt[4]{64a^5b^3} = \sqrt[4]{16a^4} \cdot \sqrt[4]{4ab^3} = 2a\sqrt[4]{4ab^3}$

57. $3\sqrt{3} - 12\sqrt{12} = 3\sqrt{3} - 12\sqrt{4 \cdot 3} = 3\sqrt{3} - 12 \cdot 2\sqrt{3}$

$= 3\sqrt{3} - 24\sqrt{3} = -21\sqrt{3}$

58. $8\sqrt{7} + 2\sqrt{63} = 8\sqrt{7} + 2\sqrt{9 \cdot 7}$

$= 8\sqrt{7} + 6\sqrt{7} = 14\sqrt{7}$

59. $\frac{\sqrt{3}}{1+\sqrt{2}} = \frac{\sqrt{3}\left(1-\sqrt{2}\right)}{\left(1+\sqrt{2}\right)\left(1-\sqrt{2}\right)} = \frac{\sqrt{3}-\sqrt{6}}{1-2}$

$= \frac{\sqrt{3}-\sqrt{6}}{-1} = \sqrt{6} - \sqrt{3}$

60. $\frac{4+\sqrt{2}}{4-\sqrt{5}} = \frac{\left(4+\sqrt{2}\right)}{\left(4-\sqrt{5}\right)} \cdot \frac{\left(4+\sqrt{5}\right)}{\left(4+\sqrt{5}\right)}$

$= \frac{16+4\sqrt{2}+4\sqrt{5}+\sqrt{10}}{16-\left(\sqrt{5}\right)^2}$

$= \frac{16+4\sqrt{2}+4\sqrt{5}+\sqrt{10}}{11}$

61. $3x - 4(x-2) = 2x + 9$

$3x - 4x + 8 = 2x + 9$

$-x + 8 = 2x + 9$

$-1 = 3x \Rightarrow x = -\frac{1}{3}$

62. $4y + 9 = -3(1-2y) + 5$

$4y + 9 = -3 + 6y + 5$

$4y + 9 = 2 + 6y$

$-2y = -7 \Rightarrow y = \frac{7}{2}$

63. $\frac{2m}{m-3} = \frac{6}{m-3} + 4$

$2m = 6 + 4(m-3)$

$2m = 6 + 4m - 12$

$2m = 4m - 6$

$6 = 2m \Rightarrow 3 = m$

Because $m = 3$ would make the denominators of the fractions equal to 0, making the fractions undefined, the given equation has no solution.

Copyright © 2019 Pearson Education, Inc.

64. $\frac{15}{k+5} = 4 - \frac{3k}{k+5}$

Multiply both sides of the equation by the common denominator $k + 5$.

$15 = 4(k+5) - 3k$

$15 = 4k + 20 - 3k \Rightarrow k = -5$

If $k = -5$, the fractions would be undefined, so the given equation has no solution.

65. $8ax - 3 = 2x$

$8ax - 2x = 3$

$x(8a-2) = 3$

$x = \frac{3}{8a-2}$

66. $b^2x - 2x = 4b^2$

$(b^2 - 2)x = 4b^2$

$x = \frac{4b^2}{b^2 - 2}$

67. $\left|\frac{2-y}{5}\right| = 8$

$\frac{2-y}{5} = 8$ or $\frac{2-y}{5} = -8$

$5\left(\frac{2-y}{5}\right) = 5(8)$ or $5\left(\frac{2-y}{5}\right) = 5(-8)$

$2 - y = 40$ or $2 - y = -40$

$-y = 38$ or $-y = -42$

$y = -38$ or $y = 42$

The solutions are –38 and 42.

68. $|4k+1| = |6k-3|$

$4k + 1 = 6k - 3$ or $4k + 1 = -(6k - 3)$

$4k + 1 = 6k - 3$ or $4k + 1 = -6k + 3$

$-2k = -4$ or $10k = 2$

$k = 2$ or $k = \frac{1}{5}$

The solutions are 2 and $\frac{1}{5}$.

69. $(b+7)^2 = 5$

Use the square root property to solve this quadratic equation.

$b + 7 = \sqrt{5}$ or $b + 7 = -\sqrt{5}$

$b = -7 + \sqrt{5}$ or $b = -7 - \sqrt{5}$

The solutions are $-7 + \sqrt{5}$ and $-7 - \sqrt{5}$, which we abbreviate as $-7 \pm \sqrt{5}$.

70. $(2p+1)^2 = 7$

Solve by the square root property.

$2p + 1 = \sqrt{7}$ or $2p + 1 = -\sqrt{7}$

$2p = -1 + \sqrt{7}$ or $2p = -1 - \sqrt{7}$

$p = \frac{-1+\sqrt{7}}{2}$ or $p = \frac{-1-\sqrt{7}}{2}$

The solutions are $\frac{-1 \pm \sqrt{7}}{2}$.

71. $2p^2 + 3p = 2$

Write the equation in standard form and solve by factoring.

$2p^2 + 3p - 2 = 0$

$(2p-1)(p+2) = 0$

$2p - 1 = 0$ or $p + 2 = 0$

$p = \frac{1}{2}$ or $p = -2$

The solutions are $\frac{1}{2}$ and –2.

72. $2y^2 = 15 + y$

Write the equation in standard form and solve by factoring.

$2y^2 - y - 15 = 0$

$(y-3)(2y+5) = 0$

$y = 3$ or $y = -\frac{5}{2}$

The solutions are 3 and $-\frac{5}{2}$.

73. $2q^2 - 11q = 21 \Rightarrow 2q^2 - 11q - 21 = 0 \Rightarrow$

$(2q+3)(q-7) = 0$

$2q + 3 = 0$ or $q - 7 = 0$

$q = -\frac{3}{2}$ or $q = 7$

The solutions are $-\frac{3}{2}$ and 7.

Copyright © 2019 Pearson Education, Inc.

74. $3x^2+2x=16 \Rightarrow 3x^2+2x-16=0 \Rightarrow$
$(3x+8)(x-2)=0$
$3x+8=0$ or $x-2=0$
$x=-\frac{8}{3}$ or $x=2$
The solutions are $-\frac{8}{3}$ and 2.

75. $6k^4+k^2=1 \Rightarrow 6k^4+k^2-1=0$
Let $p=k^2$, so $p^2=k^4$.
$6p^2+p-1=0$
$(3p-1)(2p+1)=0$
$3p-1=0$ or $2p+1=0$
$p=\frac{1}{3}$ or $p=-\frac{1}{2}$
If $p=\frac{1}{3}$, $k^2=\frac{1}{3} \Rightarrow k=\pm\sqrt{\frac{1}{3}}=\pm\frac{\sqrt{3}}{3}$
If $p=-\frac{1}{2}$, $k^2=-\frac{1}{2}$ has no real number solution.
The solutions are $\pm\frac{\sqrt{3}}{3}$.

76. $21p^4=2+p^2 \Rightarrow 21p^4-p^2-2=0$
Let $u=p^2$; then $u^2=p^4$.
$21u^2-u-2=0$
$(3u-1)(7u+2)=0$
$3u-1=0$ or $7u+2=0$
$x=\frac{1}{3}$ or $x=-\frac{2}{7}$
$p^2=\frac{1}{3}$ or $p^2=-\frac{2}{7}$
If $x=-\frac{2}{7}$, $p^2=-\frac{2}{7}$ has no real number solution.
$p=\pm\frac{1}{\sqrt{3}}=\pm\frac{\sqrt{3}}{3}$
The solutions are $\pm\frac{\sqrt{3}}{3}$.

77. $p=\frac{E^2R}{(r+R)^2}$ for r.
$p(r+R)^2=E^2R$
$p\left(r^2+2rR+R^2\right)=E^2R$
$pr^2+2rpR+R^2p=E^2R$
$pr^2+2rpR+R^2p-E^2R=0$
Use the quadratic formula to solve for r.
$$r=\frac{-2pR\pm\sqrt{4p^2R^2-4p\left(R^2p-E^2R\right)}}{2p}$$
$$r=\frac{-2pR\pm\sqrt{4pE^2R}}{2p}=\frac{-pR\pm E\sqrt{pR}}{p}$$

78. $p=\frac{E^2R}{(r+R)^2}$ for E.
$p(r+R)^2=E^2R \Rightarrow E^2=\frac{p(r+R)^2}{R} \Rightarrow$
$$E=\pm\sqrt{\frac{p(r+R)^2}{R}}=\frac{\pm(r+R)\sqrt{pR}}{R}$$

79. $K=s(s-a)$ for s.
$K=s^2-as \Rightarrow s^2-as-K=0$
Use the quadratic formula.
$$s=\frac{a\pm\sqrt{a^2-4(-K)}}{2}=\frac{a\pm\sqrt{a^2+4K}}{2}$$

80. $kz^2-hz-t=0$ for z.
Use the quadratic formula with $a=k$, $b=-h$, and $c=-t$.
$$z=\frac{-(-h)\pm\sqrt{(-h)^2-4(k)(-t)}}{2k}$$
$$=\frac{h\pm\sqrt{h^2+4kt}}{2k}$$

81. $|27-(-55)|=|82|=82\%$

82. $|-22.5-(-2.5)|=|-20|=20\%$

83. Let x = the original price
$x-.2x=1229$
$.8x=1229$
$x=1536.25$
The original price of the laptop was \$1536.25.

Copyright © 2019 Pearson Education, Inc.

84. Let x = the original price

$x+.273x=29.83$

$1.273x=29.83$

$x=23.43$

The original price of the stock was $23.43.

85. Let $O=3.63$

$3.63=.11x+2.2$

$1.43=.11x$

$13=x$

The amount of outlays reached $3.63 trillion in the year 2013.

86. Let $O=3.96$

$3.96=.11x+2.2$

$1.76=.11x$

$16=x$

The amount of outlays reached $3.96 trillion in the year 2016.

87. Let $O=10.8$

$10.8=.2x+8$

$2.8=.2x$

$14=x$

The crude oil output reached 10.8 million barrels per day in the year 2014.

88. Let $O=11.4$

$11.4=.2x+8$

$3.4=.2x$

$17=x$

The crude oil output reached 11.4 million barrels per day in the year 2017.

89. Let $x=7$

$V=.04x^2-.47x+18.38$

$V=.04(7)^2-.47(7)+18.38$

$V=17.05$

The number of vehicles sold in July 2016 was 17.05 million.

90. Let $x=10$

$V=.04x^2-.47x+18.38$

$V=.04(10)^2-.47(10)+18.38$

$V=17.68$

The number of vehicles sold in October 2016 was 17.68 million.

91. Let $V=17.5$.

$V=.04x^2-.47x+18.38$

$17.5=.04x^2-.47x+18.38$

$0=.04x^2-.47x+.88$

$$x=\frac{.47\pm\sqrt{(-.47)^2-4(.04)(.88)}}{2(.04)}$$

≈ 2.33 or 9.41

The first full month in which sales were below 17.5 million vehicles was March 2016.

92. Using the solution to 91, the second month of the year in which sales was 17.5 million was October 2016.

93. Let $x=13$

$$F=\frac{.061(13)^2+6.09(13)+38.8}{13+1}\approx 9.16$$

The total number of flights that departed in 2013 was approximately 9.16 million.

94. Let $x=20$

$$F=\frac{.061(20)^2+6.09(20)+38.8}{20+1}\approx 8.81$$

The total number of flights that will depart in 2020 will be approximately 8.81 million.

95. Let $F=9.2$.

$$F=\frac{.061x^2+6.09x+38.8}{x+1}$$

$$9.2=\frac{.061x^2+6.09x+38.8}{x+1}$$

$9.2(x+1)=.061x^2+6.09x+38.8$

$0=.061x^2-3.11x+29.6$

$$x=\frac{3.11\pm\sqrt{(-3.11)^2-4(.061)(29.6)}}{2(.061)}$$

≈ 12.67 or 38.32

The number of flight departures fell below 9.2 million in 2013.

Copyright © 2019 Pearson Education, Inc.

96. Let $F = 8.9$.

$$F = \frac{.061x^2 + 6.09x + 38.8}{x+1}$$
$$8.9 = \frac{.061x^2 + 6.09x + 38.8}{x+1}$$
$$8.9(x+1) = .061x^2 + 6.09x + 38.8$$
$$0 = .061x^2 - 2.81x + 29.9$$
$$x = \frac{2.81 \pm \sqrt{(-2.81)^2 - 4(.061)(29.9)}}{2(.061)}$$
$$\approx 16.70 \text{ or } 29.38$$

The number of flight departures fell below 8.9 million in 2017.

97. Let $x = 12$

$R = 16.2(12)^{.26} \approx 30.91$

The total amount spent on basic research by the U.S. government in 2012 was approximately \$30.91 billion.

98. Let $x = 17$

$R = 16.2(17)^{.26} \approx 33.84$

The total amount spent on basic research by the U.S. government in 2017 was approximately \$33.84 billion.

99. Let x = the single interest rate

$$2000(.12) + 500(.07) = 2500x$$
$$275 = 2500x$$
$$.11 = x$$

Therefore, a single interest rate of 11% will yield the same results.

100. Let x = the amount of beef. Then $30 - x$ = the amount of pork.

$$2.8x + 3.25(30 - x) = 3.10(30)$$
$$2.8x + 97.5 - 3.25x = 93$$
$$-.45x = -4.5$$
$$x = 10$$

Therefore the butcher should use 10 pounds of beef and 30 – 10 = 20 pounds of pork.

101. Let $x = 112$

$$D = .047x^2 - 7.69x + 341.6$$
$$D = .047(112)^2 - 7.69(112) + 341.6$$
$$D = 69.9$$

The ratio of federal debt to GDP in 2012 was approximately 69.9%.

102. Let $D = 55\%$.

$$D = .047x^2 - 7.69x + 341.6$$
$$55 = .047x^2 - 7.69x + 341.6$$
$$0 = .047x^2 - 7.69x + 286.6$$
$$x = \frac{7.69 \pm \sqrt{(-7.69)^2 - 4(.047)(286.6)}}{2(.047)}$$
$$\approx 57.42 \text{ or } 106.20$$

The most recent time the ratio of federal debt to GDP was 55% was in 2006.

103. Let x = the width of the walk.

$$\text{The area } = (10 + 2x)(15 + 2x) - 10(15)$$
$$= 150 + 20x + 30x + 4x^2 - 150$$
$$= 4x^2 + 50x$$

To use all of the cement, solve

$$200 = 4x^2 + 50x$$
$$0 = 4x^2 + 50x - 200$$

Using the quadratic formula, we have

$$x = \frac{-50 \pm \sqrt{(50)^2 - 4(4)(-200)}}{2(4)}$$
$$= \frac{-50 \pm \sqrt{5700}}{2(4)} \approx -15.6875 \text{ or } 3.1875$$

The width cannot be negative, so the solution is approximately 3.2 feet.

104. Let x = the length of the yard. Then $160 - x$ = the width of the yard. Area is equal to length times width.

$$\text{The area } = lw$$
$$4000 = (x)(160 - x)$$
$$0 = x^2 - 160x + 4000$$

Using the quadratic formula, we have

$$x = \frac{160 \pm \sqrt{(-160)^2 - 4(1)(4000)}}{2(1)}$$
$$= \frac{160 \pm \sqrt{9600}}{2(1)} \approx 31.01 \text{ or } 128.99$$

The length is longer than the width, so the length is approximately 129 feet and the width is approximately 160 – 129 = 31 feet.

105. $v_0 = 150,\ h_0 = 0,\ h = 200$

$$200 = -16t^2 + 150t + 0 \Rightarrow$$
$$16t^2 - 150t + 200 = 0 \Rightarrow$$
$$8t^2 - 75t + 100 = 0 \Rightarrow t \approx 1.61 \text{ or } 7.77$$

The ball will reach 200 feet on its downward trip after approximately 7.77 seconds.

Copyright © 2019 Pearson Education, Inc.

106. $v_0 = 55,\ h_0 = 700,\ h = 0$
$0 = -16t^2 - 55t + 700 \Rightarrow$
$16t^2 + 55t - 700 = 0 \Rightarrow t \approx -8.55$ or 5.12
The ball will reach the ground after approximately 5.12 seconds

Case 1 Consumers Often Defy Common Sense

1. The total cost to buy and run this electric hot water tank for x years is $E = 218 + 508x$.

2. The total cost to buy and run this gas hot water tank for x years is $G = 328 + 309x$.

3. Over 10 years, the electric hot water tank costs $218 + 508(10) = 5298$ or $5298, and the gas hot water tank costs $328 + 309(10) = 3418$ or $3418. The gas hot water tank costs $1880 less over 10 years.

4. The total costs for the two hot water tanks will be equal when $218 + 508x = 328 + 309x \Rightarrow$
$199x = 110 \Rightarrow x \approx 0.55.$
The costs will be equal within the first year.

5. The total cost to buy and run this Maytag refrigerator for x years is $M = 1529.10 + 50x$.

6. The total cost to buy and run this LG refrigerator for x years is $L = 1618.20 + 44x$.

7. Over 10 years, the Maytag refrigerator costs $1529.10 + 50(10) = 2029.10$ or $2029.10, and the LG refrigerator costs $1618.20 + 44(10) = 2058.20$ or $2058.20. The LG refrigerator costs $29.10 more over 10 years

8. The total costs for the two refrigerators will be equal when $1529.10 + 50x = 1618.20 + 44x \Rightarrow$
$6x = 89.10 \Rightarrow x \approx 14.85.$
The costs will be equal in the 14th year

Copyright © 2019 Pearson Education, Inc.

Chapter 2 Graphs, Lines, and Inequalities

Section 2.1 Graphs, Lines, and Inequalities

1. (1, –2) lies in quadrant IV
(–2, 1)lies in quadrant II
(3, 4) lies in quadrant I
(–5, –6) lies in quadrant III

3. (1, –3) is a solution to $3x - y - 6 = 0$ because $3(1) - (-3) - 6 = 0$ is a true statement.

5. (3, 4) is not a solution to $(x-2)^2 + (y+2)^2 = 6$ because $(3-2)^2 + (4+2)^2 = 37$, not 6.

7. $4y + 3x = 12$
Find the y-intercept. If $x = 0$,
$4y = -3(0) + 12 \Rightarrow 4y = 12 \Rightarrow y = 3$
The y-intercept is 3.
Next find the x-intercept. If $y = 0$,
$4(0) + 3x = 12 \Rightarrow 3x = 12 \Rightarrow x = 4$
The x-intercept is 4.
Using these intercepts, graph the line.

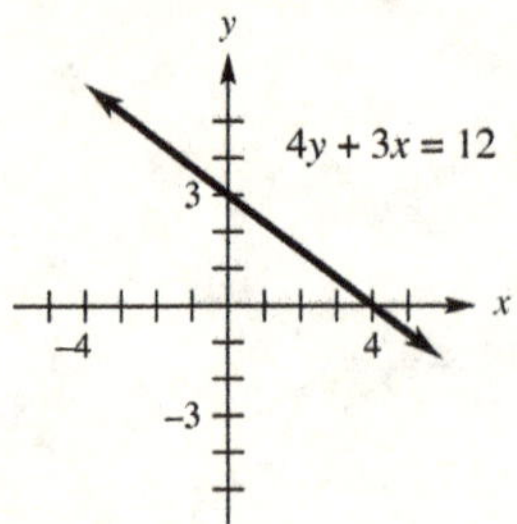

9. $8x + 3y = 12$
Find the y-intercept. If $x = 0$,
$3y = 12 \Rightarrow y = 4$
The y-intercept is 4.
Next, find the x-intercept. If $y = 0$,
$8x = 12 \Rightarrow x = \frac{12}{8} = \frac{3}{2}$
The x-intercept is $\frac{3}{2}$.
Using these intercepts, graph the line.

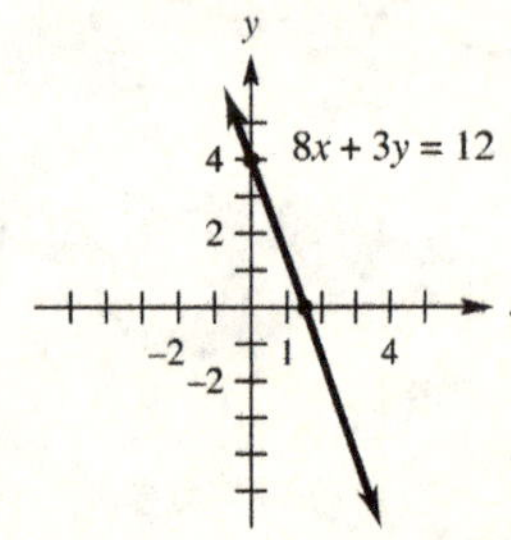

11. $x = 2y + 3$
Find the y-intercept. If $x = 0$,
$0 = 2y + 3 \Rightarrow 2y = -3 \Rightarrow y = -\frac{3}{2}$
The y-intercept is $-\frac{3}{2}$.
Next, find the x-intercept. If $y = 0$,
$x = 2(0) + 3 \Rightarrow x = 3$
The x-intercept is 3.
Using these intercepts, graph the line.

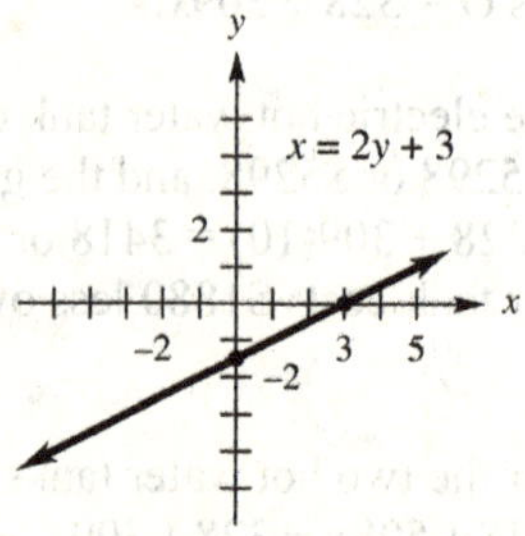

13. The x-intercepts are where the rays cross the x-axis, –2.5 and 3. The y-intercept is where the ray crosses the y-axis, 3.

15. The x-intercepts are –1 and 2. The y-intercept is –2.

17. $3x + 4y = 12$
To find the x-intercept, let $y = 0$:
$3x + 4(0) = 12 \Rightarrow 3x = 12 \Rightarrow x = 4$
The x-intercept is 4.
To find the y-intercept, let $x = 0$:
$3(0) + 4y = 12 \Rightarrow 4y = 12 \Rightarrow y = 3$
The y-intercept is 3.

19. $2x - 3y = 24$
To find the x-intercept, let $y = 0$:
$2x - 3(0) = 24 \Rightarrow 2x = 24 \Rightarrow x = 12$
The x-intercept is 12.
To find the y-intercept, let $x = 0$:
$2(0) - 3y = 24 \Rightarrow -3y = 24 \Rightarrow y = -8$
The y-intercept is –8.

21. $y = x^2 - 9$
To find the x-intercepts, let $y = 0$:
$0 = x^2 - 9 \Rightarrow x^2 = 9 \Rightarrow x = \pm\sqrt{9} = \pm 3$

Copyright © 2019 Pearson Education, Inc.

23. $y = x^2 + x - 20$

To find the x-intercepts, let $y = 0$:

$0 = x^2 + x - 20 \Rightarrow 0 = (x+5)(x-4) \Rightarrow$

$x + 5 = 0 \Rightarrow x = -5$ or $x - 4 = 0 \Rightarrow x = 4$

The x-intercepts are –5 and 4.

To find the y-intercept, let $x = 0$:

$y = 0^2 + 0 - 20 = -20$

The y-intercept is –20.

25. $y = 2x^2 - 5x + 7$

To find the x-intercepts, let $y = 0$:

$0 = 2x^2 - 5x + 7$

This equation does not have real solutions, so there are no x-intercepts.

To find the y-intercept, let $x = 0$:

$y = 2(0)^2 - 5(0) + 7 = 7$

The y-intercept is 7.

27. $y = x^2$

x-intercept: $0 = x^2 \Rightarrow x = 0$

y-intercept: $y = 0$

x	y
–2	4
–1	1
0	0
1	1
2	4

29. $y = x^2 - 3$

x-intercepts: $0 = x^2 - 3 \Rightarrow x^2 = 3 \Rightarrow x = \pm\sqrt{3}$

y-intercepts: $y = 0^2 - 3 = -3$

x	y
–3	6
–1	–2
0	–3
1	–2
3	6

31. $y = x^2 - 6x + 5$

x-intercept:

$0 = x^2 - 6x + 5 \Rightarrow 0 = (x-1)(x-5) \Rightarrow$

$x - 1 = 0 \Rightarrow x = 1$ or $x - 5 = 0 \Rightarrow x = 5$

y-intercept: $y = (0)^2 - 6(0) + 5 = 5$

x	y
–2	21
–1	12
0	5
1	0
2	–3

33. $y = x^3 + 1$

x-intercept:

$0 = x^3 + 1 \Rightarrow x^3 = -1 \Rightarrow x = \sqrt[3]{-1} = -1$

y-intercept: $y = 0^3 + 1 = 1$

x	y
–2	–7
–1	0
0	1
1	2
2	9

35. $y = \sqrt{x+4}$

x-intercept: $0 = \sqrt{x+4} \Rightarrow 0 = x + 4 \Rightarrow x = -4$

y-intercept: $y = \sqrt{0+4} = \sqrt{4} = 2$

x	y
–2	$\sqrt{2} \approx 1.4$
–1	$\sqrt{3} \approx 1.7$
0	2
2	$\sqrt{6} \approx 2.4$
5	3

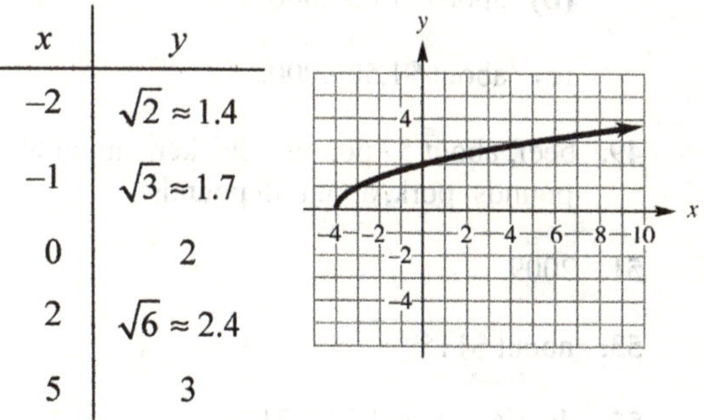

Copyright © 2019 Pearson Education, Inc.

37.

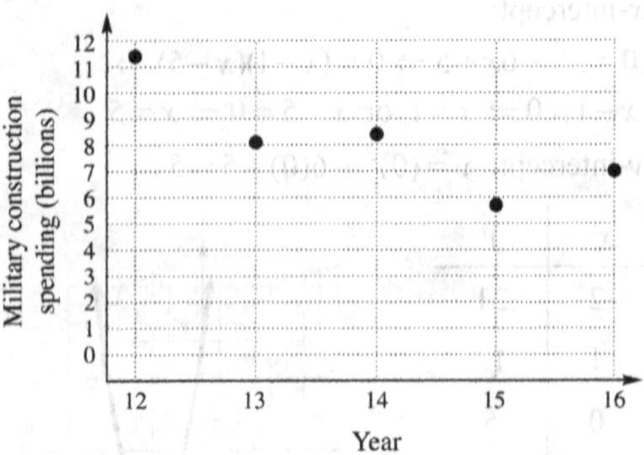

39.

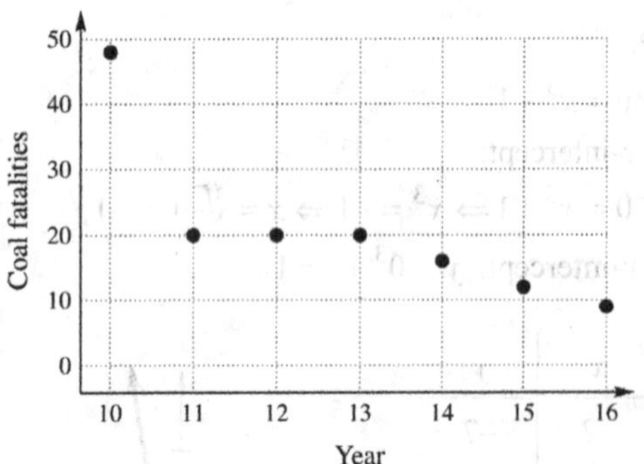

41. 1999

43. 2011

45. **(a)** about $1,250,000

(b) about $1,750,000

(c) about $4,250,000

47. **(a)** about $500,000

(b) about $1,000,000.

(c) about $1,500,000.

49. beef, about 57 pounds; chicken, about 81 pounds; pork, about 45 pounds

51. 2009

53. about $115

55. days 6, 12, and 14 – 21

57. Day 19; about $117.50

59. Day 2; about $117.00

61. about $8.50

63. $y = x^2 + x + 1$

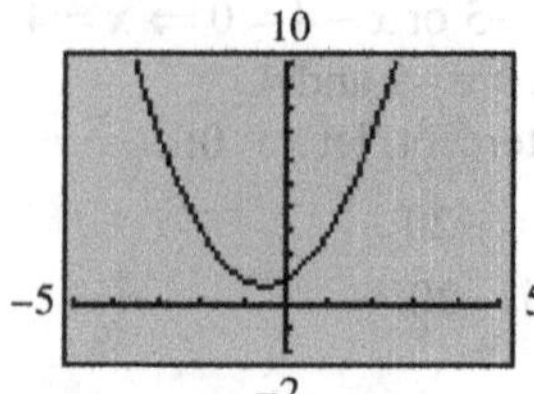

65. $y = (x-3)^3$

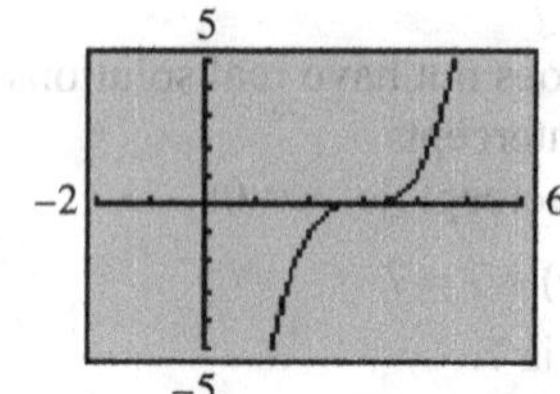

67. $y = x^3 - 3x^2 + x - 1$

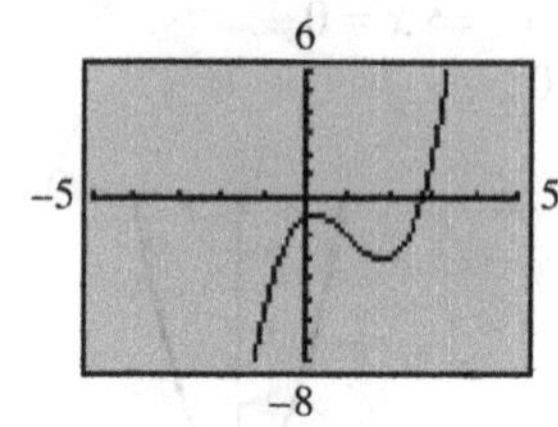

69. $y = x^4 - 2x^3 + 2x$

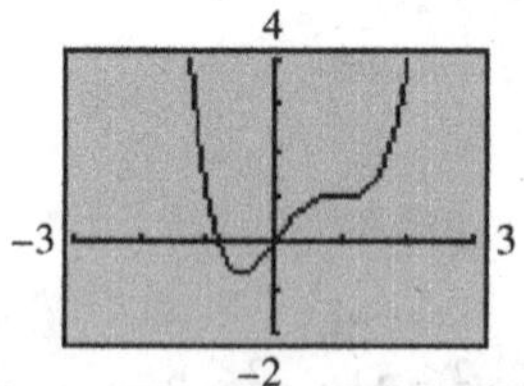

The "flat" part of the graph near $x = 1$ looks like a horizontal line segment, but it is not. The y values increase slightly as you trace along the segment from left to right.

71. $x \approx -1.1038$

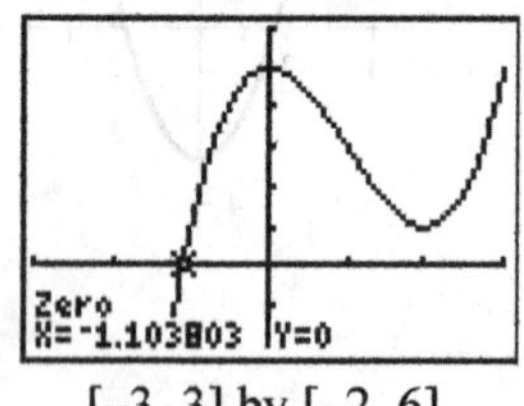

[–3, 3] by [–2, 6]

Copyright © 2019 Pearson Education, Inc.

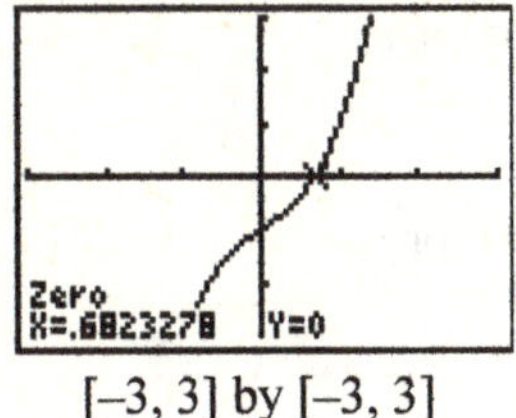

[–3, 3] by [–3, 3]

73. $x \approx 2.1017$

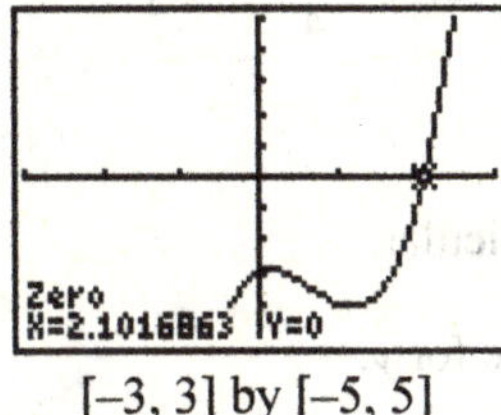

[–3, 3] by [–5, 5]

75. $x \approx -1.7521$

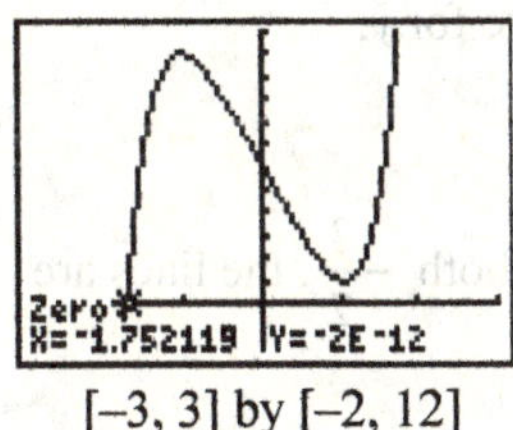

[–3, 3] by [–2, 12]

Section 2.2 Equations of Lines

1. Through (2, 5) and (0, 8)

$$\text{slope} = \frac{\Delta y}{\Delta x} = \frac{8-5}{0-2} = \frac{3}{-2} = -\frac{3}{2}$$

3. Through (–4, 14) and (3, 0)

$$\text{slope} = \frac{14-0}{-4-3} = \frac{14}{-7} = -2$$

5. Through the origin and (–4, 10); the origin has coordinate (0, 0).

$$\text{slope} = \frac{10-0}{-4-0} = \frac{10}{-4} = -\frac{5}{2}$$

7. Through (–1, 4) and (–1, 6)

$$\text{slope} = \frac{6-4}{-1-(-1)} = \frac{2}{0}, \text{ not defined}$$

The slope is undefined.

9. $b = 5, m = 4$

$y = mx + b$

$y = 4x + 5$

11. $b = 1.5, m = -2.3$

$y = mx + b$

$y = -2.3x + 1.5$

13. $b = 4,\ m = -\frac{3}{4}$

$y = mx + b$

$y = -\frac{3}{4}x + 4$

15. $2x - y = 9$

Rewrite in slope-intercept form.

$-y = -2x + 9$

$y = 2x - 9$

$m = 2, b = -9.$

17. $6x = 2y + 4$

Rewrite in slope-intercept form.

$2y = 6x - 4 \Rightarrow y = 3x - 2$

$m = 3, b = -2.$

19. $6x - 9y = 16$

Write in slope-intercept form.

$-9y = -6x + 16$

$9y = 6x - 16$

$y = \frac{2}{3}x - \frac{16}{9}$

$m = \frac{2}{3},\ b = -\frac{16}{9}.$

21. $2x - 3y = 0$

Rewrite in slope-intercept form.

$3y = 2x \Rightarrow y = \frac{2}{3}x$

$m = \frac{2}{3},\ b = 0.$

23. $x = y - 5$

Rewrite in slope-intercept form.

$y = x + 5$

$m = 1, b = 5$

25. **(a)** Largest value of slope is at C.

(b) Smallest value of slope is at B.

(c) Largest absolute value is at B

(d) Closest to 0 is at D

27. $2x - y = -2$

Find the x-intercept by setting $y = 0$ and solving for x: $2x - 0 = -2 \Rightarrow 2x = -2 \Rightarrow x = -1$

Find the y-intercept by setting $x = 0$ and solving for y: $2(0) - y = -2 \Rightarrow -y = -2 \Rightarrow y = 2$

Use the points (–1, 0) and (0, 2) to sketch the graph:

Copyright © 2019 Pearson Education, Inc.

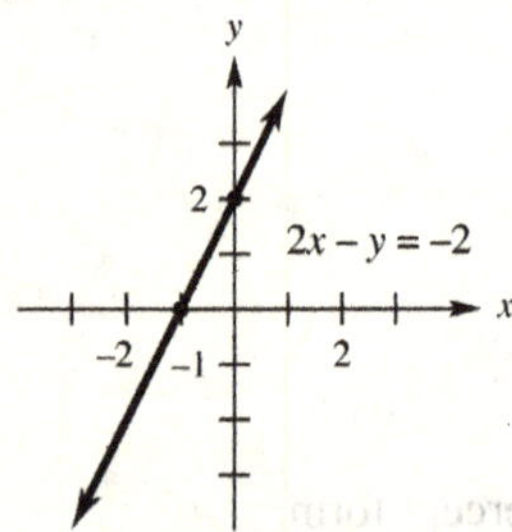

29. $2x + 3y = 4$
Find the x-intercept by setting $y = 0$ and solving for x: $2x + 3(0) = 4 \Rightarrow 2x = 4 \Rightarrow x = 2$
Find the y-intercept by setting $x = 0$ and solving for y: $2(0) + 3y = 4 \Rightarrow 3y = 4 \Rightarrow y = \frac{4}{3}$

Use the points (2, 0) and $\left(0, \frac{4}{3}\right)$ to sketch the graph:

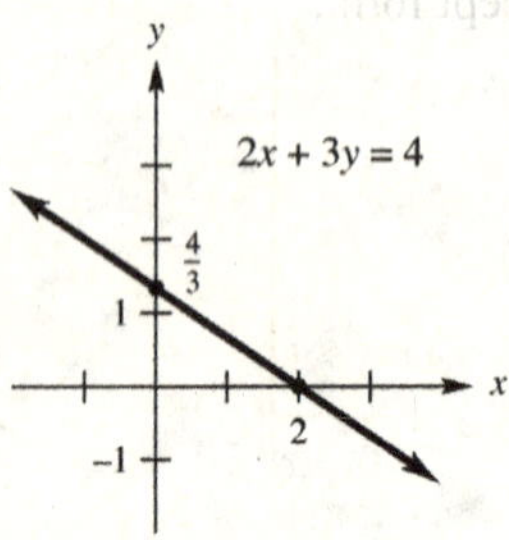

31. $4x - 5y = 2$
Find the x-intercept, by setting $y = 0$ and solving for x:

$4x - 5(0) = 2 \Rightarrow 4x = 2 \Rightarrow x = \frac{1}{2}$

Find the y-intercept by setting $x = 0$ and solving for y:

$4(0) - 5y = 2 \Rightarrow -5y = 2 \Rightarrow y = -\frac{2}{5}$

Use the points $\left(\frac{1}{2}, 0\right)$ and $\left(0, -\frac{2}{5}\right)$ to sketch the graph:

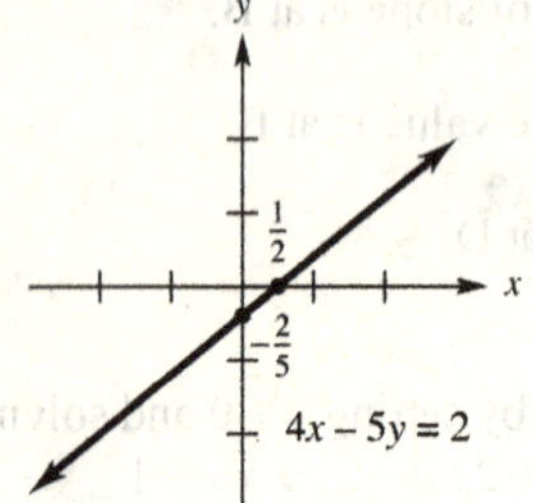

33. For $4x - 3y = 6$, solve for y.
$y = \frac{4}{3}x - 2$
For $3x + 4y = 8$, solve for y.
$y = -\frac{3}{4}x + 2$

The two slopes are $\frac{4}{3}$ and $-\frac{3}{4}$. Since
$\left(\frac{4}{3}\right)\left(-\frac{3}{4}\right) = -1,$
the lines are perpendicular.

35. For $3x + 2y = 8$, solve for y.
$y = -\frac{3}{2}x + 4$
For $6y = 5 - 9x$, solve for y.
$y = -\frac{3}{2}x + \frac{5}{6}$

Since the slopes are both $-\frac{3}{2}$, the lines are parallel.

37. For $4x = 2y + 3$, solve for y.
$y = 2x - \frac{3}{2}$
For $2y = 2x + 3$, solve for y.
$y = x + \frac{3}{2}$
Since the two slopes are 2 and 1, the lines are neither parallel nor perpendicular.

39. Triangle with vertices (9, 6), (–1, 2) and (1, –3).

a. Slope of side between vertices (9, 6) and (–1, 2):
$m = \frac{6-2}{9-(-1)} = \frac{4}{10} = \frac{2}{5}$
Slope of side between vertices (–1, 2) and (1, –3):
$m = \frac{2-(-3)}{-1-1} = \frac{5}{-2} = -\frac{5}{2}$
Slope of side between vertices (1, –3) and (9, 6):
$m = \frac{-3-6}{1-9} = \frac{-9}{-8} = \frac{9}{8}$

Copyright © 2019 Pearson Education, Inc.

b. The sides with slopes $\frac{2}{5}$ and $-\frac{5}{2}$ are perpendicular, because $\frac{2}{5}\left(-\frac{5}{2}\right)=-1$. Thus, the triangle is a right triangle.

41. Use point-slope form with $(x_1, y_1)=(-3, 2)$, $m=-\frac{2}{3}$

$$y-y_1=m(x-x_1)$$
$$y-2=-\frac{2}{3}(x-(-3))$$
$$y-2=-\frac{2}{3}(x+3)$$
$$y-2=-\frac{2}{3}x-2$$
$$y=-\frac{2}{3}x$$

43. $(x_1, y_1)=(2, 3)$, $m=3$

$$y-y_1=m(x-x_1)$$
$$y-3=3(x-2)$$
$$y-3=3x-6$$
$$y=3x-3$$

45. $(x_1, y_1)=(10, 1)$, $m=0$

$$y-y_1=m(x-x_1)$$
$$y-1=0(x-10)$$
$$y-1=0\Rightarrow y=1$$

47. Since the slope is undefined, the equation is that of a vertical line through (–2, 12).
$x=-2$

49. Through (–1, 1) and (2, 7)
Find the slope.

$$m=\frac{7-1}{2-(-1)}=\frac{6}{3}=2$$

Use the point-slope form with $(2, 7)=(x_1, y_1)$.

$$y-y_1=m(x-x_1)$$
$$y-7=2(x-2)$$
$$y-7=2x-4$$
$$y=2x+3$$

51. Through (1, 2) and (3, 9)
Find the slope.

$$m=\frac{9-2}{3-1}=\frac{7}{2}$$

Use the point-slope form with $(1, 2)=(x_1, y_1)$.

$$y-y_1=m(x-x_1)$$
$$y-2=\frac{7}{2}(x-1)$$
$$y-2=\frac{7}{2}x-\frac{7}{2}$$
$$y=\frac{7}{2}x-\frac{3}{2}$$
$$2y=7x-3$$

53. Through the origin with slope 5.
$(x_1, y_1)=(0, 0)$; $m=5$

$$y-y_1=m(x-x_1)$$
$$y-0=5(x-0)\Rightarrow y=5x$$

55. Through (6, 8) and vertical.
A vertical line has undefined slope.
$(x_1, y_1)=(6, 8)$
$x=6$

57. Through (3, 4) and parallel to $4x-2y=5$.
Find the slope of the given line because a line parallel to the line has the same slope.
$(x_1, y_1)=(3, 4)$

$$4x=2y+5$$
$$2y=4x-5$$
$$y=2x-\frac{5}{2} \quad m=2$$
$$y-y_1=m(x-x_1)$$
$$y-4=2(x-3)$$
$$y-4=2x-6$$
$$y=2x-2$$

59. x-intercept 6; y-intercept –6
Through the points (6, 0) and (0, –6).

$$m=\frac{0-(-6)}{6-0}=\frac{6}{6}=1$$

$(x_1, y_1)=(6, 0)$

$$y-y_1=m(x-x_1)$$
$$y-0=1(x-6)\Rightarrow y=x-6$$

Copyright © 2019 Pearson Education, Inc.

61. Through (–1, 3) and perpendicular to the line through (0, 1) and (2, 3).
The slope of the given line is
$m_1 = \frac{1-3}{0-2} = \frac{-2}{-2} = 1$, so the slope of a line perpendicular to the line is $m_2 = \frac{-1}{1} = -1$.

$$(x_1, y_1) = (-1, 3)$$
$$y - y_1 = m(x - x_1)$$
$$y - 3 = -1(x - (-1))$$
$$y - 3 = -x - 1$$
$$y = -x + 2$$

63. Let cost $x = 15{,}965$ and life: 12 years. Find D.
$$D = \left(\frac{1}{n}\right)x = \frac{1}{12}(15{,}965) \approx 1330.42$$
The depreciation is \$1330.42 per year.

65. Let cost $x = \$201{,}457$; life: 30 years
$$D = \left(\frac{1}{n}\right)x = \frac{1}{30}(201{,}457) \approx 6715.23$$
The depreciation is \$6715.23 per year.

67. $x = 10$
$$y = -3.9(10) + 27.3 = -11.7$$
There was a decrease of 11.7% for Miller Lite beer between 2007 and 2010.

69. $y = -32$
$$-32 = -3.9x + 27.3$$
$$-59.3 = -3.9x$$
$$15.21 \approx x$$
The first full year that there will be a decrease of 32% in Miller Lite beer shipments was 2016.

71. $x = 7$
$$y = 5.6(7) + 52 = 91.2$$
There were 91.2 million malaria cases in 2007.

73. $y = 150$
$$150 = 5.6x + 52$$
$$98 = 5.6x$$
$$17.5 = x$$
The first full year that there will be 150 million global malaria cases will be 2018.

75. $y = -.0135(2016) + 40.32 \approx 13.104$
The model predicts that the time for the men's 5000-m run will be about 13.104 minutes in the 2016 Olympics. This was a difference of .044 minutes from the actual time.

77. a. $(x_1, y_1) = (11, 5)$ and $(x_2, y_2) = (15, 23)$
Find the slope.
$$m = \frac{23-5}{15-11} = \frac{18}{4} = 4.5$$
$$y - 5 = 4.5(x - 11)$$
$$y - 5 = 4.5x - 49.5$$
$$y = 4.5x - 44.5$$

b. The year 2013 corresponds to $x = 2013 - 2000 = 13$.
$$y = 4.5(13) - 44.5 = 14$$
In 2013, Michael Kors had 14% of the luxury handbag market.

c. $y = 30$
$$30 = 4.5x - 44.5$$
$$74.5 = 4.5x$$
$$16.56 \approx x$$
The first full year that Michael Kors will had 30% of the luxury handbag market was 2017.

79. a. $(x_1, y_1) = (10, 3500)$ and $(x_2, y_2) = (15, 5000)$
Find the slope.
$$m = \frac{5000-3500}{15-10} = \frac{1500}{5} = 300$$
$$y - 3500 = 300(x - 10)$$
$$y - 3500 = 300x - 3000$$
$$y = 300x + 500$$

b. The year 2014 corresponds to $x = 2014 - 2000 = 14$.
$$y = 300(14) + 500 = 4700$$
In 2014, the average annual cost to employees for employer family health coverage was \$4700.

c. $y = 6000$
$$6000 = 300x + 500$$
$$5500 = 300x$$
$$18.33 \approx x$$
The first full year that the average annual cost to employees for employer family health coverage will exceed \$6000 will be 2019.

Copyright © 2019 Pearson Education, Inc.

Section 2.3 Linear Models

1. a. Let (x_1, y_1) be (32, 0) and (x_2, y_2) be (68, 20).
Find the slope.
$$m = \frac{20-0}{68-32} = \frac{20}{36} = \frac{5}{9}$$
Use the point-slope form with (32, 0).
$$y - 0 = \frac{5}{9}(x-32) \Rightarrow y = \frac{5}{9}(x-32)$$

b. Let $x = 50$.
$$y = \frac{5}{9}(50-32) = \frac{5}{9}(18) = 10°\text{C}$$
Let $x = 75$.
$$y = \frac{5}{9}(75-32) = \frac{5}{9}(43) \approx 23.89°\text{C}$$

c. $C = -10°\text{C}$
$$-10 = \frac{5}{9}(F-32)$$
$$-90 = 5F - 160$$
$$70 = 5F \Rightarrow F = 14°$$

d. $F = -20°\text{F}$
$$C = \frac{5}{9}(-20-32) = \frac{5}{9}(-52) \approx -28.9°C$$

3. F = 867°
$$C = \frac{5}{9}(867-32) = \frac{5}{9}(835) \approx 463.89°C$$

5. Let $(x_1, y_1) = (11, 224.9)$ and $(x_2, y_2) = (16, 240.0)$. Find the slope.
$$m = \frac{240.0-224.9}{16-11} = 3.02$$
$$y - 224.9 = 3.02(x-11)$$
$$y - 224.9 = 3.02x - 33.22$$
$$y = 3.02x + 191.68$$
To estimate the CPI in 2012, let $x = 12$.
$$y = 3.02(12) + 191.68 \approx 227.92$$
To estimate the CPI in 2015, let $x = 15$.
$$y = 3.02(15) + 191.68 \approx 236.98$$

7. Let $(x_1, y_1) = (5, 12.3)$ and $(x_2, y_2) = (14, 13.3)$. Find the slope.
$$m = \frac{13.3-12.3}{14-5} = \frac{1}{9} \approx .11$$
$$y - 12.3 = .11(x-5)$$
$$y - 12.3 = .11x - .55$$
$$y = .11x + 11.75$$
To estimate the number of employees working in educational services in 2012, let $x = 2012 - 2000 = 12$.
$$y = .11(12) + 11.75 = 13.08$$
The number of employees was estimated to be 13.08 million in 2012.

9. Find the slope of the line.
$(x_1, y_1) = (50, 320)$
$(x_2, y_2) = (80, 440)$
$$m = \frac{y_2 - y_1}{x_2 - x_1} = \frac{440-320}{80-50} = \frac{120}{30} = 4$$
Each mile per hour increase in the speed of the bat will make the ball travel 4 more feet.

11. a. $y = .07x + .83$

Data Point (x, y)	Model Point $(x, \hat{y})$	Residual $y - \hat{y}$	Squared Residual $(y - \hat{y})^2$
(6, .7)	(6,1.25)	–.55	.3025
(8, 1.22)	(8, 1.39)	–.17	.0289
(10, 1.53)	(10, 1.53)	0	0
(12,2.02)	(12, 1.67)	.35	.1225
(14,1.81)	(14,1.81)	0	0

$y = .2x - .38$

Data Point (x, y)	Model Point $(x, \hat{y})$	Residual $y - \hat{y}$	Squared Residual $(y - \hat{y})^2$
(6, .7)	(6,.82)	–.12	.0144
(8, 1.22)	(8, 1.22)	0	0
(10, 1.53)	(10, 1.62)	–.09	.0081
(12,2.02)	(12, 2.02)	0	0
(14,1.81)	(14,2.42)	–.61	.3721

Sum of the residuals for model 1 = –.37
Sum of the residuals for model 2 = –.82

Copyright © 2019 Pearson Education, Inc.

b. Sum of the squares of the residuals for model 1 = .4539
Sum of the squares of the residuals for model 2 = .3946

c. Model 2 is the better fit.

13. Plot the points.

[10, 16] by [3500, 4500]

Visually, a straight line looks to be a good model for the data.

```
LinReg
 y=ax+b
 a=-170.8
 b=6320
 r²=.9998190412
 r=-.9999095165
```

The coefficient of correlation is $r \approx -.9999$, which indicates that the regression line is a good fit for the data.

15. a. Using a graphing calculator, the regression-line model is $y = -170.8x + 6320$.

b. The year 2018 corresponds to $x = 18$. Using the regression-line model generated by a graphing calculator, we have $y = -170.8(18) + 6320 \approx 3246$, or about 3246 credit unions.

17. a. Using a graphing calculator, the regression-line model is
$y = .8x + 14.2$.

b. The third quarter of 2017 corresponds to $x = 11$. Using the regression-line model generated by a graphing calculator, we have $y = .8(11) + 14.2 = 23$, or about $23 billion in revenue.

19. a. Using a graphing calculator, the regression line model for amount of construction spending on highways and streets (in billions of dollars) is given by
$y = -1.829x + 98.4$.

b. Let $x = 12$ (December 2016).
$y = -1.829(12) + 98.4 \approx 76.5$ billion
The total amount of construction spending on highways and streets was about $76.5 billion in December 2016.

21. a. Using a graphing calculator, the regression-line model is $y = .295x + 48$.

b. $y = .295(10) + 48 \approx 50.95$. There was about $50.95 billion in revenue for the second quarter of 2017.

c. Let $y = 52$.
$$52 = .295x + 48$$
$$4 = .295x$$
$$x \approx 13.56$$
There will be $52 billion in revenue during the second quarter of 2018.

d. Using a graphing calculator, the coefficient of correlation is about .795.

Section 2.4 Linear Inequalities

1. Use brackets if you want to include the endpoint, and parentheses if you want to exclude it.

3. $-8k \le 32$

Multiply both sides of the inequality by $-\frac{1}{8}$. Since this is a negative number, change the direction of the inequality symbol.

$$-\frac{1}{8}(-8k) \ge -\frac{1}{8}(32) \Rightarrow k \ge -4$$

The solution is $[-4, \infty)$.

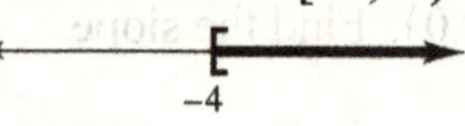

5. $-2b > 0$

Multiply both sides by $-\frac{1}{2}$.

$$-2b > 0 \Rightarrow -\frac{1}{2}(-2b) < -\frac{1}{2}(0) \Rightarrow b < 0$$

The solution is $(-\infty, 0)$. To graph this solution, put a parenthesis at 0 and draw an arrow extending to the left.

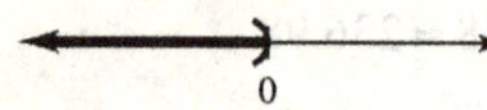

Copyright © 2019 Pearson Education, Inc.

7. $3x+4\le 14$

Subtract 4 from both sides.

$3x+4-4\le 14-4 \Rightarrow 3x\le 10$

Multiply each side by $\frac{1}{3}$.

$\frac{1}{3}(3x)\le \frac{1}{3}(10)\Rightarrow x\le \frac{10}{3}$

The solution is $\left(-\infty, \frac{10}{3}\right]$.

$\frac{10}{3}$

For exercises 9–26, we give the solutions without additional explanation.

9.

$$-5-p\ge 3$$
$$-5+5-p\ge 3+5$$
$$-p\ge 8$$
$$(-1)(-p)\le(-1)(8)$$
$$p\le -8$$

The solution is $(-\infty, -8]$.

11. $7m-5<2m+10$

$5m-5<10$

$5m<15$

$\frac{1}{5}(5m)<\frac{1}{5}(15)$

$m<3$

The solution is $(-\infty, 3)$.

3

13. $m-(4+2m)+3<2m+2$

$m-4-2m+3<2m+2$

$-1-m<2m+2$

$-m-2m<2+1$

$-3m<3$

$-\frac{1}{3}(-3m)>-\frac{1}{3}(3)$

$m>-1$

The solution is $(-1, \infty)$.

-1

15. $-2(3y-8)\ge 5(4y-2)$

$-6y+16\ge 20y-10$

$16+10\ge 20y+6y$

$26\ge 26y$

$1\ge y$ or $y\le 1$

The solution is $(-\infty, 1]$.

1

17. $3p-1<6p+2(p-1)$

$3p-1<6p+2p-2$

$-1+2<6p+2p-3p$

$1<5p$

$\frac{1}{5}<p$ or $p>\frac{1}{5}$

The solution is $\left(\frac{1}{5}, \infty\right)$.

$\frac{1}{5}$

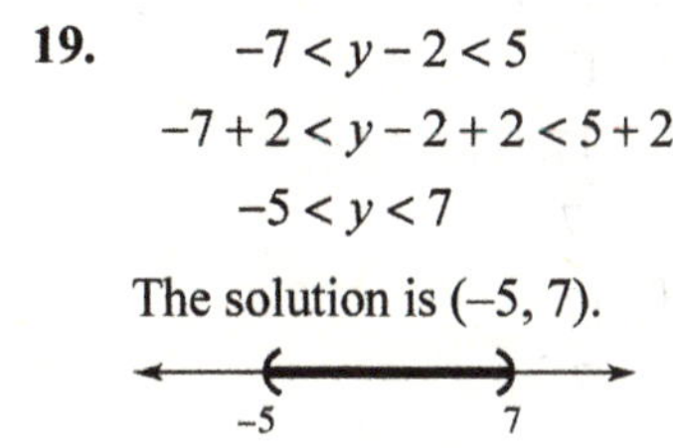

19. $-7<y-2<5$

$-7+2<y-2+2<5+2$

$-5<y<7$

The solution is $(-5, 7)$.

-5 $\quad$ 7

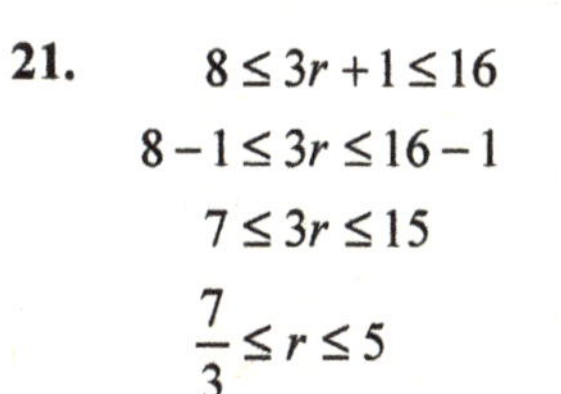

21. $8\le 3r+1\le 16$

$8-1\le 3r\le 16-1$

$7\le 3r\le 15$

$\frac{7}{3}\le r\le 5$

The solution is $\left[\frac{7}{3}, 5\right]$.

$\frac{7}{3}$ $\quad$ 5

Copyright © 2019 Pearson Education, Inc.

23.
$$-4 \le \frac{2k-1}{3} \le 2$$
$$-4(3) \le 3\left(\frac{2k-1}{3}\right) \le 2(3)$$
$$-12 \le 2k-1 \le 6$$
$$-12+1 \le 2k \le 6+1$$
$$-11 \le 2k \le 7$$
$$-\frac{11}{2} \le k \le \frac{7}{2}$$

The solution is $\left[-\frac{11}{2}, \frac{7}{2}\right]$.

$-\frac{11}{2}$ $\frac{7}{2}$

25.
$$\frac{3}{5}(2p+3) \ge \frac{1}{10}(5p+1)$$
$$10 \cdot \frac{3}{5}(2p+3) \ge 10 \cdot \frac{1}{10}(5p+1)$$
$$6(2p+3) \ge 5p+1$$
$$12p+18 \ge 5p+1$$
$$7p \ge -17$$
$$p \ge -\frac{17}{7}$$

The solution is $\left[-\frac{17}{7}, \infty\right)$.

$-\frac{17}{7}$

27. $x \ge 2$

29. $-3 < x \le 5$

31. $C = 50x + 6000;\ R = 65x$
To at least break even, $R \ge C$.
$65x \ge 50x + 6000$
$15x \ge 6000 \Rightarrow x \ge 400$
The number of units of wire must be in the interval $[400, \infty)$.

33. $C = 85x + 1000;\ R = 105x$
$R \ge C$
$105x \ge 85x + 1000$
$20x \ge 1000$
$$x \ge \frac{1000}{20} \Rightarrow x \ge 50$$
x must be in the interval $[50, \infty)$.

35. $C = 1000x + 5000;\ R = 900x$
$R \ge C$
$900x \ge 1000x + 5000$
$-100x \ge 5000$
$$x \le \frac{5000}{-100} \Rightarrow x \le -50$$
It is impossible to break even.

37. $|p| > 7$
$p < -7$ or $p > 7$
The solution is $(-\infty, -7)$ or $(7, \infty)$.

-7 7

39. $|r| \le 5 \Rightarrow -5 \le r \le 5$
The solution is $[-5, 5]$.

-5 5

41. $|b| > -5$
The absolute value of a number is always nonnegative. Therefore, $|b| > -5$ is always true, so the solution is the set of all real numbers.

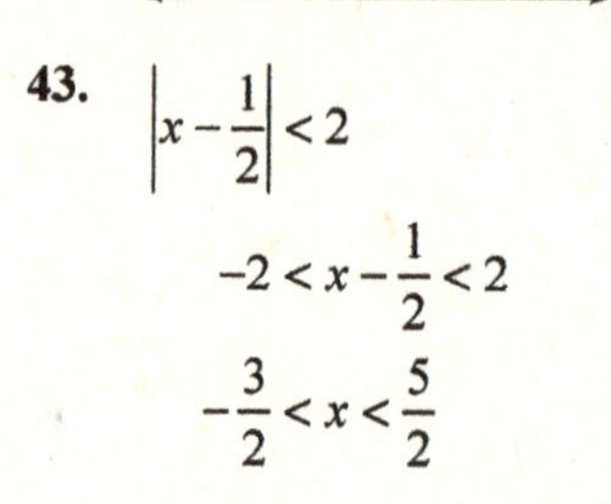

43.
$$\left|x - \frac{1}{2}\right| < 2$$
$$-2 < x - \frac{1}{2} < 2$$
$$-\frac{3}{2} < x < \frac{5}{2}$$

The solution is $\left(-\frac{3}{2}, \frac{5}{2}\right)$.

$-\frac{3}{2}$ $\frac{5}{2}$

45. $|8b+5| \ge 7$
$8b+5 \le -7$ or $8b+5 \ge 7$
$8b \le -12$ or $8b \ge 2$
$b \le -\frac{3}{2}$ or $b \ge \frac{1}{4}$

The solution is $\left(-\infty, -\frac{3}{2}\right]$ or $\left[\frac{1}{4}, \infty\right)$.

$-\frac{3}{2}$ $\frac{1}{4}$

Copyright © 2019 Pearson Education, Inc.

47. $|T-83|\le 7$
$$-7\le T-83\le 7$$
$$76\le T\le 90$$

49. $|T-61|\le 21$
$$-21\le T-61\le 21$$
$$40\le T\le 82$$

51. $|P-35.5|\le 3$
$$|P-35.5|\le \pm 3\Rightarrow$$
$$-3\le P-35.5\le 3\Rightarrow$$
$$32.5\le P\le 38.5$$

53. $35\le B\le 43$

55. The seven income ranges are:
$$0<x\le 9325$$
$$9325<x\le 37{,}950$$
$$37{,}950<x\le 91{,}900$$
$$91{,}900<x\le 191{,}650$$
$$191{,}650<x\le 416{,}700$$
$$416{,}700<x\le 418{,}400$$
$$x>418{,}400$$

Section 2.5 Polynomial and Rational Inequalities

1. $(x+4)(2x-3)\le 0$
Solve the corresponding equation.
$(x+4)(2x-3)=0$
$x+4=0$ or $2x-3=0$
$x=-4$ $\quad x=\frac{3}{2}$
Note that because the inequality symbol is "≤," –4 and $\frac{3}{2}$ are solutions of the original inequality. These numbers separate the number line into three regions.

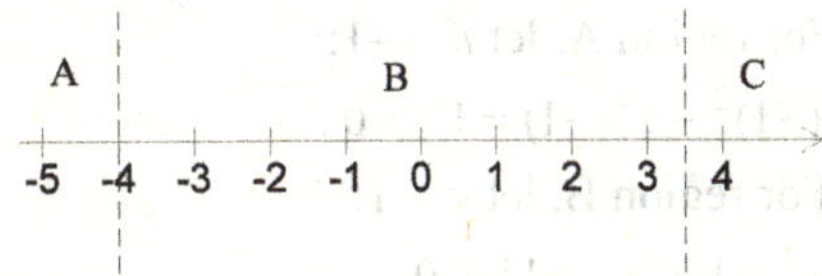

In region A, let $x=-6$:
$(-6+4)[2(-6)-3]=30>0$.
In region B, let $x=0$:
$(0+4)[2(0)-3]=-12<0$.
In region C, let $x=2$:
$(2+4)[2(2)-3]=6>0$.
The only region where $(x+4)(2x-3)$ is negative is region B, so the solution is $\left[-4,\frac{3}{2}\right]$. To graph this solution, put brackets at –4 and $\frac{3}{2}$ and draw a line segment between these two endpoints.

3. $r^2+4r>-3$
Solve the corresponding equation.
$$r^2+4r=-3$$
$$r^2+4r+3=0$$
$$(r+1)(r+3)=0$$
$$r+1=0 \quad\text{or}\quad r+3=0$$
$$r=-1 \quad\text{or}\quad r=-3$$
Note that because the inequality symbol is ">," –1 and –3 are not solutions of the original inequality.

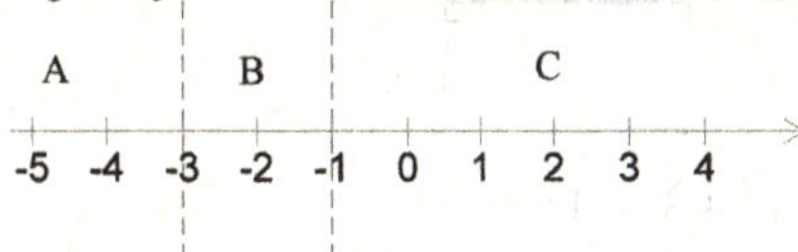

In region A, let $r=-4$:
$(-4)^2+4(-4)=0>-3$.
In region B, let $r=-2$:
$(-2)^2+4(-2)=-4<-3$.
In region C, let $r=0$:
$0^2+4(0)=0>-3$.
The solution is $(-\infty,-3)$ or $(-1,\infty)$.
To graph the solution, put a parenthesis at –3 and draw a ray extending to the left, and put a parenthesis at -1 and draw a ray extending to the right.

Copyright © 2019 Pearson Education, Inc.

5. $4m^2 + 7m - 2 \le 0$
Solve the corresponding equation.
$4m^2 + 7m - 2 = 0$
$(4m - 1)(m + 2) = 0$
$4m - 1 = 0$ or $m + 2 = 0$
$m = \frac{1}{4}$ or $m = -2$

Because the inequality symbol is "≤" $\frac{1}{4}$ and –2 are solutions of the original inequality.

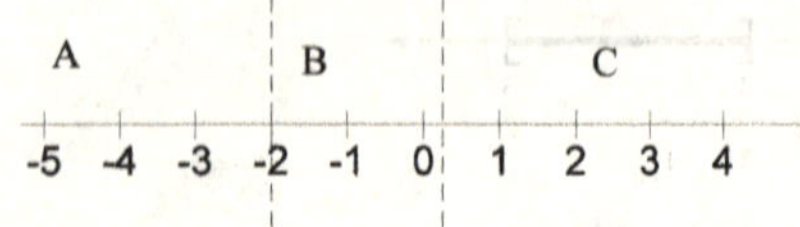

In region A, let $m = -3$:
$4(-3)^2 + 7(-3) - 2 = 13 > 0$.
In region B, let $m = 0$:
$4(0)^2 + 7(0) - 2 = -2 < 0$.
In region C, let $m = 1$:
$4(1)^2 + 7(1) - 2 = 9 > 0$.
The solution is $\left[-2, \frac{1}{4}\right]$.

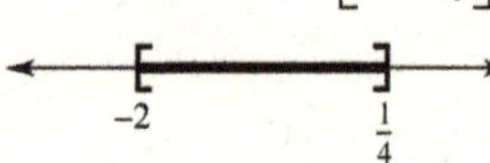

7. $4x^2 + 3x - 1 > 0$
Solve the corresponding equation.
$4x^2 + 3x - 1 = 0$
$(4x - 1)(x + 1) = 0$
$4x - 1 = 0$ or $x + 1 = 0$
$x = \frac{1}{4}$ or $x = -1$

Note that $\frac{1}{4}$ and –1 are not solutions of the original inequality.

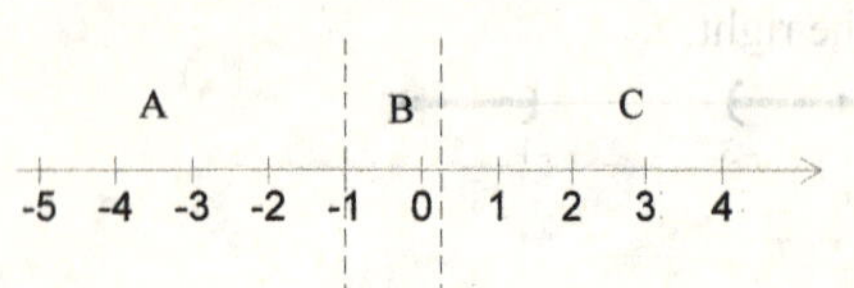

In region A, let $x = -2$:
$4(-2)^2 + 3(-2) - 1 = 9 > 0$.
In region B, let $x = 0$:
$4(0)^2 + 3(0) - 1 = -1 < 0$.
In region C, let $x = 1$:
$4(1)^2 + 3(1) - 1 = 6 > 0$.

The solution is $(-\infty, -1)$ or $\left(\frac{1}{4}, \infty\right)$.

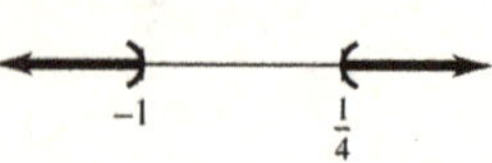

9. $x^2 \le 36$
Solve the corresponding equation.
$x^2 = 36 \Rightarrow x = \pm 6$

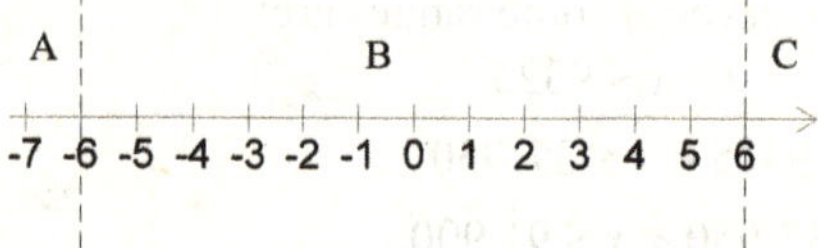

For region A, let $x = -7$: $(-7)^2 = 49 > 36$.
For region B, let $x = 0$: $0^2 = 0 < 36$.
For region C, let $x = 7$: $7^2 = 49 > 36$.
Both endpoints are included. The solution is $[-6, 6]$.

–6 6

11. $p^2 - 16p > 0$
Solve the corresponding equation.
$p^2 - 16p = 0 \Rightarrow p(p - 16) = 0 \Rightarrow$
$p = 0$ or $p = 16$
Since the inequality is ">", 0 and 16 are not solutions of the original inequality.

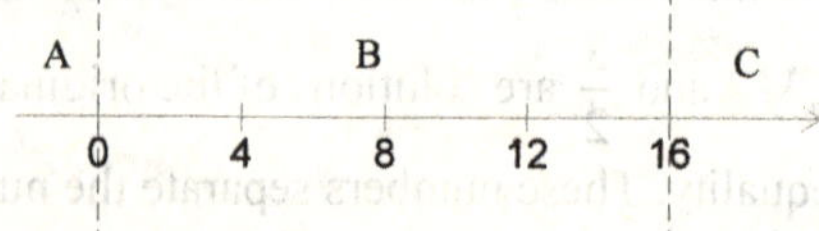

For region A, let $p = -1$:
$(-1)^2 - 16(-1) = 17 > 0$.
For region B, let $p = 1$:
$1^2 - 16(1) = -15 < 0$.
For region C, let $p = 17$:
$17^2 - 16(17) = 17 > 0$.
The solution is $(-\infty, 0)$ or $(16, \infty)$.

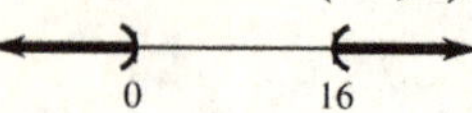

Copyright © 2019 Pearson Education, Inc.

13. $x^3 - 9x \geq 0$
Solve the corresponding equation.
$$x^3 - 9x = 0$$
$$x\left(x^2 - 9\right) = 0$$
$$x(x + 3)(x - 3) = 0$$
$x = 0$ or $x = -3$ or $x = 3$
Note that 0, –3, and 3 are all solutions of the original inequality.

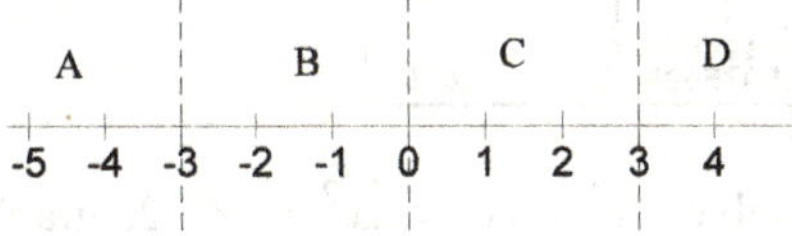

In region A, let $x = -4$:
$(-4)^3 - 9(-4) = -28 < 0$.
In region B, let $x = -1$:
$(-1)^3 - 9(-1) = 8 > 0$.
In region C, let $x = 1$:
$(1)^3 - 9(1) = -8 < 0$
In region D, let $x = 4$:
$4^3 - 9(4) = 28 > 0$.
The solution is $[-3, 0]$ or $[3, \infty)$.

15. $(x + 7)(x + 2)(x - 2) \geq 0$
Solve the corresponding equation.
$$(x + 7)(x + 2)(x - 2) = 0$$
$x + 7 = 0$ or $x + 2 = 0$ or $x - 2 = 0$
$x = -7$ or $x = -2$ or $x = 2$
Note that –7, –2 and 2 are all solutions of the original inequality.

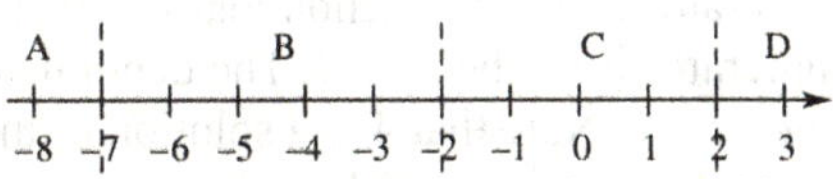

In region A, let $x = -8$:
$(-8 + 7)(-8 + 2)(-8 - 2) = -60 < 0$
In region B, let $x = -4$:
$(-4 + 7)(-4 + 2)(-4 - 2) = 36 > 0$
In region C, let $x = 0$:
$(0 + 7)(0 + 2)(0 - 2) = -28 < 0$
In region D, let $x = 3$:
$(3 + 7)(3 + 2)(3 - 2) = 50 > 0$
The solution is $[-7, -2]$ or $[2, \infty)$.

17. $(x + 5)\left(x^2 - 2x - 3\right) < 0$
Solve the corresponding equation.
$$(x + 5)\left(x^2 - 2x - 3\right) = 0$$
$$(x + 5)(x + 1)(x - 3) = 0$$
$x + 5 = 0$ or $x + 1 = 0$ or $x - 3 = 0$
$x = -5$ or $x = -1$ or $x = 3$
Note that –5, –1 and 3 are not solutions of the original inequality.

A B C D
-7 -6 -5 -4 -3 -2 -1 0 1 2 3 4

In region A, let $x = -6$:
$(-6 + 5)\left[(-6)^2 - 2(-6) - 3\right] = (-1)(45)$
$= -45 < 0$
In region B, let $x = -2$:
$(-2 + 5)\left[(-2)^2 - 2(-2) - 3\right] = 3(5) = 15 > 0$
In region C, let $x = 0$:
$(0 + 5)\left[(0)^2 - 2(0) - 3\right] = 5(-3) = -15 < 0$
In region D, let $x = 4$:
$(4 + 5)\left[(4)^2 - 2(4) - 3\right] = 9(5) = 45 > 0$
The solution is $(-\infty, -5)$ or $(-1, 3)$.

19. $6k^3 - 5k^2 < 4k \Rightarrow 6k^3 - 5k^2 - 4k < 0$
Solve the corresponding equation.
$$6k^3 - 5k^2 - 4k = 0$$
$$k\left(6k^2 - 5k - 4\right) = 0$$
$$k(3k - 4)(2k + 1) = 0$$
$k = 0$ or $k = \frac{4}{3}$ or $k = -\frac{1}{2}$
Note that 0, $\frac{4}{3}$, and $-\frac{1}{2}$ are not solutions of the original inequality.

A B C D
-5 -4 -3 -2 -1 0 1 2 3 4

Copyright © 2019 Pearson Education, Inc.

In region A, let $k = -1$:

$6(-1)^3 - 5(-1)^2 - 4(-1) = -7 < 0$

In region B, let $k = -\frac{1}{4}$:

$6\left(-\frac{1}{4}\right)^3 - 5\left(-\frac{1}{4}\right)^2 - 4\left(-\frac{1}{4}\right) = \frac{19}{32} > 0;$

In region C, let $k = 1$:

$6(1)^3 - 5(1)^2 - 4(1) = -3 < 0$

In region D, let $k = 10$:

$6(10)^3 - 5(10)^2 - 4(10) = 5460$

The given inequality is true in regions A and C.

The solution is $\left(-\infty, -\frac{1}{2}\right)$ or $\left(0, \frac{4}{3}\right)$.

21. The inequality $p^2 < 16$ should be rewritten as $p^2 - 16 < 0$ and solved by the method shown in this section for solving quadratic inequalities. This method will lead to the correct solution $(-4, 4)$. The student's method and solution are incorrect.

23. To solve $.5x^2 - 1.2x < .2$, write the inequality as $.5x^2 - 1.2x - .2 < 0$. Graph the equation $y = .5x^2 - 1.2x - .2$. Enter this equation as y_1 and use $-4 \le x \le 6$ and $-5 \le y \le 5$. On the CALC menu, use "zero" to find the x-values where the graph crosses the x-axis. These values are $x = -.1565$ and $x = 2.5565$. The graph is below the x-axis between these two values. The solution of the inequality is $(-.1565, 2.5565)$.

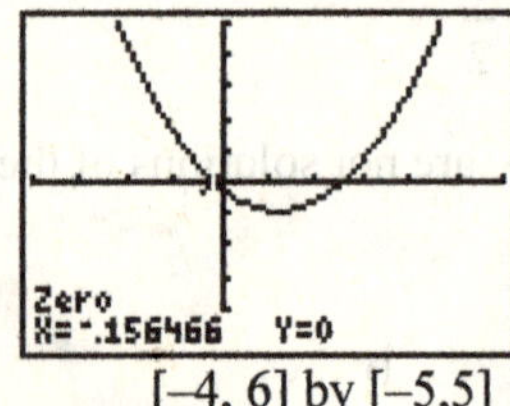

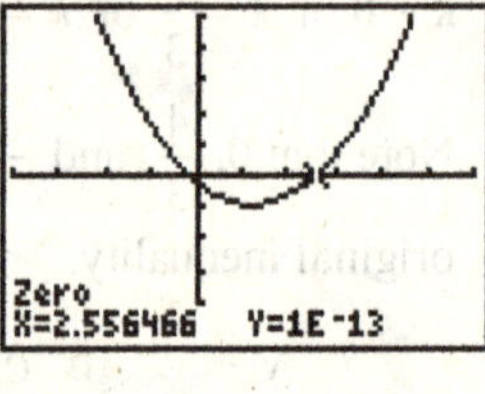

[-4, 6] by [-5,5]

25. To solve $x^3 - 2x^2 - 5x + 7 \ge 2x + 1$, graph $y_1 = x^3 - 2x^2 - 5x + 7$ and $y_2 = 2x + 1$ in the window $[-5, 5]$ by $[-10, 10]$. On the CALC menu, use "intersect" to find the x-values where the graphs intersect. These values are $x = -2.2635$, $x = .7556$ and $x = 3.5079$. The graph of y_1 is above the graph of y_2 for $[-2.2635, .7556]$ or $[3.5079, \infty)$.

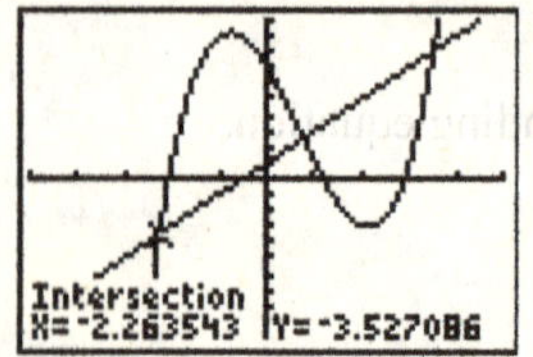

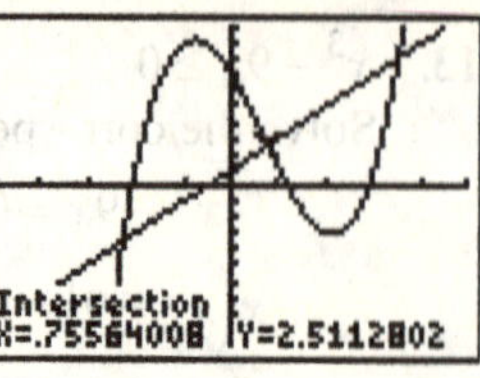

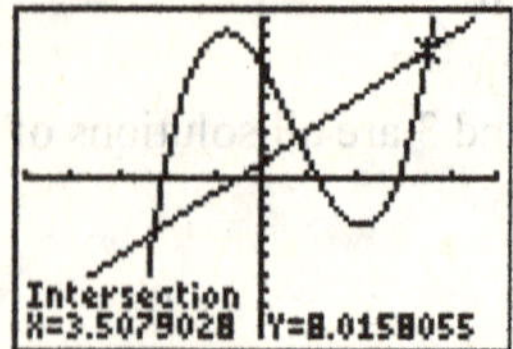

27. To solve $2x^4 + 3x^3 < 2x^2 + 4x - 2$, graph $y_1 = 2x^4 + 3x^3$ and $y_2 = 2x^2 + 4x - 2$ in the window $[-2, 2]$ by $[-5, 5]$. On the CALC menu, use "intersect" to find the x-values where the graphs intersect. These values are $x = .5$ and $x = .8393$. The graph of y_1 is below the graph of y_2 to the right of .5 and to the left of .8393. The solution of the inequality is $(.5, .8393)$.

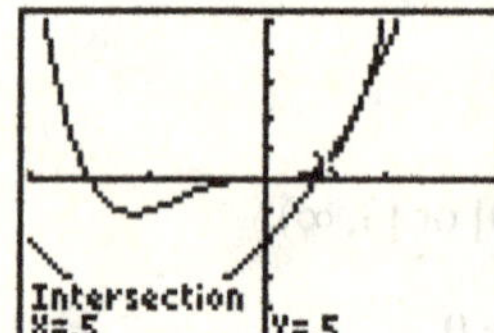

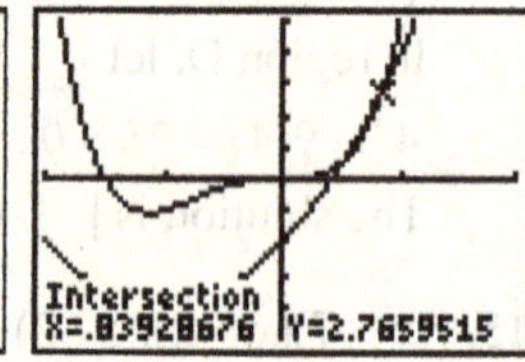

29. $\frac{r-4}{r-1} \ge 0$

Solve the corresponding equation.

$\frac{r-4}{r-1} = 0$

The quotient can change sign only when the numerator is 0 or the denominator is 0. The numerator is 0 when $r = 4$. The denominator is 0 when $r = 1$. Note that 4 is a solution of the original inequality, but 1 is not.

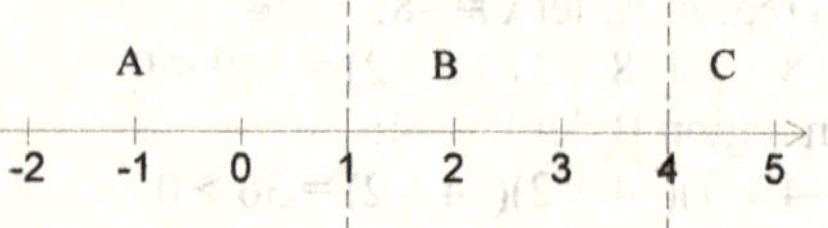

Copyright © 2019 Pearson Education, Inc.

In region A, let $r = 0$:
$$\frac{0-4}{0-1} = 4 > 0.$$
In region B, let $r = 2$:
$$\frac{2-4}{2-1} = -2 < 0.$$
In region C, let $r = 5$:
$$\frac{5-3}{5-1} = \frac{1}{4} > 0.$$
The given inequality is true in regions A and C, so the solution is $(-\infty, 1)$ or $[4, \infty)$.

31. $\frac{a-2}{a-5} < -1$

Solve the corresponding equation.
$$\frac{a-2}{a-5} = -1$$
$$\frac{a-2}{a-5} + 1 = 0$$
$$\frac{a-2}{a-5} + \frac{a-5}{a-5} = 0$$
$$\frac{2a-7}{a-5} = 0$$

The numerator is 0 when $a = \frac{7}{2}$. The denominator is 0 when $a = 5$. Note that $\frac{7}{2}$ and 5 are not solutions of the original inequality.

A B C

-3 -2 -1 0 1 2 3 4 5 6 7 8

In region A, let $a = 0$:
$$\frac{0-2}{0-5} = \frac{2}{5} > -1.$$
In region B, let $a = 4$:
$$\frac{4-2}{4-5} = \frac{2}{-1} = -2 < -1.$$
In region C, let $a = 10$:
$$\frac{10-2}{10-5} = \frac{8}{5} > -1.$$

The solution is $\left(\frac{7}{2}, 5\right)$.

33. $\frac{1}{p-2} < \frac{1}{3}$

Solve the corresponding equation.
$$\frac{1}{p-2} = \frac{1}{3}$$
$$\frac{1}{p-2} - \frac{1}{3} = 0$$
$$\frac{3-(p-2)}{3(p-2)} = 0$$
$$\frac{3-p+2}{3(p-2)} = 0$$
$$\frac{5-p}{3(p-2)} = 0$$
The numerator is 0 when $p = 5$. The denominator is 0 when $p = 2$. Note that 2 and 5 are not solutions of the original inequality.

A B C

-2 -1 0 1 2 3 4 5 6

In region A, let $p = 0$: $\frac{1}{0-2} = -\frac{1}{2} < \frac{1}{3}$.

In region B, let $p = 3$: $\frac{1}{3-2} = 1 > \frac{1}{3}$.

In region C, let $p = 6$: $\frac{1}{6-2} = \frac{1}{4} < \frac{1}{3}$.

The solution is $(-\infty, 2)$ or $(5, \infty)$.

35. $\frac{5}{p+1} > \frac{12}{p+1}$

Solve the corresponding equation.
$$\frac{5}{p+1} = \frac{12}{p+1}$$
$$\frac{5}{p+1} - \frac{12}{p+1} = 0$$
$$\frac{-7}{p+1} = 0$$
The numerator is never 0. The denominator is 0 when $p = -1$. Therefore, in this case, we separate the number line into only two regions.

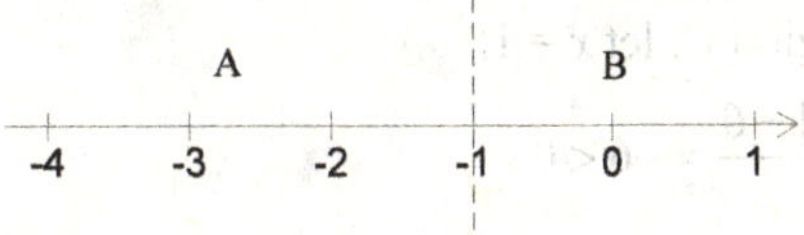

Copyright © 2019 Pearson Education, Inc.

In region A, let $p = -2$:

$\frac{5}{-2+1} = -5$

$\frac{12}{-2+1} = -12$

$-5 > -12$

In region B, let $p = 0$:

$\frac{5}{0+1} = 5$

$\frac{12}{0+1} = 12$

$12 > 5$

Therefore, the given inequality is true in region A. The only endpoint, −1, is not included because the symbol is ">." Therefore, the solution is $(-\infty, -1)$.

37. $\frac{x^2 - x - 6}{x} < 0$

Solve the corresponding equation.

$\frac{x^2 - x - 6}{x} = 0$

$x^2 - x - 6 = 0$ or $x = 0$

$(x-3)(x+2) = 0$ or $x = 0$

$x - 3 = 0$ or $x + 2 = 0$ or $x = 0$

$x = 3$ or $x = -2$ or $x = 0$

Note that −2, 0 and 3 are not solutions of the original inequality.

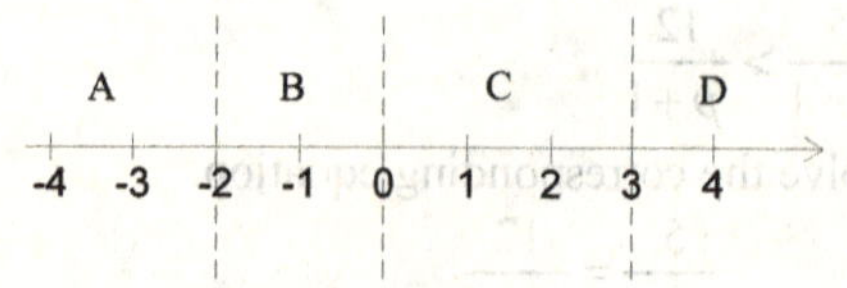

In region A, let $x = -3$:

$\frac{(-3)^2 - (-3) - 6}{-3} = \frac{9+3-6}{-3} = \frac{6}{-3} = -2 < 0.$

In region B, let $x = -1$:

$\frac{(-1)^2 - (-1) - 6}{-1} = \frac{1+1-6}{-1} = \frac{-4}{-1} = 4 > 0$

In region C, let $x = 1$:

$\frac{1^2 - 1 - 6}{1} = -6 < 0.$

In region D, let $x = 4$:

$\frac{4^2 - 4 - 6}{4} = \frac{16-10}{4} = \frac{6}{4} = \frac{3}{2} > 0.$

The solution is $(-\infty, -2)$ or $(0, 3)$.

39. To solve $\frac{2x^2 + x - 1}{x^2 - 4x + 4} \le 0$, break the inequality into two inequalities $2x^2 + x - 1 \le 0$ and $x^2 - 4x + 4 \le 0$. Graph the equations $y = 2x^2 + x - 1$ and $y = x^2 - 4x + 4$. Enter these equations as y_1 and y_2, and use $-3 < x < 3$ and $-2 < y < 2$. On the CALC menu, use "zero" to find the x-values where the graphs cross the x-axis. These values for y_1 are $x = -1$ and $x = .5$. The graph of y_1 is below the x-axis to the right of −1 and to the left of .5. The graph of y_2 is never below the x-axis. The solution of the inequality is $[-1, .5]$.

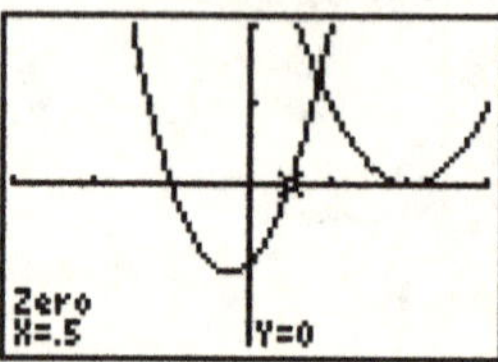

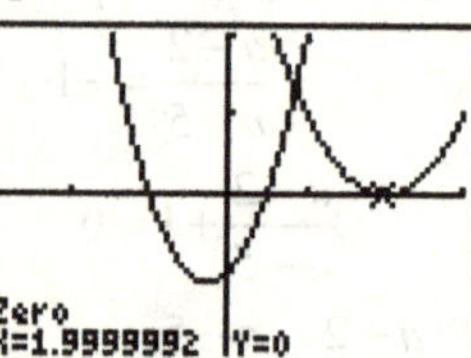

41. $P = 2x^2 - 12x - 32$

The company makes a profit when $2x^2 - 12x - 32 > 0$.

Solve the corresponding equation.

$2x^2 - 12x - 32 = 0$

$2(x^2 - 6x - 16) = 0$

$(x+2)(x-8) = 0 \Rightarrow x = -2$ or $x = 8$

The test regions are $A(-\infty, -2)$, $B(-2, 8)$, and $C(8, \infty)$. Region A makes no sense in this context, so we ignore this. Test a number from regions B and C in the original inequality.

For region B, let $x = 0$.

$2(0)^2 - 12(0) - 32 = -32 < 0$

For region C, let $x = 10$.

$2(10)^2 - 12(10) - 32 = 48 > 0$

The numbers in region C satisfy the inequality. The company makes a profit when the amount spent on advertising in hundreds of thousands of dollars is in the interval $(8, \infty)$.

Copyright © 2019 Pearson Education, Inc.

43. $P = x^2 + 300x - 18,000$
The complex makes a profit when
$x^2 + 300x - 18,000 > 0$.
Solve the corresponding equation.
$0 = x^2 + 300x - 18,000$
$$x = \frac{-300 \pm \sqrt{(300)^2 - 4(1)(-18,000)}}{2(1)}$$
$x \approx 51.25$ or $x \approx -351.25$
We only consider positive values of x because x represents the number of apartments rented.
The test regions are $A(0, 52)$ and $B(52, 200)$.
In region A, let $x = 1$:
$(1)^2 + 300(1) - 18,000 = -17,699 < 0$.
In region B, let $x = 100$:
$(100)^2 + 300(100) - 18,000 = 22,000 > 0$.
The complex makes a profit when the number of units rented is between 52 and 200, inclusive, or when x is in the interval [52, 200].

Chapter 2 Review Exercises

1. $y = x^2 - 2x - 5$
$(-2, 3)$:
$(-2)^2 - 2(-2) - 5 = 4 + 4 - 5 = 3$
$(0, -5)$
$(0)^2 - 2(0) - 5 = 0 - 0 - 5 = -5$
$(2, -3)$:
$(2)^2 - 2(2) - 5 = 4 - 4 - 5 = -5 \neq -3$
$(3, -2)$:
$(3)^2 - 2(3) - 5 = 9 - 6 - 5 = -2$
$(4, 3)$:
$(4)^2 - 2(4) - 5 = 16 - 8 - 5 = 3$
$(7, 2)$:
$(7)^2 - 2(7) - 5 = 49 - 14 - 5 = 30 \neq 2$
Solutions are (–2, 3), (0, –5), (3, –2), (4, 3).

2. $x - y = 5$
$(-2, 3): -2 - 3 = -5 \neq 5$
$(0, -5): 0 - (-5) = 0 + 5 = 5$
$(2, -3): 2 - (-3) = 2 + 3 = 5$
$(3, -2): 3 - (-2) = 3 + 2 = 5$
$(4, 3): 4 - 3 = 1 \neq 5$
$(7, 2): 7 - 2 = 5$
Solutions are (0, –5), (2, –3), (3, –2), (7, 2).

3. $5x - 3y = 15$
First, we find the y-intercept. If $x = 0$, $y = -5$, so the y-intercept is –5. Next we find the x-intercept. If $y = 0$, $x = 3$, so the x-intercept is 3. Using these intercepts, we graph the line.

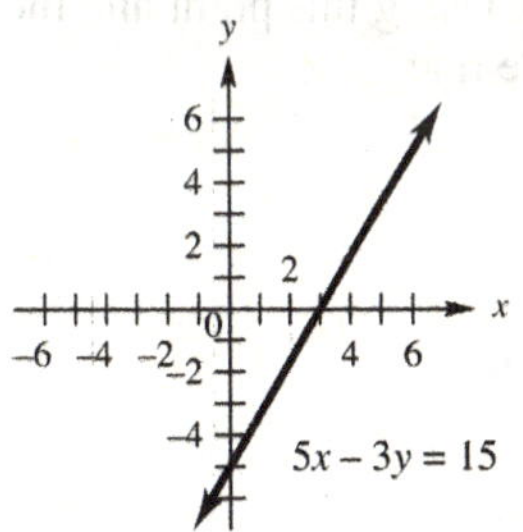

4. $2x + 7y - 21 = 0$
First we find the y-intercept. If $x = 0$, $y = 3$, so the y-intercept is 3. Next we find the x-intercept. If $y = 0$, $x = \frac{21}{2}$, so the x-intercept is $\frac{21}{2}$. Using these intercepts, we graph the line.

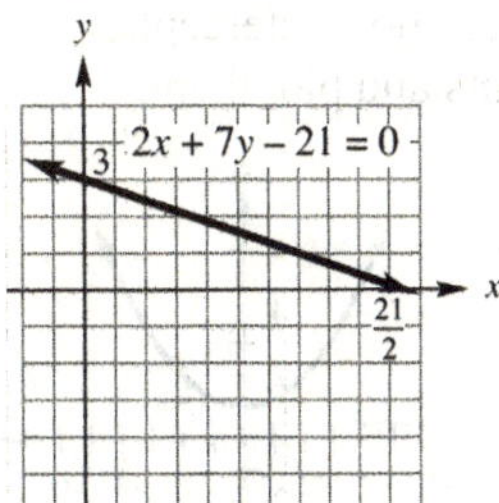

5. $y + 3 = 0$
The equation may be rewritten as $y = -3$.
The graph of $y = -3$ is a horizontal line with y-intercept of –3.

Copyright © 2019 Pearson Education, Inc.

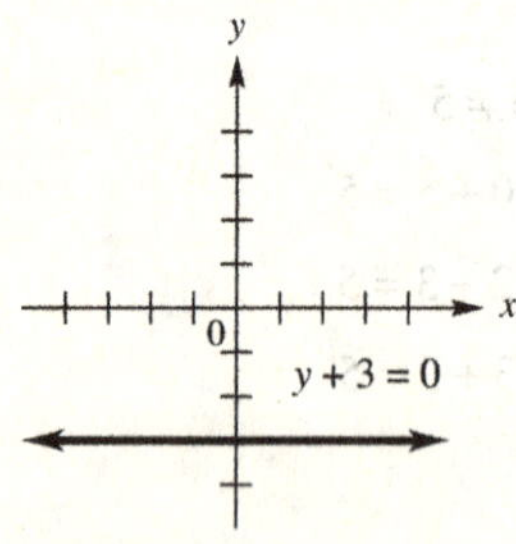

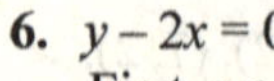

6. $y - 2x = 0$
First, we find the y-intercept. If $x = 0$, $y = 0$, so the y-intercept is 0. Since the line passes through the origin, the x-intercept is also 0. We find another point on the line by arbitrarily choosing a value for x. Let $x = 2$. Then $y - 2(2) = 0$, or $y = 4$. The point with coordinates (2, 4) is on the line. Using this point and the origin, we graph the line.

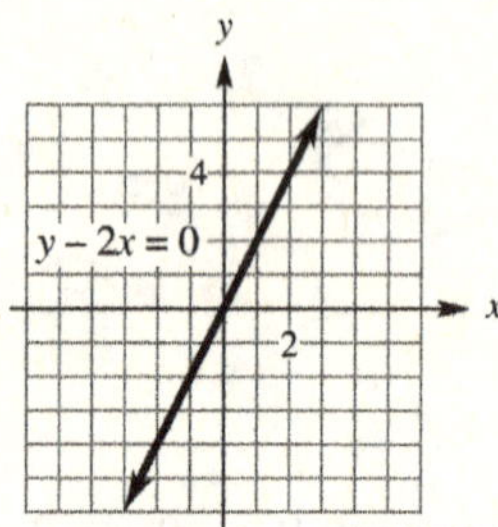

7. $y = .25x^2 + 1$
First we find the y-intercept. If $x = 0$, $y = .25(0)^2 + 1 = 1$, so the y-intercept is 1. Next we find the x-intercepts. If $y = 0$, $0 = .25x^2 + 1 \Rightarrow .25x^2 = -1 \Rightarrow x = \sqrt{-4}$, not a real number. There are no x-intercepts.
Make a table of points and plot them.

x	$.25x^2 + 1$
−4	5
−2	2
0	1
2	2
4	5

8. $y = \sqrt{x+4}$
Make a table of points and plot them.

x	$\sqrt{x+4}$
−4	0
−3	1
0	2
5	3

9. a. The temperature was over 55° from about 11:30 a.m. to about 7:30 p.m.

b. The temperature was below 40° from midnight until about 5 a.m., and after about 10:30 p.m.

10. At noon in Bratenahl the temperature was about 57°. The temperature in Greenville is 57° when the temperature in Bratenahl is 50°, or at about 10:30 a.m. and 8:30 p.m.

11. Answers vary. A possible answer is "rise over run".

12. Through (−1, 3) and (2, 6)

$$\text{slope} = \frac{\Delta y}{\Delta x} = \frac{6-3}{2-(-1)} = \frac{3}{3} = 1$$

13. Through (4, −5) and (1, 4)

$$\text{slope} = \frac{-5-4}{4-1} = \frac{-9}{3} = -3$$

14. Through (8, −3) and the origin
The coordinates of the origin are (0, 0).

$$\text{slope} = \frac{-3-0}{8-0} = -\frac{3}{8}$$

15. Through (8, 2) and (0, 4)

$$\text{slope} = \frac{4-2}{0-8} = \frac{2}{-8} = -\frac{1}{4}$$

Copyright © 2019 Pearson Education, Inc.

In exercises 16 and 17, we give the solution by rewriting the equation in slope-intercept form. Alternatively, the solution can be obtained by determining two points on the line and then using the definition of slope.

16. $3x + 5y = 25$
First we solve for y.

$5y = -3x + 25 \Rightarrow y = -\frac{3}{5}x + 5$

When the equation is written in slope-intercept form, the coefficient of x gives the slope. The slope is $-\frac{3}{5}$.

17. $6x - 2y = 7$
First we solve for y.

$6x - 2y = 7 \Rightarrow 6x - 7 = 2y \Rightarrow 3x - \frac{7}{2} = y$

The coefficient of x gives the slope, so the slope is 3.

18. $x - 2 = 0$
The graph of $x - 2 = 0$ is a vertical line. Therefore, the slope is undefined.

19. $y = -4$
The graph of $y = -4$ is a horizontal line. Therefore, the slope is 0.

20. Parallel to $3x + 8y = 0$
First, find the slope of the given line by solving for y.

$8y = -3x \Rightarrow y = -\frac{3}{8}x$

The slope is the coefficient of x, $-\frac{3}{8}$. A line parallel to this line has the same slope, so the slope of the parallel line is also $-\frac{3}{8}$.

21. Perpendicular to $x = 3y$
First, find the slope of the given line by solving for y: $y = \frac{1}{3}x$

The slope of this line is the coefficient of x, $\frac{1}{3}$. The slope of a line perpendicular to this line is the negative reciprocal of this slope, so the slope of the perpendicular line is -3.

22. Through (0, 5) with $m = -\frac{2}{3}$

Since $m = -\frac{2}{3} = \frac{-2}{3}$, we start at the point with coordinates (0, 5) and move 2 units down and 3 units to the right to obtain a second point on the line. Using these two points, we graph the line.

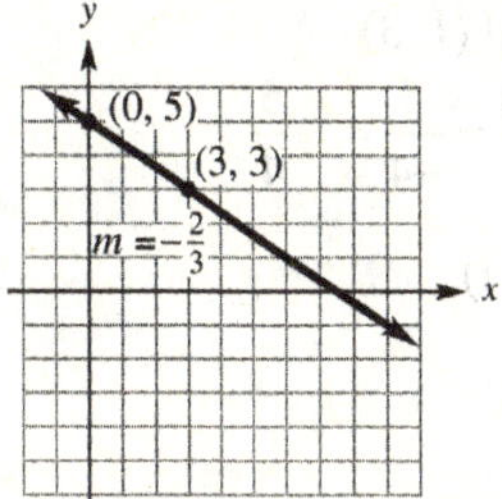

23. Through (–4, 1) with m = 3
Since $m = 3 = \frac{3}{1}$, we start at the point with coordinates (–4, 1) and move 3 units up and 1 unit to the right to obtain a second point on the line. Using these two points, we graph the line.

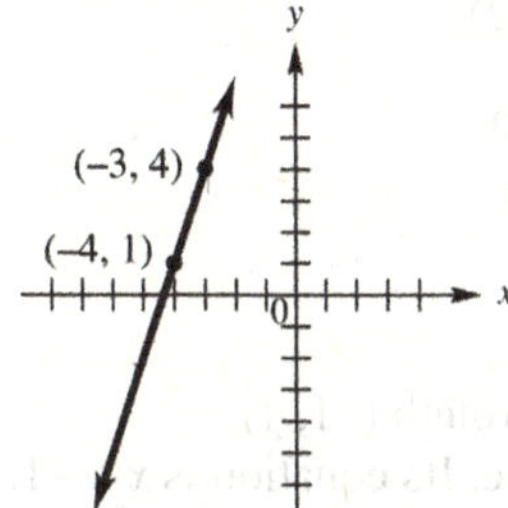

24. Answers vary. One example is:
You need two points; one point and the slope; the y-intercept and the slope.

25. Through (5, –1), slope $\frac{2}{3}$
Use the point slope form with $x_1 = 5$, $y_1 = -1$, and $m = \frac{2}{3}$.

$$y - y_1 = m(x - x_1)$$
$$y - (-1) = \frac{2}{3}(x - 5)$$
$$y + 1 = \frac{2}{3}x - \frac{10}{3}$$

Multiplying by 3 gives
$$3y + 3 = 2x - 10$$
$$3y = 2x - 13$$

Copyright © 2019 Pearson Education, Inc.

26. Through (8, 0), $m = -\frac{1}{4}$

$$y - 0 = -\frac{1}{4}(x - 8)$$
$$4y = -1(x - 8)$$
$$4y = -x + 8$$

27. Through (5, –2) and (1, 3)

$$m = \frac{3-(-2)}{1-5} = \frac{5}{-4} = -\frac{5}{4}$$
$$y - 3 = -\frac{5}{4}(x - 1)$$
$$4(y - 3) = -5(x - 1)$$
$$4y - 12 = -5x + 5$$
$$4y = -5x + 17$$

28. (2, –3) and (–3, 4)

$$m = \frac{-3-4}{2-(-3)} = -\frac{7}{5}$$
$$y - (-3) = -\frac{7}{5}(x - 2)$$
$$5(y + 3) = -7(x - 2)$$
$$5y + 15 = -7x + 14$$
$$5y = -7x - 1$$

29. Undefined slope, through (–1, 4)
This is a vertical line. Its equation is $x = -1$.

30. Slope 0, (–2, 5)
This is a horizontal line. Its equation is $y = 5$.

31. *x*-intercept –3, *y*-intercept 5
Use the points (–3, 0) and (0, 5).

$$m = \frac{5-0}{0-(-3)} = \frac{5}{3}$$
$$y = \frac{5}{3}x + 5$$
$$3y = 3\left(\tfrac{5}{3}x + 5\right)$$
$$3y = 5x + 15$$

32. *x*-intercept 3, *y*-intercept 2.
Use the points (3, 0) and (0, 2).

$$m = \frac{2-0}{0-3} = -\frac{2}{3}$$
$$y = -\frac{2}{3}x + 2$$
$$3y = 3\left(-\frac{2}{3}x + 2\right)$$
$$3y = -2x + 6$$
$$2x + 3y = 6$$

The answer is (d).

33. **a.** Let (x_1, y_1) be (10, .8) and (x_2, y_2) be (15, 1.2). Find the slope.

$$m = \frac{1.2-.8}{15-10} = \frac{.4}{5} = .08$$
$$y - .8 = .08(x - 10)$$
$$y - .8 = .08x - .8$$
$$y = .08x$$

b. The slope is positive because the total owed on federal student loans is increasing.

c. The year 2017 corresponds to $x = 17$.

$$y = .08(17) = 1.36$$

If the linear trend continued, there was $1.36 trillion owed on federal student loans in 2017.

34. **a.** Let (x_1, y_1) be (11, 175) and (x_2, y_2) be (16, 337). Find the slope.

$$m = \frac{337-175}{16-11} = \frac{162}{5} = 32.4$$
$$y - 175 = 32.4(x - 11)$$
$$y - 175 = 32.4x - 356.4$$
$$y = 32.4x - 181.4$$

b. The slope is positive because the number of bunts per plate appearance is increasing.

c. The year 2017 corresponds to $x = 17$.

$$y = 32.4(17) - 181.4 = 369.4$$

If the linear trend continued, there was 1 bunt for every 369.4 plate appearances in 2017.

Copyright © 2019 Pearson Education, Inc.

35. a. Let (x_1, y_1) be $(5, 46326)$ and (x_2, y_2) be $(10, 49276)$.
Find the slope.

$$m = \frac{49,276 - 46,326}{10 - 5} = \frac{2,950}{5} = 590$$

$$y - 46,326 = 590(x - 5)$$
$$y - 46,326 = 590x - 2,950$$
$$y = 590x + 43,376$$

b. Using a graphing calculator, the least squares regression line is $y = 964.6x + 40,833$.

c. The year 2015 corresponds to $x = 15$. Using the two-point model, we have $y = 590(15) + 43,376 = 52,226$.
Using the regression model, we have $y = 964.6(15) + 40,833 = 55,302$.
The two-point model is off by $4,290, while the regression model is off by $1,214, therefore the least-squares approximation is a better estimate.

d. The year 2018 corresponds to $x = 18$.
Regression model:
$y = 964.6(18) + 40,833 \approx 58,196$
Thus, the median household income in the year 2018 will be about $58,196.

36. a. Let (x_1, y_1) be $(8, 299.6)$ and (x_2, y_2) be $(13, 334.5)$.
Find the slope.

$$m = \frac{334.5 - 299.6}{13 - 8} = \frac{34.9}{5} = 6.98$$

$$y - 299.6 = 6.98(x - 8)$$
$$y - 299.6 = 6.98x - 55.84$$
$$y = 6.98x + 243.76$$

b. Using a graphing calculator, the least squares regression line is $y = 9.52x + 213.3$.

c. The year 2014 corresponds to $x = 14$. Using the two-point model, we have $y = 6.98(14) + 243.76 \approx \$341$ billion.
Using the regression model, we have $y = 9.52(14) + 213.3 \approx \$347$ billion.
The two-point model is off by $17.4, while the regression model is off by $11.4, therefore the least-squares approximation is a better estimate.

d. The year 2019 corresponds to $x = 19$.
Regression model:
$y = 9.52(19) + 213.3 \approx 394$
Thus, the total amount in charitable giving in the year 2019 will be about $394 billion.

37. a. The least-squares regression line is $y = 93.1x + 2034.8$.

b.

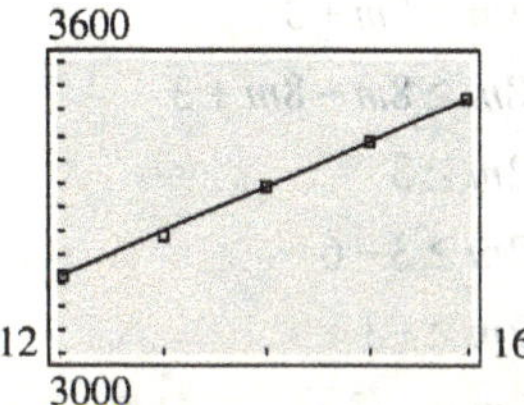

c. Yes; the line appears to fit.

d. The correlation coefficient is .9993. This indicates that the line is a good fit.

38. a. The least-squares regression line is $y = 254.3x + 5589$.

b.

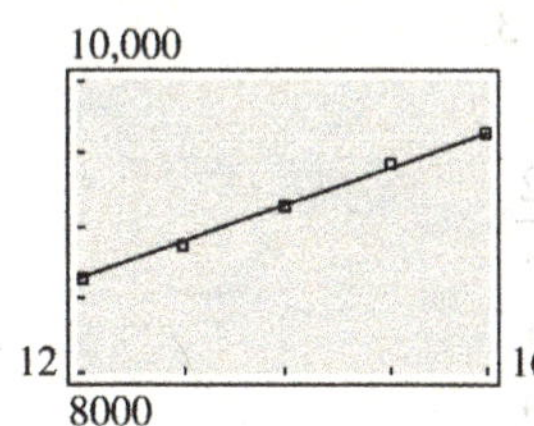

c. Yes; the line appears to fit.

d. The correlation coefficient is .9996. This indicates that the line is a good fit.

39.

$$-6x + 3 < 2x$$
$$-6x + 6x + 3 < 2x + 6x$$
$$3 < 8x$$
$$\frac{3}{8} < \frac{8x}{8}$$
$$\frac{3}{8} < x \text{ or } x > \frac{3}{8}$$

The solution is $\left(\frac{3}{8}, \infty\right)$.

Copyright © 2019 Pearson Education, Inc.

40. $12z \ge 5z - 7$

$12z - 5z \ge 5z - 5z - 7$

$7z \ge -7$

$\frac{7z}{7} \ge \frac{-7}{7}$

$z \ge -1$

The solution is $[-1, \infty)$.

41. $2(3 - 2m) \ge 8m + 3$

$6 - 4m \ge 8m + 3$

$6 - 4m - 8m \ge 8m - 8m + 3$

$6 - 12m \ge 3$

$6 - 6 - 12m \ge 3 - 6$

$-12m \ge -3$

$\frac{-12m}{-12} \le \frac{-3}{-12}$

$m \le \frac{1}{4}$

The solution is $\left(-\infty, \frac{1}{4}\right]$.

42. $6p - 5 > -(2p + 3)$

$6p - 5 > -2p - 3$

$8p - 5 > -3$

$8p > 2$

$\frac{8p}{8} > \frac{2}{8}$

$p > \frac{1}{4}$

The solution is $\left(\frac{1}{4}, \infty\right)$.

43. $-3 \le 4x - 1 \le 7$

$-2 \le 4x \le 8$

$-\frac{1}{2} \le x \le 2$

The solution is $\left[-\frac{1}{2}, 2\right]$.

44. $0 \le 3 - 2a \le 15$

$0 - 3 \le 3 - 3 - 2a \le 15 - 3$

$-3 \le -2a \le 12$

$\frac{-3}{-2} \ge \frac{-2a}{-2} \ge \frac{12}{-2}$

$\frac{3}{2} \ge a \ge -6$

The solution is $\left[-6, \frac{3}{2}\right]$.

45. $|b| \le 8 \Rightarrow -8 \le b \le 8$

The solution is $[-8, 8]$.

46. $|a| > 7 \Rightarrow a < -7$ or $a > 7$

The solution is $(-\infty, -7)$ or $(7, \infty)$.

47. $|2x - 7| \ge 3$

$2x - 7 \le -3$ or $2x - 7 \ge 3$

$2x \le 4$ or $2x \ge 10$

$x \le 2$ or $x \ge 5$

The solution is $(-\infty, 2]$ or $[5, \infty)$.

48. $|4m + 9| \le 16$

$-16 \le 4m + 9 \le 16$

$-25 \le 4m \le 7$

$-\frac{25}{4} \le m \le \frac{7}{4}$

The solution is $\left[-\frac{25}{4}, \frac{7}{4}\right]$.

49. $|5k + 2| - 3 \le 4$

$|5k + 2| \le 7$

$-7 \le 5k + 2 \le 7$

$-9 \le 5k \le 5$

$-\frac{9}{5} \le k \le 1$

The solution is $\left[-\frac{9}{5}, 1\right]$.

Copyright © 2019 Pearson Education, Inc.

50. $|3z-5|+2\geq 10$

$|3x-5|\geq 8$

$3z-5\leq -8$ or $3z-5\geq 8$

$3z\leq -3$ or $3z\geq 13$

$z\leq -1$ or $z\geq \dfrac{13}{3}$

The solution is $(-\infty, -1]$ or $\left[\dfrac{13}{3}, \infty\right)$.

51. The inequalities that represent the weight of pumpkin that he will not use are $x<2$ or $x>10$. This is equivalent to the following inequalities:

$x-6<2-6$ or $x-6>10-6$

$x-6<-4$ or $x-6>4$

$|x-6|>4$

Choose answer option (d).

52. Let x = the price of the snow thrower

$|x-600|\leq 55$

53. $45+8x\leq 65$

$8x\leq 20$

$x\leq 2.5$

Due to the fact that cell phone companies sell data by the whole gigabyte, the first plan is cheaper if you use less than or equal to 2 gigabytes.

54. Let m = number of miles driven. The rate for the second rental company is $95+.2m$. We want to determine when the second company is cheaper than the first.

$125>95+.2m\Rightarrow 30>.2m\Rightarrow 150>m$

The second company is cheaper than the first company when the number of miles driven is less than 150.

55. $r^2+r-6<0$

Solve the corresponding equation.

$r^2+r-6=0\Rightarrow (r+3)(r-2)=0\Rightarrow$

$r=-3$ or $r=2$

A B C

-4 -2 0 2 4

For region A, test −4:

$(-4)^2+(-4)-6=6>0$.

For region B, test 0:

$0^2+0-6=-6<0$.

For region C, test 3:

$3^2+3-6=6>0$.

The solution is $(-3, 2)$.

56. $y^2+4y-5\geq 0$

Solve the corresponding equation.

$y^2+4y-5=0\Rightarrow (y+5)(y-1)=0$

$y=-5$ or $y=1$

A B C

-6 -4 -2 0 2

For region A, test −6:

$(-6)^2+4(-6)-5=7>0$.

For region B, test 0:

$0^2+4(0)-5=-5<0$.

For region C, test 2:

$2^2+4(2)-5=7>0$.

Both endpoints are included because the inequality symbol is "≥." The solution is $(-\infty, -5]$ or $[1, \infty)$.

57. $2z^2+7z\geq 15$

Solve the corresponding equation.

$2z^2+7z=15$

$2z^2+7z-15=0$

$(2z-3)(z+5)=0$

$z=\dfrac{3}{2}$ or $z=-5$

These numbers are solutions of the inequality because the inequality symbol is "≥."

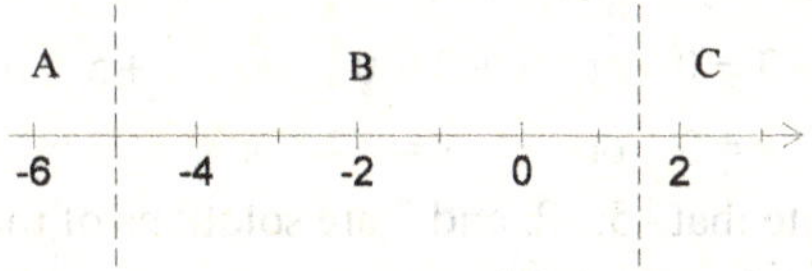

Copyright © 2019 Pearson Education, Inc.

For region A, test –6:
$2(-6)^2+7(-6)=30>15$.
For region B, test 0:
$2\cdot 0^2+7\cdot 0=0<15$.
For region C, test 2:
$2\cdot 2^2+7\cdot 2=22>15$.
The solution is $(-\infty, -5]$ or $\left[\frac{3}{2}, \infty\right)$.

58. $3k^2 \le k+14$
Solve the corresponding equation.

$$3k^2 = k+14$$
$$3k^2-k-14=0$$
$$(3k-7)(k+2)=0$$
$$k=\frac{7}{3} \text{ or } k=-2$$

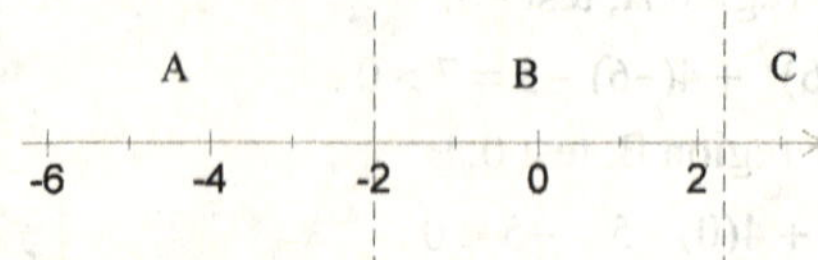

For region A, test –3:
$3(-3)^2=27,\ -3+14=11 \Rightarrow 27>11$
For region B, test 0:
$3(0)^2=0,\ 0+14=14 \Rightarrow 0<14$
For region C, test 3:
$3(3)^2=27,\ 3+14=17 \Rightarrow 27>17$
The given inequality is true in region B and at both endpoints, so the solution is $\left[-2, \frac{7}{3}\right]$.

59. $(x-3)\left(x^2+7x+10\right) \le 0$
Solve the corresponding equation.

$$(x-3)\left(x^2+7x+10\right)=0$$
$$(x-3)(x+2)(x+5)=0$$
$$x-3=0 \text{ or } x+2=0 \text{ or } x+5=0$$
$$x=3 \text{ or } x=-2 \text{ or } x=-5$$

Note that –5, –2, and 3 are solutions of the original inequality.

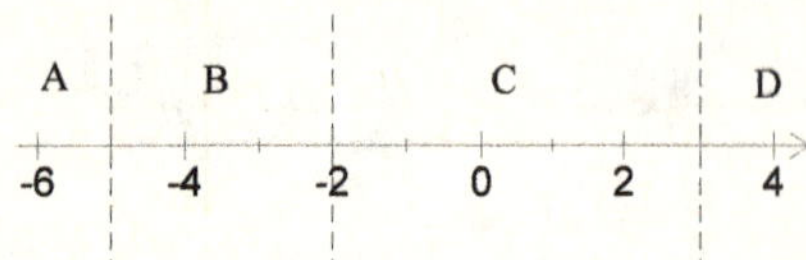

In region A, let $x=-6$:
$(-6-3)\left((-6)^2+7(-6)+10\right)$
$=-9(36-42+10)=-9(4)=-36<0$
In region B, let $x=-3$:
$(-3-3)\left((-3)^2+7(-3)+10\right)$
$=-6(9-21+10)=-6(-2)=12>0$.
In region C, let $x=0$:
$(0-3)\left(0^2+7(0)+10\right)=-3(10)=-30<0$.
In region D, let $x=4$:
$(4-3)\left(4^2+7(4)+10\right)$
$=1(16+28+10)=54>0$.
The solution is $(-\infty, -5]$ or $[-2, 3]$.

60. $(x+4)\left(x^2-1\right) \ge 0$
Solve the corresponding equation.

$$(x+4)\left(x^2-1\right)=0$$
$$(x+4)(x+1)(x-1)=0$$
$$x+4=0 \text{ or } x+1=0 \text{ or } x-1=0$$
$$x=-4 \text{ or } x=-1 \text{ or } x=1$$

Note that –4, –1, and 1 are solutions of the original inequality.

In region A, let $x=-5$:
$(-5+4)\left((-5)^2-1\right)=-1(24)=-24\le 0$.
In region B, let $x=-2$:
$(-2+4)\left((-2)^2-1\right)=2(3)=6>0$.
In region C, let $x=0$:
$(0+4)\left(0^2-1\right)=4(-1)=-4<0$.
In region D, let $x=2$:
$(2+4)\left(2^2-1\right)=6(3)=18>0$.
The solution is $[-4, -1]$ or $[1, \infty)$.

Copyright © 2019 Pearson Education, Inc.

61. $\frac{m+2}{m} \le 0$

Solve the corresponding equation $\frac{m+2}{m} = 0$.

The quotient changes sign when
$m+2=0$ or $m=0$

$m=-2$ or $m=0$

-2 is a solution of the inequality, but the inequality is undefined when $m = 0$, so the endpoint 0 must be excluded.

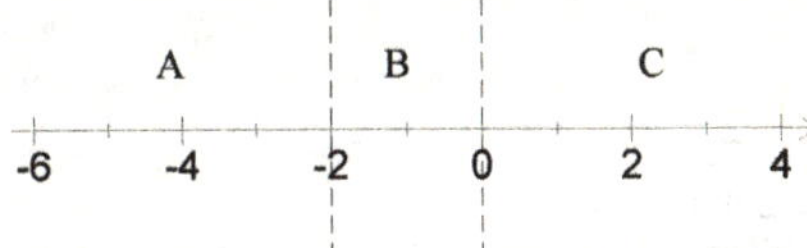

For region A, test -3:

$\frac{-3+2}{-3} = \frac{1}{3} > 0.$

For region B, test -1:

$\frac{-1+2}{-1} = -1 < 0.$

For region C, test 1:

$\frac{1+2}{1} = 3 > 0.$

The solution is $[-2, 0)$.

62. $\frac{q-4}{q+3} > 0$

Solve the corresponding equation $\frac{q-4}{q+3} = 0$.

The numerator is 0 when $q = 4$. The denominator is 0 when $q = -3$.

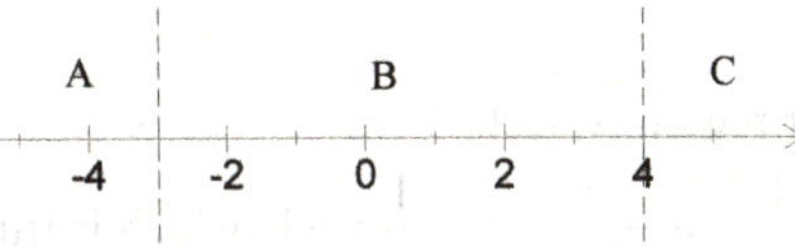

For region A, test -4:

$\frac{-4-4}{-4+3} = \frac{-8}{-1} = 8 > 0.$

For region B, test 0:

$\frac{0-4}{0+3} = -\frac{4}{3} < 0.$

For region C, test 5:

$\frac{5-4}{5+3} = \frac{1}{8} > 0.$

The inequality is true in regions A and C, and both endpoints are excluded. Therefore, the solution is $(-\infty, -3)$ or $(4, \infty)$.

63. $\frac{5}{p+1} > 2$

Solve the corresponding equation.

$$\frac{5}{p+1} = 2$$

$$\frac{5}{p+1} - 2 = 0$$

$$\frac{5-2(p+1)}{p+1} = 0$$

$$\frac{3-2p}{p+1} = 0$$

The numerator is 0 when $p = \frac{3}{2}$. The denominator is 0 when $p = -1$.
Neither of these numbers is a solution of the inequality.

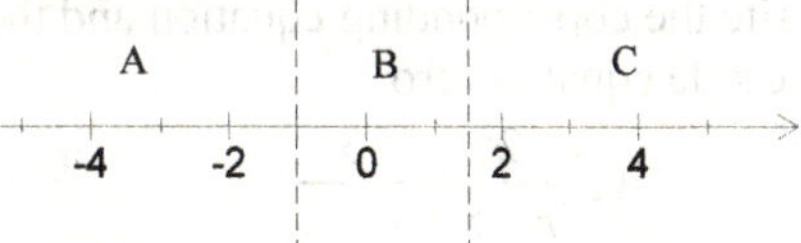

In region A, test -2:

$\frac{5}{-2+1} = -5 < 2.$

In region B, test 0:

$\frac{5}{0+1} = 5 > 2.$

In region C, test 2:

$\frac{5}{2+1} = \frac{5}{3} < 2.$

The solution is $\left(-1, \frac{3}{2}\right)$.

64. $\frac{6}{a-2} \le -3$

Solve the corresponding equation.

$$\frac{6}{a-2} = -3$$

$$\frac{6}{a-2} + 3 = 0$$

$$\frac{6+3(a-2)}{a-2} = 0$$

$$\frac{3a}{a-2} = 0$$

The numerator is 0 when $a = 0$. The denominator is 0 when $a = 2$.

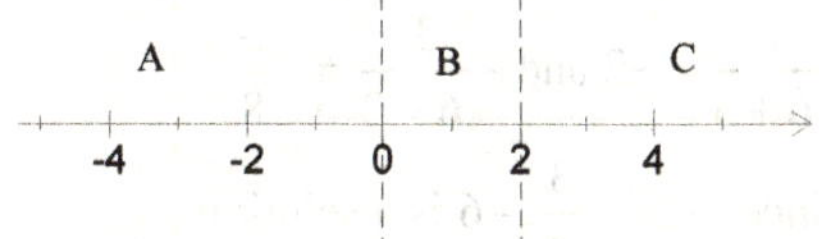

Copyright © 2019 Pearson Education, Inc.

For region A, test –1:

$\frac{6}{-1-2} = -2 \geq -3.$

For region B, test 1:

$\frac{6}{1-2} = -6 \leq -3.$

For region C, test 3:

$\frac{6}{3-2} = 6 \geq -3.$

The given inequality is true in region B. The endpoint 0 is included because the inequality symbol is "≤." However, the endpoint 2 must be excluded because it makes the denominator 0. The solution is [0, 2).

65. $\frac{2}{r+5} \leq \frac{3}{r-2}$

Write the corresponding equation and then set one side equal to zero.

$$\frac{2}{r+5} = \frac{3}{r-2}$$

$$\frac{2}{r+5} - \frac{3}{r-2} = 0$$

$$\frac{2(r-2)-3(r+5)}{(r+5)(r-2)} = 0$$

$$\frac{2r-4-3r-15}{(r+5)(r-2)} = 0$$

$$\frac{-r-19}{(r+5)(r-2)} = 0$$

The numerator is 0 when $r = -19$. The denominator is 0 when $r = -5$ or $r = 2$. –19 is a solution of the inequality, but the inequality is undefined when r = –5 or r = 2.

A B C D
-20 -16 -12 -8 -4 0 4

For region A, test –20:

$\frac{2}{-20+5} = -\frac{2}{15} \approx -.13$ and

$\frac{3}{-20-2} = -\frac{3}{22} \approx -.14$

Since –.13 > –.14, –20 is not a solution of the inequality.

For region B, test –6:

$\frac{2}{-6+5} = -2$ and $\frac{3}{-6-2} = -\frac{3}{8}.$

Since $-2 < -\frac{3}{8}$, –6 is a solution.

For region C, test 0: $\frac{2}{0+5} = \frac{2}{5}$ and $\frac{3}{0-2} = -\frac{3}{2}.$

Since $\frac{2}{5} > -\frac{3}{2}$, 0 is not a solution.

For region D, test 3: $\frac{2}{3+5} = \frac{1}{4}$ and $\frac{3}{3-2} = 3.$

Since $\frac{1}{4} < 3$, 3 is a solution. The solution is [–19, –5) or (2, ∞).

66. $\frac{1}{z-1} > \frac{2}{z+1}$

Write the corresponding equation and then set one side equal to zero.

$$\frac{1}{z-1} = \frac{2}{z+1}$$

$$\frac{1}{z-1} - \frac{2}{z+1} = 0$$

$$\frac{(z+1)-2(z-1)}{(z-1)(z+1)} = 0$$

$$\frac{3-z}{(z-1)(z+1)} = 0$$

The numerator is 0 when $z = 3$. The denominator is 0 when $z = 1$ and when $z = -1$. These three numbers, –1, 1, and 3, separate the number line into four regions.

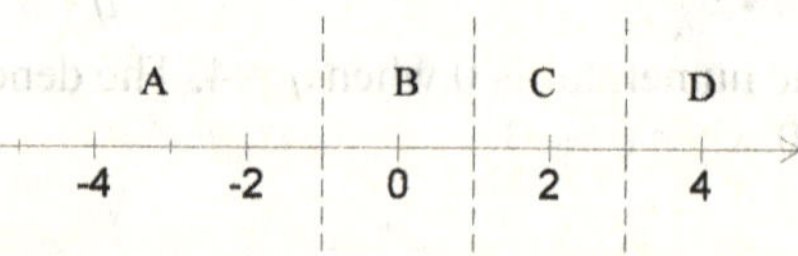

For region A, test –3.

$\frac{1}{-3-1} > \frac{2}{-3+1} \Rightarrow -\frac{1}{4} > -1$, which is true.

For region B, test 0.

$\frac{1}{0-1} > \frac{2}{0+1} \Rightarrow -1 > 2$, which is false.

For region C, test 2.

$\frac{1}{2-1} > \frac{2}{2+1} \Rightarrow 1 > \frac{2}{3}$, which is true.

For region D, test 4.

$\frac{1}{4-1} > \frac{2}{4+1} \Rightarrow \frac{1}{3} > \frac{2}{5}$, which is false.

Thus, the solution is (–∞, –1) or (1, 3).

Copyright © 2019 Pearson Education, Inc.

Case 2 Using Extrapolation for Prediction

1.

```
LinReg
 y=ax+b
 a=-.765952381
 b=82.32928571
 r²=.0399034604
 r=-.1997585053
```

This verifies the regression equation.

2.

Quarter	Table value	Predicted value	Residual
1	91.77	81.83	9.94
2	80.03	80.76	–.73
3	84.82	79.99	4.83
4	65.36	79.22	–13.86
5	67.96	78.45	–10.49
6	76.54	77.68	–1.11
7	75.81	76.91	–1.10
8	88.77	76.14	12.63

3. See the table in problem 2 for residual values.

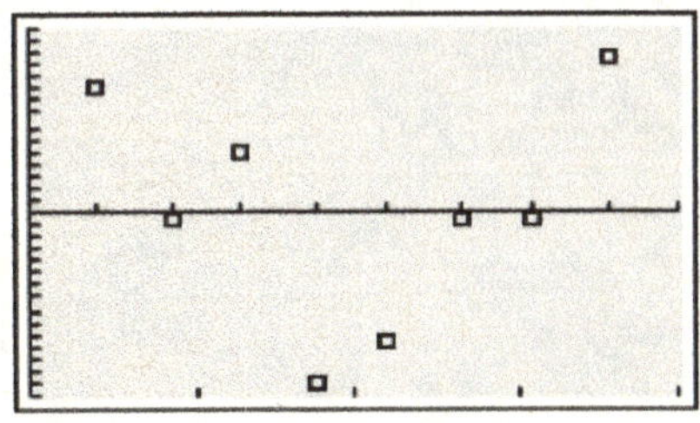

4. No; because the residuals have a nonlinear U-shape.

5. Since $x = 0$ corresponds to the year 1900, enter the following data into a computing device.

x	y
70	3.40
75	4.73
80	6.85
85	8.74
90	10.20
95	11.65
100	14.02
105	16.13
110	19.06
115	21.03

Then determine the least squares regression line.

```
LinReg
 y=ax+b
 a=.3911151515
 b=-24.59715152
 r²=.9918612299
 r=.9959223011
```

The model is verified.

6. The year 2012 corresponds to $x = 112$.
$y = .3911(112) - 24.6 \approx 19.20.$
According to the model, the hourly wage in 2012 was about \$19.20, about \$4.23 too high.

7. The year 1960 corresponds to $x = 60$.
$y = .3911(60) - 24.6 \approx -1.134.$
The model gives the hourly wage as a negative amount, which is clearly not appropriate.

Copyright © 2019 Pearson Education, Inc.

8.

Year ($x = 0$ is 1900)	Table value	Predicted value	Residual
70	3.40	2.78	.62
75	4.73	4.73	0
80	6.85	6.69	.16
85	8.74	8.64	.10
90	10.20	10.60	–.40
95	11.65	12.56	–.91
100	14.02	14.51	–.49
105	16.13	16.47	.34
110	19.06	18.42	.64
115	21.03	20.38	.65

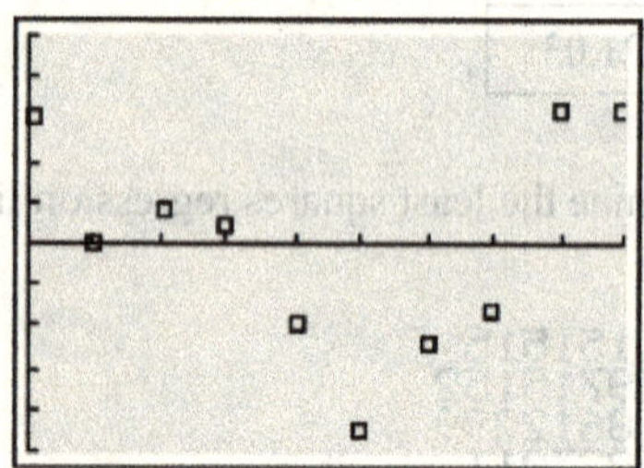

9. You'll get 0 slope and 0 intercept, because the residual represents the vertical distance from the data point to the regression line. Since r is very close to 1, the data points lie very close to the regression line.

```
LinReg
 y=ax+b
 a=.0040848485
 b=-.3068484848
 r²=.0135199307
 r=.116275237
```

Copyright © 2019 Pearson Education, Inc.

Chapter 3 Functions and Graphs

Section 3.1 Functions

1.

x	3	2	1	0	−1	−2	−3
y	9	4	1	0	1	4	9

This rule defines y as a function of x because each value of x determines one and only one value of y.

3.

x	9	4	1	0	1	4	9
y	3	2	1	0	−1	−2	−3

The rule does not define y as a function of x because some values of x determine two values for y. For example, when $x = 9$, $y = 3$ and $y = -3$.

5. The rule $y = x^3$ defines y as a function of x because each value of x determines one and only one value of y.

7. The rule $x = |y+2|$ does not define y as a function of x because some values of x determine two values for y. For example, if $x = 4$, $4 = |y+2| \Rightarrow$

$$y+2=4 \quad \text{or} \quad y+2=-4$$
$$y=2 \quad \text{or} \quad y=-6$$

9. The rule $y = \dfrac{-1}{x-1}$ defines y as a function of x because each value of x determines one and only one value of y (or none).

11. $f(x) = 4x - 1$
The domain of f is all real numbers since x may take on any real-number value. Therefore, the domain is $(-\infty, \infty)$.

13. $f(x) = x^4 - 1$
The domain of f is all real numbers since x may take on any real-number value. Therefore, the domain is $(-\infty, \infty)$.

15. $f(x) = \sqrt{-x} + 3$
In order to have $\sqrt{-x}$ be a real number, we must have $-x \geq 0$ or $x \leq 0$. Thus, the domain is all nonpositive real numbers, or $(-\infty, 0]$.

17. $g(x) = \dfrac{1}{x-2}$
Since the denominator cannot be zero, $x \neq 2$. Thus, the domain is all real numbers except 2, which in interval notation is written $(-\infty, 2)$ or $(2, \infty)$.

19. $g(x) = \dfrac{x^2+4}{x^2-4}$
Solve $x^2 - 4 = 0$ and exclude the solutions from the domain because the solutions make the denominator equal to 0.
$$x^2 - 4 = 0$$
$$(x-2)(x+2) = 0$$
$$x-2=0 \quad \text{or} \quad x+2=0$$
$$x=2 \quad \text{or} \quad x=-2$$
The domain of g is all real numbers except $x = \pm 2$ since x cannot take on $x = \pm 2$. Therefore, the domain is $(-\infty, -2)$ or $(-2, 2)$ or $(2, \infty)$.

21. $h(x) = \dfrac{\sqrt{x+4}}{x^2+x-12}$
Solve $x^2 + x - 12 = 0$ and exclude the solutions from the domain since x cannot take on these numbers.
$$x^2 + x - 12 = 0$$
$$(x-3)(x+4) = 0$$
$$x-3=0 \quad \text{or} \quad x+4=0$$
$$x=3 \quad \text{or} \quad x=-4$$
For $\sqrt{x+4}$, $x+4$ must be positive or 0. That is $x+4 \geq 0$ or $x \geq -4$. Therefore, the domain is x such that $x \neq 3$ and $x \geq -4$.

23. $f(x) = 8$
For any value of x, the value of $f(x)$ will always be 8. (This is a constant function).

a. $f(4) = 8$

b. $f(-3) = 8$

c. $f(2.7) = 8$

d. $f(-4.9) = 8$

Copyright © 2019 Pearson Education, Inc.

25. $f(x)=2x^2+4x$

a. $f(4)=2(4^2)+4(4)$
$=2(16)+16$
$=32+16=48$

b. $f(-3)=2(-3)^2+4(-3)$
$=2(9)+(-12)$
$=18-12=6$

c. $f(2.7)=2(2.7)^2+4(2.7)$
$=2(7.29)+10.8$
$=14.58+10.8=25.38$

d. $f(-4.9)=2(-4.9)^2+4(-4.9)$
$=2(24.01)+(-19.6)$
$=48.02-19.6=28.42$

27. $f(x)=\sqrt{5-x}$

a. $f(4)=\sqrt{5-4}=\sqrt{1}=1$

b. $f(-3)=\sqrt{5-(-3)}=\sqrt{8}$

c. $f(2.7)=\sqrt{5-2.7}=\sqrt{2.3}$

d. $f(-4.9)=\sqrt{5-(-4.9)}=\sqrt{9.9}$

29. $f(x)=\left|x^2-6x-4\right|$

a. $f(4)=\left|4^2-6(4)-4\right|$
$f(4)=|16-24-4|=|-12|=12$

b. $f(-3)=\left|(-3)^2-6(-3)-4\right|$
$=|9+18-4|=23$

c. $f(2.7)=\left|(2.7)^2-6(2.7)-4\right|$
$=|7.29-16.2-4|=12.91$

d. $f(-4.9)=\left|(-4.9)^2-6(-4.9)-4\right|$
$=|24.01+29.4-4|$
$=|49.41|=49.41$

31. $f(x)=\dfrac{\sqrt{x-1}}{x^2-1}$

a. $f(4)=\dfrac{\sqrt{4-1}}{4^2-1}=\dfrac{\sqrt{3}}{15}$

b. $f(-3)=\dfrac{\sqrt{-3-1}}{(-3)^2-1}=\dfrac{\sqrt{-4}}{8}$ not defined

c. $f(2.7)=\dfrac{\sqrt{2.7-1}}{2.7^2-1}=\dfrac{\sqrt{1.7}}{6.29}$

d. $f(-4.9)=\dfrac{\sqrt{-4.9-1}}{(-4.9)^2-1}$
$=\dfrac{\sqrt{-5.9}}{23.01}$ not defined

33. $f(x)=\begin{cases}-2x+4 & \text{if } x\le 1\\ 3 & \text{if } 1<x<4\\ x+1 & \text{if } x\ge 4\end{cases}$

a. $f(4)=4+1=5$

b. $f(-3)=-2(-3)+4=10$

c. $f(2.7)=3$

d. $f(-4.9)=-2(-4.9)+4=13.8$

35. $f(x)=6-x$

a. $f(p)=6-p$

b. $f(-r)=6-(-r)$
$=6+r$

c. $f(m+3)=6-(m+3)$
$=6-m-3$
$=3-m$

37. $f(x)=\sqrt{4-x}$

a. $f(p)=\sqrt{4-p}\ \ (p\le 4)$

b. $f(-r)=\sqrt{4-(-r)}$
$=\sqrt{4+r}\ \ (r\ge -4)$

c. $f(m+3)=\sqrt{4-(m+3)}$
$=\sqrt{4-m-3}$
$=\sqrt{1-m}\ \ (m\le 1)$

Copyright © 2019 Pearson Education, Inc.

39. $f(x) = x^2 - 1$

a. $f(p) = p^2 - 1$

b. $f(-r) = (-r)^2 - 1 = r^2 - 1$

c. $f(m+3) = (m+3)^2 - 1$
$= m^2 + 6m + 9 - 1$
$= m^2 + 6m + 8$

41. $f(x) = \dfrac{3}{x-1}$

a. $f(p) = \dfrac{3}{p-1}$ $(p \neq 1)$

b. $f(-r) = \dfrac{3}{-r-1}$ $(r \neq -1)$

c. $f(m+3) = \dfrac{3}{(m+3)-1} = \dfrac{3}{m+2}$ $(m \neq -2)$

43. $f(x) = 2x - 4$

$$\frac{f(x+h)-f(x)}{h} = \frac{[2(x+h)-4]-(2x-4)}{h} = \frac{2x+2h-4-2x+4}{h} = \frac{2h}{h} = 2$$

45. $f(x) = x^2 + 1$

$$\frac{f(x+h)-f(x)}{h} = \frac{(x+h)^2+1-\left(x^2+1\right)}{h} = \frac{x^2+2hx+h^2+1-x^2-1}{h} = \frac{2hx+h^2}{h} = 2x+h$$

47. $f(x) = 589x + 8858$

a. $f(13) = 589(13) + 8858 = 16{,}515$
In 2013, the GDP was about \$16,515 billion.

b. $f(15) = 589(15) + 8858 = 17{,}693$
In 2015, the GDP was about \$17,693 billion.

c. $f(16) = 589(16) + 8858 = 18{,}282$
In 2016, the GDP was about \$18,282 billion.

49. $f(x) = .8x^2 - 17x + 119$

a. $f(12) = .8(12)^2 - 17(12) + 119$
$= 30.20$
At the start of 2012, the closing price of Microsoft was \$30.20.

b. $f(14) = .8(14)^2 - 17(14) + 119$
$= 37.80$
At the start of 2014, the closing price of Microsoft was \$37.80.

c. $f(16) = .8(16)^2 - 17(16) + 119$
$= 51.80$
At the start of 2016, the closing price of Microsoft was \$51.80.

51. $T(x) =$

$$\begin{cases} .02x & \text{if } 0 \le x \le 500 \\ 10 + .04(x-500) & \text{if } 500 < x \le 3000 \\ 110 + .05(x-3000) & \text{if } x > 3000 \end{cases}$$

a. $T(100) = .02(100) = \$2.00$
A person with a taxable income of \$100 pays \$2 in state income tax.

b. $T(2500) = 10 + .04(2500 - 500) = \$90.00$
A person with a taxable income of \$2500 pays \$90 in state income tax.

c. $T(103{,}000) = 110 + .05(103{,}000 - 3000)$
$= \$5110.00$
A person with a taxable income of \$103,000 pays \$5110 in state income tax.

53. $f(x) = 21x^{.0565}$

a. $f(10) = 21(10)^{.0565} \approx 23.9$
In 2010, the mean age of a woman in the United States when het first child is born was about 23.9 years.

Copyright © 2019 Pearson Education, Inc.

b. $f(12) = 21(12)^{.0565} \approx 24.2$
In 2012, the mean age of a woman in the United States when het first child is born was about 24.2 years.

c. $f(14) = 21(14)^{.0565} \approx 24.4$
In 2014, the mean age of a woman in the United States when het first child is born was about 24.4 years

55. $f(t) = 2050 - 500t$

57. Fixed costs = \$1800
Each bag costs \$.50 to produce, and sells for \$1.20.

a. $c(x) = 1800 + .5x$

b. $r(x) = 1.2x$

c. $p(x) = r(x) - c(x)$
$p(x) = 1.2x - (1800 + .5x)$
$p(x) = .7x - 1800$

59. $f(x) = 28.2x^3 - 208x^2 + 507x + 2348$

The years 2010–2014 correspond to $x = 0$ through $x = 4$.

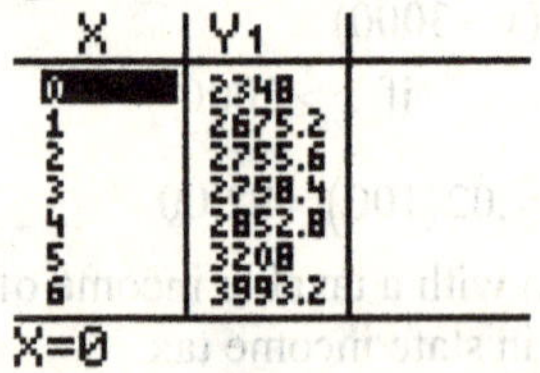

X	Y1
0	2348
1	2675.2
2	2755.6
3	2758.4
4	2852.8
5	3208
6	3993.2

X=0

Section 3.2 Graphs of Functions

1. $f(x) = -.5x + 2$
The graph is a straight line with slope −.5 and y-intercept 2.

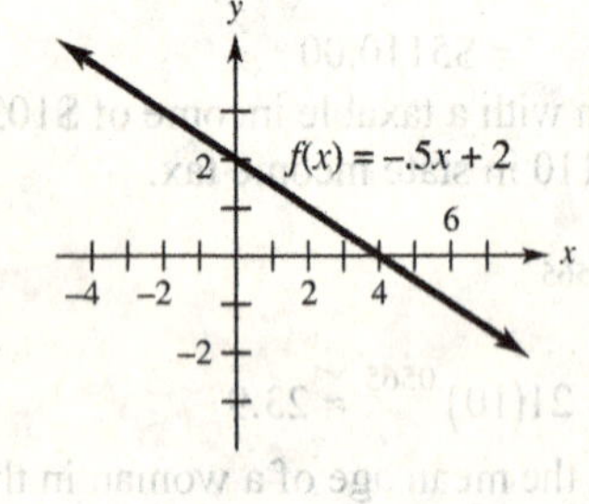

3. $f(x) = \begin{cases} x+3 & \text{if } x \le 1 \\ 4 & \text{if } x > 1 \end{cases}$

Graph the line $y = x + 3$ for $x \le 1$.
Graph the horizontal line $y = 4$ for $x > 1$.

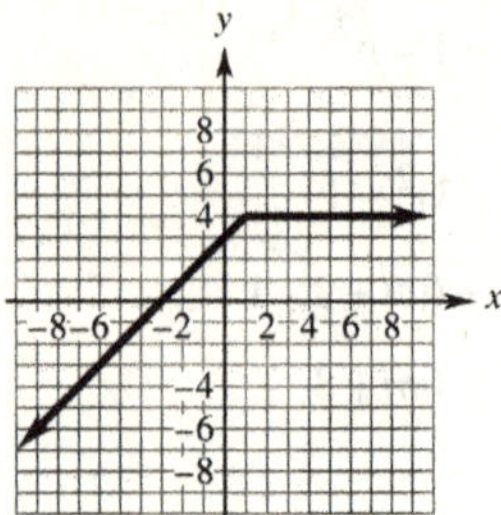

$f(x) = \begin{cases} x+3 & \text{if } x \le 1 \\ 4 & \text{if } x > 1 \end{cases}$

5. $y = \begin{cases} 4-x & \text{if } x \le 0 \\ 3x+4 & \text{if } x > 0 \end{cases}$

Graph the line $y = 4 - x$ for $x \le 0$.
Graph the line $y = 3x + 3$ for $x > 0$.

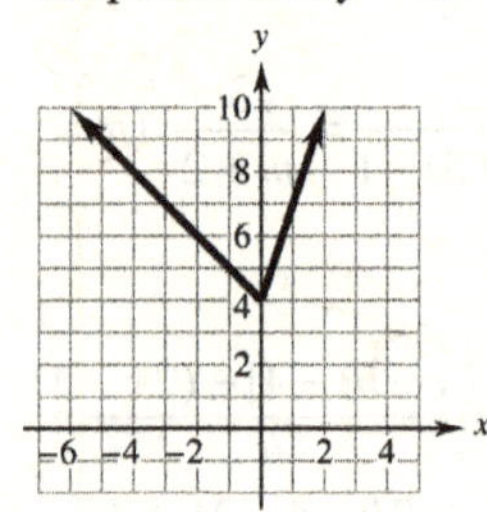

$y = \begin{cases} 4-x & \text{if } x \le 0 \\ 3x+4 & \text{if } x > 0 \end{cases}$

7. $f(x) = \begin{cases} |x| & \text{if } x < 2 \\ -2x & \text{if } x \ge 2 \end{cases}$

Rewrite the function as

$f(x) = \begin{cases} -x & \text{if } x < 0 \\ x & \text{if } 0 \le x < 2 \\ -2x & \text{if } x \ge 2 \end{cases}$

Graph the line $y = -x$ for $x \le 0$.
Graph the line $y = x$ for $0 < x \le 2$.
Graph the line $y = -2x$ for $x \ge 2$.

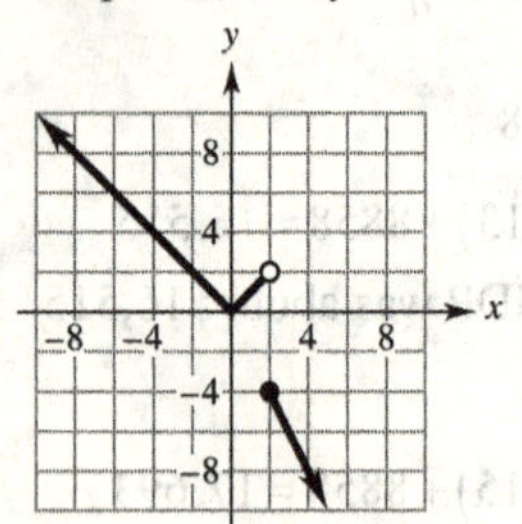

$f(x) = \begin{cases} |x| & \text{if } x < 2 \\ -2x & \text{if } x \ge 2 \end{cases}$

Copyright © 2019 Pearson Education, Inc.

9. $f(x) = |x-4|$

Using the definition of absolute values gives

$$f(x) = \begin{cases} x-4 & \text{if } x-4 \geq 0 \\ -(x-4) & \text{if } x-4 < 0 \end{cases} \text{ or}$$

$$f(x) = \begin{cases} x-4 & \text{if } x \geq 4 \\ -x+4 & \text{if } x < 4 \end{cases}$$

We graph the line $y = x - 4$ with slope 1 and y-intercept –4 for $x \geq 4$. We graph the line $y = -x + 4$ with slope –1 and y-intercept 4 for $x < 4$. Note that these partial lines meet at the point (4, 0).

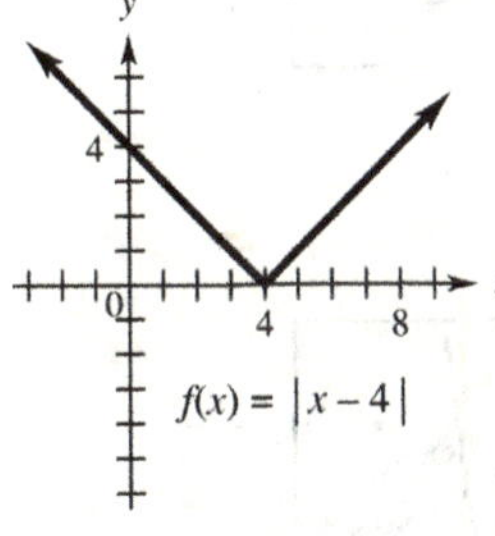

11. $f(x) = |3-3x|$

$$f(x) = \begin{cases} 3-3x & \text{if } 3-3x \geq 0 \\ -(3-3x) & \text{if } 3-3x < 0 \end{cases} \text{ or}$$

$$f(x) = \begin{cases} 3-3x & \text{if } x \leq 1 \\ -3+3x & \text{if } x > 1 \end{cases}$$

Graph the line $y = 3 - 3x$ for $x \leq 1$.
Graph the line $y = 3x - 3$ for $x > 1$.
The graph consists of two lines that meet at (1, 0).

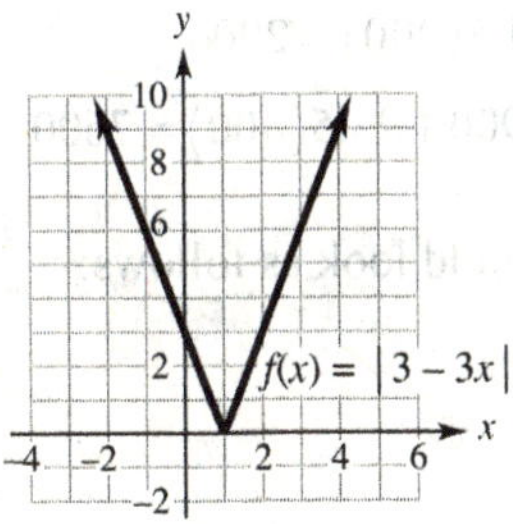

13. $y = -|x-1|$

$$y = \begin{cases} -(x-1) & \text{if } x-1 \geq 0 \\ -[-(x-1)] & \text{if } x-1 < 0 \end{cases} \text{ or}$$

$$y = \begin{cases} -x+1 & \text{if } x \geq 1 \\ x-1 & \text{if } x < 1 \end{cases}$$

Graph the line $y = -x + 1$ for $x \geq 1$.
Graph the line $y = x - 1$ for $x < 1$.
The graph consists of two lines that meet at (1, 0).

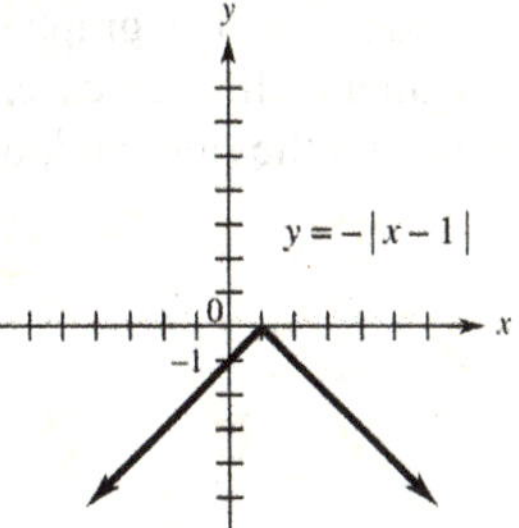

15. $y = |x-2| + 3$

$$y = \begin{cases} x+1 & \text{if } x \geq 2 \\ -x+5 & \text{if } x < 2 \end{cases}$$

Graph the line $y = x + 1$ for $x \geq 2$.
Graph the line $y = -x + 5$ for $x < 2$.
The graph consists of two lines that meet at (2, 3).

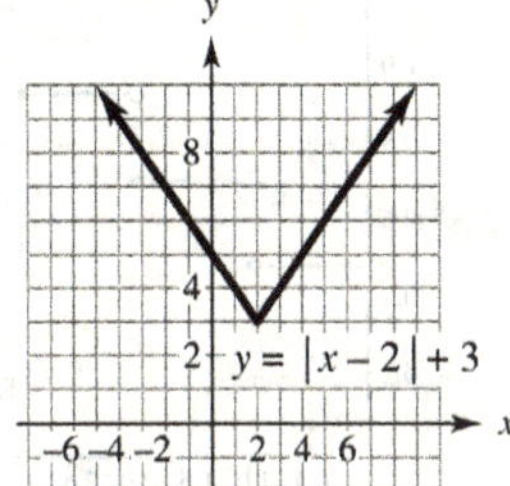

17. $f(x) = [x-3]$

For x in the interval [0, 1), the value of $[x-3] = -3$. For x in the interval [1, 2), the value of $[x-3] = -2$. For x in the interval [2, 3), the value of $[x-3] = -1$, and so on. The graph consists of a series of line segments. In each case, the left endpoint is included, and the right endpoint is excluded.

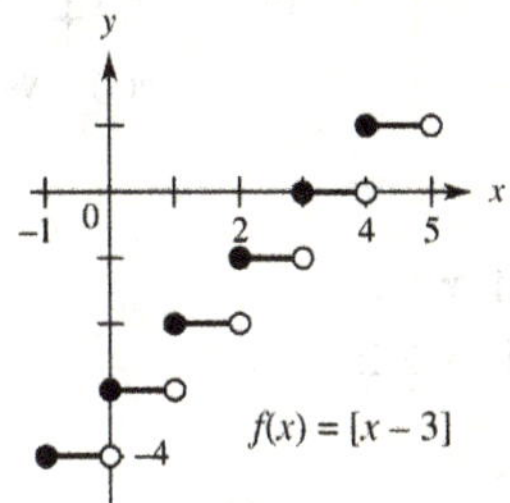

Copyright © 2019 Pearson Education, Inc.

19. $g(x) = [-x]$
For x in the interval (0, 1], the value of $[-x] = -1$. For x in the interval (1, 2], the value of $[-x] = -2$. For x in the interval (2, 3], the value of $[-x] = -3$, and so on. The graph consists of a series of line segments. In each case, the left endpoint is excluded, and the right endpoint is included.

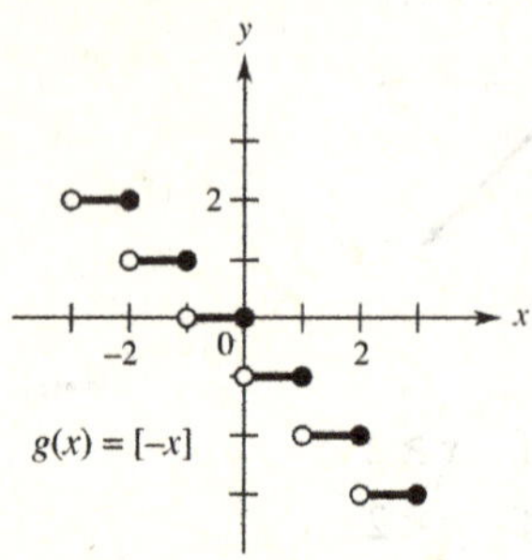

21. $f(x)$ = the price to mail a first-class letter weighing x ounces.

x	$p(x)$
(0, 1]	\$0.49
(1, 2]	\$0.71
(2, 3]	\$0.93
(3, 3.5]	\$1.15

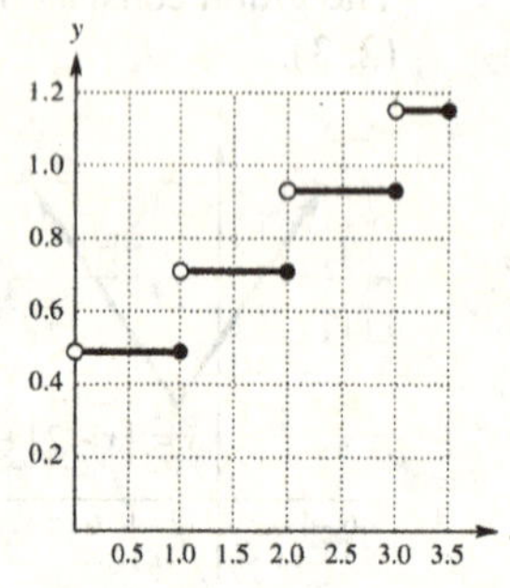

23. $g(x) = \sqrt{-x}$
Make a table of values and plot the corresponding points. The function is not defined for $x > 0$.

x	$g(x) = \sqrt{-x}$
−9	3
−4	2
−1	1
0	0

25. $f(x) = \begin{cases} x^2 & \text{if } x < 2 \\ -2x + 2 & \text{if } x \geq 2 \end{cases}$

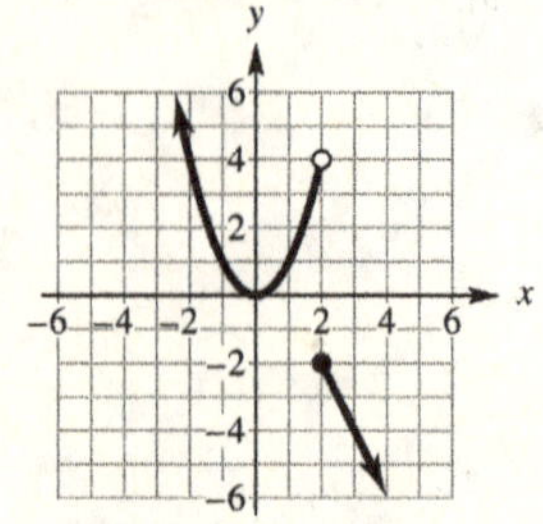

27. Every vertical line intersects this graph in at most one point, so this is the graph of a function.

29. A vertical line intersects the graph in more than one point, so this is not the graph of a function.

31. Every vertical line intersects this graph in at most one point, so this is the graph of a function.

33.

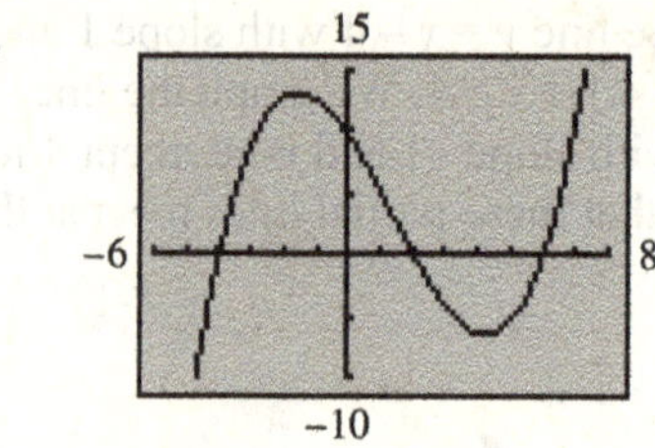

35.

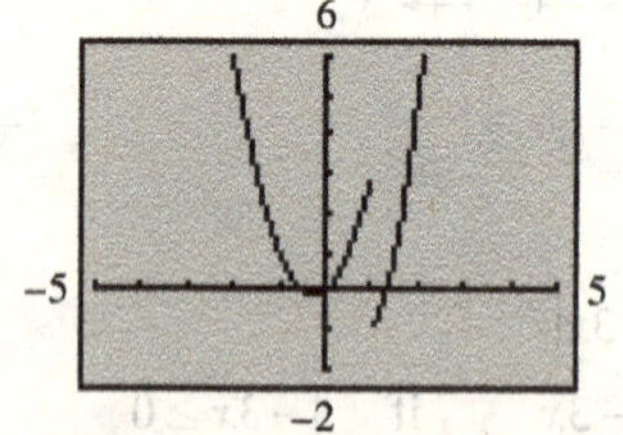

The endpoints are (1, −1), which is on the graph, and (1, 3), which is not on the graph.

37. $f(x) = \begin{cases} 14.5x & \text{if } 0 \leq x \leq 200 \\ 2900 + 23.5(x - 200) & \text{if } x > 200 \end{cases}$

a. $f(0) = 14.5(0) = 0$

$f(200) = 14.5(200) = 2900$

$f(400) = 2900 + 23.5(400) = 7600$

b. The graph would look as follows:

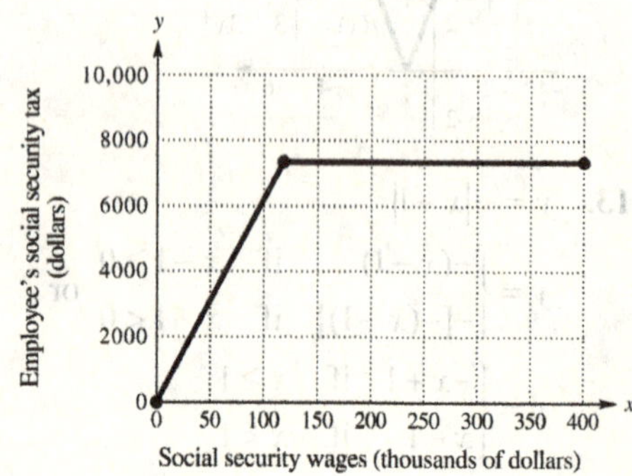

Copyright © 2019 Pearson Education, Inc.

39. The graph would look as follows:

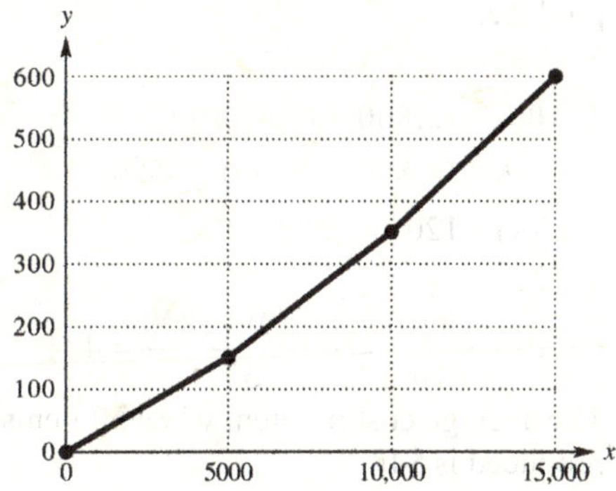

41. a. According to thegraph, the adult smoking rate in 2015 was approximately 15%.

b. $m = \dfrac{15-21}{15-5} = \dfrac{-6}{10} = -.6$

$y - 21 = -.6(x-5)$

$y - 21 = -.6x + 3$

$y = -.6x + 24$

c. For the year 2000, $x = 0$,

$y = -.6(0) + 24 = 24$

For the year 2017, $x = 17$,

$y = -.6(17) + 24 = 13.8$

The adult smoking rates in 2000 and 2017 were 24% and 13.8% respectively.

d. Let $y = 12$, $12 = -.6x + 24$

$-24 = -.6x$

$20 = x$

If the trend continues, the goal will be met in the year 2020.

43. a. The largest change in CPI for transportation occurred in 2011.

b. The largest change in CPI for apparel occurred in 2012.

c. The CPI for all items were approximately zero in the years 2009 and 2015.

45. a. In 2015, shareholders' equity in the corporation was about 290 – 170 = 120, or about $120 billion.

b. The domain represents the years from 2011 to 2015 or [2011, 2015].

c. An estimate of the range for liability is about [40, 170]

47. a. The function values in the years 2010 and 2015 were 44 cents and 49 cents respectively.

b. The figure has vertical line segments, which can't be part of the graph of a function. To make the figure into the graph of *f*, delete the vertical line segments, then for each horizontal segment of the graph, put a closed dot on the left end and an open-circle dot on the right end (or vice versa, depending on when the cost change occurred).

49. First convert 75 minutes to 1.25 hours, and 15 minutes to .25 hours. Let *x* represent the number of hours to rent the van. To calculate $C(x)$, subtract 1.25 from *x*, then divide this number by .25 and round to the next integer. Finally, multiply this result by 5 and add $19.99.

a. 2 hours

$\dfrac{2-1.25}{.25} = 3$

$5(3) + 19.99 = 34.99$

It costs $34.99 to rent a van for 2 hours.

b. 1.5 hours

$\dfrac{1.5-1.25}{.25} = 1$

$5(1) + 19.99 = 24.99$

It costs $24.99 to rent a van for 1.5 hours.

c. 3.5 hours

$\dfrac{3.5-1.25}{.25} = 9$

$5(9) + 19.99 = 64.99$

It costs $64.99 to rent a van for 3.5 hours.

d. 4 hours

$\dfrac{4-1.25}{.25} = 11$

$5(11) + 19.99 = 74.99$

It costs $74.99 to rent a van for 4 hours.

Copyright © 2019 Pearson Education, Inc.

e.

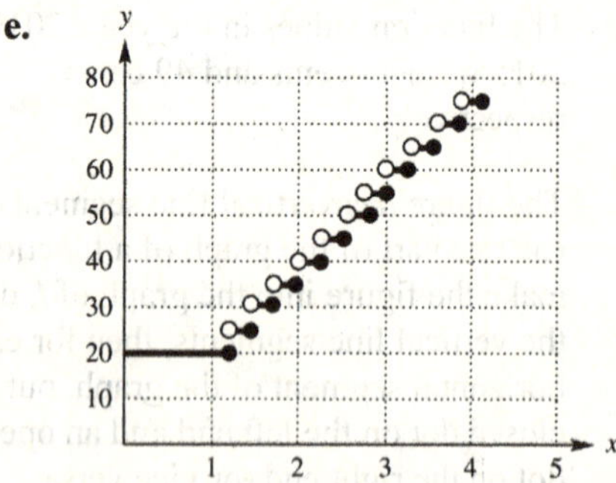

Note that for each step, the left endpoint is not included and the right point is included.

51. Make table of values:

Time	t	$g(t)$	Slope
midnight	0	1,000,000	increasing
noon	12	?	decreasing
4:00 P.M.	16	?	increasing
9:00 P.M.	21	?	vertical

There are many correct answers, including the following:

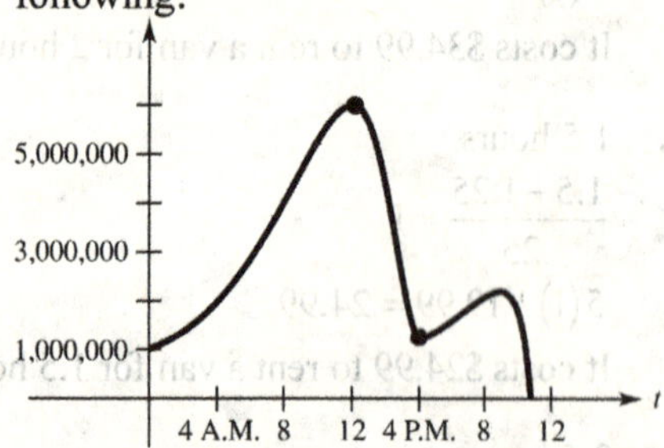

Section 3.3 Applications of Linear Functions

1. The marginal cost is \$5, while the fixed cost is \$25. Let $C(x)$ be the cost of renting a saw for x hours. Then $C(x) = 5x + 25$.

3. Let x = the number of half-hours and $C(x)$ = the total cost in dollar for x hours. Then $C(x) = 2.5x + 8$

5. Fixed cost, \$200, 50 items cost \$2000 to produce. Since the fixed cost is \$200,

$C(x) = mx + 200$

$C(50) = 2000.$

Therefore,

$2000 = m(50) + 200 \Rightarrow 50m = 1800 \Rightarrow m = 36$

$C(x) = 36x + 200$

7. Marginal cost, \$120; 100 items cost \$15,800 to produce.

$C(x) = mx + b,\ m = 120$

$C(100) = 15,800 = 120(100) + b$

$b = 15,800 - 12,000 = 3800$

$C(x) = 120x + 3800$

9. $\overline{C}(50) = \dfrac{C(50)}{50} = \dfrac{12(50) + 1800}{50} = 48$

The average cost per item when 50 items are produced is \$48.

$\overline{C}(500) = \dfrac{C(500)}{500} = \dfrac{12(500) + 1800}{500} = 15.60$

The average cost per item when 500 items are produced is \$15.60.

$\overline{C}(1000) = \dfrac{C(1000)}{1000} = \dfrac{12(1000) + 1800}{1000} = 13.80$

The average cost per item when 1000 items are produced is \$13.80.

11. $\overline{C}(200) = \dfrac{C(200)}{200} = \dfrac{6.5(200) + 9800}{200} = 55.50$

The average cost per item when 200 items are produced is \$55.50.

$\overline{C}(2000) = \dfrac{C(2000)}{2000} = \dfrac{6.5(2000) + 9800}{2000} = \$11.40$

The average cost per item when 2000 items are produced is \$11.40.

$\overline{C}(5000) = \dfrac{C(5000)}{5000} = \dfrac{6.5(5000) + 9800}{5000} = \$8.46$

The average cost per item when 5000 items are produced is \$8.46.

13. a. Let $(x_1, y_1) = (0, 25035)$ and let $(x_2, y_2) = (4, 14115)$.

$m = \dfrac{25,035 - 14,115}{0 - 4} = -2730$

The y-intercept is 25,035, so the equation is $f(x) = -2730x + 25,035$.

b. $f(5) = -2730(5) + 25,035 = 11,385$

The car will be worthe \$11,385.

c. Because the slope is -2730, the car is depreciating at a rate of \$2730 per year.

Copyright © 2019 Pearson Education, Inc.

15. a. Let $(x_1, y_1) = (0,\ 120{,}000)$ and $(x_2, y_2) = (8,\ 25{,}000)$.

$$m = \frac{25{,}000 - 120{,}000}{8 - 0} = -11{,}875$$

$$y - y_1 = m(x - x_1)$$
$$y - 120{,}000 = -11{,}875(x - 0)$$
$$y - 120{,}000 = -11{,}875x$$
$$y = -11{,}875x + 120{,}000$$
$$f(x) = -11{,}875x + 120{,}000$$

b. The domain ranges from 0 to 8 years, that is [0, 8].

c. $f(6) = -11{,}875(6) + 120{,}000 = 48{,}750$
The machine will be worth $48,750 in 6 years.

17. a. The fixed costs are
$C(0) = \$42.5(0) + 80{,}000 = \$80{,}000$

b. The slope of $C(x) = 42.5x + 80{,}000$ is 42.5, so the marginal cost is $42.50.

c. The cost of producing 1000 books is
$C(1000) = 42.5(1000) + 80{,}000 = \$122{,}500.$
The cost of producing 32,000 books is
$$C(32{,}000) = 42.5(32{,}000) + 80{,}000 = \$1{,}440{,}000.$$

d. The average cost per book when 1000 are produced is:
$$\overline{C}(1000) = \frac{C(1000)}{1000} = \frac{42.5(1000) + 80{,}000}{1000} = \$122.50$$
The average cost per book when 32,000 are produced is,
$$\overline{C}(32{,}000) = \frac{C(32{,}000)}{32{,}000} = \frac{42.5(32{,}000) + 80{,}000}{32{,}000} = \$45$$

19. a. Let (x_1, y_1) be (50, 58) and (x_2, y_2) be (250, 142).
Find the slope.
$$m = \frac{142 - 58}{250 - 50} = \frac{84}{200} = .42$$
Use the point-slope form with (50, 58).
$$y - 58 = .42(x - 50)$$
$$y = .42x + 37$$
$$C(x) = .42x + 37$$

b. The fixed costs are the constant added to the cost function or $37.

c. The total cost of producing 1000 tacos is
$C(1000) = .42(1000) + 37 = \$457.$

d. The total cost of producing 1001 tacos is
$C(1001) = .42(1001) + 37 = \$457.42.$

e. The marginal cost of the 1001st cup is
$C(1001) - C(1000) = 457.42 - 457 = \$.42,$
which is about 42¢.

f. The slope of $C(x) = .42x + 37$ is .42, so the marginal cost is $.42 or 42 cents.

21. Each customer pays $5.12 + 1.87x$ per month, where x is the number of thousands of gallons of water used. Then,
$$R(x) = 550{,}000(5.12) + 1.87x = 2{,}816{,}000 + 1.87x.$$

23. a. $C(x) = 10x + 750$

b. $R(x) = 35x$

c. $P(x) = R(x) - C(x) = 35x - (10x + 750) = 25x - 750$

d. The profit on 100 items is,
$P(100) = 25(100) - 750 = \$1750$

25. a. $C(x) = 18x + 300$

b. $R(x) = 28x$

c. $P(x) = R(x) - C(x) = 28x - (18x + 300) = 10x - 300$

d. The profit on 100 items is,
$P(100) = 10(100) - 300 = \$700$

Copyright © 2019 Pearson Education, Inc.

27. Assume that p represents the unit sales price for each of the years 2016 and 2017. Then $R(x) = px$ is the total revenue from the sale of x items. Assume that $C_0(x) = m_0x + b_0$ and $C_1(x) = m_1x + b_1$ give the total cost of x items in 2012 and 2013, respectively. The unit variable costs are m_0 and m_1 for 2016 and 2017, respectively, while b_0 and b_1 are the respective total fixed costs. Then the profit-volume functions for 2016 and 2017 are respectively
$P_0(x) = R(x) - C_0(x)$
$= px - (m_0x + b_0) = x(p - m_0) - b_0,$
and $P_1(x) = x(p - m_1) - b_1$.
From the graphs we see that the y-intercept of P_1 is greater than that of P_0 so that $-b_1 > -b_0$, or $b_0 > b_1$. Therefore, the total fixed costs have decreased from 2016 to 2017. Since from the graph we see that the slope of P_1 is less than that of P_0, it follows that $p - m_1 < p - m_0$, which means that $-m_1 < -m_0$, or $m_1 > m_0$. Therefore, the unit variable costs have increased from 2016 to 2017. The answer is (a).

29. $2x - y = 7$ and $y = 8 - 3x$
Solve the first equation for y:
$2x - y = 7 \Rightarrow y = 2x - 7$.
Set the two equations equal and solve for x.
$2x - 7 = 8 - 3x \Rightarrow 5x = 15 \Rightarrow x = 3$
Substitute x into one equation to find y.
$y = 8 - 3(3) = -1$
The lines intersect at $(3, -1)$.

31. $y = 3x - 7$ and $y = 7x + 4$
Set the two equations equal and solve for x.
$3x - 7 = 7x + 4 \Rightarrow -4x = 11 \Rightarrow x = -\frac{11}{4}$
Substitute x into one equation to find y.
$y = 3\left(-\frac{11}{4}\right) - 7 = -\frac{33}{4} - \frac{28}{4} = -\frac{61}{4}$
The lines intersect at $\left(-\frac{11}{4}, -\frac{61}{4}\right)$.

33. a. The break-even point occurs when $R(x) = C(x)$.
$125x = 100x + 5000 \Rightarrow 25x = 5000 \Rightarrow$
$x = 200$
The break-even point is $x = 200$ or 200,000 policies.

b. To graph the revenue function, graph $y = 125x$. If $x = 0$, $y = 0$. If $x = 20$, $y = 2500$. Use the points (0, 0) and (20, 2500) to graph the line. To graph the cost function, graph $y = 100x + 5000$. If $x = 0$, $y = 5000$. If $x = 30$, $y = 8000$. Use the points (0, 5000) and (30, 8000) to graph the line.

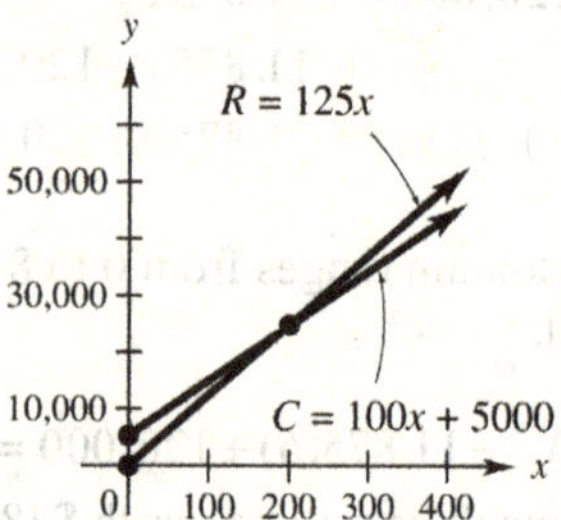

c. From the graph, when $x = 100$, cost is \$15,000 and revenue is \$12,500.

35. Given $R(x) = .21x$ and $P(x) = .084x - 1.5$

a. Cost equals revenue – profit, so
$C(x) = R(x) - P(x) = .21x - (.084x - 1.5)$
$= .21x - .084x + 1.5 = .126x + 1.5$

b. $C(7) = .126(7) + 1.5 = 2.382$
The cost of producing 7 units is \$2.382 million.

c. At the break-even point, profit $P(x) = 0$, so $C(x) = R(x)$
$.126x + 1.5 = .21x \Rightarrow 1.5 = .084x \Rightarrow$
$x \approx 17.857$
The break-even point occurs at about 17.857 units.

37. $C(x) = 80x + 7000$; $R(x) = 95x$
$95x = 80x + 7000 \Rightarrow 15x = 7000 \Rightarrow x \approx 467$
The break-even point is about 467 units. Do not produce the item since $467 > 400$.

39. $C(x) = 125x + 42{,}000$; $R(x) = 165.5x$
$125x + 42{,}000 = 165.5x \Rightarrow 42{,}000 = 40.5x \Rightarrow$
$x \approx 1037$
The break-even point is about 1037 units. Produce the item since $1037 < 2000$.

41. The percentage of female workers in both countires was the same in 2009 at about 51%.

Copyright © 2019 Pearson Education, Inc.

43. a.

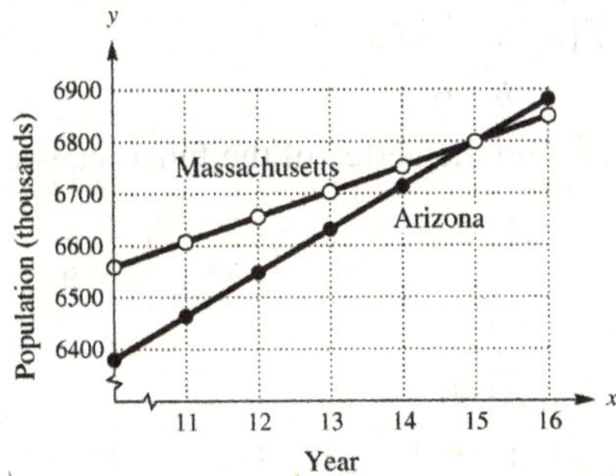

b. Yes, the graphs do intersect in this window.

c. The population of these 2 states were the same in the year 2015.

45. On the supply curve, when $q = 20, p = 140$. The point (20, 140) is on the graph. When 20 items are supplied, the price is \$140.

47. The two curves intersect at the point (10, 120). The equilibrium supply and equilibrium demand are both $q = 10$ or 10 items.

49. $p = 16 - \frac{5}{4}q$

a. If $q = 0, p = 16$, so for a demand of 0 units, the price is \$16.

b. If $q = 4$, $p = 16 - \frac{5}{4}(4) = 16 - 5 = 11$, so for a demand of 4 units, the price is \$11.

c. If $q = 8$, $p = 16 - \frac{5}{4}(8) = 16 - 10 = 6$, so for a demand of 8 units, the price is \$6.

d. From (c), if $p = 6, q = 8$, so at a price of \$6, the demand is 8 units.

e. From (b), if $p = 11, q = 4$, so at a price of \$11, the demand is 4 units.

f. From (a), if $p = 16, q = 0$, so at a price of \$16, the demand is 0 units.

g.

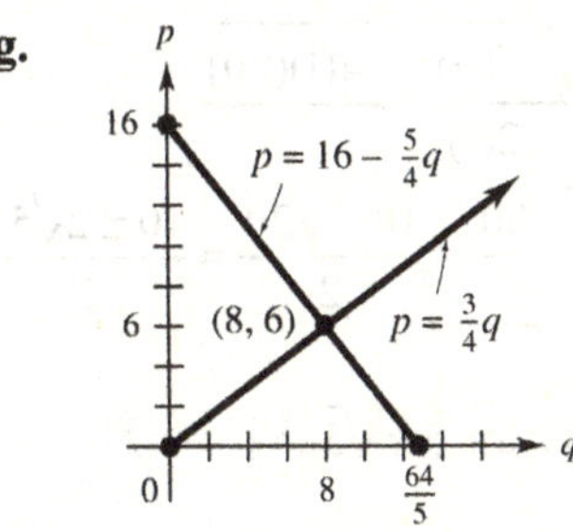

h. $p = \frac{3}{4}q$

If $p = 0, q = 0$, so at a price of \$0, the supply is 0 units.

i. If $p = 10$,

$10 = \frac{3}{4}q \Rightarrow \frac{4}{3}(10) = \frac{4}{3}\left(\frac{3}{4}\right)q \Rightarrow \frac{40}{3} = q$

When the price is \$10, the supply is $\frac{40}{3}$ units.

j. If $p = 20$,

$20 = \frac{3}{4}q \Rightarrow \frac{4}{3}(20) = \frac{4}{3}\left(\frac{3}{4}q\right) \Rightarrow \frac{80}{3} = q$

When the price is \$20, the supply is $\frac{80}{3}$ units.

k. See (g).

l. The two graphs intersect at the point (8, 6). The equilibrium supply is 8 units.

m. The equilibrium price is \$6.

51. a.

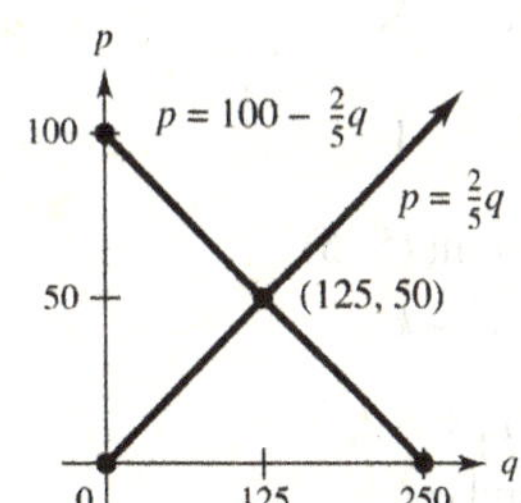

b. The two graphs intersect at the point (125, 50). The equilibrium demand is 125 units.

c. The equilibrium price is 50 cents.

d. $40 = \frac{2}{5}q \Rightarrow q = \frac{5}{2}(40) \Rightarrow$

$q = 100$

Since the value is less than the equilibrium demand value, there will be a shortage of supply.

53. Total cost increases when more items are made (because it includes the cost of all previously made items), so the graph cannot move downward. No; the average cost can decrease as more items are made, so its graph may move downward.

Copyright © 2019 Pearson Education, Inc.

Section 3.4 Quadratic Functions and Applications

1. $f(x) = x^2 - 3x - 12$
This function is written in the form $y = ax^2 + bx + c$ with a = 1, b = –3 and the parabola opens upward since $a = 1 > 0$.

3. $h(x) = -3x^2 + 14x + 1$
The parabola opens downward since $a = -3 < 0$.

5. $f(x) = 2(x-5)^2 + 7$
This function is written in the form $y = a(x-h)^2 + k$ with $a = 2$, $h = 5$, and $k = 7$. The vertex of the parabola, (h, k), is (5, 7). The parabola opens upward since $a = 2 > 0$.

7. $h(x) = -4(x+1)^2 - 9$
This function is written in the form $y = a(x-h)^2 + k$ with $a = -4$, $h = -1$, and $k = -9$. The vertex of the parabola, (h, k), is (–1, –9). The parabola opens downward since $a = -4 < 0$.

9. I **11.** K

13. J **15.** F

17. vertex (1, 2); point (5, 6)
$f(x) = a(x-h)^2 + k$
$f(x) = a(x-1)^2 + 2$
Use (5, 6) to find a.
$6 = a(5-1)^2 + 2 \Rightarrow 4 = 16a \Rightarrow a = \frac{1}{4}$
$f(x) = \frac{1}{4}(x-1)^2 + 2$

19. vertex (–1, –2); point (1, 2)
$f(x) = a(x-h)^2 + k$
$f(x) = a(x+1)^2 - 2$
Use (1, 2) to find a.
$2 = a(1+1)^2 - 2 \Rightarrow 4 = 4a \Rightarrow a = 1$
$f(x) = (x+1)^2 - 2$

21. $f(x) = x^2 + 6x - 3$
$a = 1, b = 6$
To find the vertex of the function, use the vertex formula $x = -\frac{b}{2a}$ and $y = f\left(-\frac{b}{2a}\right)$.
$x = -\frac{6}{2(1)} = -3$
$f(-3) = (-3)^2 + 6(-3) - 3 = -12$
The vertex is (–3, –12).

23. $f(x) = 3x^2 - 12x + 5$
$a = 3, b = -12$
To find the vertex of the function, use the vertex formula $x = -\frac{b}{2a}$ and $y = f\left(-\frac{b}{2a}\right)$.
$x = -\frac{-12}{2(3)} = 2$
$f(2) = 3(2)^2 - 12(2) + 5 = -7$
The vertex is (2, –7).

25. $f(x) = 3(x-2)^2 - 3$
To find the x-intercepts, set $f(x) = 0$ and then solve for x.
$0 = 3(x-2)^2 - 3 \Rightarrow 3(x-2)^2 = 3 \Rightarrow$
$(x-2)^2 = 1 \Rightarrow x - 2 = \pm 1 \Rightarrow x = 1$ or $x = 3$.
The x-intercepts are (1, 0) and (3, 0).
Let $x = 0$ to find the y-intercept.
$f(0) = 3(0-2)^2 - 3 = 9$
The y-intercept is 9.

27. $g(x) = x^2 - 10x + 20$
Set each variable equal to 0.
If $x = 0$,
$g(0) = 0^2 - 10(0) + 20 = 20$.
The y-intercept is 20.
If $y = 0$, $g(x) = y = 0$, so $0 = x^2 - 10x + 20$.
Use the quadratic formula.

$$x = \frac{-(-10) \pm \sqrt{(-10)^2 - 4(1)(20)}}{2(1)}$$

$$= \frac{10 \pm \sqrt{100 - 80}}{2} = \frac{10 \pm \sqrt{20}}{2} = \frac{10 \pm 2\sqrt{5}}{2}$$

$$= 5 \pm \sqrt{5}$$

The x-intercepts are $5 \pm \sqrt{5}$.

Copyright © 2019 Pearson Education, Inc.

29. $f(x)=(x+2)^2 \Rightarrow y=1(x+2)^2+0$
The vertex is (–2, 0). The axis is $x=-2$.

x	–1	0	1
y	1	4	9

Plot these points and use the axis of symmetry to find corresponding points on the other side of the axis. Connect the points with a smooth curve.

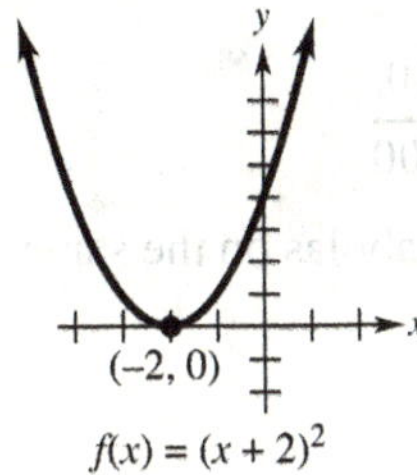

$f(x)=(x+2)^2$

In exercises 31–32, we complete the square to write the equation in the form $f(x)=a(x-h)^2+k$. Alternatively, we can use the vertex formula, $h=-\frac{b}{2a}$, $k=f(h)$, to find the vertex.

31. $f(x)=x^2-4x+6$
$=(x^2-4x+4)+(6-4)$
$=(x-2)^2+2$
The vertex is (2, 2). The axis is $x=2$.

x	3	4	5
y	3	6	11

Use the axis of symmetry to find corresponding points on the other side of the axis.

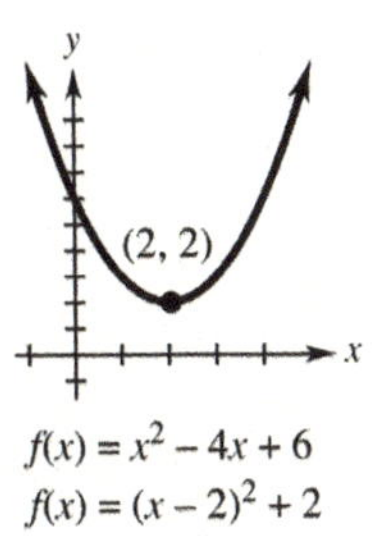

$f(x)=x^2-4x+6$
$f(x)=(x-2)^2+2$

33. **a.** To find the vertex of the function, use the vertex formula $x=-\frac{b}{2a}$ and $y=f\left(-\frac{b}{2a}\right)$.

$x=-\frac{13}{2(-.5)}=13$

$f(13)=-.5(13)^2+13(13)-57=27.5$
The vertex is (13, 27.5).

b. The revenue was a maximum in the year 2013 because the x coordinate of the vertex represents the year after 2000.

c. The maximum revenue is about \$28 billion because the y coordinate of the vertex represents revenue.

35. **a.** $y=-x^2+20x-60=-(x^2-20x)-60$
$=-(x^2-20x+100)-60+100$
$=-(x-10)^2+40$
The graph of this parabola opens downward, so the maximum occurs at the vertex, (10, 40). The maximum firing rate will be reached in 10 milliseconds.

b. When $x=10$, $y=40$, so the maximum firing rate is 40 responses/millisec.

37. $f(x)=.0328x^2-3.55x+115$
Let x be the age of the driver. Because the graph is a parabola which opens upward, the rate is lowest at the vertex. We must find the x-coordinate of the vertex.
$a=.0328$, $b=-3.55$
$x=-\frac{b}{2a}=-\frac{-3.55}{2(.0328)}\approx 54.12$
The rate is lowest at about age 54.

39. Given supply: $p=\frac{1}{5}q^2$, and

demand: $p=-\frac{1}{5}q^2+40$.

a. $10=-\frac{1}{5}q^2+40 \Rightarrow -30=-\frac{1}{5}q^2 \Rightarrow$
$150=q^2 \Rightarrow q\approx 12$
About 12 books are demanded.

Copyright © 2019 Pearson Education, Inc.

b. $20 = -\frac{1}{5}q^2 + 40 \Rightarrow -20 = -\frac{1}{5}q^2 \Rightarrow$
$100 = q^2 \Rightarrow q = 10$
Ten books are demanded.

c. $30 = -\frac{1}{5}q^2 + 40 \Rightarrow -10 = -\frac{1}{5}q^2 \Rightarrow$
$50 = q^2 \Rightarrow q \approx 7$
About 7 books are demanded.

d. $40 = -\frac{1}{5}q^2 + 40 \Rightarrow 0 = -\frac{1}{5}q^2 \Rightarrow q = 0$
No books are demanded.

e. $5 = \frac{1}{5}q^2 \Rightarrow 25 = q^2 \Rightarrow 5 = q$
5 books are supplied.

f. $10 = \frac{1}{5}q^2 \Rightarrow 50 = q^2 \Rightarrow 7 \approx q$
About 7 books are supplied.

g. $20 = \frac{1}{5}q^2 \Rightarrow 100 = q^2 \Rightarrow 10 = q$
10 books are supplied.

h. $30 = \frac{1}{5}q^2 \Rightarrow 150 = q^2 \Rightarrow 12 \approx q$
About 12 books are supplied.

i. Using the values from the tables, plot points and graph the functions.

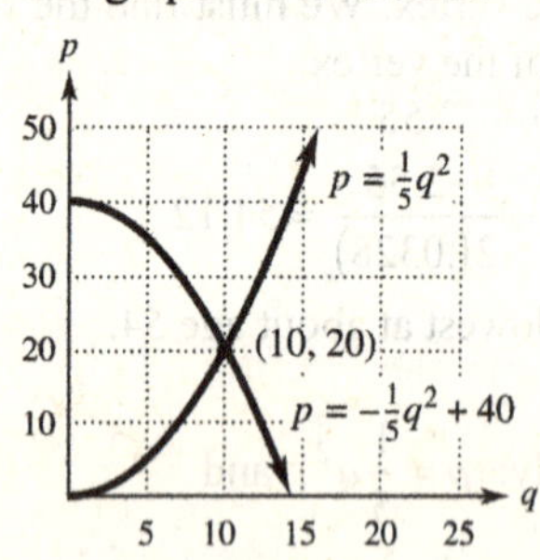

41. a. $p = 640 - 5(0)^2 = 640$
When 0 widgets are demanded, the price is \$640.

b. $p = 640 - 5(5)^2 = 515$
When 5 hundred widgets are demanded, the price is \$515.

c. $p = 640 - 5(10)^2 = 140$
When 10 hundred widgets are demanded, the price is \$140.

d. $p = 640 - 5q^2$

q	0	4	8	10
p	640	560	320	140

$p = 5q^2$

q	0	4	8	10
p	0	80	320	500

Graph these two parabolas on the same axes.

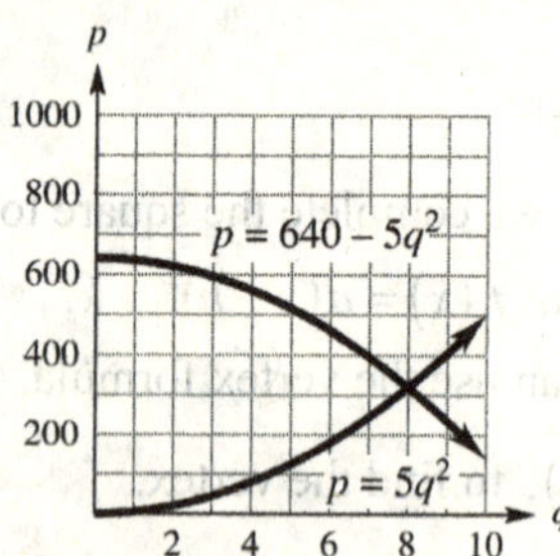

e. supply = demand $\Rightarrow$
$5q^2 = 640 - 5q^2 \Rightarrow 10q^2 = 640 \Rightarrow$
$q^2 = 640 \Rightarrow q = \pm 8$
A negative value of q is not meaningful. The equilibrium supply is 8 hundreds or 800 units.

f. If $q = 8,\ p = 5(8)^2 = 320$.
The equilibrium cost is \$320.

Copyright © 2019 Pearson Education, Inc.

43. Set $p = 45q$ and $p = -q^2 + 10{,}000$ equal.

$45q = -q^2 + 10{,}000$

Write this quadratic equation in standard form and solve using the quadratic formula.

$q^2 + 45q - 10{,}000 = 0$

$$q = \frac{-45 \pm \sqrt{45^2 - 4(1)(-10{,}000)}}{2(1)}$$

$$= \frac{-45 \pm \sqrt{42025}}{2} = \frac{-45 \pm 205}{2}$$

$$q = \frac{-45 + 205}{2} = 80 \text{ or } q = \frac{-45 - 205}{2} = -125$$

Since q cannot be negative, $q = 80$.
Since $p = 45(80) = 3600$ and $p = -(80)^2 + 10{,}000 = 3600$, the equilibrium quantity is $q = 80$ units and the equilibrium price is $p = \$3600$.

45. Set $p = q^2 + 20q$ and $p = -2q^2 + 10q + 3000$ equal.

$q^2 + 20q = -2q^2 + 10q + 3000$

Write this quadratic equation in standard form and solve using the quadratic formula.

$3q^2 + 10q - 3000 = 0$

$$q = \frac{-10 \pm \sqrt{10^2 - 4(3)(-3000)}}{2(3)}$$

$$= \frac{-10 \pm \sqrt{36{,}100}}{6} = \frac{-10 \pm 190}{6}$$

$$q = \frac{-10 - 190}{6} = \frac{-200}{6} = -\frac{100}{3} \text{ or}$$

$$q = \frac{-10 + 190}{6} = \frac{180}{6} = 30$$

Since q cannot be negative, $q = 30$.
Since $p = 30^2 + 20(30) = 1500$ and $p = -2(30)^2 + 10(30) + 3000 = 1500$, the equilibrium quantity is $q = 30$ units and the equilibrium price is $p = \$1500$.

47. Set the revenue function equal to the cost function and solve for x.

$200x - x^2 = 70x + 2200$

$0 = x^2 - 130x + 2200$

By the quadratic formula,

$$x = \frac{-(-130) \pm \sqrt{(-130)^2 - 4(1)(2200)}}{2(1)}$$

$$= \frac{130 \pm \sqrt{8100}}{2}$$

$x = 20$ or $x = 110$

Since these equations are only valid for $0 \le x \le 100$, 20 is the number of units needed to break even.

49. Set the revenue function equal to the cost function and solve for x.

$400x - 2x^2 = -x^2 + 200x + 1900$

$-x^2 + 200x - 1900 = 0$

By the quadratic formula,

$$x = \frac{-200 \pm \sqrt{(200)^2 - 4(-1)(-1900)}}{2(-1)}$$

$$= \frac{-200 \pm \sqrt{32400}}{-2}$$

$x = 10$ or $x = 190$

Since these equations are only valid for $0 \le x \le 100$, 10 is the number of units needed to break even.

51. **a.** If R is the revenue function,
$R(x) = xp = x(84 - 2x) = 84x - 2x^2$.

b. To find the vertex of the function, use the vertex formula

$$x = -\frac{b}{2a} \text{ and } y = f\left(-\frac{b}{2a}\right).$$

$$x = -\frac{84}{2(-2)} = 21$$

$f(21) = 84(21) - 2(21)^2 = 882$

The vertex is at (21, 882). The number of units that need to be sold that lead to maximum revenue is $x = 21$.

c. From the vertex, the maximum revenue is \$882.

Copyright © 2019 Pearson Education, Inc.

53. a. If R is the revenue function,
$R(x) = xp = x(300 - 10x) = 300x - 10x^2$.

b. To find the vertex of the function, use the vertex formula
$x = -\frac{b}{2a}$ and $y = f\left(-\frac{b}{2a}\right)$.
$x = -\frac{300}{2(-10)} = 15$
$f(15) = 300(15) - 10(15)^2 = 2250$
The vertex is at (15, 2250). The number of nights that need to be rented that lead to maximum revenue is $x = 15$.

c. From the vertex, the maximum revenue is $2250.

55. a. Let x be the number of unsold seats. Then the number of people flying is $100 - x$, and the price per ticket is $200 + 4x$. The total revenue is
$R(x) = (200 + 4x)(100 - x)$
$= 20,000 - 200x + 400x - 4x^2$
$R(x) = 20,000 + 200x - 4x^2$

b. $R(x) = -4\left(x^2 - 50x\right) + 20,000$
$= -4\left(x^2 - 50x + 625 - 625\right) + 20,000$
$= -4(x - 25)^2 + 2500 + 20,000$
$R(x) = -4(x - 25)^2 + 22,500$
The vertex is (25, 22,500).

x	5	15	25
$R(x)$	20,900	22,100	22,500
x	35	45	
$R(x)$	22,100	20,900	

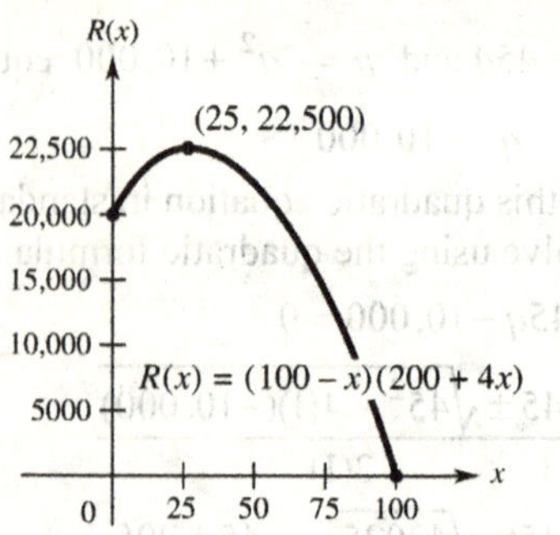

c. The maximum revenue occurs at the vertex, when 25 seats are unsold.

d. $R(25) = 22,500$, so the maximum revenue is $22,500.

57. Let x represent the number of weeks she should wait. Let R represent her revenue in dollars per hog.
$R(x) = (90 + 5x)(.88 - .02x)$
$= 79.2 - 1.8x + 4.4x - .1x^2$
$= \left(-1.x^2 + 2.6x\right) + 79.2$
$= -1.\left(x^2 - 26x\right) + 79.2$
$= -.1\left(x^2 - 26x + 169 - 169\right) + 79.2$

$R(x) = -.1\left(x^2 - 26x + 169\right) + 16.9 + 79.2$
$= -1.(x - 13)^2 + 96.1$
The vertex is (13, 96.1). She should wait 13 weeks and will receive $96.10/hog.

59. a. The vertex is given as (0, 5674), so the equation is of the form
$f(x) = a(x - 0)^2 + 5674$. The year 2007 corresponds to $x = 7$. Using the point (7, 9008), we have
$9008 = a(7 - 0)^2 + 5674 \Rightarrow 3334 = 49a \Rightarrow$
$a = \frac{3334}{49} \approx 68.$
The equation is $g(x) = 68x^2 + 5674$.

b. The year 2008 corresponds to $x = 8$.
$f(8) = 68(8)^2 + 5674 \approx 10,026.$
According to the model, the federal debt in 2008 was about $10,026 billion. Yes, this is a reasonable estimate based on the actual debt.

Copyright © 2019 Pearson Education, Inc.

61. a. The vertex is given as (13, 545), so the equation is of the form $f(x) = a(x-13)^2 + 545$. The year 2015 corresponds to $x = 15$. Using the point (15, 395), we have

$$395 = a(15-13)^2 + 545 \Rightarrow$$
$$-150 = 4a \Rightarrow$$
$$a = \frac{-150}{4} = -37.5.$$

The equation is
$$f(x) = -37.5(x-13)^2 + 545.$$

b. The year 2011 corresponds to $x = 11$.
$$f(11) = -37.5(11-13)^2 + 545 = 395.$$
According to the model, the price of soybean meal in 2011 was \$395. Yes, this is a reasonable estimate based on the actual price.

63.
```
QuadReg
 y=ax²+bx+c
 a=29.91883117
 b=305.8993506
 c=5580.909091
 R²=.9954821227
```

The quadratic regression is
$$f(x) = 29.92x^2 + 305.90x + 5580.91.$$

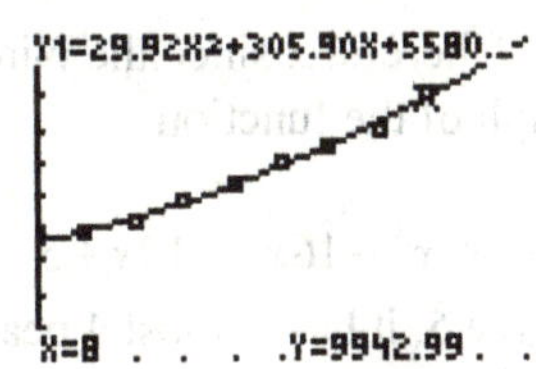

The estimate for 2008 is about \$9943 billion which is reasonably close to the answer calculated in exercise 59.

65.
```
QuadReg
 y=ax²+bx+c
 a=-39.71428571
 b=1032.371429
 c=-6151.685714
 R²=.9860472831
```

The quadratic regression is
$$f(x) = -39.71x^2 + 1032.37x - 6151.69.$$

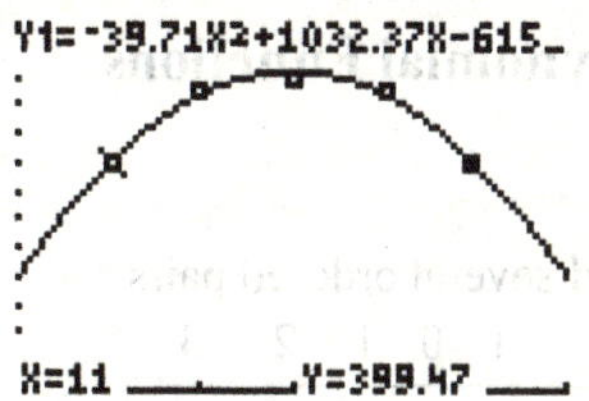

The quadratic regression estimates that the price of soybeans was about \$400 in 2011 which is reasonably close to the model in exercise 61.

67. $R(x) = 400x - 2x^2$; $C(x) = 200x + 2000$

a.
$$R(x) = C(x)$$
$$400x - 2x^2 = 200x + 2000$$
$$0 = 2x^2 - 200x + 2000$$
$$0 = x^2 - 100x + 1000$$
$$x = \frac{100 \pm \sqrt{(-100)^2 - 4(1)(1000)}}{2(1)}$$
$$= \frac{100 \pm \sqrt{6000}}{2} = \frac{100 \pm 20\sqrt{15}}{2}$$
$$= 50 \pm 10\sqrt{15}$$
$$x \approx 88.7 \text{ and } 11.3$$

The break-even point occurs when $x \approx 11.3$ and $x \approx 88.7$.

b.
$$P(x) = R(x) - C(x)$$
$$= (400x - 2x^2) - (200x + 2000)$$
$$= -2x^2 + 200x - 2000$$
$$= -2(x^2 - 100x + 2500) - 2000 + 5000$$
$$= -2(x-50)^2 + 3000$$

The vertex is (50, 3000). Profit is maximum when $x = 50$.

c. From the vertex, the maximum profit is \$3000.

d. Since the break-even points are 88.7 and 11.3, a loss will occur if $x < 11.3$ or $x > 88.7$.

e. A profit will occur for $11.3 < x < 88.7$.

Copyright © 2019 Pearson Education, Inc.

Section 3.5 Polynomial Functions

1. $f(x) = x^4$
First we find several ordered pairs.

x	−3	−2	−1	0	1	2	3
y	81	16	1	0	1	16	81

Plot these ordered pairs and draw a smooth curve through them.

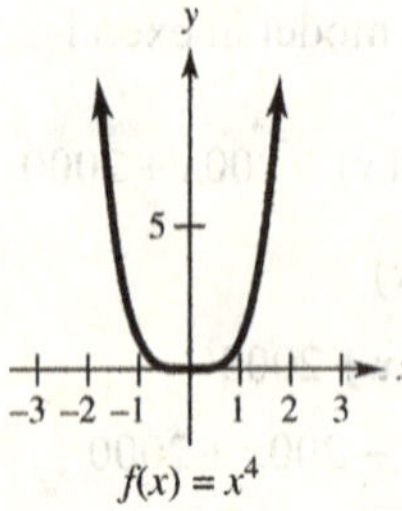

3. $h(x) = -.2x^5$
First find several ordered pairs

x	−3	−2	−1	0	1	2	3
y	48.6	6.4	.2	0	−.2	−6.4	−48.6

To graph the function, plot these ordered pairs, and draw a smooth curve through them.

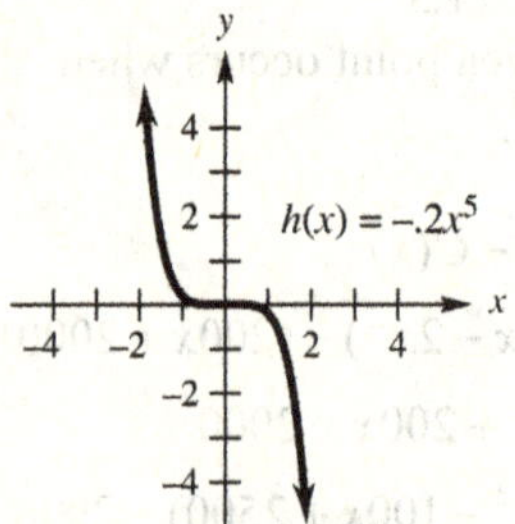

5.a. Yes. The graph could possibly be the graph of some polynomial function because it moves sharply away from the x-axis at the far left and far right.

b. No. The total number of peaks and valleys on the graph of a polynomial function of degree n is at most $n - 1$. Since the graph has 2 peaks and 2 valleys, this could not be the graph of a polynomial function of degree 3.

c. No. Since the graph has 2 peaks and 2 valleys, it could not be the graph of a polynomial function of degree 4.

d. Yes. Since the graph has 2 peaks and 2 valleys and the opposite ends of the graph move sharply away from the x-axis in the opposite direction, it could be the graph of a polynomial function of degree 5.

7. a. Yes. The graph could possibly be the graph of some polynomial function because it moves sharply away from the x-axis as $|x|$ gets large.

b. No. Since it has 1 peak and 2 valleys, it could not be the graph of a polynomial function of degree 3 (since $n - 1 = 3 - 1 = 2$).

c. Yes. Since it has 1 peak and 2 valleys, it could be the graph of a polynomial function of degree 4 (since $n - 1 = 4 - 1 = 3$).

d. No. The graph could not be the graph of a polynomial function of degree 5 because the opposite ends of the graph move sharply away from the x-axis in the same direction, indicating an even degree.

9. $f(x) = x^3 - 7x - 9$
Since $f(x)$ is degree 3, it has at most 2 peaks and valleys. When $|x|$ is large, the graph resembles the graph of x^3. Therefore, since the y-intercept is −9, D is the graph of the function.

11. $f(x) = x^4 - 5x^2 + 7.$
Since $f(x)$ is degree 4, it has at most 3 peaks and valleys. When $|x|$ is large, the graph resembles the graph of x^4. Therefore, since the y-intercept is 7, B is the graph of the function.

13. $f(x) = -x^5 + 4x^4 + x^3 - 16x^2 + 12x + 5$
Since $f(x)$ is degree 5, it has at most 4 peaks and valleys. When $|x|$ is large, the graph resembles the graph of $-x^5$. Therefore, since the y-intercept is 5, A is the graph of the function.

15. $f(x) = (x+3)(x-4)(x+1)$
First, find x-intercepts by setting $f(x) = 0$ and solving for x.
$f(x) = (x+3)(x-4)(x+1) = 0 \Rightarrow$
$x = -3, 4, -1$
These three numbers divide the x-axis into four regions. Choose an x-value in each region as a test number and compute the function value.

Copyright © 2019 Pearson Education, Inc.

Region	Test No.	Value of $f(x)$	Graph
$x<-3$	–5	–72	below x-axis
$-3<x<-1$	–2	6	above x-axis
$-1<x<4$	2	–30	below x-axis
$4<x$	5	48	above x-axis

Use this information, the x-intercepts, and the fact that the graph can have a total of at most 2 peaks and valleys to sketch the graph.

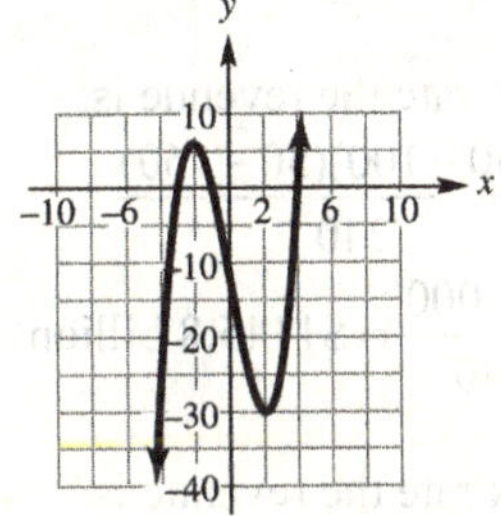

17. $f(x)=x^2(x+3)(x-1)$

First, find x-intercepts by setting $f(x)=0$ and solving for x.

$f(x)=x^2(x+3)(x-1)=0 \Rightarrow$

$x=0,\ -3,\ 1$

These three numbers divide the x-axis into four regions. Choose an x-value in each region as a test number and compute the function value.

Region	Test No.	Value of $f(x)$	Graph
$x<-3$	–4	80	above x-axis
$-3<x<0$	–1	–4	below x-axis
$0<x<1$	0.5	–.4375	below x-axis
$1<x$	2	20	above x-axis

Use this information, the x-intercepts, and the fact that the graph can have a total of at most 3 peaks and valleys to sketch the graph.

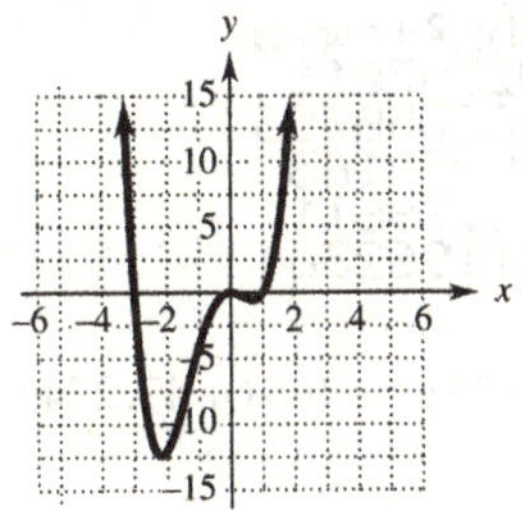

19. $f(x)=x^3-x^2-20x$

First, find x-intercepts by setting $f(x)=0$ and solving for x.

$f(x)=0=x^3-x^2-20x=x(x^2-x-20)$

$=x(x-5)(x+4) \Rightarrow x=0,\ 5,\ -4$

These three numbers divide the x-axis into four regions. Choose an x-value in each region as a test number and compute the function value.

Region	Test No.	Value of $f(x)$	Graph
$x<-4$	–5	–50	below x-axis
$-4<x<0$	–1	18	above x-axis
$0<x<5$	1	–20	below x-axis
$5<x$	10	700	above x-axis

Use this information, the x-intercepts, and the fact that the graph can have a total of at most 2 peaks and valleys to sketch the graph.

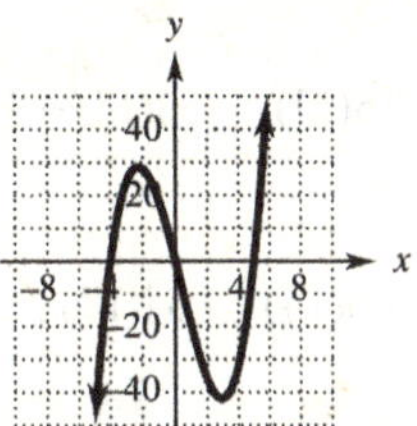

21. **a.** $R(4)=.19(4)^3-5.2(4)^2+44(4)-40$

$R(4)\approx 65$

In 2004, Home Depot revenue was about \$65 billion.

b. $R(6)=.19(6)^3-5.2(6)^2+44(6)-40$

$R(6)\approx 78$

$R(11)=.19(11)^3-5.2(11)^2+44(11)-40$

$R(11)\approx 68$

$R(14)=.19(14)^3-5.2(14)^2+44(14)-40$

$R(14)\approx 78$

The Home Depot revenue for 2006, 2011, and 2014 was about \$78 billion, about \$68 billion, and about \$78 billion respectivly.

Copyright © 2019 Pearson Education, Inc.

c.

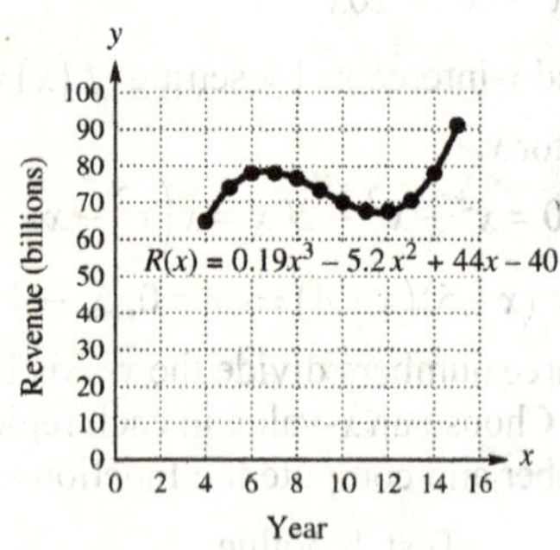

d. Based on the graph, revenue fluctuates.

23. **a.** $C(4) = .13(4)^3 - 3.6(4)^2 + 31(4) - 34$

$C(4) \approx 41$

In 2004, Home Depot cost was about $41 billion.

b. $C(6) = .13(6)^3 - 3.6(6)^2 + 31(6) - 34$

$C(6) \approx 50$

$C(11) = .13(11)^3 - 3.6(11)^2 + 31(11) - 34$

$C(11) \approx 44$

$C(14) = .13(14)^3 - 3.6(14)^2 + 31(14) - 34$

$C(14) \approx 51$

The Home Depot cost for 2006, 2011, and 2014 was about $50 billion, about $44 billion, and about $51 billion respectivly.

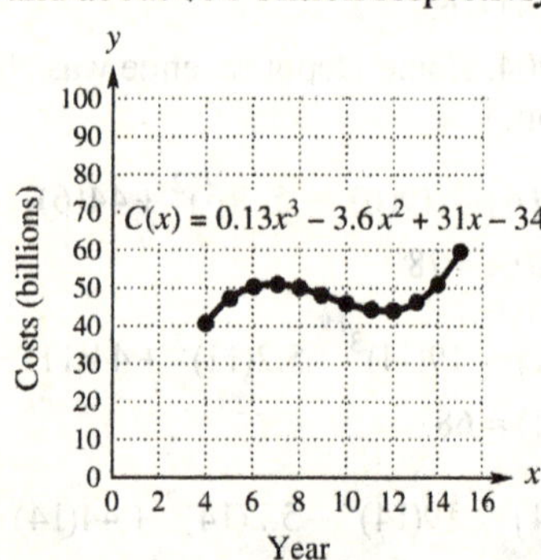

c.

d. Based on the graph, costs fluctuates.

25. $P(x) = R(x) - C(x)$

$= (.19x^3 - 5.2x^2 + 44x - 40)$

$-(.13x^3 - 3.6x^2 + 31x - 34)$

$P(x) = .06x^3 - 1.6x^2 + 13x - 6$

$P(14) = .06(14)^3 - 1.6(14)^2 + 13(14) - 6$

$P(14) \approx 27$

The Home Depot profit in 2014 was about $27 billion.

27. **a.** For a 20% tax rate the revenue is,

$$f(20) = \frac{20(20-100)(20-160)}{240}$$

$$= \frac{224{,}000}{240} \approx \$933.33 \text{ billion}$$

b. For a 40% tax rate the revenue is,

$$f(40) = \frac{40(40-100)(40-160)}{240}$$

$$= \frac{288{,}000}{240} = \$1200 \text{ billion}$$

c. For a 50% tax rate the revenue is,

$$f(50) = \frac{50(50-100)(50-160)}{240}$$

$$= \frac{275{,}000}{240} \approx \$1145.8 \text{ billion}$$

d. For a 70% tax rate the revenue is,

$$f(70) = \frac{70(70-100)(70-160)}{240}$$

$$= \frac{189{,}000}{240} = \$787.5 \text{ billion}$$

e.

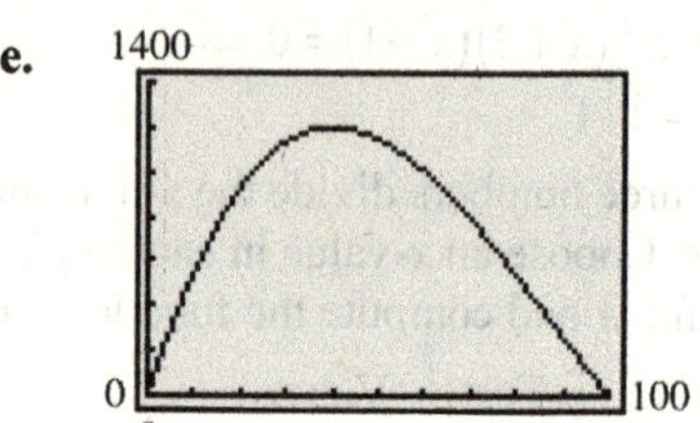

29. **a.**

b.

```
CubicReg
y=ax³+bx²+cx+d
a=.9702397011
b=-32.66750753
c=464.8859163
d=47273.79611
R²=.9911962304
```

$f(x) = x^3 - 33x^2 + 465x + 47{,}274$

Copyright © 2019 Pearson Education, Inc.

c.

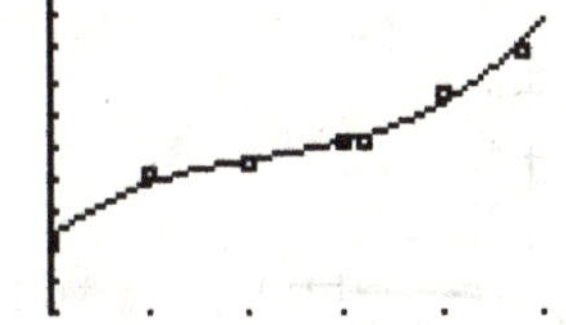

Yes; the graph fits reasonably well.

d.

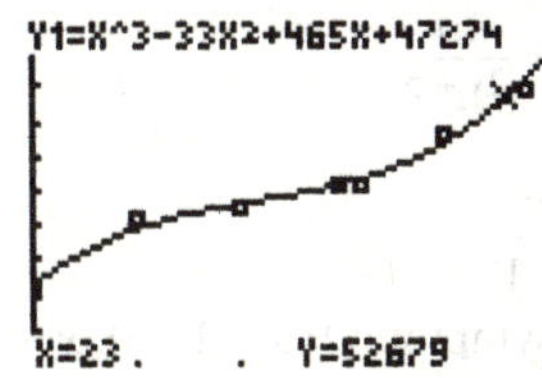

According to the model, the enrollment will be about 52,679,000 in 2023.

31. a.

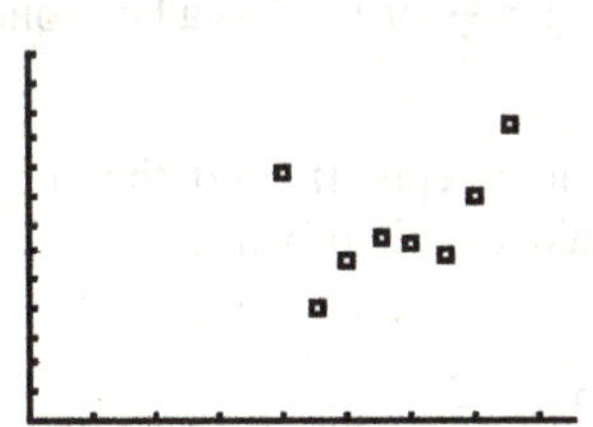

b.

```
QuarticReg
 y=ax⁴+bx³+…+e
 a=.5634469697
 b=-26.12310606
 c=449.8229167
 d=-3403.767587
↓e=9593.992965
```

$$f(x) = .5634x^4 - 26.123x^3 + 449.82x^2 - 3403.8x + 9594$$

c.

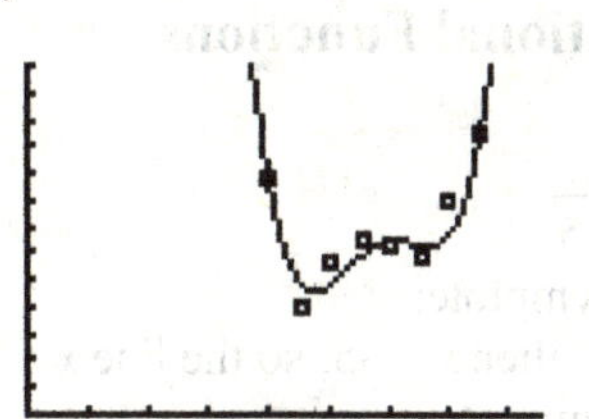

Yes; the graph fits reasonably well.

d.

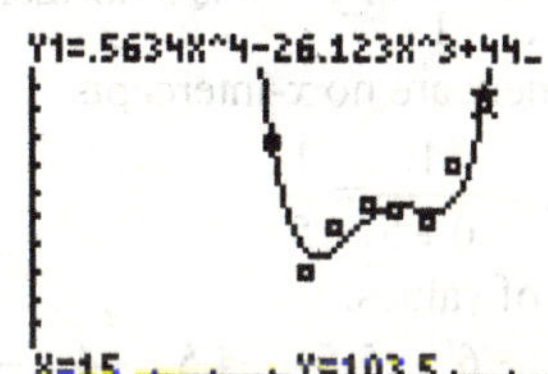

According to the model, the consumer confidence index was 103.5 in 2015. Yes, this value is close to the actual index value.

33. To graph $g(x) = x^3 - 3x^2 - 4x - 5$, enter the function as Y_1. There are many possible viewing windows that will show the complete graph. One such window is $-3 \le x \le 5$ and $-20 \le y \le 5$.

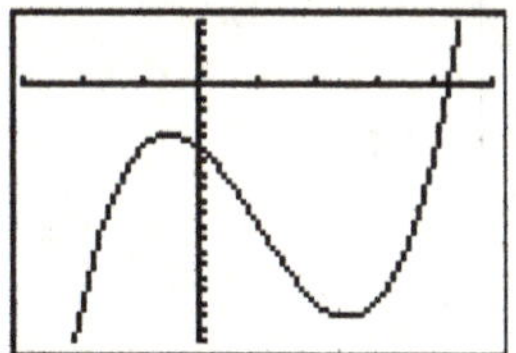

35. To graph $f(x) = 2x^5 - 3.5x^4 - 10x^3 + 5x^2 + 12x + 6$, enter the function as Y_1. There are many possible viewing windows that will show the complete graph. One such window is $-3 \le x \le 4$ and $-35 \le y \le 20$.

Copyright © 2019 Pearson Education, Inc.

Section 3.6 Rational Functions

1. $f(x)=\dfrac{1}{x+5}$
Vertical asymptote:
If $x+5=0$, then $x=-5$, so the line $x=-5$ is a vertical asymptote.
Horizontal asymptote: $y=\dfrac{1}{x+5}=\dfrac{0x+1}{1x+5}$

The line $y=\dfrac{a}{c}=\dfrac{0}{1}$ or $y=0$ is a horizontal asymptote. There are no x-intercepts.
y-intercept: $y=\dfrac{1}{0+5}=\dfrac{1}{5}$
Make a table of values.

x	−8	−7	−6	−5.5	−4.5	−4	−3	−2
y	$-\frac{1}{3}$	$-\frac{1}{2}$	−1	−2	2	1	$\frac{1}{2}$	$\frac{1}{3}$

Using these points and the asymptotes, we graph the function.

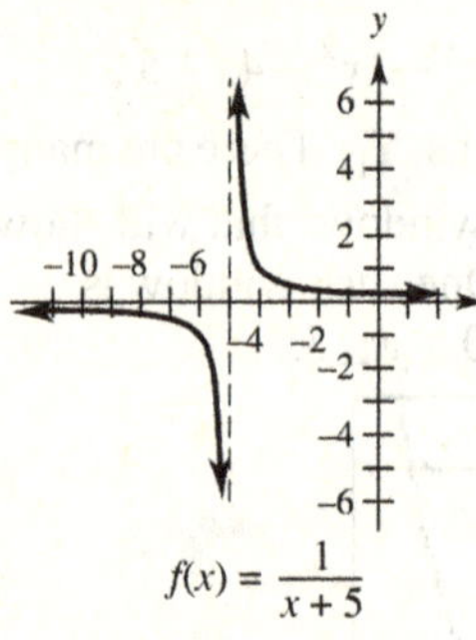

3. $f(x)=\dfrac{-3}{2x+5}$
Vertical asymptote:
If $2x+5=0$, $x=-\dfrac{5}{2}$, so the line $x=-\dfrac{5}{2}$ is a vertical asymptote.
Horizontal asymptote: $y=-\dfrac{3}{2x+5}=\dfrac{0x-3}{2x+5}$

The line $y=\dfrac{0}{2}$ or $y=0$ is a horizontal asymptote. There are no x-intercepts. If $x=0$, then $y=\dfrac{-3}{2(0)+5}=-\dfrac{3}{5}$.
Make a table of values.

x	−5	−4	−3	−2	−1	0
y	$\frac{3}{5}$	1	3	−3	−1	$-\frac{3}{5}$

Using these points and the information above, we graph the function.

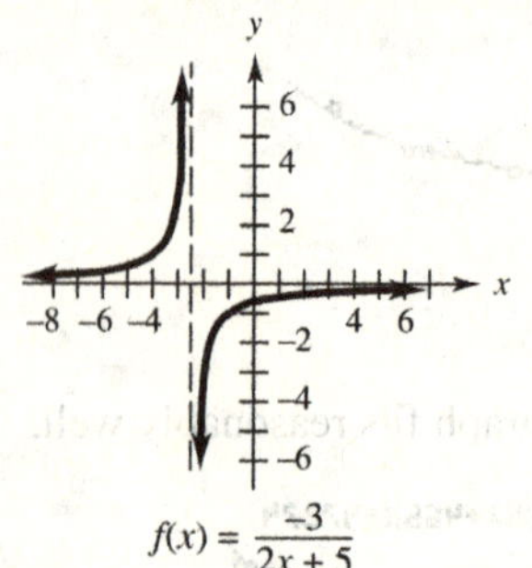

5. $f(x)=\dfrac{3x}{x-1}$
Vertical asymptote: If $x-1=0$, $x=1$, so the line $x=1$ is a vertical asymptote.
Horizontal asymptote: $y=\dfrac{3x+0}{1x-1}$

The line $y=\dfrac{3}{1}$ or $y=3$ is a horizontal asymptote.
Find the intercepts. If $x=0$, then $y=f(x)=0$.
Next, make a table of values.

x	−3	−2	−1	0	2	3	4	5
y	$\frac{9}{4}$	2	$\frac{3}{2}$	0	6	$\frac{9}{2}$	4	$\frac{15}{4}$

Using these points and the information above, we graph the function.

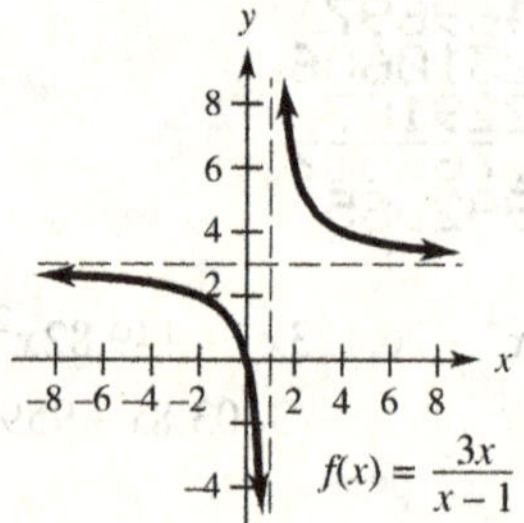

Copyright © 2019 Pearson Education, Inc.

7. $f(x)=\dfrac{x-3}{x+5}$
To find the vertical asymptote, set the denominator equal to 0 and solve.
$x+5=0 \Rightarrow x=-5$
The line $x=-5$ is a vertical asymptote.
To find the horizontal asymptote, write the equation as $y=\dfrac{1x-3}{1x+5}$.
The line $y=\dfrac{1}{1}$ or $y=1$ is a horizontal asymptote. Find the intercepts. If $x=0$, then $y=\dfrac{0-3}{0+5}=-\dfrac{3}{5}$. The y-intercept is $-\dfrac{3}{5}$. If $y=0$, then $0=\dfrac{x-3}{x+5} \Rightarrow 0=x-3 \Rightarrow 3=x$.

The x-intercept is 3. Make a table of values.

x	−8	−7	−6	−5.5	−4.5	−4	−3	−2
y	$\frac{11}{3}$	5	9	17	−15	−7	−3	$-\frac{5}{3}$

Using these points and the information above, we graph the function.

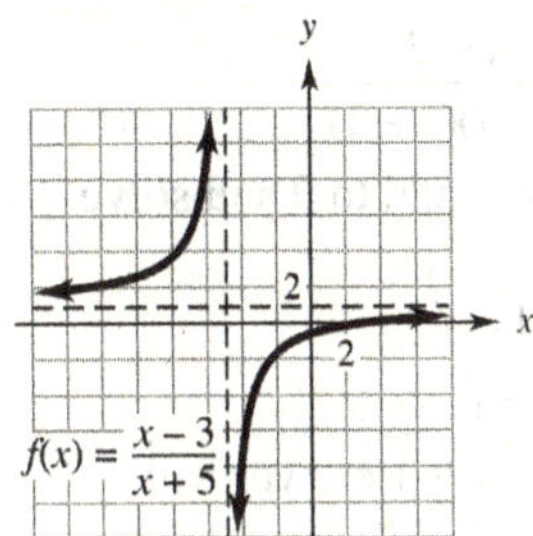

9. $f(x)=\dfrac{2-x}{x-3}$
Vertical asymptote:
If $x-3=0$, $x=3$, so the line $x=3$ is a vertical asymptote. Horizontal asymptote: $y=\dfrac{-1x+2}{1x-3}$
The line $y=-\dfrac{1}{1}$ or $y=-1$ is a horizontal asymptote. Find the intercepts. If $x=0$, then $f(x)=y=-\dfrac{2}{3}$. If $y=0$, then $x=2$. Now, make a table of values.

x	−1	0	1	2	4	5	6	7
y	$-\frac{3}{4}$	$-\frac{2}{3}$	$-\frac{1}{2}$	0	−2	$-\frac{3}{2}$	$-\frac{4}{3}$	$-\frac{5}{4}$

Using these points and the information above, we graph the function.

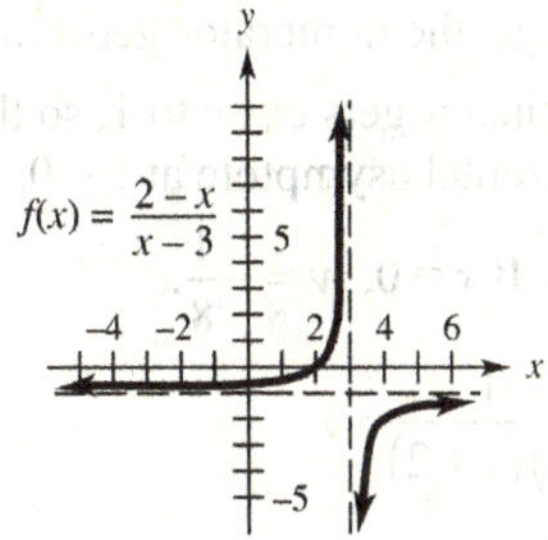

11. $f(x)=\dfrac{4x-8}{8x+1}$
To find the vertical asymptote, set the denominator equal to 0 and solve.
$8x+1=0 \Rightarrow x=-\dfrac{1}{8}$
The horizontal asymptote is $y=\dfrac{4}{8}=\dfrac{1}{2}$. Find the intercepts. If $x=0$, then $y=-8$. If $y=0$, then $0=\dfrac{4x-8}{8x+1} \Rightarrow 4x-8=0 \Rightarrow x=2$. Now make a table of values.

x	−6	−2	−1	0	1	2	3	4
y	$\frac{32}{47}$	$\frac{16}{15}$	$\frac{12}{7}$	−8	$-\frac{4}{9}$	0	$\frac{4}{25}$	$\frac{8}{33}$

Using these points and the asymptotes, graph the function.

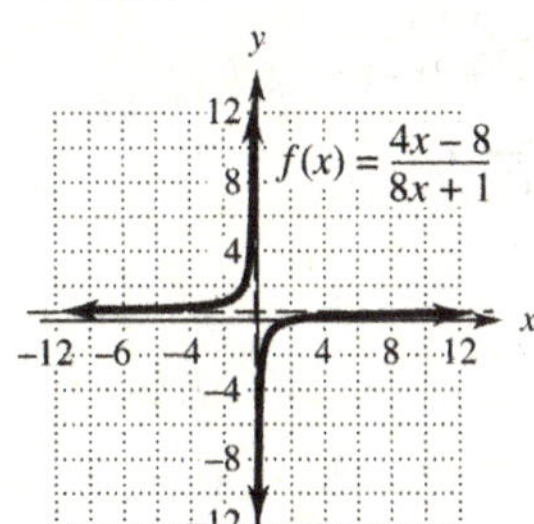

13. $h(x)=\dfrac{x+1}{x^2+2x-8}=\dfrac{x+1}{(x-4)(x+2)}$
Set the factors in the denominator equal to 0 and solve. $x-4=0 \Rightarrow x=4$; $x+2=0 \Rightarrow x=-2$.
To find horizontal asymptote, divide the numerator and the denominator by x^2.

$$\frac{\frac{x}{x^2}+\frac{1}{x^2}}{\frac{x^2}{x^2}+\frac{2x}{x^2}-\frac{8}{x^2}}=\frac{\frac{1}{x}+\frac{1}{x^2}}{1+\frac{2x}{x^2}-\frac{8}{x^2}}$$

Copyright © 2019 Pearson Education, Inc.

As $|x|$ gets very large, the numerator gets close to 0 and the denominator gets close to 1, so the function has a horizontal asymptote at $y = 0$.

Find the intercepts. If $x = 0$, $y = -\frac{1}{8}$.

If $y = 0$, $0 = \frac{x+1}{(x-4)(x+2)} \Rightarrow$
$x+1=0 \Rightarrow x=-1$.

Now, make a table of values.

x	−10	−8	−6	−4
y	$-\frac{1}{8}$	$-\frac{7}{40}$	$-\frac{5}{16}$	undefined

x	−3	−2	−1	0	1
y	$\frac{2}{5}$	$\frac{1}{8}$	0	$-\frac{1}{8}$	$-\frac{2}{5}$

x	2	4	6	8	10
y	undefined	$\frac{5}{16}$	$\frac{7}{40}$	$\frac{1}{8}$	$\frac{11}{112}$

Using these points and the asymptotes, graph the function.

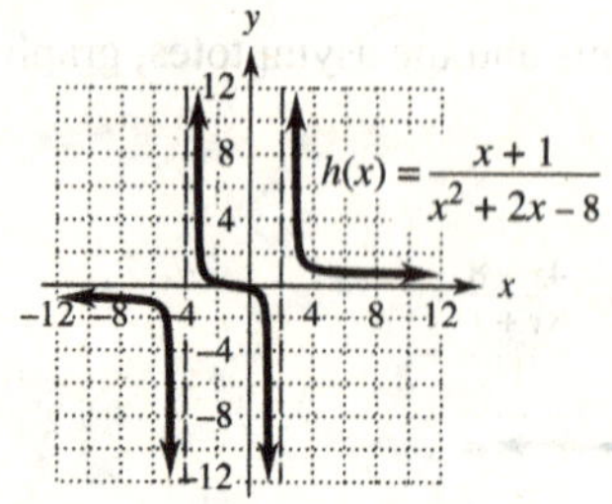

15. $f(x) = \frac{x^2+4}{x^2-4}$
To find the vertical asymptotes, set the denominator equal to 0 and solve.
$x^2 - 4 = 0 \Rightarrow (x-2)(x+2) = 0 \Rightarrow x = \pm 2$
To find horizontal asymptotes, divide the numerator and the denominator by x^2.

$$\frac{\frac{x^2}{x^2}+\frac{4}{x^2}}{\frac{x^2}{x^2}-\frac{4}{x^2}} = \frac{1+\frac{4}{x^2}}{1-\frac{4}{x^2}}$$

As $|x|$ gets very large, the function approaches 1. So the function has a horizontal asymptote at $y = 1$. Find the intercepts.

If $x = 0$, $\frac{0+4}{0-4} = -1$, so the y-intercept is −1.

If $y = 0$, $0 = \frac{x^2+4}{x^2-4} \Rightarrow x^2 = -4$ is undefined, so there is no x-intercept. Now make a table of values.

x	−6	−4	−3	−1	0	1	3	4
y	$1\frac{1}{4}$	$1\frac{2}{3}$	2.6	$-1\frac{2}{3}$	−1	$-1\frac{2}{3}$	$2\frac{3}{5}$	$1\frac{2}{3}$

Use these points and the information above, graph the function.

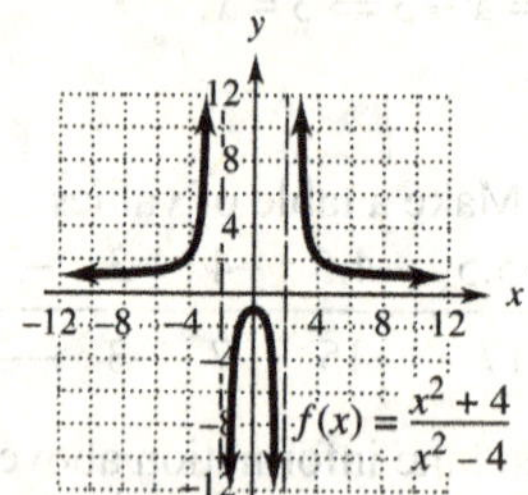

17. $g(x) = \frac{x+2}{x^2-1} = \frac{x+2}{(x+1)(x-1)}$
Set the denominator equal to 0 and solve.
$(x+1)(x-1) = 0$
$x+1=0$ or $x-1=0$
$x=-1$ or $x=1$
The lines $x = -1$ and $x = 1$ are vertical asymptotes.

19. $g(x) = \frac{x^2+2x}{x^2-4x-5} = \frac{x(x+2)}{(x-5)(x+1)}$
If $(x-5)(x+1) = 0$, then $x = 5$ or $x = -1$. The lines $x = 5$ and $x = -1$ are vertical asymptotes.

21. **a.** $C(x) = 100 + 5x$

b. $\overline{C}(x) = \frac{C(x)}{x} = \frac{100+5x}{x}$

c. $\overline{C}(5) = \frac{100+5(5)}{5} = 25$ When producing 5 pairs, the average cost is \$25 per pair.

d. $\overline{C}(50) = \frac{100+5(50)}{50} = 7$ When producing 50 pairs, the average cost is \$7 per pair.

Copyright © 2019 Pearson Education, Inc.

e. The horizontal asymptote will be $\frac{5x}{x}=5$. The average cost approaches $5 as more pairs are produced.

23. $f(x)=\dfrac{5x}{2+x}$

a. $f(0)=\dfrac{5(0)}{2+0}=0$

$f(2)=\dfrac{5(2)}{2+2}=2.5$

$f(8)=\dfrac{5(8)}{2+8}=4$

$f(78)=\dfrac{5(7)}{2+78}=4.875$

In the presence of no food, there will be no growth; with 2 units of food, the growth rate will be 2.5%; with 8 units of food, the growth rate will be 4%; with 78 units of food, the growth rate will be 4.875%.

b. The horizontal asymptote will be $\frac{5x}{x}=5$ or $y=5$.

c. No; there are no vertical asymptotes.

d.

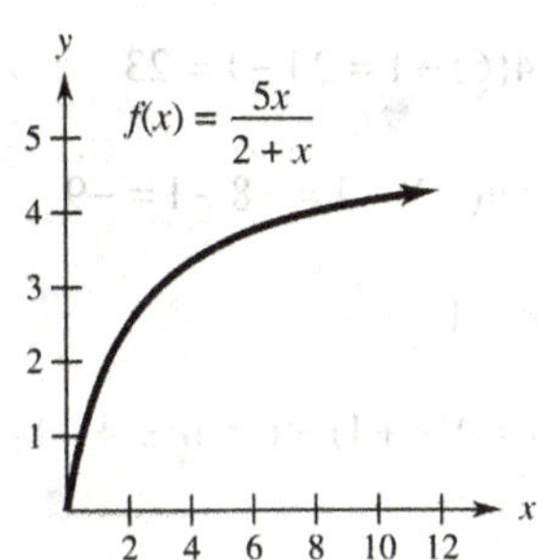

e. The maximum possible growth rate.

25. $f(x)=\dfrac{4.3x}{100-x}$

a. $f(50)=\dfrac{4.3(50)}{100-50} \Rightarrow = 4.3$ or $4300

b. $f(70)=\dfrac{4.3(70)}{100-70}=10.03333$ or $10,033.33

c. $f(80)=\dfrac{4.3(80)}{100-80}=17.2$ or $17,200

d. $f(90)=\dfrac{4.3(90)}{100-90}=38.7$ or $38,700

e. $f(95)=\dfrac{4.3(95)}{100-95}=81.7$ or $81,700

f. $f(98)=\dfrac{4.3(98)}{100-98}=210.7$ or $210,700

g. $f(99)=\dfrac{4.3(99)}{100-99}=425.7$ or $425,700

h. Since $f(100)$ is undefined, all the pollutant cannot be removed according to his model.

i.

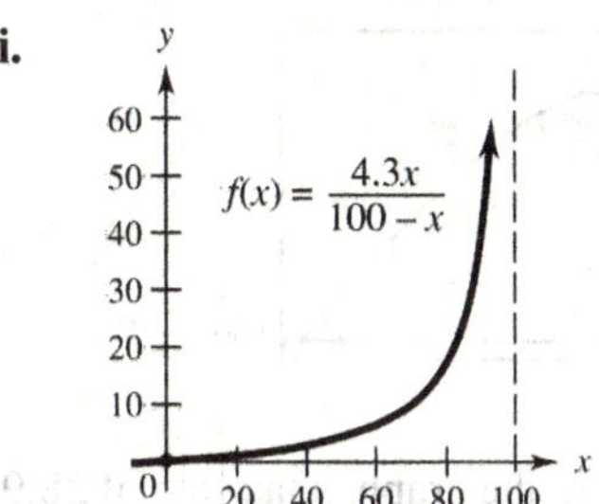

27. $y=\dfrac{125,000-25x}{125+2x}$

If $x=0$, $y=1000$. If $y=0$, $x=5000$.

$x=-62.5$ is a vertical asymptote and $y=\frac{-25}{2}$ or $y=-\frac{25}{2}$ is a horizontal asymptote. Use the intercepts and asymptotes to graph the quadrant I portion of the graph.

(*continued next page*)

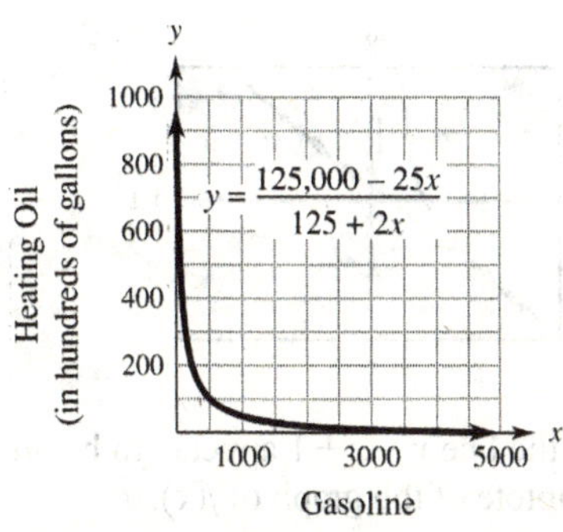

From the intercepts, the maximum value of y is 1000 which represents 100,000 gallons of oil per day and the maximum value of x is 5000 which represents 500,000 gallons of gasoline per day.

Copyright © 2019 Pearson Education, Inc.

29. a. Let $x = 16$.

$$f(16) = \frac{300(16) - 3(16)^2}{10(16) + 200} = \$11.2 \text{ billion}$$

b. Let $x = 25$.

$$f(25) = \frac{300(25) - 3(25)^2}{10(25) + 200} = \$12.5 \text{ billion}$$

c. Let $x = 40$.

$$f(40) = \frac{300(40) - 3(40)^2}{10(40) + 200} = \$12 \text{ billion}$$

d. Let $x = 55$.

$$f(55) = \frac{300(55) - 3(55)^2}{10(55) + 200} = \$9.9 \text{ billion}$$

e.

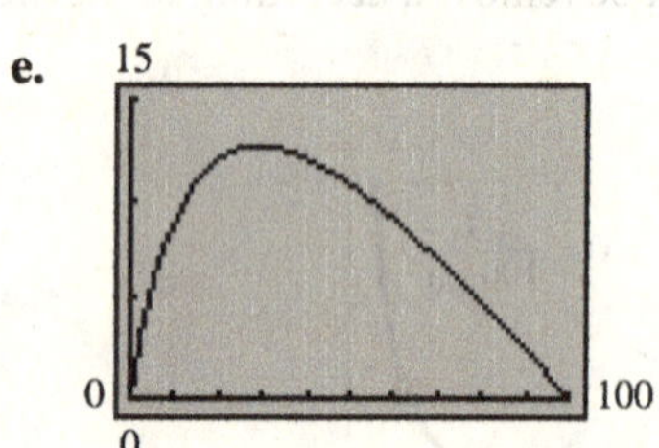

f. According to the graph, a tax rate of 28.9% produces the maximum revenue of \$12.6 billion.

31. a.

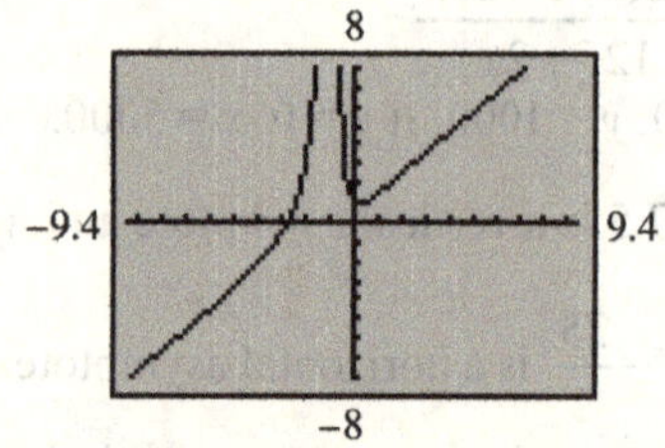

b. The graph does not appear to have a horizontal asymptote, although it does seem to have a non-horizontal straight line that is an asymptote.

c.

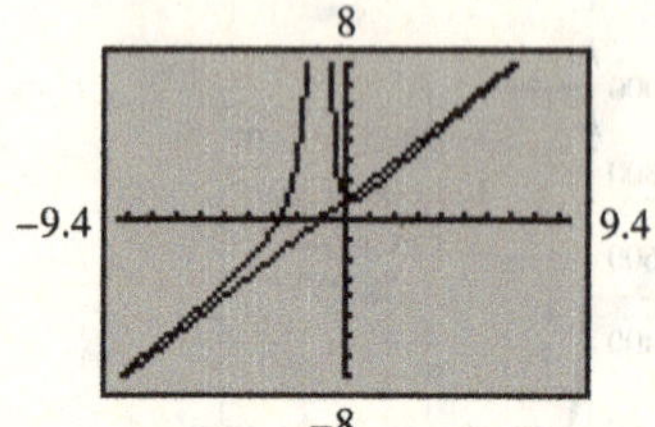

Yes, the line $y = x + 1$ appears to be an asymptote of the graph of $f(x)$.

Chapter 3 Review Exercises

1. This rule does not define y as a function of x because the x-value of 2 corresponds to both 5 and –5.

2. This rule defines y as a function of x because each value of x determines one and only one value of y.

3. $y = \sqrt{x}$
This rule defines y as a function of x because each value of x determines one and only one value of y.

4. $x = |y|$
This rule does not define y as a function of x. A given value of x may define two values for y. For example, if $x = 4$, we have $4 = |y|$, so $y = 4$ and $y = -4$.

5. $x = y^2 + 1$
This rule does not define y as a function of x. A given value of x may define two values for y. For example, if $x = 10$, we have $10 = y^2 + 1$, or $y^2 = 9$, so $y = \pm 3$.

6. $y = 5x - 2$
This rule defines y as a function of x because each value of x determines one and only one value of y.

7. $f(x) = 4x - 1$

   a. $f(6) = 4(6) - 1 = 24 - 1 = 23$

   b. $f(-2) = 4(-2) - 1 = -8 - 1 = -9$

   c. $f(p) = 4p - 1$

   d. $f(r + 1) = 4(r + 1) - 1 = 4r + 4 - 1 = 4r + 3$

8. $f(x) = 3 - 4x$

   a. $f(6) = 3 - 4(6) = -21$

   b. $f(-2) = 3 - 4(-2) = 3 + 8 = 11$

   c. $f(p) = 3 - 4p$

   d. $f(r + 1) = 3 - 4(r + 1) = -1 - 4r$

Copyright © 2019 Pearson Education, Inc.

9. $f(x) = -x^2 + 2x - 4$

a. $f(6) = -6^2 + 2(6) - 4 = -36 + 12 - 4 = -28$

b. $f(-2) = -(-2)^2 + 2(-2) - 4$
$= -4 - 4 - 4 = -12$

c. $f(p) = -p^2 + 2p - 4$

d. $f(r+1) = -(r+1)^2 + 2(r+1) - 4$
$= -(r^2 + 2r + 1) + 2r + 2 - 4$
$= -r^2 - 2r - 1 + 2r + 2 - 4$
$= -r^2 - 3$

10. $f(x) = 8 - x - x^2$

a. $f(6) = 8 - 6 - 36 = -34$

b. $f(-2) = 8 - (-2) - (-2)^2 = 8 + 2 - 4 = 6$

c. $f(p) = 8 - p - p^2$

d. $f(r+1) = 8 - (r+1) - (r+1)^2$
$= 8 - r - 1 - (r^2 + 2r + 1)$
$= 8 - r - 1 - r^2 - 2r - 1$
$= -r^2 - 3r + 6$

11. $f(x) = 5x - 3$ and $g(x) = -x^2 + 4x$

a. $f(-2) = 5(-2) - 3 = -10 - 3 = -13$

b. $g(3) = -3^2 + 4(3) = -9 + 12 = 3$

c. $g(-k) = -(-k)^2 + 4(-k) = -k^2 - 4k$

d. $g(3m) = -(3m)^2 + 4(3m) = -9m^2 + 12m$

e. $g(k-5) = -(k-5)^2 + 4(k-5)$
$= -(k^2 - 10k + 25) + 4k - 20$
$= -k^2 + 10k - 25 + 4k - 20$
$= -k^2 + 14k - 45$

f. $f(3-p) = 5(3-p) - 3 = 15 - 5p - 3$
$= 12 - 5p$

12. $f(x) = x^2 + x + 1$

a. $f(3) = 3^2 + 3 + 1 = 13$

b. $f(1) = 1^2 + 1 + 1 = 3$

c. $f(4) = 4^2 + 4 + 1 = 21$

d. No, $f(3) + f(1) = 13 + 3 = 16$, while $f(3+1) = f(4) = 21$.

13. $f(x) = |x| - 3$

$$y = \begin{cases} x - 3 & \text{if } x \geq 0 \\ -x - 3 & \text{if } x < 0 \end{cases}$$

We graph the line $y = x - 3$ for $x \geq 0$. We graph the line $y = -x - 3$ for $x < 0$.

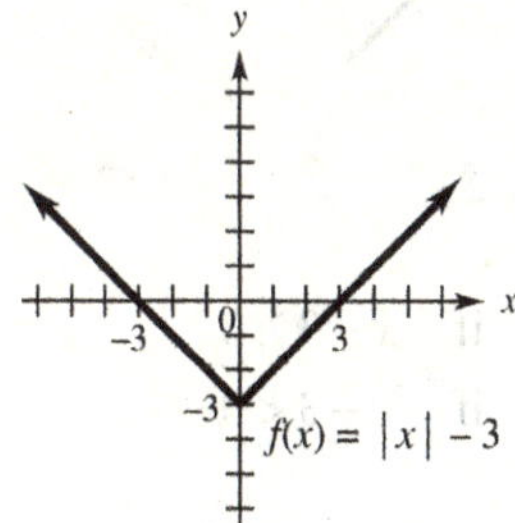

14. $f(x) = -|x| - 2$

$$y = \begin{cases} -x - 2 & \text{for } x \geq 0 \\ -(-x) - 2 & \text{for } x < 0 \end{cases}$$

or

$$y = \begin{cases} -x - 2 & \text{for } x \geq 0 \\ x - 2 & \text{for } x < 0 \end{cases}$$

We graph the line $y = -x - 2$ for $x \geq 0$. We graph the line $y = x - 2$ for $x < 0$.

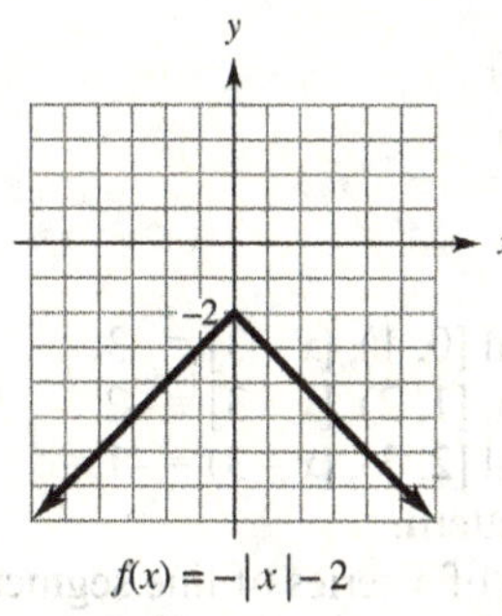

Copyright © 2019 Pearson Education, Inc.

15. $f(x) = -|x+1| + 3$

$$y = \begin{cases} -(x+1)+3 & \text{if } x+1 \geq 0 \\ -[-(x+1)]+3 & \text{if } x+1 < 0 \end{cases}$$

or

$$y = \begin{cases} -x+2 & \text{if } x \geq -1 \\ x+4 & \text{if } x < -1 \end{cases}$$

We graph the line $y = -x + 2$ for $x \geq -1$. We graph the line $y = x + 4$ for $x < -1$.

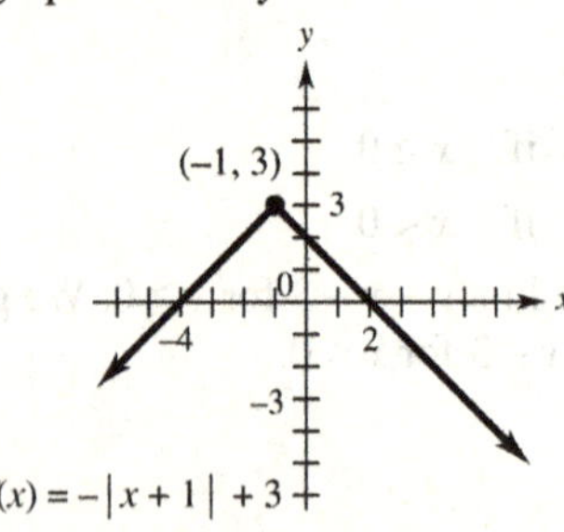

16. $f(x) = 2|x-3| - 4$

$$y = \begin{cases} 2(x-3)-4 & \text{if } x-3 \geq 0 \\ -2(x-3)-4 & \text{if } x-3 < 0 \end{cases}$$

or

$$y = \begin{cases} 2x-10 & \text{if } x \geq 3 \\ -2x+2 & \text{if } x < 3 \end{cases}$$

We graph the line $y = 2x - 10$ for $x \geq 3$. We graph the line $y = -2x + 2$ for $x < 3$.

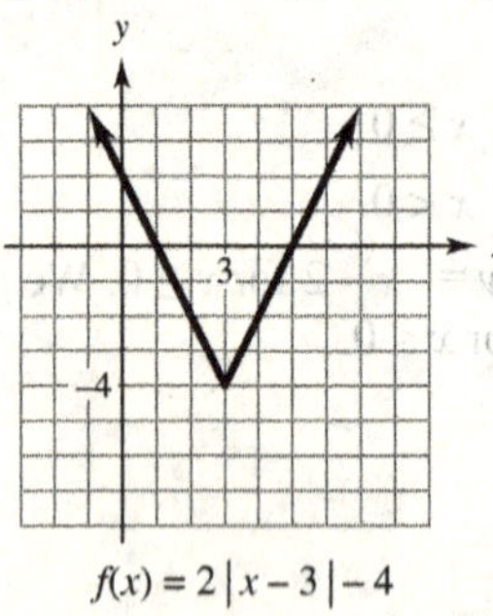

17. $f(x) = [x-3]$

For x in the interval $[0, 1)$, $[x - 3] = -3$.
For x in the interval $[1, 2)$, $[x - 3] = -2$.
For x in the interval $[2, 3)$, $[x - 3) = -1$.
Continue in this pattern.
The graph consists of a series of line segments. In each case the left endpoint is included, and the right endpoint is excluded.

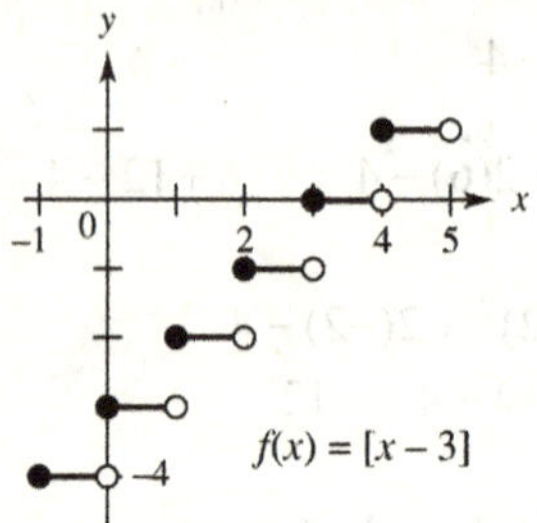

18. $f(x) = \left[\frac{1}{2}x - 2\right]$

$f(x)$ is a step function with breaks in the graph whenever x is an even integer.
If $x \in [-2, 0)$, $f(x) = -3$.
If $x \in [0, 2)$, $f(x) = -2$.
If $x \in [2, 4)$, $f(x) = -1$.
If $x \in [4, 6)$, $f(x) = 0$.
If $x \in [6, 8)$, $f(x) = 1$.

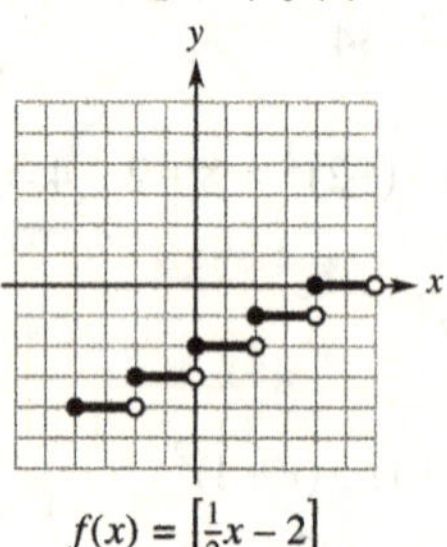

19. $f(x) = \begin{cases} -4x+2 & \text{if } x \leq 1 \\ 3x-5 & \text{if } x > 1 \end{cases}$

For $x \leq 1$, graph the line $y = -4x + 2$ using the two points $(0, 2)$ and $(1, -2)$. For $x > 1$, graph the line $y = 3x - 5$ using the two points $(1, -2)$ and $(4, 7)$. Note that the two lines meet at their common endpoint $(1, -2)$.

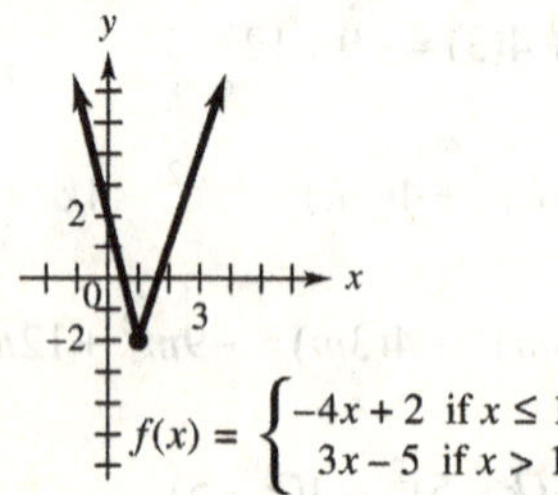

20. $f(x) = \begin{cases} 3x+1 & \text{if } x < 2 \\ -x+4 & \text{if } x \geq 2 \end{cases}$

For $x < 2$, plot the following points to graph the line $y = 3x + 1$.

x	0	1	2
y	1	4	7

Copyright © 2019 Pearson Education, Inc.

For $x \geq 2$, plot the following points to graph the line $y = -x + 4$.

x	2	3	4
y	2	1	0

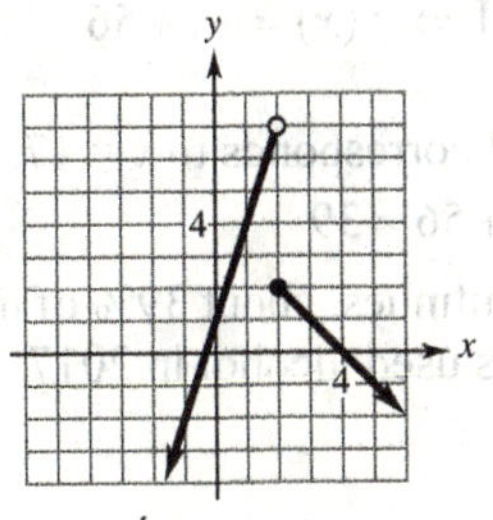

$f(x) = \begin{cases} 3x+1 & \text{if } x < 2 \\ -x+4 & \text{if } x \geq 2 \end{cases}$

21. $f(x) = \begin{cases} x^2 & \text{if } x \leq 0 \\ x & \text{if } x > 0 \end{cases}$

For $x \leq 0$, plot the following points to graph the line $y = x^2$.

x	−2	−1	0
y	4	1	0

For $x > 0$, plot the following points to graph the line $y = x$.

x	0	1	2
y	0	1	2

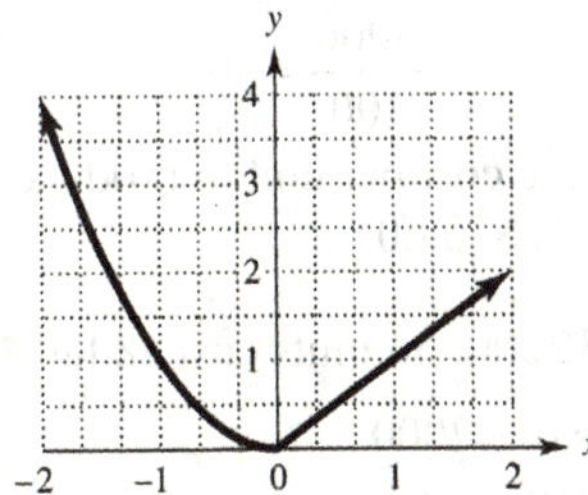

22. $f(x) = \begin{cases} |x| & \text{if } x < 3 \\ 6-x & \text{if } x \geq 3 \end{cases}$

For $x < 3$, graph $y = |x|$ using points from the following table.

x	−3	−2	−1	0	1	2	2.9
y	3	2	1	0	1	2	2.9

For $x \geq 3$, graph the line $y = 6 - x$ using the two points (3, 3) and (6, 0).

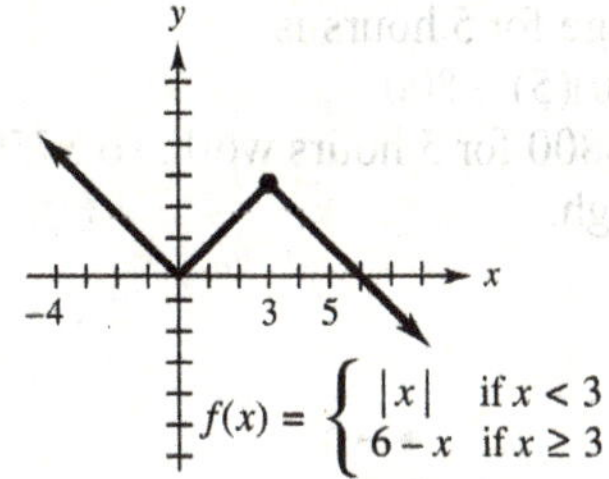

$f(x) = \begin{cases} |x| & \text{if } x < 3 \\ 6-x & \text{if } x \geq 3 \end{cases}$

23. $h(x) = \sqrt{x} + 2$

Note that x must be greater than or equal to 0. Use the table of values to draw the graph.

x	0	1	4	9
$h(x)$	2	3	4	5

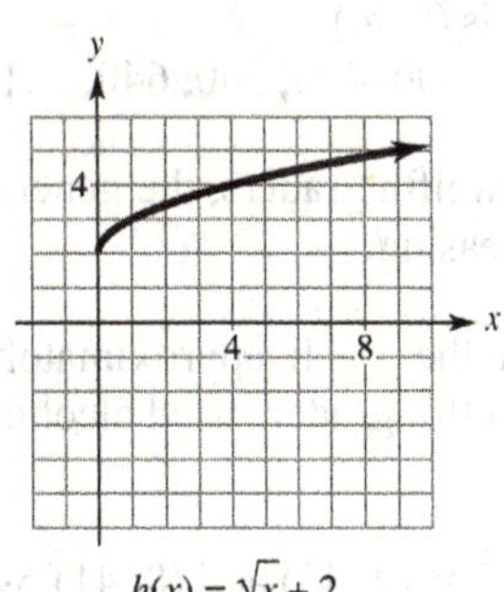

$h(x) = \sqrt{x} + 2$

24. $f(x) = \sqrt{x^2}$

Use the table of values to draw the graph.

x	−4	−3	−2	−1	0	1	2	3	4	5
$f(x)$	4	3	2	1	0	1	2	3	4	5

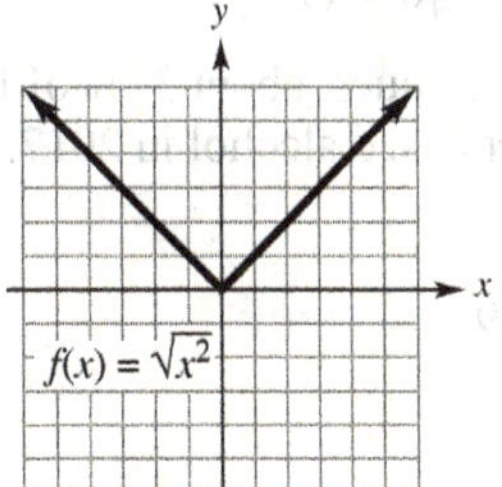

$f(x) = \sqrt{x^2}$

25. a.

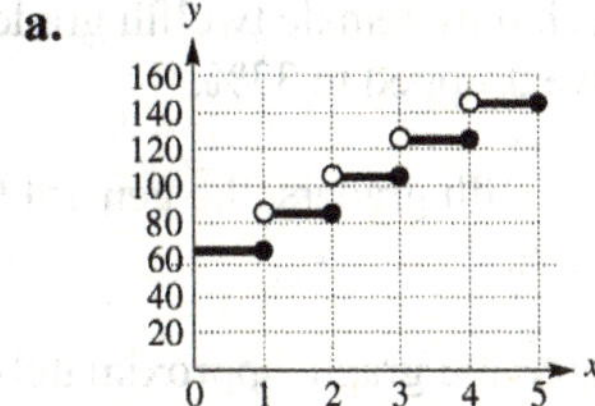

b. The domain is $(0, \infty)$.
The range is {65, 85, 105, 125, ...}.

c. If he can spend no more than \$90, he can rent the washer for at most 2 days.

Copyright © 2019 Pearson Education, Inc.

26. **a.** The charge for 5 hours is $400 + (80)(5) = 800$. It costs \$800 for 5 hours work, so \$750 is not enough.

b.

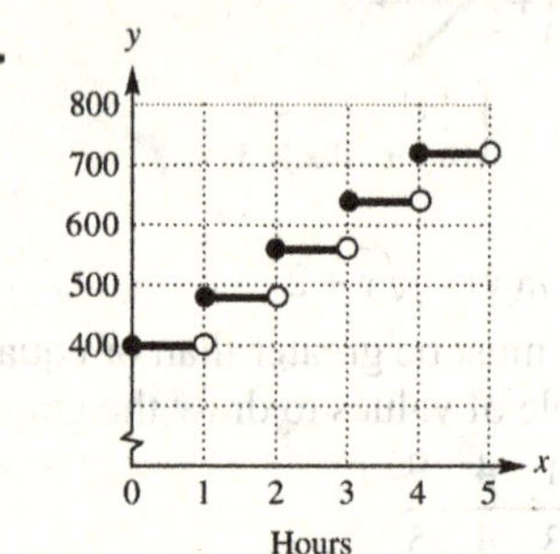

c. The domain is $(0, \infty)$ The range is $\{400, 480, 560, 640, \ldots\}$.

27. **a.** For female twelfth graders, the general trend is decreasing.

b. According to the graph, approximately 45% of female twelfth graders used alcohol in 2004.

c. Using the points (4, 45) and (8, 41), we have $m = \dfrac{41-45}{8-4} = -1$.

$y-45 = -1(x-4) \Rightarrow$

$y-45 = -x+4 \Rightarrow f(x) = -x+49$

d. The year 2017 corresponds to $x = 17$.

$f(17) = -17+49 = 32$

If the trend continues, about 32% of female twelfth graders used alcohol in 2017.

e. Let $y = 33$.

$33 = -x+49$

$-16 = -x$

$16 = x$

If the trend continued, in the year 2016 the use of alcohol by female twelfth graders would have dropped to 33%.

28. **a.** For male twelfth graders, the general trend is decreasing.

b. According to the graph, approximately 55% of male twelfth graders used alcohol in 2001.

c. Using the points (1, 55) and (12, 44), we have $m = \dfrac{44-55}{12-1} = -1$.

$y-55 = -1(x-1) \Rightarrow$

$y-55 = -x+1 \Rightarrow f(x) = -x+56$

d. The year 2017 corresponds to $x = 17$.

$f(17) = -17+56 = 39$

If the trend continues, about 39% of male twelfth graders used alcohol in 2017.

e. Let $y = 40$.

$40 = -x+56$

$-16 = -x$

$16 = x$

If the trend continued, in the year 2016 the use of alcohol by male twelfth graders would have dropped to 40%

29. Eight units cost \$300; fixed cost is \$60.

a. $C(x) = mx + b$, $b = 60$ and $C(8) = 300$

$C(x) = mx + 60 \Rightarrow 300 = m(8) + 60 \Rightarrow$

$8m = 240 \Rightarrow m = 30$

$C(x) = 30x + 60$

b. The marginal cost is given by the slope, so the marginal cost is \$30.

c. Average cost $= \dfrac{C(100)}{100} = \dfrac{30(100)+60}{100}$

$= \dfrac{3060}{100} = 30.6$

The average cost per unit to produce 100 units is \$30.60.

30. Fixed cost is \$2000; 36 units cost \$8480.

a. $C(x) = mx + 2000$

$C(36) = 8480 = 36m + 2000$

$m = \dfrac{8480-2000}{36} = 180$

$C(x) = 180x + 2000$

b. Since $m = 180$, the marginal cost is \$180.

c. Average cost $= \dfrac{C(100)}{100} = \dfrac{180(100)+2000}{100}$

$= \dfrac{20{,}000}{100} = 200$

The average cost per unit to produce 100 units is \$200.

Copyright © 2019 Pearson Education, Inc.

31. Twelve units cost $445; 50 units cost $1585.

a. $C(x) = mx + b$
Use the two points (12, 445) and (50, 1585).
$$m = \frac{1585-445}{50-12} = \frac{1140}{38} = 30$$
$$y-1585 = 30(x-50) \Rightarrow$$
$$y-1585 = 30x-1500 \Rightarrow y = 30x+85$$
$$C(x) = 30x+85$$

b. The slope is 30, so the marginal cost is $30.

c. $$\text{Average cost} = \frac{C(100)}{100} = \frac{30(100)+85}{100} = \frac{3085}{100} = 30.85$$
The average cost per unit to produce 100 units is $30.85.

32. Thirty units cost $1500; 120 units cost $5640.
Use the points (30, 1500) and (120, 5640).

a. $$m = \frac{5640-1500}{120-30} = \frac{4140}{90} = 46$$
$$C(x) = 46x+b \Rightarrow C(30) = 1500 \Rightarrow$$
$$46(30)+b = 1500 \Rightarrow b = 120$$
$$C(x) = 46x+120$$

b. The slope is 46, so the marginal cost is $46.

c. $$\text{Average cost} = \frac{C(100)}{100} = \frac{46(100)+120}{100} = \frac{4720}{100} = 47.20$$
The average cost per item to produce 100 units is $47.20.

33. a. The fixed cost is $18,000.

b. Revenue = (Price per item) × (Number of items)
$R(x) = 28x$

c. Set the cost function equal to the revenue function and solve for x.
$$24x+18,000 = 28x$$
$$-4x = -18,000$$
$$x = 4500 \text{ cartridges}$$

d. Let $x = 4500$.
$R(4500) = 28(4500) = \$126,000$

34. a. The fixed cost is $2,300,000.

b. Revenue = (Price per item) × (Number of items)
$R(x) = 450x$

c. Set the cost function equal to the revenue function and solve for x.
$$325x+2,300,000 = 450x$$
$$-125x = -2,300,000$$
$$x = 18,400 \text{ printers}$$

d. Let $x = 18,400$.
$R(18,400) = 450(18,400) = \$8,280,000$

35. $-.5q+30.95 = .3q+2.15 \Rightarrow -.8q = -28.8 \Rightarrow$
$q = 36$
$-.5(36) + 30.95 = \$12.95$ per month
The equilibrium quantity is 36 million subscribers at an equilibrium price of $12.95 per month.

36. $.0015q+1 = -.0025q+64.36 \Rightarrow$
$.004q = 63.36 \Rightarrow q = 15,840$
$p = .0015(15,840) + 1 = 24.76$
The equilibrium quantity is 15,840 prescriptions at an equilibrium price of $24.76.

37. $f(x) = 3(x-2)^2+6$
Here $a = 3$, $h = 2$, and $k = 6$. The vertex is (2, 6). The parabola opens upward since $a = 3 > 0$.

38. $f(x) = 2(x+3)^2-5$
Here $a = 2$, $h = -3$, and $k = -5$. The vertex is (–3, –5). The parabola opens upward since $a = 2 > 0$.

39. $g(x) = -4(x+1)^2+8$
$a = -4$, $h = -1$, $k = 8$. The vertex is (–1, 8). The parabola opens downward since $a = -4 < 0$.

40. $g(x) = -5(x-4)^2-6$
$a = -5$, $h = 4$, $k = -6$. The vertex is (4, –6). The parabola opens downward since $a = -5 < 0$.

Copyright © 2019 Pearson Education, Inc.

41. $f(x) = x^2 - 9$

First, locate the vertex of the parabola and find the axis of the parabola.

$y = 1(x-0)^2 - 9$

The vertex is at (0, –9). The axis is the line $x = 0$. Make a table of values to find points on one side of the axis and then use the axis of symmetry to find the corresponding points on the other side.

x	1	2	3
y	–8	–5	0

Connect the points with a smooth curve.

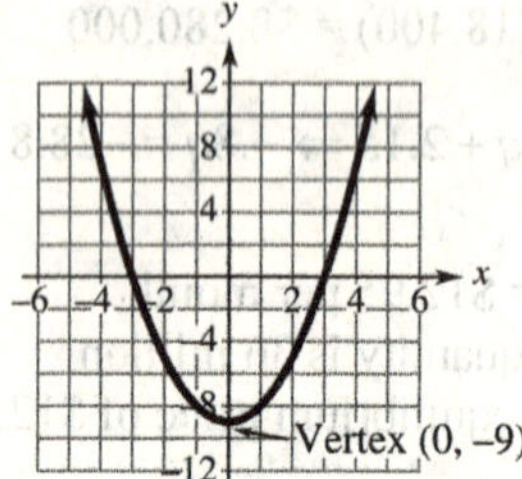

42. $f(x) = 5 - 2x^2 = -2(x-0)^2 + 5$

The vertex is (0, 5). The axis is the line $x = 0$. Make a table of values to find points on one side of the axis and then use the axis of symmetry to find the corresponding points on the other side.

x	–3	–2	–1	0
y	–13	–3	3	5

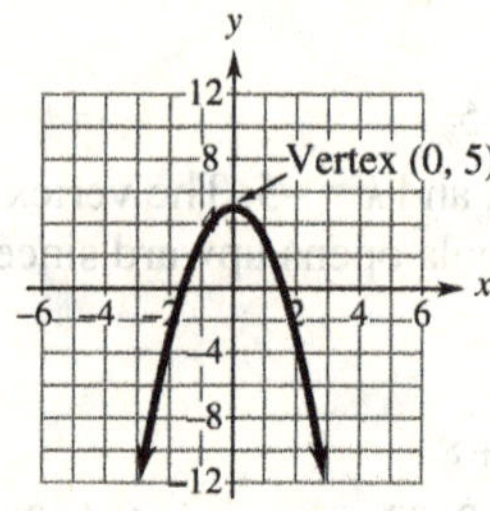

43. $f(x) = x^2 + 2x - 6$

We must complete the square to find the vertex.

$f(x) = (x^2 + 2x + 1) - 6 - 1 = (x+1)^2 - 7$

The vertex is (–1, –7). The axis is the line $x = -1$. Make a table of values to find points on one side of the axis and then use the axis of symmetry to find the corresponding points on the other side.

x	–4	–3	–2	–1
y	2	–3	–6	–7

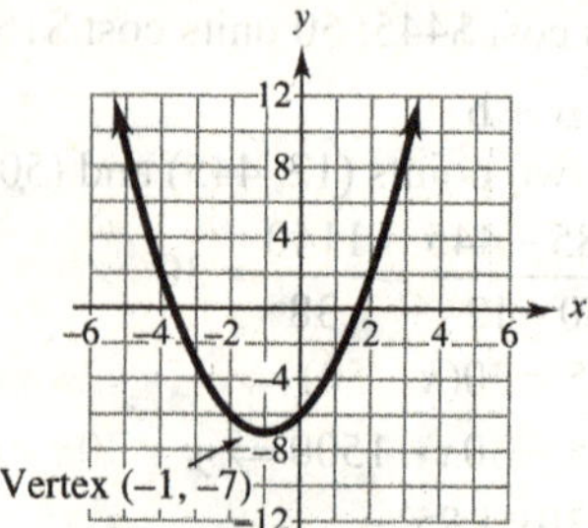

44. $f(x) = -x^2 + 8x - 1$

We must complete the square to find the vertex.

$$\begin{aligned} f(x) &= -(x^2 - 8x) - 1 \\ &= -(x^2 - 8x + 16 - 16) - 1 \\ &= -(x^2 - 8x + 16) - 1 + 16 \\ &= -(x-4)^2 + 15 \end{aligned}$$

The vertex is (4, 15). The axis is the line $x = 4$. Make a table of values to find points on one side of the axis and then use the axis of symmetry to find the corresponding points on the other side.

x	0	1	2	3	4
y	–1	6	11	14	15

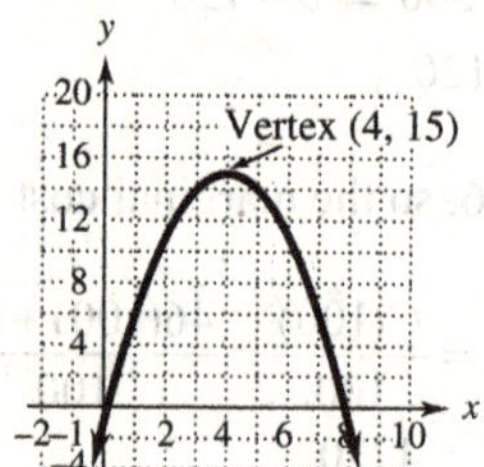

45. $f(x) = -x^2 - 6x + 5$

We must complete the square to find the vertex.

$$\begin{aligned} f(x) &= -(x^2 + 6x) + 5 = -(x^2 + 6x + 9 - 9) + 5 \\ &= -(x^2 + 6x + 9) + 5 + 9 = -(x+3)^2 + 14 \end{aligned}$$

The vertex is (–3, 14). The axis is the line $x = -3$. Make a table of values to find points on one side of the axis and then use the axis of symmetry to find the corresponding points on the other side.

x	–6	–5	–4	–3
y	5	10	13	14

Copyright © 2019 Pearson Education, Inc.

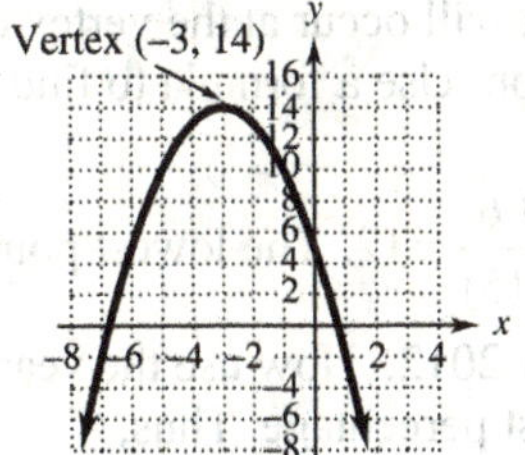

46. $f(x) = 5x^2 + 20x - 2$

We must complete the square to find the vertex.

$$f(x) = 5(x^2 + 4x) - 2 = 5(x^2 + 4x + 4 - 4) - 2$$
$$= 5(x^2 + 4x + 4) - 20 - 2 = 5(x+2)^2 - 22$$

The vertex is (–2, –22). The axis is the line $x = -2$. Make a table of values to find points on one side of the axis and then use the axis of symmetry to find the corresponding points on the other side.

x	–5	–4	–3	–2
y	23	–2	–17	–22

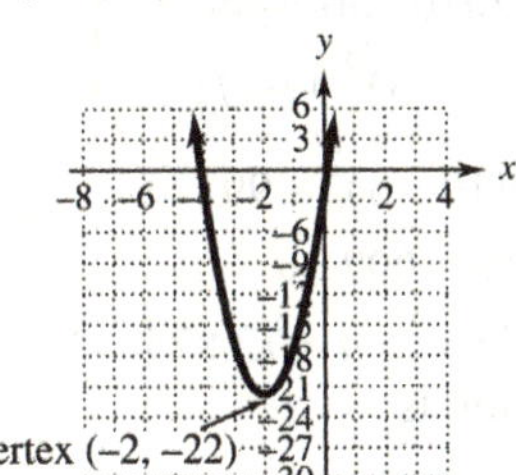

47. $f(x) = 2x^2 - 12x + 10$

We must complete the square to find the vertex.

$$f(x) = 2(x^2 - 6x) + 10 = 2(x^2 - 6x + 9 - 9) + 10$$
$$= 2(x^2 - 6x + 9) - 18 + 10 = 2(x-3)^2 - 8$$

The vertex is (3, –8). The axis is the line $x = 3$. Make a table of values to find points on one side of the axis and then use the axis of symmetry to find the corresponding points on the other side.

x	0	1	2	3
y	10	0	–6	–8

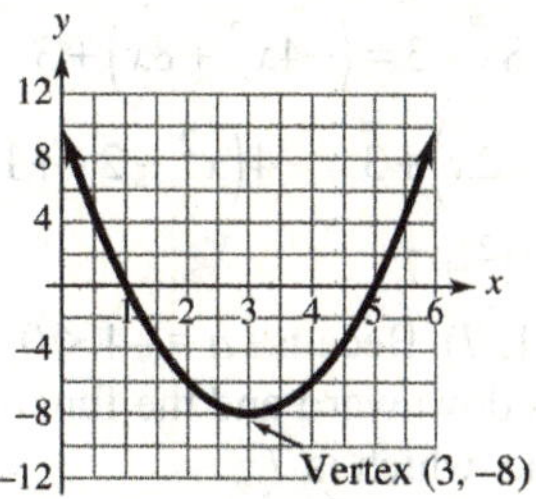

48. $f(x) = -3x^2 - 12x - 2$

We must complete the square to find the vertex.

$$f(x) = -3(x^2 + 4x) - 2$$
$$= -3(x^2 + 4x + 4 - 4) - 2$$
$$= -3(x^2 + 4x + 4) + 12 - 2$$
$$= -3(x+2)^2 + 10$$

The vertex is (–2, 10). The axis is the line $x = -2$. Make a table of values to find points on one side of the axis and then use the axis of symmetry to find the corresponding points on the other side.

x	–5	–4	–3	–2
y	–17	–2	7	10

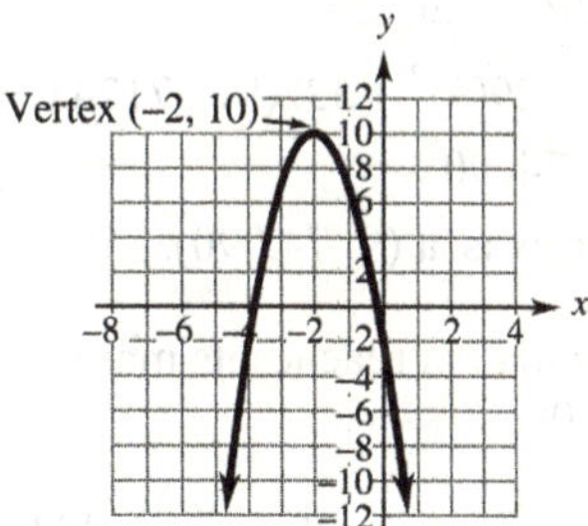

49. $f(x) = x^2 + 6x - 2 = (x^2 + 6x + 9 - 9) - 2$
$$= (x^2 + 6x + 9) - 9 - 2 = (x+3)^2 - 11$$

The vertex is (–3, –11). Because $a = 1 > 0$, the parabola opens upward and the function has a minimum value. This is the y-value of the vertex, which is –11.

50. $f(x) = x^2 + 4x + 5 = (x^2 + 4x + 4) - 4 + 5$
$$== (x+2)^2 + 1$$

The vertex is (–2, 1). Because $a = 1 > 0$, the parabola opens upward and the function has a minimum value, which is 1.

Copyright © 2019 Pearson Education, Inc.

51. $g(x) = -4x^2 + 8x + 3 = \left(-4x^2 + 8x\right) + 3$
$= -4\left(x^2 - 2x\right) + 3 = -4\left(x^2 - 2x + 1\right) + 3 + 4$
$= -4(x-1)^2 + 7$
The vertex is (1, 7). Because $a = -4 < 0$, the parabola opens downward and the function has a maximum value, which is 7.

52. $g(x) = -3x^2 - 6x + 3 = \left(-3x^2 - 6x\right) + 3$
$= -3\left(x^2 + 2x\right) + 3$
$= -3\left(x^2 + 2x + 1\right) + 3 + 3 = -3(x+1)^2 + 6$
The vertex is (–1, 6). The parabola opens downward and the function has a maximum value, which is 6.

53. **a.** $f(0) = -29(0)^2 + 522(0) + 71741$
$f(0) = 71741$
The population in the year 2000 was 71,741,000.

b. The maximum population is the vertex.
$x = -\frac{b}{2a} = -\frac{522}{2(-29)} = 9$
$f(x) = -29(9)^2 + 522(9) + 71741$
$f(x) = 74090$
The vertex is at (9, 74,090).

c. The population reaches its maximum in the year 2009.

d. The maximum population would be 74,090,000.

54. **a.** $f(0) = -.2(0)^2 + 10.4(0) + 1263$
$f(0) = 1263$
The population in the year 2000 was 1263 million.

b. The maximum population is the vertex.
$x = -\frac{b}{2a} = -\frac{10.4}{2(-.2)} = 26$
$f(x) = -.2(26)^2 + 10.4(26) + 1263$
$f(x) = 1398.2$
The vertex is at (26, 1398.2).

c. The population reaches its maximum in the year 2026.

d. The maximum population would be 1398.2 million.

55. The lowest point will occur at the vertex of the quadratic function. Use a formula to find the x coordinate:
$x = -\frac{b}{2a} = -\frac{-3.6}{2(.15)} = 12.$ The lowest point will occur in the year 2012. Now use the year 2012 to find the lowest percentage. Thus,
$f(12) = .15(12)^2 - 3.6(12) + 25$
≈ 3.4
The lowest percentage will be about 3.4% and occur in the year 2012.

56. The peak will occur at the vertex of the quadratic function. Use a formula to find the x coordinate:
$x = -\frac{b}{2a} = -\frac{16.9}{2(-.65)} = 13.$ The peak will occur in the year 2013. Now use the year 2013 to find the peak price. Thus,
$f(13) = -.65(13)^2 + 16.9(13) - 98$
≈ 11.85
The peak price will be about $11.85and occurred in the year 2013.

57. **a.** The year 2010 corresponds to $x = 10$.
$f(x) = a(x-6)^2 + 997 \Rightarrow$
$2163 = a(10-6)^2 + 997 \Rightarrow$
$2163 = 16a + 997 \Rightarrow$
$16a = 1166 \Rightarrow a = 72.875$
The function is
$f(x) = 72.875(x-6)^2 + 997.$

b. The year 2016 corresponds to $x = 16$.
$f(16) = 72.875(16-6)^2 + 997 \approx 8284.5$
According to the model, Netflix revenue was $8284.5 million in 2016.

58. **a.** The year 2010 corresponds to $x = 10$.
$f(x) = a(x-6)^2 + 627 \Rightarrow$
$1357 = a(10-6)^2 + 627 \Rightarrow$
$1357 = 16a + 627 \Rightarrow$
$16a = 730 \Rightarrow a = 45.625$
The function is
$f(x) = 45.625(x-6)^2 + 627.$

b. The year 2016 corresponds to $x = 16$.
$f(16) = 45.625(16-6)^2 + 627 = 5189.5$
According to the model, Netflix costs were $5189.5 million in 2016.

Copyright © 2019 Pearson Education, Inc.

59. a.

```
QuadReg
 y=ax²+bx+c
 a=64.51893939
 b=-724.255303
 c=3046.586364
 R²=.9954978011
```

$R(x) = 65x^2 - 724x + 3047$

b. Yes;

```
Y1=65X²-724X+3047
X=16        Y=8103
```

c. The year 2016 corresponds to $x = 16$.

$R(16) = 65(16)^2 - 724(16) + 3047$

≈ 8103

According to the model, Netflix revenue was $8103 million.

60. a.

```
QuadReg
 y=ax²+bx+c
 a=44.28030303
 b=-489.680303
 c=1979.627273
 R²=.9961060641
```

$C(x) = 44x^2 - 490x + 1980$

b. Yes;

```
Y1=44X²-490X+1980
X=16        Y=5404
```

c. The year 2016 corresponds to $x = 16$.

$C(16) = 44(16)^2 - 490(16) + 1980$

≈ 5404

According to the model, Netflix costs were $5404 million in 2016.

61. $g(x) = x^3 - 4x = x(x^2 - 4) = x(x-2)(x+2)$

The x-intercepts are 0, –2, and 2.
These three numbers divide the x-axis into four regions. Choose an x-value in each region as a test number and compute the function value.

Region	Test No.	Value of $f(x)$	Graph
$x < -2$	–3	–15	below x-axis
$-2 < x < 0$	–1	3	above x-axis
$0 < x < 2$	1	–3	below x-axis
$2 < x$	3	15	above x-axis

Use this information, the x-intercepts, and the fact that the graph can have a total of at most 2 peaks and valleys to sketch the graph.

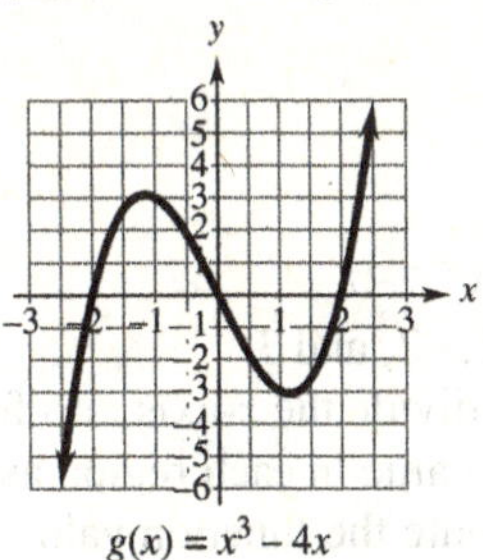

$g(x) = x^3 - 4x$

62. $f(x) = x^4 - 5$

First we find several ordered pairs.

x	–2	–1.5	–1	0	1	1.5	2
y	11	$\frac{1}{16}$	–4	–5	–4	$\frac{1}{16}$	11

Plot these ordered pairs and draw a smooth curve through them.

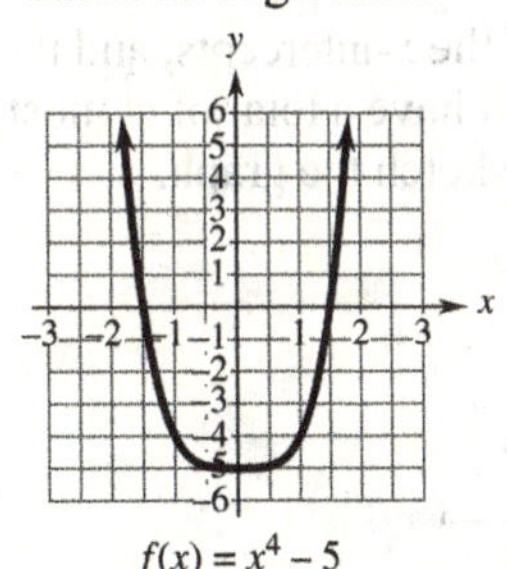

$f(x) = x^4 - 5$

63. $f(x) = x(x-4)(x+1)$

The x-intercepts are 0, 4, and –1. These three numbers divide the x-axis into four regions. Choose an x-value in each region as a test number and compute the function value.

Copyright © 2019 Pearson Education, Inc.

Region	Test No.	Value of $f(x)$	Graph
$x<-1$	-2	-12	below x-axis
$-1<x<0$	$-.5$	1.125	above x-axis
$0<x<4$	1	-6	below x-axis
$4<x$	5	30	above x-axis

Use this information, the x-intercepts, and the fact that the graph can have a total of at most 2 peaks and valleys to sketch the graph.

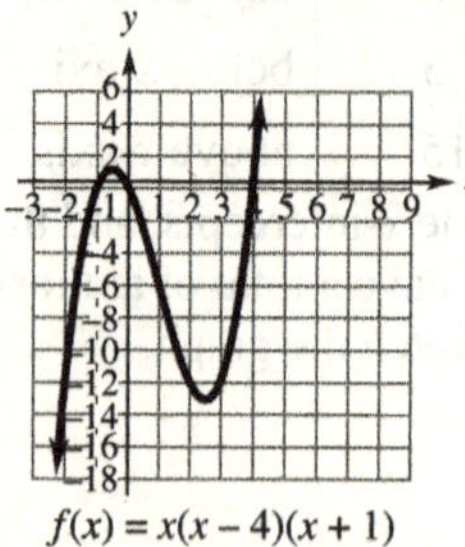

$f(x) = x(x-4)(x+1)$

64. $f(x)=(x-1)(x+2)(x-3)$

The x-intercepts are 1, –2, and 3.

These three numbers divide the x-axis into four regions. Choose an x-value in each region as a test number and compute the function value.

Region	Test No.	Value of $f(x)$	Graph
$x<-2$	-3	-24	below x-axis
$-2<x<1$	0	6	above x-axis
$1<x<3$	2	-4	below x-axis
$3<x$	4	18	above x-axis

Use this information, the x-intercepts, and the fact that the graph can have a total of at most 2 peaks and valleys to sketch the graph.

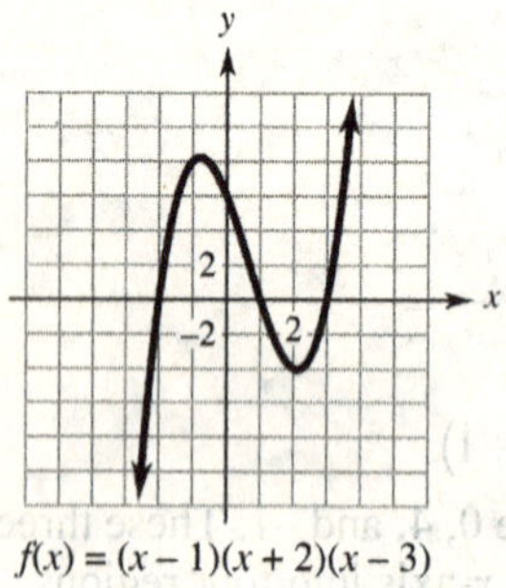

$f(x) = (x-1)(x+2)(x-3)$

65. $f(x)=x^4-5x^2-6=(x^2-6)(x^2+1)$

First, find x-intercepts by setting $f(x) = 0$ and solving for x. Note that there is no real solution for $x^2+1=0$.

$(x^2-6)(x^2+1)=0 \Rightarrow x^2=6 \Rightarrow$

$x=\pm\sqrt{6}\approx \pm 2.4$

These two numbers divide the x-axis into three regions. Choose a point in each region as a test point and compute the function value.

Region	Test No.	Value of $f(x)$	Graph
$x<-\sqrt{6}$	-3	30	above x-axis
$-\sqrt{6}<x<\sqrt{6}$	0	-6	below x-axis
$\sqrt{6}<x$	3	30	above x-axis

Use this information, the x-intercepts, and the fact that the graph can have a total of at most 3 peaks and valleys to sketch the graph.

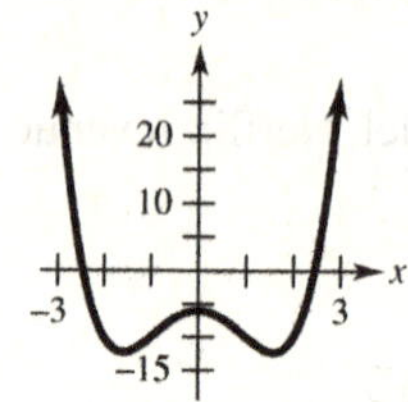

$f(x) = x^4-5x^2-6$

66. $f(x)=x^4-7x^2-8=(x^2-8)(x^2+1)$
$=(x-\sqrt{8})(x+\sqrt{8})(x^2+1)$

The x-intercepts are $-\sqrt{8}$ and $\sqrt{8}$.

(Note that $x^2+1\neq 0$ for any x.) These two numbers divide the x-axis into three regions. Choose a point in each region as a test point and compute the function value.

Region	Test No.	Value of $f(x)$	Graph
$x<-\sqrt{8}$	-3	10	above x-axis
$-\sqrt{8}<x<\sqrt{8}$	0	-8	below x-axis
$\sqrt{8}<x$	3	10	above x-axis

Use this information, the x-intercepts, and the fact that the graph can have a total of at most 3 peaks and valleys to sketch the graph.

Copyright © 2019 Pearson Education, Inc.

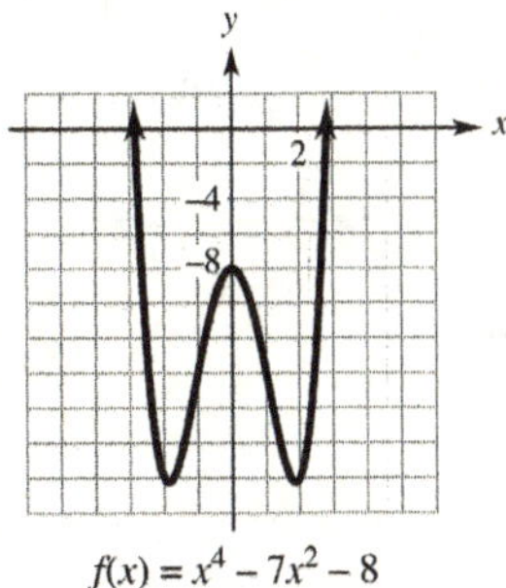

$f(x) = x^4 - 7x^2 - 8$

67. a.

```
CubicReg
 y=ax³+bx²+cx+d
 a=-.1944444444
 b=6.575757576
 c=-69.94191919
 d=256.2878788
 R²=.9746424242
```

$$R(x) = -.194x^3 + 6.576x^2 - 69.942x + 256.288$$

b. Yes;

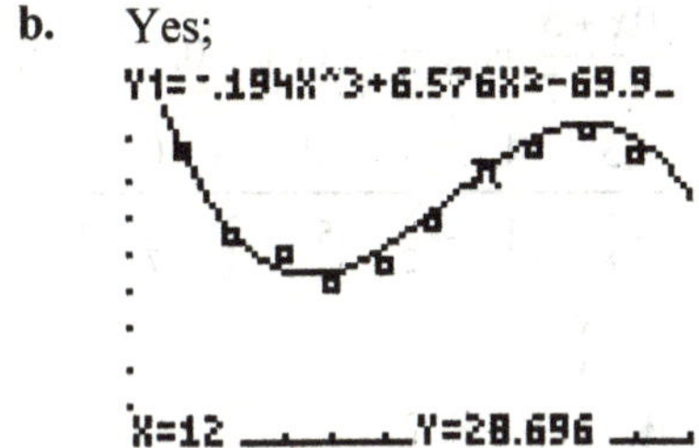

c. The year 2012 corresponds to $x = 12$.

$R(12) \approx 28.696$

According to the model, Forward Industries revenue was $28.696 million in 2012. Yes, it is close to the actual revenue.

68. a.

```
CubicReg
 y=ax³+bx²+cx+d
 a=-.1534576535
 b=5.136946387
 c=-53.90190365
 d=195.1680653
 R²=.9359676733
```

$$C(x) = -.153x^3 + 5.137x^2 - 53.902x + 195.168$$

b. Yes;

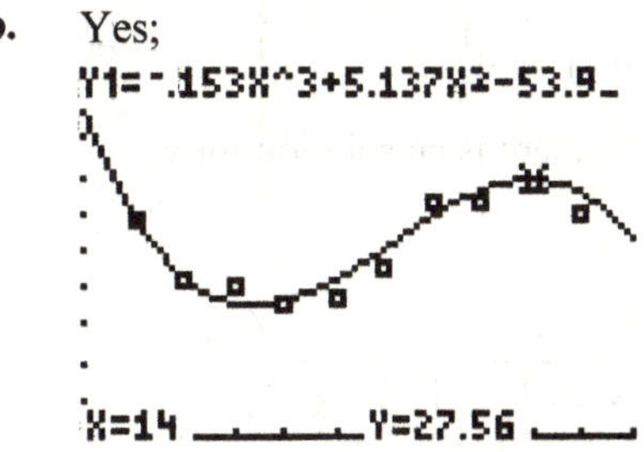

c. The year 2014 corresponds to $x = 14$.

$C(14) \approx 27.56$

According to the model, Forward Industries costs were $27.56 million in 2014. Yes, it is close to the actual costs.

69. $P(x) = R(x) - C(x)$

$$= (-.194x^3 + 6.576x^2 - 69.942x + 256.288) - (-.153x^3 + 5.137x^2 - 53.902x + 195.168)$$

$$P(x) = -.041x^3 + 1.439x^2 - 16.040x + 61.120$$

$$P(16) = -.041(16)^3 + 1.439(16)^2 - 16.040(16) + 61.120$$

$$P(16) \approx 4.928$$

The Forward Industries profit in 2016 was about $4.928 million.

70. $P(x) = R(x) - C(x)$

$$= (65x^2 - 724x + 3047) - (44x^2 - 490x + 1980)$$

$$P(x) = 21x^2 - 234x + 1067$$

$$P(16) = 21(16)^2 - 234(16) + 1067$$

$$P(16) \approx 2699$$

The Netflix profit in 2016 was about $2699 million.

71. $f(x) = \dfrac{1}{x-3}$

Find the vertical asymptote by setting the denominator equal to zero and solving for x.

$x - 3 = 0 \Rightarrow x = 3$

The line $x = 3$ is a vertical asymptote. The horizontal asymptote of the function in the form $y = \dfrac{0x+1}{1x-3}$ is $y = \dfrac{0}{1}$ or $y = 0$.

Copyright © 2019 Pearson Education, Inc.

If $x = 0$, $y = -\frac{1}{3}$.

If $y = 0$, there is no solution for x.

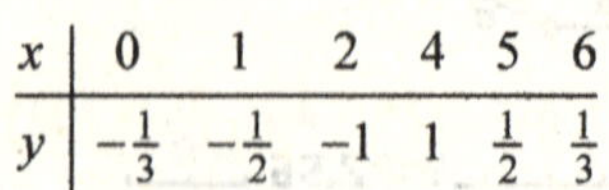

x	0	1	2	4	5	6
y	$-\frac{1}{3}$	$-\frac{1}{2}$	-1	1	$\frac{1}{2}$	$\frac{1}{3}$

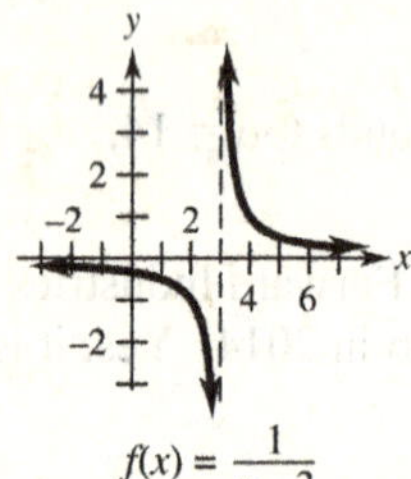

$f(x) = \frac{1}{x-3}$

72. $f(x) = \frac{-2}{x+4}$

Vertical asymptote: $x = -4$

Horizontal asymptote: $y = 0$

x	-8	-6	-5	$-\frac{9}{2}$	$-\frac{7}{2}$	-3	-2	0
y	$\frac{1}{2}$	1	2	4	-4	-2	-1	$-\frac{1}{2}$

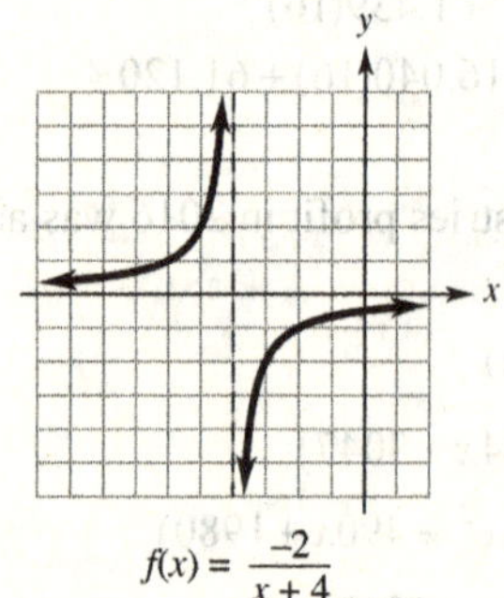

$f(x) = \frac{-2}{x+4}$

73. $f(x) = \frac{-3}{2x-4}$

Find the vertical asymptote by setting the denominator equal to zero and solving for x.

$2x - 4 = 0 \Rightarrow x = 2$

The line $x = 2$ is a vertical asymptote. The horizontal asymptote of the function in the form

$y = \frac{0x-3}{2x-4}$ is $y = \frac{0}{2}$ or $y = 0$.

If $x = 0$, $y = \frac{3}{4}$.

If $y = 0$, there is no solution for x.

x	-1	0	1	3	4	5
y	$\frac{1}{2}$	$\frac{3}{4}$	$\frac{3}{2}$	$-\frac{3}{2}$	$-\frac{3}{4}$	$-\frac{1}{2}$

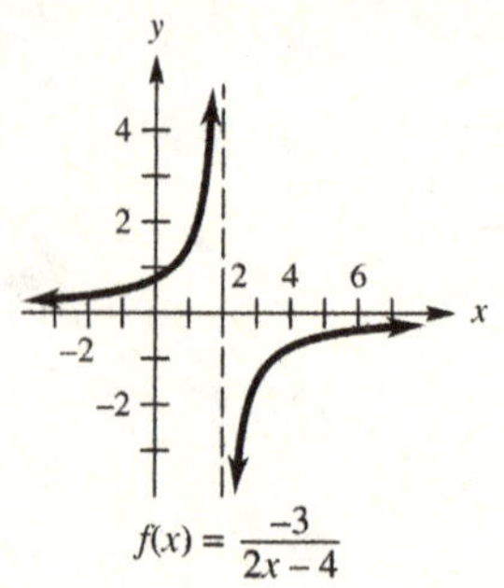

$f(x) = \frac{-3}{2x-4}$

74. $f(x) = \frac{5}{3x+7}$

Find the vertical asymptote by setting the denominator equal to zero and solving for x.

$3x + 7 = 0 \Rightarrow x = -\frac{7}{3}$

Vertical asymptote: $x = -\frac{7}{3}$

The horizontal asymptote of the function in the form $y = \frac{0x+5}{3x+7}$ is $y = \frac{0}{3}$ or $y = 0$.

x	-6	-5	-4	-3	-2	-1	0	1
y	$-\frac{5}{11}$	$-\frac{5}{8}$	-1	$-\frac{5}{2}$	5	$\frac{5}{4}$	$\frac{5}{7}$	$\frac{1}{2}$

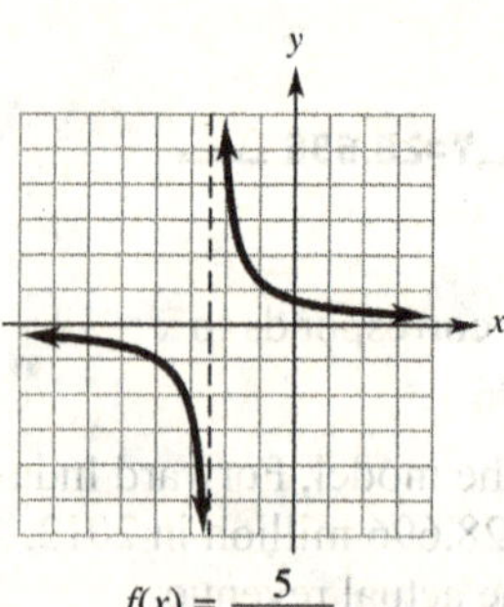

$f(x) = \frac{5}{3x+7}$

75. $g(x) = \frac{5x-2}{4x^2-4x-3}$

Find the vertical asymptotes by setting the denominator equal to zero and solving for x.

$4x^2 - 4x - 3 = 0 \Rightarrow (2x-3)(2x+1) = 0 \Rightarrow$

$2x - 3 = 0$ or $2x + 1 = 0$

$2x = 3$ or $2x = -1$

$x = \frac{3}{2}$ or $x = -\frac{1}{2}$

The lines $x = \frac{3}{2}$ and $x = -\frac{1}{2}$ are vertical asymptotes.

Copyright © 2019 Pearson Education, Inc.

Divide the numerator and denominator by x^2 to find the horizontal asymptotes.

$$\frac{\frac{5x}{x^2}-\frac{2}{x^2}}{\frac{4x^2}{x^2}-\frac{4x}{x^2}-\frac{3}{x^2}}=\frac{\frac{5}{x}-\frac{2}{x^2}}{4-\frac{4}{x}-\frac{3}{x^2}}$$

As $|x|$ gets very large, the numerator gets close to 0. The function has a horizontal asymptote at $y = 0$. Find the intercepts.

If $x = 0$, $y=\frac{-2}{-3}=\frac{2}{3}$. The y-intercept is $\frac{2}{3}$.

If $y = 0$, $0=\frac{5x-2}{4x^2-4x-3}\Rightarrow 5x=2\Rightarrow x=\frac{2}{5}$

The x-intercept is $\frac{2}{5}$.

x	-5	-3	-1	$-.3$	0	$\frac{2}{5}$
y	$-.23$	$-.38$	-1.4	2.4	$\frac{2}{3}$	0

x	1	$1\frac{1}{4}$	$1\frac{3}{4}$	2	4
y	-1	-2.4	3	1.6	$.4$

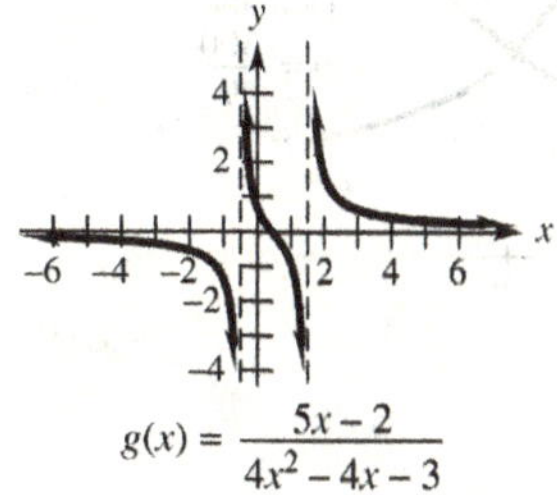

76. $g(x)=\frac{x^2}{x^2-1}$

Find the vertical asymptotes by setting the denominator equal to zero and solving for x.

$x^2-1=0\Rightarrow x^2=1\Rightarrow x=\pm 1$

The lines $x = 1$ and $x = -1$ are vertical asymptotes.

Divide the numerator and denominator by x^2 to find the horizontal asymptotes.

$$\frac{\frac{x^2}{x^2}}{\frac{x^2}{x^2}-\frac{1}{x^2}}=\frac{1}{1-\frac{1}{x^2}}$$

As $|x|$ gets very large, the function approaches 1. The function has a horizontal asymptote at $y = 1$.

Find the intercepts.

If $x = 0$, $y = 0$. The y-intercept is 0.

If $y = 0$, $0=\frac{x^2}{x^2-1}\Rightarrow x=0$

The x-intercept is 0.

x	-3	-2	$-\frac{3}{2}$	$-\frac{2}{3}$	$-\frac{1}{2}$	0
y	$\frac{9}{8}$	$\frac{4}{3}$	$\frac{9}{5}$	$-\frac{4}{5}$	$-\frac{1}{3}$	0

x	$\frac{1}{2}$	$\frac{2}{3}$	$\frac{3}{2}$	2	3
y	$-\frac{1}{3}$	$-\frac{4}{5}$	$\frac{9}{5}$	$\frac{4}{3}$	$\frac{9}{8}$

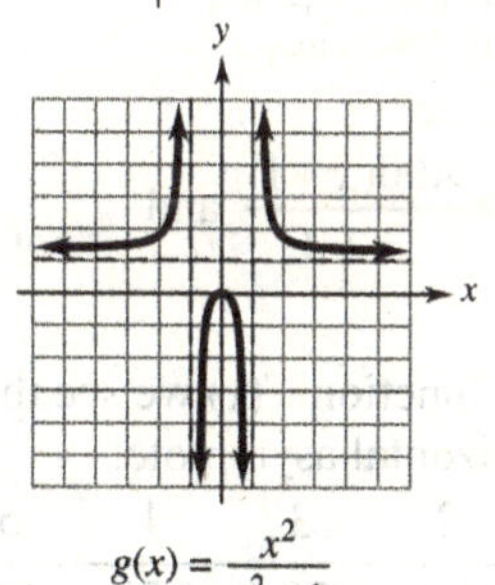

77. $A(x)=\frac{650}{2x+40}$

a. $A(10)=\frac{650}{2(10)+40}\approx 10.83$

The average cost per carton to produce 10 cartons is about \$10.83.

b. $A(50)=\frac{650}{2(50)+40}\approx 4.64$

The average cost per carton to produce 50 cartons is about \$4.64.

c. $A(70)=\frac{650}{2(70)+40}\approx 3.61$

The average cost per carton to produce 70 cartons is about \$3.61.

d. $A(100)=\frac{650}{2(100)+40}\approx 2.71$

The average cost per carton to produce 100 cartons is about \$2.71.

Copyright © 2019 Pearson Education, Inc.

e. We see that $y = 0$ (the x-axis) is a horizontal asymptote.

x	0	10	30	50	70	100
y	16.25	10.83	6.50	4.64	3.61	2.71

Using these points and the horizontal asymptote, we graph the function.

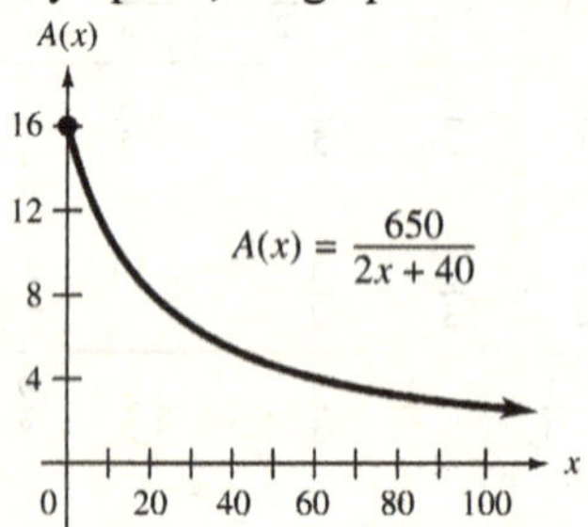

78. $C(x) = \frac{400x + 400}{x + 4} = \frac{400(x + 1)}{x + 4}$ and
$R(x) = 100x$

a. In graphing the function $C(x)$ we see that $y = 400$ is a horizontal asymptote.

x	0	1	2	3	4	5
y	100	160	200	228.6	250	266.7

Using the asymptote and these points we graph the function $C(x)$. The graph of the function $R(x)$ is a line with a slope of 100 and a y-intercept of 0.

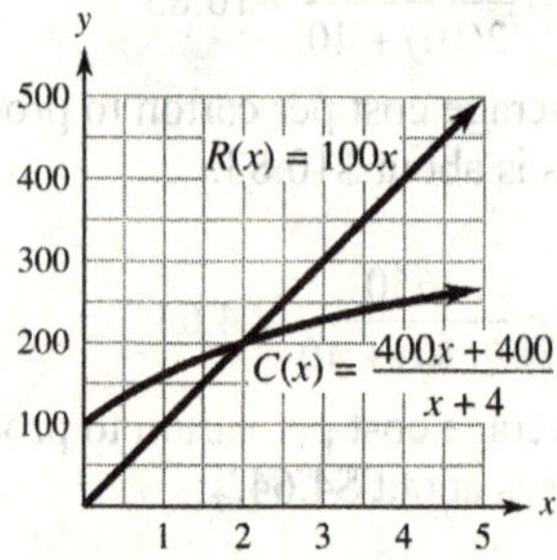

b. To find the break-even point, set $C(x) = R(x)$ and solve for x.

$\frac{400(x + 1)}{x + 4} = 100x$

Divide both sides by 100.

$\frac{4(x + 1)}{x + 4} = x \Rightarrow 4x + 4 = x^2 + 4x \Rightarrow$

$x^2 = 4 \Rightarrow x = \pm 2$

Reject –2 as a solution because the number of hundreds of units cannot be negative. $x = 2$ represents 200 units, so the break-even point is 200 units.

c. $P(1)$ represents a loss. From the graph we see that $C(1) > R(1)$. If the cost is greater than revenue, there is a loss.

d. $P(4)$ represents a profit. From the graph we see that $R(4) > C(4)$. If revenue is greater than cost, there is a profit.

79. Supply: $p = \frac{q^2}{4} + 25$

Demand: $p = \frac{500}{q}$

q	0	5	10	15
Supply	25	31.25	50	81.25
Demand	No value	100	50	33.3

The equilibrium point is (10, 50).

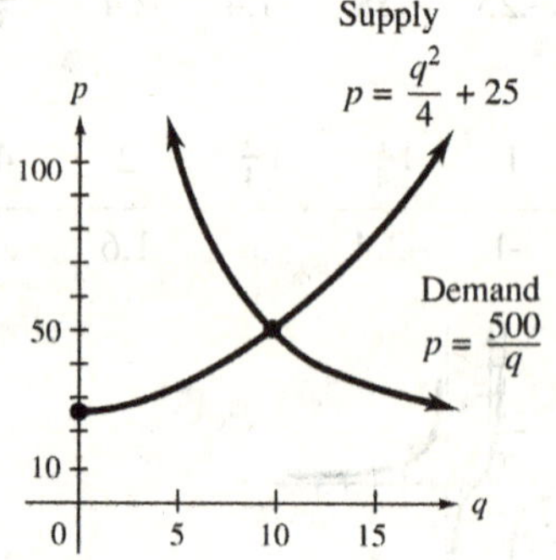

80. $y = \frac{9.2x}{106 - x}$

a. If $x = 50$, $y = \frac{9.2(50)}{106 - 50} \approx 8.2$ or about $8200.

b. If $x = 98$, $y = \frac{9.2(98)}{106 - 98} \approx 112.7$ or about $112,700

c. If $y = \$22,000$, then we have

$22 = \frac{9.2x}{106 - x} \Rightarrow 22(106 - x) = 9.2x \Rightarrow$

$2332 - 22x = 9.2x \Rightarrow 2332 = 31.2x \Rightarrow$

$x \approx 74.7$

About 75% of the pollutant can be removed for $22,000.

Copyright © 2019 Pearson Education, Inc.

Case 3 Maximizing Profit

1. The maximum profit occurs at the vertex.

$$x = -\frac{b}{2a} = -\frac{5.50}{2(-.005)} = 550$$

$$f(550) = -.005(550)^2 + 5.50(550) - 520$$
$$f(550) = 992.5$$

The vertex is at (550, 992.5). Therefore, 550 washing machine loads will lead to a maximum profit of $992.50.

2. No, the values are not the same. The owner should prefer to maximize profit.

3. Based on this information, let $x = 550$ and find the price:

$$p = -.005x + 6$$
$$p = -.005(550) + 6$$
$$p = 3.25$$

The owner should charge $3.25 per load to maximize the profit.

a. The maximum profit occurs at the vertex.

$$x = -\frac{b}{2a} = -\frac{2.50}{2(-.0025)} = 500$$

$$f(500) = -.0025(500)^2 + 2.50(500) - 520$$
$$f(500) = 105$$

The vertex is at (500, 105). Therefore, 500 dryer loads will lead to a maximum profit.

b. The maximum profit from the dryer loads would be $105.

c. Based on this information, let $x = 500$ and find the price:

$$p = -.0025x + 2.75$$
$$p = -.0025(500) + 2.75$$
$$p = 1.50$$

The owner should charge $1.50 per load to maximize the profit.

Case 3 Extended Project

$$m = \frac{1.00 - 1.25}{700 - 600} = \frac{-.25}{100} = -.0025$$

$$p - p_1 = m(x - x_1)$$
$$p - 1 = -.0025(x - 700)$$
$$p - 1 = -.0025 + 1.75$$
$$p = -.0025 + 2.75$$

$$R(x) = xp$$
$$R(x) = x(-.0025x + 2.75)$$
$$R(x) = -.0025x^2 + 2.75x \quad \text{and}$$
$$C(x) = 520 + .25x$$

Therefore, the profit function will be:

$$P(x) = R(x) - C(x)$$
$$P(x) = (-.0025x^2 + 2.75x) - (520 + .25x)$$
$$P(x) = -.0025x^2 + 2.50x - 520$$

Copyright © 2019 Pearson Education, Inc.

Chapter 4 Exponential and Logarithmic Functions

Section 4.1 Exponential Functions

1. $f(x)=6^x$

   This function is exponential, because the variable is in the exponent and the base is a positive constant other than 1.

3. $h(x)=4x^2-x+5$

   This function is quadratic because it is a polynomial function of degree 2.

5. $f(x)=675(1.055^x)$

   This function is exponential because the variable is in the exponent and the base is a positive constant other than 1.

7. $f(x)=.6^x$

   a. The equation describing this function has the format $f(x)=a^x$ with $0<a<1$, so the graph lies entirely above the x-axis and falls from left to right. It falls relatively steeply until it reaches the y-intercept 1 and then falls slowly, with the positive x-axis as a horizontal asymptote.

   b. $f(0)=.8^0=1;\ (0,\ 1)$

   $f(1)=.6^1=.6;\ (1,\ .6)$

9. $h(x)=2^{.5x}$

   a. The base of this exponential function is greater than 1, so the graph lies entirely above the x-axis and rises from left to right. The negative x-axis is a horizontal asymptote. The graph rises slowly until it reaches the y-intercept 1 and then rises quite steeply.

   b. $h(0)=2^{.5(0)}=2^0=1;\ (0,\ 1)$

   $h(1)=2^{.5(1)}=2^{.5};\ (1,\ 2^{.5})=(1,\ \sqrt{2})$

11. $f(x)=e^{-x}$

   a. $e^{-x}=\left(\frac{1}{e}\right)^x$, so this exponential function is of the form $f(x)=a^x$ with $0<a<1$. The graph lies entirely above the x-axis and falls from left to right. It falls very steeply until it reaches the y-intercept 1 and then falls slowly, with the positive x-axis as a horizontal asymptote.

   b. $f(0)=e^{-0}=1;\ (0,1)$

   $f(1)=e^{-1}=\frac{1}{e};\ \left(1,\frac{1}{e}\right)\approx(1,.367879)$

13. $f(x)=3^x$

   Construct a table of ordered pairs, plot the points, and connect them to form a smooth curve.

x	−2	−1	0	1	2
y	$\frac{1}{9}$	$\frac{1}{3}$	1	3	9

   This graph crosses the y-axis at 1 and it is always above the x-axis.

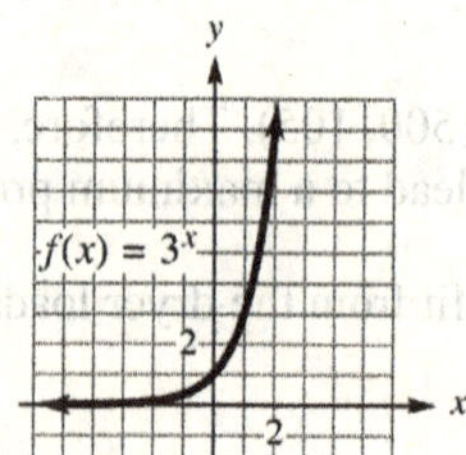

15. $f(x)=2^{x/2}$

   Construct a table of ordered pairs.

x	−4	−2	0	2	4
y	$\frac{1}{4}$	$\frac{1}{2}$	1	2	4

   This graph crosses the y-axis at 1 and it is always above the x-axis.

Copyright © 2019 Pearson Education, Inc.

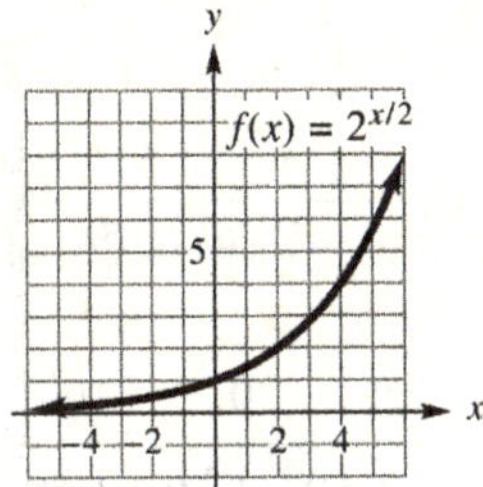

17. $f(x)=\left(\frac{1}{5}\right)^x$

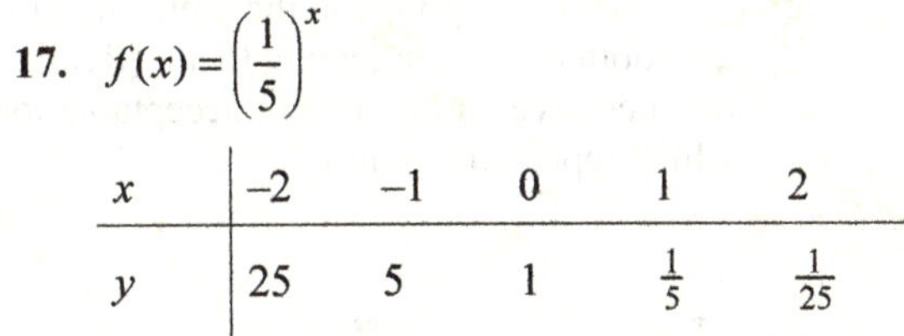

x	-2	-1	0	1	2
y	25	5	1	$\frac{1}{5}$	$\frac{1}{25}$

This graph crosses the y-axis at 1 and it is always above the x-axis.

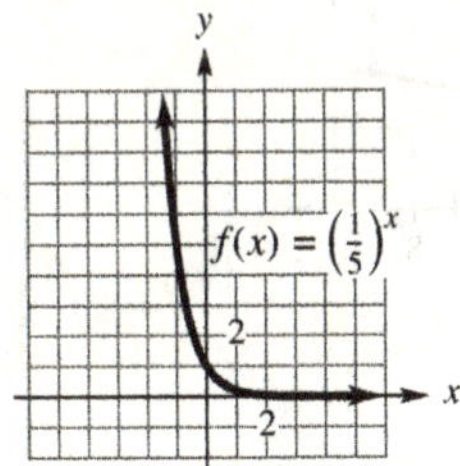

19. **a.** $f(x)=2^x$

x	-2	-1	0	1	2
y	$\frac{1}{4}$	$\frac{1}{2}$	1	2	4

b. $g(x)=2^{x+3}$

x	-2	-1	0	1
y	2	4	8	16

c. $h(x)=2^{x-4}$

x	-1	0	1	2	3	4
y	$\frac{1}{32}$	$\frac{1}{16}$	$\frac{1}{8}$	$\frac{1}{4}$	$\frac{1}{2}$	1

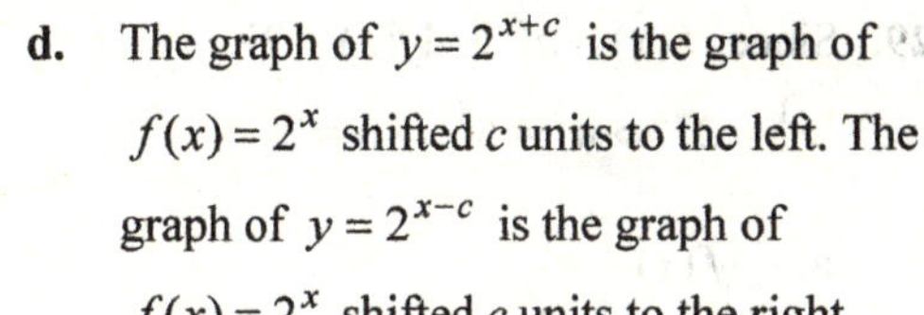

d. The graph of $y=2^{x+c}$ is the graph of $f(x)=2^x$ shifted c units to the left. The graph of $y=2^{x-c}$ is the graph of $f(x)=2^x$ shifted c units to the right.

21. $y=a^x$

$y=2.3^x$ is A

23. $y=a^x$

$y=.75^x$ is C

25. $y=a^x$

$y=.31^x$ is E

27. **a.** $a>1$

b. Domain: $(-\infty, \infty)$
Range: $(0, \infty)$

c.

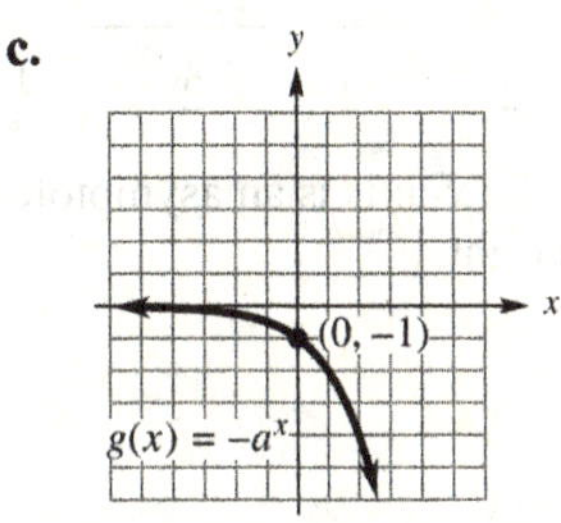

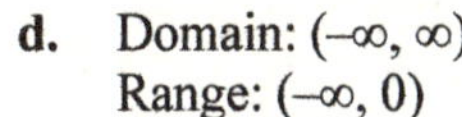

d. Domain: $(-\infty, \infty)$
Range: $(-\infty, 0)$

e.

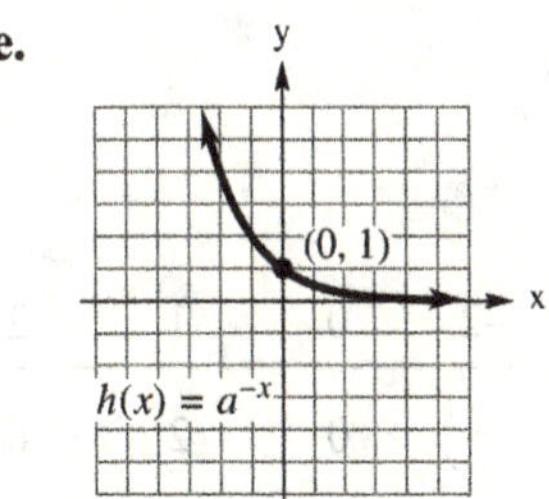

f. Domain: $(-\infty, \infty)$
Range: $(0, \infty)$

Copyright © 2019 Pearson Education, Inc.

29. Since $f(x) = a^x$ and $f(3) = 27$,

$a^3 = 27 \Rightarrow a^3 = 3^3 \Rightarrow a = 3$

a. $f(x) = 3^x$
$f(1) = 3^1 = 3$

b. $f(x) = 3^x$
$f(-1) = 3^{-1} = \frac{1}{3}$

c. $f(x) = 3^x$
$f(2) = 3^2 = 9$

d. $f(x) = 3^x$
$f(0) = 3^0 = 1$

31. $f(x) = 2^{-x^2+2}$

x	−2	−1	0	1	2
y	$\frac{1}{4}$	2	4	2	$\frac{1}{4}$

For this graph, the x-axis is an asymptote on the left and on the right.

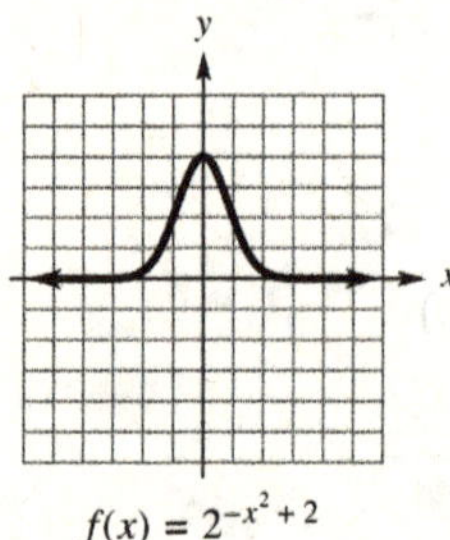

33. $f(x) = x \cdot 2^x$

x	−4	−2	0	1	2
y	$-\frac{1}{4}$	$-\frac{1}{2}$	0	2	8

This graph passes through the origin, and it has the negative x-axis as a horizontal asymptote towards the left.

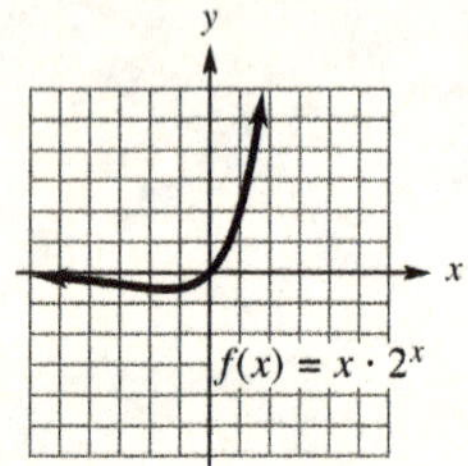

35. $y = (1.06)^t$

a.

t	0	1	2	3	4	5
y	1	1.06	1.12	1.19	1.26	1.34

t	6	7	8	9	10
y	1.42	1.50	1.59	1.69	1.79

b. Plot the eleven points in the table and draw a smooth curve connecting them. (Realize that negative values are not acceptable for t, which represents time.)

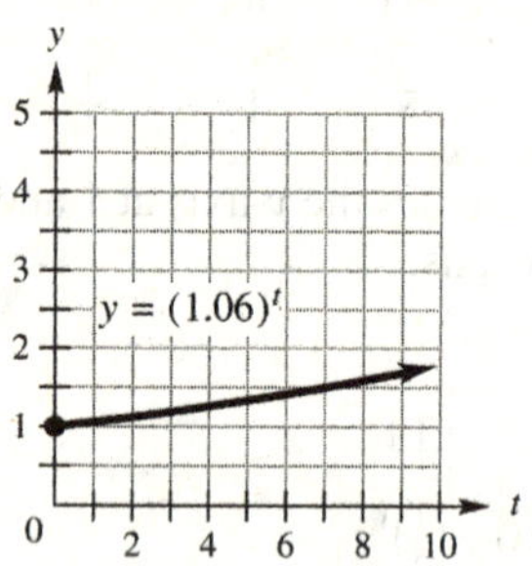

37. $y = (.97)^t$

a. When $t = 10, y = (.97)^{10} \approx .74$.
Let x represent the unknown cost.
$.74x = 105,000$
$x = \frac{105,000}{.74}$
$x \approx 141,892$
The house will cost about $141,892.

b. When $t = 8$, $y = (.97)^8 \approx .78$.
Let x represent the unknown cost.
$.78x = 50 \Rightarrow x = \frac{50}{.78} \Rightarrow x \approx 64.10$
The book will cost about $64.10.

39. $f(x) = 315e^{.056x}$
$x = 7$ corresponds to 2007.

a. $x = 2012 - 2000 = 12$
$f(12) = 315e^{.056(12)} \approx 616.8$
The assets for Prudential Financial, Inc were approximately $616.8 billion in 2012.

b. $x = 2015 - 2000 = 15$
$f(15) = 315e^{.056(15)} \approx 729.7$
The assets for Prudential Financial, Inc were approximately $729.7 billion in 2015.

Copyright © 2019 Pearson Education, Inc.

c. $x = 2018 - 2000 = 18$

$f(18) = 315e^{.056(18)} \approx 863.1$

The assets for Prudential Financial, Inc will be approximately \$863.1 billion in 2018.

41. $W(x) = 2^{-x/24{,}360}$

a. $W(1000) = 2^{-1000/24{,}360} \approx .97$

After 1000 years, about .97 kg will be left.

b. $W(10{,}000) = 2^{-10{,}000/24{,}360} \approx .75$

After 10,000 years, about .75 kg will be left.

c. $W(15{,}000) = 2^{-15{,}000/24{,}360} \approx .65$

After 15,000 years, about .65 kg will be left.

d. $W(24{,}360) = 2^{-24{,}360/24{,}360} \approx .5$

It will take 24,360 years for the one kilogram to decay to half its original weight.

43. If $C = \$244{,}000$, $n = 12$, and $r = .15$, then

$$S = C(1-r)^n = 244{,}000(1-.15)^{12}$$
$$= 244{,}000(.85)^{12} \approx \$34{,}706.99$$

The scrap value is about \$34,706.99.

45. $P(t) = 4.834\left(1.01^{(t-1980)}\right)$

a. $t = 2015$

$$P(2015) = 4.834\left(1.01^{(2015-1980)}\right)$$
$$= 4.834(1.01^{35}) \approx 6.85$$

In 2015, the population was about 6.85 billion.

b. $t = 2020$

$$P(2020) = 4.834\left(1.01^{(2020-1980)}\right)$$
$$= 4.834(1.01^{40}) \approx 7.20$$

In 2020, the population will be about 7.20 billion.

c. $t = 2040$

$$P(2040) = 4.834\left(1.01^{(2040-1980)}\right)$$
$$= 4.834(1.01^{60}) \approx 8.78$$

In 2040, the population will be about 8.78 billion.

d. Answers will vary.

For Exercises 47 – 50, use the following equations:

$$f(x) = .00011\left(1.102^x\right), \quad g(x) = .0132\left(1.067^x\right)$$

47. $x = 0$ corresponds to 1900.

$x = 2015 - 1900 = 115$

$f(115) = .00011\left(1.102^{115}\right) \approx 7.8$

$x = 2025 - 1900 = 125$

$f(125) = .00011\left(1.102^{125}\right) \approx 20.6$

According to the model, in 2015, the GDP of China was about \$7.8 trillion, and the GDP of China will be about \$20.6 trillion in 2025.

49. Graph $Y_1 = .00011(1.102)^x$ and $Y_2 = 25$ in the window [60, 150] by [0, 30], then find the intersection.

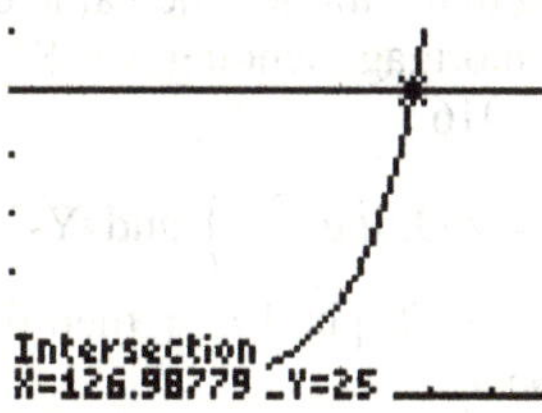

The GDP of China will surpass \$25 trillion in the year 2027.

Copyright © 2019 Pearson Education, Inc.

51. a. $x = 2012 - 2000 = 12$

$A(12) = .34\left(1.19^{12}\right) \approx 2.74$

$x = 2016 - 2000 = 16$

$A(16) = .34\left(1.19^{16}\right) \approx 5.50$

According to the model, the amount of money in passively – managed mutual funds was $2.74 trillion and $5.50 trillion in 2012 and 2016 respectively.

b. Graph $Y_1 = .34\left(1.19^x\right)$ and $Y_2 = 6.5$ in the window [0, 20] by [0, 8], then find the intersection.

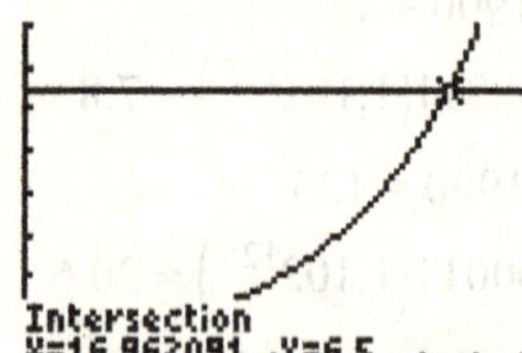

The amount of passively – managed mutual funds will exceed $6.5 trillion in 2017.

53. a. $x = 2016 - 2000 = 16$

$g(16) = 243.5\left(e^{-.25x}\right) \approx 4.46$

According to the model, the value of subprime mortgage lending was $4.46 billion in 2016.

b. Graph $Y_1 = 243.5\left(e^{-.25x}\right)$ and $Y_2 = 3$ in the window [0, 25] by [0, 5], then find the intersection.

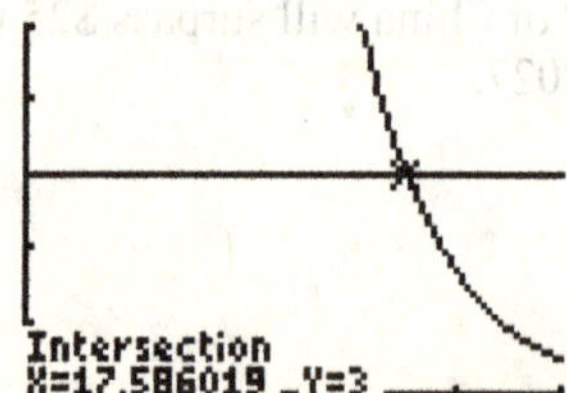

The value of subprime mortgage lending will fall below $3 billion in 2018.

Section 4.2 Applications of Exponential Functions

1. $B(t) = 800\left(.9898^t\right)$

a. $B(6) = 800\left(.9898^6\right) \approx 752.27$

Your balance after 6 months is $752.27.

b. $B(12) = 800\left(.9898^{12}\right) \approx 707.39$

Your balance after 1 year is $707.39.

c. $B(60) = 800\left(.9898^{60}\right) \approx 432.45$

Your balance after 5 years is $432.45.

d. $B(96) = 800\left(.9898^{96}\right) \approx 298.98$

Your balance after 8 years is $298.98.

e. Answers vary. Your balance will take years to reach 0, so never pay the minimum amount.

3. $g(t) = 14.66\left(1.042^t\right)$, where $t = 10$ corresponds to 2010.

a. $t = 2015 - 2000 = 15$

$g(15) = 14.66\left(1.042^{15}\right) \approx 27.17$

According to the model, in 2015, the amount of dry natural gas energy generated in the United States was about 27.17 trillion cubic feet.

b. $t = 2022 - 2000 = 22$

$g(22) = 14.66\left(1.042^{22}\right) \approx 36.24$

According to the model, in 2022, the amount of dry natural gas energy generated in the United States will be about 36.24 trillion cubic feet.

5. a. Let $f(t)$ represent the average asking price for office space rent in San Francisco in the year t after 2004.
The values of $f(t)$ at $t = 0$ and $t = 2015 - 2004 = 11$ are given, that is, $f(0) = 30.48$ and $f(11) = 66.71$. Solving the first of these equations for y_0 in $f(t) = y_0 b^t$:

$f(0) = 30.48 \Rightarrow y_0 b^0 = 30.48 \Rightarrow$
$y_0 = 30.48$

The model has the form $f(t) = 30.48b^t$.
Solving the second equation, $f(11) = 66.71$, for b:

$f(11) = 66.71 \Rightarrow 30.48b^{11} = 66.71 \Rightarrow$

$$b^{11} = \frac{66.71}{30.48} \Rightarrow b = \left(\frac{66.71}{30.48}\right)^{1/11} \approx 1.074$$

The model is $f(t) = 30.48(1.074)^t$.

Copyright © 2019 Pearson Education, Inc.

b. $t = 2016 - 2004 = 12$

$f(12) = 30.48(1.074)^{12} \approx 71.79$

According to the model, the average asking office space rent per square foot in San Francisco was about \$71.79 in 2016.

c. Graph $Y_1 = 30.48(1.074)^x$ and $Y_2 = 45$ in the window [0, 20] by [1000, 3000], then find the intersection.

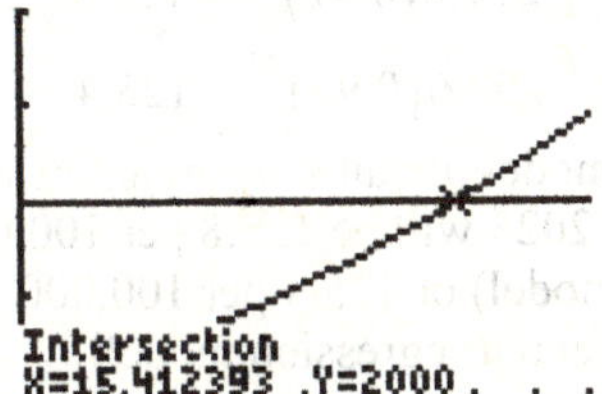

According to the model, the average asking price for office space rent per square foot in San Francisco exceeded \$45 in 2010.

7. a. Let $f(t)$ represent the expenditures of video and audio equipment, computers, and related services in billions in year t after 1990. The values of $f(t)$ at $t = 0$ and $t = 2015 - 1990 = 25$ are given, that is, $f(0) = 9.6$ and $f(25) = 109.2$. Solve the first of these equations for y_0 in $f(t) = y_0b^t$.

$f(0) = 9.6 \Rightarrow y_0b^0 = 9.6 \Rightarrow y_0 = 9.6$

Now solve the second equation $f(25) = 109.2$ for b.

$f(25) = 109.2 \Rightarrow 9.6b^{25} = 109.2 \Rightarrow$

$b^{25} = \dfrac{109.2}{9.6} \Rightarrow b = \left(\dfrac{109.2}{9.6}\right)^{1/25} \approx 1.102$

The model is $f(t) = 9.6(1.102^t)$.

b. $t = 2014 - 1990 = 24$

$f(24) = 9.6(1.102^{24}) \approx 98.8$

According to the model, expenditures were about \$98.8 billion in 2014.

c. $t = 2020 - 1990 = 30$

$f(30) = 9.6(1.102^{30}) \approx 176.9$

According to the model, expenditures will be about \$176.9 billion in 2020.

9. a. Let $t = 0$ correspond to 2000. Using the data for 2000 and 2015 to find a function of the form $f(t) = y_0b^t$, first solve for y_0:

$f(0) = 17 \Rightarrow y_0b^0 = 17 \Rightarrow y_0 = 17$

The function has the form $f(t) = 17b^t$.

$t = 2015 - 2000 = 15$

Now solve for b.

$f(15) = 199 \Rightarrow 17b^{15} = 199 \Rightarrow$

$b = \left(\dfrac{199}{17}\right)^{1/15} \approx 1.178$

Thus the function is $f(t) = 17(1.178)^t$.

Using exponential regression on a graphing calculator, the function produced is $g(t) = 17.4(1.181^t)$.

```
ExpReg
 y=a*b^x
 a=17.36283303
 b=1.181113756
 r²=.9966464141
 r=.9983217989
```

b. Two-point model:

$t = 2018 - 2000 = 18$

$f(18) = 17(1.178)^{18} \approx 324.4$

In 2018, the total Chinese Bank assets will be about 324.4 trillion yuan.

$t = 2021 - 2000 = 21$

$f(21) = 17(1.178)^{21} \approx 530.3$

In 2021, the total Chinese Bank assets will be about 530.3 trillion yuan.

Exponential regression model:

$g(18) = 17.4(1.181)^{18} \approx 347.6$

In 2018, the total Chinese Bank assets will be about 347.6 trillion yuan.

$g(21) = 17.4(1.181)^{21} \approx 572.5$

In 2021, the total Chinese Bank assets will be about 572.5 trillion yuan.

c. Two point model:

Graph $Y_1 = 17(1.178)^x$ and $Y_2 = 450$ in the window [0, 25] by [0, 500], then find the intersection.

Copyright © 2019 Pearson Education, Inc.

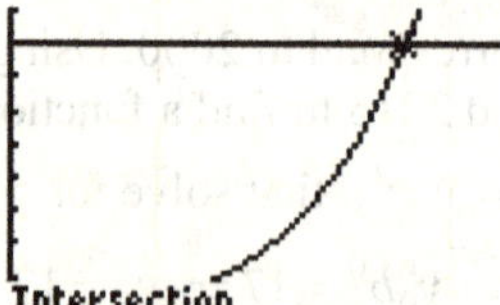

Intersection
X=19.998001 Y=450

According to the two-point model, the Chinese Bank assets will exceed 450 trillion yuan in 2020.

Exponential regression model:

Graph $Y_1 = 17.4(1.181^x)$ and $Y_2 = 450$ in the window [0, 20] by [0, 500], then find the intersection.

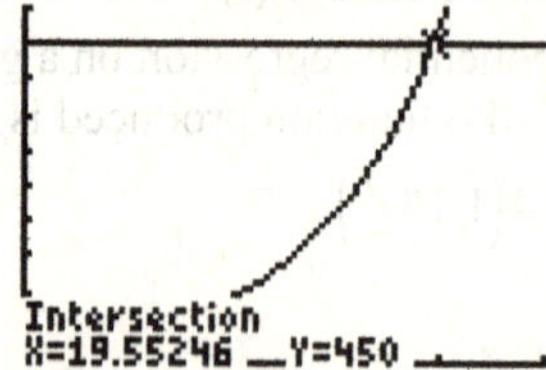

Intersection
X=19.55246 Y=450

According to the exponential regression model, the Chinese Bank assets will exceed 450 trillion yuan in 2020.

11. a. Let $f(t)$ be the death rate per 100,000 population in the year t after 2000.

Two-point model:

$f(0) = 257.6 \Rightarrow y_0 b^0 = 257.6 \Rightarrow$
$y_0 = 257.6$

$t = 2014 - 2000 = 14$

$f(14) = 167.0 \Rightarrow 257.6b^{14} = 167.0 \Rightarrow$

$$b^{14} = \frac{167.0}{257.6} \Rightarrow b = \left(\frac{167.0}{257.6}\right)^{1/14} \approx 0.97$$

$f(t) = 257.6(0.97)^t$, where $t = 0$ corresponds to 2000.

Exponential regression model:

```
ExpReg
 y=a*b^x
 a=252.6155691
 b=.9679901205
 r²=.970707696
 r=-.9852449929
```

$g(t) = 252.6(0.97)^t$

b. $t = 2019 - 2000 = 19$

$f(19) = 257.6(0.97)^{19} \approx 144.4$

$g(19) = 252.6(0.97)^{19} \approx 141.6$

If the model remains accurate, the death-rate in 2019 will be 144.4 per 100,000 (two-point model) or 141.6 per 100,000 (exponential regression model).

$t = 2023 - 2000 = 23$

$f(23) = 257.6(0.97)^{23} \approx 127.8$

$g(23) = 252.6(0.97)^{23} \approx 125.4$

If the model remains accurate, the death-rate in 2023 will be 127.8 per 100,000 (two-point model) or 125.4 per 100,000 (exponential regression model).

c. Two point model:

Graph $Y_1 = 257.6(0.97)^x$ and $Y_2 = 100$ in the window [20, 40] by [0, 125], then find the intersection.

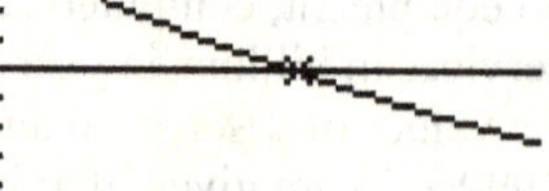

Intersection
X=31.06574 Y=100

According to the two-point model, the death rate from heart disease will drop to 100 per 100,000 population in the year 2032.

Exponential regression model:

Graph $Y_1 = 252.6(0.97)^x$ and $Y_2 = 100$ in the window [20, 40] by [0, 125], then find the intersection.

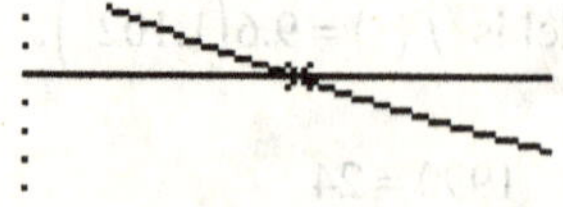

Intersection
X=30.42223 Y=100

According to the exponential regression model, the death rate from heart disease will drop to 100 per 100,000 population in the year 2031.

Copyright © 2019 Pearson Education, Inc.

13. $P(t) = 25 - 25e^{-.3t}$

a. $P(1) = 25 - 25e^{-.3(1)} \approx 6$

b. $P(8) = 25 - 25e^{-.3(8)} \approx 23$

c. The maximum number of items that can be produced is 25.

15. $F(t) = T_0 + Cb^t$, where T_0 is the temperature of the constant environment. At $t = 0$, $F(0) = 100$. So,

$100 = T_0 + Cb^0 \Rightarrow 100 = -18 + Cb^0 \Rightarrow$
$100 = -18 + C(1) \Rightarrow C = 118$. The function then becomes $F(t) = -18 + 118b^t$. At $t = 24$, $F(24) = 50$.

So $50 = -18 + 118b^{24} \Rightarrow$

$$68 = 118b^{24} \Rightarrow b^{24} = \frac{68}{118} \Rightarrow$$

$$b = \left(\frac{68}{118}\right)^{1/24} \approx 0.9773$$

The function now becomes
$F(t) = -18 + 118\left(0.9773^t\right)$. At $t = 76$,

$F(76) = -18 + 118\left(0.9773^{76}\right) \approx 2.6$. So after 76 minutes, the temperature of the water will be about 2.6° C.

17. a. Let $t = 2006 - 2000 = 6$, then

$$g(6) = \frac{49.8}{1 + 52.8e^{-.47(6)}} = \frac{49.8}{1 + 52.8e^{-2.82}} \approx 12.0$$

Let $t = 2015 - 2000 = 15$, then

$$g(15) = \frac{49.8}{1 + 52.8e^{-.47(15)}} = \frac{49.8}{1 + 52.8e^{-7.05}} \approx 47.6$$

According to the model, in the years 2006 and 2015, the percentage of Chinese residents that used the internet was 12.0% and 47.6% respectively.

b. Let $t = 2050 - 2000 = 50$, then

$$g(50) = \frac{49.8}{1 + 52.8e^{-.47(50)}} = \frac{49.8}{1 + 52.8e^{-23.5}} \approx 49.8$$

According to the model, the percentage of Chinese residents using the internet will not reach 50% by the year 2050.

19. $f(x) = \dfrac{120.9}{1 + 3.42e^{-.23x}}$, $x = 0$ corresponds to 2000.

a. $x = 2000 - 2000 = 0$

$$f(0) = \frac{120.9}{1 + 3.42e^{-.23(0)}} \approx 27.4$$

The function estimates the amount of money the U.S. government spent on food and nutrition assistance was about $27.4 billion in 2000.

$x = 2015 - 2000 = 15$

$$f(15) = \frac{120.9}{1 + 3.42e^{-.23(1)}} \approx 109.1$$

The function estimates the amount of money the U.S. government spent on food and nutrition assistance was about $109.1 billion in 2015.

b. To determine the year in which revenues will exceed $100 billion, graph $y_1 = \dfrac{120.9}{1 + 3.42e^{-.23x}}$ and $y_2 = 100$ on the same screen. On the CALC menu, select "intersect" and compute the point of intersection.

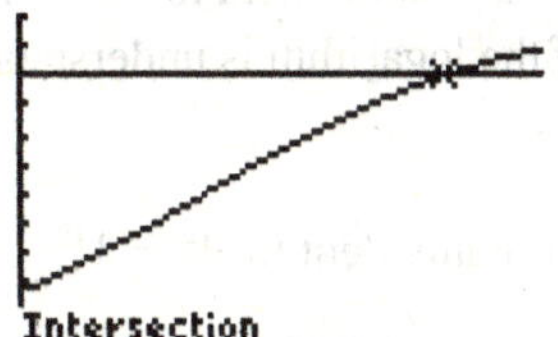

The coordinates of this point are (12.15, 100) so the revenues will exceed $100 billion in the year 2013.

Copyright © 2019 Pearson Education, Inc.

21. $L(x) = \frac{67.9}{1+.014e^{.132x}}$, $x = 0$ corresponds to 2000.

a. $x = 2005 - 2000 = 5$

$g(5) = \frac{67.9}{1+.014e^{.132(5)}} \approx 66.1$

The function estimates the percentage of labor force participation was about 66.1% in 2005.

$x = 2015 - 2000 = 15$

$g(15) = \frac{67.9}{1+.014e^{.132(15)}} \approx 61.6$

The function estimates the percentage of labor force participation was about 61.6% in 2015.

b. To graph $g(x) = \frac{67.9}{1+.014e^{.132x}}$, enter this as y_1 and use [5, 15] by [60, 70]

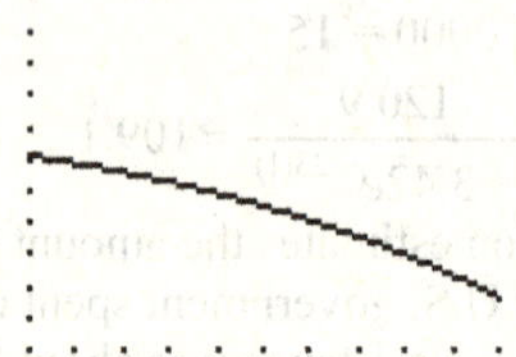

Section 4.3 Logarithmic Functions

1. $y = \log_a x$ means $\underline{x = a^y}$.

3. It is missing the value that equals b^y. If that value is x, it should read $y = \log_b x$.

5. $\log 100{,}000 = 5$ is equivalent to $10^5 = 100{,}000$. (The base of the logarithm is understood to be 10.)

7. $\log_9 81 = 2$ is equivalent to $9^2 = 81$.

9. $10^{1.9823} = 96$ means $\log 96 = 1.9823$.

11. $3^{-2} = \frac{1}{9}$ is equivalent to $\log_3\left(\frac{1}{9}\right) = -2$.

13. $\log 1000 = \log_{10} 10^3 = 3$

15. $\log_6 36 = \log_6 6^2 = 2$

17. $\log_4 64 = \log_4 4^3 = 3$

19. $\log_2 \frac{1}{4} = \log_2 2^{-2} = -2$

21. $\ln\sqrt{e} = \ln e^{1/2} = \frac{1}{2}$

23. $\ln e^{8.77} = 8.77$

25. $\log 53 \approx 1.724$

27. $\ln .0068 \approx -4.991$

29. Answers will vary. Possible answer: $\log_a 1 = 0$ because $a^0 = 1$ for any valid base a.

31. $\log 6 + \log 8 - \log 2 = \log\frac{6(8)}{2} = \log 24$

33. $2\ln 5 - \frac{1}{2}\ln 25 = \ln 5^2 - \ln 25^{1/2} = \ln 25 - \ln 5$

$= \ln\frac{25}{5} = \ln 5$

35. $2\log u + 3\log w - 6\log v$

$= \log u^2 + \log w^3 - \log v^6 = \log\left(\frac{u^2w^3}{v^6}\right)$

37. $2\ln(x+2) - \ln(x+3) = \ln(x+2)^2 - \ln(x+3)$

$= \ln\left(\frac{(x+2)^2}{x+3}\right)$

39. $\ln\sqrt{6m^4n^2} = \ln(6m^4n^2)^{1/2} = \frac{1}{2}\ln 6m^4n^2$

$= \frac{1}{2}(\ln 6 + \ln m^4 + \ln n^2)$

$= \frac{1}{2}(\ln 6 + 4\ln m + 2\ln n)$

$= \frac{1}{2}\ln 6 + 2\ln m + \ln n$

41. $\log\frac{\sqrt{xz}}{z^3} = \log\frac{(xz)^{1/2}}{z^3}$

$= \log(xz)^{1/2} - \log z^3$

$= \frac{1}{2}\log(xz) - 3\log z$

$= \frac{1}{2}\log x + \frac{1}{2}\log z - 3\log z$

$= \frac{1}{2}\log x - \frac{5}{2}\log z$

Copyright © 2019 Pearson Education, Inc.

43. $\ln(x^2y^5) = \ln x^2 + \ln y^5$
$= 2\ln x + 5\ln y$
$= 2u + 5v$

45. $\ln\left(\frac{x^3}{y^2}\right) = \ln x^3 - \ln y^2 = 3\ln x - 2\ln y$
$= 3u - 2v$

47. $\log_6 384 = \frac{\ln 384}{\ln 6} \approx 3.32112$

49. $\log_{35} 5646 = \frac{\ln 5646}{\ln 35} \approx 2.429777$

51. $\log(b + c) = \log b + \log c$
Consider the values $b = 1$ and $c = 2$.
Then $\log(b + c)$ becomes
$\log(1 + 2) = \log 3 \approx .4771$,
while $\log b + \log c$ becomes
$\log 1 + \log 2 \approx .3010$.
Many other choices for the b and c values would demonstrate just as clearly that the statement $\log(b + c) = \log b + \log c$ is generally false.

53. $y = \ln(x + 2)$
We must have $x + 2 > 0$, so the domain is $x > -2$.

x	y
−1.99	−4.6
−1.5	−.7
−1	0
0	.7
2	1.4
4	1.8

Connect these points with a smooth curve.

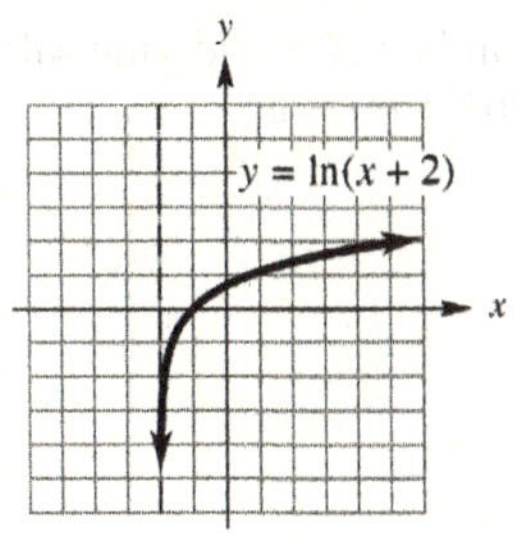

55. $y = \log(x - 3)$
We must have $x - 3 > 0$, so the domain is $x > 3$.

x	y
3.01	−2
3.5	−.30
4	0
6	.48
8	.70

Connect these points with a smooth curve.

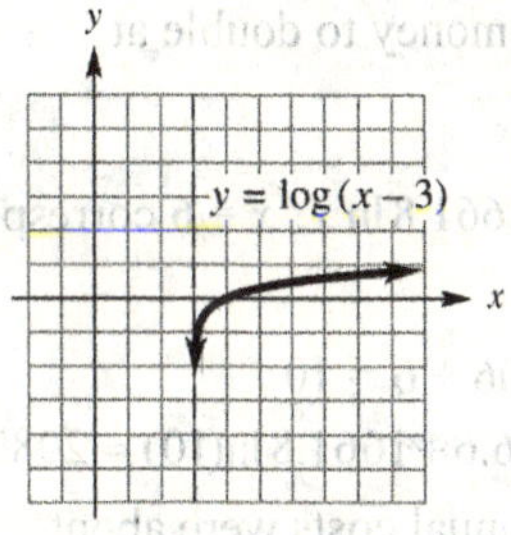

57. Answers will vary. Possible answer:

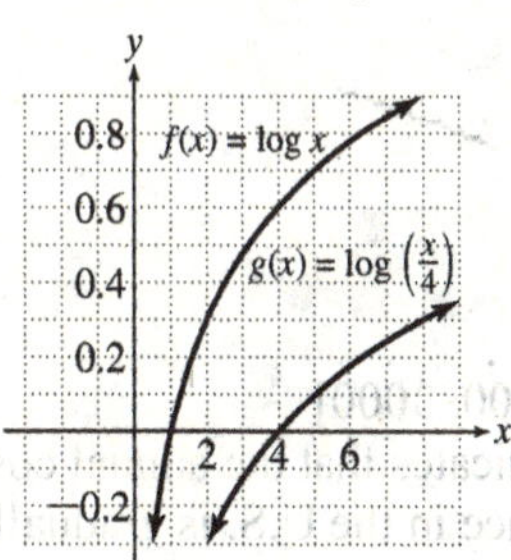

$\log\left(\frac{x}{4}\right) = \log x - \log 4$

$g(x)$ equals log 4 subtracted from $f(x)$

59. $\ln 2.75 = 1.0116009$
$e^{1.0116009} = 2.75$

61. $D(r) = \frac{\ln 2}{\ln(1 + r)}$

a. $D(4\%) = D(.04) = \frac{\ln 2}{\ln(1 + .04)} \approx 17.67$
It takes 17.67 years to double.

b. $D(8\%) = D(.08) = \frac{\ln 2}{\ln(1 + .08)} \approx 9.01$
It takes 9.01 years to double.

Copyright © 2019 Pearson Education, Inc.

c. $D(18\%) = D(.18) = \dfrac{\ln 2}{\ln(1+.18)} \approx 4.19$

It takes 4.19 years to double.

d. $D(36\%) = D(.36) = \dfrac{\ln 2}{\ln(1+.36)} \approx 2.25$

It takes 2.25 years to double.

e. $17.67 \approx 18 = \dfrac{72}{4}$; $9.01 \approx 9 = \dfrac{72}{8}$;

$4.19 \approx 4 = \dfrac{72}{18}$; $2.25 \approx 2 = \dfrac{72}{36}$

The pattern is that it takes about 72/*k* years for money to double at *k*% interest.

63. $g(x) = -1736.6 + 1661.8 \ln x$, $x = 6$ corresponds to 2006.

a. $x = 2010 - 2006 + 6 = 10$

$g(10) = -1736.6 + 1661.8 \ln(10) \approx 2089.84$

In 2010, the annual costs were about $2089.84.

b.

[6, 15] by [1000, 3000]

c. The graph indicates that the annual costs for health insurance in the U.S. is gradually increasing as time goes on.

65. a. $x = 2009 - 2006 + 6 = 9$

$f(9) = 310.4 + 40.7 \ln 9 \approx 399.83$

In 2009, the average expenditures were approximately 399.83.

$x = 2014 - 2006 + 6 = 14$

$f(14) = 310.4 + 40.7 \ln 14 \approx 417.81$

In 2014, the average expenditures were approximately 417.81.

b.

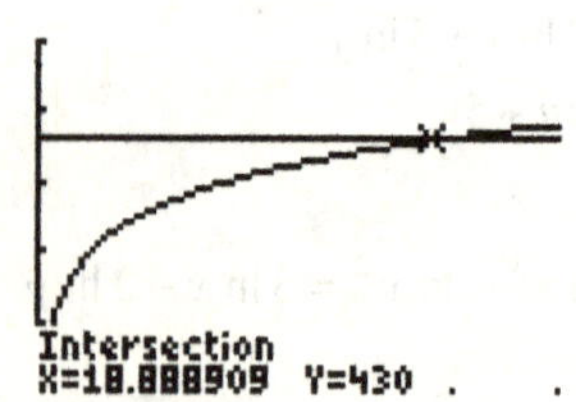

[0, 25] by [250, 500]

According to the model, the average annual expenditures will exceed $430 in 2019.

67. a. $x = 2005 - 2002 + 2 = 5$

$f(5) = .185 + 1.209 \ln 5 \approx 2.1$

In 2005, the amount spent on border patrol was approximately $2.1 billion.

$x = 2016 - 2002 + 2 = 16$

$f(16) = .185 + 1.209 \ln 16 \approx 3.5$

In 2016, the amount spent on border patrol was approximately $3.5 billion.

b.

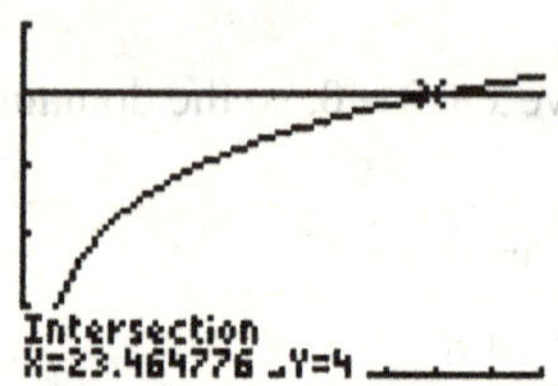

[0, 30] by [0, 5]

According to the model, the amount spent on border patrol will exceed $4.0 billion in 2024.

69. a. $x = 2015 - 2011 + 11 = 15$

$g(15) = -875 + 399 \ln 15 \approx 205.5$

In 2015, the number of worldwide sales for iPhones was 205.5 million.

Copyright © 2019 Pearson Education, Inc.

b.

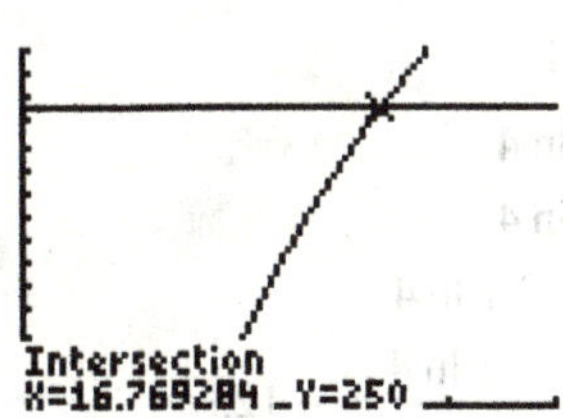

[0, 25] by [0, 300]

According to the model, the number of worldwide sales for iPhones exceeded 250 million in 2017.

Section 4.4 Logarithmic and Exponential Equations

1. $\ln(x+3) = \ln(2x-5) \Rightarrow x+3 = 2x-5$
$x = 8$

3. $\ln(3x+1) - \ln(5+x) = \ln 2$

$$\ln\frac{3x+1}{5+x} = \ln 2$$
$$\frac{3x+1}{5+x} = 2$$
$$3x+1 = 10+2x$$
$$x = 9$$

5. $2\ln(x-3) = \ln(x+5) + \ln 4$

$$\ln(x-3)^2 = \ln[4(x+5)]$$
$$(x-3)^2 = 4(x+5)$$
$$x^2 - 6x + 9 = 4x + 20$$
$$x^2 - 10x - 11 = 0$$
$$(x-11)(x+1) = 0$$
$$x = 11 \text{ or } x = -1$$

Since −1 is not in the domain of ln (x − 3), the only solution is 11.

7. $\log_3(6x-2) = 2 \Rightarrow 6x-2 = 3^2 \Rightarrow$
$6x - 2 = 9 \Rightarrow 6x = 11 \Rightarrow x = \frac{11}{6}$

9. $\log x - \log(x+4) = -1 \Rightarrow \log\left(\frac{x}{x+4}\right) = -1 \Rightarrow$
$\frac{x}{x+4} = 10^{-1} \Rightarrow \frac{x}{x+4} = \frac{1}{10} \Rightarrow 10x = x+4 \Rightarrow$
$9x = 4 \Rightarrow x = \frac{4}{9}$

11. $\log_3(y+2) = \log_3(y-7) + \log_3 4$

$$\log_3(y+2) = \log_3[4(y-7)]$$
$$y+2 = 4(y-7)$$
$$y+2 = 4y-28$$
$$30 = 3y$$
$$y = 10$$

13. $\ln(x+9) - \ln x = 1 \Rightarrow \ln\left(\frac{x+9}{x}\right) = 1 \Rightarrow$
$\frac{x+9}{x} = e^1 \Rightarrow x+9 = ex \Rightarrow x - ex = -9 \Rightarrow$
$(1-e)x = -9 \Rightarrow x = -\frac{-9}{e-1} \approx 5.2378$

15. $\log x + \log(x-9) = 1$

$$\log[x(x-9)] = 1$$
$$x(x-9) = 10^1$$
$$x^2 - 9x = 10$$
$$x^2 - 9x - 10 = 0$$
$$(x-10)(x+1) = 0$$
$$x - 10 = 0 \text{ or } x + 1 = 0$$
$$x = 10 \text{ or } x = -1$$

$x = -1$ is not in the domains of log x or log $(x - 9)$, so the only solution is $x = 10$.

17. $\log(3+b) = \log(4c-1) \Rightarrow 3+b = 4c-1 \Rightarrow$
$4 + b = 4c \Rightarrow \frac{4+b}{4} = c \Rightarrow c = \frac{4+b}{4}$

19. $2 - b = \log(6c+5) \Rightarrow 6c + 5 = 10^{2-b} \Rightarrow$
$6c = 10^{2-b} - 5 \Rightarrow c = \frac{10^{2-b} - 5}{6}$

21. Answers will vary. Sample answer:
Not necessarily. For example, the solution to the example $\log(x+103) = 2$ is $x = -3$. Negative answers must be rejected when they are not in the domain of any of the terms in the equation.

Copyright © 2019 Pearson Education, Inc.

23. $2^{x-1}=8 \Rightarrow 2^{x-1}=2^3 \Rightarrow x-1=3 \Rightarrow x=4$

25. $25^{-3x}=3125 \Rightarrow (5^2)^{-3x}=5^5 \Rightarrow$
$5^{-6x}=5^5 \Rightarrow -6x=5 \Rightarrow x=-\dfrac{5}{6}$

27. $6^{-x} \Rightarrow 36^{x+6}=6^{-x}=\left(6^2\right)^{x+6} \Rightarrow$
$6^{-x}=6^{2(x+6)} \Rightarrow -x=2(x+6) \Rightarrow$
$-x=2x+12 \Rightarrow -3x=12 \Rightarrow x=-4$

29. $\left(\dfrac{3}{4}\right)^x=\dfrac{16}{9} \Rightarrow \left(\dfrac{3}{4}\right)^x=\left(\dfrac{4}{3}\right)^2 \Rightarrow$
$\left(\dfrac{3}{4}\right)^x=\left(\dfrac{3}{4}\right)^{-2} \Rightarrow x=-2$

31. $2^x=5$
Take natural logarithms of both sides.
$\ln 2^x=\ln 5$
$x\ln 2=\ln 5$
$x=\dfrac{\ln 5}{\ln 2}\approx\dfrac{1.6094}{.6931}\approx 2.3219$

33.
$$\begin{aligned}2^x&=3^{x-1}\\ \ln 2^x&=\ln 3^{x-1}\\ x\ln 2&=(x-1)(\ln 3)\\ x\ln 2&=x\ln 3-1\ln 3\\ x\ln 2-x\ln 3&=-\ln 3\\ (\ln 2-\ln 3)x&=-\ln 3\\ x&=\frac{-\ln 3}{\ln 2-\ln 3}\approx 2.710\end{aligned}$$

35.
$$\begin{aligned}3^{1-2x}&=5^{x+5}\\ \ln 3^{1-2x}&=\ln 5^{x+5}\\ (1-2x)(\ln 3)&=(x+5)(\ln 5)\\ \ln 3-2x\ln 3&=x\ln 5+5\ln 5\\ \ln 3-5\ln 5&=x\ln 5+2x\ln 3\\ \ln 3-5\ln 5&=(\ln 5+2\ln 3)x\\ \frac{\ln 3-5\ln 5}{\ln 5+2\ln 3}&=x\\ x&\approx-1.825\end{aligned}$$

37. $e^{3x}=6 \Rightarrow \ln e^{3x}=\ln 6 \Rightarrow$
$3x=\ln 6 \Rightarrow x=\dfrac{\ln 6}{3}\approx .597253$

39.
$$\begin{aligned}2e^{5a+2}&=8\\ e^{5a+2}&=4\\ \ln e^{5a+2}&=\ln 4\\ 5a+2&=\ln 4\\ 5a&=-2+\ln 4\\ a&=\frac{-2+\ln 4}{5}\approx-.123\end{aligned}$$

41.
$$\begin{aligned}10^{4c-3}&=d\\ \log 10^{4c-3}&=\log d\\ 4c-3&=\log d\\ 4c&=\log d+3\\ c&=\frac{\log d+3}{4}\end{aligned}$$

43. $e^{2c-1}=b \Rightarrow \ln e^{2c-1}=\ln b \Rightarrow$
$2c-1=\ln b \Rightarrow 2c=\ln b+1 \Rightarrow c=\dfrac{\ln b+1}{2}$

45.
$$\begin{aligned}\log_7(r+3)+\log_7(r-3)&=1\\ \log_7\left[(r+3)(r-3)\right]&=1\\ (r+3)(r-3)&=7^1\\ r^2-9&=7\\ r^2&=16 \Rightarrow r=\pm 4\end{aligned}$$
Since –4 is not in the domain of $\log_7(r-3)$, 4 is the only solution.

47. $\log_3(a-3)=1+\log_3(a+1) \Rightarrow$
$\log_3(a-3)-\log_3(a+1)=1 \Rightarrow$
$\log_3\dfrac{a-3}{a+1}=1 \Rightarrow \dfrac{a-3}{a+1}=3^1 \Rightarrow$
$a-3=3a+3 \Rightarrow -6=2a \Rightarrow a=-3$
Since –3 is not in the domain of $\log_3(a-3)$ nor in that of $\log_3(a+1)$, there is no solution.

49.
$$\begin{aligned}\log_2\sqrt{2y^2}-1&=\frac{3}{2}\\ \log_2\sqrt{2y^2}&=\frac{5}{2}\\ \sqrt{2y^2}&=2^{5/2}\\ 2y^2&=2^5=32\\ y^2&=16 \Rightarrow y=\pm 4\end{aligned}$$

Copyright © 2019 Pearson Education, Inc.

51. $\log_2(\log_3 x) = 1 \Rightarrow \log_3 x = 2^1 = 2 \Rightarrow$
$x = 3^2 = 9$

53. $5^{-2x} = \frac{1}{25} \Rightarrow 5^{-2x} = 5^{-2} \Rightarrow -2x = -2 \Rightarrow x = 1$

55. $2^{|x|} = 16 \Rightarrow 2^{|x|} = 2^4 \Rightarrow |x| = 4 \Rightarrow x = \pm 4$

57.
$$2^{x^2-1} = 10$$
$$\ln 2^{x^2-1} = \ln 10$$
$$(x^2-1)\ln 2 = \ln 10$$
$$x^2 - 1 = \frac{\ln 10}{\ln 2}$$
$$x^2 - 1 \approx \frac{2.3026}{.6931}$$
$$x^2 - 1 \approx 3.3219$$
$$x^2 = 4.3219$$
$$x = \pm\sqrt{4.3219} \approx \pm 2.0789$$

59. $2(e^x + 1) = 10 \Rightarrow e^x + 1 = 5 \Rightarrow e^x = 4 \Rightarrow$
$\ln e^x = \ln 4 \Rightarrow x = \ln 4 \Rightarrow x \approx 1.386$

61. Answers will vary. Possible answer:
Since $x^2 \geq 0$ for every x, $x^2 + 1 \geq 1$ for every x.
Hence $4^{x^2+1} \geq 4^1 = 4$ for every x.

63. $f(x) = .227(1.17)^x$,
$x = 5$ corresponds to 2005.

a. $f(x) = 1.5 \Rightarrow 1.5 = .227(1.17)^x \Rightarrow$
$\frac{1.5}{.227} = 1.17^x \Rightarrow$
$\ln\left(\frac{1.5}{.227}\right) = x \ln 1.17 \Rightarrow$
$$x = \frac{\ln\left(\frac{1.5}{.227}\right)}{\ln 1.17} \approx 12.03$$
The amount of foreign earnings kept overseas by S&P 500 companies exceeded \$1.5 trillion in 2013.

b. $f(x) = 2.5 \Rightarrow 2.5 = .227(1.17)^x \Rightarrow$
$\frac{2.5}{.227} = 1.17^x \Rightarrow$
$\ln\left(\frac{2.5}{.227}\right) = x \ln 1.17 \Rightarrow$
$$x = \frac{\ln\left(\frac{2.5}{.227}\right)}{\ln 1.17} \approx 15.28$$
The amount of foreign earnings kept overseas by S&P 500 companies exceeded \$2.5 trillion in 2016.

65. $h(x) = 23.5(1.08)^x$,
$x = 5$ corresponds to 1995.

a. $x = 2013 - 1995 + 5 = 23$
$h(23) = 23.5(1.08)^{23} \approx 138.0$
The amount for total expenditures in 2013 was approximately \$138.0 billion.

b. $h(x) = 110 \Rightarrow 110 = 23.5(1.08)^x \Rightarrow$
$\frac{110}{23.5} = 1.08^x \Rightarrow \ln\left(\frac{110}{23.5}\right) = x \ln 1.08 \Rightarrow$
$$x = \frac{\ln\left(\frac{110}{23.5}\right)}{\ln 1.08} \approx 20.06$$
The amount for total expenditures on benefits for U.S. veterans exceeded \$110 billion in 2011.

67. $f(x) = -335.5 + 193.1 \ln x$,
$x = 6$ corresponds to 2006.

a. $f(x) = 250 \Rightarrow 250 = -335.5 + 193.1 \ln x \Rightarrow$
$585.5 = 193.1 \ln x \Rightarrow \frac{585.5}{193.1} = \ln x \Rightarrow$
$x = e^{585.5/193.1} \approx 20.74$
The annual U.S. power generation from wind will exceed 250 million megawatts-hours in 2021.

b. $f(x) = 300 \Rightarrow 300 = -335.5 + 193.1 \ln x \Rightarrow$
$635.5 = 193.1 \ln x \Rightarrow \frac{635.5}{193.1} = \ln x \Rightarrow$
$x = e^{635.5/193.1} \approx 26.87$
The annual U.S. power generation from wind will exceed 300 million megawatts-hours in 2027.

Copyright © 2019 Pearson Education, Inc.

69. $g(x) = -5.3 + 3.54 \ln x,$

$x = 7$ corresponds to 2007.

a. $x = 2010 - 2007 + 7 = 10$

$g(10) = -5.3 + 3.54 \ln(10) \approx 2.85$

The earnings per share in 2010 were approximately \$2.85.

b. $g(x) = 3.25 \Rightarrow 3.25 = -5.3 + 3.54 \ln x \Rightarrow$

$8.55 = 3.54 \ln x \Rightarrow \frac{8.55}{3.54} = \ln x \Rightarrow$

$x = e^{8.55/3.54} \approx 11.2$

The earnings per share exceeded \$3.25 in 2012.

71. $f(x) = 41.6e^{.088x},$

$x = 7$ corresponds to 2007.

a. $x = 2011 - 2007 + 7 = 11$

$f(11) = 41.6e^{.088(11)} \approx 109.5$

The total revenue for CVS Health in 2011 was approximately \$109.5 billion.

b. $f(x) = 140 \Rightarrow 140 = 41.6e^{.088x} \Rightarrow$

$\frac{140}{41.6} = e^{.088x} \Rightarrow \ln\left(\frac{140}{41.6}\right) = .088x \Rightarrow$

$x = \frac{\ln\left(\frac{140}{41.6}\right)}{.088} \approx 13.8$

The total revenue for CVS Health exceeded \$140 billion in 2014.

73. $g(x) = 22e^{-.23x},$

$g(x) = 5.0 \Rightarrow 5.0 = 22e^{-.23x} \Rightarrow$

$\frac{5.0}{22} = e^{-.23x} \Rightarrow \ln\left(\frac{5.0}{22}\right) = -.23x \Rightarrow$

$x = \frac{\ln\left(\frac{5.0}{22}\right)}{-.23} \approx 6.44$

The stock price to earnings ratio for Twitter was lower than 5.0 in the seventh quarter after the first quarter of 2015 or the third quarter of 2016.

75. $C(t) = 25e^{-.14t}$

a. $C(0) = 25e^{-.14(0)} = 25e^0 = 25$

Initially, 25 g of cobalt was present.

b. We determine the half-life by finding a value of t such that $C(t) = \left(\frac{1}{2}\right)(25) = 12.5.$

$12.5 = 25e^{-.14t} \Rightarrow \frac{1}{2} = e^{-.14t} \Rightarrow$

$\ln\frac{1}{2} = \ln e^{-.14t} \Rightarrow \ln\frac{1}{2} = -.14t \Rightarrow$

$t = \frac{\ln\frac{1}{2}}{-.14} = \frac{\ln .5}{-.14} \approx 4.95$

The half-life of cobalt is about 4.95 years.

77. $y = y_0\left(.5^{t/5730}\right)$

36% lost means 64% remains. Since 64% = .64, replace y with $.64y_0$ and solve for t.

$.64y_0 = y_0\left(.5^{t/5730}\right) \Rightarrow .64 = \left(.5^{t/5730}\right) \Rightarrow$

$\ln .64 = \ln\left(.5^{t/5730}\right) \Rightarrow \ln .64 = \frac{t}{5730}\ln .5 \Rightarrow$

$t = \frac{5730 \ln .64}{\ln .5} \approx 3689.3$

The ivory is about 3689 years old.

79. Use the formula for the Richter scale given in Example 9 in this section of the textbook.

$R(i) = \log\left(\frac{i}{i_0}\right)$

a. $R(i) = 7.9 = \log\left(\frac{i}{i_0}\right) \Rightarrow$

$10^{7.9} = \frac{i}{i_0} \Rightarrow i = 10^{7.9} i_0 \approx 79{,}432{,}823 i_0$

b. $R(i) = 5.4 = \log\left(\frac{i}{i_0}\right) \Rightarrow$

$10^{5.4} = \frac{i}{i_0} \Rightarrow i = 10^{5.4} i_0 \approx 251{,}189 i_0$

c. The China earthquake was

$\frac{10^{7.9}}{10^{5.4}} = 10^{7.9-5.4} = 10^{2.5} \approx 316.23$

times stronger than the Los Angeles earthquake.

81. $D(i) = 10 \cdot \log\left(\frac{i}{i_0}\right)$

a. $D(115i_0) = 10\log\left(\frac{115i_0}{i_0}\right)$

$= 10\log 115 \approx 21$

Copyright © 2019 Pearson Education, Inc.

b. $D(10^{10} i_0) = 10\log\left(\dfrac{10^{10} i_0}{i_0}\right)$
$= 10\cdot(10\log 10)$
$= 10\cdot(10\cdot 1) = 100$

c. $D(31,600,000,000 i_0)$
$= 10\log\left(\dfrac{31,600,000,000 i_0}{i_0}\right)$
$= 10\log 31,600,000,000 \approx 105$

d. $D(895,000,000,000 i_0)$
$= 10\log\left(\dfrac{895,000,000,000 i_0}{i_0}\right)$
$= 10\log(895,000,000,000) \approx 120$

e. $D(109,000,000,000,000 i_0)$
$= 10\log\left(\dfrac{109,000,000,000,000 i_0}{i_0}\right)$
$= 10\log(109,000,000,000,000) \approx 140$

Chapter 4 Review Exercises

1. $y = a^{x+2}$ is (c).

2. $y = a^x + 2$ is (a).

3. $y = -a^x + 2$ is (d).

4. $y = a^{-x} + 2$ is (b).

5. $0 < a < 1$

6. Domain of f: $(-\infty, \infty)$ or all real numbers.

7. Range of f: $(0, \infty)$ or all positive real numbers.

8. $f(0) = 1$

9. $f(x) = 4^x$

x	−2	−1	0	1	2
y	$\frac{1}{16}$	$\frac{1}{4}$	1	4	16

The negative x-axis is a horizontal asymptote for the graph.

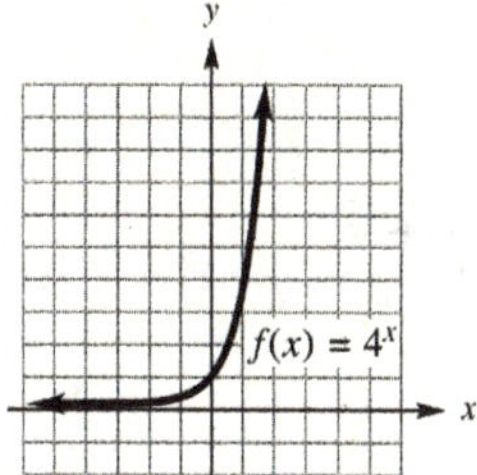

10. $g(x) = 4^{-x}$

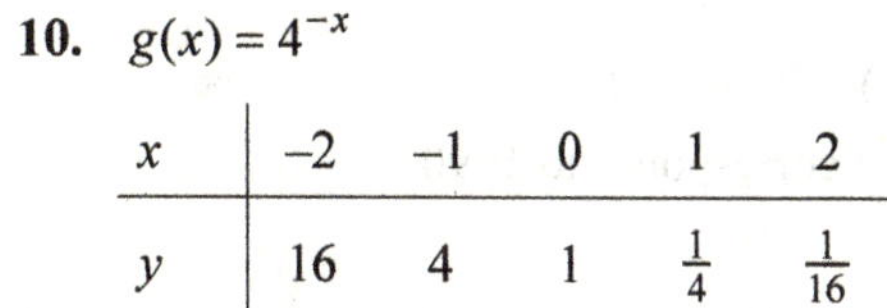

x	−2	−1	0	1	2
y	16	4	1	$\frac{1}{4}$	$\frac{1}{16}$

The positive x-axis is a horizontal asymptote for the graph.

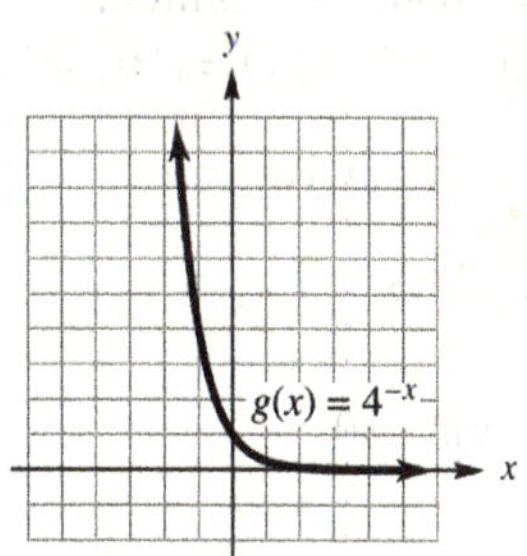

11. $f(x) = \ln x + 5$

x	.01	.1	1	2	4	6	8
y	.4	2.7	5	5.7	6.4	6.8	7.1

The negative y-axis is a vertical asymptote for the graph.

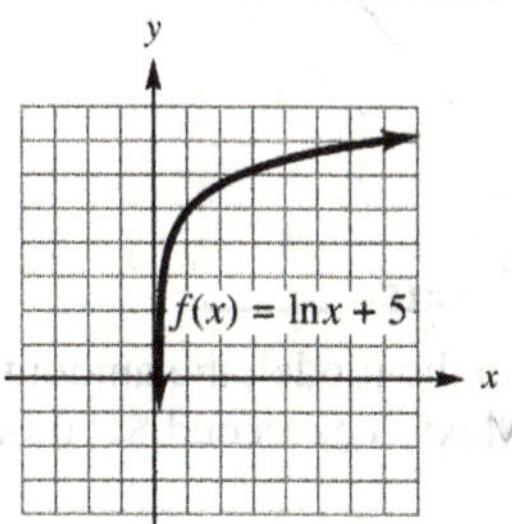

12. $g(x) = \log x - 3$

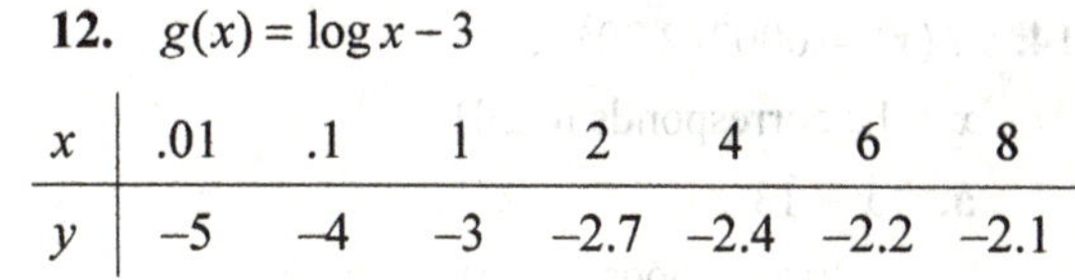

x	.01	.1	1	2	4	6	8
y	−5	−4	−3	−2.7	−2.4	−2.2	−2.1

The negative y-axis is a vertical asymptote for the graph.

Copyright © 2019 Pearson Education, Inc.

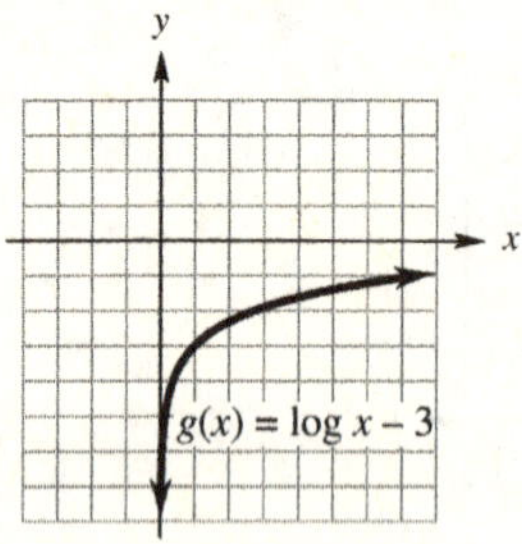

13. $g(x) = 27.1(1.10)^x$,
$x = 0$ corresponds to 1990.

a. $x = 25$

$g(25) = 27.1(1.10)^{25} \approx 294$

In 2015, the amount of U.S. exports to Mexico was about $294 billion.

b. To determine x for $g(x) = 150$, solve:

$$g(x) = 150 = 27.1(1.10)^x \Rightarrow$$

$$\frac{150}{27.1} = (1.10)^x \Rightarrow$$

$$\ln\left(\frac{150}{27.1}\right) = x \ln 1.10 \Rightarrow$$

$$x = \frac{\ln\left(\frac{150}{27.1}\right)}{\ln 1.10} \approx 17.95$$

Alternatively, graph $Y_1 = 27.1(1.10)^x$ and $Y_2 = 150$ in the window [0, 25] by [0, 175], then find the intersection.

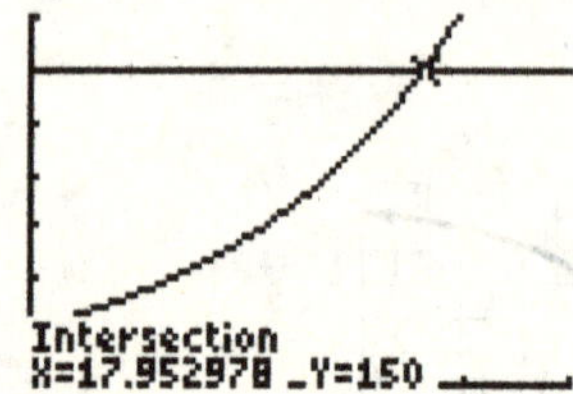

According to the model, the amount of U.S. exports to Mexico exceeded $150 billion in 2008.

14. $f(x) = .0007(2.29)^x$,
$x = 12$ corresponds to 2012.

a. $x = 13$

$f(13) = .0007(2.29)^{13} \approx 33.3$

In 2013, the percentage gain in share price for the cruise line Royal Caribbean was 33.3%.

b. To determine x for $f(x) = 100$, solve:

$$f(x) = 100 = .0007(2.29)^x \Rightarrow$$

$$\frac{100}{.0007} = (2.29)^x \Rightarrow$$

$$\ln\left(\frac{100}{.0007}\right) = x \ln 2.29 \Rightarrow$$

$$x = \frac{\ln\left(\frac{100}{.0007}\right)}{\ln 2.29} \approx 14.33$$

Alternatively, graph $Y_1 = .0007(2.29)^x$ and $Y_2 = 100$ in the window [0, 20] by [0, 125], then find the intersection.

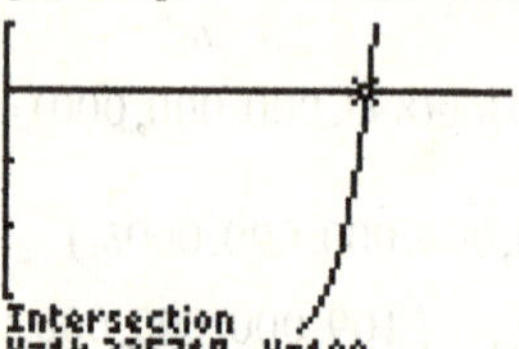

According to the model, the percent gain in share price for Royal Caribbean was greater than 100% in 2015.

15. $10^{2.53148} = 340$ is equivalent to $\log 340 = 2.53148$.

16. $5^4 = 625$ is equivalent to $\log_5 625 = 4$.

17. $e^{3.8067} = 45$ is equivalent to $\ln 45 = 3.8067$.

18. $7^{1/2} = \sqrt{7}$ is equivalent to $\log_7 \sqrt{7} = \frac{1}{2}$.

19. $\log 10{,}000 = 4$ is equivalent to $10^4 = 10{,}000$.

20. $\log 26.3 = 1.4200$ is equivalent to $10^{1.4200} = 26.3$.

21. $\ln 81.1 = 4.3957$ is equivalent to $e^{4.3957} = 81.1$.

22. $\log_2 4096 = 12$ is equivalent to $2^{12} = 4096$.

23. $\ln e^5 = 5$ because $\ln e^k = k$ for every real number k.

24. $\log \sqrt[3]{10} = \log_{10} 10^{1/3} = \frac{1}{3}$ because $\log_a a^y = y$ for every positive real number y.

Copyright © 2019 Pearson Education, Inc.

25. $10^{\log 8.9} = 8.9$ because $a^{\log_a x} = x$ for every positive real number x.

26. $\ln e^{3t^2} = 3t^2$ because $\ln e^j = j$ for every real number j.

27. Let $x = \log_8 2$.

$\log_8 2 = x \Rightarrow 8^x = 2 \Rightarrow \left(2^3\right)^x = 2 \Rightarrow 2^{3x} = 2 \Rightarrow$

$3x = 1 \Rightarrow x = \frac{1}{3}$

Therefore, $\log_8 2 = \frac{1}{3}$.

28. Let $x = \log_8 32$.

$\log_8 32 = x \Rightarrow 8^x = 32 \Rightarrow \left(2^3\right)^x = 2^5 \Rightarrow$

$2^{3x} = 2^5 \Rightarrow 3x = 5 \Rightarrow x = \frac{5}{3}$

Therefore, $\log_8 32 = \frac{5}{3}$.

29. $\log 4x + \log 5x^5 = \log(4x \cdot 5x^5) = \log\left(20x^6\right)$

30. $$\begin{aligned} 4\log u - 5\log u^6 &= \log u^4 - \log\left(u^6\right)^5 \\ &= \log u^4 - \log u^{30} \\ &= \log\left(\frac{u^4}{u^{30}}\right) = \log\left(\frac{1}{u^{26}}\right) \end{aligned}$$

31. $3\log b - 2\log c = \log b^3 - \log c^2 = \log\left(\frac{b^3}{c^2}\right)$

32. $$\begin{aligned} 7\ln x - 3(\ln x^3 + 5\ln x) &= 7\ln x - 3\ln x^3 - 15\ln x \\ &= -3\ln x^3 - 8\ln x \\ &= \ln\left(x^3\right)^{-3} + \ln x^{-8} \\ &= \ln\left(x^{-9}\right) + \ln x^{-8} \\ &= \ln\left(x^{-9} \cdot x^{-8}\right) = \ln x^{-17} \\ &= \ln\left(\frac{1}{x^{17}}\right) \end{aligned}$$

33. $\ln(m+8) - \ln m = \ln 3 \Rightarrow$

$\ln\frac{m+8}{m} = \ln 3 \Rightarrow \frac{m+8}{m} = 3 \Rightarrow 3m = m + 8 \Rightarrow$

$2m = 8 \Rightarrow m = 4$

34. $2\ln(y+1) = \ln(y^2 - 1) + \ln 5$

$\ln(y+1)^2 = \ln\left[(y^2 - 1)(5)\right]$

$(y+1)^2 = 5(y^2 - 1)$

$$\begin{aligned} y^2 + 2y + 1 &= 5y^2 - 5 \\ 0 &= 4y^2 - 2y - 6 \\ 0 &= 2(2y^2 - y - 3) \\ 0 &= 2(2y - 3)(y + 1) \\ y &= \frac{3}{2} \text{ or } y = -1 \end{aligned}$$

Since -1 is not in the domain of $\ln(y + 1)$ nor in that of $\ln(y^2 - 1)$, the only solution is $\frac{3}{2}$.

35. $\log(m+3) = 2 \Rightarrow m + 3 = 10^2 \Rightarrow$

$m + 3 = 100 \Rightarrow m = 97$

36. $\log x^3 = 2 \Rightarrow x^3 = 10^2 \Rightarrow x^3 = 100 \Rightarrow$

$x = \sqrt[3]{100} \approx 4.642$

37. $\log_2(3k+1) = 4 \Rightarrow 3k + 1 = 2^4 \Rightarrow$

$3k + 1 = 16 \Rightarrow 3k = 15 \Rightarrow k = 5$

38. $\log_5\left(\frac{5z}{z-2}\right) = 2 \Rightarrow 5^2 = \frac{5z}{z-2} \Rightarrow$

$25(z-2) = 5z \Rightarrow 25z - 50 = 5z \Rightarrow$

$20z = 50 \Rightarrow z = \frac{50}{20} = 2.5$

39. $\log x + \log(x-3) = 1 \Rightarrow \log\left[x(x-3)\right] = 1 \Rightarrow$

$x(x-3) = 10^1 \Rightarrow x^2 - 3x = 10 \Rightarrow$

$x^2 - 3x - 10 = 0 \Rightarrow (x-5)(x+2) = 0 \Rightarrow$

$x = 5$ or $x = -2$

Since -2 is not in the domain of $\log x$ nor in that of $\log(x - 3)$, the only solution is 5.

40. $$\begin{aligned} \log_2 r + \log_2(r-2) &= 3 \\ \log_2\left[r(r-2)\right] &= 3 \\ r(r-2) &= 2^3 \\ r^2 - 2r - 8 &= 0 \\ (r-4)(r+2) = 0 &\Rightarrow r = 4 \text{ or } r = -2 \end{aligned}$$

Since -2 is not in the domain of $\log_2 r$ nor in that of $\log_2(r-2)$, the only solution is 4.

Copyright © 2019 Pearson Education, Inc.

41. $2^{3x} = \frac{1}{64} \Rightarrow 2^{3x} = 2^{-6} \Rightarrow 3x = -6 \Rightarrow x = -2$

42. $\left(\frac{9}{16}\right)^x = \frac{3}{4} \Rightarrow \left[\left(\frac{3}{4}\right)^2\right]^x = \left(\frac{3}{4}\right)^1 \Rightarrow$

$\left(\frac{3}{4}\right)^{2x} = \left(\frac{3}{4}\right)^1 \Rightarrow 2x = 1 \Rightarrow x = \frac{1}{2}$

43. $9^{2y+1} = 27^y \Rightarrow (3^2)^{2y+1} = (3^3)^y \Rightarrow$

$3^{4y+2} = 3^{3y} \Rightarrow 4y + 2 = 3y \Rightarrow y = -2$

44. $\frac{1}{2} = \left(\frac{b}{4}\right)^{1/4} \Rightarrow \left(\frac{1}{2}\right)^4 = \left[\left(\frac{b}{4}\right)^{1/4}\right]^4 \Rightarrow$

$\frac{1}{16} = \frac{b}{4} \Rightarrow 16b = 4 \Rightarrow b = \frac{4}{16} = \frac{1}{4}$

45. $8^p = 19 \Rightarrow \ln 8^p = \ln 19 \Rightarrow p \ln 8 = \ln 19 \Rightarrow$

$p = \frac{\ln 19}{\ln 8} \Rightarrow p \approx \frac{2.9444}{2.0794} \Rightarrow p \approx 1.416$

46. $3^z = 11 \Rightarrow \ln 3^z = \ln 11 \Rightarrow z \ln 3 = \ln 11 \Rightarrow$

$z = \frac{\ln 11}{\ln 3} \Rightarrow z \approx 2.183$

47. $5 \cdot 2^{-m} = 35 \Rightarrow 2^{-m} = 7 \Rightarrow \ln(2^{-m}) = \ln 7 \Rightarrow$

$-m \ln 2 = \ln 7 \Rightarrow -m = \frac{\ln 7}{\ln 2} \Rightarrow$

$-m = \frac{1.9459}{.6931} \Rightarrow m \approx -2.807$

48. $2 \cdot 15^{-k} = 18 \Rightarrow 15^{-k} = 9 \Rightarrow \ln 15^{-k} = \ln 9 \Rightarrow$

$-k \ln 15 = \ln 9 \Rightarrow k = -\frac{\ln 9}{\ln 15} \Rightarrow k \approx -.811$

49. $e^{-5-2x} = 5 \Rightarrow \ln e^{-5-2x} = \ln 5 \Rightarrow$

$-5 - 2x = \ln 5 \Rightarrow -5 - 2x \approx 1.6094 \Rightarrow$

$-2x = 6.6094 \Rightarrow x \approx -3.305$

50. $e^{3x-1} = 12 \Rightarrow \ln e^{3x-1} = \ln 12 \Rightarrow 3x - 1 = \ln 12 \Rightarrow$

$x = \frac{1 + \ln 12}{3} \approx 1.162$

51.

$$\begin{aligned} 6^{2-m} &= 2^{3m+1} \\ \ln 6^{2-m} &= \ln 2^{3m+1} \\ (2-m)\ln 6 &= (3m+1)\ln 2 \\ 2\ln 6 - m\ln 6 &= 3m\ln 2 + \ln 2 \\ 2\ln 6 - \ln 2 &= 3m\ln 2 + m\ln 6 \\ 2\ln 6 - \ln 2 &= (3\ln 2 + \ln 6)m \\ m &= \frac{2\ln 6 - \ln 2}{3\ln 2 + \ln 6} \approx .747 \end{aligned}$$

52.

$$\begin{aligned} 5^{3r-1} &= 6^{2r+5} \\ \ln 5^{3r-1} &= \ln 6^{2r+5} \\ (3r-1)\ln 5 &= (2r+5)\ln 6 \\ 3r\ln 5 - \ln 5 &= 2r\ln 6 + 5\ln 6 \\ (3\ln 5 - 2\ln 6)r &= 5\ln 6 + \ln 5 \\ r &= \frac{5\ln 6 + \ln 5}{3\ln 5 - 2\ln 6} \approx 8.490 \end{aligned}$$

53.

$$\begin{aligned} (1 + .003)^k &= 1.089 \\ 1.003^k &= 1.089 \\ \ln 1.003^k &= \ln 1.089 \\ k\ln 1.003 &= \ln 1.089 \\ k &= \frac{\ln 1.089}{\ln 1.003} \approx 28.463 \end{aligned}$$

54.

$$\begin{aligned} (1 + .094)^z &= 2.387 \\ 1.094^z &= 2.387 \\ \ln 1.094^z &= \ln 2.387 \\ z\ln 1.094 &= \ln 2.387 \\ z &= \frac{\ln 2.387}{\ln 1.094} \approx 9.684 \end{aligned}$$

55. $y = 4e^{.03t}$

a. The population will double, so $y = 2 \cdot 4$. The answer is C.

b. The population will be 12 thousand, so $y = 12$. The answer is A.

c. $t = 6$ The answer is D.

d. $t = 6(12) = 72$ The answer is B.

Copyright © 2019 Pearson Education, Inc.

56. $y = 2e^{.02t}$

a. The population will triple, so $y = 3 \cdot 2$.
The answer is B.

b. The population will be 3 million, so $y = 3$.
The answer is D.

c. $t = 3$
The answer is C.

d. $t = \frac{4}{12} = \frac{1}{3}$
The answer is A.

57. $g(x) = 846e^{-.034x}$,
$x = 0$ corresponds to 1990.

a. $x = 2012 - 1990 = 22$
$g(22) = 846e^{-.034(22)} \approx 400$
The total number of murders in Chicago in 2012 was approximately 400.

b. $g(x) = 600 \Rightarrow 600 = 846e^{-.034x} \Rightarrow$
$\frac{600}{846} = e^{-.034x} \Rightarrow \ln\left(\frac{600}{846}\right) = -.034x \Rightarrow$
$x = \frac{\ln\left(\frac{600}{846}\right)}{-.034} \approx 10.11$
The total number of murders in Chicago exceeded 600 in 2001.

58. $A(t) = 10e^{-.00495t}$

a. $A(0) = 10e^{-.00495(0)} = 10e^0 = 10$
The amount of polonium present initially was 10 g.

b. We determine the half-life by finding a value of t such that $A(t) = \left(\frac{1}{2}\right)(10) = 5$.

$5 = 10e^{-.00495(t)}$

$\frac{1}{2} = e^{-.00495t}$

$\ln\frac{1}{2} = \ln e^{-.00495t}$

$\ln\frac{1}{2} = -.00495t$

$t = \frac{\ln .5}{-.00495} \approx 140$

The half-life of polonium is about 140 days.

c. Find t such that $A(t) = 3$.

$10e^{-.00495(t)} = 3$

$e^{-.00495(t)} = .3$

$\ln e^{-.00495(t)} = \ln .3$

$-.00495t = \ln .3$

$t = \frac{\ln .3}{-.00495} \approx 243$

It will take about 243 days for the polonium to decay to 3 g.

59. $g(x) = \frac{124}{1 + .118e^{.371x}}$,
$x = 1$ corresponds to the first quarter of 2014.

a. $x = 5$
$g(5) = \frac{124}{1 + .118e^{.371(5)}} \approx 70.69$
The future price of crude oil for the first quarter of 2015 was \$70.69.

b. $g(x) = 50 \Rightarrow 50 = \frac{124}{1 + .118e^{.371x}} \Rightarrow$
$50\left(1 + .118e^{.371x}\right) = 124 \Rightarrow$
$50 + 5.9e^{.371x} = 124 \Rightarrow$
$5.9e^{.371x} = 74 \Rightarrow$
$\ln\left(\frac{74}{5.9}\right) = .371x \Rightarrow$
$x = \frac{\ln\left(\frac{74}{5.9}\right)}{.371} \approx 6.82$
The future price of crude oil fell below \$50 in the seventh quarter after the first quarter of 2014 or the third quarter of 2015.

60. $h(x) = \frac{170}{1 + .009e^{.342x}}$,
$x = 11$ corresponds to 2011.

a. $x = 2016 - 2011 + 11 = 16$
$h(16) = \frac{170}{1 + .009e^{.342(16)}} \approx 54.11$
The future price of coal in 2016 was \$54.11 per metric ton.

Copyright © 2019 Pearson Education, Inc.

b. $h(x) = 90 \Rightarrow 90 = \dfrac{170}{1 + .009e^{.342x}} \Rightarrow$

$90\left(1 + .009e^{.342x}\right) = 170 \Rightarrow$

$90 + .81e^{.342x} = 170 \Rightarrow$

$.81e^{.342x} = 80 \Rightarrow$

$\ln\left(\dfrac{80}{.81}\right) = .342x \Rightarrow$

$x = \dfrac{\ln\left(\dfrac{80}{.81}\right)}{.342} \approx 13.43$

The future price of coal below \$90 per metric ton in 2014.

61. a. $i(7.8) = i_0 10^x = i_0 10^{7.8}$

$i(7.8) = 63{,}095{,}734.45 i_0$

b. $i(6) = i_0 10^6 = 1{,}000{,}000 i_0$

c. $\dfrac{i(7.8)}{i(6)} = \dfrac{63{,}095{,}734.45 i_0}{1{,}000{,}000 i_0} \approx 63.1$

The earthquake in Ecuador was about 63.1 times more intense than the earthquake in Mexico.

62. For earthquakes, increasing the ground motion by a factor of 10^k increases the Richter magnitude by k units. In this problem, the second earthquake, with ground motion $1000 = 10^3$ times greater than the first earthquake, will measure 4.6 + 3 = 7.6 on the Richter scale.

63. $F(t) = T_0 + Ce^{-kt}$

$T_0 = 50, F(0) = 50 + Ce^{-k(0)} = 300, C = 250$

$F(t) = 50 + 250e^{-kt}$

$F(4) = 175 = 50 + 250e^{-k(4)}$

$125 = 250e^{-k(4)}$

$\dfrac{125}{250} = e^{-k(4)} \Rightarrow \ln\left(\dfrac{125}{250}\right) = -k(4) \Rightarrow k \approx .1733$

$F(12) = 50 + 250e^{-.1733(12)} \approx 81.25$

The temperature after 12 minutes is 81.25° C.

64. $F(t) = T_0 + Ce^{-kt}, T_0 = 18$

$F(0) = 3.4 = 18 + C \Rightarrow C = -14.6$

$F(30) = 18 - 14.6e^{-k(30)} = 7.2$

$-14.6e^{-k(30)} = 7.2 - 18$

$e^{-k(30)} = \dfrac{7.2 - 18}{-14.6} \approx 0.739726$

$k = \dfrac{\ln 0.739726}{-30} \approx 0.01$

$F(t) = 18 - 14.6e^{-(0.01)t} = 10 \Rightarrow$

$-14.6e^{-(0.01)t} = 10 - 18 \Rightarrow$

$e^{-(0.01)t} = \dfrac{-8}{-14.6} \Rightarrow t \approx \dfrac{\ln 0.5479}{-0.01} \approx 60$

It will thaw to 10°C in 1 hour.

65. a. $f(0) = 25, f(5) = 33$

$f(x) = a(b)^x \Rightarrow f(0) = 25 = a(b)^0 \Rightarrow$
$a = 25$

$f(5) = a(b)^5 \Rightarrow 33 = 25(b)^5 \Rightarrow$

$b^5 = \dfrac{33}{25} \Rightarrow b = \left(\dfrac{33}{25}\right)^{1/5} \Rightarrow b \approx 1.057$

The function is $f(x) = 25(1.057)^x$.

b.

```
ExpReg
 y=a*b^x
 a=25.08285532
 b=1.052468062
 r²=.9692183643
 r=.9844888848
```

$g(x) = 25.08(1.052)^x$

c. $x = 6$ for 2017

$f(6) = 25(1.057)^6 \approx 34.9$

$g(6) = 25.08(1.052)^6 \approx 34.0$

Using the two-point model, the gross profit for Fed Ex was about \$34.9 billion. Using the exponential regression equation, the profit was about \$34.0 billion.

Copyright © 2019 Pearson Education, Inc.

d. $f(x) = 40 = 25(1.057)^x \Rightarrow$

$\frac{40}{25} = (1.057)^x \Rightarrow$

$\ln(1.6) = x \ln 1.057 \Rightarrow$

$x = \frac{\ln(1.6)}{\ln 1.057} \approx 8.48$

$g(x) = 40 = 25.08(1.052)^x \Rightarrow$

$\frac{40}{25.08} = 1.052^x \Rightarrow$

$\ln\left(\frac{40}{25.08}\right) = x \ln 1.052 \Rightarrow$

$x = \frac{\ln\left(\frac{40}{25.08}\right)}{\ln 1.052} \approx 9.21$

The two point model predicts that gross profit for Fed Ex will exceed $40 billion in 2020, while the regression model predicts this will happen in 2021.

66. a. At $x = 0$, $f(0) = 10.0$ and at $x = 11$, $f(11) = 15.9$.

To find $f(x) = a(b^x)$, first solve for a:

$f(0) = 10.0 \Rightarrow a(b^0) = 10.0 \Rightarrow a = 10$

Now solve for b:

$f(11) = 15.9 \Rightarrow 10b^{11} = 15.9 \Rightarrow$

$b^{11} = \frac{15.9}{10} \Rightarrow b = \left(\frac{15.9}{10}\right)^{1/11} \approx 1.043$

The model becomes $f(x) = 10(1.043^x)$

b.
```
ExpReg
 y=a*b^x
 a=9.472667099
 b=1.048004306
 r²=.8988571639
 r=.9480807792
```

$h(x) = 9.47(1.048^x)$

c. $x = 12$

$f(12) = 10(1.043^{12}) \approx 16.6$

$h(12) = 9.47(1.048^{12}) \approx 16.6$

Both models estimate the number of unique visitors using the Starbucks app were approximately 16.6 million in June 2015.

d. Two-point model:

$10(1.043^x) = 18 \Rightarrow 1.043^x = \frac{18}{10} \Rightarrow$

$x \ln 1.043 = \ln(1.8) \Rightarrow$

$x = \frac{\ln(1.8)}{\ln 1.043} \approx 13.96$

Regression model:

$9.47(1.048^x) = 18$

$1.048^x = \frac{18}{9.47}$

$x \ln(1.048) = \ln\left(\frac{18}{9.47}\right)$

$x = \frac{\ln\left(\frac{18}{9.47}\right)}{\ln 1.048} \approx 13.70$

Both models predict that the Starbucks app exceeded 18 million unique users in August 2015.

67. a. $f(1) = 4.9$, $f(12) = 6.6$

$f(x) = a + b \ln x \Rightarrow$

$f(1) = 4.9 = a + b \ln 1 \Rightarrow 4.9 = a$

$f(12) = 4.9 + b \ln 12 \Rightarrow 6.6 = 4.9 + b \ln 12 \Rightarrow$

$b \ln 12 = 1.7 \Rightarrow b = \frac{1.7}{\ln 12} \approx .684$

The function is $f(x) = 4.9 + .684 \ln x$.

b.
```
LnReg
 y=a+blnx
 a=5.055165823
 b=.6973463025
 r²=.4792385742
 r=.6922705932
```

$g(x) = 5.1 + .697 \ln x$

c. June 2015 corresponds to $x = 13$

$f(13) = 4.9 + .684 \ln 13 \approx 6.6$

$g(13) = 5.1 + .697 \ln 13 \approx 6.9$

According to the two-point model, the number of unique visitors to the Dunkin' Donuts app was about 6.6 million. According to the logarithmic regression, the unique users was about 6.9 million in June 2015.

Copyright © 2019 Pearson Education, Inc.

d. $f(x) = 7.0$
$4.9 + .684 \ln x = 7.0 \Rightarrow$
$.684 \ln x = 2.1 \Rightarrow \ln x = \dfrac{2.1}{.684} \Rightarrow$
$x = e^{2.1/.684} \approx 21.5$
$g(x) = 7.0$
$5.1 + .697 \ln x = 7.0 \Rightarrow$
$.697 \ln x = 1.9 \Rightarrow \ln x = \dfrac{1.9}{.697} \Rightarrow$
$x = e^{1.9/.697} \approx 16.2$
According to the two-point model, the number of unique users to the Dunkin' Donuts app exceeded 7.0 million in March 2016. According to the logarithmic regression model, the unique users exceeded 7.0 million in October 2015.

68. a. $f(1) = 5.8,\ f(7) = 12.2$
$f(x) = a + b \ln x \Rightarrow$
$f(1) = 5.8 = a + b \ln 1 \Rightarrow 5.8 = a$
$f(7) = 5.8 + b \ln 7 \Rightarrow$
$12.2 = 5.8 + b \ln 7 \Rightarrow$
$b \ln 7 = 6.4 \Rightarrow b = \dfrac{6.4}{\ln 7} \approx 3.29$
The function is $f(x) = 5.8 + 3.29 \ln x$.

b.
```
LnReg
 y=a+blnx
 a=5.820442977
 b=3.384909439
 r²=.9914761999
 r=.9957289792
```

$h(x) = 5.8 + 3.38 \ln x$

c. 2016 corresponds to $x = 8$
$f(8) = 5.8 + 3.29 \ln 8 \approx 12.6$
$h(8) = 5.8 + 3.38 \ln 8 \approx 12.8$
According to the two-point model, the common equity ratio was about 12.6% in 2016. According to the logarithmic regression, the ratio was about 12.8% in 2016.

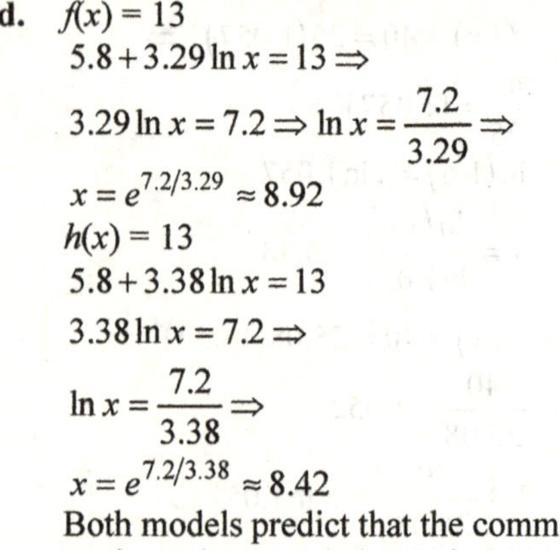

d. $f(x) = 13$
$5.8 + 3.29 \ln x = 13 \Rightarrow$
$3.29 \ln x = 7.2 \Rightarrow \ln x = \dfrac{7.2}{3.29} \Rightarrow$
$x = e^{7.2/3.29} \approx 8.92$
$h(x) = 13$
$5.8 + 3.38 \ln x = 13$
$3.38 \ln x = 7.2 \Rightarrow$
$\ln x = \dfrac{7.2}{3.38} \Rightarrow$
$x = e^{7.2/3.38} \approx 8.42$
Both models predict that the common equity ratio exceeded 13% in 2017.

Case 4 Gapminder.org

1.

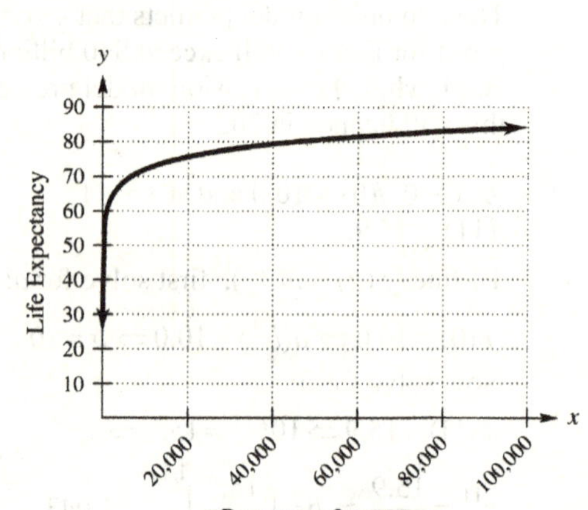

2. The graph is increasing and then leveling off.

3. $x = 4000$
$f(4000) = 23.97 + 5.221 \ln 4000 \approx 67.3$
According to the model, the life expectancy in the Central African Republic if the per person income level was \$4000 would be about 67.3 years.

4. $x = 20{,}000$
$f(20{,}000) = 23.97 + 5.221 \ln 20{,}000 \approx 75.7$
According to the model, the life expectancy in Indonesia if the per person income level was \$20,000 would be about 75.7 years.

5. $f(x) = 60 \Rightarrow 60 = 23.97 + 5.221 \ln x \Rightarrow$
$36.03 = 5.221 \ln x \Rightarrow \dfrac{36.03}{5.221} = \ln x \Rightarrow$
$x = e^{36.03/5.221} \approx 993$
According to the model, if there is a life expectancy of 60 years, the per person income level would be about \$993.

Copyright © 2019 Pearson Education, Inc.

6. $f(x)=75 \Rightarrow 75=23.97+5.221\ln x \Rightarrow$

$51.03=5.221\ln x \Rightarrow \dfrac{51.03}{5.221}=\ln x \Rightarrow$

$x=e^{51.03/5.221}\approx 17{,}571$

According to the model, if there is a life expectancy of 75 years, the per person income level would be about $17,571.

7.

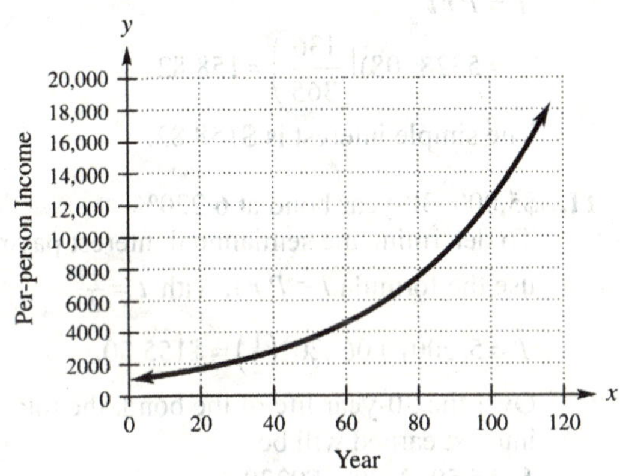

8. The graph is increasing in an exponential fashion.

9. $x=1950-1900=50$

$g(50)=1037(1.025)^{50}\approx 3564$

According to the model, the per person income for Brazil in 1950 was about $3564.

$x=2000-1900=100$

$g(100)=1037(1.025)^{100}\approx 12251$

According to the model, the per person income for Brazil in 2000 was about $12,251.

10. $1037(1.025^x)=4000 \Rightarrow 1.025^x=\dfrac{4000}{1037}\Rightarrow$

$x\ln 1.025=\ln\left(\dfrac{4000}{1037}\right)\Rightarrow$

$x=\dfrac{\ln\left(\frac{4000}{1037}\right)}{\ln 1.025}\approx 54.67$

According to the model, the per person income for Brazil exceeded $4000 in the year 1955.

11.

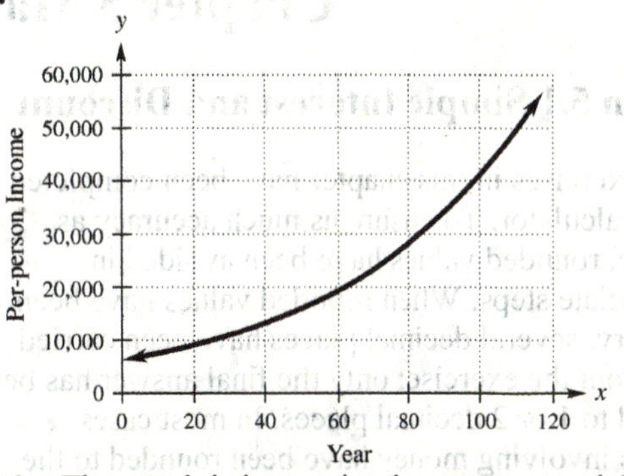

12. The graph is increasing in an exponential fashion.

13. $x=1960-1900=60$

$h(60)=6260(1.019)^{60}\approx 19{,}365$

According to the model, the per person income for the U.S. in 1960 was about $19,365.

$x=2010-1900=110$

$h(110)=6260(1.019)^{110}\approx 49{,}629$

According to the model, the per person income for the U.S. in 2010 was about $49,629.

14. $6260(1.019^x)=30{,}000\Rightarrow$

$1.019^x=\dfrac{30{,}000}{6260}\Rightarrow$

$x\ln 1.019=\ln\left(\dfrac{30{,}000}{6260}\right)\Rightarrow$

$x=\dfrac{\ln\left(\frac{30{,}000}{6260}\right)}{\ln 1.019}\approx 83.26$

According to the model, the per person income for the U.S. exceeded $30,000 in the year 1984.

Copyright © 2019 Pearson Education, Inc.

Chapter 5 Mathematics of Finance

Section 5.1 Simple Interest and Discount

Note: Exercises in this chapter have been completed with a calculator. To ensure as much accuracy as possible, rounded values have been avoided in intermediate steps. When rounded values have been necessary, several decimal places have been carried throughout the exercise; only the final answer has been rounded to 1 or 2 decimal places. In most cases, answers involving money have been rounded to the nearest cent. Students who use rounded intermediate values should expect their final answers to differ from the answers given here. Depending on the magnitude of the numbers used in the exercise, the difference could be a few pennies or several thousand dollars.

1. The factors are time and interest rate.

3. \$2850 at 7% for 8 months

$P = 2850, r = .07, \text{ and } t = \frac{8}{12}.$

$I = Prt$

$= 2850(.07)\left(\frac{8}{12}\right) = 133.00$

The simple interest is \$133.00.

5. \$3650 at 6.5% for 11 months

$P = 3650, r = .065, t = \frac{11}{12}$

$I = Prt$

$= 3650(.065)\left(\frac{11}{12}\right) = 217.48$

The simple interest is \$217.48.

7. \$2830 at 8.9% for 125 days

$P = 2830, \text{r} = .089, t = \frac{125}{365}$

$I = Prt$

$= 2830(.089)\left(\frac{125}{365}\right) = 86.26$

The simple interest is \$86.26.

9. \$5328 at 8%; loan made on August 16 is due December 30.
The duration of this loan is
$(31 - 16) + 30 + 31 + 30 + 30 = 136$ days.

$P = 5328, r = .08, t = \frac{136}{365}$

$I = Prt$

$= 5328(.08)\left(\frac{136}{365}\right) = 158.82$

The simple interest is \$158.82.

11. \$5,000, 30-year bond at 6.220%
To determine the semiannual interest payment, use the formula $I = P\,r\,t$, with $t = \frac{1}{2}$.

$I = 5,000(0.06220)\left(\frac{1}{2}\right) = \$155.50.$

Over the 30-year life of the bond, the total interest earned will be
$\$155.50 \cdot 2 \cdot 30 = \$9330.00.$

13. \$12,500, 10-year bond at 4.100%
To determine the semiannual interest payment, use the formula $I = P\,r\,t$, with $t = \frac{1}{2}$.

$I = 12,500(0.04100)\left(\frac{1}{2}\right) = \$256.25.$

Over the 10-year life of the bond, the total interest earned will be
$\$256.25 \cdot 2 \cdot 10 = \$5125.00.$

15. \$6,500, 5-year bond at 1.200%
To determine the semiannual interest payment, use the formula $I = P\,r\,t$, with $t = \frac{1}{2}$.

$I = 6,500(0.01200)\left(\frac{1}{2}\right) = \$39.00.$

Over the 5-year life of the bond, the total interest earned will be $\$39.00 \cdot 2 \cdot 5 = \$390.00.$

17. \$12,000 loan at 3.5% for three months
Three months is $\frac{1}{4}$ of a year.

$A = P + I = P + Prt$

$= 12,000 + 12,000(.035)\left(\frac{1}{4}\right) = \$12,105$

19. \$6500 loan at 5.25% for eight months.
Eight months is $\frac{2}{3}$ of a year.

$A = P + I = P + Prt$

$= 6500 + 6500(.0525)\left(\frac{2}{3}\right) \approx \$6727.50$

Copyright © 2019 Pearson Education, Inc.

21. Answers vary. Sample answer: The present value is the amount of money that can be deposited today to yield some larger amount in the future.

23. \$15,000 for 9 months; money earns 6%. Use the formula for present value with

$A = 15{,}000,\ t = \frac{9}{12},\ r = .06$

$$P = \frac{A}{1+rt} = \frac{15{,}000}{1+(.06)\left(\frac{9}{12}\right)} = \frac{15{,}000}{1.045} = \$14{,}354.07$$

The present value is \$14,354.07

25. \$15,402 for 120 days; money earns 6.3%.

$A = 15{,}402,\ t = \frac{120}{365}\ r = .063$

$$P = \frac{A}{1+rt} = \frac{15{,}402}{1+(.063)\left(\frac{120}{365}\right)} = \frac{15{,}402}{1.0207} = \$15{,}089.46$$

The present value is \$15,089.46

27. Three-month \$20,000 T-bill with discount rate .27%

a. Three months is $\frac{1}{4}$ of a year.

$\text{Discount} = Prt = 20000 \cdot .0027 \cdot \frac{1}{4} = \$13.50$

The price of the T-bill is the face value minus the discount.

$20{,}000 - 13.50 = \$19{,}986.50$

b. $I = Prt \Rightarrow 13.50 = 19{,}986.50(r)\left(\frac{1}{4}\right) \Rightarrow$

$r \approx .0027018$

The actual interest rate is about .27018%.

29. Six-month \$15,500 T-bill with discount rate .36%

a. Six months is $\frac{1}{2}$ of a year.

$\text{Discount} = Prt$

$= 15{,}500 \cdot .0036 \cdot \frac{1}{2} = \$27.90$

The price of the T-bill is the face value minus the discount.

$15{,}500 - 27.90 = \$15{,}472.10$

b. $I = Prt \Rightarrow 27.90 = 15{,}472.10(r)\left(\frac{1}{2}\right) \Rightarrow$

$r \approx .0036065$

The actual interest rate is about .36065%.

31. Three-month \$20,000 T-bill with discount rate 5.07%

a. Three months is $\frac{1}{4}$ of a year.

$\text{Discount} = Prt$

$= 20000 \cdot .0507 \cdot \frac{1}{4}$

$= \$253.50$

The price of the T-bill is the face value minus the discount.

$20{,}000 - 253.50 = \$19{,}746.50$

b. $I = Prt \Rightarrow 253.50 = 19{,}746.50(r)\left(\frac{1}{4}\right) \Rightarrow$

$r \approx .051351$

The actual interest rate is about 5.1351%.

33. Six-month \$15,500 T-bill with discount rate 5.10%

a. Six months is $\frac{1}{2}$ of a year.

$\text{Discount} = Prt$

$= 15{,}500 \cdot .0510 \cdot \frac{1}{2} = \$395.25$

The price of the T-bill is the face value minus the discount.

$15{,}500 - 395.25 = \$15{,}104.75$

b. $I = Prt \Rightarrow 395.25 = 15{,}104.75(r)\left(\frac{1}{2}\right) \Rightarrow$

$r \approx .052335$

The actual interest rate is about 5.2335%.

35. $P = 650{,}000,\ r = .052,\ t = \frac{30}{365}$

$$A = P(1+rt) = 650{,}000\left[1+(.052)\left(\frac{30}{365}\right)\right] = 652{,}778.08$$

The total amount paid was \$652,778.08.

37. $P = 3000,\ r = .015,\ t = 2$

$$A = P(1+rt) = 3000[1+(.015)(2)] = 3090.00$$

After 2 years, the value of the account will be \$3090.00.

39. $\text{Interest} = 67{,}359.39 - 67{,}081.20 = 278.19$

$P = 67{,}081.20,\ t = \frac{1}{12}$

$I = Prt$

$$r = \frac{I}{Pt} = \frac{278.19}{(67{,}081.20)\left(\frac{1}{12}\right)} \approx .050$$

The interest rate was about 5.0%.

Copyright © 2019 Pearson Education, Inc.

41. Interest $= 7675 - 7500 = 175$, $P = 7500$, $r = .07$

$I = Prt \Rightarrow 175 = 7500(.07)t \Rightarrow$

$t = \dfrac{175}{7500(.07)} = \dfrac{1}{3}$

The time period for the loan was $\frac{1}{3}$ of a year, or four months.

43. Want present value of \$6000 in 10 months at 3.6% interest.

$A = 6000$, $r = .036$, $t = \dfrac{10}{12}$

$P = \dfrac{A}{1+rt} = \dfrac{6000}{1+(.036)\left(\dfrac{10}{12}\right)} = 5825.24$

The present value would be \$5825.24.

45. $A = 50(1704) = 85{,}200$, $r = .0342$, $t = \dfrac{7}{12}$

$P = \dfrac{A}{1+rt} = \dfrac{85{,}200}{1+(.0342)\left(\frac{7}{12}\right)}$

$= \dfrac{85{,}200}{1.01995} = 83{,}533.51$

The firm could deposit \$83,533.51.

47. $P = 4000$, $t = \frac{1}{2}$

Discount $= 4000 - 3930 = 70$

Find the discount rate as follows:

$70 = 4000(r)\left(\frac{1}{2}\right) \Rightarrow r = .035$

The discount rate is 3.5%

49. $P = 7600$, $t = \frac{1}{4}$, discount $= 80.75$

Find the discount rate as follows:

$80.75 = 7600(r)\left(\frac{1}{4}\right) \Rightarrow r = .0425$

The discount rate is 4.25%

51. Interest $= (24 - 22) + .50 = 2.50$, $P = 22$, $t = 1$

$I = Prt \Rightarrow 2.50 = 22(r)(1) \Rightarrow$

$r = \dfrac{2.50}{22(1)} \approx .114$

The simple interest rate is about 11.4%.

53. You will receive \$700 from your tax preparer if you take the offer.

$60 = 760(r)\left(\frac{28}{365}\right) \Rightarrow r \approx 1.0291$

The actual interest rate is about 102.91%!

55. First find the maturity value of the loan, the amount that the contractor must repay.

$A = P(1+rt) \Rightarrow$

$A = 13{,}500\left(1+.09\left(\frac{9}{12}\right)\right) \approx 14{,}411.25$

In six months, the bank will receive \$14,411.25. Since the bank wants a 10% return, compute the present value of this amount at 10% for six months.

$P = \dfrac{A}{1+rt} = \dfrac{14{,}411.25}{1+.1\left(\frac{1}{2}\right)} \approx 13{,}725$

The bank pays the plumber \$13,725 and receives \$14,411.25. This is enough for the plumber to pay a bill for \$13,650.

57. y_1 is the future value after t years of \$100 invested at 8% simple interest. y_2 is the future value after t years of \$200 invested at 3%.

a. $y_1 = 100(.08)t + 100 = 8t + 100$

$y_2 = 200(.03)t + 200 = 6t + 200$

b. The graph of y_1 is a line with slope 8 and y-intercept 100. The graph of y_2 is a line with slope 6 and y-intercept 200.

c.

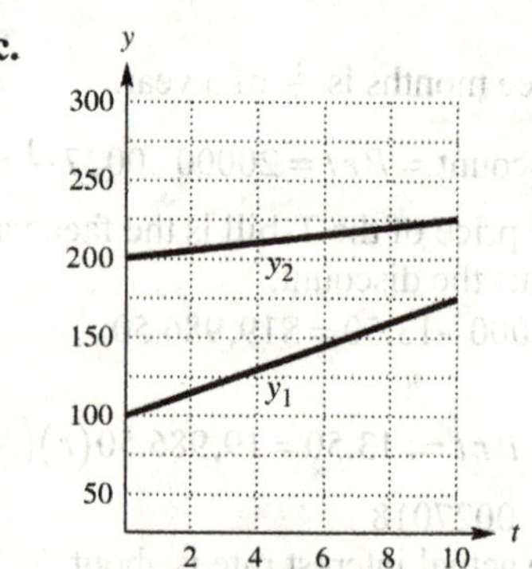

d. The y-intercept of each graph indicates the amount invested. The slope of the graph is the annual amount of interest paid.

Section 5.2 Compound Interest

1. r is the interest rate per year, while i is the interest rate per compounding period. t is the number of years, while n is the number of compounding periods.

3. The interest rate and number of compounding periods determine the amount of interest earned on a fixed principal.

5. The present value is the amount of money to be invested to earn a certain amount of return.

Copyright © 2019 Pearson Education, Inc.

7. \$470 at 8% compounded monthly for 12 years

$P = 470,\ i = \dfrac{8\%}{12} = .00667,\ n = 12(12) = 144$

$A = P(1+i)^n = 470(1.00667)^{144} \approx 1223.59$

The compound amount is \$1223.59. The interest earned is $I = 1223.59 - 470 = \$753.59$.

9. \$6500 at 4.5% compounded quarterly for 8 years

$P = 6500,\ i = \dfrac{4.5\%}{4} = .01125,\ n = 8(4) = 32$

$A = P(1+i)^n = 6500(1.01125)^{32} \approx 9297.93$

The compound amount is \$9297.93. The interest earned is $I = 9297.93 - 6500 = \$2797.93$.

11. \$10,000 at 1.11% compounded daily for 1 yr

$P = 10{,}000,\ i = .0111,\ n = 365$

Find the compound amount and then amount of interest.

$$A = P(1+\tfrac{i}{n})^{nt}$$
$$= 10{,}000(1+\tfrac{.0111}{365})^{(365)(1)}$$
$$\approx 10{,}111.62$$

The compound amount is \$10,111.62. The amount of interest earned is \$10,111.62 – \$10,000 = \$111.62.

13. \$15,000 at 1.00% compounded monthly for 3 years

Find the compound amount and then amount of interest.

$P = 15{,}000,\ i = .01,\ n = 12$

$$A = P(1+\tfrac{i}{n})^{nt}$$
$$= 15{,}000(1+\tfrac{.01}{12})^{(12)(3)}$$
$$\approx 15{,}456.62$$

The amount of interest earned is \$15,456.62 – \$15000 = \$456.62.

14. \$60,000 at .95% compounded monthly for 2 years

Find the compound amount and then amount of interest.

$P = 60{,}000,\ i = .0095,\ n = 12$

$$A = P(1+\tfrac{i}{n})^{nt}$$
$$= 60{,}000(1+\tfrac{.0095}{12})^{(12)(2)}$$
$$\approx 61{,}150.44$$

The amount of interest earned is \$61,150.44 – \$60,000 = \$1,150.44.

15. \$75,000 at .90% compounded quarterly for 4 years

$P = 75{,}000,\ i = .0090,\ n = 4$

$$A = P(1+\tfrac{i}{n})^{nt}$$
$$= 75{,}000(1+\tfrac{.0090}{4})^{(4)(4)}$$
$$\approx 77{,}746.04$$

The amount of interest earned is \$77,746.04 – \$75,000 = \$2,746.04.

17. $P = 3000,\ A = 3606,\ n = 5$. Solve for *i*.

$$P(1+i)^n = A \Rightarrow 3000(1+i)^5 = 3606 \Rightarrow$$
$$(1+i)^5 = \frac{3606}{3000} \Rightarrow 1+i = \sqrt[5]{\frac{3606}{3000}} \Rightarrow$$
$$i = \sqrt[5]{\frac{3606}{3000}} - 1 \approx .0375$$

The interest rate is about 3.75%

19. $P = 8500,\ A = 12{,}161,\ n = 7$. Solve for *i*.

$$P(1+i)^n = A \Rightarrow 8500(1+i)^7 = 12{,}161 \Rightarrow$$
$$(1+i)^7 = \frac{12{,}161}{8500} \Rightarrow \sqrt[7]{(1+i)^7} = \sqrt[7]{\frac{12{,}161}{8500}} \Rightarrow$$
$$1+i = \sqrt[7]{\frac{12{,}161}{8500}} \Rightarrow i = \sqrt[7]{\frac{12{,}161}{8500}} - 1 \approx 0.0525$$

The interest rate is about 5.25%.

21. \$20,000 at 3.5% compounded continuously for 5 years

$P = 20{,}000,\ r = .035$

$$A = Pe^{rt}$$
$$= 20{,}000e^{(.035)(5)}$$
$$\approx 23{,}824.92$$

The amount of interest earned is \$23,824.92 – \$20,000 = \$3824.92.

23. \$30,000 at 1.8% compounded continuously for 3 years

$P = 30{,}000,\ r = .018$

$$A = Pe^{rt}$$
$$= 30{,}000e^{(.018)(3)}$$
$$\approx 31{,}664.54$$

The amount of interest earned is \$31,664.54 – 30,000 = \$1664.54.

25. $P = 4630,\ r = .052,\ n = 2(15) = 30$

$$A = P\left(1+\frac{i}{n}\right)^{nt} = 4630\left(1+\frac{.052}{2}\right)^{30} \approx 10{,}000$$

The face value is \$10,000.

Copyright © 2019 Pearson Education, Inc.

For exercises 27–29, we use the compound interest formula. Interest is paid twice per year.

27. $P = 9992, r = .035, t = 2(20) = 40$

$$A = P\left(1+\frac{i}{n}\right)^n = 9992\left(1+\frac{.035}{2}\right)^{40} \approx 20{,}000$$

The face value is \$20,000.

29. 4% compounded semiannually
Use the formula for effective rate with $r = .04$ and $m = 2$.

$$r_E = \left(1+\frac{r}{m}\right)^m - 1 = \left(1+\frac{.04}{2}\right)^2 - 1$$
$$= (1.02)^2 - 1 = 1.0404 - 1 = .0404$$

The effective rate is 4.04%.

31. 5% compounded quarterly
$r = .05, m = 4$

$$r_E = \left(1+\frac{r}{m}\right)^m - 1 = \left(1+\frac{.05}{4}\right)^4 - 1 \approx .05095$$

The effective rate is 5.095%.

33. \$12,000 at 5% compounded annually for 6 yr
Use the present value formula for compound interest with
$A = 12{,}000$, $i = .05$, and $n = 6$.

$$P = A(1+i)^{-n} = 12{,}000(1.05)^{-6} \approx 8954.58$$

The present value is \$8954.58.

35. \$17,230 at 4% compounded quarterly for 10 yr

$$A = 17{,}230,\ i = \frac{.04}{4} = .01,\ n = 10(4) = 40$$

$$P = A(1+i)^{-n} = 17{,}230(1.01)^{-40} \approx 11{,}572.58$$

The present value is \$11,572,58.

37. $A = 5000,\ i = \frac{.035}{2},\ n = 5(2) = 10$

$$P = \frac{A}{(1+i)^n} = \frac{5000}{\left(1+\frac{.035}{2}\right)^{10}} \approx 4203.64$$

A fair price would be \$4203.64.

39. $A = 20{,}000,\ i = \frac{.047}{2},\ n = 15(2) = 30$

$$P = \frac{A}{(1+i)^n} = \frac{20{,}000}{\left(1+\frac{.047}{2}\right)^{30}} \approx 9963.10$$

A fair price would be \$9963.10.

41. $A = 5000, i = .0162, n = 4$

$$P = \frac{A}{(1+i)^n} = \frac{5000}{(1+.0162)^4} \approx 4688.71$$

The \$5000 item costs about \$4688.71 years prior.

43. $A = 500, i = .036, n = 2$

$$P = \frac{A}{(1+i)^n} = \frac{500}{(1+.036)^2} \approx 465.85$$

The \$500 item costs about \$465.85 2 years prior.

45. $A = 1210,\ i = \frac{.08}{4} = .02,\ n = 5(4) = 20$

$$P = A(1+i)^{-n} = 1210(1.02)^{-20} \approx 814.30$$

Since this amount is less than \$1000, "\$1000 now" is greater.

47. $P = 10{,}000$
Money market fund:

$$i = \frac{.058}{365},\ n = 365(2) = 730$$

$$A = P(1+i)^n = 10{,}000\left(1+\frac{.058}{365}\right)^{730}$$
$$\approx 11{,}229.86$$

$\text{Interest} = 11{,}229.86 - 10{,}000 = 1229.86$
Treasury note (recall that treasury notes pay simple interest–see section 5.1):
$A = P(1+rt)$
$A = 10{,}000(1+.06(2)) \approx 11{,}200$
$\text{Interest} = 11{,}200 - 10{,}000 = 1200$
The treasury note pays the most interest.

49. $P = 4510,\ r = \frac{.061}{2},\ n = 2(20)$

$$A = P(1+r)^n = 4510\left(1+\frac{.061}{2}\right)^{40} \approx 15{,}000.18$$

$$P = 5809,\ r = \frac{.048}{2},\ n = 2(20)$$

$$A = P(1+r)^n = 5809\left(1+\frac{.048}{2}\right)^{40} \approx 15{,}000.29$$

The 4.8% bond pays slightly more than the 6.1% bond.

Copyright © 2019 Pearson Education, Inc.

51. $A = 2.9, i = \frac{.05}{12}, n = 5(12) = 60$

Use the formula for present value with compound interest.

$$P = A(1+i)^{-n} \approx 2.9\left(1+\frac{.05}{12}\right)^{-60} \approx 2.259696$$

The company should invest about $2,259,696 now.

53. $A = 23,500, i = .0375, n = 3$

Use the formula for present value with compound interest.

$$P = A(1+i)^{-n}$$
$$= 23,500(1+.0375)^{-3} \approx 21,042.80$$

The car would have cost about $21,043 three years ago.

55. To find the number of years it will take $1 to inflate to $2, use the formula for compound amount with $A = 2$, $P = 1$, and $i = .03$.

$$A = P(1+i)^n \Rightarrow 2 = 1(1.03)^n \Rightarrow 2 = (1.03)^n \Rightarrow$$
$$\ln 2 = n \ln 1.03 \Rightarrow n = \frac{\ln 2}{\ln 1.03} \approx 23.4$$

Prices will double in about 23.4 yr.

57. Find the number of years, it will take for $1 to inflate to $2 using the formula for compound amount with $A = 2$, $P = 1$, $i = .05$

$$A = P(1+i)^n \Rightarrow 2 = 1(1.05)^n \Rightarrow 2 = (1.05)^n \Rightarrow$$
$$\ln 2 = n \ln 1.05 \Rightarrow n = \frac{\ln 2}{\ln 1.05} \approx 14.21$$

Prices will double in about 14.2 yr.

59. To find the number of years it will take for a demand of 1 unit of electricity to increase to a demand of 2 units, use the formula for compound amount with $A = 2$, $P = 1$, and $i = .06$.

$$A = P(1+i)^n \Rightarrow 2 = 1(1.06)^n \Rightarrow 2 = (1.06)^n \Rightarrow$$
$$\ln 2 = n \ln 1.06 \Rightarrow n = \frac{\ln 2}{\ln 1.06} \approx 11.90$$

The electric utilities will need to double their generating capacity in about 11.9 yr.

61. a. $P = 16,000 - 30 = 15,970,\ i = \frac{.055}{12},\ n = 12$

Use the formula for compound amount.

$$A = P(1+i)^n = 15,970\left(1+\frac{.055}{12}\right)^{12}$$
$$\approx 16,870.83$$

After the annual charges,

$$16,870.83 - .0125(16,870.83) \approx \$16,659.95$$

b. Note that the amount in part (a) can be written in terms of P.

$$A = P\left(1+\frac{.055}{12}\right)^{12}(.9875) \approx \$16,659.95$$

Therefore, the amount in the account after 7 years can be found by repeatedly multiplying P by the factors $\left(1+\frac{.055}{12}\right)^{12}(.9875)$. So, by commutatively of multiplication this amount is

$$A = P\left(1+\frac{.055}{12}\right)^{12(7)}(.9875)^7 \approx \$21,472.67.$$

63. The company must pay tax on what they could have received in interest, based on the present value of the three-year note, which is .75(200,000) = $150,000. At the end of the first year (2015), the value of the note would be $150,000 + 10% interest

= 150,000 + (.1)(150,000) = $165,000.

So the interest for the second year (2016) would be (.1)(165,000) = $16,500.

Section 5.3 Annuities, Future Value, and Sinking Funds

1. $1 + 1.05 + 1.05^2 + 1.05^3 + \cdots + 1.05^{14}$

$$= \frac{1.05^{15} - 1}{1.05 - 1} \approx 21.5786$$

In Exercises 3–8, use the formula

$$S = R \cdot s_{\overline{n}|i} \text{ or } S = R\left[\frac{(1+i)^n - 1}{i}\right].$$

3. $R = 2,000,\ i = \frac{.0110}{12} = .000917,,\ n = 3(12) = 36$

$$S = 2,000\left[\frac{(1.000917)^{36} - 1}{.000917}\right] \approx 73,167.09$$

The future value is $73,167.09.

Copyright © 2019 Pearson Education, Inc.

5. $R = 750,\ i = \dfrac{.0105}{2} = .00525,\ n = 5(2) = 10$

$$S = 750\left[\frac{(1.00525)^{10} - 1}{.00525}\right] \approx 7{,}679.69$$

The future value is \$7,679.69.

7. $R = 2500,\ i = \dfrac{.01}{4} = .0025,\ n = 10(4) = 40$

$$S = 2500\left[\frac{(1.0025)^{40} - 1}{.0025}\right] \approx 105{,}033.01$$

The future value is \$105,033.01.

9. $R = 400,\ i = \dfrac{.04}{12},\ n = 12(10) = 120$

$$S = 400\left[\frac{\left(1 + \frac{.04}{12}\right)^{120} - 1}{\frac{.04}{12}}\right] \approx 58{,}899.92$$

At the end of 10 years, there will be about \$58,899.92 in the account. Using the compound amount formula, the future value of this money is $58{,}899.92\left(1 + \frac{.06}{12}\right)^{120} \approx 107{,}162.32$, for $i = \dfrac{.06}{12}$, and $n = 12(10) = 120$. Now compute the amount in the account for $R = 600,\ i = \dfrac{.06}{12}$, and $n = 12(10) = 120$.

$$S = 600\left[\frac{\left(1 + \frac{.06}{12}\right)^{120} - 1}{\frac{.06}{12}}\right] \approx 98{,}327.61.$$

$98{,}327.61 + 107{,}162.32 = 205{,}489.93$

There will be about \$205,490 in the account at the end of 20 years.

11. $R = 1000,\ i = \dfrac{.042}{4},\ n = 4(10) = 40$

$$S = 1000\left[\frac{\left(1 + \frac{.042}{4}\right)^{40} - 1}{\frac{.042}{4}}\right] \approx 49{,}393.58$$

At the end of 10 years, there will be about \$49,393.58 in the account. Using the compound amount formula, the future value of this money is $49{,}393.58\left(1 + \frac{.074}{4}\right)^{60} \approx 148{,}365.36$, for $i = \dfrac{.074}{4}$, and $n = 4(15) = 60$. Now compute the amount in the account for $R = 1500,\ i = \dfrac{.074}{4}$, and $n = 4(15) = 60$.

$$S = 1500\left[\frac{\left(1 + \frac{.074}{4}\right)^{60} - 1}{\frac{.074}{4}}\right] \approx 162{,}465.22.$$

$148{,}365.36 + 162{,}465.22 = 310{,}830.58$

There will be about \$310,831 in the account at the end of 25 years.

13. $S = 11{,}000,\ i = \dfrac{.05}{2} = .025,\ n = 6(2) = 12$

$$11{,}000 = R\left[\frac{(1.025)^{12} - 1}{.025}\right]$$

$$11{,}000 \approx R(13.79555)$$

$$R \approx 797.36$$

The periodic payment is \$797.36.

15. $S = 50{,}000,\ i = \dfrac{.08}{4} = .02,\ n = \left(2\frac{1}{2}\right)(4) = 10$

$$50{,}000 = R\left[\frac{(1.02)^{10} - 1}{.02}\right]$$

$$50{,}000 \approx R(10.94972)$$

$$R \approx 4566.33$$

The periodic payment is \$4566.33.

17. $S = 6000,\ i = \dfrac{.06}{12}, = .005,\ n = 3(12) = 36$

$$6000 = R\left[\frac{(1.005)^{36} - 1}{.005}\right]$$

$$6000 \approx R(39.33610)$$

$$R \approx 152.53$$

The periodic payment is \$152.53.

Copyright © 2019 Pearson Education, Inc.

For exercises 19–22, we use the future value formula,
$$S = R\left[\frac{(1+i)^n - 1}{i}\right].$$

19. $R = 3940, n = 10, S = 50{,}000, i = r$
$$50{,}000 = 3940\left[\frac{(1+i)^{10} - 1}{i}\right]$$
Graphing each side of the equation, we find that the intersection is $x \approx .0519$

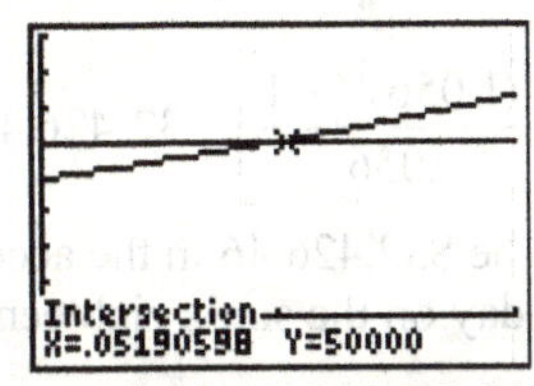

[0, .1] by [–10,000, 50,000]

An interest rate of about 5.19% is needed.

21. $R = 1675, n = 5(4) = 20, S = 38{,}000, i = \frac{r}{4}$
$$38{,}000 = 1675\left[\frac{\left(1+\frac{r}{4}\right)^{20} - 1}{\frac{r}{4}}\right]$$
Graphing each side of the equation, we find that the intersection is $x \approx .05223$

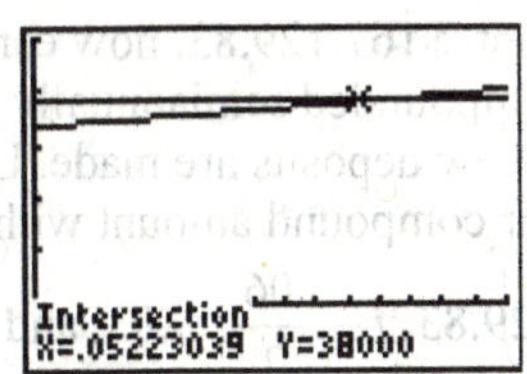

[0, .075] by [–10,000, 50,000]

An interest rate of about 5.223% is needed.

23. A sinking fund is a fund that is set up to receive periodic payments. Businesses and corporations use sinking funds to repay bond issues or to provide for replacement of fixed assets.

25. $R = 500, i = .0025, n = 10$
$$S = 500\left[\frac{(1.0025)^{11} - 1}{.0025}\right] - 500 \approx 5069.27$$
The future value is $5069.27.

27. $R = 1{,}600, i = \frac{.0018}{12} = .00015, n = 11(12) = 132$
$$S = 1{,}600\left[\frac{(1.00015)^{133} - 1}{.00015}\right] - 16{,}000$$
$\approx 213{,}320.59$

The future value is $213,320.59.

29. $R = 1000, i = \frac{.0015}{2} = .00075, n = 9(2) = 18$
$$S = 1000\left[\frac{(1.00075)^{19} - 1}{.00075}\right] - 1000$$
$\approx 18{,}128.80$

The future value is $18,128.80.

31. $R = 200, i = \frac{.0030}{4} = .00075, n = 7(4) = 28$
$$S = 200\left[\frac{(1.00075)^{29} - 1}{.00075}\right] - 200$$
≈ 5661.31

The future value is $5661.31.

33. $S = 8000, n = 3(4) = 12, i = \frac{.044}{4}$
$$8000 = R\left[\frac{\left(1+\frac{.044}{4}\right)^{13} - 1}{\frac{.044}{4}}\right] - R \Rightarrow$$
$$8000 = R\left(\frac{\left(1+\frac{.044}{4}\right)^{13} - 1}{\frac{.044}{4}} - 1\right) \Rightarrow$$
$$R = \frac{8000}{\frac{\left(1+\frac{.044}{4}\right)^{13} - 1}{\frac{.044}{4}} - 1} \approx 620.46$$
The payment should be $620.46.

35. $S = 55{,}000, n = 12(12) = 144, i = \frac{.057}{12}$
$$55{,}000 = R\left[\frac{\left(1+\frac{.057}{12}\right)^{145} - 1}{\frac{.057}{12}}\right] - R \Rightarrow$$
$$55{,}000 = R\left(\frac{\left(1+\frac{.057}{12}\right)^{145} - 1}{\frac{.057}{12}} - 1\right) \Rightarrow$$
$$R = \frac{55{,}000}{\frac{\left(1+\frac{.057}{12}\right)^{145} - 1}{\frac{.057}{12}} - 1} \approx 265.71$$
The payment should be $265.71.

Copyright © 2019 Pearson Education, Inc.

37. $R = 400,\ i = \dfrac{.0085}{12},\ n = 30(12) = 360$

Use the formula for future value of an ordinary annuity.

$$S = R \cdot s_{\overline{n}|i}$$

$$= 400\left[\frac{\left(1+\frac{.0085}{12}\right)^{360} - 1}{\frac{.0085}{12}}\right] \approx 163{,}959.61$$

After 30 years, the amount in the account will be about \$163,959.61.

39. $R = 12{,}000,\ i = .06,\ n = (9)(12) = 108$

a. Use the formula for future value of an ordinary annuity.

$$S = R \cdot s_{\overline{n}|i}$$

$$= 12{,}000\left[\frac{(1+.06)^9 - 1}{.06}\right] \approx 137{,}895.79$$

She will have \$137,895.79 on deposit after 9 years.

b. $i = .05$

$$S = 12{,}000\left[\frac{(1+.05)^9 - 1}{.05}\right] \approx 132{,}318.77$$

She will have \$132,18.77 on deposit after 9 years.

c. $137{,}895.79 - 132{,}318.77 = 5{,}577.02$

She would lose \$5,577.02.

41. $R = 1200,\ i = \dfrac{.072}{4},\ n = 10(4) = 40$

a. Use the formula for future value of an ordinary annuity.

$$S = R \cdot s_{\overline{n}|i}$$

$$= 1200\left[\frac{\left(1+\frac{.072}{4}\right)^{40} - 1}{\frac{.072}{4}}\right] \approx 69{,}421.34$$

She will have \$69,421.34 in the account when she is 60.

b. $A = P(1+i)^n = 69{,}421.34\left(1+\dfrac{.072}{4}\right)^{4(5)}$

$\approx 99{,}185.58$

She will have \$99,185.58 for retirement at age 65.

43. This may be considered an annuity due, since payments are made at the beginning of each year, starting with the day the son is born. However, a payment should not be subtracted at the end, since a nineteenth payment is made on his eighteenth birthday. Thus, the future value is given by

$$S = R\left[\frac{(1+i)^{n+1} - 1}{i}\right],$$

where $R = 1000$, $i = .056$, and $n = 18$. Therefore,

$$S = 1000\left[\frac{(1.056)^{19} - 1}{.056}\right] = 32{,}426.46$$

There will be \$32,426.46 in the account at the end of the day on the son's eighteenth birthday.

45. For the first 12 years, we have an annuity due. To find the amount in this account after 12 years, use the formula for the future value of an annuity due with $R = 10{,}000$, $i = .05$, and $n = 12$.

$$S = R\left[\frac{(1+i)^{n+1} - 1}{i}\right] - R$$

$$= 10{,}000\left[\frac{(1.05)^{13} - 1}{.05}\right] - 10{,}000$$

$\approx 167{,}129.83$

This amount, \$167,129.83, now earns 6% interest compounded semiannually for another 9 yr, but no new deposits are made. Use the formula for compound amount with

$P = 167{,}129.83,\ i = \dfrac{.06}{2} = .03$, and

$n = 9(2) = 18.$

$A = P(1+i)^n = 167{,}129.83(1.03)^{18}$

$= 284{,}527.35.$

The final amount on deposit after 21 yr is \$284,527.35.

47. $S = 10{,}000,\ n = 8(4) = 32$

a. $i = \dfrac{.05}{4} = .0125$

$$S = R \cdot s_{\overline{n}|i}$$

$$10{,}000 = R\left[\frac{(1.0125)^{32} - 1}{.0125}\right] \Rightarrow$$

$$10{,}000 \approx R(39.05044) \Rightarrow R = \$256.08$$

He should deposit \$256.08 at the end of each quarter.

Copyright © 2019 Pearson Education, Inc.

b. $i = \frac{.058}{4} = .0145$

$S = R \cdot s_{\overline{n}|i}$

$10,000 = R\left[\frac{(1.0145)^{32} - 1}{.0145}\right]$

$10,000 \approx R(40.35398) \Rightarrow R = \$247.81$

He should deposit $247.81 quarterly.

49. $S = 24,000,\ i = \frac{.05}{4} = .0125,\ n = 6(4) = 24$

$S = R \cdot s_{\overline{n}|i}$

$24,000 = R\left[\frac{(1.0125)^{24} - 1}{.0125}\right] \Rightarrow$

$R = \frac{24,000}{\frac{(1.0125)^{24} - 1}{.0125}} \approx 863.68$

She should deposit $863.68 at the end of each quarter.

51. a. $P = 60,000,\ r = .08,\ t = \frac{1}{4}$

$I = Prt = 60,000(.08)\left(\frac{1}{4}\right) = \$1200$

Each quarterly interest payment is $1200.

b. $S = 60,000,\ i = \frac{.06}{2} = .03,$

$n = 7(2) = 14$

$S = R \cdot s_{\overline{n}|i}$

$60,000 = R\left[\frac{(1.03)^{14} - 1}{.03}\right]$

$60,000 \approx R(17.08632)$

$R = \$3511.58$

The amount of each payment is $3511.58.

53. $S = 147,126,\ R = 300,\ n = 12(20),\ i = \frac{r}{12}$

$S = R\left[\frac{(1+i)^n - 1}{i}\right]$

$147,126 = 300\left[\frac{\left(1+\frac{r}{12}\right)^{12(20)} - 1}{\frac{r}{12}}\right]$

Graph each side of this equation. The intersection is (.065, 147,126).

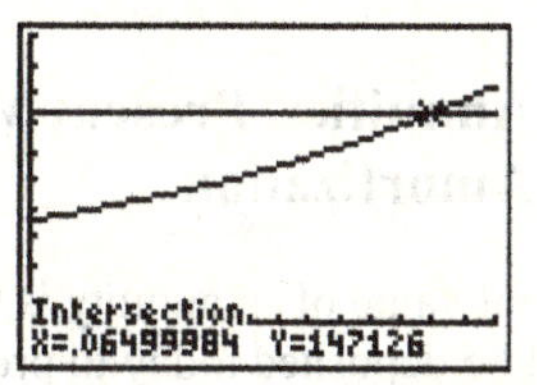

[0, .075] by [–20,000, 200,000]

She received an interest rate of 6.5%.

55. a. Answers will vary.

b. Joe contributed (600)(40) = $24,000 to his retirement.

c. For the first ten years, Joe has an ordinary annuity with $R = 600,\ i = \frac{.081}{4}$, and $n = 10(4) = 40$. Use the formula for the future value of an ordinary annuity.

$S = \left[\frac{(1+i)^n - 1}{i}\right] = 600\left[\frac{\left(1+\frac{.081}{4}\right)^{40} - 1}{\frac{.081}{4}}\right]$

$= 36,438.25$

He makes no further deposits, but leaves this amount in the account at the same interest rate. Use the formula for compound amount with $P = 36,438.25,\ i = \frac{.081}{4}$, and $n = 35(4) = 140$ to find the value of this amount after 35 years.

$A = P(1+i)^n$

$= 36,438.25\left(1+\frac{.081}{4}\right)^{140} \approx 603,229$

After 45 years, there will be about $603,229 in Joe's account.

d. Sarah contributed (600)(4)(35) = $84,000 to her retirement.

e. Sarah has an ordinary annuity for 35 years with $R = 600,\ i = \frac{.081}{4}$, and $n = 35(4) = 140$.

$S = \left[\frac{(1+i)^n - 1}{i}\right] = 600\left[\frac{\left(1+\frac{.081}{4}\right)^{140} - 1}{\frac{.081}{4}}\right]$

$\approx 460,884$

Sarah will have about $460,884 in her account after 45 years.

Copyright © 2019 Pearson Education, Inc.

Section 5.4 Annuities, Present Value, and Amortization

1. The present value of an annuity is the amount that must be deposited today to provide all the payments for the term of the annuity. The future value of an annuity is the final sum in an account receiving annuity payments after all payments are made.

In Exercises 2–6, use the formula

$$P = R\left[\frac{1-(1+i)^{-n}}{i}\right].$$

3. $R = 1400$, $i = .06$, $n = 8$

$$P = 1400\left[\frac{1-(1.06)^{-8}}{.06}\right] \approx 8693.71$$

The present value is \$8693.71.

5. $R = 50{,}000$, $i = \frac{.05}{4} = .0125$, $n = 10(4) = 40$

$$P = 50{,}000\left[\frac{1-(1.0125)^{-40}}{.0125}\right] \approx 1{,}566{,}346.66$$

The present value is \$1,566,346.66.

7. $R = 650$, $i = \frac{.049}{4}$, $n = 5(4) = 20$

$$P = 650\left[\frac{1-\left(1+\frac{.049}{4}\right)^{-20}}{\frac{.049}{4}}\right] \approx 11{,}468.10$$

About \$11,468.10 is needed to fund the withdrawals.

9. $R = 425$, $i = \frac{.061}{12}$, $n = 10(12) = 120$

$$P = 425\left[\frac{1-\left(1+\frac{.061}{12}\right)^{-120}}{\frac{.061}{12}}\right] \approx 38{,}108.61$$

About \$38,108.61 is needed to fund the withdrawals.

11. $P = 90{,}000$, $i = \frac{.049}{12}$, $n = 22(12) = 264$

$$90{,}000 = R\left[\frac{1-\left(1+\frac{.049}{12}\right)^{-264}}{\frac{.049}{12}}\right] \Rightarrow$$

$$R = \frac{90{,}000}{\frac{1-\left(1+\frac{.049}{12}\right)^{-264}}{\frac{.049}{12}}} \approx 557.68$$

The monthly payment will be \$557.68.

13. $P = 275{,}000$, $i = \frac{.06}{4} = .015$, $n = 18(4) = 72$

$$275{,}000 = R\left[\frac{1-(1+.015)^{-72}}{.015}\right] \Rightarrow$$

$$R = \frac{275{,}000}{\frac{1-(1+.015)^{-72}}{.015}} \approx 6272.14$$

The monthly payment will be \$6272.14.

15. $P = 2500$, $i = \frac{.08}{4} = .02$, $n = 6$

$$R = 2500\left[\frac{.02}{1-(1.02)^{-6}}\right] \approx 446.31$$

Quarterly payments of \$446.31 are required to amortize this loan.

17. $P = 90{,}000$, $i = \frac{.07}{1} = .07$, $n = 12$

$$R = 90{,}000\left[\frac{.07}{1-(1.07)^{-12}}\right] \approx 11{,}331.18$$

Annual payments of \$11,331.18 are required to amortize this loan.

19. $P = 7400$, $i = \frac{.082}{2} = .041$, $n = 18$

$$R = 7400\left[\frac{.041}{1-(1.041)^{-18}}\right] \approx 589.31$$

Semiannual payments of \$589.31 are required to amortize this loan.

Copyright © 2019 Pearson Education, Inc.

For exercises 21–28, we use the amortization payment formula $R = \frac{Pi}{1-(1+i)^{-n}}$.

21. $P = 225{,}000$, $i = \frac{.0368}{12}$, $n = 30(12) = 360$

$$R = 225{,}000\left[\frac{\frac{.0368}{12}}{1-\left(1+\frac{.0368}{12}\right)^{-360}}\right] \approx 1033.09$$

Monthly payments of \$1033.09 are required to amortize this loan.

23. $P = 140{,}000$, $i = \frac{.0263}{12}$, $n = 15(12) = 180$

$$R = 140{,}000\left[\frac{\frac{.0263}{12}}{1-\left(1+\frac{.0263}{12}\right)^{-180}}\right] \approx 942.10$$

Monthly payments of \$942.10 are required to amortize this loan.

25. $P = 26{,}799$, $i = \frac{.0374}{12}$, $n = 6(12) = 72$

$$R = 26{,}799\left[\frac{\frac{.0374}{12}}{1-\left(1+\frac{.0374}{12}\right)^{-72}}\right] \approx 416.11$$

After 2 years (24 payments), the remaining 48 payments can be thought of as annuity. The present value of this annuity is the remaining balance.

So, we use the present value formula with $R = 16.11$, $i = \frac{.0374}{12}$, $n = 48$

$$P = 416.11\left[\frac{1-\left(1+\frac{.0374}{12}\right)^{-48}}{\frac{.0374}{12}}\right] \approx 18{,}524.33$$

The monthly payment is \$416.11 and the balance remaining after two years is about \$18,524.33.

27. $P = 210{,}000$, $i = \frac{.0363}{12}$, $n = 30(12) = 360$

$$R = 210{,}000\left[\frac{\frac{.0363}{12}}{1-\left(1+\frac{.0363}{12}\right)^{-360}}\right] \approx 958.30$$

After 12 years (144 payments), the remaining 216 payments can be thought of as annuity. The present value of this annuity is the remaining balance. So, we use the present value formula with $R = 958.30$, $i = \frac{.0363}{12}$, $n = 216$

$$P = 958.30\left[\frac{1-\left(1+\frac{.0363}{12}\right)^{-216}}{\frac{.0363}{12}}\right] \approx 151{,}811.44$$

The monthly payment is \$958.30 and the balance remaining after 12 years is about \$151,811.44.

29. Locate the table entry that is in the row labeled Payment Number 5 and in the column labeled Interest for Period. Note that \$6.80 of the fifth payment is interest.

31. To find how much interest is paid in the first 5 months of the loan, add the first five nonzero entries in the column labeled Interest for Period.

$$\begin{array}{r} \$10.00 \\ 9.21 \\ 8.42 \\ 7.61 \\ +6.80 \\ \hline \$42.04 \end{array}$$

In the first 5 months of the loan, \$42.04 is paid in interest.

For exercises 43–46, we use the formula for the present value of an annuity due, $S = R + R\left[\frac{1-(1+i)^{-(n-1)}}{i}\right]$

33. The yearly payment for \$57.6 million is $\frac{57{,}600{,}000}{30} = 1{,}920{,}000.$

$R = 1{,}920{,}000$, $i = .051$, $n = 30$

$$S = 1{,}920{,}000 + 1{,}920{,}000\left[\frac{1-(1.051)^{-(30-1)}}{.051}\right]$$

$$\approx 30{,}669{,}881$$

The cash value is about \$30,669,881.

The figure shows the solution using the TVM on a TI-84 Plus calculator.

Copyright © 2019 Pearson Education, Inc.

35. The yearly payment for $41.6 million is $\frac{41,600,000}{30} = 1,386,666.67.$

$R = 1,386,666.67$, $i = .04735$, $n = 30$

$$S = 1,386,666 + 1,386,666\left[\frac{1-(1.04735)^{-29}}{.04735}\right]$$

$$\approx 23,016,361$$

The cash value is about $23,016,361. Using the TVM on a TI-84 Plus calculator will verify the answer.

37. The monthly payments form an annuity with $R = 30$, $i = .0125$, and $n = 36$.

a. The price of the stereo system is $600 plus the present value of the annuity.

$$P = 600 + 30\left[\frac{1-(1.0125)^{-36}}{.0125}\right]$$
$$= 600 + 865.42$$
$$= 1465.42$$

The stereo costs $1465.42.

b. The total amount paid is

$$600 + 30(36) = 600 + 1080 = \$1680$$

$$\text{Interest} = 1680.00 - 1465.42 = 214.58$$

The total amount of interest paid is $214.58.

39. $P = 15,000,\ i = \frac{.10}{2} = .05,\ n = 4(2) = 8$

$$R = \frac{P}{a_{\overline{n}|i}} = \frac{15,000}{\left[\frac{1-(1.05)^{-8}}{.05}\right]} \approx \frac{15,000}{6.46321} = \$2320.83$$

Each payment is $2320.83.

41. $R = \frac{35,000}{\left[\frac{1-(1+\frac{.0743}{12})^{-120}}{\frac{.0743}{12}}\right]} \approx 414.18$

Total payments = 10(12)(414.18) = 49,701.60
Interest = 49,701.60 − 35,000 = 14,701.60
The monthly payments are $414.18 and the total interest paid is $14,701.60.

43. a. $P = 14,000 + 7200 - 1200 = 20,000$

$i = \frac{.12}{2} = .06,\ n = 5(2) = 10$

$$R = \frac{20,000}{\left[\frac{1-(1.06)^{-10}}{.06}\right]} \approx 2717.36$$

Each payment is $2717.36.

b. There were 2 payments left at the time she decided to pay off the loan.

45. a. $P = 257,000 - .2(257,000) = 205,600$

$i = \frac{.039}{12} = .00325, n = 30(12) = 360$

$$R = \frac{205,600}{\left[\frac{1-(1.00325)^{-360}}{.00325}\right]} \approx 969.75$$

Their monthly payments are $969.75.

b. There are 360 − 60 = 300 payments left to be made. This can be thought of as an annuity consisting of 300 payments of $969.75 at 3.9% interest

$\frac{.039}{12} = .00325$ per period.

$$969.75\left[\frac{1-(1.00325)^{-300}}{.00325}\right] = 185,658.15$$

Their loan balance is $185,658.15.

c. Total payments = (360)(969.75) = 349,110
Interest = 349,110 − 205,600 = 143,510
The total interest paid is $143,510.

Copyright © 2019 Pearson Education, Inc.

47. We can think of these as two ordinary annuities. First, we determine the future value of her retirement account with $R = 400$, $i = \frac{.07}{12}$, and $n = 45(12) = 540$.

$$S = R\left[\frac{(1+i)^n - 1}{i}\right] = 400\left[\frac{\left(1+\frac{.07}{12}\right)^{540} - 1}{\frac{.07}{12}}\right]$$

$$\approx 1{,}517{,}038$$

Now we compute the amount of each withdrawal, using $P = 1{,}517{,}038$, $i = \frac{.05}{12}$, and $n = 30(12) = 360$.

$$P = R\left[\frac{1-(1+i)^{-n}}{i}\right] \Rightarrow$$

$$1{,}517{,}038 = R\left[\frac{1-\left(1+\frac{.05}{12}\right)^{-360}}{\frac{.05}{12}}\right] \Rightarrow$$

$$R = \frac{1{,}517{,}038}{\frac{1-\left(1+\frac{.05}{12}\right)^{-360}}{\frac{.05}{12}}} \approx 8143.79$$

Her maximum monthly withdrawal is \$8143.79.

49. Amount needed for retirement:

$$2500 = \frac{P}{\left[\frac{1-(1+\frac{.071}{12})^{-120}}{\frac{.071}{12}}\right]} \Rightarrow P = 214{,}363.15$$

Monthly contribution to acquire \$214,363.15:

$$214{,}363.15 = R\left[\frac{(1+\frac{.071}{12})^{240} - 1}{\frac{.071}{12}}\right]$$

$$R = 406.53$$

She must make monthly payments of \$406.53.

51. Amount needed for retirement:

$$3400 = \frac{P}{\left[\frac{1-(1+\frac{.082}{12})^{-120}}{\frac{.082}{12}}\right]} \Rightarrow P = 277{,}807.34$$

Amount needed after 120 equal contributions (i.e., how much money is needed to earn interest for 10 years with no contributions to amount to \$277,807.34):

$$277{,}807.34 = P\left[1+\frac{.082}{12}\right]^{10(12)}$$

$$P = 122{,}696.87$$

Monthly contributions to acquire \$122,696.87:

$$122{,}696.87 = R\left[\frac{(1+\frac{.082}{12})^{120} - 1}{\frac{.082}{12}}\right]$$

$$R = 663.22$$

He needs to make monthly contributions of \$663.22.

53. $P = 35{,}000$, $i = \frac{.052}{12} = .004333$, $n = 6(12) = 72$

a. $$R = \frac{P}{a_{\overline{n}|i}} = \frac{35{,}000}{\left[\frac{1-(1.004333)^{-72}}{.004333}\right]}$$

$$\approx \frac{35{,}000}{61.7365} \approx 566.93$$

The amount of each monthly payment is \$566.93.

b. Jessica will pay a total of $72(\$566.93) = \$40{,}818.96$. The price of the car is \$35,000, so the total amount of interest she will pay is $\$40{,}818.96 - \$35{,}000 = \$5{,}818.96$

c. Through the use of a spreadsheet, if Jessica paid an additional \$100 per month toward the principal, she would pay off the loan in 60 months.

d. Through the use of a spreadsheet, Jessica would pay back a total of \$39,795.04 with the additional \$100 added per month. Since the price of the car is \$35,000, the total amount of interest she will pay is $\$39{,}795.04 - \$35{,}000 = \$4{,}795.04$.

Copyright © 2019 Pearson Education, Inc.

In Exercises 55–58, the computer program "Explorations in Finite Mathematics" has been used to prepare the amortization schedules. Answers found using other computer programs or calculators may differ slightly.

55. $P = 4000,\ i = .08,\ n = 4$

Each annual payment is

$$R = \frac{P}{a_{\overline{n}|i}} = \frac{4000}{\left[\frac{1-(1.08)^{-4}}{.08}\right]} = \$1207.68$$

(although the last payment may differ slightly).

The interest for each period is 8% of the principal at the end of the previous period.

The portion to principal for each period is the difference between the amount of the payment and the interest for the period.

The principal at the end of each new period is obtained by subtracting the new portion to principal from the principal at the end of the previous period.

Repeat these steps four times to obtain the following amortization schedule.

Payment Number	Amount of Payment	Interest for Period	Portion to Principal	Principal at End of Period
0	------	------	------	$4000.00
1	$1207.68	$320.00	$887.68	3112.32
2	1207.68	248.99	958.69	2153.63
3	1207.68	172.29	1035.39	1118.24
4	1207.70	89.46	1118.24	0

57. $P = 8(1048) - 1200 = 7184,\ i = \frac{.12}{12} = .01,\ n = 4(12) = 48$

Each monthly payment is

$$R = \frac{P}{a_{\overline{n}|i}} = \frac{7184}{\left[\frac{1-(1.01)^{-48}}{.01}\right]} = \$189.18.$$

The interest for the first period is .01(7184) = $71.84.

The portion to principal for the first period is $189.18 – 71.84 = $117.34.

The principal at the end of the first period is $7184 – 117.34 = $7066.66.

Repeat these steps to construct three more rows of the table, which will look as follows.

Payment Number	Amount of Payment	Interest for Period	Portion to Principal	Principal at End of Period
0	------	------	------	$7184.00
1	$189.18	$71.84	$117.34	7066.66
2	189.18	70.67	118.51	6948.15
3	189.18	69.48	119.70	6828.45
4	189.18	68.28	120.90	6707.55

Chapter 5 Review Exercises

1. $P = 4902,\ r = .065,\ t = \frac{11}{12}$

$$I = Prt = 4902(.065)\left(\frac{11}{12}\right) \approx 292.08$$

The simple interest is $292.08.

2. $P = 42{,}368,\ r = .0922,\ t = \frac{5}{12}$

$$I = Prt = (42{,}368)(.0922)\left(\frac{5}{12}\right) \approx 1627.64$$

The simple interest is $1627.64.

Copyright © 2019 Pearson Education, Inc.

3. $P = 3478,\ r = .074,\ t = \dfrac{88}{365}$

$I = Prt = 3478(.074)\left(\dfrac{88}{365}\right) \approx 62.05$

The simple interest is \$62.05.

4. $P = 2390,\ r = .087,\ t = \dfrac{[(31-3)+30+28]}{365} = \dfrac{86}{365}$

$I = Prt = (2390)(.087)\left(\dfrac{86}{365}\right) \approx 48.99$

The simple interest is \$48.99.

5. $I = Prt = 12{,}000 \cdot .0475 \cdot \dfrac{1}{2} = 285$

The semiannual simple interest payment is \$285. The total interest earned over the life of the loan is (\$285)(12) = \$3420.

6. $I = Prt = 20{,}000 \cdot .0525 \cdot \dfrac{1}{2} = 525$

The semiannual simple interest payment is \$525. The total interest earned over the life of the loan is (\$525)(18) = \$9450.

7. $A = P(1+rt) = 7750\left(1+.068\cdot\frac{4}{12}\right) = \$7925.67$

The maturity value is \$7925.67.

8. $A = P(1+rt) = 15{,}600\left(1+.082\cdot\frac{9}{12}\right)$
$= \$16{,}559.40$

The maturity value is \$16,559.40.

9. $A = 3250,\ r = .0587,\ t = \dfrac{11}{12}$

$P = \dfrac{A}{1+rt} = \dfrac{3250}{1+.0587\left(\frac{11}{12}\right)} \approx 3084.05$

The present value is \$3,084.05.

10. $A = 459.57,\ r = .055,\ t = \dfrac{7}{12}$

$P = \dfrac{A}{1+rt} = \dfrac{459.57}{1+.055\left(\frac{7}{12}\right)} \approx 445.28$

The present value is \$445.28.

11. $A = 80{,}612,\ r = .0677,\ t = \dfrac{128}{365}$

$P = \dfrac{A}{1+rt} = \dfrac{80{,}612}{1+(.0677)\left(\frac{128}{365}\right)} \approx 78{,}742.54$

The present value is \$78,742.54.

12. Discount $= Prt = 7000 \cdot .035 \cdot \frac{9}{12} = \$183.75$

The price of the T-bill is the face value – the discount. $7000 - 183.75 = \$6816.25$

13. Discount $= Prt = 10{,}000 \cdot .04 \cdot \frac{6}{12} = \$200$

The price of the T-bill is the face value – the discount. $10{,}000 - 200 = \$9800$

Now compute the actual interest rate:

$I = Prt \Rightarrow 200 = 980(r)\left(\frac{1}{2}\right) \Rightarrow$

$r \approx .040816$

The actual interest rate is about 4.082%.

14. Answers vary. Possible answer: Compound interest produces more interest because interest is paid on the previously earned interest as well as on the original principal.

15. $P = 2800,\ i = .06,\ n = 12$

$A = P(1+i)^n = 2800(1.06)^{12} \approx 5634.15$

The compound amount is \$5634.15. The amount of interest earned is
\$5634.15 – \$2800 = \$2834.15.

16. $P = 57{,}809.34,\ i = \dfrac{.04}{4} = .01,\ n = 6(4) = 24$

$A = P(1+i)^n = 57{,}809.34(1.01)^{24} \approx 73{,}402.52$

The compound amount is \$73,402.52.
The amount of interest earned is
\$73,402.52 – \$57,809.34 = \$15,593.18.

17. $P = 12{,}903.45,\ i = \dfrac{.0637}{4} = .015925,\ n = 29$

$A = P(1+i)^n$
$= 12{,}903.45(1.015925)^{29} \approx \$20{,}402.98$

The compound amount is \$20,402.98. The amount of interest earned is
\$20,402.98 – \$12,903.45 = \$7499.53.

18. $P = 4677.23,\ i = \dfrac{.0457}{12},\ n = 32$

$A = P(1+i)^n$

$= 4677.23\left(1+\dfrac{.0457}{12}\right)^{32} \approx 5282.19$

The compound amount is \$5282.19. The amount of interest earned is
\$5282.19 – \$4677.23 = \$604.96.

Copyright © 2019 Pearson Education, Inc.

19. $P = 22{,}000,\ i = \frac{.055}{4},\ n = 6(4) = 24$

$A = P(1+i)^n$

$= 22{,}000\left(1+\frac{.055}{4}\right)^{24} \approx 30{,}532.58$

The compound amount is \$30,532.52. The amount of interest earned is \$30,532.52 − \$22,000 = \$8532.58.

20. $P = 2975,\ i = \frac{.047}{12},\ n = 4(12) = 48$

$A = P(1+i)^n$

$= 2975\left(1+\frac{.047}{12}\right)^{48} \approx 3589.01$

The compound amount is \$3580.01. The amount of interest earned is \$3589.01 − \$2975 = \$614.01.

21. $P = 12{,}366,\ i = \frac{.039}{2},\ n = 5(2) = 10$

$A = 12{,}366\left(1+\frac{.039}{2}\right)^{10} \approx 15{,}000$

The face value is \$15,000.

22. $P = 11{,}575,\ i = \frac{.052}{2},\ n = 15(2) = 30$

$A = 11{,}575\left(1+\frac{.052}{2}\right)^{30} \approx 25{,}000$

The face value is \$25,000.

23. Use the formula for effective rate with $r = .05$ and $m = 2$.

$r_E = \left(1+\frac{r}{m}\right)^m - 1 = \left(1+\frac{.05}{2}\right)^2 - 1$

$= .050625$

The effective rate is 5.0625%.

24. Use the formula for effective rate with $r = .065$ and $m = 365$.

$r_E = \left(1+\frac{r}{m}\right)^m - 1 = \left(1+\frac{.065}{365}\right)^{365} - 1$

$= .067153$

The effective rate is 6.7153%.

25. $A = 42{,}000,\ i = \frac{.12}{12} = .01,\ n = 7(12) = 84$

$P = \frac{A}{(1+i)^n} = \frac{42{,}000}{(1.01)^{84}} \approx 18{,}207.65$

The present value is \$18,207.65.

26. $A = 17{,}650,\ i = \frac{.08}{4} = .02,\ n = 4(4) = 16$

$P = \frac{A}{(1+i)^n} = \frac{17{,}650}{(1.02)^{16}} \approx 12{,}857.07$

The present value is \$12,857.07.

27. $A = 1347.89,\ i = \frac{.062}{2} = .031,$

$n = (3.5)(2) = 7$

$P = \frac{A}{(1+i)^n} = \frac{1347.89}{(1.031)^7} \approx 1088.54$

The present value is \$1088.54.

28. $A = 2388.90,\ i = \frac{.0575}{12},\ n = 44$

$P = \frac{A}{(1+i)^n} = \frac{2388.90}{\left(1+\frac{.0575}{12}\right)^{44}} \approx 1935.77$

The present value is \$1935.77.

29. $P = \frac{A}{(1+i)^n} = \frac{15{,}000}{\left(1+\frac{.044}{2}\right)^{2(10)}} \approx 9706.74$

A purchaser should be willing to pay about \$9706.74.

30. $P = \frac{A}{(1+i)^n} = \frac{30{,}000}{\left(1+\frac{.062}{2}\right)^{2(25)}} \approx 6519.11$

A purchaser should be willing to pay about \$6519.11.

31. The future value of an annuity is the balance in the account at the end of the term.

32. $R = 1288,\ i = .07,\ n = 14$

$S = R \cdot s_{\overline{n}|i} = 1288\left[\frac{(1.07)^{14} - 1}{.07}\right] \approx 29{,}045.03$

The future value of this ordinary annuity is \$29,045.03.

33. $R = 4000,\ i = \frac{.06}{4} = .015,\ n = 8(4) = 32$

$S = R \cdot s_{\overline{n}|i} = 4000\left[\frac{(1.015)^{32} - 1}{.015}\right]$

$\approx 162{,}753.15$

The future value of this ordinary annuity is \$162,753.15.

Copyright © 2019 Pearson Education, Inc.

34. $R = 233,\ i = \frac{.06}{12} = .005,\ n = 4(12) = 48$

$$S = R \cdot s_{\overline{n}|i} = 233\left[\frac{(1.005)^{48} - 1}{.005}\right] \approx 12{,}604.79$$

The future value of this ordinary annuity is $12,604.79.

35. $R = 672,\ i = \frac{.05}{4} = .0125,\ n = 7(4) = 28$

Because deposits are made at the beginning of each time period, this is an annuity due.

$$S = R \cdot s_{\overline{n+1}|i} - R = 672\left[\frac{(1.0125)^{29} - 1}{.0125}\right] - 672$$
$$\approx 22{,}643.29$$

The future value of this annuity due is $22,643.29.

36. $R = 11{,}900,\ i = \frac{.07}{12} = .005833,\ n = 13$

$$S = R \cdot s_{\overline{n+1}|i} - R$$
$$= 11{,}900\left[\frac{1.0058333^{14} - 1}{.0058333}\right] - 11{,}900$$
$$\approx 161{,}166.70$$

The future value of this annuity due is $161,166.70.

37. Answers vary. Possible answer: The purpose of a sinking fund is to receive funds deposited regularly for some goal.

38. $S = 6500,\ i = .05,\ n = 6$

$$S = R \cdot s_{\overline{n}|i} \Rightarrow R = \frac{S}{s_{\overline{n}|i}} = \frac{6500}{\left[\frac{(1.05)^6 - 1}{.05}\right]} \approx 955.61$$

The amount of each payment into this sinking fund is $955.61.

39. $S = 57{,}000,\ i = \frac{.06}{2} = .03,\ n = \left(8\frac{1}{2}\right)(2) = 17$

$$R = \frac{S}{s_{\overline{n}|i}} = \frac{57{,}000}{\left[\frac{(1.03)^{17} - 1}{.03}\right]} \approx 2619.29$$

The amount of each payment is $2619.29.

40. $S = 233{,}188,\ i = \frac{.057}{4} = .01425,$

$$n = \left(7\frac{3}{4}\right)(4) = 31$$

$$R = \frac{S}{s_{\overline{n}|i}} = \frac{233{,}188}{\left[\frac{(1.01425)^{31} - 1}{.01425}\right]} \approx 6035.27$$

The amount of each payment is $6035.27.

41. $S = 56{,}788,\ i = \frac{.0612}{12} = .0051,$

$$n = \left(4\frac{1}{2}\right)(12) = 54$$

$$R = \frac{S}{s_{\overline{n}|i}} = \frac{56{,}788}{\left[\frac{(1.0051)^{54} - 1}{.0051}\right]} \approx 916.12$$

The amount of each payment is $916.12.

42. $R = 850,\ i = .05,\ n = 4$

$$P = R \cdot a_{\overline{n}|i} = 850\left[\frac{1 - (1.05)^{-4}}{.05}\right] \approx 3014.06$$

The present value of this ordinary annuity is $3014.06.

43. $R = 1500,\ i = \frac{.08}{4} = .02,\ n = 7(4) = 28$

$$P = R \cdot a_{\overline{n}|i} = 1500\left[\frac{1 - (1.02)^{-28}}{.02}\right] \approx 31{,}921.91$$

The present value is $31,921.91.

44. $R = 4210,\ i = \frac{.056}{2} = .028,\ n = 8(2) = 16$

$$P = R \cdot a_{\overline{n}|i} = 4210\left[\frac{1 - (1.028)^{-16}}{.028}\right]$$
$$\approx \$53{,}699.94$$

The present value is $53,699.94.

45. $R = 877.34,\ i = \frac{.064}{12} \approx .0053,\ n = 17$

$$P = R \cdot a_{\overline{n}|i} = 877.34\left[\frac{1 - \left(1 + \frac{.064}{12}\right)^{-17}}{\frac{.064}{12}}\right]$$
$$\approx 14{,}222.42$$

The present value is $14,222.42.

Copyright © 2019 Pearson Education, Inc.

For exercises 46–48, recall that an annuity in which the payments are made at the end of each period is an ordinary annuity. We are seeking the present value.

46. $R = 800,\ i = \dfrac{.046}{4},\ n = 4(4) = 16$

$$P = 800\left[\frac{1-\left(1+\frac{.046}{4}\right)^{-16}}{\frac{.046}{4}}\right] \approx 11{,}630.63$$

About \$11,630.63 is needed to fund the withdrawals.

47. $R = 1500,\ i = \dfrac{.058}{12},\ n = 10(12) = 120$

$$P = 1500\left[\frac{1-\left(1+\frac{.058}{12}\right)^{-120}}{\frac{.058}{12}}\right] \approx 136{,}340.32$$

About \$136,340.32 is needed to fund the withdrawals.

48. $R = 3000,\ i = .062,\ n = 15$

$$P = 3000\left[\frac{1-(1.062)^{-15}}{.062}\right] \approx 28{,}759.74$$

About \$28,759.74 is needed to fund the withdrawals.

49. $P = 150{,}000,\ i = \dfrac{.051}{12},\ n = 15(12) = 180$

$$150{,}000 = R\left[\frac{1-\left(1+\frac{.051}{12}\right)^{-180}}{\frac{.051}{12}}\right] \Rightarrow$$

$$R = \frac{150{,}000}{\frac{1-\left(1+\frac{.051}{12}\right)^{-180}}{\frac{.051}{12}}} \approx 1194.02$$

The monthly payment will be \$1194.02.

50. $P = 25{,}000,\ i = \dfrac{.049}{4},\ n = 8(4) = 32$

$$25{,}000 = R\left[\frac{1-\left(1+\frac{.049}{4}\right)^{-32}}{\frac{.049}{4}}\right] \Rightarrow$$

$$R = \frac{25{,}000}{\frac{1-\left(1+\frac{.049}{4}\right)^{-32}}{\frac{.049}{4}}} \approx 949.07$$

The monthly payment will be \$949.07.

51. Find the future value of an annuity with $R = 4200,\ i = .045,$ and $n = 12.$

$$S = R\left[\frac{(1+i)^n - 1}{i}\right] = 4200\left[\frac{(1.045)^{12} - 1}{.045}\right]$$
$$\approx 64{,}948.93$$

Now find the present value of \$64,948.93 at 4.5% compounded annually for 12 years.

$A = 64{,}948.93,\ i = .045,\ n = 12$

$A = P(1+i)^n \Rightarrow 64{,}948.93 = P(1.045)^{12} \Rightarrow$

$$P = \frac{64{,}948.93}{(1.045)^{12}} \approx 38{,}298.04$$

The required lump sum is \$38,298.04.

52. The interest that is paid each half-year is $I = Prt = 24{,}000 \cdot .05 \cdot \dfrac{1}{2} = \$600.$ So, we have a two-part investment, an annuity that pays \$600 every six months for 6 years and the \$24,000 face value of the bond which will be paid when the bond matures 6 years from now. The purchaser should be willing to pay the present value of each part of the investment, assuming 6.5% interest compounded semiannually.

Present value of annuity

$R = 600,\ i = \dfrac{.065}{2} = .0325,\ n = 6(2) = 12$

$$P = 600\left[\frac{1-(1.0325)^{-12}}{.0325}\right] \approx 5884.25$$

Present value of \$24,000 in 6 years

$P = A(1+i)^{-n}$

$P = 24{,}000(1.0325)^{-12} \approx 16{,}350.48$

The purchaser should be willing to pay \$5884.25 + \$16,350.48 = \$22,234.73, or about \$22,235.

53. $P = 32{,}000,\ i = \dfrac{.084}{4} = .021,\ n = 10$

$$P = R \cdot a_{\overline{n}|i} \Rightarrow R = \frac{P}{a_{\overline{n}|i}} = \frac{32{,}000}{\left[\frac{1-(1.021)^{-10}}{.021}\right]}$$
$$\approx 3581.11$$

Quarterly payments of \$3581.11 are necessary to amortize this loan.

Copyright © 2019 Pearson Education, Inc.

54. $P = 5607,\ i = \dfrac{.076}{12},\ n = 32$

$$R = \frac{P}{a_{\overline{n}|i}} = \frac{5607}{\left[\dfrac{1-\left(1+\frac{.076}{12}\right)^{-32}}{\frac{.076}{12}}\right]} \approx 194.13$$

Monthly payments of $194.13 are needed to amortize this loan.

55. $P = 95,000,\ i = \dfrac{.0367}{12},\ n = 30(12) = 360$

$$R = \frac{P}{a_{\overline{n}|i}} = \frac{95,000}{\left[\dfrac{1-\left(1+\frac{.0367}{12}\right)^{-360}}{\frac{.0367}{12}}\right]} \approx 435.66$$

The monthly payment for this mortgage is $435.66.

56. $P = 167,000,\ i = \dfrac{.0291}{12},\ n = 15(12) = 180$

$$R = \frac{P}{a_{\overline{n}|i}} = \frac{167,000}{\left[\dfrac{1-\left(1+\frac{.0291}{12}\right)^{-180}}{\frac{.0291}{12}}\right]} \approx 1146.06$$

The monthly payment for this mortgage is $1146.06.

57. After 5 years of payments, the remaining 25 years of payments can be thought of as annuity. The present value of this annuity is the remaining balance. So, we use the present value formula with $R = 435.66,\ i = \dfrac{.0367}{12},\ n = 300$

$$P = 435.66\left[\frac{1-\left(1+\frac{.0367}{12}\right)^{-300}}{\frac{.0367}{12}}\right] \approx 85,459.20$$

The balance remaining after five years is about $85,459.20.

58. After 7.5 years (90 payments), the remaining 90 payments can be thought of as annuity. The present value of this annuity is the remaining balance. So, we use the present value formula with $R = 1146.06,\ i = \dfrac{.0291}{12},\ n = 90$

$$P = 1146.06\left[\frac{1-\left(1+\frac{.0291}{12}\right)^{-90}}{\frac{.0291}{12}}\right] \approx 92,565.32$$

The balance remaining after 7.5 years of payments is about $92,565.32.

59. Locate the entry of the table that is in the row labeled "Payment Number 7" and in the column labeled "Interest for Period," to observe that $45.04 of the seventh payment is interest.

60. Locate the entry of the table that is in the row labeled "Payment Number 4" and in the column labeled "Portion to Principal," to observe that $104.14 of the forth payment is used to reduce the debt.

61. To find out how much interest is paid in the first 6 months of the loan, add the first six nonzero entries in the column labeled "Interest for Period."

$$\begin{array}{r} \$47.00 \\ 46.68 \\ 46.35 \\ 46.03 \\ 45.70 \\ +\ 45.37 \\ \hline \$277.13 \end{array}$$

62. To find out how much the debt has been reduced by the end of the first 8 months, subtract the balance at the end of the eighth payment from the original value.
$\$15,000 - 14,165.59 = \$834.41$

63. a. $\dfrac{429,600,000}{30} = 14,320,000$

Each yearly payment would have been $14,320,000.

b. We use the formula for the present value of an annuity due, $S = R + R\left[\dfrac{1-(1+i)^{-(n-1)}}{i}\right]$.

$R = 14,320,000,\ i = .0358,\ n = 30$

$$S = 14,320,000 + 14,320,000\left[\frac{1-(1.0358)^{-29}}{.0358}\right] \approx 270,088,972.46$$

The cash value is about $270,088,972.46.

64. $P = 15,000,\ i = \dfrac{.06}{2} = .03,\ n = 7.5(2) = 15$

$$I = P(1+i)^n - P$$
$$= 15,000(1+.03)^{15} - 15,000 \approx 8369.51$$

Copyright © 2019 Pearson Education, Inc.

65. Starting at age 23:

$P = 500,\ i = \frac{.05}{4} = .0125,$

$n = (65 - 23)4 = 42(4) = 168$

$A = P(1+i)^n = 500(1.0125)^{168} \approx 4030.28$

Starting at age 40:

$t = (65 - 40)4 = 25(4) = 100$

$A = 500(1.0125)^{100} \approx 1731.70$

$4030.28 - 1731.70 = 2298.58$

He will have $2298.58 more if he invests now.

66. $r_F = \left(1 + \frac{r}{m}\right)^m - 1;\ r_E = \left(1 + \frac{r}{m}\right)^m - 1$

Frontenac Bank:

$r_F = \left(1 + \frac{.0394}{4}\right)^4 - 1 \approx .04$

The effective rate is about 4.00%.

E*TRADE Bank:

$r_E = \left(1 + \frac{.0393}{365}\right)^{365} - 1 \approx .0401$

The effective rate is about 4.01%.

E*TRADE Bank paid a higher effective rate.

67. $R = 3200,\ i = \frac{.068}{2} = .034,\ n = (3.5)(2) = 7$

Use the formula for the future value of an ordinary annuity.

$$S = R \cdot s_{\overline{n}|i} = 3200\left[\frac{(1.034)^7 - 1}{.034}\right] \approx 24{,}818.76$$

The final amount in the account will be $24,818.76.

The interest earned will be $24,818.76 – 7($3200) = $2418.76.

68. $S = 625{,}000{,}000,\ i = \frac{.075}{12} = .00625,$

$n = 20(12) = 240$

Use the formula for future value of an ordinary annuity to find the periodic payment.

$S = R \cdot s_{\overline{n}|i} \Rightarrow$

$$R = \frac{S}{s_{\overline{n}|i}} = \frac{625{,}000{,}000}{\left[\frac{(1.00625)^{240} - 1}{.00625}\right]} \approx 1{,}128{,}707.46$$

The firm should invest $1,128,707.46 monthly.

69. $S = 2{,}000{,}000,\ i = \frac{.055}{12},\ n = 10(12) = 120$

Use the formula for future value of an annuity.

$$R = S\left[\frac{i}{(1+i)^n - 1}\right]$$

$$\approx 2{,}000{,}000\left[\frac{\frac{.055}{12}}{(1 + \frac{.055}{12})^{120} - 1}\right] \approx 12{,}538.59$$

She would have to put $12,538.59 into her savings account every month.

70. Age 55: $P = 9000,\ i = \frac{.08}{1} = .08,\ n = 20(1) = 20$

$$S = 9000\left[\frac{(1.08)^{20} - 1}{.08}\right] \approx 411{,}857.68$$

Age 65: $P = 18{,}000,\ i = \frac{.08}{1} = .08,\ n = 10(1) = 10$

$$S = 18{,}000\left[\frac{(1.08)^{10} - 1}{.08}\right] \approx 260{,}758.12$$

If the pension starts at age 55, the total is $411,857.68, whereas, if the pension starts at age 65, the total is $260,758.12. Thus, taking the option at age 55 produces the larger amount.

71. $A = 7500,\ i = \frac{.10}{2} = .05,\ n = 3(2) = 6$

$A = P(1+i)^n$

$$P = \frac{A}{(1+i)^n} = \frac{7500}{(1.05)^6} \approx 5596.62$$

The required lump sum is $5596.62.

72. $P = 15{,}000,\ i = \frac{.072}{12} = .006,\ n = 36$

This is an ordinary annuity.

$P = R \cdot a_{\overline{n}|i} \Rightarrow$

$$R = \frac{P}{a_{\overline{n}|i}} = \frac{15{,}000}{\left[\frac{1-(1.006)^{-36}}{.006}\right]} \approx 464.53$$

The amount of each payment will be $464.53.

73. $P = 40{,}000,\ i = \frac{.09}{2} = .045,\ n = 8(2) = 16$

This is an ordinary annuity.

$$R = \frac{P}{a_{\overline{n}|i}} = \frac{40{,}000}{\left[\frac{1-(1.045)^{-16}}{.045}\right]} \approx \$3560.61$$

The amount of each payment will be $3560.61.

Copyright © 2019 Pearson Education, Inc.

74. Amount of loan
$= \$210,000 - \$42,000 = \$168,000$

a. Use the formula for amortization payments with $P = 168,000$, $i = \frac{.0375}{12}$, and $n = 30(12) = 360$.

$$R = \frac{Pi}{1-(1+i)^{-n}} = \frac{168,000(\frac{.0375}{12})}{1-(1+\frac{.0375}{12})^{-360}} \approx 778.03$$

The monthly payment for this mortgage is \$778.03.

b. To find the amount of the first payment that goes to interest, use $I = Prt$ with $P = 168,000$, $i = .003125$, and $t = 1$.

$$I = (168,000)(.003125)(1) = 525.00$$

Of the first payment, \$525.00 is interest.

c. Using method 1, since 180 payments were made, there are 180 remaining payments. The present value is

$$778.03\left[\frac{1-(1.003125)^{-180}}{.003125}\right] \approx 106,986.52,$$

so the remaining balance is \$106,986.52. Using method 2, since 180 payments were already made, we have

$$778.03\left[\frac{1-(1.003125)^{-180}}{.003125}\right] \approx 106,986.52.$$

They still owe \$106,986.52 after 15 years of payments.

d. Amount of money received
= Selling price–Current mortgage balance
Using method 1, the amount received is
\$255,000 – \$106,986.52
= \$148,013.48.
Using method 2, the amount received is
\$255,000 – \$106,986.52
= \$148,013.48.

75. This is an ordinary annuity. We are seeking the future value.

$$R = 325,\ i = \frac{.068}{12},\ n = 20(12) = 240$$

$$S = 325\left[\frac{\left(1+\frac{.068}{12}\right)^{240}-1}{\frac{.068}{12}}\right] \approx 165,249.03$$

There will be about \$165,249.03 in the account at the end of 20 years.

76. Total amount after 60 deposits:

$$R = 400,\ i = \frac{.031}{12},\ n = 60$$

$$S = 400\left[\frac{(1+\frac{.031}{12})^{60}-1}{\frac{.031}{12}}\right] = 25,923.81$$

Total amount after next 96 months:

$$A = 25,923.81(1+\tfrac{.031}{12})^{96} = 33,209.70$$

Total of 60 \$525 deposits:

$$R = 525,\ i = \frac{.031}{12},\ n = 60$$

$$S = 525\left[\frac{(1+\frac{.031}{12})^{60}-1}{\frac{.031}{12}}\right] \approx 34,025.00$$

Gene's account balance is
\$33,209.71 + \$34,025.00 = \$67,234.71.

77. $P = 50,000$, $i = .042$, $n = 10$

$$A = P(1+i)^n = 50,000(1.042)^{10} \approx \$75,447.91$$

At the end of 10 years, the value of the death benefit will be \$75,447.91. This balance of \$75,447.91 is to be paid out in 120 equal monthly payments, with $i = \frac{.025}{12} = .0020833$. The full balance will be paid out, so we use the amortization formula. Let X be the amount of each payment.

$$X = \frac{75,447.91}{a_{\overline{120}|.0020833}} = \frac{75,447.91}{\left[\frac{1-(1.0020833)^{-120}}{.0020833}\right]} \approx \$711.25$$

Copyright © 2019 Pearson Education, Inc.

78. Amount needed for retirement:

$$75,000 = \frac{x}{\left[\frac{1-(1.101)^{-20}}{.101}\right]}$$

$$x = 634,183.19$$

Annual contribution to acquire \$634,183.19:

$$634,183.19 = R\left[\frac{(1.101)^{30}-1}{.101}\right]$$

$$R \approx 3783.01$$

She will have to make annual deposits of \$3783.01.

Case 5 Continuous Compounding

1. $PMT = 150,\ i = \frac{.06}{12},\ n = 30(12) = 360$

$$FV = PMT\left[\frac{(1+i)^n - 1}{i}\right]$$

$$= 150\left[\frac{(1+.005)^{360} - 1}{.005}\right]$$

$$\approx 150,677.26$$

The future value of Shelia's IRA would be approximately \$150,677.26.

2. $PMT = 300,\ i = \frac{.12}{12},\ n = 25(12) = 300$

$$FV = PMT\left[\frac{(1+i)^n - 1}{i}\right]$$

$$= 300\left[\frac{(1+.01)^{300} - 1}{.01}\right]$$

$$\approx 563,653.99$$

The future value of Hector's IRA would be approximately \$563,653.99.

3. a. Using the rule of 72 and the 6% interest rate Sheila was receiving, it would take 12 years from when she stopped making deposits for her money to double or 55 + 12 = 67 years old.

b. The balance would be twice the value from exercise 1 or 2 (150,677.26) = \$301,354.52.

4. a. Using the rule of 72 and the 12% interest rate Hector was receiving, it would take 6 years from when he stopped making deposits for his money to double or 55 + 6 = 61 years old.

b. The balance would be twice the value from exercise 2 or 2 (563,653.99) = \$1,127,307.98.

5. Using the rule of 72 and a 6% interest rate, it would take 12 years for the money to double. Therefore, the money will double 3 times in 36 years. The value at the end of the 36 years will be (10,000)(2)(2)(2) = \$80,000.

6. Using the rule of 72 and an 8% interest rate, it would take 9 years for the money to double. Therefore, the money will double 4 times in 36 years. The value at the end of the 36 years will be (10,000)(2)(2)(2)(2) = \$160,000.

Copyright © 2019 Pearson Education, Inc.

Chapter 6 Systems of Linear Equations and Matrices

Section 6.1 Systems of Two Linear Equations in Two Variables

1. Is (–1, 3) a solution of
$2x + y = 1$ (1)
$-3x + 2y = 9$ (2)
Check (–1, 3) in equation (1).
$2(-1) + 3 = 1$?
$-2 + 3 = 1$? True
Check (–1, 3) in equation (2).
$-3(-1) + 2(3) = 9$?
$3 + 6 = 9$? True
The ordered pair (–1, 3) is a solution of this system of equations.

3. $3x - y = 1$ (1)
$x + 2y = -9$ (2)
Solve for y from equation (1) to get $y = 3x - 1$.
Then substitute $y = 3x - 1$
into (2) $x + 2(3x - 1) = -9$. Solve for x.
$x + 2(3x - 1) = -9 \Rightarrow x + 6x - 2 = -9 \Rightarrow$
$7x - 2 = -9 \Rightarrow 7x = -7 \Rightarrow x = -1$
Substitute $x = -1$ into $y = 3x - 1$ to solve for y.
$y = 3(-1) - 1 \Rightarrow y = -4$.
The solution set is (–1, –4).

5. $3x - 2y = 4$ (1)
$2x + y = -1$ (2)
Solve for y from equation 2 to get $y = -1 - 2x$.
Substitute $y = -1 - 2x$ into equation (1)
$3x - 2(-1 - 2x) = 4 \Rightarrow 3x + 2 + 4x = 4 \Rightarrow$
$7x + 2 = 4 \Rightarrow 7x = 2 \Rightarrow x = \frac{2}{7}$
Substitute $x = \frac{2}{7}$ into $y = -1 - 2x$ to solve for y.
$$y = -1 - 2\left(\frac{2}{7}\right) = -\frac{11}{7}$$
The solution set is $\left(\frac{2}{7}, -\frac{11}{7}\right)$.

7. $x - 2y = 5$ (1)
$2x + y = 3$ (2)
Multiply equation (2) by 2 and add the result to equation (1).
$$\begin{array}{ll} x - 2y = 5 & (1) \\ \underline{4x + 2y = 6} & \\ 5x \quad = 11 \Rightarrow x = \frac{11}{5} & \end{array}$$
Substitute $\frac{11}{5}$ for x in equation (1) to get
$\frac{11}{5} - 2y = 5 \Rightarrow -2y = \frac{14}{5} \Rightarrow y = -\frac{7}{5}$.
The solution of the system is $\left(\frac{11}{5}, -\frac{7}{5}\right)$.

9. $2x - 2y = 12$ (1)
$-2x + 3y = 10$ (2)
Add the two equations to get $y = 22$.
Substitute 22 for y in equation (1) to get
$2x - 2(22) = 12 \Rightarrow 2x - 44 = 12 \Rightarrow$
$2x = 56 \Rightarrow x = 28$
The solution of the system is (28, 22).

11. $x + 3y = -1$ (1)
$2x - y = 5$ (2)
Multiply equation (2) by 3, then add the result to equation (1)
$$\begin{array}{ll} x + 3y = -1 & (1) \\ \underline{6x - 3y = 15} & (2) \\ 7x \quad = 14 \Rightarrow x = 2 & \end{array}$$
Substitute 2 for x in equation (1) and solve for y.
$2 + 3y = -1 \Rightarrow 3y = -3 \Rightarrow y = -1$
The solution of the system is (2, –1).

13. $2x + 3y = 15$ (1)
$8x + 12y = 40$ (2)
Multiply equation (1) by –4 and add the result to equation (1).
$$\begin{array}{ll} -8x - 12y = -60 & \\ \underline{8x + 12y = 40} & (2) \\ 0 = -20 & \text{False} \end{array}$$
The system is inconsistent. There is no solution.

Copyright © 2019 Pearson Education, Inc.

15. $2x - 8y = 2$ (1)
$3x - 12y = 3$ (2)

Multiply equation (1) by 3 and equation (2) by –2 to get

$$6x - 24y = 6 \quad (3)$$
$$-6x + 24y = -6. \quad (4)$$

Add the two equations.

$$\begin{array}{r} 6x - 24y = 6 \quad (3) \\ -6x + 24y = -6 \quad (4) \\ \hline 0 = 0 \end{array}$$

The system is dependent, so there are an infinite number of solutions. To solve the system in terms of one of the parameters, say y, solve equation (1) for x in terms of y.

$2x - 8y = 2 \Rightarrow 2x = 8y + 2 \Rightarrow x = 4y + 1$

The solution to the system is the infinite set of pairs of the form $(4y + 1, y)$ for any real number y.

Note: the system could also be solved in terms of x. The ordered pairs would all have the form $\left(x, \frac{1}{4}x - \frac{1}{4}\right)$.

17. $\frac{x}{2} + \frac{y}{3} = 8$ (1)

$\frac{2x}{3} + \frac{3y}{2} = 17$ (2)

Clear the fractions by multiplying equation (1) by 6 and equation (2) by 6.

$3x + 2y = 48$ (3)
$4x + 9y = 102$ (4)

Now, multiply equation (3) by 4 and equation (4) by –3, then add the resulting equations and solve for y

$$\begin{array}{rr} 12x + 8y = 192 & (5) \\ -12x - 27y = -306 & (6) \\ \hline -19y = -114 \Rightarrow y = 6 & \end{array}$$

Solve for x by substituting 6 for y into equation (1). $\frac{x}{2} + \frac{6}{3} = 8 \Rightarrow \frac{x}{2} + 2 = 8 \Rightarrow \frac{x}{2} = 6 \Rightarrow x = 12$

The solution of the system is (12, 6).

19. Multiply the second equation by 5 and then add the equations:

$$\begin{array}{r} 10x + 13y = 1125 \\ -10x + 35y = 2475 \\ \hline 48y = 3600 \Rightarrow y = 75 \text{ million} \end{array}$$

Substitute 75 for y in the first equation and solve for x.

$10x + 13(75) = 1125 \Rightarrow$
$x = 15$

In 2015, the number of millennials and boomers were each 75 million.

21. $-x + 150y = 1710$ (1)
$-41x + 300y = 2430$ (2)

We are asked to find the year in which the two states have the same population, which is represented by the x-coordinate of the solution. Multiply equation (1) by –2, then add the result to equation (2) and solve for x.

$$\begin{array}{r} 2x - 300y = -3420 \\ -41x + 300y = 2430 \\ \hline -39x = -990 \Rightarrow x \approx 25.38 \end{array}$$

$2000 + 25.38 = 2025.38$

Thus, the two states will have the same population in 2025.

23. $-25x + 6y = 20$ (1)
$15x + 2y = 322$ (2)

We are asked to find when the interest in LED lighting will surpass CFL lighting, i.e., the year. This is represented by the x-coordinate of the solution. Multiply equation (2) by –3 and then add the result to equation (1) to solve for x.

$$\begin{array}{r} -25x + 6y = 20 \\ -45x - 6y = -966 \\ \hline -70x = -946 \Rightarrow x \approx 13.51 \end{array}$$

$2000 + 13.51 = 2013.51$

According to the model, the interest in LED lighting surpassed CFL lighting 2014.

Copyright © 2019 Pearson Education, Inc.

25. $-7x+16y=1070$ (1)
$-5x+10y=759$ (2)

We are asked to find if the number of women in the workforce will be the same as the number of men, i.e, the year. This is represented by the x-coordinate of the solution. Multiply equation (1) by 10 and equation (2) by –16 and then add the results to solve for x.

$$\begin{array}{l} -70x+160y=10,700 \\ \underline{80x-160y=-12,144} \\ 10x \quad =-1444 \Rightarrow x \approx -144.4 \end{array}$$

According to the model, the number of women in the workforce will not be the same as the number of men during the time period.

27. Let x = the number of adults and let y = the number of children.

$x+\ y=200$ (1)
$8x+5y=1435$ (2)

Solve equation (1) for y, then substitute that expression into equation (2) and solve for x.

$x+y=200 \Rightarrow y=200-x$
$8x+5(200-x)=1435 \Rightarrow$
$8x+1000-5x=1435 \Rightarrow 3x=435 \Rightarrow x=145$

Substitute x = 145 into equation (1) and solve for y.

$145+y=200 \Rightarrow y=55$

There were 145 adults and 55 children.

29. We use the formula rate × time = distance. Let x = the speed of the plane and let y = the speed of the wind.

$5(x+y)=3000 \Rightarrow x+y=600$
$6(x-y)=3000 \Rightarrow x-y=500$

Add the equations and solve for x.

$2x=1100 \Rightarrow x=550$

Substitute x = 550 into the first equation and solve for y.

$550+y=600 \Rightarrow y=50$

The plane's speed is 550 mph and the wind speed is 50 mph.

31. a. To find the equation of the line through (1, 2) and (3, 4), first find the slope.

$$m=\frac{4-2}{3-1}=\frac{2}{2}=1$$

Use the point-slope form with $(x_1, y_1)=(1,2)$.

$y-y_1=m(x-x_1) \Rightarrow y-2=1(x-1) \Rightarrow$
$y-2=x-1 \Rightarrow y=x+1$

An alternate solution can be written using a system of equations. Using the form $y = mx + b$, the line through (1, 2) and (3, 4) would yield the system of equations.

$2=m(1)+b$ (1)
$4=m(3)+b.$ (2)

This simplifies to

$m+b=2$ (1)
$3m+b=4.$ (2)

This system can be solved by multiplying equation (1) by –1 and adding the result to equation (2).

$$\begin{array}{l} -m-b=-2 \\ \underline{3m+b=4} \\ 2m=2 \Rightarrow m=1 \end{array}$$

Substitute 1 for m in equation (1).

$m+b=2 \Rightarrow 1+b=2 \Rightarrow b=1$

So, the equation is $y = x + 1$.

b. To find the equation of the line with slope $m = 3$ through (–1, 1), use the form $y = mx + b$. Since $m = 3$, the equation becomes $y = 3x + b$.

Substitute –1 for x and 1 for y.

$1=3(-1)+b \Rightarrow 1=-3+b \Rightarrow 4=b$

The equation is $y = 3x + 4$.

c. To find a point on both lines, solve the system

$y=x+1$ (1)
$y=3x+4$ (2)

Equate the expressions for y, then solve.

$$x+1=3x+4 \Rightarrow -2x=3 \Rightarrow x=-\frac{3}{2}$$

Substitute $-\frac{3}{2}$ for x in equation (1).

$$y=x+1 \Rightarrow y=-\frac{3}{2}+1 \Rightarrow y=-\frac{1}{2}$$

The point $\left(-\frac{3}{2}, -\frac{1}{2}\right)$ is on both lines.

Section 6.2 Larger Systems of Linear Equations

1. $x \qquad -3z=2$
$2x-4y+5z=1$
$5x-8y+7z=6$
$3x-4y+2z=3$

Copyright © 2019 Pearson Education, Inc.

3. $x+2y+4z=3$
$x \quad +2z=0$
$x+2y+\frac{1}{2}z=\frac{3}{2}$

5. $x+y+2z+3w=1$
$-y-\ z-2w=-1$
$3x+y+4z+5w=2$

7. $x+12y-3z+4w=10$
$2y+3z\ +w=4$
$-z \quad =-7$
$6y-2z-3w=0$

9. $x+3y-4z+2w=1$ (1)
$y+z-w=4$ (2)
$2z+2w=-6$ (3)
$3w=9$ (4)

Divide (4) by 3 to obtain $w = 3$. Substitute $w = 3$ into (3) and solve for z.
$2z+2(3)=-6 \Rightarrow 2z=-12 \Rightarrow z=-6$
Substitute $w = 3$ and $z = -6$ into (2) to solve for y.
$y+(-6)-3=4 \Rightarrow y=13$
Substitute $y = 13$, $w = 3$, and $z = -6$ into (1) to solve for x.
$x+3(13)-4(-6)+2(3)=1 \Rightarrow x=-68$
The solution set is $(-68, 13, -6, 3)$.

11. $2x+2y-4z+\ w=-5$ (1)
$3y+4z-\ w=0$ (2)
$2z-7w=-6$ (3)
$5w=15$ (4)

Divide (4) by 5 to obtain $w = 3$. Substitute $w = 3$ into (3) to solve for z.
$$2z-7(3)=-6 \Rightarrow 2z-21=-6 \Rightarrow z=\frac{15}{2}$$
Substitute $w = 3$ and $z=\frac{15}{2}$ into (2) to solve for y.
$$3y+4\left(\frac{15}{2}\right)-3=0 \Rightarrow 3y+30=3 \Rightarrow$$
$$3y=-27 \Rightarrow y=-9$$
Substitute $y = -9$, $z=\frac{15}{2}$, and $w = 3$ into (1) to solve for x.
$$2x+2(-9)-4\left(\frac{15}{2}\right)+3=-5 \Rightarrow$$
$$2x-18-30+3=-5 \Rightarrow 2x-45=-5 \Rightarrow$$
$$2x=40 \Rightarrow x=20$$
The solution set is $\left(20, -9, \frac{15}{2}, 3\right)$.

13. The augmented matrix for the system
$2x+y+z=3$
$3x-4y+2z=-5$
$x+y+z=2$
is $\left[\begin{array}{ccc|c} 2 & 1 & 1 & 3 \\ 3 & -4 & 2 & -5 \\ 1 & 1 & 1 & 2 \end{array}\right]$.

15. The system of equations for the augmented matrix
$\left[\begin{array}{ccc|c} 2 & 3 & 8 & 20 \\ 1 & 4 & 6 & 12 \\ 0 & 3 & 5 & 10 \end{array}\right]$ is
$2x+3y+8z=20$
$x+4y+6z=12$
$3y+5z=10.$

17. $\left[\begin{array}{ccc|c} 5 & -1 & 4 & 8 \\ 2 & 2 & 9 & 3 \\ 1 & 7 & -3 & 5 \end{array}\right]$
Interchange R_1 and R_3 to obtain
$\left[\begin{array}{ccc|c} 1 & 7 & -3 & 5 \\ 2 & 2 & 9 & 3 \\ 5 & -1 & 4 & 8 \end{array}\right]$

19. $\left[\begin{array}{ccc|c} 2 & 5 & 1 & -1 \\ -4 & 0 & 4 & 6 \\ 6 & 0 & 8 & -4 \end{array}\right]$
Replace R_3 by $\frac{1}{4}R_3$ to obtain
$\left[\begin{array}{ccc|c} 2 & 5 & 1 & -1 \\ -4 & 0 & 4 & 6 \\ \frac{3}{2} & 0 & 2 & -1 \end{array}\right]$.

Copyright © 2019 Pearson Education, Inc.

21. $\left[\begin{array}{cccc|c} -4 & -3 & 1 & -1 & 2 \\ 8 & 2 & 5 & 0 & 6 \\ 0 & -2 & 9 & 4 & 5 \end{array}\right]$

Replace R_2 with $2R_1 + R_2$ to obtain

$\left[\begin{array}{cccc|c} -4 & -3 & 1 & -1 & 2 \\ 0 & -4 & 7 & -2 & 10 \\ 0 & -2 & 9 & 4 & 5 \end{array}\right]$.

23. $\left[\begin{array}{cccc|c} 1 & 0 & 0 & 0 & \frac{3}{2} \\ 0 & 1 & 0 & 0 & 17 \\ 0 & 0 & 1 & 0 & -5 \\ 0 & 0 & 0 & 1 & 0 \end{array}\right]$

The solution of the system is $\left(\frac{3}{2}, 17, -5, 0\right)$.

25. $\left[\begin{array}{ccc|c} 1 & 0 & 0 & -5 \\ 0 & 1 & 0 & 9 \\ 0 & 0 & 0 & 1 \end{array}\right]$

The last row, which represents $0 = 1$, indicates that the system has no solution.

27. $\left[\begin{array}{cccc|c} 1 & 0 & 0 & 1 & 12 \\ 0 & 1 & 0 & 2 & -3 \\ 0 & 0 & 1 & 0 & -5 \\ 0 & 0 & 0 & 0 & 0 \end{array}\right]$

The last row, which represents $0 = 0$, indicates that the system has no unique solution. The linear system associated with this matrix is

$x + w = 12 \quad (1)$
$y + 2w = -3 \quad (2)$
$z = 5 \quad (3)$

To put the solution in terms of w, solve for x in equation (1) and y in equation (2).
The solution is $(12 - w, -3 - 2w, -5, w)$ for any real number w.

29. $x + 2y \quad = 0 \quad (1)$
$y - z = 2 \quad (2)$
$x + y + z = -2 \quad (3)$

Multiply equation (3) by –1, then add the result to equation (1).

$x + 2y \quad = 0$
$-x - y - z = 2$
$y - z = 2$

The system now becomes

$y - z = 2 \quad (4)$
$y - z = 2 \quad (2)$
$x + y + z = -2 \quad (3)$

Multiply equation (4) by –1, then add the result to equation (2).

$y - z = 2$
$-y + z = -2$
$0 = 0 \quad (5)$

Since equation (5) is true, the system is dependent.

31. $x + 2y + 4z = 6 \quad (1)$
$y + z = 1 \quad (2)$
$x + 3y + 5z = 10 \quad (3)$

Multiply equation (3) by –1 and add the result to equation (1).

$x + 2y + 4z = 6$
$-x - 3y - 5z = -10$
$-y - z = -4$

The system now becomes

$-y - z = -4 \quad (1)$
$y + z = 1 \quad (2)$
$x + 3y + 5z = 10 \quad (4)$

Add equation (2) to equation (4).

$y + z = 1 \quad (2)$
$-y - z = -4 \quad (4)$
$0 = -3 \quad (5)$

Since equation (5) is false, the system is inconsistent.

Copyright © 2019 Pearson Education, Inc.

33. $a-3b-2c=-3$ (1)
$3a+2b-c=12$ (2)
$-a-b+4c=3$ (3)

Multiply equation (1) by –3 and add the result to equation (2). Also, add equation (1) to equation (3).

The new system is

$a-3b-2c=-3$ (1)
$11b+5c=21$ (4)
$-4b+2c=0.$ (5)

Multiply equation (4) by $\frac{1}{11}$. This gives

$a-3b-2c=-3$ (1)
$b+\frac{5}{11}c=\frac{21}{11}$ (6)
$-4b+2c=0.$ (5)

Multiply equation (6) by 4 and add the result to equation (5). This gives

$a-3b-2c=-3$ (1)
$b+\frac{5}{11}c=\frac{21}{11}$ (6)
$\frac{42}{11}c=\frac{84}{11}.$ (7)

Multiply equation (7) by $\frac{11}{42}$ to get

$a-3b-2c=-3$ (1)
$b+\frac{5}{11}c=\frac{21}{11}$ (6)
$c=2.$ (8)

Substitute 2 for c in equation (6) to get

$b+\frac{10}{11}=\frac{21}{11}$, or $b=\frac{11}{11}=1$.

Substitute 2 for c and 1 for b in equation (1) to get $a=4$. The system is independent.

35. The augmented matrix is

$$\left[\begin{array}{ccc|c} -1 & 3 & 2 & 0 \\ 2 & -1 & -1 & 3 \\ 1 & 2 & 3 & 0 \end{array}\right].$$

Now use the matrix method to solve the system.

$$\left[\begin{array}{ccc|c} 1 & -3 & -2 & 0 \\ 2 & -1 & -1 & 3 \\ 1 & 2 & 3 & 0 \end{array}\right]\begin{array}{l} -R_1 \\ \\ \\ \end{array}$$

$$\left[\begin{array}{ccc|c} 1 & -3 & -2 & 0 \\ 0 & 5 & 3 & 3 \\ 1 & 2 & 3 & 0 \end{array}\right]\begin{array}{l} \\ -2R_1+R_2 \\ \\ \end{array}$$

$$\left[\begin{array}{ccc|c} 1 & -3 & -2 & 0 \\ 0 & 5 & 3 & 3 \\ 0 & -5 & -5 & 0 \end{array}\right]\begin{array}{l} \\ \\ -R_3+R_1 \end{array}$$

$$\left[\begin{array}{ccc|c} 1 & -3 & -2 & 0 \\ 0 & 5 & 3 & 3 \\ 0 & 0 & -2 & 3 \end{array}\right]\begin{array}{l} \\ \\ R_2+R_3 \end{array}$$

$$\left[\begin{array}{ccc|c} 1 & -3 & -2 & 0 \\ 0 & 1 & \frac{3}{5} & \frac{3}{5} \\ 0 & 0 & 1 & -\frac{3}{2} \end{array}\right]\begin{array}{l} \\ \frac{1}{5}R_2 \\ -\frac{1}{2}R_3 \end{array}$$

The system is now in row echelon form, and

$z=-\frac{3}{2}$

Using back substitution, we find that

$y+\frac{3}{5}\left(-\frac{3}{2}\right)=\frac{3}{5}\Rightarrow y=\frac{3}{2}$

Now solve for x.

$-x-3\left(\frac{3}{2}\right)-2\left(-\frac{3}{2}\right)=0\Rightarrow x=\frac{3}{2}$

The solution set is $\left(\frac{3}{2},\frac{3}{2},-\frac{3}{2}\right)$.

Copyright © 2019 Pearson Education, Inc.

37. The augmented matrix is

$$\left[\begin{array}{rrr|r} 1 & -2 & 4 & 6 \\ 1 & 2 & 13 & 6 \\ -2 & 6 & -1 & -10 \end{array}\right].$$

Now use the matrix method to solve the system.

$$\left[\begin{array}{rrr|r} 1 & -2 & 4 & 6 \\ 0 & -4 & -9 & 0 \\ 0 & 2 & 7 & 2 \end{array}\right] \begin{array}{l} \\ R_1 - R_2 \\ 2R_1 + R_3 \end{array}$$

$$\left[\begin{array}{rrr|r} 1 & -2 & 4 & 6 \\ 0 & -4 & -9 & 0 \\ 0 & 0 & 5 & 4 \end{array}\right] \begin{array}{l} \\ \\ R_2 + 2R_3 \end{array}$$

$$\left[\begin{array}{rrr|r} 1 & -2 & 4 & 6 \\ 0 & 1 & \frac{9}{4} & 0 \\ 0 & 0 & 1 & \frac{4}{5} \end{array}\right] \begin{array}{l} \\ -\frac{1}{4}R_2 \\ \frac{1}{5}R_3 \end{array}$$

The system is now in row echelon form, and $z = \frac{4}{5}$. Using back substitution, we find that

$$y + \frac{9}{4}\left(\frac{4}{5}\right) = 0 \Rightarrow y = -\frac{9}{5}$$

$$x - 2\left(-\frac{9}{5}\right) + 4\left(\frac{4}{5}\right) = 6 \Rightarrow x = -\frac{4}{5}$$

The solution set for the system is $(-.8, -1.8, .8)$.

39. Note that we have four equations in three unknowns. The augmented matrix is

$$\left[\begin{array}{rrr|r} 1 & 1 & 1 & 200 \\ 1 & -2 & 0 & 0 \\ 2 & 3 & 5 & 600 \\ 2 & -1 & 1 & 200 \end{array}\right]$$

Use the matrix method to solve the system.

$$\left[\begin{array}{rrr|r} 1 & 1 & 1 & 200 \\ 0 & 3 & 1 & 200 \\ 0 & 4 & 4 & 400 \\ 2 & -1 & 1 & 200 \end{array}\right] \begin{array}{l} \\ R_1 - R_2 \\ R_3 - R_4 \\ \\ \end{array}$$

$$\left[\begin{array}{rrr|r} 1 & 1 & 1 & 200 \\ 0 & 3 & 1 & 200 \\ 0 & 4 & 4 & 400 \\ 0 & 3 & 1 & 200 \end{array}\right] \begin{array}{l} \\ \\ \\ 2R_1 - R_4 \end{array}$$

$$\left[\begin{array}{rrr|r} 1 & 1 & 1 & 200 \\ 0 & 3 & 1 & 200 \\ 0 & 1 & 1 & 100 \\ 0 & 0 & 0 & 0 \end{array}\right] \begin{array}{l} \\ \\ \frac{1}{4}R_3 \\ R_2 - R_4 \end{array}$$

$$\left[\begin{array}{rrr|r} 1 & 1 & 1 & 200 \\ 0 & 3 & 1 & 200 \\ 0 & 0 & 2 & 100 \\ 0 & 0 & 0 & 0 \end{array}\right] \begin{array}{l} \\ \\ 3R_3 - R_2 \\ \\ \end{array}$$

$$\left[\begin{array}{rrr|r} 1 & 1 & 1 & 200 \\ 0 & 1 & \frac{1}{3} & \frac{200}{3} \\ 0 & 0 & 1 & 50 \\ 0 & 0 & 0 & 0 \end{array}\right] \begin{array}{l} \\ \frac{1}{3}R_2 \\ \frac{1}{2}R_3 \\ \\ \end{array}$$

The system is now in row echelon form, and $z = 50$. Use back-substitution to solve for x and y.

$$y + \frac{1}{3}(50) = \frac{200}{3} \Rightarrow y = 50$$

$$x + 1(50) + 1(50) = 200 \Rightarrow x = 100$$

The solution to the system is $(100, 50, 50)$.

41.
$$\begin{aligned} x + y + z &= 5 \\ 2x + y - z &= 2 \\ x - y + z &= -2 \end{aligned}$$

The augmented matrix is

$$\left[\begin{array}{rrr|r} 1 & 1 & 1 & 5 \\ 2 & 1 & -1 & 2 \\ 1 & -1 & 1 & -2 \end{array}\right]$$

Now use the matrix method to solve the system:

$$\left[\begin{array}{rrr|r} 1 & 1 & 1 & 5 \\ 0 & -1 & -3 & -8 \\ 0 & -2 & 0 & -7 \end{array}\right] \begin{array}{l} \\ -2R_1 + R_2 \\ -R_1 + R_3 \end{array}$$

$$\left[\begin{array}{rrr|r} 1 & 1 & 1 & 5 \\ 0 & -1 & -3 & -8 \\ 0 & 0 & 6 & 9 \end{array}\right] \begin{array}{l} \\ \\ -2R_2 + R_3 \end{array}$$

$$\left[\begin{array}{rrr|r} 1 & 1 & 1 & 5 \\ 0 & 1 & 3 & 8 \\ 0 & 0 & 1 & \frac{3}{2} \end{array}\right] \begin{array}{l} \\ -R_2 \\ \frac{1}{6}R_3 \end{array}$$

The system is now in row echelon form and $z = \frac{3}{2}$. Use back-substitution to solve for x and y.

$$y + 3\left(\frac{3}{2}\right) = 8 \Rightarrow y = \frac{7}{2}$$

$$x + \frac{7}{2} + \frac{3}{2} = 5 \Rightarrow x = 0$$

The solution set is $(0, 3.5, 1.5)$.

Copyright © 2019 Pearson Education, Inc.

43. $x+2y+z=5$
$2x+y-3z=-2$
$3x+y+4z=-5$

The matrix is

$$\left[\begin{array}{ccc|c} 1 & 2 & 1 & 5 \\ 2 & 1 & -3 & -2 \\ 3 & 1 & 4 & -5 \end{array}\right]$$

$$\left[\begin{array}{ccc|c} 1 & 2 & 1 & 5 \\ 0 & -3 & -5 & -12 \\ 0 & -5 & 1 & -20 \end{array}\right]\begin{array}{l} \\ -2R_1+R_2 \\ -3R_1+R_3 \end{array}$$

$$\left[\begin{array}{ccc|c} 1 & 2 & 1 & 5 \\ 0 & 1 & \frac{5}{3} & 4 \\ 0 & -5 & 1 & -20 \end{array}\right]\begin{array}{l} \\ -\frac{1}{3}R_2 \\ \\ \end{array}$$

$$\left[\begin{array}{ccc|c} 1 & 0 & -\frac{7}{3} & -3 \\ 0 & 1 & \frac{5}{3} & 4 \\ 0 & 0 & \frac{28}{3} & 0 \end{array}\right]\begin{array}{l} -2R_2+R_1 \\ \\ 5R_2+R_3 \end{array}$$

$$\left[\begin{array}{ccc|c} 1 & 0 & -\frac{7}{3} & -3 \\ 0 & 1 & \frac{5}{3} & 4 \\ 0 & 0 & 1 & 0 \end{array}\right]\begin{array}{l} \\ \\ \frac{3}{28}R_3 \end{array}$$

$$\left[\begin{array}{ccc|c} 1 & 0 & 0 & -3 \\ 0 & 1 & 0 & 4 \\ 0 & 0 & 1 & 0 \end{array}\right]\begin{array}{l} \frac{7}{3}R_3+R_1 \\ -\frac{5}{3}R_3+R_2 \\ \\ \end{array}$$

The solution is (–3, 4, 0).

45. $x+3y-6z=7$
$2x-y+2z=0$
$x+y+2z=-1$

The matrix is

$$\left[\begin{array}{ccc|c} 1 & 3 & -6 & 7 \\ 2 & -1 & 2 & 0 \\ 1 & 1 & 2 & -1 \end{array}\right]$$

$$\left[\begin{array}{ccc|c} 1 & 3 & -6 & 7 \\ 0 & -7 & 14 & -14 \\ 0 & -2 & 8 & -8 \end{array}\right]\begin{array}{l} \\ -2R_1+R_2 \\ -1R_1+R_3 \end{array}$$

$$\left[\begin{array}{ccc|c} 1 & 3 & -6 & 7 \\ 0 & 1 & -2 & 2 \\ 0 & -2 & 8 & -8 \end{array}\right]\begin{array}{l} \\ -\frac{1}{7}R_2 \\ \\ \end{array}$$

$$\left[\begin{array}{ccc|c} 1 & 0 & 0 & 1 \\ 0 & 1 & -2 & 2 \\ 0 & 0 & 4 & -4 \end{array}\right]\begin{array}{l} -3R_2+R_1 \\ \\ 2R_2+R_3 \end{array}$$

$$\left[\begin{array}{ccc|c} 1 & 0 & 0 & 1 \\ 0 & 1 & -2 & 2 \\ 0 & 0 & 1 & -1 \end{array}\right]\begin{array}{l} \\ \\ \frac{1}{4}R_3 \end{array}$$

$$\left[\begin{array}{ccc|c} 1 & 0 & 0 & 1 \\ 0 & 1 & 0 & 0 \\ 0 & 0 & 1 & -1 \end{array}\right]\begin{array}{l} \\ 2R_3+R_2 \\ \\ \end{array}$$

The solution is (1, 0, –1).

47. $x-2y+4z=9$
$x+y+13z=6$
$-2x+6y-z=-10$

The augmented matrix is

$$\left[\begin{array}{ccc|c} 1 & -2 & 4 & 9 \\ 1 & 1 & 13 & 6 \\ -2 & 6 & -1 & -10 \end{array}\right]$$

$$\left[\begin{array}{ccc|c} 1 & -2 & 4 & 9 \\ 0 & 3 & 9 & -3 \\ 0 & 2 & 7 & 8 \end{array}\right]\begin{array}{l} \\ R_2-R_1 \\ 2R_1+R_3 \end{array}$$

$$\left[\begin{array}{ccc|c} 1 & -2 & 4 & 9 \\ 0 & 1 & 3 & -1 \\ 0 & 2 & 7 & 8 \end{array}\right]\begin{array}{l} \\ \frac{1}{3}R_2 \\ \\ \end{array}$$

$$\left[\begin{array}{ccc|c} 1 & -2 & 4 & 9 \\ 0 & 1 & 3 & -1 \\ 0 & 0 & 1 & 10 \end{array}\right]\begin{array}{l} \\ \\ R_3-2R_2 \end{array}$$

$$\left[\begin{array}{ccc|c} 1 & -2 & 0 & -31 \\ 0 & 1 & 0 & -31 \\ 0 & 0 & 1 & 10 \end{array}\right]\begin{array}{l} R_1-4R_3 \\ R_2-3R_3 \\ \\ \end{array}$$

$$\left[\begin{array}{ccc|c} 1 & 0 & 0 & -93 \\ 0 & 1 & 0 & -31 \\ 0 & 0 & 1 & 10 \end{array}\right]\begin{array}{l} R_1+2R_2 \\ \\ \\ \end{array}$$

The solution is (–93, –31, 10).

Copyright © 2019 Pearson Education, Inc.

49. $x+3y+4z=14$
$2x-3y+2z=10$
$3x-\ y+\ z=9$
$4x+2y+5z=9$

The augmented matrix is

$$\left[\begin{array}{ccc|c} 1 & 3 & 4 & 14 \\ 2 & -3 & 2 & 10 \\ 3 & -1 & 1 & 9 \\ 4 & 2 & 5 & 9 \end{array}\right]$$

Now use the matrix method to solve the system.

$$\left[\begin{array}{ccc|c} 1 & 3 & 4 & 14 \\ 0 & -9 & -6 & -18 \\ 0 & -10 & -11 & -33 \\ 0 & -10 & -11 & -47 \end{array}\right]\begin{array}{l} \\ -2R_1+R_2 \\ -3R_1+R_3 \\ -4R_1+R_4 \end{array}$$

$$\left[\begin{array}{ccc|c} 1 & 3 & 4 & 14 \\ 0 & -9 & -6 & -18 \\ 0 & -10 & -11 & -33 \\ 0 & 0 & 0 & -14 \end{array}\right]\begin{array}{l} \\ \\ \\ -R_3+R_4 \end{array}$$

The last row is equivalent to the equation $0=-14$, which is false. Therefore the system is inconsistent and has no solution.

51. $4x-4y+12z=28$
$3x-2y-\ z=36$

The augmented matrix is

$$\left[\begin{array}{ccc|c} 4 & -4 & 12 & 28 \\ 3 & -2 & -1 & 36 \end{array}\right]$$

Now use the matrix method to solve the system:

$$\left[\begin{array}{ccc|c} 1 & -1 & 3 & 7 \\ 3 & -2 & -1 & 36 \end{array}\right]\begin{array}{l} .25R_1 \\ \\ \end{array}$$

$$\left[\begin{array}{ccc|c} 1 & -1 & 3 & 7 \\ 0 & 1 & -10 & 15 \end{array}\right]\begin{array}{l} \\ R_2-3R_1 \end{array}$$

$$\left[\begin{array}{ccc|c} 1 & 0 & -7 & 22 \\ 0 & 1 & -10 & 15 \end{array}\right]\begin{array}{l} R_1+R_2 \\ \\ \end{array}$$

Since there are only two equations, the system has an infinite number of solutions. Solve the equation represented by the first row of the matrix for x in terms of the parameter z.
$x-7z=22 \Rightarrow x=22+7z$
Now solve the equation represented by the second row of the matrix for y in terms of the parameter z. $y-10z=15 \Rightarrow x=15+10z$
The solution is $(22+7z, 15+10z,\ z)$ for any real number z.

53. $-8x-\ 9y=11$
$24x+34y=2$
$16x+11y=-57$

The matrix is

$$\left[\begin{array}{cc|c} -8 & -9 & 11 \\ 24 & 34 & 2 \\ 16 & 11 & -57 \end{array}\right]$$

$$\left[\begin{array}{cc|c} -8 & -9 & 11 \\ 0 & 7 & 35 \\ 0 & -7 & -35 \end{array}\right]\begin{array}{l} \\ 3R_1+R_2 \\ 2R_1+R_3 \end{array}$$

$$\left[\begin{array}{cc|c} 1 & \frac{9}{8} & -\frac{11}{8} \\ 0 & 1 & 5 \\ 0 & 1 & 5 \end{array}\right]\begin{array}{l} -\frac{1}{8}R_1 \\ \frac{1}{7}R_2 \\ -\frac{1}{7}R_3 \end{array}$$

$$\left[\begin{array}{cc|c} 1 & 0 & -7 \\ 0 & 1 & 5 \\ 0 & 0 & 0 \end{array}\right]\begin{array}{l} -\frac{9}{8}R_2+R_1 \\ \\ -1R_2+R_3 \end{array}$$

The solution is $(-7, 5)$.

55. $x-\ 2y+z=5$
$2x+\ y\ -z=2$
$-2x+4y-2z=2$

The matrix is

$$\left[\begin{array}{ccc|c} 1 & -2 & 1 & 5 \\ 2 & 1 & -1 & 2 \\ -2 & 4 & -2 & 2 \end{array}\right]$$

$$\left[\begin{array}{ccc|c} 1 & -2 & 1 & 5 \\ 0 & 5 & -3 & -8 \\ 0 & 0 & 0 & 12 \end{array}\right]\begin{array}{l} \\ -2R_1+R_2 \\ 2R_1+R_3 \end{array}$$

The third row represents the equation $0=12$, which is false. The system is inconsistent and has no solution.

Copyright © 2019 Pearson Education, Inc.

57. $x+y-z=-20$
$2x-y+z=11$

The matrix is

$$\left[\begin{array}{ccc|c}1&1&-1&-20\\2&-1&1&11\end{array}\right]$$

$$\left[\begin{array}{ccc|c}1&1&-1&-20\\0&-3&3&51\end{array}\right]\begin{array}{l}\\-2R_1+R_2\end{array}$$

$$\left[\begin{array}{ccc|c}1&1&-1&-20\\0&1&-1&-17\end{array}\right]\begin{array}{l}\\-\frac{1}{3}R_2\end{array}$$

$$\left[\begin{array}{ccc|c}1&0&0&-3\\0&1&-1&-17\end{array}\right]\begin{array}{l}-R_2+R_1\\ \end{array}$$

This last matrix is the augmented matrix for the system

$x=-3$ (1)
$y-z=-17$ (2)

Solve equation (2) for y in terms of z. $y=z-17$. The solution is all ordered triples of the form $(-3, z-17, z)$ for any real number z.

59. $2x+y+3z-2w=-6$
$4x+3y+z-w=-2$
$x+y+z+w=-5$
$-2x-2y-2z+2w=-10$

The matrix is

$$\left[\begin{array}{cccc|c}2&1&3&-2&-6\\4&3&1&-1&-2\\1&1&1&1&-5\\-2&-2&-2&2&-10\end{array}\right]$$

Interchange R_1 and R_3.

$$\left[\begin{array}{cccc|c}1&1&1&1&-5\\4&3&1&-1&-2\\2&1&3&-2&-6\\-2&-2&-2&2&-10\end{array}\right]$$

$$\left[\begin{array}{cccc|c}1&1&1&1&-5\\0&-1&-3&-5&18\\0&-1&1&-4&4\\0&0&0&4&-20\end{array}\right]\begin{array}{l}\\-4R_1+R_2\\-2R_1+R_3\\2R_1+R_4\end{array}$$

$$\left[\begin{array}{cccc|c}1&1&1&1&-5\\0&1&3&5&-18\\0&-1&1&-4&4\\0&0&0&1&-5\end{array}\right]\begin{array}{l}\\-R_2\\ \\ \frac{1}{4}R_4\end{array}$$

$$\left[\begin{array}{cccc|c}1&0&-2&-4&13\\0&1&3&5&-18\\0&0&4&1&-14\\0&0&0&1&-5\end{array}\right]\begin{array}{l}-R_2+R_1\\ \\R_2+R_3\\ \end{array}$$

$$\left[\begin{array}{cccc|c}1&0&-2&-4&13\\0&1&3&5&-18\\0&0&1&\frac{1}{4}&-\frac{7}{2}\\0&0&0&1&-5\end{array}\right]\begin{array}{l}\\ \\ \frac{1}{4}R_3\\ \end{array}$$

$$\left[\begin{array}{cccc|c}1&0&-2&-4&13\\0&1&3&5&-18\\0&0&1&0&-\frac{9}{4}\\0&0&0&1&-5\end{array}\right]\begin{array}{l}\\ \\ -\frac{1}{4}R_4+R_3\\ \end{array}$$

$$\left[\begin{array}{cccc|c}1&0&-2&0&-7\\0&1&3&0&7\\0&0&1&0&-\frac{9}{4}\\0&0&0&1&-5\end{array}\right]\begin{array}{l}4R_4+R_1\\-5R_4+R_2\\ \\ \end{array}$$

$$\left[\begin{array}{cccc|c}1&0&0&0&-\frac{23}{2}\\0&1&0&0&\frac{55}{4}\\0&0&1&0&-\frac{9}{4}\\0&0&0&1&-5\end{array}\right]\begin{array}{l}2R_3+R_1\\-3R_3+R_2\\ \\ \end{array}$$

The solution is $(-11.5, 13.75, -2.25, -5)$.

61. $x+2y-z=3$
$3x+y+w=4$
$2x-y+z+w=2$

The matrix is

$$\left[\begin{array}{cccc|c}1&2&-1&0&3\\3&1&0&1&4\\2&-1&1&1&2\end{array}\right]$$

$$\left[\begin{array}{cccc|c}1&2&-1&0&3\\0&-5&3&1&-5\\0&-5&3&1&-4\end{array}\right]\begin{array}{l}\\-3R_1+R_2\\-2R_1+R_3\end{array}$$

$$\left[\begin{array}{cccc|c}1&2&-1&0&3\\0&-5&3&1&-5\\0&0&0&0&1\end{array}\right]\begin{array}{l}\\ \\-1R_2+R_3\end{array}$$

Since the last row yields a false statement, there is no solution.

63. $2x+3y+z=9$
$4x+y-3z=-7$
$6x+2y-4z=-8$

The matrix is

$$\left[\begin{array}{ccc|c} 2 & 3 & 1 & 9 \\ 4 & 1 & -3 & -7 \\ 6 & 2 & -4 & -8 \end{array}\right]$$

$$\left[\begin{array}{ccc|c} 1 & \frac{3}{2} & \frac{1}{2} & \frac{9}{2} \\ 4 & 1 & -3 & -7 \\ 6 & 2 & -4 & -8 \end{array}\right] \tfrac{1}{2}R_1$$

$$\left[\begin{array}{ccc|c} 1 & \frac{3}{2} & \frac{1}{2} & \frac{9}{2} \\ 0 & -5 & -5 & -25 \\ 0 & -7 & -7 & -35 \end{array}\right] \begin{array}{l} \\ -4R_1+R_2 \\ -6R_1+R_3 \end{array}$$

$$\left[\begin{array}{ccc|c} 1 & \frac{3}{2} & \frac{1}{2} & \frac{9}{2} \\ 0 & 1 & 1 & 5 \\ 0 & 1 & 1 & 5 \end{array}\right] \begin{array}{l} \\ -\frac{1}{5}R_2 \\ -\frac{1}{7}R_3 \end{array}$$

$$\left[\begin{array}{ccc|c} 1 & 0 & -1 & -3 \\ 0 & 1 & 1 & 5 \\ 0 & 0 & 0 & 0 \end{array}\right] \begin{array}{l} -\frac{3}{2}R_2+R_1 \\ \\ -R_2+R_3 \end{array}$$

Since the last row is all zeros, the system has no unique solution. Rows 1 and 2 can be written as the system

$x-z=-3$ (1)
$y+z=5.$ (2)

Let z be arbitrary, so

$x=z-3$ (1)
$y=-z+5$ (2)

The solution is $(z-3, -z+5, z)$ for any real number z.

65. We are given the system

$x+20y=112$
$-x+10y=38$.
$y=5$

The associated matrix is

$$\left[\begin{array}{cc|c} 1 & 20 & 112 \\ -1 & 10 & 38 \\ 0 & 1 & 5 \end{array}\right].$$

$$\left[\begin{array}{cc|c} 1 & 0 & 12 \\ -1 & 0 & -12 \\ 0 & 1 & 5 \end{array}\right] \begin{array}{l} -20R_3+R_1 \\ -10R_3+R_2 \\ \\ \end{array}$$

$$\left[\begin{array}{cc|c} 1 & 0 & 12 \\ 1 & 0 & 12 \\ 0 & 1 & 5 \end{array}\right] \begin{array}{l} \\ -R_2 \\ \\ \end{array}$$

Thus, the solution is (12, 5).
All three counties had the same population in 2010 + 12 = 2012. The population was 5000.

67. $x+1.25y+.25z=457.5$
$x+\ .6y+\ .4z=390$
$3.16x+3.48y+\ .4z=1297.2$

The associated matrix is

$$\left[\begin{array}{ccc|c} 1 & 1.25 & .25 & 457.5 \\ 1 & .6 & .4 & 390 \\ 3.16 & 3.48 & .4 & 1297.2 \end{array}\right].$$

$$\left[\begin{array}{ccc|c} 1 & 1.25 & .25 & 457.5 \\ 0 & .65 & -.15 & 67.5 \\ 0 & -.47 & -.39 & -148.5 \end{array}\right] \begin{array}{l} \\ -R_2+R_1 \\ -3.16R_1+R_3 \end{array}$$

$$\left[\begin{array}{ccc|c} 1 & 1.25 & .25 & 457.5 \\ 0 & 1 & -\frac{3}{13} & \frac{1350}{13} \\ 0 & -.47 & -.39 & -148.5 \end{array}\right] \begin{array}{l} \\ \frac{1}{.65}R_2 \\ \\ \end{array}$$

At this point, it is probably easier to convert to fractions.

$$\left[\begin{array}{ccc|c} 1 & \frac{5}{4} & \frac{1}{4} & \frac{4575}{10} \\ 0 & 1 & -\frac{3}{13} & \frac{1350}{13} \\ 0 & -\frac{47}{100} & -\frac{39}{100} & -\frac{1485}{10} \end{array}\right]$$

$$\left[\begin{array}{ccc|c} 1 & \frac{5}{4} & \frac{1}{4} & \frac{4575}{10} \\ 0 & 1 & -\frac{3}{13} & \frac{1350}{13} \\ 0 & 0 & -\frac{162}{325} & -\frac{1296}{13} \end{array}\right] \begin{array}{l} \\ \\ \frac{47}{100}R_2+R_3 \end{array}$$

Copyright © 2019 Pearson Education, Inc.

$$\left[\begin{array}{ccc|c} 1 & 0 & \frac{7}{13} & \frac{4260}{13} \\ 0 & 1 & -\frac{3}{13} & \frac{1350}{13} \\ 0 & 0 & 1 & 200 \end{array}\right] \begin{array}{l} -\frac{5}{4}R_2 + R_1 \\ \\ -\frac{325}{162}R_3 \end{array}$$

$$\left[\begin{array}{ccc|c} 1 & 0 & 0 & 220 \\ 0 & 1 & 0 & 150 \\ 0 & 0 & 1 & 200 \end{array}\right] \begin{array}{l} -\frac{7}{13}R_3 + R_1 \\ \frac{3}{13}R_3 + R_2 \\ \\ \end{array}$$

The TI-84 Plus calculator solution is shown below.

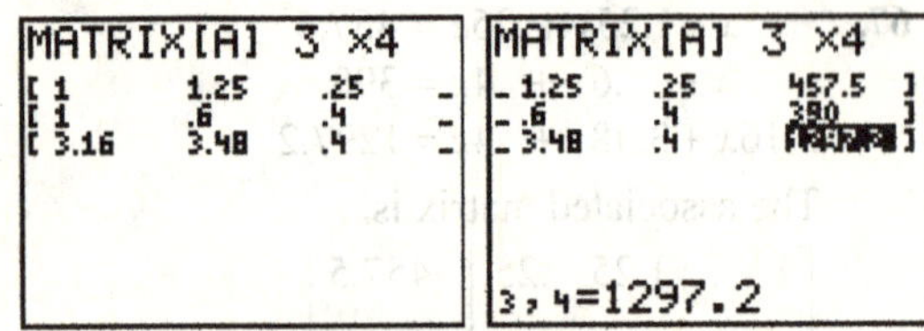

Left three columns of A right three columns of A

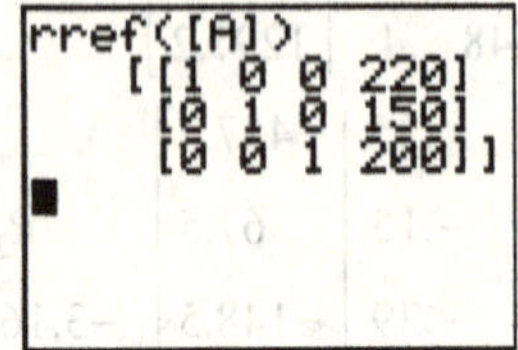

There were 220 adults, 150 teenagers, and 200 preteen children present.

69.

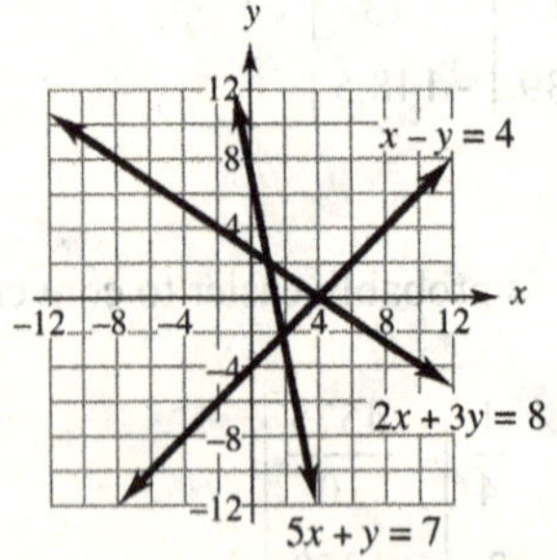

Answers vary. Sample answer: The three lines do not have one common intersection point.

71. $y = ax^2 + bx + c$

Since (2, 3), (–1, 0), and (–2, 2) satisfy the above equation, the following system of equations can be obtained.

$3 = a(2)^2 + b(2) + c$ (1)
$0 = a(-1)^2 + b(-1) + c$ (2)
$2 = a(-2)^2 + b(-2) + c$ (3)

The system simplifies to

$4a + 2b + c = 3$ (1)
$a - b + c = 0$ (2)
$4a - 2b + c = 2.$ (3)

Interchange equations (2) and (1).

$a - b + c = 0$ (2)
$4a + 2b + c = 3$ (1)
$4a - 2b + c = 2$ (3)

Eliminate a in equations (1) and (3).

$a - b + c = 0$ (2)
$6b - 3c = 3$ (4)
$2b - 3c = 2$ (5)

Multiply equation (4) by $\frac{1}{6}$.

$a - b + c = 0$ (2)
$b - \frac{1}{2}c = \frac{1}{2}$ (6)
$2b - 3c = 2$ (5)

Eliminate b in equation (5).

$a - b + c = 0$ (2)
$b - \frac{1}{2}c = \frac{1}{2}$ (6)
$-2c = 1$ (7)

Solve for c in equation (7).

$a - b + c = 0$ (2)
$b - \frac{1}{2}c = \frac{1}{2}$ (6)
$c = -\frac{1}{2}$ (8)

Use back-substitution in equations (6) and (2) to solve for b and then a. thus,

$a = \frac{3}{4}$, $b = \frac{1}{4}$, and $c = -\frac{1}{2}$. Therefore, the equation is $y = .75x^2 + .25x - .5$.

Copyright © 2019 Pearson Education, Inc.

Section 6.3 Applications of Systems of Linear Equations

For these exercises, we show just one of the various solution methods that could be used.

1. Let x = the number of minutes and let y = the number of miles. We have the system

$$.2x+1.4y=10 \ (1)$$
$$x=3y \ (2)$$

We substitute equation (2) into equation (1) and solve for the variable y.

$$.2(3y)+1.4y=10$$
$$.6y+1.4y=10$$
$$2y=10$$
$$y=5$$

Now take the value for y and solve for x using equation (2). We get $x = 3(5) = 15$. Therefore, your trip was 15 minutes and covered 5 miles.

3. Let x = the number of euros and let y = the number of pounds. We have the system

$$1.2x+1.4y=1000 \ (1)$$
$$x=3y \quad (2)$$

We substitute equation (2) into equation (1) and solve for the variable y.

$$1.2(3y)+1.4y=1000$$
$$3.6y+1.4y=1000$$
$$5y=1000$$
$$y=200$$

Now take the value for y and solve for x using equation (2). We get $x = 3(200) = 600$. Therefore, Abigail should change her $1000 and receive 600 euros and 200 pounds.

5. Let x = the number of Coca-Cola shares and let y = the number of Intel Shares. We have the system: $45x+32y=25,000$ (1)

$$90x+48y=42,000 \ (2)$$

Multiply equation (1) by –1, then add the result to equation (2)

$$-90x-64y=-50,000$$
$$\underline{90x+48y=42,000}$$
$$-16y=-8,000 \Rightarrow y=500$$

Substitute 500 for y in equation (1) and solve for x.

$$45x+32(500)=25,000$$
$$45x+16,000=25,000$$
$$45x=9000$$
$$x=200$$

The Portfolio has 200 shares of Coca-Cola and 500 shares of Intel.

7. a. Let x = the number of free-throws made, let y = the number of regular baskets made, z = the number of three-point baskets made. We have the system: $x+2y+3z=31$ (1)

$$y+z=12 \ (2)$$
$$x=4 \quad (3)$$

b. Simplifying the system in part a yields:

$$2y+3z=27 \ (4)$$
$$y+z=12 \ (5)$$

Multiply equation (5) by –2, then add the result to equation (4):

$$2y+3z=27$$
$$\underline{-2y-2z=-24}$$
$$z=3$$

Substitute 3 for z in equation (5) and solve for y.

$$y+z=12$$
$$y+3=12$$
$$y=9$$

The center for the basketball team made 4 free-throws, 9 regular baskets, and 3 three-point baskets in the game.

9. From example 1, let x = the number of vans, let y = the number of small trucks, and let z = the number of large trucks. We have the system

$$\begin{aligned} x+ \quad y+ \quad z &= 175 && (1)\\ 25,000x+30,000y+40,000z &= 5,000,000 && (2)\\ x- \quad 2y \quad &= 0 && (3)\end{aligned}$$

We divide equation (2) by 5000, so the system becomes

$$\begin{aligned} x+ y+ z &= 175\\ 5x+6y+8z &= 1000\\ x-2y &= 0\end{aligned}$$

Form the augmented matrix and transform it into reduced row echelon form.

$$\left[\begin{array}{ccc|c} 1 & 1 & 1 & 175\\ 5 & 6 & 8 & 1000\\ 1 & -2 & 0 & 0\end{array}\right]$$

$$\left[\begin{array}{ccc|c} 1 & 1 & 1 & 175\\ 0 & 1 & 3 & 125\\ 0 & 3 & 1 & 175\end{array}\right] \begin{array}{l} \\ -5R_1+R_2 \\ -R_3+R_1\end{array}$$

$$\left[\begin{array}{ccc|c} 1 & 1 & 1 & 175\\ 0 & 1 & 3 & 125\\ 0 & 0 & -8 & -200\end{array}\right] \begin{array}{l} \\ \\ -3R_2+R_3\end{array}$$

Copyright © 2019 Pearson Education, Inc.

$$\left[\begin{array}{ccc|c} 1 & 1 & 1 & 175 \\ 0 & 1 & 3 & 125 \\ 0 & 0 & 1 & 25 \end{array}\right] -\tfrac{5}{31}R_3$$

$$\left[\begin{array}{ccc|c} 1 & 1 & 0 & 150 \\ 0 & 1 & 0 & 50 \\ 0 & 0 & 1 & 25 \end{array}\right] \begin{array}{l} -R_3+R_1 \\ -3R_3+R_2 \\ \\ \end{array}$$

$$\left[\begin{array}{ccc|c} 1 & 0 & 0 & 100 \\ 0 & 1 & 0 & 50 \\ 0 & 0 & 1 & 25 \end{array}\right] \begin{array}{l} -R_3+R_1 \\ \\ \\ \end{array}$$

The company should buy 100 vans, 50 small trucks, and 25 large trucks.

11. From example 3, let x = the number of units of corn, let y = the number of units of soybeans, and let z = the number of units of cottonseed. We have the system

$$\begin{aligned} 10x+20y+30z &= 1800 \\ 30x+20y+40z &= 2400 \\ 20x+40y+25z &= 2200 \end{aligned}$$

Form the augmented matrix and transform it into reduced row echelon form.

$$\left[\begin{array}{ccc|c} 10 & 20 & 30 & 1800 \\ 30 & 20 & 40 & 2400 \\ 20 & 40 & 25 & 2200 \end{array}\right]$$

$$\left[\begin{array}{ccc|c} 10 & 20 & 30 & 1800 \\ 0 & 40 & 50 & 3000 \\ 0 & 0 & 35 & 1400 \end{array}\right] \begin{array}{l} \\ 3R_1-R_2 \\ 2R_1-R_3 \end{array}$$

$$\left[\begin{array}{ccc|c} 1 & 2 & 3 & 180 \\ 0 & 40 & 50 & 3000 \\ 0 & 0 & 1 & 40 \end{array}\right] \begin{array}{l} \tfrac{1}{10}R_1 \\ \\ \tfrac{1}{35}R_3 \end{array}$$

$$\left[\begin{array}{ccc|c} 1 & 2 & 0 & 60 \\ 0 & 40 & 0 & 1000 \\ 0 & 0 & 1 & 40 \end{array}\right] \begin{array}{l} -3R_3+R_1 \\ -50R_3+R_2 \\ \\ \end{array}$$

$$\left[\begin{array}{ccc|c} 1 & 2 & 0 & 60 \\ 0 & 1 & 0 & 25 \\ 0 & 0 & 1 & 40 \end{array}\right] \begin{array}{l} \\ \tfrac{1}{40}R_2 \\ \\ \end{array}$$

$$\left[\begin{array}{ccc|c} 1 & 0 & 0 & 10 \\ 0 & 1 & 0 & 25 \\ 0 & 0 & 1 & 40 \end{array}\right] \begin{array}{l} -2R_2+R_1 \\ \\ \\ \end{array}$$

The feed should contain 10 units of corn, 25 units of soybeans, and 40 units of cottonseed.

13. Let x = the amount charged for a shirt, let y = the amount charged for a pair of slack, and let z = the amount charged for a sport coat. We have the system

$$\begin{aligned} 3x+\ y \qquad &= 10.96 \\ 7x+2y+z &= 30.40 \\ 4x+ \qquad z &= 14.45 \end{aligned}$$

which has the augmented matrix

$$\left[\begin{array}{ccc|c} 3 & 1 & 0 & 10.96 \\ 7 & 2 & 1 & 30.40 \\ 4 & 0 & 1 & 14.45 \end{array}\right]$$

$$\left[\begin{array}{ccc|c} 4 & 0 & 1 & 14.45 \\ 7 & 2 & 1 & 30.40 \\ 3 & 1 & 0 & 10.96 \end{array}\right] \text{ interchange } R_1 \text{ and } R_3$$

$$\left[\begin{array}{ccc|c} 1 & 0 & \frac{1}{4} & 3.6125 \\ 7 & 2 & 1 & 30.40 \\ 3 & 1 & 0 & 10.96 \end{array}\right] \begin{array}{l} \tfrac{1}{4}R_1 \\ \\ \\ \end{array}$$

$$\left[\begin{array}{ccc|c} 1 & 0 & \frac{1}{4} & 3.6125 \\ 0 & 2 & -\frac{3}{4} & 5.1125 \\ 0 & 1 & -\frac{3}{4} & .1225 \end{array}\right] \begin{array}{l} \\ -7R_1+R_2 \\ -3R_1+R_3 \end{array}$$

$$\left[\begin{array}{ccc|c} 1 & 0 & \frac{1}{4} & 3.6125 \\ 0 & 1 & 0 & 4.99 \\ 0 & 1 & -\frac{3}{4} & .1225 \end{array}\right] \begin{array}{l} \\ R_2-R_3 \\ \\ \end{array}$$

$$\left[\begin{array}{ccc|c} 1 & 0 & \frac{1}{4} & 3.6125 \\ 0 & 1 & 0 & 4.99 \\ 0 & 0 & 1 & 6.49 \end{array}\right] \begin{array}{l} \\ \\ \tfrac{4}{3}(R_2-R_3) \end{array}$$

$$\left[\begin{array}{ccc|c} 1 & 0 & 0 & 1.99 \\ 0 & 1 & 0 & 4.99 \\ 0 & 0 & 1 & 6.49 \end{array}\right] \begin{array}{l} R_1-\tfrac{1}{4}R_2 \\ \\ \\ \end{array}$$

Shirts cost \$1.99 each, slacks cost \$4.99 each, and sports coats cost \$6.49 each.

Copyright © 2019 Pearson Education, Inc.

15. Let x = the number of adult tickets, let y = the number of teenager tickets, let z = the number of preteen tickets. Then we have

$$x+\ y+\ z=570 \quad (1)$$
$$5x+3y+2z=1950 \quad (2).$$
$$\frac{3}{4}z=y \quad (3)$$

Using substitution, we have

$$x+\frac{3}{4}z+z=570 \quad (4)$$
$$5x+3\left(\frac{3}{4}z\right)+2z=1950 \quad (5)$$

Clear the fractions and combine like terms.

$$4x+7z=2280 \quad (6)$$
$$20x+17z=7800 \quad (7)$$

Multiply (6) by –5, then add the result to (7) and solve for z.

$$\begin{aligned} -20x-35z &= -11400 \\ 20x+17z &= 7800 \\ \hline -18z &= -3600 \Rightarrow z=200 \end{aligned}$$

Substitute $z = 200$ into equation (6) and solve for x, and into equation (3) to solve for y.

$$4x+7(200)=2280 \Rightarrow 4x=880 \Rightarrow x=220$$
$$y=\frac{3}{4}(200)=150$$

There were 220 adult tickets, 150 teenager tickets, and 200 preteen tickets sold.

17. Let x = the amount invested in AAA bonds; let y = the amount invested in A bonds; and let z = the amount invested in B bonds.

a. Solve the system

$$x+y+z=25,000 \quad (1)$$
$$.06x+.07y+.1z=1810 \quad (2)$$
$$x=2z. \quad (3)$$

Equation (3) should be rewritten so that the system becomes

$$x+y+z=25,000 \quad (1)$$
$$.06x+.07y+.1z=1810 \quad (2)$$
$$x-2z=0. \quad (3)$$

The matrix is

$$\left[\begin{array}{ccc|c} 1 & 1 & 1 & 25,000 \\ .06 & .07 & .01 & 1810 \\ 1 & 0 & -2 & 0 \end{array}\right]$$

$$\left[\begin{array}{ccc|c} 1 & 1 & 1 & 25,000 \\ 6 & 7 & 10 & 181,000 \\ 1 & 0 & -2 & 0 \end{array}\right] \begin{matrix} \\ 100R_2 \\ \\ \end{matrix}$$

$$\left[\begin{array}{ccc|c} 1 & 1 & 1 & 25,000 \\ 0 & 1 & 4 & 31,000 \\ 0 & -1 & -3 & -25,000 \end{array}\right] \begin{matrix} \\ -6R_1+R_2 \\ -R_1+R_3 \end{matrix}$$

$$\left[\begin{array}{ccc|c} 1 & 0 & -3 & -6000 \\ 0 & 1 & 4 & 31,000 \\ 0 & 0 & 1 & 6000 \end{array}\right] \begin{matrix} -R_2+R_1 \\ \\ R_2+R_3 \end{matrix}$$

$$\left[\begin{array}{ccc|c} 1 & 0 & 0 & 12,000 \\ 0 & 1 & 0 & 7000 \\ 0 & 0 & 1 & 6000 \end{array}\right] \begin{matrix} 3R_3+R_1 \\ -4R_3+R_2 \\ \\ \end{matrix}$$

The client should invest \$12,000 in AAA bonds at 6%, \$7000 in A bonds at 7%, and \$6000 in B bonds at 10%.

b. The new system is

$$x+y+z=30,000$$
$$.06x+.07y+.1z=2150$$
$$x-2z=0.$$

The matrix is

$$\left[\begin{array}{ccc|c} 1 & 1 & 1 & 30,000 \\ .06 & .07 & .1 & 2150 \\ 1 & 0 & -2 & 0 \end{array}\right]$$

$$\left[\begin{array}{ccc|c} 1 & 1 & 1 & 30,000 \\ 6 & 7 & 10 & 215,000 \\ 1 & 0 & -2 & 0 \end{array}\right] \begin{matrix} \\ 100R_2 \\ \\ \end{matrix}$$

$$\left[\begin{array}{ccc|c} 1 & 1 & 1 & 30,000 \\ 0 & 1 & 4 & 35,000 \\ 0 & -1 & -3 & -30,000 \end{array}\right] \begin{matrix} \\ -6R_1+R_2 \\ -R_1+R_3 \end{matrix}$$

$$\left[\begin{array}{ccc|c} 1 & 0 & -3 & -5000 \\ 0 & 1 & 4 & 35,000 \\ 0 & 0 & 1 & 5000 \end{array}\right] \begin{matrix} -R_2+R_1 \\ \\ R_2+R_3 \end{matrix}$$

$$\left[\begin{array}{ccc|c} 1 & 0 & 0 & 10,000 \\ 0 & 1 & 0 & 15,000 \\ 0 & 0 & 1 & 5000 \end{array}\right] \begin{matrix} 3R_3+R_1 \\ -4R_3+R_2 \\ \\ \end{matrix}$$

The client should invest \$10,000 in AAA bonds at 6%, \$15,000 in A bonds at 7%, and \$5000 in B bonds at 10%.

Copyright © 2019 Pearson Education, Inc.

c. The new system is

$$\begin{aligned} x+y+z &= 40,000 \\ .06x+.07y+.1z &= 2900 \\ x-2z &= 0. \end{aligned}$$

The matrix is

$$\left[\begin{array}{ccc|c} 1 & 1 & 1 & 40,000 \\ .06 & .07 & .1 & 2900 \\ 1 & 0 & -2 & 0 \end{array}\right]$$

$$\left[\begin{array}{ccc|c} 1 & 1 & 1 & 40,000 \\ 6 & 7 & 10 & 290,000 \\ 1 & 0 & -2 & 0 \end{array}\right] \begin{matrix} \\ 100R_2 \\ \\ \end{matrix}$$

$$\left[\begin{array}{ccc|c} 1 & 1 & 1 & 40,000 \\ 0 & 1 & 4 & 50,000 \\ 0 & -1 & -3 & -40,000 \end{array}\right] \begin{matrix} \\ -6R_1+R_2 \\ -R_1+R_3 \end{matrix}$$

$$\left[\begin{array}{ccc|c} 1 & 0 & -3 & -10,000 \\ 0 & 1 & 4 & 50,000 \\ 0 & 0 & 1 & 10,000 \end{array}\right] \begin{matrix} -R_2+R_1 \\ \\ R_2+R_3 \end{matrix}$$

$$\left[\begin{array}{ccc|c} 1 & 0 & 0 & 20,000 \\ 0 & 1 & 0 & 10,000 \\ 0 & 0 & 1 & 10,000 \end{array}\right] \begin{matrix} 3R_3+R_1 \\ -4R_3+R_2 \\ \\ \end{matrix}$$

The client should invest $20,000 in AAA bonds at 6%, $10,000 in A bonds at 7%, and $10,000 in B bonds at 10%.

19. Let x = amount in mutual fund
let y = amount in corporate bonds
let z = amount in fast food franchise

$x+y+z=x+2x+z=3x+z=70,000$

$.02x+.06(2x)+.1z=.14x+.1z=4800$

The matrix is

$$\left[\begin{array}{cc|c} 3 & 1 & 70,000 \\ .14 & .1 & 4800 \end{array}\right]$$

$$\left[\begin{array}{cc|c} 3 & 1 & 70,000 \\ 14 & 10 & 480,000 \end{array}\right]$$

$$\left[\begin{array}{cc|c} 1 & \frac{1}{3} & \frac{70,000}{3} \\ 14 & 10 & 480,000 \end{array}\right] \begin{matrix} \frac{1}{3}R_1 \\ \\ \end{matrix}$$

$$\left[\begin{array}{cc|c} 1 & \frac{1}{3} & \frac{70,000}{3} \\ 0 & \frac{16}{3} & \frac{460,000}{3} \end{array}\right] \begin{matrix} \\ -14R_1+R_2 \end{matrix}$$

$$\left[\begin{array}{cc|c} 1 & \frac{1}{3} & \frac{70,000}{3} \\ 0 & 1 & 28,750 \end{array}\right] \begin{matrix} \\ \frac{3}{16}R_2 \end{matrix}$$

$$\left[\begin{array}{cc|c} 1 & 0 & 13,750 \\ 0 & 1 & 28,750 \end{array}\right] \begin{matrix} -\frac{1}{3}R_2+R_1 \\ \\ \end{matrix}$$

Therefore, she should invest $13,750 in the mutual fund, $27,500 in bonds, and $28,750 in the food franchise.

21. Let x = the number of pounds of pretzels, y = the number of pounds of dried fruit, and z = the number of pounds of nuts.
The system of equations is

$$\begin{aligned} x+y+z &= 140 \\ 3x+4y+8z &= 6(140) \\ x &= 2y. \end{aligned}$$

The system simplifies to

$$\begin{aligned} x+y+z &= 140 \\ 3x+4y+8z &= 840 \\ x-2y &= 0. \end{aligned}$$

The matrix is

$$\left[\begin{array}{ccc|c} 1 & 1 & 1 & 140 \\ 3 & 4 & 8 & 840 \\ 1 & -2 & 0 & 0 \end{array}\right]$$

$$\left[\begin{array}{ccc|c} 1 & 1 & 1 & 140 \\ 0 & 1 & 5 & 420 \\ 0 & -3 & -1 & -140 \end{array}\right] \begin{matrix} \\ -3R_1+R_2 \\ -R_1+R_3 \end{matrix}$$

$$\left[\begin{array}{ccc|c} 1 & 0 & -4 & -280 \\ 0 & 1 & 5 & 420 \\ 0 & 0 & 14 & 1120 \end{array}\right] \begin{matrix} -R_2+R_1 \\ \\ 3R_2+R_3 \end{matrix}$$

$$\left[\begin{array}{ccc|c} 1 & 0 & -4 & -280 \\ 0 & 1 & 5 & 420 \\ 0 & 0 & 1 & 80 \end{array}\right] \begin{matrix} \\ \\ \frac{1}{14}R_3 \end{matrix}$$

$$\left[\begin{array}{ccc|c} 1 & 0 & 0 & 40 \\ 0 & 1 & 0 & 20 \\ 0 & 0 & 1 & 80 \end{array}\right] \begin{matrix} 4R_3+R_1 \\ -5R_3+R_2 \\ \\ \end{matrix}$$

Use 40 lb of pretzels, 20 lb of dried fruit, and 80 lb of nuts.

23. Let x = the number of cases of Brand A,
let y = the number of cases of Brand B,
let z = the number of cases of Brand C, and
let w = the number of cases of Brand D.
The system of equations is

$$\begin{aligned} 25x+50y+75z+100w &= 1200 \\ 30x+30y+30z+60w &= 600 \\ 30x+20y+20z+30w &= 400. \end{aligned}$$

Copyright © 2019 Pearson Education, Inc.

This system can be simplified to

$$\begin{aligned} x+2x+3z+4w&=48\\ x+y+z+2w&=20\\ 3x+2y+2z+3w&=40. \end{aligned}$$

The matrix is

$$\left[\begin{array}{cccc|c} 1&2&3&4&48\\ 1&1&1&2&20\\ 3&2&2&3&40 \end{array}\right]$$

$$\left[\begin{array}{cccc|c} 1&2&3&4&48\\ 0&-1&-2&-2&-28\\ 0&-4&-7&-9&-104 \end{array}\right]\begin{array}{l} \\ -R_1+R_2\\ -3R_1+R_3 \end{array}$$

$$\left[\begin{array}{cccc|c} 1&2&3&4&48\\ 0&1&2&2&28\\ 0&-4&-7&-9&-104 \end{array}\right]\begin{array}{l} \\ -R_2\\ \\ \end{array}$$

$$\left[\begin{array}{cccc|c} 1&0&-1&0&-8\\ 0&1&2&2&28\\ 0&0&1&-1&8 \end{array}\right]\begin{array}{l} -2R_2+R_1\\ \\ 4R_2+R_3 \end{array}$$

$$\left[\begin{array}{cccc|c} 1&0&0&-1&0\\ 0&1&0&4&12\\ 0&0&1&-1&8 \end{array}\right]\begin{array}{l} R_3+R_1\\ -2R_3+R_2\\ \\ \end{array}$$

This last matrix is the augmented matrix for the dependent system

$$\begin{aligned} x-w&=0 \quad (1)\\ y+4w&=12 \quad (2)\\ z-w&=8 \quad (3) \end{aligned}$$

Solve equations (1), (2), and (3) in terms of w.

$$\begin{aligned} x&=w \quad (1)\\ y&=-4w+12 \quad (2)\\ z&=w+8 \quad (3) \end{aligned}$$

The solution is all points of the form $(w, -4w+12, w+8, w)$. But in order for y to be positive, w has to be 0, 1, 2, or 3. So, there are four possibilities, as follows:

0 cases of A, 12 cases of B, 8 cases of C, 0 cases of D;
1 case of A, 8 cases of B, 9 cases of C, 1 case of D;
2 cases of A, 4 cases of B, 10 cases of C, 2 cases of D;
3 cases of A, 0 cases of B, 11 cases of C, 3 cases of D.

25. **a.** $f(x)=ax^2+bx+c$

We have the system

$$\begin{aligned} 163&=a(15)^2+b(15)+c\\ 187&=a(35)^2+b(35)+c\,.\\ 192&=a(40)^2+b(40)+c \end{aligned}$$

This becomes

$$\begin{aligned} 225a+15b+c&=163\\ 1225a+35b+c&=187\,.\\ 1600a+40b+c&=192 \end{aligned}$$

The augmented matrix is

$$\left[\begin{array}{ccc|c} 225&15&1&163\\ 1225&35&1&187\\ 1600&40&1&192 \end{array}\right].$$

$$\left[\begin{array}{ccc|c} 1375&25&0&29\\ 375&5&0&5\\ 1600&40&1&192 \end{array}\right]\begin{array}{l} -R_1+R_3\\ -R_2+R_3\\ \\ \end{array}$$

$$\left[\begin{array}{ccc|c} 1375&25&0&29\\ 75&1&0&1\\ 1600&40&1&192 \end{array}\right]\begin{array}{l} \\ \frac{1}{5}R_2\\ \\ \end{array}$$

$$\left[\begin{array}{ccc|c} -500&0&0&4\\ 75&1&0&1\\ -1400&0&1&152 \end{array}\right]\begin{array}{l} -25R_2+R_1\\ \\ R_3-40R_2 \end{array}$$

$$\left[\begin{array}{ccc|c} 1&0&0&-0.008\\ 75&1&0&1\\ -1400&0&1&152 \end{array}\right]\begin{array}{l} -\frac{1}{500}R_1\\ \\ \\ \end{array}$$

$$\left[\begin{array}{ccc|c} 1&0&0&-0.008\\ 0&1&0&1.6\\ 0&0&1&140.8 \end{array}\right]\begin{array}{l} \\ -75R_1+R_2\\ 1400R_1+R_3 \end{array}$$

Thus, $a=-0.008$, $b=1.6$, and $c=140.8$.

The equation is

$f(x)=-0.008x^2+1.6x+140.8.$

b. The year 2020 corresponds to $x=20$.

$$\begin{aligned} f(20)&=-0.008(20)^2+1.6(20)+140.8\\ &=169.6 \end{aligned}$$

The U.S. female population will be about 169.6 million in 2020.

Copyright © 2019 Pearson Education, Inc.

27. a. An equation $y = ax^2 + bx + c$ is sought.
When $x = 6$, $y = 2.80$. Therefore
$36a + 6b + c = 2.80$.
When $x = 8$, $y = 2.48$. Therefore,
$64a + 8b + c = 2.48$.
When $x = 10$, $y = 2.24$. Therefore,
$100a + 10b + c = 2.24$.
Thus, the system of equations is

$$\begin{aligned} 36a+6b+c&=2.80 \quad (1)\\ 64a+8b+c&=2.48 \quad (2)\\ 100a+10b+c&=2.24. \quad (3)\end{aligned}$$

Solve the system by elimination. Subtract equation (1) from equation (2) to obtain
$28a + 2b = -.32$,
and subtract equation (2) from equation (3) to obtain
$36a + 2b = -.24$.
Now the system is

$$\begin{aligned} 28a + 2b &= -.32 \quad (4)\\ 36a + 2b &= -.24 \quad (5)\end{aligned}$$

Subtract equation (5) from equation (4).
$-8a = -.08$
Thus, $a = .01$. Substituting into
$28a + 2b = -.32$ gives

$$\begin{aligned} 28(.01) + 2b &= -.32\\ 2b &= -.60\\ b &= -.30.\end{aligned}$$

Substituting into equation (1) gives

$$\begin{aligned} 36(.01) + 6(-.30) + c &= 2.80\\ .36 - 1.80 + c &= 2.80\\ -1.44 + c &= 2.80\\ c &= 4.24.\end{aligned}$$

Thus, the equation is
$y = .01x^2 - .3x + 4.24$.

b. Write $y = .01x^2 - .3x + 4.24$ in the form
$y = a(x-h)^2 + k$.

$$\begin{aligned} y &= .01(x^2 - 30x) + 4.24\\ y &= .01(x^2 - 30x + 225) + 4.24 - 2.25\\ y &= .01(x-15)^2 + 1.99\end{aligned}$$

The minimum value of y is 1.99, occurring when $x = 15$. Thus, 15 platters should be fired at one time to minimize the fuel cost. The minimum fuel cost is \$1.99.

29. Let x = number of cups of Hearty Chicken Rotini
let y = number of cups of Hearty Chicken, and
let z = number of cups of Chunky Chicken Noodle
The system of equations is

$$\begin{aligned} 100x + 130y + 110z &= 1710\\ 960x + 480y + 890z &= 11{,}580\\ 7x + 9y + 8z &= 121\end{aligned}$$

After simplifying the equations, the matrix is

$$\left[\begin{array}{ccc|c} 10 & 13 & 11 & 171\\ 96 & 48 & 89 & 1158\\ 7 & 9 & 8 & 121\end{array}\right]$$

$$\left[\begin{array}{ccc|c} 1 & \frac{13}{10} & \frac{11}{10} & \frac{171}{10}\\ 96 & 48 & 89 & 1158\\ 7 & 9 & 8 & 121\end{array}\right] \begin{array}{l} \frac{1}{10}R_1\\ \\ \\ \end{array}$$

$$\left[\begin{array}{ccc|c} 1 & \frac{13}{10} & \frac{11}{10} & \frac{171}{10}\\ 0 & -\frac{384}{5} & -\frac{83}{5} & -\frac{2418}{5}\\ 0 & -\frac{1}{10} & \frac{3}{10} & \frac{13}{10}\end{array}\right] \begin{array}{l} \\ -96R_1 + R_3\\ -7R_1 + R_3\end{array}$$

$$\left[\begin{array}{ccc|c} 1 & \frac{13}{10} & \frac{11}{10} & \frac{171}{10}\\ 0 & -\frac{384}{5} & -\frac{83}{5} & -\frac{2418}{5}\\ 0 & 1 & -3 & -13\end{array}\right] \begin{array}{l} \\ \\ -10R_3\end{array}$$

Interchange R_3 and R_2.

$$\left[\begin{array}{ccc|c} 1 & \frac{13}{10} & \frac{11}{10} & \frac{171}{10}\\ 0 & 1 & -3 & -13\\ 0 & -\frac{384}{5} & -\frac{83}{5} & -\frac{2418}{5}\end{array}\right]$$

$$\left[\begin{array}{ccc|c} 1 & 0 & 5 & 34\\ 0 & 1 & -3 & -13\\ 0 & 0 & -247 & -1482\end{array}\right] \begin{array}{l} -\frac{13}{10}R_2 + R_1\\ \\ \frac{384}{5}R_2 + R_3\end{array}$$

$$\left[\begin{array}{ccc|c} 1 & 0 & 5 & 34\\ 0 & 1 & -3 & -13\\ 0 & 0 & 1 & 6\end{array}\right] \begin{array}{l} \\ \\ -\frac{1}{247}R_3\end{array}$$

$$\left[\begin{array}{ccc|c} 1 & 0 & 0 & 4\\ 0 & 1 & 0 & 5\\ 0 & 0 & 1 & 6\end{array}\right] \begin{array}{l} R_1 - 5R_3\\ R_2 + 3R_3\\ \\ \end{array}$$

Therefore, use 4 cups of Hearty Chicken Rotini, 5 cups of Hearty Chicken, and 6 cups of Chunky Chicken Noodle.

$$\text{Serving size} = \frac{4+5+6}{10} = 1.5 \text{ cups}$$

Copyright © 2019 Pearson Education, Inc.

Section 6.4 Basic Matrix Operations

1. $\begin{bmatrix} 7 & -8 & 4 \\ 0 & 13 & 9 \end{bmatrix}$ is a 2×3 matrix.

Its additive inverse is $\begin{bmatrix} -7 & 8 & -4 \\ 0 & -13 & -9 \end{bmatrix}$.

3. $\begin{bmatrix} -3 & 0 & 11 \\ 1 & \frac{1}{4} & -7 \\ 5 & -3 & 9 \end{bmatrix}$ is a 3×3 square matrix.

Its additive inverse is $\begin{bmatrix} 3 & 0 & -11 \\ -1 & -\frac{1}{4} & 7 \\ -5 & 3 & -9 \end{bmatrix}$.

5. $\begin{bmatrix} 7 \\ 11 \end{bmatrix}$ is a 2×1 column matrix.

Its additive inverse is $\begin{bmatrix} -7 \\ -11 \end{bmatrix}$.

7. If $A + B = A$, then B must be a zero matrix. Because A is a 5×3 matrix and only matrices of the same size can be added, B must also be 5×3. Therefore, B is a 5×3 zero matrix.

9.
$$\begin{bmatrix} 1 & 2 & 7 & -1 \\ 8 & 0 & 2 & -4 \end{bmatrix}+\begin{bmatrix} -8 & 12 & -5 & 5 \\ -2 & -3 & 0 & 0 \end{bmatrix}$$
$$=\begin{bmatrix} 1+(-8) & 2+12 & 7+(-5) & -1+5 \\ 8+(-2) & 0+(-3) & 2+0 & -4+0 \end{bmatrix}$$
$$=\begin{bmatrix} -7 & 14 & 2 & 4 \\ 6 & -3 & 2 & -4 \end{bmatrix}$$

11.
$$\begin{bmatrix} -1 & -5 & 9 \\ 2 & 2 & 3 \end{bmatrix}+\begin{bmatrix} 4 & 4 & -7 \\ 1 & -1 & 2 \end{bmatrix}$$
$$=\begin{bmatrix} -1+4 & -5+4 & 9+(-7) \\ 2+1 & 2+(-1) & 3+2 \end{bmatrix}$$
$$=\begin{bmatrix} 3 & -1 & 2 \\ 3 & 1 & 5 \end{bmatrix}$$

13.
$$\begin{bmatrix} 0 & -2 \\ 1 & 9 \\ -5 & -9 \end{bmatrix}-\begin{bmatrix} 8 & 6 \\ 9 & 17 \\ 3 & -1 \end{bmatrix}=\begin{bmatrix} 0-8 & -2-6 \\ 1-9 & 9-17 \\ -5-3 & -9-(-1) \end{bmatrix}$$
$$=\begin{bmatrix} -8 & -8 \\ -8 & -8 \\ -8 & -8 \end{bmatrix}$$

15.
$$\begin{bmatrix} 9 & 1 \\ 0 & -3 \\ 4 & 10 \end{bmatrix}-\begin{bmatrix} 1 & 9 & -4 \\ -1 & 1 & 0 \end{bmatrix}$$

This subtraction cannot be performed since the matrices have different sizes.

17. $2A=2\begin{bmatrix} -2 & 0 \\ 5 & 3 \end{bmatrix}=\begin{bmatrix} -4 & 0 \\ 10 & 6 \end{bmatrix}$

19. $-4B=-4\begin{bmatrix} 0 & 2 \\ 4 & -6 \end{bmatrix}=\begin{bmatrix} 0 & -8 \\ -16 & 24 \end{bmatrix}$

21.
$$-4A+5B=-4\begin{bmatrix} -2 & 0 \\ 5 & 3 \end{bmatrix}+5\begin{bmatrix} 0 & 2 \\ 4 & -6 \end{bmatrix}$$
$$=\begin{bmatrix} 8 & 0 \\ -20 & -12 \end{bmatrix}+\begin{bmatrix} 0 & 10 \\ 20 & -30 \end{bmatrix}$$
$$=\begin{bmatrix} 8 & 10 \\ 0 & -42 \end{bmatrix}$$

23. $A=\begin{bmatrix} 1 & -2 \\ 4 & 3 \end{bmatrix}$, $B=\begin{bmatrix} 2 & -1 \\ 0 & 5 \end{bmatrix}$
$$2A+3B=2\begin{bmatrix} 1 & -2 \\ 4 & 3 \end{bmatrix}+3\begin{bmatrix} 2 & -1 \\ 0 & 5 \end{bmatrix}$$
$$=\begin{bmatrix} 2 & -4 \\ 8 & 6 \end{bmatrix}+\begin{bmatrix} 6 & -3 \\ 0 & 15 \end{bmatrix}=\begin{bmatrix} 8 & -7 \\ 8 & 21 \end{bmatrix}$$

If $2X = 2A + 3B$, then

$2X=\begin{bmatrix} 8 & -7 \\ 8 & 21 \end{bmatrix}$ and $X=\begin{bmatrix} 4 & -\frac{7}{2} \\ 4 & \frac{21}{2} \end{bmatrix}$.

25. $X+T=\begin{bmatrix} x & y \\ z & w \end{bmatrix}+\begin{bmatrix} r & s \\ t & u \end{bmatrix}=\begin{bmatrix} x+r & y+s \\ z+t & w+u \end{bmatrix}$,

which is another 2×2 matrix.

Copyright © 2019 Pearson Education, Inc.

27. Show that $X+(T+P)=(X+T)+P$. On the left-hand side, the sum of $T+P$ is obtained first, and then $X+(T+P)$. This gives the matrix

$$\begin{bmatrix} x+(r+m) & y+(s+n) \\ z+(t+p) & w+(u+q) \end{bmatrix}.$$

For the right-hand side, first the sum $X+T$ is obtained, and then $(X+T)+P$. This gives the matrix

$$\begin{bmatrix} (x+r)+m & (y+s)+n \\ (z+t)+p & (w+u)+q \end{bmatrix}.$$

Comparing corresponding elements shows that they are equal by the associative property of addition of real numbers. Thus, $X+(T+P)=(X+T)+P$.

29. Show that $P+O=P$.

$$P+O=\begin{bmatrix} m & n \\ p & q \end{bmatrix}+\begin{bmatrix} 0 & 0 \\ 0 & 0 \end{bmatrix}=\begin{bmatrix} m+0 & n+0 \\ p+0 & q+0 \end{bmatrix}$$
$$=\begin{bmatrix} m & n \\ p & q \end{bmatrix}=P$$

31. Several possible answers, including:

	basketball	hockey	football	baseball
percent of no shows	16	16	20	18
lost revenue per fan ($)	18.20	18.25	19	15.40
lost annual revenue (millions $)	22.7	35.8	51.9	96.3

33.

	2010	2016
heart	2874	4154
lung	1759	1470
liver	15,394	14,794
kidney	83,919	100,467

35. a.

	Bread	Milk	Peanut butter	Cold cuts
I	88	48	16	112
II	105	72	21	147
III	60	40	0	50

b. For store I,
$1.25(88)=110$, $1.25(48)=60$, $1.25(16)=20$, $1.25(112)=140$.

For store II,
$\frac{4}{3}(105)=140$, $\frac{4}{3}(72)=96$, $\frac{4}{3}(21)=28$, $\frac{4}{3}(147)=196$.

For store III,
$1.10(60)=66$, $1.10(40)=44$, $1.10(0)=0$, $1.10(50)=55$.

The new matrix is

$$\begin{bmatrix} 110 & 60 & 20 & 140 \\ 140 & 96 & 28 & 196 \\ 66 & 44 & 0 & 55 \end{bmatrix}.$$

Copyright © 2019 Pearson Education, Inc.

c. To find the total sales, add the matrices from parts (a) and (b).

$$\begin{bmatrix} 88 & 48 & 16 & 112 \\ 105 & 72 & 21 & 147 \\ 60 & 40 & 0 & 50 \end{bmatrix} + \begin{bmatrix} 110 & 60 & 20 & 140 \\ 140 & 96 & 28 & 196 \\ 66 & 44 & 0 & 55 \end{bmatrix} = \begin{bmatrix} 198 & 108 & 36 & 252 \\ 245 & 168 & 49 & 343 \\ 126 & 84 & 0 & 105 \end{bmatrix}$$

Section 6.5 Matrix Products and Inverses

1. AB is 2×2. A has 2 rows, B has 2 columns. BA is 2×2. B has 2 rows, A has 2 columns.

3. AB is 3×3. A has 3 rows, B has 3 columns. BA is 5×5. B has 5 rows, A has 5 columns.

5. AB does not exist because the number of columns of A is not the same as the number of rows of B. BA is 3×2. B has 3 rows, A has 2 columns.

7. Columns; rows

9. $$\begin{bmatrix} 1 & 2 \\ 3 & 4 \end{bmatrix}\begin{bmatrix} -1 \\ 3 \end{bmatrix} = \begin{bmatrix} 1(-1)+2(3) \\ 3(-1)+4(3) \end{bmatrix} = \begin{bmatrix} 5 \\ 9 \end{bmatrix}$$

11. $$\begin{bmatrix} 2 & 2 & -1 \\ 5 & 0 & 1 \end{bmatrix}\begin{bmatrix} 0 & -2 \\ -1 & 5 \\ 0 & 2 \end{bmatrix} = \begin{bmatrix} (2)(0)+(2)(-1)+(-1)(0) & (2)(-2)+(2)(5)+(-1)(2) \\ (5)(0)+(0)(-1)+(1)(0) & (5)(-2)+(0)(5)+(1)(2) \end{bmatrix} = \begin{bmatrix} -2 & 4 \\ 0 & -8 \end{bmatrix}$$

13. $$\begin{bmatrix} -4 & 1 \\ 2 & -3 \end{bmatrix}\begin{bmatrix} 1 & 0 \\ 0 & 1 \end{bmatrix} = \begin{bmatrix} (-4)(1)+(1)(0) & (-4)(0)+(1)(1) \\ (2)(1)+(-3)(0) & (2)(0)+(-3)(1) \end{bmatrix} = \begin{bmatrix} -4 & 1 \\ 2 & -3 \end{bmatrix}$$

15. $$\begin{bmatrix} 1 & 0 & 0 \\ 0 & 1 & 0 \\ 0 & 0 & 1 \end{bmatrix}\begin{bmatrix} 3 & -5 & 7 \\ -2 & 1 & 6 \\ 0 & -3 & 4 \end{bmatrix} = \begin{bmatrix} (1)(3)+(0)(-2)+(0)(0) & (1)(-5)+(0)(1)+(0)(-3) & (1)(7)+(0)(6)+(0)(4) \\ (0)(3)+(1)(-2)+(0)(0) & (0)(-5)+(1)(1)+(0)(-3) & (0)(7)+(1)(6)+(0)(4) \\ (0)(3)+(0)(-2)+(1)(0) & (0)(-5)+(0)(1)+(1)(-3) & (0)(7)+(0)(6)+(1)(4) \end{bmatrix}$$
$$= \begin{bmatrix} 3 & -5 & 7 \\ -2 & 1 & 6 \\ 0 & -3 & 4 \end{bmatrix}$$

17. $$\begin{bmatrix} 1 & 2 & 3 \\ 4 & 0 & 6 \\ 7 & 8 & 9 \end{bmatrix}\begin{bmatrix} -1 & 4 \\ 7 & 0 \\ 1 & 2 \end{bmatrix} = \begin{bmatrix} (1)(-1)+(2)(7)+(3)(1) & (1)(4)+(2)(0)+(3)(2) \\ (4)(-1)+(0)(7)+(6)(1) & (4)(4)+(0)(0)+(6)(2) \\ (7)(-1)+(8)(7)+(9)(1) & (7)(4)+(8)(0)+(9)(2) \end{bmatrix} = \begin{bmatrix} 16 & 10 \\ 2 & 28 \\ 58 & 46 \end{bmatrix}$$

19. $$AB = \begin{bmatrix} -3 & -9 \\ 2 & 6 \end{bmatrix}\begin{bmatrix} 4 & 6 \\ 2 & 3 \end{bmatrix} = \begin{bmatrix} -30 & -45 \\ 20 & 30 \end{bmatrix}, \quad BA = \begin{bmatrix} 4 & 6 \\ 2 & 3 \end{bmatrix}\begin{bmatrix} -3 & -9 \\ 2 & 6 \end{bmatrix} = \begin{bmatrix} 0 & 0 \\ 0 & 0 \end{bmatrix}$$

Since $\begin{bmatrix} -30 & -45 \\ 20 & 30 \end{bmatrix} \neq \begin{bmatrix} 0 & 0 \\ 0 & 0 \end{bmatrix}$, $AB \neq BA$.

Therefore, matrix multiplication is not commutative.

Copyright © 2019 Pearson Education, Inc.

21.

$$A+B=\begin{bmatrix}-3 & -9\\ 2 & 6\end{bmatrix}+\begin{bmatrix}4 & 6\\ 2 & 3\end{bmatrix}=\begin{bmatrix}1 & -3\\ 4 & 9\end{bmatrix}$$

$$A-B=\begin{bmatrix}-3 & -9\\ 2 & 6\end{bmatrix}-\begin{bmatrix}4 & 6\\ 2 & 3\end{bmatrix}=\begin{bmatrix}-7 & -15\\ 0 & 3\end{bmatrix}$$

$$(A+B)(A-B)=\begin{bmatrix}1 & -3\\ 4 & 9\end{bmatrix}\begin{bmatrix}-7 & -15\\ 0 & 3\end{bmatrix}=\begin{bmatrix}-7 & -24\\ -28 & -33\end{bmatrix}$$

$$A^2=\begin{bmatrix}-3 & -9\\ 2 & 6\end{bmatrix}\begin{bmatrix}-3 & -9\\ 2 & 6\end{bmatrix}=\begin{bmatrix}-9 & -27\\ 6 & 18\end{bmatrix}$$

$$B^2=\begin{bmatrix}4 & 6\\ 2 & 3\end{bmatrix}\begin{bmatrix}4 & 6\\ 2 & 3\end{bmatrix}=\begin{bmatrix}28 & 42\\ 14 & 21\end{bmatrix}$$

$$A^2-B^2=\begin{bmatrix}-9 & -27\\ 6 & 18\end{bmatrix}-\begin{bmatrix}28 & 42\\ 14 & 21\end{bmatrix}=\begin{bmatrix}-37 & -69\\ -8 & -3\end{bmatrix}$$

Since $\begin{bmatrix}-7 & -24\\ -28 & -33\end{bmatrix}\neq\begin{bmatrix}-37 & -69\\ -8 & -3\end{bmatrix}$, $(A+B)(A-B)\neq A^2-B^2$.

23. Verify that $(PX)T = P(XT)$.

$$(PX)T=\left(\begin{bmatrix}m & n\\ p & q\end{bmatrix}\begin{bmatrix}x & y\\ z & w\end{bmatrix}\right)\begin{bmatrix}r & s\\ t & u\end{bmatrix}=\begin{bmatrix}mx+nz & my+nw\\ px+qz & py+qw\end{bmatrix}\begin{bmatrix}r & s\\ t & u\end{bmatrix}$$

$$=\begin{bmatrix}(mx+nz)r+(my+nw)t & (mx+nz)s+(my+nw)u\\ (px+qz)r+(py+qw)t & (px+qz)s+(py+qw)u\end{bmatrix}$$

$$=\begin{bmatrix}mxr+nzr+myt+nwt & mxs+nzs+myu+nwu\\ pxr+qzr+pyt+qwt & pxs+qzs+pyu+qwu\end{bmatrix}$$ Distributive property for real numbers

$$P(XT)=\begin{bmatrix}m & n\\ p & q\end{bmatrix}\left(\begin{bmatrix}x & y\\ z & w\end{bmatrix}\begin{bmatrix}r & s\\ t & u\end{bmatrix}\right)=\begin{bmatrix}m & n\\ p & q\end{bmatrix}\begin{bmatrix}xr+yt & xs+yu\\ zr+wt & zs+wu\end{bmatrix}$$

$$=\begin{bmatrix}m(xr+yt)+n(zr+wt) & m(xs+yu)+n(zs+wu)\\ p(xr+yt)+q(zr+wt) & p(xs+yu)+q(zs+wu)\end{bmatrix}$$

$$=\begin{bmatrix}mxr+myt+nzr+nwt & mxs+myu+nzs+nwu\\ pxr+pyt+qzr+qwt & pxs+pyu+qzs+qwu\end{bmatrix}$$ Distributive property for real numbers

$$=\begin{bmatrix}mxr+nzr+myt+nwt & mxs+nzs+myu+nwu\\ pxr+qzr+pyt+qwt & pxs+qzs+pyu+qwu\end{bmatrix}$$ Commutative property for real numbers

Thus, $(PX)T = P(XT)$.

25. Verify $k(X+T) = kX + kT$ for any real number k.

$$k(X+T)=k\left[\begin{bmatrix}x & y\\ z & w\end{bmatrix}+\begin{bmatrix}r & s\\ t & u\end{bmatrix}\right]=k\begin{bmatrix}x+r & y+s\\ z+t & w+u\end{bmatrix}=\begin{bmatrix}k(x+r) & k(y+s)\\ k(z+t) & k(w+u)\end{bmatrix}=\begin{bmatrix}kx+kr & ky+ks\\ kz+kt & kw+ku\end{bmatrix}$$

$$=\begin{bmatrix}kx & ky\\ kz & kw\end{bmatrix}+\begin{bmatrix}kr & ks\\ kt & ku\end{bmatrix}=k\begin{bmatrix}x & y\\ z & w\end{bmatrix}+k\begin{bmatrix}r & s\\ t & u\end{bmatrix}=kX+kT$$

Copyright © 2019 Pearson Education, Inc.

27. $\begin{bmatrix} 5 & 2 \\ 3 & -1 \end{bmatrix}\begin{bmatrix} -1 & 2 \\ 3 & -4 \end{bmatrix} = \begin{bmatrix} 1 & 2 \\ -6 & 10 \end{bmatrix} \neq I$, so the given matrices are not inverses of each other.

29. $\begin{bmatrix} 3 & -1 \\ -4 & 2 \end{bmatrix}\begin{bmatrix} 1 & \frac{1}{2} \\ 2 & \frac{3}{2} \end{bmatrix} = \begin{bmatrix} 1 & 0 \\ 0 & 1 \end{bmatrix} = I$

$\begin{bmatrix} 1 & \frac{1}{2} \\ 2 & \frac{3}{2} \end{bmatrix}\begin{bmatrix} 3 & -1 \\ -4 & 2 \end{bmatrix} = \begin{bmatrix} 1 & 0 \\ 0 & 1 \end{bmatrix} = I$

Therefore, the given matrices are inverses of each other.

31. $\begin{bmatrix} 1 & 1 & 1 \\ 2 & 3 & 0 \\ 1 & 2 & 1 \end{bmatrix}\begin{bmatrix} 1.5 & .5 & -1.5 \\ -1 & 0 & 1 \\ .5 & -.5 & .5 \end{bmatrix} = \begin{bmatrix} 1 & 0 & 0 \\ 0 & 1 & 0 \\ 0 & 0 & 1 \end{bmatrix} = I$

$\begin{bmatrix} 1.5 & .5 & -1.5 \\ -1 & 0 & 1 \\ .5 & -.5 & .5 \end{bmatrix}\begin{bmatrix} 1 & 1 & 1 \\ 2 & 3 & 0 \\ 1 & 2 & 1 \end{bmatrix} = \begin{bmatrix} 1 & 0 & 0 \\ 0 & 1 & 0 \\ 0 & 0 & 1 \end{bmatrix} = I$

The given matrices are inverses of each other.

33. To find the inverse of $\begin{bmatrix} 1 & 2 \\ 3 & 6 \end{bmatrix}$, write the augmented matrix $[A \mid I]$.

$\left[\begin{array}{cc|cc} 1 & 2 & 1 & 0 \\ 3 & 6 & 0 & 1 \end{array}\right]$

$\left[\begin{array}{cc|cc} 1 & 2 & 1 & 0 \\ 0 & 0 & -3 & 1 \end{array}\right] \; -3R_1 + R_2$

Since there is no way to continue the transformation, the given matrix has no inverse.

35. To find the inverse of $\begin{bmatrix} -1 & 2 \\ 1 & -1 \end{bmatrix}$, write the augmented matrix $[A \mid I]$.

$\left[\begin{array}{cc|cc} -1 & 2 & 1 & 0 \\ 1 & -1 & 0 & 1 \end{array}\right]$

$\left[\begin{array}{cc|cc} -1 & 2 & 1 & 0 \\ 0 & 1 & 1 & 1 \end{array}\right] \; R_1 + R_2$

$\left[\begin{array}{cc|cc} -1 & 0 & -1 & -2 \\ 0 & 1 & 1 & 1 \end{array}\right] \; -2R_2 + R_1$

$\left[\begin{array}{cc|cc} 1 & 0 & 1 & 2 \\ 0 & 1 & 1 & 1 \end{array}\right] \; -R_1$

The inverse is $\begin{bmatrix} 1 & 2 \\ 1 & 1 \end{bmatrix}$.

37. To find the inverse of $\begin{bmatrix} 2 & -3 \\ -1 & 2 \end{bmatrix}$, write the augmented matrix $[A \mid I]$.

$\left[\begin{array}{cc|cc} 2 & -3 & 1 & 0 \\ -1 & 2 & 0 & 1 \end{array}\right]$.

$\left[\begin{array}{cc|cc} 1 & -1 & 1 & 1 \\ -1 & 2 & 0 & 1 \end{array}\right] \; R_1 + R_2$

$\left[\begin{array}{cc|cc} 1 & -1 & 1 & 1 \\ 0 & 1 & 1 & 2 \end{array}\right] \; R_1 + R_2$

$\left[\begin{array}{cc|cc} 1 & 0 & 2 & 3 \\ 0 & 1 & 1 & 2 \end{array}\right] \; R_1 + R_2$

The inverse is $\begin{bmatrix} 2 & 3 \\ 1 & 2 \end{bmatrix}$.

39. To find the inverse of $\begin{bmatrix} 1 & -1 & 0 \\ -1 & 2 & 3 \\ 1 & 0 & 2 \end{bmatrix}$, write the augmented matrix $[A \mid I]$.

$\left[\begin{array}{ccc|ccc} 1 & -1 & 0 & 1 & 0 & 0 \\ -1 & 2 & 3 & 0 & 1 & 0 \\ 1 & 0 & 2 & 0 & 0 & 1 \end{array}\right]$

$\left[\begin{array}{ccc|ccc} 1 & -1 & 0 & 1 & 0 & 0 \\ 0 & 1 & 3 & 1 & 1 & 0 \\ 0 & -1 & -2 & 1 & 0 & -1 \end{array}\right] \; \begin{matrix} \\ R_1 + R_2 \\ R_1 - R_3 \end{matrix}$

$\left[\begin{array}{ccc|ccc} 1 & 0 & 3 & 2 & 1 & 0 \\ 0 & 1 & 3 & 1 & 1 & 0 \\ 0 & 0 & 1 & 2 & 1 & -1 \end{array}\right] \; \begin{matrix} R_1 + R_2 \\ \\ R_2 + R_3 \end{matrix}$

$\left[\begin{array}{ccc|ccc} 1 & 0 & 0 & -4 & -2 & 3 \\ 0 & 1 & 0 & -5 & -2 & 3 \\ 0 & 0 & 1 & 2 & 1 & -1 \end{array}\right] \; \begin{matrix} R_1 - 3R_3 \\ R_2 - 3R_3 \\ \end{matrix}$

The inverse of the given matrix is

$\begin{bmatrix} -4 & -2 & 3 \\ -5 & -2 & 3 \\ 2 & 1 & -1 \end{bmatrix}$.

Copyright © 2019 Pearson Education, Inc.

41. To find the inverse of $\begin{bmatrix} 1 & -1 & 4 \\ 0 & 1 & 3 \\ 2 & -3 & 4 \end{bmatrix}$,

write the augmented matrix $[A \mid I]$.

$$\left[\begin{array}{ccc|ccc} 1 & -1 & 4 & 1 & 0 & 0 \\ 0 & 1 & 3 & 0 & 1 & 0 \\ 2 & -3 & 4 & 0 & 0 & 1 \end{array}\right]$$

$$\left[\begin{array}{ccc|ccc} 1 & 0 & 7 & 1 & 1 & 0 \\ 0 & 1 & 3 & 0 & 1 & 0 \\ 0 & -1 & -4 & -2 & 0 & 1 \end{array}\right] \begin{array}{l} R_1 + R_2 \\ \\ -2R_1 + R_3 \end{array}$$

$$\left[\begin{array}{ccc|ccc} 1 & 0 & 7 & 1 & 1 & 0 \\ 0 & 1 & 3 & 0 & 1 & 0 \\ 0 & 0 & 1 & 2 & -1 & -1 \end{array}\right] \begin{array}{l} \\ \\ -(R_2 + R_3) \end{array}$$

$$\left[\begin{array}{ccc|ccc} 1 & 0 & 0 & -13 & 8 & 7 \\ 0 & 1 & 0 & -6 & 4 & 3 \\ 0 & 0 & 1 & 2 & -1 & -1 \end{array}\right] \begin{array}{l} -7R_3 + R_1 \\ -3R_3 + R_2 \\ \\ \end{array}$$

The inverse of the given matrix is

$$\begin{bmatrix} -13 & 8 & 7 \\ -6 & 4 & 3 \\ 2 & -1 & -1 \end{bmatrix}$$

43. To find the inverse of $\begin{bmatrix} 1 & 2 & 0 \\ 3 & -1 & 2 \\ -2 & 3 & -2 \end{bmatrix}$,

write the augmented matrix $[A \mid I]$.

$$\left[\begin{array}{ccc|ccc} 1 & 2 & 0 & 1 & 0 & 0 \\ 3 & -1 & 2 & 0 & 1 & 0 \\ -2 & 3 & -2 & 0 & 0 & 1 \end{array}\right]$$

$$\left[\begin{array}{ccc|ccc} 1 & 2 & 0 & 1 & 0 & 0 \\ 0 & -7 & 2 & -3 & 1 & 0 \\ 0 & 7 & -2 & 2 & 0 & 1 \end{array}\right] \begin{array}{l} \\ -3R_1 + R_2 \\ 2R_1 + R_3 \end{array}$$

$$\left[\begin{array}{ccc|ccc} 1 & 2 & 0 & 1 & 0 & 0 \\ 0 & -7 & 2 & -3 & 1 & 0 \\ 0 & 0 & 0 & -1 & 1 & 1 \end{array}\right] \begin{array}{l} \\ \\ R_2 + R_3 \end{array}$$

Since there is no way to continue the desired transformation, the given matrix has no inverse.

45. $A = \begin{bmatrix} 1 & 2 & 3 \\ 1 & 4 & 2 \\ 0 & 1 & -1 \end{bmatrix}$

```
[A]-1
  [[6   -5 8 ]
   [-1  1  -1]
   [-1  1  -2]]
```

47. Use a graphing calculator to find the inverse of

$$\begin{bmatrix} 1 & 0 & -2 & 0 \\ -2 & 1 & 2 & 2 \\ 3 & -1 & -2 & -3 \\ 0 & 1 & 4 & 1 \end{bmatrix}.$$

Using a graphing calculator, we find that the inverse of the given matrix is

$$\begin{bmatrix} \frac{1}{2} & -1 & -\frac{1}{2} & \frac{1}{2} \\ \frac{1}{2} & 4 & \frac{5}{2} & -\frac{1}{2} \\ -\frac{1}{4} & -\frac{1}{2} & -\frac{1}{4} & \frac{1}{4} \\ \frac{1}{2} & -2 & -\frac{3}{2} & \frac{1}{2} \end{bmatrix}.$$

49. a. $AB = \begin{bmatrix} 6 & 2 \\ 1 & 3 \end{bmatrix}\begin{bmatrix} 8 & 7 \\ 4 & 5 \end{bmatrix} = \begin{bmatrix} \$56 & \$52 \\ \$20 & \$22 \end{bmatrix}$

b. Row 1 represents you, and row 2 represents your roommate.

c. Column 1 represents Timesaver, and column 2 represents Wheels.

d. TimeSaver is the total of column 1; \$56 + \$20 = \$76. Wheels is the total of column 2; \$52 + \$22 = \$74.

51. a. $AB = \begin{bmatrix} 5 & 25 \\ 1 & 4 \end{bmatrix}\begin{bmatrix} 41 & 6 \\ 20 & 2 \end{bmatrix} = \begin{bmatrix} 705 & 80 \\ 121 & 14 \end{bmatrix}$

b. Row 1 represents the restaurant and row 2 the food truck. Column 1 represents total salary and column 2 total benefits (in thousands) for all employees at those locations.

c. The total salary for all employees at restaurants and the Food Trucks were about \$705,000 and \$121,00 respectively . The total cost of benefits for all employees at the restaurant and food truck was about \$80,000 and \$14,000 respectively.

Copyright © 2019 Pearson Education, Inc.

53. **a.** No; the number of columns of P is not the same as the number of rows of S.

b. $$SP = \begin{bmatrix} 8 & 7 & 2 \\ 4 & 5 & 2 \\ 6 & 9 & 4 \end{bmatrix}\begin{bmatrix} 250 \\ 300 \\ 350 \end{bmatrix} = \begin{bmatrix} 4800 \\ 3200 \\ 5600 \end{bmatrix}$$

The matrix represents the revenue in each state.

c. The total revenue is the sum of all values in the column matrix from part b; \$4800 + \$3200 + \$5600 = \$13,600.

55. **a.** $$A = \begin{bmatrix} 195 & 143 & 1225 & 1341 \\ 210 & 141 & 1387 & 1388 \end{bmatrix}$$

b. $$B = \begin{bmatrix} .016 & .006 \\ .011 & .014 \\ .023 & .008 \\ .013 & .007 \end{bmatrix}$$

c. $$AB = \begin{bmatrix} 195 & 143 & 1225 & 1341 \\ 210 & 141 & 1387 & 1388 \end{bmatrix}\begin{bmatrix} .016 & .006 \\ .011 & .014 \\ .023 & .008 \\ .013 & .007 \end{bmatrix} = \begin{bmatrix} 50 & 22 \\ 55 & 24 \end{bmatrix}$$

d. The rows correspond to years 2010 and 2020. The entries in column 1 give the total number of births (in millions) in those years and column 2 the total deaths (in millions).

e. The total number of deaths in these four countries combined in 2010 was about 22 million. The projected number of births in these four countries in 2020 will be about 55 million.

Copyright © 2019 Pearson Education, Inc.

Section 6.6 Applications of Matrices

1. The solution to the matrix equation $AX = B$ is $X = A^{-1}B$.

$$A = \begin{bmatrix} 1 & -1 \\ 5 & 6 \end{bmatrix}, B = \begin{bmatrix} 4 \\ -2 \end{bmatrix}$$

Use a graphing calculator or row operations on $[A \mid I]$ to find A^{-1}.

$$\left[\begin{array}{cc|cc} 1 & -1 & 1 & 0 \\ 5 & 6 & 0 & 1 \end{array}\right]$$

$$\left[\begin{array}{cc|cc} 1 & -1 & 1 & 0 \\ 0 & 11 & -5 & 1 \end{array}\right] -5R_1 + R_2$$

$$\left[\begin{array}{cc|cc} 1 & -1 & 1 & 0 \\ 0 & 1 & -\frac{5}{11} & \frac{1}{11} \end{array}\right] \frac{1}{11}R_2$$

$$\left[\begin{array}{cc|cc} 1 & 0 & \frac{6}{11} & \frac{1}{11} \\ 0 & 1 & -\frac{5}{11} & \frac{1}{11} \end{array}\right] R_1 + R_2$$

$$A^{-1} = \begin{bmatrix} \frac{6}{11} & \frac{1}{11} \\ -\frac{5}{11} & \frac{1}{11} \end{bmatrix}$$

$$X = A^{-1}B = \begin{bmatrix} \frac{6}{11} & \frac{1}{11} \\ -\frac{5}{11} & \frac{1}{11} \end{bmatrix}\begin{bmatrix} 4 \\ -2 \end{bmatrix} = \begin{bmatrix} 2 \\ -2 \end{bmatrix}.$$

3. The solution to the matrix equation $AX = B$ is $X = A^{-1}B$.

$$A = \begin{bmatrix} 3 & 1 \\ 4 & 2 \end{bmatrix}, B = \begin{bmatrix} 3 & 4 \\ 5 & 6 \end{bmatrix}.$$

Use a graphing calculator or row operations on $[A \mid I]$ to find A^{-1}.

$$\left[\begin{array}{cc|cc} 3 & 1 & 1 & 0 \\ 4 & 2 & 0 & 1 \end{array}\right]$$

$$\left[\begin{array}{cc|cc} 1 & \frac{1}{3} & \frac{1}{3} & 0 \\ 4 & 2 & 0 & 1 \end{array}\right] \frac{1}{3}R_1$$

$$\left[\begin{array}{cc|cc} 1 & \frac{1}{3} & \frac{1}{3} & 0 \\ 0 & \frac{2}{3} & -\frac{4}{3} & 1 \end{array}\right] R_2 - 4R_1$$

$$\left[\begin{array}{cc|cc} 1 & \frac{1}{3} & \frac{1}{3} & 0 \\ 0 & 1 & -2 & \frac{3}{2} \end{array}\right] \frac{3}{2}R_2$$

$$\left[\begin{array}{cc|cc} 1 & 0 & 1 & -\frac{1}{2} \\ 0 & 1 & -2 & \frac{3}{2} \end{array}\right] R_1 - \frac{1}{3}R_2$$

$$A^{-1} = \begin{bmatrix} 1 & -\frac{1}{2} \\ -2 & \frac{3}{2} \end{bmatrix}$$

$$X = A^{-1}B = \begin{bmatrix} 1 & -\frac{1}{2} \\ -2 & \frac{3}{2} \end{bmatrix}\begin{bmatrix} 3 & 4 \\ 5 & 6 \end{bmatrix} = \begin{bmatrix} \frac{1}{2} & 1 \\ \frac{3}{2} & 1 \end{bmatrix}$$

5. The solution to the matrix equation $AX = B$ is $X = A^{-1}B$.

$$A = \begin{bmatrix} 2 & 1 & 0 \\ -4 & -1 & 3 \\ 3 & 1 & -2 \end{bmatrix}, B = \begin{bmatrix} 1 \\ 4 \\ 0 \end{bmatrix}.$$

Use a graphing calculator or row operations on $[A \mid I]$ to find that

$$A^{-1} = \begin{bmatrix} 1 & -2 & -3 \\ -1 & 4 & 6 \\ 1 & -1 & -2 \end{bmatrix}.$$

$$X = A^{-1}B = \begin{bmatrix} 1 & -2 & -3 \\ -1 & 4 & 6 \\ 1 & -1 & -2 \end{bmatrix}\begin{bmatrix} 1 \\ 4 \\ 0 \end{bmatrix} = \begin{bmatrix} -7 \\ 15 \\ -3 \end{bmatrix}.$$

7.
$$x + 2y + 3z = 10$$
$$2x + 3y + 2z = 6$$
$$-x - 2y - 4z = -1$$

has coefficient matrix

$$A = \begin{bmatrix} 1 & 2 & 3 \\ 2 & 3 & 2 \\ -1 & -2 & -4 \end{bmatrix}.$$

Find A^{-1}:

$$[A \mid I] = \left[\begin{array}{ccc|ccc} 1 & 2 & 3 & 1 & 0 & 0 \\ 2 & 3 & 2 & 0 & 1 & 0 \\ -1 & -2 & -4 & 0 & 0 & 1 \end{array}\right]$$

$$\left[\begin{array}{ccc|ccc} 1 & 2 & 3 & 1 & 0 & 0 \\ 0 & -1 & -4 & -2 & 1 & 0 \\ 0 & 0 & -1 & 1 & 0 & 1 \end{array}\right] \begin{array}{l} \\ -2R_1 + R_2 \\ R_1 + R_3 \end{array}$$

$$\left[\begin{array}{ccc|ccc} 1 & 0 & -5 & -3 & 2 & 0 \\ 0 & 1 & 4 & 2 & -1 & 0 \\ 0 & 0 & 1 & -1 & 0 & -1 \end{array}\right] \begin{array}{l} 2R_2 + R_1 \\ -1R_2 \\ -1R_3 \end{array}$$

$$\left[\begin{array}{ccc|ccc} 1 & 0 & 0 & -8 & 2 & -5 \\ 0 & 1 & 0 & 6 & -1 & 4 \\ 0 & 0 & 1 & -1 & 0 & -1 \end{array}\right] \begin{array}{l} 5R_3 + R_1 \\ -4R_3 + R_2 \\ \end{array}$$

Copyright © 2019 Pearson Education, Inc.

$$A^{-1} = \begin{bmatrix} -8 & 2 & -5 \\ 6 & -1 & 4 \\ -1 & 0 & -1 \end{bmatrix}$$

$$X = A^{-1}B = \begin{bmatrix} -8 & 2 & -5 \\ 6 & -1 & 4 \\ -1 & 0 & -1 \end{bmatrix} \begin{bmatrix} 10 \\ 6 \\ -1 \end{bmatrix} = \begin{bmatrix} -63 \\ 50 \\ -9 \end{bmatrix}$$

The solution is (–63, 50, –9).

9. $x + 4y + 3z = -12$
$x - 3y - 2z = 0$
$2x + 5y + 4z = 7$
has the coefficient matrix

$$A = \begin{bmatrix} 1 & 4 & 3 \\ 1 & -3 & -2 \\ 2 & 5 & 4 \end{bmatrix}.$$

From Exercise 41 of Section 6.5,

$$A^{-1} = \begin{bmatrix} 2 & 1 & -1 \\ 8 & 2 & -5 \\ -11 & -3 & 7 \end{bmatrix}.$$

$$X = A^{-1}B = \begin{bmatrix} 2 & 1 & -1 \\ 8 & 2 & -5 \\ -11 & -3 & 7 \end{bmatrix} \begin{bmatrix} -12 \\ 0 \\ 7 \end{bmatrix} = \begin{bmatrix} -31 \\ -131 \\ 181 \end{bmatrix}.$$

The solution is (–31, –131, 181).

11. $x + 2y + 3z = 4$
$x + 4y + 2z = 8$
$y - z = -4$
has coefficient matrix

$$A = \begin{bmatrix} 1 & 2 & 3 \\ 1 & 4 & 2 \\ 0 & 1 & -1 \end{bmatrix}.$$

From Exercise 45 of Section 6.5,

$$A^{-1} = \begin{bmatrix} 6 & -5 & 8 \\ -1 & 1 & -1 \\ -1 & 1 & -2 \end{bmatrix}.$$

$$X = A^{-1}B = \begin{bmatrix} 6 & -5 & 8 \\ -1 & 1 & -1 \\ -1 & 1 & -2 \end{bmatrix} \begin{bmatrix} 4 \\ 8 \\ -4 \end{bmatrix} = \begin{bmatrix} -48 \\ 8 \\ 12 \end{bmatrix}$$

The solution is (–48, 8, 12).

13. $x + y + 2w = 3$
$2x - y + z - w = 3$
$3x + 3y + 2z - 2w = 5$
$x + 2y + z = 3$
The coefficient matrix is

$$\begin{bmatrix} 1 & 1 & 0 & 2 \\ 2 & -1 & 1 & -1 \\ 3 & 3 & 2 & -2 \\ 1 & 2 & 1 & 0 \end{bmatrix}.$$

The inverse of this coefficient matrix was calculated in Exercise 48 of Section 6.5. Use that result to obtain

$$\begin{bmatrix} x \\ y \\ z \\ w \end{bmatrix} = \begin{bmatrix} \frac{1}{2} & 0 & \frac{1}{2} & -1 \\ \frac{1}{10} & -\frac{2}{5} & \frac{3}{10} & -\frac{1}{5} \\ -\frac{7}{10} & \frac{4}{5} & -\frac{11}{10} & \frac{12}{5} \\ \frac{1}{5} & \frac{1}{5} & -\frac{2}{5} & \frac{3}{5} \end{bmatrix} \begin{bmatrix} 3 \\ 3 \\ 5 \\ 3 \end{bmatrix} = \begin{bmatrix} 1 \\ 0 \\ 2 \\ 1 \end{bmatrix}.$$

The solution is (1, 0, 2, 1).

15. Let x = the amount of sales, and let y = the number of transactions.
The system of equations is $.029x + .30y = 291$
$.03x + .25y = 295$

Then, we have

$$A = \begin{bmatrix} .029 & .30 \\ .03 & .25 \end{bmatrix} \text{ and } B = \begin{bmatrix} 291 \\ 295 \end{bmatrix}.$$

Using a calculator, we find that the inverse is

$$\begin{bmatrix} -\frac{1000}{7} & \frac{1200}{7} \\ \frac{120}{7} & -\frac{116}{7} \end{bmatrix}.$$

$$X = A^{-1}B = \begin{bmatrix} -\frac{1000}{7} & \frac{1200}{7} \\ \frac{120}{7} & -\frac{116}{7} \end{bmatrix} \begin{bmatrix} 291 \\ 295 \end{bmatrix} = \begin{bmatrix} 9000 \\ 100 \end{bmatrix}$$

Therefore, there was $9,000 in sales and 100 transactions for the month.

17. Let x = the number of buffets, let y = the number of chairs, and let z = the number of tables. The information can be summarized in a table.

	Cutting hours	Assembly hours	Finishing hours
Buffets	$15x$	$20x$	$5x$
Chairs	$5y$	$8y$	$5y$
Tables	$10z$	$6z$	$6z$
Total	4900	6600	3900

Copyright © 2019 Pearson Education, Inc.

The system of equations is

$$15x+5y+10z=4900$$
$$20x+8y+6z=6600$$
$$5x+5y+6z=3900$$

Then, we have

$$A=\begin{bmatrix}15 & 5 & 10\\20 & 8 & 6\\5 & 5 & 6\end{bmatrix} \text{ and } B=\begin{bmatrix}4900\\6600\\3900\end{bmatrix}.$$

Using a calculator, we find that the inverse is

$$\begin{bmatrix}\frac{3}{70} & \frac{1}{21} & -\frac{5}{42}\\-\frac{3}{14} & \frac{2}{21} & \frac{11}{42}\\\frac{1}{7} & -\frac{5}{42} & \frac{1}{21}\end{bmatrix}.$$

$$X=A^{-1}B=\begin{bmatrix}\frac{3}{70} & \frac{1}{21} & -\frac{5}{42}\\-\frac{3}{14} & \frac{2}{21} & \frac{11}{42}\\\frac{1}{7} & -\frac{5}{42} & \frac{1}{21}\end{bmatrix}\begin{bmatrix}4900\\6600\\3900\end{bmatrix}=\begin{bmatrix}60\\600\\100\end{bmatrix}$$

Therefore, 60 buffets, 600 chairs, and 100 tables should be produced each week.

19. Let x = the wholesale price of jeans, let y = the wholesale price of jackets, let z = the wholesale price of sweaters, and let w = the wholesale price of shirts. The system is

$$3000x+3000y+2200z+4200w=507,650$$
$$2700x+2500y+2100z+4300w=459,075$$
$$5000x+2000y+1400z+7500w=541,225$$
$$7000x+1800y+600z+8000w=571,500$$

Store the following matrix as matrix A.

$$\left[\begin{array}{cccc|c}3000 & 3000 & 2200 & 4200 & 507,650\\2700 & 2500 & 2100 & 4300 & 459,075\\5000 & 2000 & 1400 & 7500 & 541,225\\7000 & 1800 & 600 & 8000 & 571,500\end{array}\right]$$

Using row operations, transform this matrix to obtain

$$\left[\begin{array}{cccc|c}1 & 0 & 0 & 0 & 34.5\\0 & 1 & 0 & 0 & 72\\0 & 0 & 1 & 0 & 44\\0 & 0 & 0 & 1 & 21.75\end{array}\right]$$

Therefore, the wholesale price for jeans is \$34.50, jacket is \$72, sweater is \$44, and a shirt is \$21.75.

21. The following solution presupposes the use of a TI-82 graphing calculator. Similar results can be obtained from other graphing calculators. Let x = number of bacterium 1, let y = number of bacterium 2, and let z = number of bacterium 3.

	Food		
	I	II	III
Bacterium 1	1.3	1.3	2.3
Bacterium 2	1.1	2.4	3.7
Bacterium 3	8.1	2.9	5.1
Totals	16,000	28,000	44,000

The data in the table produce the system of equations

$$1.3x+1.1y+8.1z=16,000$$
$$1.3x+2.4y+2.9z=28,000$$
$$2.3x+3.7y+5.1z=44,000.$$

We store the following matrix as matrix A:

$$\left[\begin{array}{ccc|c}1.3 & 1.1 & 8.1 & 16,000\\1.3 & 2.4 & 2.9 & 28,000\\2.3 & 3.7 & 5.1 & 44,000\end{array}\right].$$

Using row operations, we transform this to obtain

$$\left[\begin{array}{ccc|c}1 & 0 & 0 & 2339.74359\\0 & 1 & 0 & 10,128.20513\\0 & 0 & 1 & 224.3589744\end{array}\right]$$

Thus, 2340 of the first species, 10,128 of the second species, and 224 of the third species can be maintained.

23. $A=\begin{bmatrix}.1 & .03\\.07 & .6\end{bmatrix}, D=\begin{bmatrix}5\\10\end{bmatrix}$

First calculate $I-A$.

$$I-A=\begin{bmatrix}.9 & -.03\\-.07 & .4\end{bmatrix}$$

Use row operations to find the inverse of $I-A$.

$$(I-A)^{-1}=\begin{bmatrix}1.118 & .084\\.196 & 2.515\end{bmatrix}$$

Since $X=(I-A)^{-1}D$,

$$X=\begin{bmatrix}1.118 & .084\\.196 & 2.515\end{bmatrix}\begin{bmatrix}5\\10\end{bmatrix}=\begin{bmatrix}6.43\\26.12\end{bmatrix}.$$

Copyright © 2019 Pearson Education, Inc.

25. $A=\begin{bmatrix} .25 & .08 \\ .33 & .11 \end{bmatrix}, D=\begin{bmatrix} 690 \\ 920 \end{bmatrix}$

$(I-A)^{-1}=\begin{bmatrix} 1.388 & 0.1248 \\ 0.5147 & 1.1698 \end{bmatrix}$

$X=(I-A)^{-1}D=\begin{bmatrix} 1.388 & 0.1248 \\ 0.5147 & 1.1699 \end{bmatrix}\begin{bmatrix} 690 \\ 920 \end{bmatrix}$

$=\begin{bmatrix} 1073 \\ 1431 \end{bmatrix}$

Produce 1073 metric tons of wheat, and 1431 metric tons of oil.

27. $A=\begin{bmatrix} \frac{1}{4} & \frac{1}{6} \\ \frac{1}{2} & 0 \end{bmatrix}, I-A=\begin{bmatrix} \frac{3}{4} & -\frac{1}{6} \\ -\frac{1}{2} & 1 \end{bmatrix}$

$(I-A)^{-1}=\begin{bmatrix} \frac{3}{2} & \frac{1}{4} \\ \frac{3}{4} & \frac{9}{8} \end{bmatrix}$

a. $X=\begin{bmatrix} \frac{3}{2} & \frac{1}{4} \\ \frac{3}{4} & \frac{9}{8} \end{bmatrix}\begin{bmatrix} 1 \\ 1 \end{bmatrix}=\begin{bmatrix} \frac{7}{4} \\ \frac{15}{8} \end{bmatrix}$

Thus, $\frac{7}{4}$ bushes of yams and $\frac{15}{8}\approx 2$ pigs should be produced.

b. $X=\begin{bmatrix} \frac{3}{2} & \frac{1}{4} \\ \frac{3}{4} & \frac{9}{8} \end{bmatrix}\begin{bmatrix} 100 \\ 70 \end{bmatrix}=\begin{bmatrix} 167.5 \\ 153.75 \end{bmatrix}$

Thus, 167.5 bushels of yams and 153.75 ≈ 154 pigs should be produced.

29. First, write the input-output matrix A.

$A=\begin{bmatrix} .4 & .6 \\ .5 & .25 \end{bmatrix}=\begin{bmatrix} \frac{2}{5} & \frac{3}{5} \\ \frac{1}{2} & \frac{1}{4} \end{bmatrix}$

$I-A=\begin{bmatrix} 1 & 0 \\ 0 & 1 \end{bmatrix}-\begin{bmatrix} \frac{2}{5} & \frac{3}{5} \\ \frac{1}{2} & \frac{1}{4} \end{bmatrix}=\begin{bmatrix} \frac{3}{5} & -\frac{3}{5} \\ -\frac{1}{2} & \frac{3}{4} \end{bmatrix}$

Form $[I-A \mid I]$.

$\left[\begin{array}{cc|cc} \frac{3}{5} & -\frac{3}{5} & 1 & 0 \\ -\frac{1}{2} & \frac{3}{4} & 0 & 1 \end{array}\right]$

$\left[\begin{array}{cc|cc} 1 & -1 & \frac{5}{3} & 0 \\ -\frac{1}{2} & \frac{3}{4} & 0 & 1 \end{array}\right]\frac{5}{3}R_1$

$\left[\begin{array}{cc|cc} 1 & -1 & \frac{5}{3} & 0 \\ 0 & \frac{1}{4} & \frac{5}{6} & 1 \end{array}\right]\frac{1}{2}R_1+R_2$

$\left[\begin{array}{cc|cc} 1 & -1 & \frac{5}{3} & 0 \\ 0 & 1 & \frac{10}{3} & 4 \end{array}\right]4R_2$

$\left[\begin{array}{cc|cc} 1 & 0 & 5 & 4 \\ 0 & 1 & \frac{10}{3} & 4 \end{array}\right]R_2+R_1$

Thus, $(I-A)^{-1}=\begin{bmatrix} 5 & 4 \\ \frac{10}{3} & 4 \end{bmatrix}$

Since $D=\begin{bmatrix} 15 \text{ million} \\ 12 \text{ million} \end{bmatrix}$ and $X=(I-A)^{-1}D$,

$X=\begin{bmatrix} 5 & 4 \\ \frac{10}{3} & 4 \end{bmatrix}\begin{bmatrix} 15 \text{ million} \\ 12 \text{ million} \end{bmatrix}=\begin{bmatrix} 123 \text{ million} \\ 98 \text{ million} \end{bmatrix}.$

The output should be \$123 million of electricity and \$98 million of gas.

Copyright © 2019 Pearson Education, Inc.

31. a. From the input-output matrix, we see that .40 unit of agriculture, .12 unit of manufacturing, and 3.60 units of households are required for the manufacturing sector to produce 1 unit.

b. $A = \begin{bmatrix} .25 & .40 & .133 \\ .14 & .12 & .100 \\ .80 & 3.60 & .133 \end{bmatrix}$

$I - A = \begin{bmatrix} .75 & -.40 & -.133 \\ -.14 & .88 & -.100 \\ -.80 & -3.60 & .867 \end{bmatrix}$

Next, calculate $(I - A)^{-1}$.

$(I - A)^{-1} \approx \begin{bmatrix} 6.61 & 13.53 & 2.57 \\ 3.3 & 8.91 & 1.53 \\ 19.8 & 49.5 & 9.9 \end{bmatrix}$

$X = (I - A)^{-1} D$

$\approx \begin{bmatrix} 6.61 & 13.53 & 2.57 \\ 3.3 & 8.91 & 1.53 \\ 19.8 & 49.5 & 9.9 \end{bmatrix} \begin{bmatrix} 35 \\ 38 \\ 40 \end{bmatrix} \approx \begin{bmatrix} 848 \\ 516 \\ 2970 \end{bmatrix}$

Therefore, 848 units of agriculture, 516 units of manufacturing, and 2970 units of households need to be produced.

c. Since .25 units of agriculture are used in producing each unit of agriculture, (848)(.25) = 212 units are used. Since .4 units of agriculture are used in producing each unit of manufacturing, (.4)(516) = 206.4 units are used. Since .133 units of agriculture are used in producing each unit of households, (.133)(2970) = 395.01. Thus, 212 + 206 + 395 = 813 units of agriculture are used in the production process.

33. a. The energy sector requires .017 unit of manufacturing and .216 unit of energy to produce one unit.

b. We are given the input-output matrix A and the production matrix X.

$A = \begin{bmatrix} .293 & 0 & 0 \\ .014 & .207 & .017 \\ .044 & .010 & .216 \end{bmatrix}, \ X = \begin{bmatrix} 175{,}000 \\ 22{,}000 \\ 12{,}000 \end{bmatrix}$

$D = (I - A)X$

$= \begin{bmatrix} .707 & 0 & 0 \\ -.014 & .793 & -.017 \\ .044 & -.010 & .784 \end{bmatrix} \begin{bmatrix} 175{,}000 \\ 22{,}000 \\ 12{,}000 \end{bmatrix}$

$= \begin{bmatrix} 123{,}725 \\ 14{,}792 \\ 1488 \end{bmatrix}$

Therefore, about 123,725,000 pounds of agriculture, 14,792,000 pounds of manufacturing, and 1,488,000 pounds of energy is available to satisfy demand.

c. $X = (I - A)^{-1} D$

$\approx \begin{bmatrix} 1.414 & 0 & 0 \\ .027 & 1.261 & .027 \\ .080 & .016 & 1.276 \end{bmatrix} \begin{bmatrix} 138{,}213 \\ 17{,}597 \\ 1786 \end{bmatrix}$

$\approx \begin{bmatrix} 195{,}492 \\ 25{,}933 \\ 13{,}580 \end{bmatrix}$

Therefore, about 195,492,000 pounds of agriculture, 25,933,000 pounds of manufacturing, and 13,580,000 pounds of energy should be produced.

Copyright © 2019 Pearson Education, Inc.

35. $A = \begin{bmatrix} .1045 & .0428 & .0029 & .0031 \\ .0826 & .1087 & .0584 & .0321 \\ .0867 & .1019 & .2032 & .3555 \\ .6253 & .3448 & .6106 & .0798 \end{bmatrix}, D = \begin{bmatrix} 450 \\ 300 \\ 125 \\ 100 \end{bmatrix}$

$$(I-A)^{-1} \approx \begin{bmatrix} 1.133 & .062 & .019 & .013 \\ .202 & 1.19 & .171 & .108 \\ .748 & .536 & 1.87 & .742 \\ 1.343 & .845 & 1.315 & 1.63 \end{bmatrix}$$

$$X = (I-A)^{-1} D$$

$$X \approx \begin{bmatrix} 1.133 & .062 & .019 & .013 \\ .202 & 1.19 & .171 & .108 \\ .748 & .536 & 1.87 & .742 \\ 1.343 & .845 & 1.315 & 1.63 \end{bmatrix} \begin{bmatrix} 450 \\ 300 \\ 125 \\ 100 \end{bmatrix} \approx \begin{bmatrix} 532 \\ 481 \\ 805 \\ 1185 \end{bmatrix}$$

This means $532 million of natural resources, $481 million of manufacturing, $805 million of trade and services, and $1185 million of personal consumption.

37. The message *Anne is home* broken into groups of 2 letters would have the following matrices:

$$\begin{bmatrix} 1 \\ 14 \end{bmatrix}, \begin{bmatrix} 14 \\ 5 \end{bmatrix}, \begin{bmatrix} 27 \\ 9 \end{bmatrix}, \begin{bmatrix} 19 \\ 27 \end{bmatrix}, \begin{bmatrix} 8 \\ 15 \end{bmatrix}, \begin{bmatrix} 13 \\ 5 \end{bmatrix}.$$

Multiply $M = \begin{bmatrix} 1 & 3 \\ 2 & 7 \end{bmatrix}$ by each matrix in the code to obtain:

$$\begin{bmatrix} 43 \\ 100 \end{bmatrix}, \begin{bmatrix} 29 \\ 63 \end{bmatrix}, \begin{bmatrix} 54 \\ 117 \end{bmatrix}, \begin{bmatrix} 100 \\ 227 \end{bmatrix}, \begin{bmatrix} 53 \\ 121 \end{bmatrix}, \begin{bmatrix} 28 \\ 61 \end{bmatrix}.$$

39. $\begin{bmatrix} 57 \\ 129 \end{bmatrix}, \begin{bmatrix} 37 \\ 79 \end{bmatrix}.$

Multiply $M^{-1} = \begin{bmatrix} 7 & -3 \\ -2 & 1 \end{bmatrix}$ by each matrix in the code to obtain:

$$\begin{bmatrix} 12 \\ 15 \end{bmatrix}, \begin{bmatrix} 22 \\ 5 \end{bmatrix}$$

The message is: *Love.*

41. $A = \begin{bmatrix} 0 & 1 & 2 & 2 \\ 1 & 0 & 1 & 0 \\ 2 & 1 & 0 & 1 \\ 2 & 0 & 1 & 0 \end{bmatrix}$

$$A^2 = \begin{bmatrix} 9 & 2 & 3 & 2 \\ 2 & 2 & 2 & 3 \\ 3 & 2 & 6 & 4 \\ 2 & 3 & 4 & 5 \end{bmatrix}$$

a. The number of ways to travel from city 1 to city 3 by passing through exactly one city is the entry in row 1, column 3 of A^2, which is 3.

b. The number of ways to travel from city 2 to city 4 by passing through exactly one city is the entry in row 2, column 4 of A^2, which is 3.

c. The number of ways to travel from city 1 to city 3 by passing through at most one city is the sum of the entries in row 1, column 3 of A and A^2, which is 2 + 3 or 5.

d. The number of ways to travel from city 2 to city 4 by passing through at most one city is the sum of the entries in row 2, column 4 of A and A^2, which is 0 + 3 or 3.

43. a.

$$C = \begin{array}{c|cccc} & d & r & c & m \\ \hline d & 0 & 1 & 1 & 1 \\ r & 0 & 0 & 0 & 1 \\ c & 0 & 1 & 0 & 1 \\ m & 0 & 0 & 0 & 0 \end{array}$$

b. $C^2 = \begin{bmatrix} 0 & 1 & 1 & 1 \\ 0 & 0 & 0 & 1 \\ 0 & 1 & 0 & 1 \\ 0 & 0 & 0 & 0 \end{bmatrix} \begin{bmatrix} 0 & 1 & 1 & 1 \\ 0 & 0 & 0 & 1 \\ 0 & 1 & 0 & 1 \\ 0 & 0 & 0 & 0 \end{bmatrix}$

$$= \begin{bmatrix} 0 & 1 & 0 & 2 \\ 0 & 0 & 0 & 0 \\ 0 & 0 & 0 & 1 \\ 0 & 0 & 0 & 0 \end{bmatrix}$$

C^2 gives the number of food sources once removed from the feeder. Thus, since dogs eat rats and rats eat mice, mice are an indirect as well as a direct food source of dogs.

Copyright © 2019 Pearson Education, Inc.

Chapter 6 Review Exercises

1. $-5x - 3y = -3 \quad (1)$
$2x + y = 4 \quad (2)$
Multiply equation (2) by 3 and add to equation (1).

$$\begin{aligned} -5x - 3y &= -3 \\ \underline{6x + 3y} &\underline{= 12} \\ x \quad\quad &= 9 \end{aligned}$$

Substitute 9 for x in equation (2) to solve for y.
$2(9) + y = 4 \Rightarrow y = 14$
The solution is $(9, -14)$.

2. $3x - y = 8 \quad (1)$
$2x + 3y = 6 \quad (2)$
Multiply equation (1) by 3 and add to equation (2).

$$\begin{aligned} 9x - 3y &= 24 \\ \underline{2x + 3y} &\underline{= 6} \\ 11x \quad\quad &= 30 \Rightarrow x = \frac{30}{11} \end{aligned}$$

Substitute $\frac{30}{11}$ for x in equation (1) to solve for y.

$$3\left(\frac{30}{11}\right) - y = 8 \Rightarrow y = \frac{2}{11}$$

The solution is $\left(\frac{30}{11}, \frac{2}{11}\right)$.

3. $3x - 5y = 16 \quad (1)$
$2x + 3y = -2 \quad (2)$
Multiply equation (1) by 2 and equation (2) by -3 and add the results.

$$\begin{aligned} 6x - 10y &= 32 \\ \underline{-6x - 9y} &\underline{= 6} \\ -19y &= 38 \Rightarrow y = -2 \end{aligned}$$

Substitute -2 for y in equation (1) to solve for x.
$3x - 5(-2) = 16 \Rightarrow 3x = 6 \Rightarrow x = 2$
The solution is $(2, -2)$.

4. $\frac{1}{4}x - \frac{1}{3}y = -\frac{1}{4} \quad (1)$
$\frac{1}{10}x + \frac{2}{5}y = \frac{2}{5} \quad (2)$
Multiply equation (1) by 12 and equation (2) by 10 and add the results.

$$\begin{aligned} 3x - 4y &= -3 \quad (3) \\ \underline{x + 4y} &\underline{= 4} \quad (4) \\ 4x \quad\quad &= 1 \Rightarrow x = \frac{1}{4} \end{aligned}$$

Substitute $\frac{1}{4}$ for x in equation (4) to solve for y.

$$\frac{1}{4} + 4y = 4 \Rightarrow 4y = \frac{15}{4} \Rightarrow y = \frac{15}{16}$$

The solution is $\left(\frac{1}{4}, \frac{15}{16}\right)$.

5. Let x = the number of shares of the first stock, and let y = the number of shares of the second stock.
We will solve the following system.
$32x + 23y = 10,100 \quad (1)$
$1.2x + 1.4y = 540 \quad (2)$
Multiply equation (1) by 1.2 and equation (2) by -32 and add the results.

$$\begin{aligned} 38.4x + 27.6y &= 12,120 \\ \underline{-38.4x - 44.8y} &\underline{= -17,280} \\ -17.2y &= -5160 \Rightarrow y = 300 \end{aligned}$$

Substitute 300 for y in equation (2) and solve for x.
$1.2x + 1.4(300) = 540 \Rightarrow 1.2x = 120 \Rightarrow x = 100$
He should buy 100 shares of the first stock and 300 shares of the second stock.

Copyright © 2019 Pearson Education, Inc.

6. Let x = the amount invested in the first fund, and let y = the amount invested in the second fund. We will solve the system

$$.08x + .02y = 780 \quad (1)$$
$$.1x + .01y = 810 \quad (2)$$

Multiply equation (2) by –2 and add to equation (1).

$$\begin{aligned} .08x + .02y &= 780 \\ -.2x - .02y &= -1620 \\ \hline -.12x \qquad &= -840 \Rightarrow x = 7000 \end{aligned}$$

Substitute 7000 for x in equation (1) and solve for y.

$.08(7000) + .02y = 780 \Rightarrow .02y = 220 \Rightarrow$
$y = 11{,}000$

Maria has $7000 invested in the first fund and $11,000 in the second.

7.
$$x - 2y = 1 \quad (1)$$
$$4x + 4y = 2 \quad (2)$$
$$10x + 8y = 4 \quad (3)$$

Multiply equation (1) by 2 and add to equation (2).

$$\begin{aligned} 2x - 4y &= 2 \\ 4x + 4y &= 2 \\ \hline 6x \qquad &= 4 \Rightarrow x = \frac{2}{3} \end{aligned}$$

Substitute $\frac{2}{3}$ for x in equation (1) to solve for y.

$$\frac{2}{3} - 2y = 1 \Rightarrow -2y = \frac{1}{3} \Rightarrow y = -\frac{1}{6}$$

The ordered pair $\left(\frac{2}{3}, -\frac{1}{6}\right)$ satisfies equations (1) and (2), but not equation (3). The system is inconsistent; there is no solution.

8.
$$4x - y - 2z = 4 \quad (1)$$
$$x - y - \frac{1}{2}z = 1 \quad (2)$$
$$2x - y - z = 8 \quad (3)$$

Multiply equation (2) by –4 and add to equation (1).

$$\begin{aligned} 4x - y - 2z &= 4 \quad (1) \\ -4x + 4y + 2z &= -4 \quad (4) \\ \hline 3y &= 0 \Rightarrow y = 0 \end{aligned}$$

Substitute 0 for y in equations (1) and (3).

$$4x - 2z = 4 \quad (5)$$
$$2x - z = 8 \quad (6)$$

Multiply equation (6) by –2 and add to equation (5).

$$\begin{aligned} 4x - 2z &= 4 \quad (5) \\ -4x + 2z &= -16 \quad (7) \\ \hline 0 &= -12 \quad (8) \end{aligned}$$

Since equation (8) is false, this system is inconsistent; there is no solution.

9.
$$3x + y - z = 3 \quad (1)$$
$$x + 2z = 6 \quad (2)$$
$$-3x - y + 2z = 9 \quad (3)$$

Add equations (1) and (3).

$$\begin{aligned} 3x + y - z &= 3 \\ -3x - y + 2z &= 9 \\ \hline z &= 12 \end{aligned}$$

Substitute 12 for z in equation (2) and solve for x.

$x + 2(12) = 6 \Rightarrow x = -18$

Substitute –18 for x and 12 for z in equation (1) and solve for y.

$3(-18) + y - 12 = 3 \Rightarrow y = 69$

The solution is (–18, 69, 12).

Copyright © 2019 Pearson Education, Inc.

10. $x+y-4z=0 \quad (1)$
$2x+y-3z=2 \quad (2)$

Multiply equation (1) by −1 and add to equation (2).

$$\begin{array}{rl} -x-y+4z=0 & (3) \\ 2x+y-3z=2 & (2) \\ \hline x \quad\quad +z=2 & (4) \end{array}$$

Solve equation (4) for x in terms of z.
$x+z=2 \Rightarrow x=2-z$

Substitute $2-z$ for x in equation (1) and solve for y in terms of z.

$$(2-z)+y-4z=0$$
$$2+y-5z=0$$
$$y=5z-2$$

The system is dependent, and the solution is all ordered triples of the form $(2-z, 5z-2, z)$ for any real number z.

We solve exercises 11–16 using matrix methods.

11. $x+z=-3$
$y-z=6$
$2x+3z=5$

The augmented matrix is

$$\left[\begin{array}{ccc|c} 1 & 0 & 1 & -3 \\ 0 & 1 & -1 & 6 \\ 2 & 0 & 3 & 5 \end{array}\right]$$

$$\left[\begin{array}{ccc|c} 1 & 0 & 1 & -3 \\ 0 & 1 & -1 & 6 \\ 0 & 0 & 1 & 11 \end{array}\right] -2R_1+R_3$$

$$\left[\begin{array}{ccc|c} 1 & 0 & 0 & -14 \\ 0 & 1 & 0 & 17 \\ 0 & 0 & 1 & 11 \end{array}\right] \begin{array}{l} -1R_3+R_1 \\ R_3+R_2 \\ \\ \end{array}$$

The solution is (−14, 17, 11).

12. $2x+3y+4z=8$
$-x+y-2z=-9$
$2x+2y+6z=16$

The augmented matrix is

$$\left[\begin{array}{ccc|c} 2 & 3 & 4 & 8 \\ -1 & 1 & -2 & -9 \\ 2 & 2 & 6 & 16 \end{array}\right]$$

$$\left[\begin{array}{ccc|c} 2 & 3 & 4 & 8 \\ -1 & 1 & -2 & -9 \\ 1 & 1 & 3 & 8 \end{array}\right] \begin{array}{l} \\ \\ \frac{1}{2}R_3 \end{array}$$

$$\left[\begin{array}{ccc|c} 1 & 1 & 3 & 8 \\ -1 & 1 & -2 & -9 \\ 2 & 3 & 4 & 8 \end{array}\right] \text{Interchange } R_1 \text{ and } R_3$$

$$\left[\begin{array}{ccc|c} 1 & 1 & 3 & 8 \\ 0 & 2 & 1 & -1 \\ 0 & 1 & -2 & -8 \end{array}\right] \begin{array}{l} \\ R_1+R_2 \\ -2R_1+R_3 \end{array}$$

$$\left[\begin{array}{ccc|c} 1 & 0 & 5 & 16 \\ 0 & 5 & 0 & -10 \\ 0 & 1 & -2 & -8 \end{array}\right] \begin{array}{l} R_1-R_3 \\ 2R_2+R_3 \\ \\ \end{array}$$

$$\left[\begin{array}{ccc|c} 1 & 0 & 5 & 16 \\ 0 & 1 & 0 & -2 \\ 0 & 1 & -2 & -8 \end{array}\right] \begin{array}{l} \\ \frac{1}{5}R_2 \\ \\ \end{array}$$

$$\left[\begin{array}{ccc|c} 1 & 0 & 5 & 16 \\ 0 & 1 & 0 & -2 \\ 0 & 0 & -2 & -6 \end{array}\right] \begin{array}{l} \\ \\ -R_2+R_3 \end{array}$$

$$\left[\begin{array}{ccc|c} 1 & 0 & 5 & 16 \\ 0 & 1 & 0 & -2 \\ 0 & 0 & 1 & 3 \end{array}\right] \begin{array}{l} \\ \\ -\frac{1}{2}R_3 \end{array}$$

$$\left[\begin{array}{ccc|c} 1 & 0 & 0 & 1 \\ 0 & 1 & 0 & -2 \\ 0 & 0 & 1 & 3 \end{array}\right] \begin{array}{l} -5R_3+R_1 \\ \\ \\ \end{array}$$

The solution is (1, −2, 3).

Copyright © 2019 Pearson Education, Inc.

13. $x-2y+5z=3$

$4x+3y-4z=1$

$3x+5y-9z=7$

The augmented matrix is

$$\left[\begin{array}{rrr|r} 1 & -2 & 5 & 3 \\ 4 & 3 & -4 & 1 \\ 3 & 5 & -9 & 7 \end{array}\right]$$

$$\left[\begin{array}{rrr|r} 1 & -2 & 5 & 3 \\ 0 & 11 & -24 & -11 \\ 0 & 11 & -24 & -2 \end{array}\right]\begin{array}{l} \\ -4R_1+R_2 \\ -3R_1+R_2 \end{array}$$

$$\left[\begin{array}{rrr|r} 1 & -2 & 5 & 3 \\ 0 & 11 & -24 & -11 \\ 0 & 0 & 0 & 9 \end{array}\right]\begin{array}{l} \\ \\ -R_2+R_3 \end{array}$$

Since row 3 has all zeros except for the last entry, there is no solution.

14. $5x-8y+z=1$

$3x-2y+4z=3$

$10x-16y+2z=3$

The augmented matrix is

$$\left[\begin{array}{rrr|r} 5 & -8 & 1 & 1 \\ 3 & -2 & 4 & 3 \\ 10 & -16 & 2 & 3 \end{array}\right]$$

$$\left[\begin{array}{rrr|r} 1 & -\frac{8}{5} & \frac{1}{5} & \frac{1}{5} \\ 3 & -2 & 4 & 3 \\ 10 & -16 & 2 & 3 \end{array}\right]\begin{array}{l} \frac{1}{5}R_1 \\ \\ \\ \end{array}$$

$$\left[\begin{array}{rrr|r} 1 & -\frac{8}{5} & \frac{1}{5} & \frac{1}{5} \\ 0 & \frac{14}{5} & \frac{17}{5} & \frac{12}{5} \\ 0 & 0 & 0 & 1 \end{array}\right]\begin{array}{l} \\ -3R_1+R_2 \\ -10R_1+R_3 \end{array}$$

Since row 3 has all zeros except for the last entry, there is no solution.

15. $x-2y+3z=4$

$2x+y-4z=3$

$-3z+4y-z=-2$

The augmented matrix is

$$\left[\begin{array}{rrr|r} 1 & -2 & 3 & 4 \\ 2 & 1 & -4 & 3 \\ -3 & 4 & -1 & -2 \end{array}\right]$$

$$\left[\begin{array}{rrr|r} 1 & -2 & 3 & 4 \\ 0 & 5 & -10 & -5 \\ 0 & -2 & 8 & 10 \end{array}\right]\begin{array}{l} \\ -2R_1+R_2 \\ 3R_1+R_3 \end{array}$$

$$\left[\begin{array}{rrr|r} 1 & -2 & 3 & 4 \\ 0 & 1 & -2 & -1 \\ 0 & -2 & 8 & 10 \end{array}\right]\begin{array}{l} \\ \frac{1}{5}R_2 \\ \\ \end{array}$$

$$\left[\begin{array}{rrr|r} 1 & 0 & -1 & 2 \\ 0 & 1 & -2 & -1 \\ 0 & 0 & 4 & 8 \end{array}\right]\begin{array}{l} 2R_2+R_1 \\ \\ 2R_2+R_3 \end{array}$$

$$\left[\begin{array}{rrr|r} 1 & 0 & -1 & 2 \\ 0 & 1 & -2 & -1 \\ 0 & 0 & 1 & 2 \end{array}\right]\begin{array}{l} \\ \\ \frac{1}{4}R_3 \end{array}$$

$$\left[\begin{array}{rrr|r} 1 & 0 & 0 & 4 \\ 0 & 1 & 0 & 3 \\ 0 & 0 & 1 & 2 \end{array}\right]\begin{array}{l} R_3+R_1 \\ 2R_3+R_2 \\ \\ \end{array}$$

The solution is (4, 3, 2).

Copyright © 2019 Pearson Education, Inc.

16. $3x+2y-6z=3$

$x+y+2z=2$

$2x+2y+5z=0$

The augmented matrix is

$$\left[\begin{array}{ccc|c} 3 & 2 & -6 & 3 \\ 1 & 1 & 2 & 2 \\ 2 & 2 & 5 & 0 \end{array}\right]$$

Interchange R_1 and R_2.

$$\left[\begin{array}{ccc|c} 1 & 1 & 2 & 2 \\ 3 & 2 & -6 & 3 \\ 2 & 2 & 5 & 0 \end{array}\right]$$

$$\left[\begin{array}{ccc|c} 1 & 1 & 2 & 2 \\ 0 & -1 & -12 & -3 \\ 0 & 0 & 1 & -4 \end{array}\right] \begin{array}{l} \\ -3R_1+R_2 \\ -2R_1+R_3 \end{array}$$

$$\left[\begin{array}{ccc|c} 1 & 1 & 2 & 2 \\ 0 & 1 & 12 & 3 \\ 0 & 0 & 1 & -4 \end{array}\right] \begin{array}{l} \\ -1R_2 \\ \\ \end{array}$$

$$\left[\begin{array}{ccc|c} 1 & 0 & -10 & -1 \\ 0 & 1 & 12 & 3 \\ 0 & 0 & 1 & -4 \end{array}\right] \begin{array}{l} -1R_2+R_1 \\ \\ \\ \end{array}$$

$$\left[\begin{array}{ccc|c} 1 & 0 & 0 & -41 \\ 0 & 1 & 0 & 51 \\ 0 & 0 & 1 & -4 \end{array}\right] \begin{array}{l} 10R_3+R_1 \\ -12R_3+R_2 \\ \\ \end{array}$$

The solution is (–41, 51, –4).

Exercises 17–20 can be solved using either the elimination method or matrix methods.

17. Let x = the number of one dollar bills, let y = the number of five dollar bills, and let z = the number of ten dollar bills.

From the given information, we have the system

$$x+y+z=35 \quad (1)$$
$$x+5y+10z=144 \quad (2)$$
$$z=y+2 \quad (3)$$

Substitute $y+2$ for z in equation (1) to get

$$x+y+y+2=35$$
$$x+2y+2=35$$
$$x+2y=33 \quad (4)$$

Substitute $y+2$ for z in equation (2) to get

$$x+5y+10(y+2)=144$$
$$x+5y+10y+20=144$$
$$x+15y=124 \quad (5)$$

Multiply equation (4) by –1 and add to equation (5).

$$-x-2y=-33$$
$$\underline{x+15y=\ 124}$$
$$13y=\ 91 \Rightarrow y=7$$

Substitute 7 for y in equation (4) and solve for x.

$x+2(7)=33 \Rightarrow x=19$

Substitute 7 for y in equation (3) and solve for z.

$z=7+2=9$

The solution is 19 ones, 7 fives, and 9 tens.

18. Organize the information into a table.

	I	II	III
Food	100	200	150
Shelter	250	0	200
Counseling	0	100	100

Let x = the number of clients from source I, let y = the number of clients from source II, and let z = the number of clients from source III.

The system is

$$100x+200y+150z=50,000$$
$$250x \qquad +200z=32,500$$
$$100y+100z=25,000$$

Using matrix methods, we have

$$\left[\begin{array}{ccc|c} 100 & 200 & 150 & 50,000 \\ 250 & 0 & 200 & 32,500 \\ 0 & 100 & 100 & 25,000 \end{array}\right]$$

$$\left[\begin{array}{ccc|c} 1 & 2 & \frac{3}{2} & 500 \\ 1 & 0 & \frac{4}{5} & 130 \\ 0 & 1 & 1 & 250 \end{array}\right] \begin{array}{l} \frac{1}{100}R_1 \\ \frac{1}{250}R_2 \\ \frac{1}{100}R_3 \end{array}$$

$$\left[\begin{array}{ccc|c} 1 & 0 & -\frac{1}{2} & 0 \\ 0 & 2 & \frac{7}{10} & 370 \\ 0 & 1 & 1 & 250 \end{array}\right] \begin{array}{l} R_1-2R_3 \\ R_1-R_2 \\ \\ \end{array}$$

$$\left[\begin{array}{ccc|c} 1 & 0 & -\frac{1}{2} & 0 \\ 0 & 2 & \frac{7}{10} & 370 \\ 0 & 0 & \frac{13}{10} & 130 \end{array}\right] \begin{array}{l} \\ \\ 2R_3-R_2 \end{array}$$

$$\left[\begin{array}{ccc|c} 1 & 0 & -\frac{1}{2} & 0 \\ 0 & 2 & \frac{7}{10} & 370 \\ 0 & 0 & 1 & 100 \end{array}\right] \begin{array}{l} \\ \\ \frac{10}{13}R_3 \end{array}$$

Copyright © 2019 Pearson Education, Inc.

$$\begin{bmatrix} 1 & 0 & 0 & 50 \\ 0 & 2 & 0 & 300 \\ 0 & 0 & 1 & 100 \end{bmatrix} \begin{matrix} R_1 + \frac{1}{2}R_3 \\ R_2 - \frac{7}{10}R_3 \\ \\ \end{matrix}$$

$$\begin{bmatrix} 1 & 0 & 0 & 50 \\ 0 & 1 & 0 & 150 \\ 0 & 0 & 1 & 100 \end{bmatrix} \begin{matrix} \\ \frac{1}{2}R_2 \\ \\ \end{matrix}$$

Thus, the agency can serve 50 clients from source I, 150 from source II, and 100 from source III.

19. Let x = the number of blankets, let y = the number of rugs, and let z = the number of skirts. Solve the following system.

$$24x + 30y + 12z = 306 \quad (1)$$
$$4x + 5y + 3z = 59 \quad (2)$$
$$15x + 18y + 9z = 201 \quad (3)$$

Simplify the system by dividing equation (1) by 6 and equation (3) by 3.

$$4x + 5y + 2z = 51 \quad (4)$$
$$4x + 5y + 3z = 59 \quad (2)$$
$$5x + 6y + 3z = 67 \quad (5)$$

Multiply equation (4) by –1 and add to equation (2).

$$-4x - 5y - 2z = -51$$
$$\underline{4x + 5y + 3z = \ 59}$$
$$z = \ 8$$

Substitute 8 for z in equation (4).

$$4x + 5y + 2(8) = 51$$
$$4x + 5y = 35 \quad (5)$$

Substitute 8 for z in equation (5).

$$5x + 6y + 3(8) = 67$$
$$5x + 6y = 43 \quad (6)$$

Multiply equation (5) by 5 and equation (6) by –4 and add the results.

$$20x + 25y = 175$$
$$\underline{-20x - 24y = -172}$$
$$y = \ 3$$

Substitute 3 for y in equation (5) and solve for x.

$$4x + 5(3) = 35 \Rightarrow 4x = 20 \Rightarrow x = 5$$

They can make 5 blankets, 3 rugs, and 8 skirts.

20. Let x = the number of chairs, and let y = number tables, and let z = number of chests.

	Construction	Painting	Packing
Chair	2	1	2
Table	4	3	3
Chest	8	6	4
Totals	2000	1400	1300

$$2x + 4y + 8z = 2000$$
$$x + 3y + 6z = 1400$$
$$2x + 3y + 4z = 1300$$

Divide the first equation by 2 and write the augmented matrix.

$$\begin{bmatrix} 1 & 2 & 4 & 1000 \\ 1 & 3 & 6 & 1400 \\ 2 & 3 & 4 & 1300 \end{bmatrix}$$

$$\begin{bmatrix} 1 & 2 & 4 & 1000 \\ 0 & 1 & 2 & 400 \\ 0 & -1 & -4 & -700 \end{bmatrix} \begin{matrix} \\ -1R_1 + R_2 \\ -2R_1 + R_3 \end{matrix}$$

$$\begin{bmatrix} 1 & 0 & 0 & 200 \\ 0 & 1 & 2 & 400 \\ 0 & 0 & -2 & -300 \end{bmatrix} \begin{matrix} -2R_2 + R_1 \\ \\ R_2 + R_3 \end{matrix}$$

$$\begin{bmatrix} 1 & 0 & 0 & 200 \\ 0 & 1 & 0 & 100 \\ 0 & 0 & -2 & -300 \end{bmatrix} \begin{matrix} \\ R_3 + R_2 \\ \\ \end{matrix}$$

$$\begin{bmatrix} 1 & 0 & 0 & 200 \\ 0 & 1 & 0 & 100 \\ 0 & 0 & 1 & 150 \end{bmatrix} \begin{matrix} \\ \\ -\frac{1}{2}R_3 \end{matrix}$$

The factory can produce 200 chairs, 100 tables, and 150 chests.

21. $\begin{bmatrix} 2 & 3 \\ 5 & 9 \end{bmatrix}$

The matrix is 2×2. Since the matrix is 2×2, it is square.

22. $\begin{bmatrix} 2 & -1 \\ 4 & 6 \\ 5 & 7 \end{bmatrix}$

The size of this matrix is 3×2.

23. $\begin{bmatrix} 12 & 4 & -8 & -1 \end{bmatrix}$

The matrix is 1×4. The matrix is a row matrix.

Copyright © 2019 Pearson Education, Inc.

24. $\begin{bmatrix} -7 & 5 & 6 & 4 \\ 3 & 2 & -1 & 2 \\ -1 & 12 & 8 & -1 \end{bmatrix}$

This is a 3×4 matrix.

25. $\begin{bmatrix} 6 & 8 & 10 \\ 5 & 3 & -2 \end{bmatrix}$

This matrix is 2×3.

26. $\begin{bmatrix} -9 \\ 15 \\ 4 \end{bmatrix}$

This matrix is 3×1. It is a column matrix.

27. As a 3 x 3 matrix, the data is

$$\begin{bmatrix} 162.45 & 164.44 & 163.64 \\ 31.88 & 31.94 & 31.86 \\ 51.66 & 52.19 & 51.95 \end{bmatrix}$$

28. $\begin{bmatrix} 8 & 8 & 8 \\ 10 & 5 & 9 \\ 7 & 10 & 7 \\ 8 & 9 & 7 \end{bmatrix}$

29. $B = \begin{bmatrix} 1 & 2 & -3 \\ 2 & 3 & 0 \\ 0 & 1 & 4 \end{bmatrix}$; $-B = \begin{bmatrix} -1 & -2 & 3 \\ -2 & -3 & 0 \\ 0 & -1 & -4 \end{bmatrix}$

30. $D = \begin{bmatrix} 6 \\ 1 \\ 0 \end{bmatrix}$; $-D = \begin{bmatrix} -6 \\ -1 \\ 0 \end{bmatrix}$

31. $A = \begin{bmatrix} 4 & 6 \\ -2 & -2 \\ 5 & 9 \end{bmatrix}, C = \begin{bmatrix} 5 & 0 \\ -1 & 3 \\ 4 & 7 \end{bmatrix}$

$A + 2C$

$$A + 2C = \begin{bmatrix} 4 & 6 \\ -2 & -2 \\ 5 & 9 \end{bmatrix} + \begin{bmatrix} 10 & 0 \\ -2 & 6 \\ 8 & 14 \end{bmatrix} = \begin{bmatrix} 14 & 6 \\ -4 & 4 \\ 13 & 23 \end{bmatrix}$$

32. $F = \begin{bmatrix} -1 & 2 \\ 6 & 7 \end{bmatrix}, G = \begin{bmatrix} 2 & 5 \\ 1 & 6 \end{bmatrix}$

$$F + 3G = \begin{bmatrix} -1 & 2 \\ 6 & 7 \end{bmatrix} + \begin{bmatrix} 6 & 15 \\ 3 & 18 \end{bmatrix} = \begin{bmatrix} 5 & 17 \\ 9 & 25 \end{bmatrix}$$

33. $B = \begin{bmatrix} 1 & 2 & -3 \\ 2 & 3 & 0 \\ 0 & 1 & 4 \end{bmatrix}, C = \begin{bmatrix} 5 & 0 \\ -1 & 3 \\ 4 & 7 \end{bmatrix}$

$2B - 5C$ is not defined, since the matrices have different sizes.

34. $D = \begin{bmatrix} 6 \\ 1 \\ 0 \end{bmatrix}, E = \begin{bmatrix} 1 & 3 & -4 \end{bmatrix}$

$D + E$ is not defined, since the matrices have different sizes.

35. $A = \begin{bmatrix} 4 & 6 \\ -2 & -2 \\ 5 & 9 \end{bmatrix}, C = \begin{bmatrix} 5 & 0 \\ -1 & 3 \\ 4 & 7 \end{bmatrix}$

$3A - 2C$

$$3A - 2C = \begin{bmatrix} 12 & 18 \\ -6 & -6 \\ 15 & 27 \end{bmatrix} - \begin{bmatrix} 10 & 0 \\ -2 & 6 \\ 8 & 14 \end{bmatrix} = \begin{bmatrix} 2 & 18 \\ -4 & -12 \\ 7 & 13 \end{bmatrix}$$

36. $G = \begin{bmatrix} 2 & 5 \\ 1 & 6 \end{bmatrix}, F = \begin{bmatrix} -1 & 2 \\ 6 & 7 \end{bmatrix}$

$$G - 2F = \begin{bmatrix} 2 & 5 \\ 1 & 6 \end{bmatrix} - \begin{bmatrix} -2 & 4 \\ 12 & 14 \end{bmatrix} = \begin{bmatrix} 4 & 1 \\ -11 & -8 \end{bmatrix}$$

Copyright © 2019 Pearson Education, Inc.

37. The 3 × 2 matrix that represents the first day's data is:

$$\begin{bmatrix} -.08 & 25.1 \\ 1.50 & 12.6 \\ .65 & 3.1 \end{bmatrix}$$

The 3 × 2 matrix that represents the next day's data is:

$$\begin{bmatrix} .10 & 27.2 \\ .63 & 12.4 \\ .47 & 2.9 \end{bmatrix}$$

The sum of these matrices is:

$$\begin{bmatrix} .02 & 52.3 \\ -.87 & 25 \\ 1.12 & 6 \end{bmatrix}$$

38. a. First shipment:

	Tulsa	New Orleans
Chicago	110,000	85,000
Dallas	73,000	108,000
Atlanta	95,000	69,000

Second shipment:

$$\begin{bmatrix} 58,000 & 40,000 \\ 33,000 & 52,000 \\ 80,000 & 30,000 \end{bmatrix}$$

b. Add the 2 matrices in part (a).

$$\begin{bmatrix} 168,000 & 125,000 \\ 106,000 & 160,000 \\ 175,000 & 99,000 \end{bmatrix}$$

39. $A = \begin{bmatrix} 4 & 6 \\ -2 & -2 \\ 5 & 9 \end{bmatrix}, G = \begin{bmatrix} 2 & 5 \\ 1 & 6 \end{bmatrix}$

$$AG = \begin{bmatrix} (4)(2)+(6)(1) & (4)(5)+(6)(6) \\ (-2)(2)+(-2)(1) & (-2)(5)+(-2)(6) \\ (5)(2)+(9)(1) & (5)(5)+(9)(6) \end{bmatrix} = \begin{bmatrix} 14 & 56 \\ -6 & -22 \\ 19 & 79 \end{bmatrix}$$

40. $E = [1 \quad 3 \quad -4], B = \begin{bmatrix} 1 & 2 & -3 \\ 2 & 3 & 0 \\ 0 & 1 & 4 \end{bmatrix}$

$EB = [7 \quad 7 \quad -19]$

41. $G = \begin{bmatrix} 2 & 5 \\ 1 & 6 \end{bmatrix}, A = \begin{bmatrix} 4 & 6 \\ -2 & -2 \\ 5 & 9 \end{bmatrix}$

GA is not defined since the number of columns in G is not equal to the number of rows in A

42. $C = \begin{bmatrix} 5 & 0 \\ -1 & 3 \\ 4 & 7 \end{bmatrix}, A = \begin{bmatrix} 4 & 6 \\ -2 & -2 \\ 5 & 9 \end{bmatrix}$

CA is not defined since the number of columns in C is not equal to the number of rows in A.

43. $A = \begin{bmatrix} 4 & 6 \\ -2 & -2 \\ 5 & 9 \end{bmatrix}, G = \begin{bmatrix} 2 & 5 \\ 1 & 6 \end{bmatrix}, F = \begin{bmatrix} -1 & 2 \\ 6 & 7 \end{bmatrix}$

$$GF = \begin{bmatrix} 2 & 5 \\ 1 & 6 \end{bmatrix}\begin{bmatrix} -1 & 2 \\ 6 & 7 \end{bmatrix} = \begin{bmatrix} 28 & 39 \\ 35 & 44 \end{bmatrix}.$$

Then, $AGF = A(GF)$

$$AGF = A(GF) = \begin{bmatrix} 4 & 6 \\ -2 & -2 \\ 5 & 9 \end{bmatrix}\begin{bmatrix} 28 & 39 \\ 35 & 44 \end{bmatrix} = \begin{bmatrix} 322 & 420 \\ -126 & -166 \\ 455 & 591 \end{bmatrix}$$

44. $E = [1 \quad 3 \quad -4], B = \begin{bmatrix} 1 & 2 & -3 \\ 2 & 3 & 0 \\ 0 & 1 & 4 \end{bmatrix}, D = \begin{bmatrix} 6 \\ 1 \\ 0 \end{bmatrix}$

From Exercise 40, $EB = [7 \quad 7 \quad -19]$.

Then, $EBD = (EB)D$

$$EBD = (EB)D = [7 \quad 7 \quad -19]\begin{bmatrix} 6 \\ 1 \\ 0 \end{bmatrix} = [49]$$

Copyright © 2019 Pearson Education, Inc.

45. There were 8000 athlete exposures for the league. The column matrix used will have values of 4 because there are an equal number and they are measured in thousands. Therefore, the matrix is

$$\begin{bmatrix} 3.54 & 1.41 \\ 1.53 & 1.57 \\ .34 & .29 \\ 7.53 & 6.21 \end{bmatrix} \cdot \begin{bmatrix} 4 \\ 4 \end{bmatrix} = \begin{bmatrix} 3.54(4)+1.41(4) \\ 1.53(4)+1.57(4) \\ .34(4)+.29(4) \\ 7.53(4)+6.21(4) \end{bmatrix} = \begin{bmatrix} 19.8 \\ 12.4 \\ 2.52 \\ 54.96 \end{bmatrix}$$

Therefore, there are about 20 head and face injuries, about 12 concussions, about 3 neck injuries, and about 55 other injuries.

46. a.

	Cutting	Shaping
Standard	$\frac{1}{4}$	$\frac{1}{2}$
Extra Large	$\frac{1}{3}$	$\frac{1}{3}$

b. $\begin{bmatrix} 48 & 66 \end{bmatrix} \begin{bmatrix} \frac{1}{4} & \frac{1}{2} \\ \frac{1}{3} & \frac{1}{3} \end{bmatrix} = \begin{bmatrix} 34 & 46 \end{bmatrix}$

The cutting machine will operate for 34 hr and the shaping machine for 46 hr.

47. a.

	Cost Per Share	Dividends Per Share
Starbucks	60.94	1.00
PepsiCo	113.22	3.22
Anh.-Busch	119.11	4.24

b.

	Stock		
	S	P	A
Number of shares	500	300	200

c. $\begin{bmatrix} 500 & 300 & 200 \end{bmatrix} \begin{bmatrix} 60.94 & 1.00 \\ 113.22 & 3.22 \\ 119.11 & 4.24 \end{bmatrix}$

$= \begin{bmatrix} 88{,}258 & 3214 \end{bmatrix}$

Total cost = \$88,258.
Total dividend = \$3,214.

48. a.

	Cost Per Share	Dividends Per Share
Apple	153.01	2.52
Texas Instr.	79.46	2.00
Microsoft	68.94	1.56

b.

	Stock		
	A	TI	M
Number of shares	1000	500	700

c. $\begin{bmatrix} 1000 & 500 & 700 \end{bmatrix} \begin{bmatrix} 153.01 & 2.52 \\ 79.46 & 2.00 \\ 68.94 & 1.56 \end{bmatrix}$

$= \begin{bmatrix} 240{,}998 & 4612 \end{bmatrix}$

Total cost = \$240,998.
Total dividend = \$4,612.

49. There are many correct answers. Here is one example.

$A = \begin{bmatrix} 3 & 0 \\ 2 & 1 \end{bmatrix}$; let $B = \begin{bmatrix} 1 & 2 \\ 3 & 4 \end{bmatrix}$.

$AB = \begin{bmatrix} 3 & 6 \\ 5 & 8 \end{bmatrix}$; $BA = \begin{bmatrix} 7 & 2 \\ 17 & 4 \end{bmatrix}$

Thus, *AB* and *BA* are both defined, and $AB \neq BA$.

50. No. $A = 4I$, so $AB = BA = 4B$.

51. Let $A = \begin{bmatrix} -2 & 2 \\ 0 & 5 \end{bmatrix}$.

$$[A \mid I] = \left[\begin{array}{cc|cc} -2 & 2 & 1 & 0 \\ 0 & 5 & 0 & 1 \end{array}\right]$$

$$= \left[\begin{array}{cc|cc} 1 & -1 & -\frac{1}{2} & 0 \\ 0 & 1 & 0 & \frac{1}{5} \end{array}\right] \begin{matrix} -\frac{1}{2}R_1 \\ \frac{1}{5}R_2 \end{matrix}$$

$$= \left[\begin{array}{cc|cc} 1 & 0 & -\frac{1}{2} & \frac{1}{5} \\ 0 & 1 & 0 & \frac{1}{5} \end{array}\right] R_2 + R_1$$

$$A^{-1} = \begin{bmatrix} -.5 & .2 \\ 0 & .2 \end{bmatrix}$$

52. Let $A = \begin{bmatrix} 3 & -1 \\ -5 & 2 \end{bmatrix}$.

$$[A \mid I] = \left[\begin{array}{cc|cc} 3 & -1 & 1 & 0 \\ -5 & 2 & 0 & 1 \end{array}\right]$$

$$\left[\begin{array}{cc|cc} 1 & -\frac{1}{3} & \frac{1}{3} & 0 \\ -5 & 2 & 0 & 1 \end{array}\right] \tfrac{1}{3}R_1$$

$$\left[\begin{array}{cc|cc} 1 & -\frac{1}{3} & \frac{1}{3} & 0 \\ 0 & \frac{1}{3} & \frac{5}{3} & 1 \end{array}\right] 5R_1 + R_2$$

$$\left[\begin{array}{cc|cc} 1 & 0 & 2 & 1 \\ 0 & 1 & 5 & 3 \end{array}\right] \begin{matrix} R_1 + R_2 \\ 3R_2 \end{matrix}$$

$$A^{-1} = \begin{bmatrix} 2 & 1 \\ 5 & 3 \end{bmatrix}$$

Copyright © 2019 Pearson Education, Inc.

53. Let $A=\begin{bmatrix}6 & 4\\3 & 2\end{bmatrix}$.

$[A\mid I]=\left[\begin{array}{cc|cc}6 & 4 & 1 & 0\\3 & 2 & 0 & 1\end{array}\right]$

$\left[\begin{array}{cc|cc}1 & \frac{2}{3} & \frac{1}{6} & 0\\3 & 2 & 0 & 1\end{array}\right]\frac{1}{6}R_1$

$\left[\begin{array}{cc|cc}1 & \frac{2}{3} & \frac{1}{6} & 0\\0 & 0 & -\frac{1}{2} & 1\end{array}\right]-3R_1+R_2$

The second row can never become $[0 \quad 1]$, so A has no inverse.

54. Let $A=\begin{bmatrix}15 & -12\\5 & -4\end{bmatrix}$.

$[A\mid I]=\left[\begin{array}{cc|cc}15 & -12 & 1 & 0\\5 & -4 & 0 & 1\end{array}\right]$

$\left[\begin{array}{cc|cc}1 & -\frac{4}{5} & \frac{1}{15} & 0\\5 & -4 & 0 & 1\end{array}\right]\frac{1}{15}R_1$

$\left[\begin{array}{cc|cc}1 & -\frac{4}{5} & \frac{1}{15} & 0\\0 & 0 & -\frac{1}{3} & 1\end{array}\right]-5R_1+R_2$

The second row can never become $[0 \quad 1]$, so A has no inverse.

55. Let $A=\begin{bmatrix}2 & 0 & 6\\1 & -1 & 0\\0 & 1 & -3\end{bmatrix}$

$[A\mid I]=\left[\begin{array}{ccc|ccc}2 & 0 & 6 & 1 & 0 & 0\\1 & -1 & 0 & 0 & 1 & 0\\0 & 1 & -3 & 0 & 0 & 1\end{array}\right]$

Interchange rows.

$\left[\begin{array}{ccc|ccc}1 & -1 & 0 & 0 & 1 & 0\\0 & 1 & -3 & 0 & 0 & 1\\2 & 0 & 6 & 1 & 0 & 0\end{array}\right]$

$\left[\begin{array}{ccc|ccc}1 & -1 & 0 & 0 & 1 & 0\\0 & 1 & -3 & 0 & 0 & 1\\0 & 2 & 6 & 1 & -2 & 0\end{array}\right]-2R_1+R_3$

$\left[\begin{array}{ccc|ccc}1 & 0 & -3 & 0 & 1 & 1\\0 & 1 & -3 & 0 & 0 & 1\\0 & 0 & 12 & 1 & -2 & -2\end{array}\right]\begin{matrix}R_2+R_1\\ \\ -2R_2+R_3\end{matrix}$

$\left[\begin{array}{ccc|ccc}1 & 0 & -3 & 0 & 1 & 1\\0 & 1 & -3 & 0 & 0 & 1\\0 & 0 & 1 & \frac{1}{12} & -\frac{1}{6} & -\frac{1}{6}\end{array}\right]\frac{1}{12}R_3$

$[I\mid B]=\left[\begin{array}{ccc|ccc}1 & 0 & 0 & \frac{1}{4} & \frac{1}{2} & \frac{1}{2}\\0 & 1 & 0 & \frac{1}{4} & -\frac{1}{2} & \frac{1}{2}\\0 & 0 & 1 & \frac{1}{12} & -\frac{1}{6} & -\frac{1}{6}\end{array}\right]\begin{matrix}3R_3+R_1\\3R_3+R_2\\ \\ \end{matrix}$

$A^{-1}=\begin{bmatrix}\frac{1}{4} & \frac{1}{2} & \frac{1}{2}\\\frac{1}{4} & -\frac{1}{2} & \frac{1}{2}\\\frac{1}{12} & -\frac{1}{6} & -\frac{1}{6}\end{bmatrix}$

56. Let $A=\begin{bmatrix}2 & -1 & 0\\1 & 0 & 2\\1 & -4 & 0\end{bmatrix}$

$\left[\begin{array}{ccc|ccc}2 & -1 & 0 & 1 & 0 & 0\\1 & 0 & 2 & 0 & 1 & 0\\1 & -4 & 0 & 0 & 0 & 1\end{array}\right]$

Interchange rows 1 and 2.

$\left[\begin{array}{ccc|ccc}1 & 0 & 2 & 0 & 1 & 0\\2 & -1 & 0 & 1 & 0 & 0\\1 & -4 & 0 & 0 & 0 & 1\end{array}\right]$

$\left[\begin{array}{ccc|ccc}1 & 0 & 2 & 0 & 1 & 0\\0 & -1 & -4 & 1 & -2 & 0\\0 & -4 & -2 & 0 & -1 & 1\end{array}\right]\begin{matrix} \\ -2R_1+R_2\\-1R_1+R_3\end{matrix}$

Multiply row 2 by –1 and continue.

$\left[\begin{array}{ccc|ccc}1 & 0 & 2 & 0 & 1 & 0\\0 & 1 & 4 & -1 & 2 & 0\\0 & 0 & 14 & -4 & 7 & 1\end{array}\right]-4R_2+R_3$

$\left[\begin{array}{ccc|ccc}1 & 0 & 2 & 0 & 1 & 0\\0 & 1 & 4 & -1 & 2 & 0\\0 & 0 & 1 & -\frac{2}{7} & \frac{1}{2} & \frac{1}{14}\end{array}\right]\frac{1}{14}R_3$

$\left[\begin{array}{ccc|ccc}1 & 0 & 0 & \frac{4}{7} & 0 & -\frac{1}{7}\\0 & 1 & 0 & \frac{1}{7} & 0 & -\frac{2}{7}\\0 & 0 & 1 & -\frac{2}{7} & \frac{1}{2} & \frac{1}{14}\end{array}\right]\begin{matrix}-2R_3+R_1\\-4R_3+R_2\\ \\ \end{matrix}$

$A^{-1}=\begin{bmatrix}\frac{4}{7} & 0 & -\frac{1}{7}\\\frac{1}{7} & 0 & -\frac{2}{7}\\-\frac{2}{7} & \frac{1}{2} & \frac{1}{14}\end{bmatrix}$

Copyright © 2019 Pearson Education, Inc.

57. Let $A = \begin{bmatrix} 2 & 3 & 5 \\ -2 & -3 & -5 \\ 1 & 4 & 2 \end{bmatrix}$.

$[A \mid I] = \left[\begin{array}{ccc|ccc} 2 & 3 & 5 & 1 & 0 & 0 \\ -2 & -3 & -5 & 0 & 1 & 0 \\ 1 & 4 & 2 & 0 & 0 & 1 \end{array}\right]$

$\left[\begin{array}{ccc|ccc} 1 & 4 & 2 & 0 & 0 & 1 \\ -2 & -3 & -5 & 0 & 1 & 0 \\ 2 & 3 & 5 & 1 & 0 & 0 \end{array}\right]$ Interchange R_1 and R_3

$\left[\begin{array}{ccc|ccc} 1 & 4 & 2 & 0 & 0 & 1 \\ -2 & -3 & -5 & 0 & 1 & 0 \\ 0 & 0 & 0 & 1 & 1 & 0 \end{array}\right] R_2 + R_3$

The third row can never become $[0 \quad 0 \quad 1]$, so A does not have an inverse.

58. Let $A = \begin{bmatrix} 1 & 3 & 6 \\ 4 & 0 & 9 \\ 5 & 15 & 30 \end{bmatrix}$

$\left[\begin{array}{ccc|ccc} 1 & 3 & 6 & 1 & 0 & 0 \\ 4 & 0 & 9 & 0 & 1 & 0 \\ 5 & 15 & 30 & 0 & 0 & 1 \end{array}\right]$

$\left[\begin{array}{ccc|ccc} 1 & 3 & 6 & 1 & 0 & 0 \\ 0 & -12 & -15 & -4 & 1 & 0 \\ 0 & 0 & 0 & -5 & 0 & 1 \end{array}\right] \begin{matrix} \\ -4R_1 + R_2 \\ -5R_1 + R_3 \end{matrix}$

The last row is all zeros, so no inverse exists.

59. Let $A = \begin{bmatrix} 1 & 3 & -2 & -1 \\ 0 & 1 & 1 & 2 \\ -1 & -1 & 1 & -1 \\ 1 & -1 & -3 & -2 \end{bmatrix}$.

Use a graphing calculator to find A^{-1}.

$A^{-1} = \begin{bmatrix} -\frac{2}{3} & -\frac{17}{3} & -\frac{14}{3} & -3 \\ \frac{1}{3} & \frac{1}{3} & \frac{1}{3} & 0 \\ -\frac{1}{3} & -\frac{10}{3} & -\frac{7}{3} & -2 \\ 0 & 2 & 1 & 1 \end{bmatrix}$

60. Let $A = \begin{bmatrix} 3 & 2 & 0 & -1 \\ 2 & 0 & 1 & 2 \\ 1 & 2 & -1 & 0 \\ 2 & -1 & 1 & 1 \end{bmatrix}$.

Use a graphing calculator to find A^{-1}.

$A^{-1} = \begin{bmatrix} 0 & -\frac{1}{3} & \frac{1}{3} & \frac{2}{3} \\ \frac{1}{3} & \frac{2}{3} & -\frac{1}{3} & -1 \\ \frac{2}{3} & 1 & -\frac{4}{3} & -\frac{4}{3} \\ -\frac{1}{3} & \frac{1}{3} & \frac{1}{3} & 0 \end{bmatrix}$

61. $F = \begin{bmatrix} -1 & 2 \\ 6 & 7 \end{bmatrix}$

Form the augmented matrix $[F \mid I]$.

$\left[\begin{array}{cc|cc} -1 & 2 & 1 & 0 \\ 6 & 7 & 0 & 1 \end{array}\right]$

$\left[\begin{array}{cc|cc} 1 & -2 & -1 & 0 \\ 6 & 7 & 0 & 1 \end{array}\right] -1R_1$

$\left[\begin{array}{cc|cc} 1 & -2 & -1 & 0 \\ 0 & 19 & 6 & 1 \end{array}\right] -6R_1 + R_2$

$\left[\begin{array}{cc|cc} 1 & -2 & -1 & 0 \\ 0 & 1 & \frac{6}{19} & \frac{1}{19} \end{array}\right] \frac{1}{19}R_2$

$\left[\begin{array}{cc|cc} 1 & 0 & -\frac{7}{19} & \frac{2}{19} \\ 0 & 1 & \frac{6}{19} & \frac{1}{19} \end{array}\right] R_1 + 2R_2$

$F^{-1} = \begin{bmatrix} -\frac{7}{19} & \frac{2}{19} \\ \frac{6}{19} & \frac{1}{19} \end{bmatrix}$

62. $G = \begin{bmatrix} 2 & 5 \\ 1 & 6 \end{bmatrix}$

Form the augmented matrix $[G \mid I]$.

$\left[\begin{array}{cc|cc} 2 & 5 & 1 & 0 \\ 1 & 6 & 0 & 1 \end{array}\right]$

Interchange R_1 and R_2.

$\left[\begin{array}{cc|cc} 1 & 6 & 0 & 1 \\ 2 & 5 & 1 & 0 \end{array}\right]$

$\left[\begin{array}{cc|cc} 1 & 6 & 0 & 1 \\ 0 & -7 & 1 & -2 \end{array}\right] -2R_1 + R_2$

$\left[\begin{array}{cc|cc} 1 & 6 & 0 & 1 \\ 0 & 1 & -\frac{1}{7} & \frac{2}{7} \end{array}\right] -\frac{1}{7}R_2$

Copyright © 2019 Pearson Education, Inc.

$$\left[\begin{array}{cc|cc} 1 & 0 & \frac{6}{7} & -\frac{5}{7} \\ 0 & 1 & -\frac{1}{7} & \frac{2}{7} \end{array}\right] -6R_2 + R_1$$

$$G^{-1} = \begin{bmatrix} \frac{6}{7} & -\frac{5}{7} \\ -\frac{1}{7} & \frac{2}{7} \end{bmatrix}$$

63. $G = \begin{bmatrix} 2 & 5 \\ 1 & 6 \end{bmatrix}, F = \begin{bmatrix} -1 & 2 \\ 6 & 7 \end{bmatrix}$

$$G - F = \begin{bmatrix} 3 & 3 \\ -5 & -1 \end{bmatrix}$$

Form the augmented matrix $[G - F \mid I]$.

$$\left[\begin{array}{cc|cc} 3 & 3 & 1 & 0 \\ -5 & -1 & 0 & 1 \end{array}\right]$$

$$\left[\begin{array}{cc|cc} 1 & 1 & \frac{1}{3} & 0 \\ -5 & -1 & 0 & 1 \end{array}\right] \frac{1}{3}R_1$$

$$\left[\begin{array}{cc|cc} 1 & 1 & \frac{1}{3} & 0 \\ 0 & 4 & \frac{5}{3} & 1 \end{array}\right] 5R_1 + R_2$$

$$\left[\begin{array}{cc|cc} 1 & 1 & \frac{1}{3} & 0 \\ 0 & 1 & \frac{5}{12} & \frac{1}{4} \end{array}\right] \frac{1}{4}R_2$$

$$\left[\begin{array}{cc|cc} 1 & 0 & -\frac{1}{12} & -\frac{1}{4} \\ 0 & 1 & \frac{5}{12} & \frac{1}{4} \end{array}\right] -R_2 + R_1$$

$$(G - F)^{-1} = \begin{bmatrix} -\frac{1}{12} & -\frac{1}{4} \\ \frac{5}{12} & \frac{1}{4} \end{bmatrix}$$

64. $F = \begin{bmatrix} -1 & 2 \\ 6 & 7 \end{bmatrix}, G = \begin{bmatrix} 2 & 5 \\ 1 & 6 \end{bmatrix}$

$$F + G = \begin{bmatrix} 1 & 7 \\ 7 & 13 \end{bmatrix}$$

Form the augmented matrix $[F + G \mid I]$.

$$\left[\begin{array}{cc|cc} 1 & 7 & 1 & 0 \\ 7 & 13 & 0 & 1 \end{array}\right]$$

$$\left[\begin{array}{cc|cc} 1 & 7 & 1 & 0 \\ 0 & -36 & -7 & 1 \end{array}\right] -7R_1 + R_2$$

$$\left[\begin{array}{cc|cc} 1 & 7 & 1 & 0 \\ 0 & 1 & \frac{7}{36} & -\frac{1}{36} \end{array}\right] -\frac{1}{36}R_2$$

$$\left[\begin{array}{cc|cc} 1 & 0 & -\frac{13}{36} & \frac{7}{36} \\ 0 & 1 & \frac{7}{36} & -\frac{1}{36} \end{array}\right] -7R_2 + R_1$$

$$(F + G)^{-1} = \begin{bmatrix} -\frac{13}{36} & \frac{7}{36} \\ \frac{7}{36} & -\frac{1}{36} \end{bmatrix}$$

65. $B = \begin{bmatrix} 1 & 2 & -3 \\ 2 & 3 & 0 \\ 0 & 1 & 4 \end{bmatrix}$

Form the augmented matrix $[B \mid I]$.

$$\left[\begin{array}{ccc|ccc} 1 & 2 & -3 & 1 & 0 & 0 \\ 2 & 3 & 0 & 0 & 1 & 0 \\ 0 & 1 & 4 & 0 & 0 & 1 \end{array}\right]$$

$$\left[\begin{array}{ccc|ccc} 1 & 0 & -11 & 1 & 0 & -2 \\ 0 & 1 & -6 & 2 & -1 & 0 \\ 0 & 1 & 4 & 0 & 0 & 1 \end{array}\right] \begin{array}{l} -2R_3 + R_1 \\ -R_2 + 2R_1 \\ \\ \end{array}$$

$$\left[\begin{array}{ccc|ccc} 1 & 0 & -11 & 1 & 0 & -2 \\ 0 & 1 & -6 & 2 & -1 & 0 \\ 0 & 0 & 1 & -\frac{1}{5} & \frac{1}{10} & \frac{1}{10} \end{array}\right] \frac{1}{10}(-R_2 + R_3)$$

$$\left[\begin{array}{ccc|ccc} 1 & 0 & 0 & -\frac{6}{5} & \frac{11}{10} & -\frac{9}{10} \\ 0 & 1 & 0 & \frac{4}{5} & -\frac{2}{5} & \frac{3}{5} \\ 0 & 0 & 1 & -\frac{1}{5} & \frac{1}{10} & \frac{1}{10} \end{array}\right] \begin{array}{l} R_1 + 11R_3 \\ R_2 + 6R_3 \\ \\ \end{array}$$

$$B^{-1} = \begin{bmatrix} -1.2 & 1.1 & -.9 \\ .8 & -.4 & .6 \\ -.2 & .1 & .1 \end{bmatrix}$$

66. Answers vary. Sample answer: There is no way to transform the given matrix into the identity matrix using row operations

67. $A = \begin{bmatrix} -3 & 4 \\ -1 & 2 \end{bmatrix}, B = \begin{bmatrix} 3 \\ -1 \end{bmatrix}$

If $AX = B$, then $X = A^{-1}B$.

If $A = \begin{bmatrix} -3 & 4 \\ -1 & 2 \end{bmatrix}$, then $A^{-1} = \begin{bmatrix} -1 & 2 \\ -\frac{1}{2} & \frac{3}{2} \end{bmatrix}$.

Thus, $X = \begin{bmatrix} -1 & 2 \\ -\frac{1}{2} & \frac{3}{2} \end{bmatrix}\begin{bmatrix} 3 \\ -1 \end{bmatrix} = \begin{bmatrix} -5 \\ -3 \end{bmatrix}$.

68. $A = \begin{bmatrix} 1 & 3 \\ -2 & 4 \end{bmatrix}, B = \begin{bmatrix} 9 \\ 6 \end{bmatrix}$

$$A^{-1} = \begin{bmatrix} \frac{2}{5} & -\frac{3}{10} \\ \frac{1}{5} & \frac{1}{10} \end{bmatrix}$$

Then $X = A^{-1}B$

$$\begin{bmatrix} \frac{2}{5} & -\frac{3}{10} \\ \frac{1}{5} & \frac{1}{10} \end{bmatrix}\begin{bmatrix} 9 \\ 6 \end{bmatrix} = \begin{bmatrix} \frac{9}{5} \\ \frac{12}{5} \end{bmatrix}.$$

Copyright © 2019 Pearson Education, Inc.

69. $A=\begin{bmatrix}1 & 0 & 2\\ -1 & 1 & 0\\ 3 & 0 & 4\end{bmatrix}$, $B=\begin{bmatrix}8\\ 4\\ -6\end{bmatrix}$

If $AX = B$, then $X = A^{-1}B$.

If $A=\begin{bmatrix}1 & 0 & 2\\ -1 & 1 & 0\\ 3 & 0 & 4\end{bmatrix}$,

then $A^{-1}=\begin{bmatrix}-2 & 0 & 1\\ -2 & 1 & 1\\ \frac{3}{2} & 0 & -\frac{1}{2}\end{bmatrix}$.

$X=\begin{bmatrix}-2 & 0 & 1\\ -2 & 1 & 1\\ \frac{3}{2} & 0 & -\frac{1}{2}\end{bmatrix}\begin{bmatrix}8\\ 4\\ -6\end{bmatrix}=\begin{bmatrix}-22\\ -18\\ 15\end{bmatrix}$.

70. $A=\begin{bmatrix}2 & 4 & 0\\ 1 & -2 & 0\\ 0 & 0 & 3\end{bmatrix}$, $B=\begin{bmatrix}72\\ -24\\ 48\end{bmatrix}$

$A^{-1}=\begin{bmatrix}\frac{1}{4} & \frac{1}{2} & 0\\ \frac{1}{8} & -\frac{1}{4} & 0\\ 0 & 0 & \frac{1}{3}\end{bmatrix}$

$X=A^{-1}B=\begin{bmatrix}\frac{1}{4} & \frac{1}{2} & 0\\ \frac{1}{8} & -\frac{1}{4} & 0\\ 0 & 0 & \frac{1}{3}\end{bmatrix}\begin{bmatrix}72\\ -24\\ 48\end{bmatrix}=\begin{bmatrix}6\\ 15\\ 16\end{bmatrix}$.

71. $x + y = -2$

$2x + 5y = 2$

The system as a matrix equation is

$\begin{bmatrix}1 & 1\\ 2 & 5\end{bmatrix}\begin{bmatrix}x\\ y\end{bmatrix}=\begin{bmatrix}-2\\ 2\end{bmatrix}$.

Let $A=\begin{bmatrix}1 & 1\\ 2 & 5\end{bmatrix}$ and $B=\begin{bmatrix}-2\\ 2\end{bmatrix}$.

$A^{-1}=\begin{bmatrix}\frac{5}{3} & -\frac{1}{3}\\ -\frac{2}{3} & \frac{1}{3}\end{bmatrix}$

$\begin{bmatrix}x\\ y\end{bmatrix}=\begin{bmatrix}\frac{5}{3} & -\frac{1}{3}\\ -\frac{2}{3} & \frac{1}{3}\end{bmatrix}\begin{bmatrix}-2\\ 2\end{bmatrix}=\begin{bmatrix}-4\\ 2\end{bmatrix}$.

The solution is (–4, 2).

72. $5x - 3y = -2$

$2x + 7y = -9$

The system as a matrix equation is

$\begin{bmatrix}5 & -3\\ 2 & 7\end{bmatrix}\begin{bmatrix}x\\ y\end{bmatrix}=\begin{bmatrix}-2\\ -9\end{bmatrix}$.

Let $A=\begin{bmatrix}5 & -3\\ 2 & 7\end{bmatrix}$ and $B=\begin{bmatrix}-2\\ -9\end{bmatrix}$.

$A^{-1}=\begin{bmatrix}\frac{7}{41} & \frac{3}{41}\\ -\frac{2}{41} & \frac{5}{41}\end{bmatrix}$

$X=A^{-1}B=\begin{bmatrix}\frac{7}{41} & \frac{3}{41}\\ -\frac{2}{41} & \frac{5}{41}\end{bmatrix}\begin{bmatrix}-2\\ -9\end{bmatrix}=\begin{bmatrix}-1\\ -1\end{bmatrix}$

The solution is (–1, –1).

73. $2x + y = 10$

$3x - 2y = 8$

The system as a matrix equation is

$\begin{bmatrix}2 & 1\\ 3 & -2\end{bmatrix}\begin{bmatrix}x\\ y\end{bmatrix}=\begin{bmatrix}10\\ 8\end{bmatrix}$.

Let $A=\begin{bmatrix}2 & 1\\ 3 & -2\end{bmatrix}$ and $B=\begin{bmatrix}10\\ 8\end{bmatrix}$.

$A^{-1}=\begin{bmatrix}\frac{2}{7} & \frac{1}{7}\\ \frac{3}{7} & -\frac{2}{7}\end{bmatrix}$

$X=\begin{bmatrix}\frac{2}{7} & \frac{1}{7}\\ \frac{3}{7} & -\frac{2}{7}\end{bmatrix}\begin{bmatrix}10\\ 8\end{bmatrix}=\begin{bmatrix}4\\ 2\end{bmatrix}$.

The solution is (4, 2).

74. $x - 2y = 7$

$3x + y = 7$

The system as a matrix equation is

$\begin{bmatrix}1 & -2\\ 3 & 1\end{bmatrix}\begin{bmatrix}x\\ y\end{bmatrix}=\begin{bmatrix}7\\ 7\end{bmatrix}$.

$A=\begin{bmatrix}1 & -2\\ 3 & 1\end{bmatrix}$, $B=\begin{bmatrix}7\\ 7\end{bmatrix}$. $A^{-1}=\begin{bmatrix}\frac{1}{7} & \frac{2}{7}\\ -\frac{3}{7} & \frac{1}{7}\end{bmatrix}$

$X=A^{-1}B=\begin{bmatrix}3\\ -2\end{bmatrix}$

The solution is (3, –2).

Copyright © 2019 Pearson Education, Inc.

75. $x + y + z = 1$
$2x - y = -2$
$3y + z = 2$

The system as a matrix equation is

$$\begin{bmatrix} 1 & 1 & 1 \\ 2 & -1 & 0 \\ 0 & 3 & 1 \end{bmatrix}\begin{bmatrix} x \\ y \\ z \end{bmatrix} = \begin{bmatrix} 1 \\ -2 \\ 2 \end{bmatrix}.$$

Let $A = \begin{bmatrix} 1 & 1 & 1 \\ 2 & -1 & 0 \\ 0 & 3 & 1 \end{bmatrix}$ and $B = \begin{bmatrix} 1 \\ -2 \\ 2 \end{bmatrix}$.

$$A^{-1} = \begin{bmatrix} -\frac{1}{3} & \frac{2}{3} & \frac{1}{3} \\ -\frac{2}{3} & \frac{1}{3} & \frac{2}{3} \\ 2 & -1 & -1 \end{bmatrix}$$

$$X = \begin{bmatrix} -\frac{1}{3} & \frac{2}{3} & \frac{1}{3} \\ -\frac{2}{3} & \frac{1}{3} & \frac{2}{3} \\ 2 & -1 & -1 \end{bmatrix}\begin{bmatrix} 1 \\ -2 \\ 2 \end{bmatrix} = \begin{bmatrix} -1 \\ 0 \\ 2 \end{bmatrix}.$$

The solution is (–1, 0, 2).

76. $x = -3$
$y + z = 6$
$2x - 3z = -9$

The system as a matrix equation is

$$\begin{bmatrix} 1 & 0 & 0 \\ 0 & 1 & 1 \\ 2 & 0 & -3 \end{bmatrix}\begin{bmatrix} x \\ y \\ z \end{bmatrix} = \begin{bmatrix} -3 \\ 6 \\ -9 \end{bmatrix}.$$

$$A = \begin{bmatrix} 1 & 0 & 0 \\ 0 & 1 & 1 \\ 2 & 0 & -3 \end{bmatrix}, B = \begin{bmatrix} -3 \\ 6 \\ -9 \end{bmatrix}$$

$$A^{-1} = \begin{bmatrix} 1 & 0 & 0 \\ -\frac{2}{3} & 1 & \frac{1}{3} \\ \frac{2}{3} & 0 & -\frac{1}{3} \end{bmatrix}$$

$$X = A^{-1}B = \begin{bmatrix} -3 \\ 5 \\ 1 \end{bmatrix}$$

The solution is (–3, 5, 1).

77. $3x - 2y + 4z = 4$
$4x + y - 5z = 2$
$-6x + 4y - 8z = -2$

The system as a matrix equation is

$$\begin{bmatrix} 3 & -2 & 4 \\ 4 & 1 & -5 \\ -6 & 4 & -8 \end{bmatrix}\begin{bmatrix} x \\ y \\ z \end{bmatrix} = \begin{bmatrix} 4 \\ 2 \\ -2 \end{bmatrix}.$$

Let $A = \begin{bmatrix} 3 & -2 & 4 \\ 4 & 1 & -5 \\ -6 & 4 & -8 \end{bmatrix}$.

Since row 3 is –2 times row 1, the matrix will have no inverse, and the system cannot be solved by this method. Another method should be used to complete the solution. Use the elimination method. Multiply equation (1) by 2 and add the result to equation (3).

$$\begin{aligned} 6x - 4y + 8z &= 8 \\ -6x + 4y - 8z &= -2 \\ \hline 0 &= 6 \end{aligned}$$

This false result indicates that the system has no solution.

78. $-2x + 3y - z = 1$
$5x - 7y + 8z = 4$
$6x - 9y + 3z = 2$

The system as a matrix equation is

$$\begin{bmatrix} -2 & 3 & -1 \\ 5 & -7 & 8 \\ 6 & -9 & 3 \end{bmatrix}\begin{bmatrix} x \\ y \\ z \end{bmatrix} = \begin{bmatrix} 1 \\ 4 \\ 2 \end{bmatrix}.$$

Let $A = \begin{bmatrix} -2 & 3 & -1 \\ 5 & -7 & 8 \\ 6 & -9 & 3 \end{bmatrix}$.

Since row 3 is –3 times row 1, the matrix will have no inverse, and the system cannot be solved by this method. Another method should be used to complete the solution. Use the elimination method. Multiply equation (1) by 3 and add the result to equation (3).

$$\begin{aligned} -6x + 9y - 3z &= 3 \\ 6x - 9y + 3z &= 2 \\ \hline 0 &= 5 \end{aligned}$$

This false result indicates that the system has no solution.

Copyright © 2019 Pearson Education, Inc.

Exercises 79–86 can be solved using either the elimination method or matrix methods.

79. Let x = the amount of the 9% wine and let y = the amount of the 14% wine.
We have the system

$$\begin{aligned} x + \quad y &= 40 \\ .09x + .14y &= .12(40). \end{aligned}$$

This system can be simplified to

$$\begin{aligned} x + \quad y &= 40 \quad (1) \\ .09x + .14y &= 4.8. \quad (2) \end{aligned}$$

Multiply equation (1) by –.09 and add to equation (2).

$$\begin{aligned} -.09x - .09y &= -3.6 \\ .09x + .14y &= 4.8 \\ \hline .05y &= 1.2 \Rightarrow y = 24 \end{aligned}$$

Substitute 24 for y in equation (1) to get $x = 16$.
The wine maker should mix 16 liters of the 9% wine and 24 liters of the 14% wine.

80. Let x = number of grams of 12 carat gold and let y = number of grams of 22 carat gold.
The system is

$$\begin{aligned} x + \quad y &= 25 \quad (1) \\ \frac{12}{24}x + \frac{22}{24}y &= \frac{15}{24}(25) \quad (2) \end{aligned}$$

Multiply equation (2) by 24.

$$\begin{aligned} x + \quad y &= 25 \quad (1) \\ 12x + 22y &= 375 \quad (3) \end{aligned}$$

Multiply equation (1) by –12 and add the result to equation (3).

$$\begin{aligned} -12x - 12y &= -300 \\ 12x + 22y &= 375 \\ \hline 10y &= 75 \Rightarrow y = 7.5 \end{aligned}$$

Substitute 7.5 for y in equation (1)
$x + 7.5 = 25 \Rightarrow x = 17.5$
The merchant should mix 17.5 grams of 12 carat gold and 7.5 grams of 22 carat gold.

81. Let x = the number of liters of 40% solution and let y = the number of liters of 60% solution.

$$\begin{aligned} x + \quad y &= 40 \\ .4x + .6y &= .45(40) = 18 \end{aligned}$$

Multiply the second equation by 5.

$$\begin{aligned} x + \quad y &= 40 \\ 2x + 3y &= 90 \end{aligned}$$

Use the Gauss-Jordan method.

$$\left[\begin{array}{cc|c} 1 & 1 & 40 \\ 2 & 3 & 90 \end{array}\right]$$

$$\left[\begin{array}{cc|c} 1 & 1 & 40 \\ 0 & 1 & 10 \end{array}\right] -2R_1 + R_2$$

$$\left[\begin{array}{cc|c} 1 & 0 & 30 \\ 0 & 1 & 10 \end{array}\right] \begin{array}{l} -R_2 + R_1 \\ \\ \end{array}$$

The chemist should use 30 liters of the 40% solution and 10 liters of the 60% solution.

82. Let x = the number of pounds of tea worth \$4.60 a pound and let y = the number of pounds of tea worth \$6.50 a pound.
The system is

$$\begin{aligned} x + \quad y &= 10 \quad (1) \\ 4.60x + 6.50y &= 5.74(10). \quad (2) \end{aligned}$$

Multiply equation (1) by –4.60 and add the result to equation (2).

$$\begin{aligned} -4.60x - 4.60y &= -46.0 \\ 4.60x + 6.50y &= 57.4 \\ \hline 1.90y &= 11.4 \Rightarrow y = 6 \end{aligned}$$

Substitute 6 for y in equation (1) and solve for x.
$x + 6 = 10 \Rightarrow x = 4$
Thus, 4 pounds of the tea worth \$4.60 a pound should be used.

Copyright © 2019 Pearson Education, Inc.

83. Let x = the number of bowls and let y = the number of plates.
After converting 8 hours to 480 minutes, we will solve the system

$$3x + 2y = 480 \quad (1)$$
$$.25x + .2y = 44 \quad (2)$$

Multiply equation (2) by –10 and add to equation (1).

$$\begin{aligned} 3x + 2y &= 480 \\ -2.5x - 2y &= -440 \\ \hline .5x \quad &= 40 \Rightarrow x = 80 \end{aligned}$$

Substitute 80 for x in equation (1) and solve for y.

$$3(80) + 2y = 480 \Rightarrow y = 120$$

The factory can produce 80 bowls and 120 plates.

84. Let x = the speed of the boat and let y = the speed of the current. The system is

$$3(x + y) = 57 \quad (1)$$
$$5(x - y) = 55. \quad (2)$$

Dividing equation (1) by 3 and equation (2) by 5 gives

$$x + y = 19 \quad (3)$$
$$x - y = 11. \quad (4)$$

Adding the above two equations gives
$2x = 30 \Rightarrow x = 15.$
Substituting into (3) gives
$15 + y = 19 \Rightarrow y = 4.$

The speed of the boat is 15 km/hr, and the speed of the current is 4 km/hr.

85. Let x = the amount invested at 8%, let y = the amount invested at $8\frac{1}{2}$%, and let z = the amount invested at 11%. The system is

$$x + y + z = 50{,}000$$
$$.08x + .085y + .11z = 4436.25$$
$$.11z = .08x + 80.$$

Write the system in the proper form.

$$x + \quad y + \quad z = 50{,}000$$
$$.08x + .085y + .11z = 4436.25$$
$$-.08x \qquad + .11z = 80$$

Write the augmented matrix of the system.

$$\left[\begin{array}{ccc|c} 1 & 1 & 1 & 50{,}000 \\ .08 & .085 & .11 & 4436.25 \\ -.08 & 0 & .11 & 80 \end{array}\right]$$

$$\left[\begin{array}{ccc|c} 1 & 1 & 1 & 50{,}000 \\ 0 & .005 & .03 & 436.25 \\ 0 & .08 & .19 & 4080 \end{array}\right] \begin{array}{l} \\ -.08R_1 + R_2 \\ .08R_1 + R_3 \end{array}$$

$$\left[\begin{array}{ccc|c} 1 & 1 & 1 & 50{,}000 \\ 0 & 1 & 6 & 87{,}250 \\ 0 & .08 & .19 & 4080 \end{array}\right] \begin{array}{l} \\ \frac{1}{.005}R_2 \\ \\ \end{array}$$

$$\left[\begin{array}{ccc|c} 1 & 0 & -5 & -37{,}250 \\ 0 & 1 & 6 & 87{,}250 \\ 0 & 0 & -.29 & -2900 \end{array}\right] \begin{array}{l} -1R_2 + R_1 \\ \\ -.08R_2 + R_3 \end{array}$$

$$\left[\begin{array}{ccc|c} 1 & 0 & -5 & -37{,}250 \\ 0 & 1 & 6 & 87{,}250 \\ 0 & 0 & 1 & 10{,}000 \end{array}\right] \begin{array}{l} \\ \\ -\frac{1}{.29}R_3 \end{array}$$

$$\left[\begin{array}{ccc|c} 1 & 0 & 0 & 12{,}750 \\ 0 & 1 & 0 & 27{,}250 \\ 0 & 0 & 1 & 10{,}000 \end{array}\right] \begin{array}{l} 5R_3 + R_1 \\ -6R_3 + R_2 \\ \\ \end{array}$$

Thus, $x = 12{,}750$, $y = 27{,}250$, and $z = 10{,}000$. Mr. Miller invested \$12,750 at 8%, \$27,250 at $8\frac{1}{2}$%, and \$10,000 at 11%.

86. Let x = the number of student tickets, let y = the number of alumni tickets, let z = the number of other adult tickets, and let w = the number of children's tickets.

$$x + y + z + w = 3750 \quad (1)$$
$$4x + 10y + 12z + 6w = 29{,}100 \quad (2)$$
$$x = 6w \quad (3)$$
$$y = \tfrac{4}{5}x \quad (4)$$

Solve equation (3) for w in terms of x.
$x = 6w \Rightarrow w = \frac{1}{6}x. \quad (5)$

Now substitute the values for y and w into equations (1) and (2)

$$x + \tfrac{4}{5}x + z + \tfrac{1}{6}x = 3750 \quad (6)$$
$$4x + 10\left(\tfrac{4}{5}x\right) + 12z + 6\left(\tfrac{1}{6}x\right) = 29{,}100 \quad (7)$$

Simplify by multiplying equation (6) by 30 and combining like terms, and combining like terms in (7).

$$59x + 30z = 112{,}500 \quad (8)$$
$$13x + 12z = 29{,}100 \quad (9)$$

Copyright © 2019 Pearson Education, Inc.

Multiply (8) by 2 and (9) by −5, then add the resulting equations and solve for x.

$$\begin{array}{rcl} 118x+60z &=& 225{,}000 \\ -65x-60z &=& -145{,}500 \\ \hline 53x &=& 79{,}500 \Rightarrow x=1500 \end{array}$$

Substitute $x = 1500$ into (3), (4), and (8) to solve for the remaining variables.

$1500 = 6w \Rightarrow w = 250$

$y = \frac{4}{5}(1500) = 1200$

$59(1500) + 30z = 112{,}500 \Rightarrow 30z = 24{,}000 \Rightarrow$

$z = 800$

There were 1500 students, 1200 alumni, 800 other adults, and 250 children at the game.

87. $A=\begin{bmatrix} 0 & \frac{1}{4} \\ \frac{1}{2} & 0 \end{bmatrix}, D=\begin{bmatrix} 2100 \\ 1400 \end{bmatrix}$

a. $I-A=\begin{bmatrix} 1 & 0 \\ 0 & 1 \end{bmatrix}-\begin{bmatrix} 0 & \frac{1}{4} \\ \frac{1}{2} & 0 \end{bmatrix}=\begin{bmatrix} 1 & -\frac{1}{4} \\ -\frac{1}{2} & 1 \end{bmatrix}$

b. $\left[\begin{array}{cc|cc} 1 & -\frac{1}{4} & 1 & 0 \\ -\frac{1}{2} & 1 & 0 & 1 \end{array}\right]$

$\left[\begin{array}{cc|cc} 1 & -\frac{1}{4} & 1 & 0 \\ 0 & \frac{7}{8} & \frac{1}{2} & 1 \end{array}\right] \frac{1}{2}R_1+R_2$

$\left[\begin{array}{cc|cc} 1 & -\frac{1}{4} & 1 & 0 \\ 0 & 1 & \frac{4}{7} & \frac{8}{7} \end{array}\right] \frac{8}{7}R_2$

$\left[\begin{array}{cc|cc} 1 & 0 & \frac{8}{7} & \frac{2}{7} \\ 0 & 1 & \frac{4}{7} & \frac{8}{7} \end{array}\right] \frac{1}{4}R_2+R_1$

$(I-A)^{-1}=\begin{bmatrix} \frac{8}{7} & \frac{2}{7} \\ \frac{4}{7} & \frac{8}{7} \end{bmatrix}.$

c. $X=(I-A)^{-1}D=\begin{bmatrix} \frac{8}{7} & \frac{2}{7} \\ \frac{4}{7} & \frac{8}{7} \end{bmatrix}\begin{bmatrix} 2100 \\ 1400 \end{bmatrix}=\begin{bmatrix} 2800 \\ 2800 \end{bmatrix}$

88. a. The input-output matrix is

$$\begin{array}{c} \\ c \\ g \end{array}\begin{array}{c} \begin{array}{cc} c & g \end{array} \\ \begin{bmatrix} 0 & \frac{1}{2} \\ \frac{2}{3} & 0 \end{bmatrix} \end{array} = A.$$

b. $I-A=\begin{bmatrix} 1 & -\frac{1}{2} \\ -\frac{2}{3} & 1 \end{bmatrix}, D=\begin{bmatrix} 400 \\ 800 \end{bmatrix}$

$(I-A)^{-1}=\begin{bmatrix} \frac{3}{2} & \frac{3}{4} \\ 1 & \frac{3}{2} \end{bmatrix}$

$X=(I-A)^{-1}D=\begin{bmatrix} 1200 \\ 1600 \end{bmatrix}$

The production required is 1200 units of cheese and 1600 units of goats.

89. Write the input-output matrix.

$$A=\begin{array}{c} \\ A \\ M \end{array}\begin{array}{c} \begin{array}{cc} A & M \end{array} \\ \begin{bmatrix} .10 & .70 \\ .40 & .20 \end{bmatrix} \end{array}=\begin{bmatrix} \frac{1}{10} & \frac{7}{10} \\ \frac{4}{10} & \frac{2}{10} \end{bmatrix}$$

$I-A=\begin{bmatrix} 1 & 0 \\ 0 & 1 \end{bmatrix}-\begin{bmatrix} \frac{1}{10} & \frac{7}{10} \\ \frac{4}{10} & \frac{2}{10} \end{bmatrix}=\begin{bmatrix} \frac{9}{10} & -\frac{7}{10} \\ -\frac{4}{10} & \frac{8}{10} \end{bmatrix}$

Find $(I-A)^{-1}$.

$\left[\begin{array}{cc|cc} \frac{9}{10} & -\frac{7}{10} & 1 & 0 \\ -\frac{4}{10} & \frac{8}{10} & 0 & 1 \end{array}\right]$

$\left[\begin{array}{cc|cc} 1 & -\frac{7}{9} & \frac{10}{9} & 0 \\ -\frac{4}{10} & \frac{8}{10} & 0 & 1 \end{array}\right] \frac{10}{9}R_1$

$\left[\begin{array}{cc|cc} 1 & -\frac{7}{9} & \frac{10}{9} & 0 \\ 0 & \frac{44}{90} & \frac{4}{9} & 1 \end{array}\right] \frac{4}{10}R_1+R_2$

$\left[\begin{array}{cc|cc} 1 & -\frac{7}{9} & \frac{10}{9} & 0 \\ 0 & 1 & \frac{10}{11} & \frac{90}{44} \end{array}\right] \frac{90}{44}R_2$

$\left[\begin{array}{cc|cc} 1 & 0 & \frac{180}{99} & \frac{70}{44} \\ 0 & 1 & \frac{10}{11} & \frac{90}{44} \end{array}\right] \frac{7}{9}R_2+R_1$

$(I-A)^{-1}=\begin{bmatrix} \frac{180}{99} & \frac{70}{44} \\ \frac{10}{11} & \frac{90}{44} \end{bmatrix}$

$X=(I-A)^{-1}D$

$X=\begin{bmatrix} \frac{180}{99} & \frac{70}{44} \\ \frac{10}{11} & \frac{90}{44} \end{bmatrix}\begin{bmatrix} 60{,}000 \\ 20{,}000 \end{bmatrix}$

$X=\begin{bmatrix} 140{,}909 \\ 95{,}455 \end{bmatrix}$ Rounded

The agriculture industry should produce \$140,909, while the manufacturing industry should produce \$95,455.

Copyright © 2019 Pearson Education, Inc.

90. a. From the input-output matrix, we know that .9 unit agriculture; .4 unit services; .02 unit mining; .9 unit manufacturing are required to produce 1 unit.

b. The input-output matrix, A, and the matrix, $I-A$, are

$$A = \begin{bmatrix} .02 & .9 & 0 & .001 \\ 0 & .4 & 0 & .06 \\ .01 & .02 & .06 & .07 \\ .25 & .9 & .9 & .4 \end{bmatrix}$$

$$I - A = \begin{bmatrix} .980 & -.900 & 0 & -.001 \\ 0 & .600 & 0 & -.060 \\ -.010 & -.020 & .940 & -.070 \\ -.250 & -.900 & -.900 & .600 \end{bmatrix}$$

Next, calculate $(I-A)^{-1}$.

$$(I-A)^{-1} \approx \begin{bmatrix} 1.079 & 1.960 & .2132 & .223 \\ .064 & 2.129 & .230 & .240 \\ .060 & .4107 & 1.242 & .186 \\ .635 & 4.627 & 2.296 & 2.398 \end{bmatrix}$$

$$D = \begin{bmatrix} 760 \\ 1600 \\ 1000 \\ 2000 \end{bmatrix}$$

$$X = (I-A)^{-1}D \approx \begin{bmatrix} 4615 \\ 4165 \\ 2317 \\ 14,979 \end{bmatrix}$$

This is about 4615 units of agriculture, 4165 units of services, 2317 units of mining, and 14,979 units of manufacturing.

c. Since .25 units of manufacturing are used in producing each unit of agriculture, $(4615)(.25) = 1153.75$ units are used. Since .9 units of manufacturing are used in producing each unit of service, $(.9)(4165) = 3748.5$ units are used. Since .9 units of manufacturing are used in producing each unit of mining, $(.9)(2317) = 2085.3$. Since .4 units of manufacturing are used in producing each unit of manufacturing, $(.4)(14,979) = 5991.6$.
Thus, $1153.75 + 3748.5 + 2085.3 + 5991.6 \approx 12,979$ units of manufacturing are used in the production process.

91. a. From the input-output matrix, we know that .4 unit agriculture; .09 unit construction, .4 unit energy, .1 unit manufacturing, .9 unit transportation are required to produce 1 unit.

b. The input-output matrix, A and the matrix, $I-A$, are

$$A = \begin{bmatrix} .18 & .017 & .4 & .005 & 0 \\ .14 & .018 & .09 & .001 & 0 \\ .9 & 0 & .4 & .06 & .002 \\ .19 & .16 & .1 & .008 & .5 \\ .14 & .25 & .9 & .4 & .12 \end{bmatrix}$$

$(I-A)$

$$= \begin{bmatrix} .82 & -.017 & -.4 & -.005 & 0 \\ -.14 & .982 & -.09 & -.001 & 0 \\ -.9 & 0 & .6 & -.06 & -.002 \\ -.19 & -.16 & -.1 & .992 & -.5 \\ -.14 & -.25 & -.9 & -.4 & .88 \end{bmatrix}$$

$$X = \begin{bmatrix} 28,067 \\ 9383 \\ 51,372 \\ 61,364 \\ 90,403 \end{bmatrix}$$

$$D = (I-A)X = \begin{bmatrix} 1999.809 \\ 599.882 \\ 1700.254 \\ 3700.378 \\ 2499.110 \end{bmatrix}$$

This represents about 2000 units of agriculture, 600 units of construction, 1700 units of energy, 3700 units of manufacturing, and 2500 units of transportation.

Copyright © 2019 Pearson Education, Inc.

92. Multiply matrix $(I-A)^{-1}$ by matrix D where

$$D=\begin{bmatrix}2400\\850\\1400\\3200\\1800\end{bmatrix}$$

$$(I-A)^{-1} \approx \begin{bmatrix}7.744 & .294 & 5.745 & .509 & .302\\ 2.319 & 1.111 & 1.890 & .167 & .099\\ 13.100 & .543 & 11.546 & 1.006 & .598\\ 14.119 & .976 & 12.001 & 2.357 & 1.367\\ 21.707 & 1.361 & 18.714 & 2.229 & 2.445\end{bmatrix}$$

$$X=(I-A)^{-1}D \approx \begin{bmatrix}29{,}049\\9869\\52{,}362\\61{,}520\\90{,}987\end{bmatrix}$$

This represents 29,049 units of agriculture, 9869 units of construction, 52,362 units of energy, 61,520 units of manufacturing, and 90,987 units of transportation.

93.

$$M=\begin{array}{c} \\ A\\B\\C\\D\end{array}\begin{array}{c}\begin{array}{cccc}A&B&C&D\end{array}\\ \begin{bmatrix}0&1&1&0\\1&0&1&1\\1&1&0&1\\0&1&1&0\end{bmatrix}\end{array}$$

a. M^2 gives the number of one-step routes between stations.

$$M^2=\begin{bmatrix}0&1&1&0\\1&0&1&1\\1&1&0&1\\0&1&1&0\end{bmatrix}\begin{bmatrix}0&1&1&0\\1&0&1&1\\1&1&0&1\\0&1&1&0\end{bmatrix}=\begin{bmatrix}2&1&1&2\\1&3&2&1\\1&2&3&1\\2&1&1&2\end{bmatrix}$$

b. $m_{14}^2=2$ so there are 2 one-stop routes between stations A and D.

c. The number of two-stop routes is given by

$$M^3=M^2\cdot M=\begin{bmatrix}2&5&5&2\\5&4&5&5\\5&5&4&5\\2&5&5&2\end{bmatrix}$$

d. $m_{14}^3=2$ so there are 2 two-stop routes between stations A and D

94.

$$M=\begin{array}{c} \\ A\\B\\C\\D\end{array}\begin{array}{c}\begin{array}{cccc}A&B&C&D\end{array}\\ \begin{bmatrix}0&1&0&1\\1&0&0&1\\0&0&0&1\\1&1&1&0\end{bmatrix}\end{array}$$

a. M^2 gives the number of one-step flights between cities.

$$M^2=\begin{bmatrix}0&1&0&1\\1&0&0&1\\0&0&0&1\\1&1&1&0\end{bmatrix}\begin{bmatrix}0&1&0&1\\1&0&0&1\\0&0&0&1\\1&1&1&0\end{bmatrix}=\begin{bmatrix}2&1&1&1\\1&2&1&1\\1&1&1&0\\1&1&0&3\end{bmatrix}$$

$m_{13}^2=1$ so there is 1 one-stop flight between cities A and C.

b. The number of direct or one-stop flights from B to C is $m_{23}^2+m_{23}=1+0=1$.

c. The number of two-stop flights is given by

$$M^3=M^2\cdot M=\begin{bmatrix}2&3&1&4\\3&2&1&4\\1&1&0&3\\4&4&3&2\end{bmatrix}$$

Copyright © 2019 Pearson Education, Inc.

95. The message "leave now" broken into groups of 2 letters would have the following matrices:

$$\begin{bmatrix}12\\5\end{bmatrix}, \begin{bmatrix}1\\22\end{bmatrix}, \begin{bmatrix}5\\27\end{bmatrix}, \begin{bmatrix}14\\15\end{bmatrix}, \begin{bmatrix}23\\27\end{bmatrix}$$

Multiply $M = \begin{bmatrix}2 & 6\\1 & 4\end{bmatrix}$ by each matrix in the code to obtain $\begin{bmatrix}54\\32\end{bmatrix}, \begin{bmatrix}134\\89\end{bmatrix}, \begin{bmatrix}172\\113\end{bmatrix}, \begin{bmatrix}118\\74\end{bmatrix}, \begin{bmatrix}208\\131\end{bmatrix}$.

96. To decode the message, use $M^{-1} = \begin{bmatrix}2 & -3\\-\frac{1}{2} & 1\end{bmatrix}$.

Case 6 Airline Route Maps

1. As the matrix A^2 indicates, the only city *not* reachable from San Antonio in a two-flight sequence is Lubbock, since this is the only city whose column has a zero in the row for San Antonio. Thus, the cities reachable by a two-flight sequence from San Antonio are all Stampede cities except Lubbock.
In the matrix A^3, there are no zero entries in the row for San Antonio, so all Stampede Air cities may be reached by a three-flight sequence.

2. El Paso—Austin—Dallas—San Antonio
El Paso—Austin—Houston—San Antonio
El Paso—Dallas—Houston—San Antonio
El Paso—Houston—Dallas—San Antonio

3. The trips between two different cities for which the matrix entries remain zero until A^3 are trips between Lubbock and Corpus Christi, and Lubbock and San Antonio. These trips each take three flights.

4. There are 5 vertices, so the adjacency matrix is a 5×5 matrix. If the vertices in the Vacation Air graph respectively correspond to Jacksonville (J), Orlando (O), West Palm Beach (W), Miami (M), and Tampa (T), then the adjacency matrix for Vacation Air is

$$\begin{matrix} & J & O & W & M & T \end{matrix}$$
$$B = \begin{bmatrix}0 & 0 & 0 & 1 & 1\\0 & 0 & 0 & 0 & 1\\0 & 0 & 0 & 0 & 1\\1 & 0 & 0 & 0 & 1\\1 & 1 & 1 & 1 & 0\end{bmatrix}.$$

5. To find the cities that may be reached by a two-flight sequence, find the matrix that represents B^2.

$$\begin{matrix} & J & O & W & M & T \end{matrix}$$
$$B^2 = \begin{bmatrix}2 & 1 & 1 & 1 & 1\\1 & 1 & 1 & 1 & 0\\1 & 1 & 1 & 1 & 0\\1 & 1 & 1 & 2 & 1\\1 & 0 & 0 & 1 & 4\end{bmatrix}.$$

This matrix shows the two-flight sequences between cities. An entry is nonzero whenever there is a two-step sequence between the cities.

Copyright © 2019 Pearson Education, Inc.

6. Add the matrices found in Exercises 4 and 5.

$$B+B^2=\begin{array}{c} \begin{matrix} J & O & W & M & T \end{matrix} \\ \begin{bmatrix} 2 & 1 & 1 & 2 & 2 \\ 1 & 1 & 1 & 1 & 1 \\ 1 & 1 & 1 & 1 & 1 \\ 2 & 1 & 1 & 2 & 2 \\ 2 & 1 & 1 & 2 & 4 \end{bmatrix} \end{array}.$$

From this matrix, we can conclude that all entries are nonzero. This means that one can fly between and two of these cities in at most two steps.

Copyright © 2019 Pearson Education, Inc.

Chapter 7 Linear Programming

Section 7.1 Graphing Linear Inequalities in Two Variables

1. $y \ge -x - 2$

Graph of $y = -x - 2$ is a solid line. The intercepts are (–2, 0) and (0, –2). Use the origin as a test point. Since $0 \ge -0 - 2$ is true, the origin will be included in the region, so shade the half-plane above the line. This matches choice F.

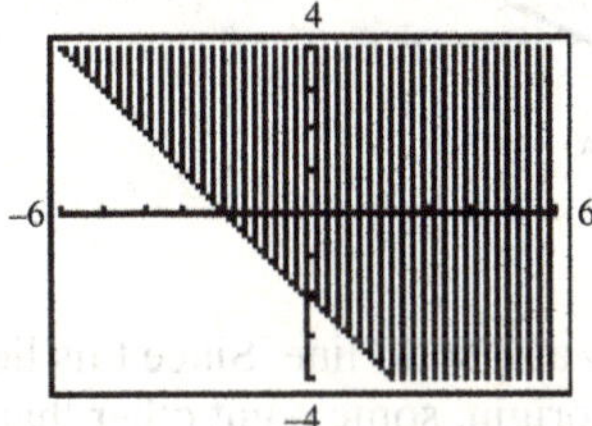

3. $y \le x + 2$

Graph $y = x + 2$ is a solid line. The intercepts are (–2, 0) and (0, 2). Use the origin as a test point. Since $0 \le 0 + 2$ is true, the origin will be included in the region, so shade the half-plane below the line. This matches choice A.

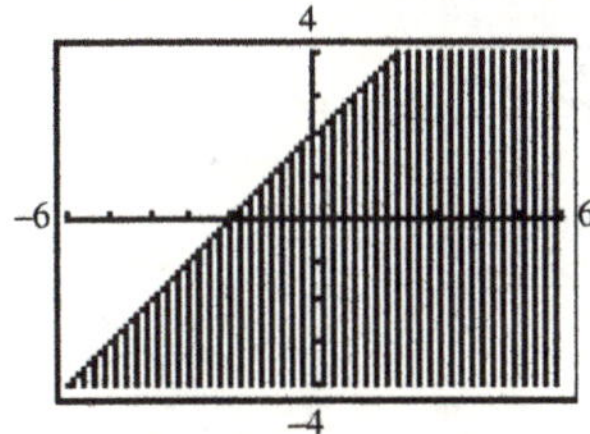

5. $6x + 4y \ge -12$

Graph $6x + 4y = -12$ is a solid line. The intercepts are (–2, 0) and (0, –3). Use the origin as a test point. Since $6(0) + 4(0) \ge -12$ is true, the origin will be included in the region, so shade the half-plane above the line. This matches choice E.

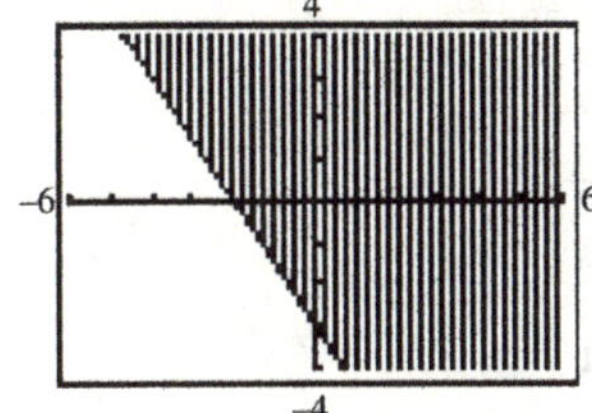

7. $y < 5 - 2x$

Graph $y = 5 - 2x$ as a dashed line. The intercepts are (0, 5) and $\left(\frac{5}{2}, 0\right)$. Test (0, 0) in the original inequality. The result, $0 < 5$, is true, so shade the region that contains (0, 0).

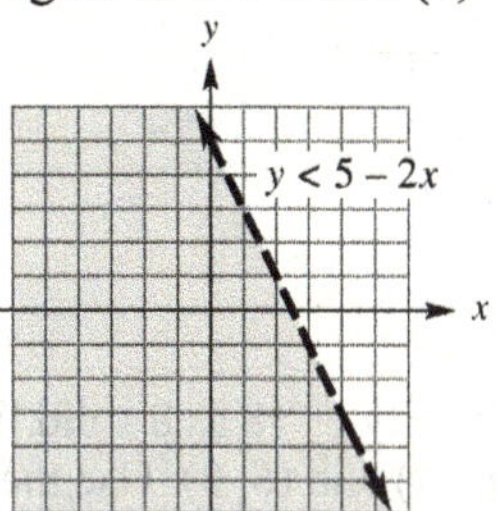

9. $x - 2y \ge 3$

Graph $x - 2y = 3$ as a solid line. The intercepts are (0, –1.5) and (3, 0). Use the origin as a test point. Since $(0) - 2(0) \ge 3$ is false, the origin will not be included in the region, so shade the half-plane below the line.

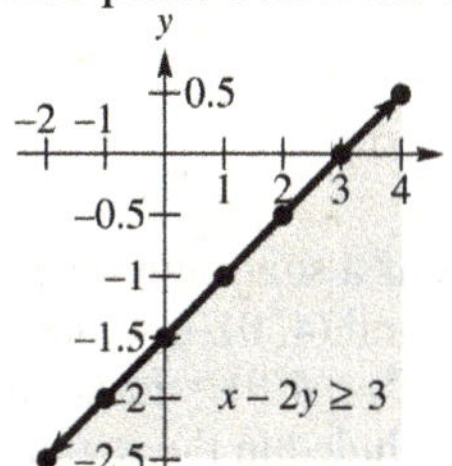

11. $2x - y \le 4$

Graph $2x - y = 4$ as a solid line. The intercepts are (0, –4) and (2, 0). Use the origin as a test point. Since $2(0) - 0 \le 4$ is true, the origin will be included in the region, so shade the half-plane above the line.

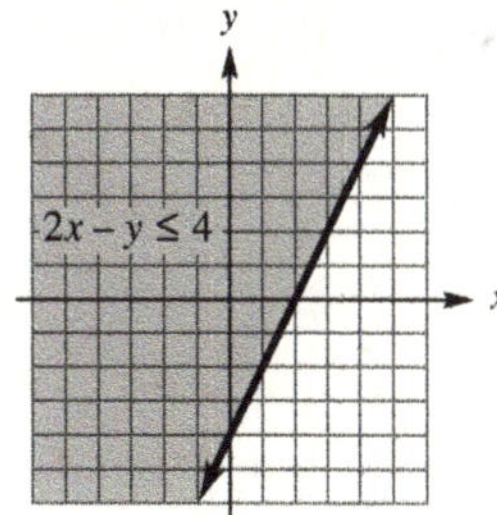

Copyright © 2019 Pearson Education, Inc.

13. $y \le -4$
Graph $y = -4$ as a solid line. $y = -4$ is a horizontal line crossing the y-axis at $(0, -4)$. Since the inequality symbol is $\le$, shade the half-plane below the line.

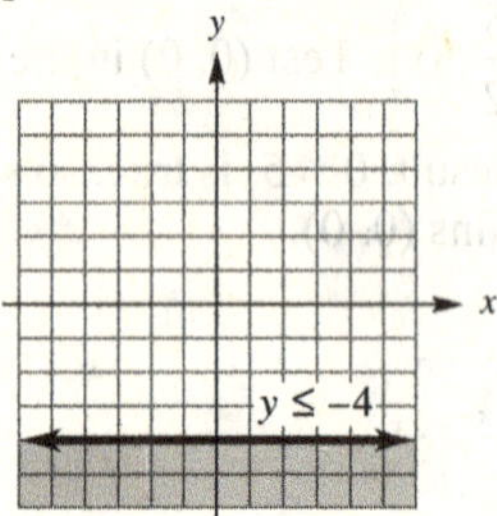

15. $3x - 2y \ge 18$
Graph the line $3x - 2y = 18$ as a solid line. The intercepts are $(0, -9)$ and $(6, 0)$. Use the origin as a test point. Since $3(0) - 2(0) \ge 18$ is false, the origin will not be included in the region, so shade the half-plane below the line.

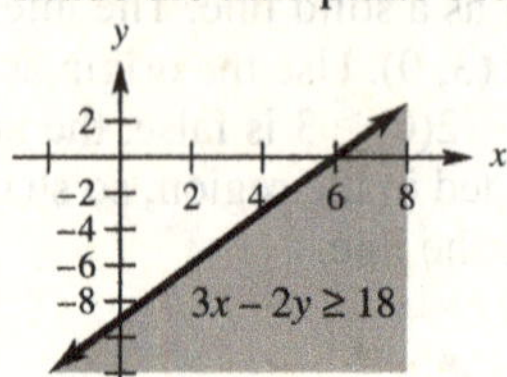

17. $3x + 4y \ge 12$
Graph $3x + 4y = 12$ as a solid line. The intercepts are $(0, 3)$ and $(4, 0)$. Use the origin as a test point. Since $3(0) + 4(0) > 12$ is false, the origin will not be included in the region, so shade the half-plane above the line.

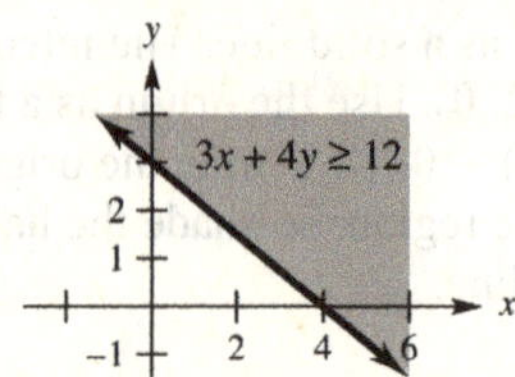

19. $2x - 4y \le 3$
Graph $2x - 4y = 3$ as a solid line. The intercepts are $\left(\frac{3}{2}, 0\right)$ and $\left(0, -\frac{3}{4}\right)$. Use the origin as a test point. $2(0) - 4(0) < 3$ is true, so the region above the line, which includes the origin, is the correct region to shade.

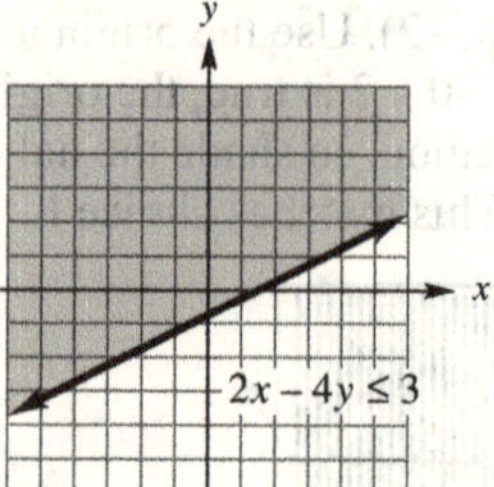

21. $x \le 5y$
Graph $x = 5y$ as a solid line. Since this line contains the origin, some point other than $(0, 0)$ must be used as a test point. The point $(1, 2)$ gives $1 \le 5(2)$ or $1 \le 10$, a true sentence. Shade the side of the line containing $(1, 2)$, that is, the side above the line.

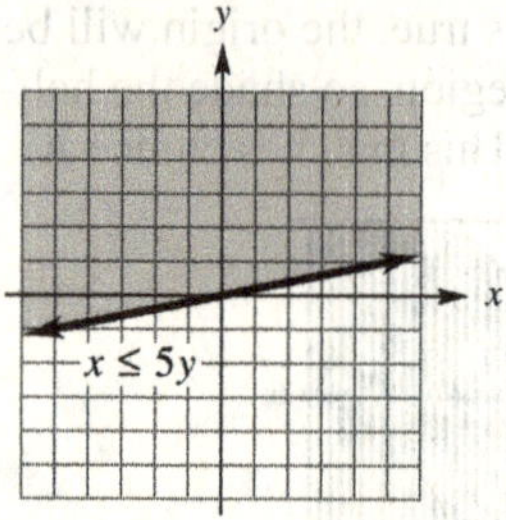

23. $-3x \le y$
Graph $y = -3x$ as a solid line. Since this line contains the origin, use some point other than $(0, 0)$ as a test point. $(1, 1)$ used as a test point gives $-3 < 1$, a true sentence. Shade the region containing $(1, 1)$, which is the region above the line.

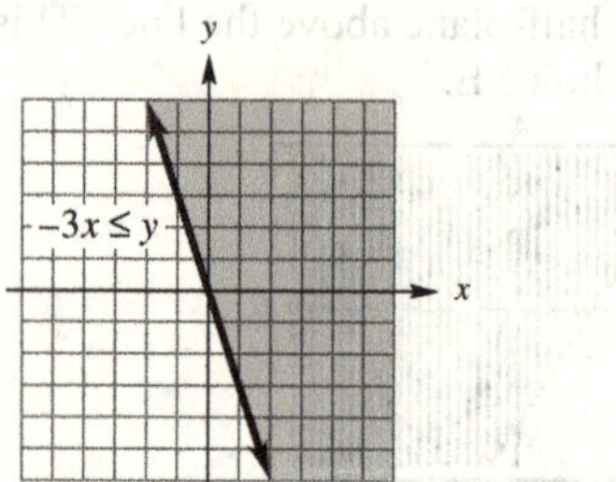

Copyright © 2019 Pearson Education, Inc.

25. $y \le x$

Graph $y = x$ as a dashed line. Since this line contains the origin, choose a point other than (0, 0) as a test point. (2, 3) gives $3 < 2$, which is false. Shade the region that does not contain the test point, that is, the region below the line.

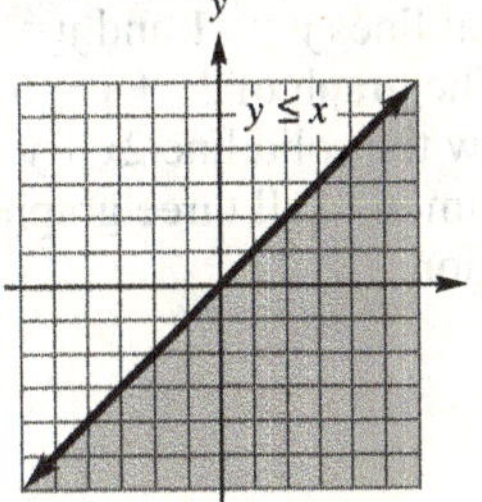

27. Answers may vary.

Possible answer: When the inequality is $<$ or $>$, the line is dashed. When the inequality is $\le$ or $\ge$, the line is solid.

29. $y \ge 3x - 6$
$y \ge -x + 1$

Graph $y \ge 3x - 6$ as the region on or above the solid line $y = 3x - 6$. Graph $y \ge -x + 1$ as the region on or above the solid line $y = -x + 1$. The feasible region is the overlap of the two half-planes.

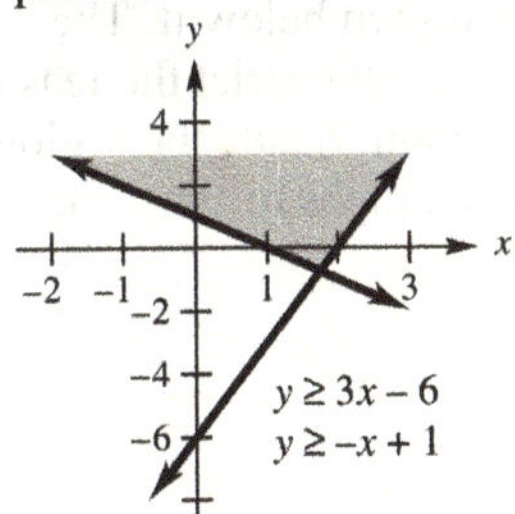

31. $2x + y \le 5$
$x + 2y \le 5$

Graph $2x + y \le 5$ as the region on or below the solid line $2x + y = 5$, which has intercepts (0, 5) and $\left(\frac{5}{2}, 0\right)$. Graph $x + 2y \le 5$ as the region on or below the solid line $x + 2y = 5$, which has intercepts $\left(0, \frac{5}{2}\right)$ and (5, 0). The feasible region is the overlap of the two half-planes.

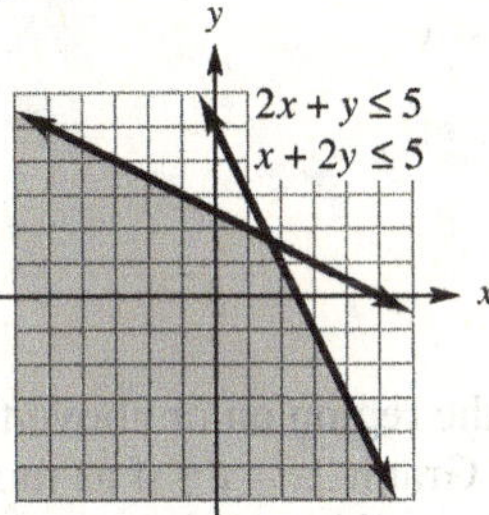

33. $2x + y \ge 8$
$4x - y \le 3$

Graph $2x + y \ge 8$ as the region on or above the dashed line $2x + y = 8$, which has intercepts (0, 8) and (4, 0). Graph $4x - y \le 3$ as the region above the solid line $4x - y = 3$, which has intercepts (0, –3) and $\left(\frac{3}{4}, 0\right)$.

The overlap of these two regions is the feasible region.

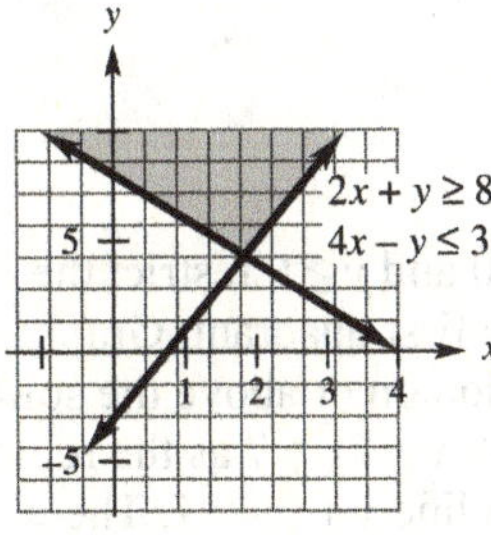

Copyright © 2019 Pearson Education, Inc.

35. $2x - y \leq 1$
$3x + y \leq 6$

Graph $2x - y \leq 1$ as the region on or above the dashed line $2x - y = 1$. Graph $3x + y \leq 6$ as the region on or below the solid line $3x + y = 6$. Shade the overlapping part of these two regions to show the feasible region.

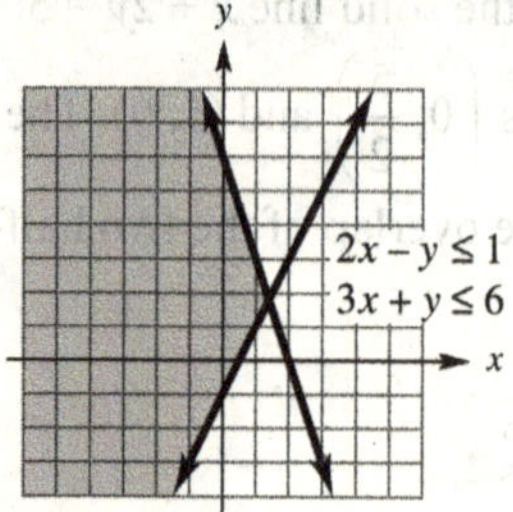

37. $-x - y \leq 5$
$2x - y \leq 4$

Graph $-x - y \leq 5$ as the region on or above the solid line $-x - y = 5$. Graph $2x - y \leq 4$ as the region on or above the solid line $2x - y = 4$. Shade the overlapping part of these two regions to show the feasible region.

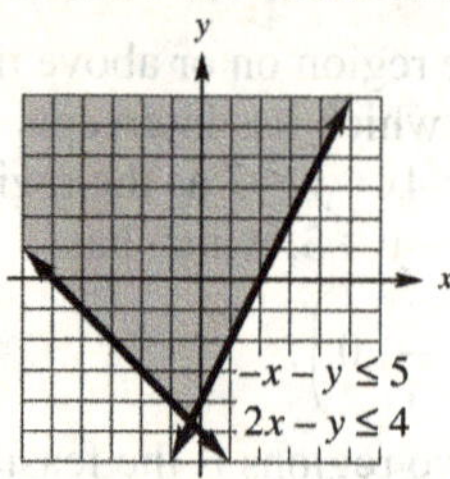

39. $3x + y \geq 6$
$x + 2y \geq 7$
$x \geq 0$
$y \geq 0$

The inequalities $x \geq 0$ and $y \geq 0$ restrict the feasible region to the first quadrant. Graph $3x + y \geq 6$ as the region on or above the solid line $3x + y = 6$. Graph $x + 2y \geq 7$ as the region on or above the solid line $x + 2y = 7$. The feasible region is the overlap of these half-planes.

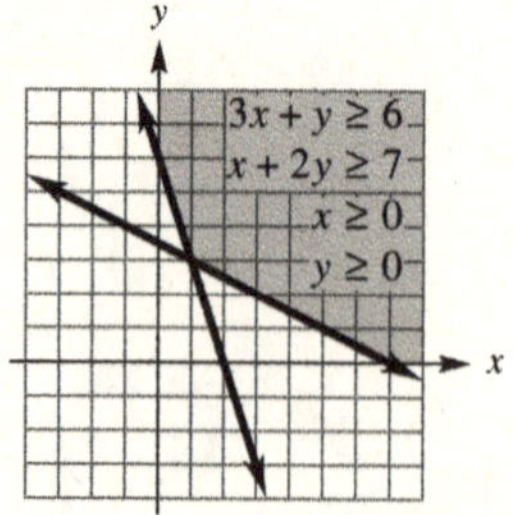

41. $-2 \leq x \leq 3$
$-1 \leq y \leq 5$
$2x + y \leq 6$

The graph of $-2 \leq x \leq 3$ is the region between the vertical line $x = -2$ and $x = 3$, including the lines. The graph of $-1 \leq y \leq 5$ is the region between the horizontal lines $y = -1$ and $y = 5$, including the lines. The graph of $2x + y \leq 6$ is the region on or below the solid line $2x + y = 6$. Shade the region common to all three graphs to show the feasible region.

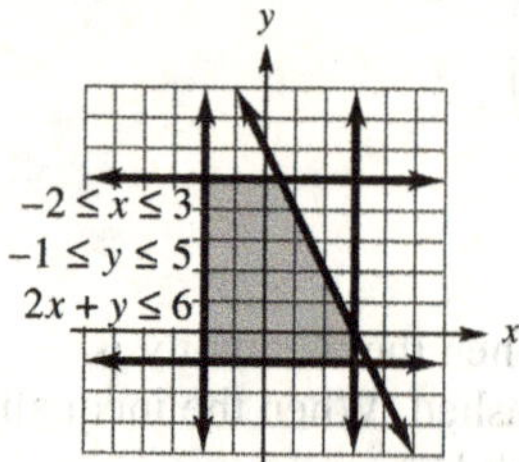

43. $2y - x \geq -5$
$y \leq 3 + x$
$x \geq 0$
$y \geq 0$

The graph of $2y - x \geq -5$ consists of the boundary line $2y - x = 5$ and the region above it. The graph of $y \leq 3 + x$ consists of the boundary line $y = 3 + x$ and the region below it. The inequalities $x \geq 0$ and $y \geq 0$ restrict the feasible region to the first quadrant. Shade the region common to all of these graphs to show the feasible region.

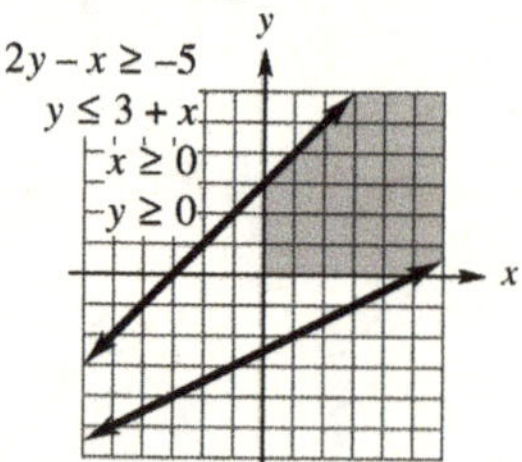

Copyright © 2019 Pearson Education, Inc.

45. $3x+4y \geq 12$
$2x-3y \leq 6$
$0 \leq y \leq 2$
$x \geq 0$

$3x + 4y \geq 12$ is the set of points on or above the solid line $3x + 4y = 12$. $2x - 3y \leq 6$ is the set of points on or above the solid line $2x - 3y = 6$. $0 \leq y \leq 2$ is the rectangular strip of points lying on or between the horizontal lines $y = 0$ and $y = 2$. $x \geq 0$ consists of all the points on and to the right of the y-axis. The feasible region is the triangular region satisfying all the inequalities.

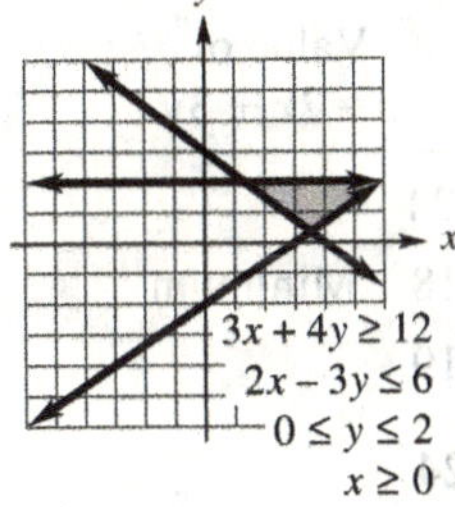

47. The shaded region lies between the horizontal lines $y = 0$ and $y = 4$, so this inequality is $0 \leq y \leq 4$. The shaded region also lies to the right of the vertical line $x = 0$, so this inequality is $x \geq 0$. The shaded region also lies below the line containing the points (3, 4) and (6, 0) which has the equation $4x + 3y = 24$.

Using the origin as a test point, we determine that
$4(0) + 3(0) \leq 24$ is true, so the inequality must be $\leq$. So the third inequality is $4x + 3y \leq 24$. So, the inequalities are
$x \geq 0$
$0 \leq y \leq 4$
$4x+3y \leq 24$

49. The feasible region is the interior of the rectangle with vertices (2, 3), (2, –1), (7, 3), and (7, –1). The x-values range from 2 to 7, not including 2 or 7. The y-values range from –1 to 3, not including –1 or 3. Thus, the system of inequalities is
$2 < x < 7$
$-1 < y < 3.$

51. The system is

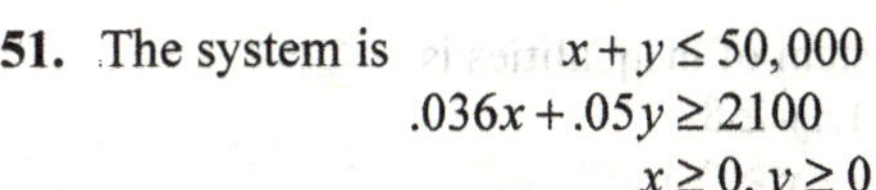

$x+y \leq 50{,}000$
$.036x+.05y \geq 2100$
$x \geq 0, y \geq 0$

This system will reduce to: $y \leq -x+50{,}000$
$y \geq -.72x+45{,}000$
$x \geq 0, y \geq 0$

The first inequality gives the region below the line that crosses the y axis at 50,000, including the points on the line. The second inequality gives the region above the line that crosses the y axis at 45,000, including the points on the line. The other inequalities are the x and y axes.

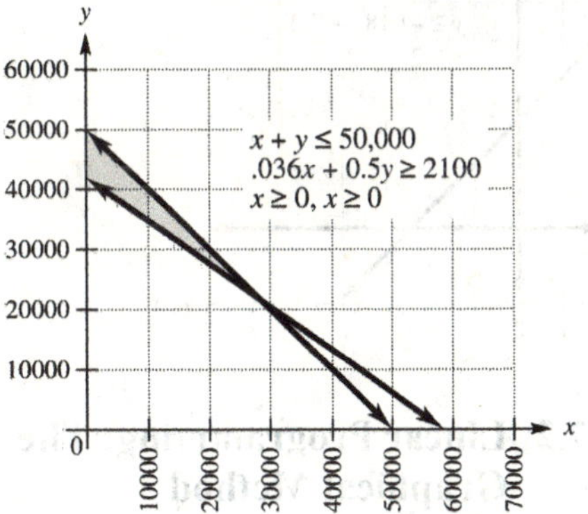

53. The system is $x+y \leq 60{,}000$
$.024x+.042y \geq 2016$
$x \geq 0, y \geq 0$

This system will reduce to: $y \leq -x+60{,}000$
$y \geq -\frac{12}{21}x+48{,}000$
$x \geq 0, y \geq 0$

The first inequality gives the region below the line that crosses the y axis at 60,000, including the points on the line. The second inequality gives the region above the line that crosses the y axis at 48,000, including the points on the line. The other inequalities are the x and y axes.

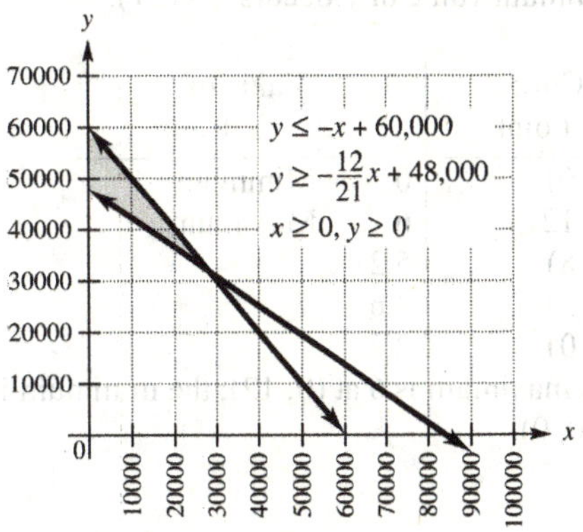

Copyright © 2019 Pearson Education, Inc.

55. The system of inequalities is
$1.3x + 1.1y \ge 8$
$x \ge 2, y \ge 2$

This system will reduce to: $y \ge -1.18x + 7.27$
$x \ge 2, y \ge 2$

The first inequality gives the region above the line that crosses the y axis at 7.27, including the points on the line. The other inequalities are a horizontal and vertical line crossing the the x and y axes at 2.

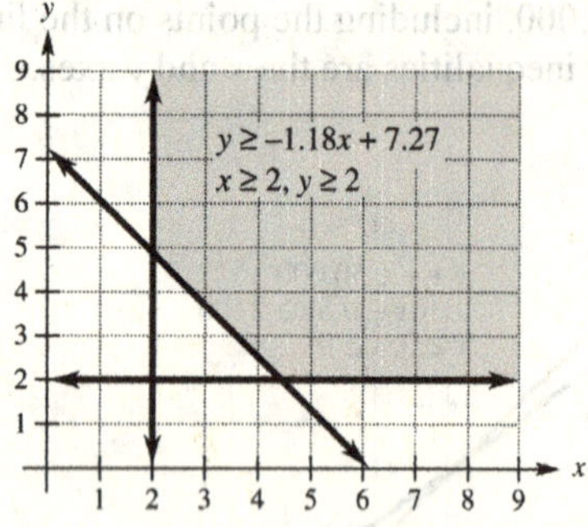

Section 7.2 Linear Programming: The Graphical Method

1. Make a table indicating the value of the objective function $z = 6x + y$ at each corner point.

Corner Point	Value of $z = 6x + y$
(1, 1)	$6(1)+1=7$ Minimum
(2, 7)	$6(2)+1(7)=19$
(5, 10)	$6(5)+10=40$ Maximum
(6, 3)	$6(6)+3=39$

The maximum value of 40 occurs at (5, 10). The minimum value of 7 occurs at (1, 1).

3.

Corner Point	Value of $z = .3x + .5y$	
(0, 0)	0	Minimum
(0, 12)	6	Maximum
(4, 8)	5.2	
(7, 3)	3.6	
(9, 0)	2.7	

The maximum is 6 at (0, 12); the minimum is 0 at (0, 0).

5. a.

Corner Point	Value of $z = x + 5y$
(0, 8)	40
(3, 4)	23
$\left(\frac{13}{2}, 2\right)$	16.5
(12, 0)	12 Minimum

The minimum is 12 at (12, 0). There is no maximum because the feasible region is unbounded.

b.

Corner Point	Value of $z = 2x + 3y$
(0, 8)	24
(3, 4)	18 Minimum
$\left(\frac{13}{2}, 2\right)$	19
(12, 0)	24

The minimum is 18 at (3, 4). There is no maximum because the feasible region is unbounded.

c.

Corner Point	Value of $z = 2x + 4y$
(0, 8)	32
(3, 4)	22
$\left(\frac{13}{2}, 2\right)$	21 Minimum
(12, 0)	24

The minimum is 21 at $\left(\frac{13}{2}, 2\right)$. There is no maximum.

d.

Corner Point	Value of $z = 4x + y$
(0, 8)	8 Minimum
(3, 4)	16
$\left(\frac{13}{2}, 2\right)$	28
(12,0)	48

The minimum is 8 at (0, 8). There is no maximum.

The minimum is 18 at (2, 4). There is no maximum value of z because the feasible region is unbounded.

Copyright © 2019 Pearson Education, Inc.

7. Maximize $z = 4x + 3y$
subject to: $2x + 3y \le 6$
$4x + y \le 6$
$x \ge 0,\ y \ge 0.$

Graph the feasible region, and identify the corner points.

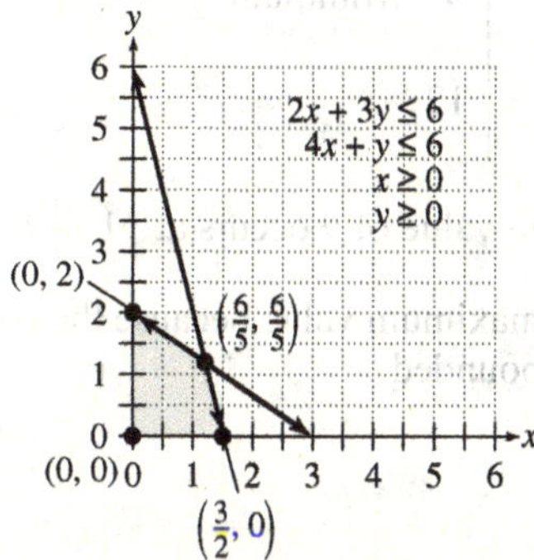

The graph shows the feasible region is bounded. The corner points are

$(0, 0), (0, 2), \left(\frac{3}{2}, 0\right), \left(\frac{6}{5}, \frac{6}{5}\right)$, which is the intersection of $2x + 3y = 6$ and $4x + y = 6$. Use the corner points to find the maximum value of the objective function.

Corner Point	Value of $z = 4x + 3y$
(0, 0)	0
(0, 2)	6
$\left(\frac{6}{5}, \frac{6}{5}\right)$	$\frac{42}{5} = 8.4$ Maximum
$\left(\frac{3}{2}, 0\right)$	6

The maximum value of $z = 4x + 3y$ is 8.4 at $x = 1.2, y = 1.2.$

9. Minimize $z = 2x + y$
subject to: $3x - y \ge 12$
$x + y \le 15$
$x \ge 2,\ y \ge 3.$

Graph the feasible region, and identify the corner points

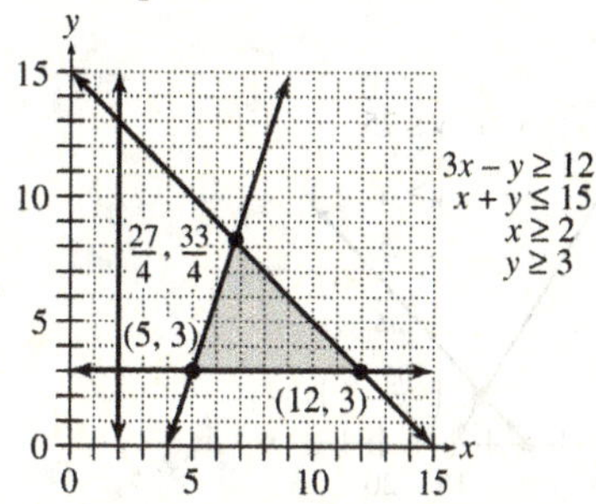

The feasible region is bounded with corner points $(5, 3), \left(\frac{27}{4}, \frac{33}{4}\right), (12, 3)$.

Corner Point	Value of $z = 2x + y$
(5, 3)	13 Minimum
$\left(\frac{27}{4}, \frac{33}{4}\right)$	$\frac{87}{4}$
(12, 3)	27

The minimum is 13 at (5, 3).

Copyright © 2019 Pearson Education, Inc.

11. Maximize $z = 5x + y$
subject to: $x - y \le 10$
$5x + 3y \le 75$
$x \ge 0,\ y \ge 0.$

Graph the feasible region, and identify the corner points.

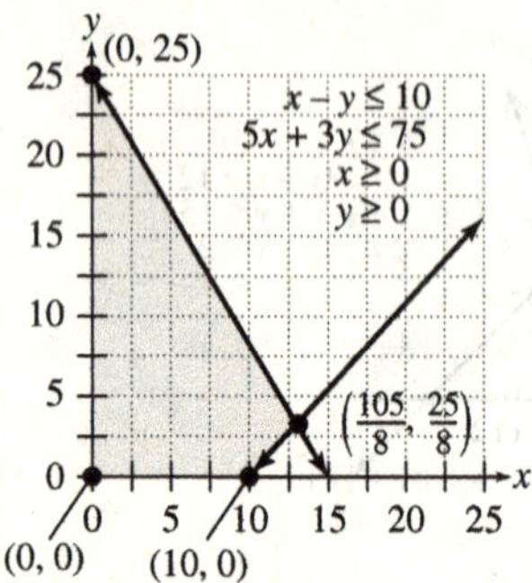

This region is bounded, with corner points $(0, 0)$, $(0, 25)$, $\left(\frac{105}{8}, \frac{25}{8}\right)$, $(10, 0)$.

Corner Point	Value of $z = 5x + y$
$(0, 0)$	0
$(0, 25)$	25
$\left(\frac{105}{8}, \frac{25}{8}\right)$	$\frac{275}{4} = 68.75$ Maximum
$(10, 0)$	50

The maximum is 68.75 at $\left(\frac{105}{8}, \frac{25}{8}\right)$.

13. $3x + 2y \ge 6$
$x + 2y \ge 4$
$x \ge 0,\ y \ge 0$

Graph the feasible region and identify the corner points.

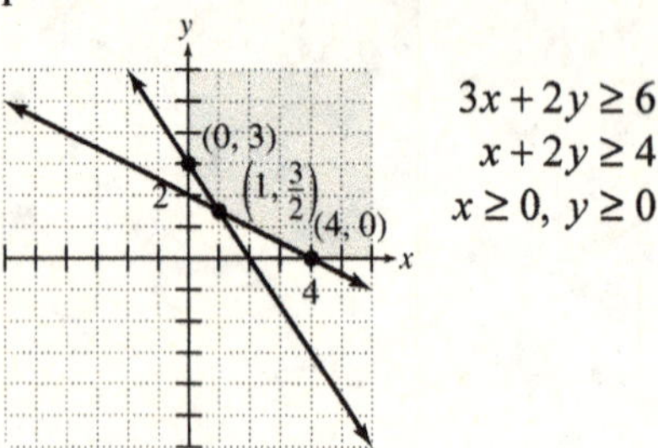

Corner Point	Value of $z = 3x + 4y$
$(0, 3)$	12
$\left(1, \frac{3}{2}\right)$	9 Minimum
$(4, 0)$	12

The minimum value of 9 occurs at $\left(1, \frac{3}{2}\right)$.

There is no maximum value because the feasible region is unbounded.

15. $x + y \le 6$
$-x + y \le 2$
$2x - y \le 8$

Graph the feasible region, and identify the corner points.

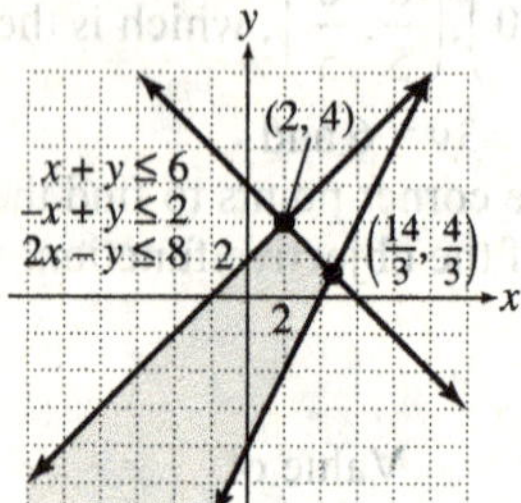

Corner Point	Value of $z = 3x + 4y$
$(2, 4)$	22 Maximum
$\left(\frac{14}{3}, \frac{4}{3}\right)$	$\frac{58}{3}$

The maximum value of z is 22 at (2, 4). There is no minimum because the feasible region is unbounded.

Copyright © 2019 Pearson Education, Inc.

17. a. 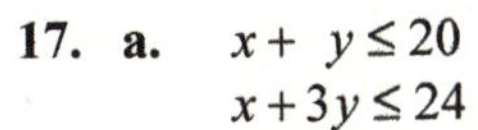

$x + y \le 20$
$x + 3y \le 24$

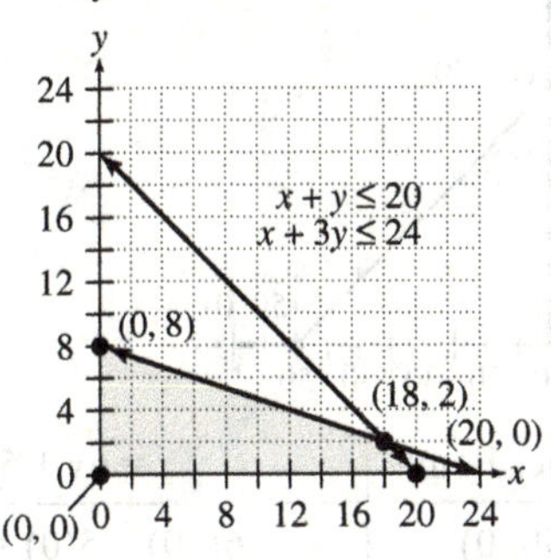

Corner Point	Value of $z = 10x + 12y$
(0, 0)	0
(0, 8)	96
(18, 2)	204 Maximum
(20, 0)	200

The maximum value of 204 occurs when $x = 18$ and $y = 2$, or at (18, 2).

b. $3x + y \le 15$
$x + 2y \le 18$

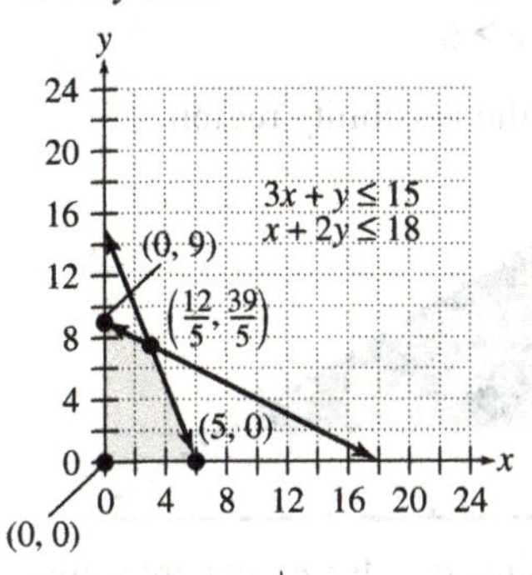

Corner Point	Value of $z = 10x + 12y$
(0, 0)	0
(0, 9)	108
$\left(\frac{12}{5}, \frac{39}{5}\right)$	$\frac{588}{5}$
(5, 0)	50

The maximum value of $\frac{588}{5}$ or $117\frac{3}{5}$ occurs when $x = \frac{12}{5}$ and $y = \frac{39}{5}$, or at $\left(\frac{12}{5}, \frac{39}{5}\right)$.

c. $x + 2y \ge 10$
$2x + y \ge 12$
$x - y \le 8$

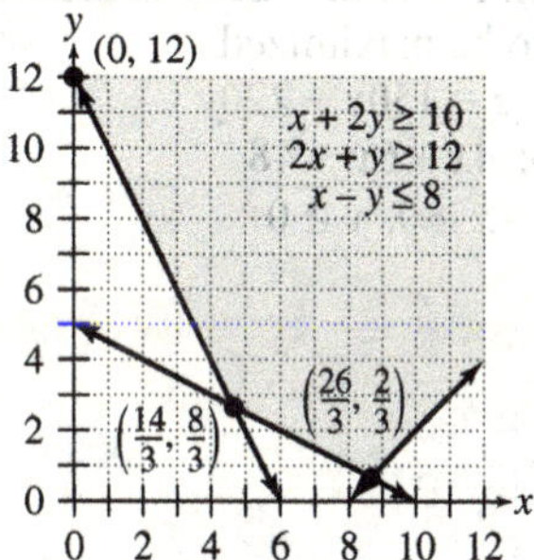

The feasible region is unbounded, so there is no maximum.

19. Answer varies. Sample response: The constraints do not describe a feasible region, i.e. there does not exist a point that satisfies all five constraints.

Section 7.3 Applications of Linear Programming

1. Let x = the number of canoes and let y = the number of rowboats. The constraints are $8x + 5y \le 110, x \ge 0, y \ge 0$.

3. Let x = the number of radio spots and let y = the number of TV ads. The constraints are $250x + 750y \le 9500, x \ge 0$ $y \ge 0$.

Copyright © 2019 Pearson Education, Inc.

5. Let x = the number of chain saws and let y = the number of wood chippers.
Assembling x chain saws at 4 hours each takes $4x$ hours while assembling y wood chippers at 6 hours each takes $6y$ hours. There are only 48 available hours, the first constraint is
$4x + 6y \leq 48$.
The number of chain saws and wood chippers assembled cannot be negative, so $x \geq 0, y \geq 0$.
Each chain saw produces a profit of \$150 and each wood chipper, \$220. If z represents total profit, then $z = 150x + 220y$ is the objective function to be maximized.
Maximize $z = 150x + 220y$
subject to: $4x + 6y \leq 48$
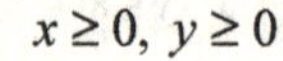
$x \geq 0,\ y \geq 0$

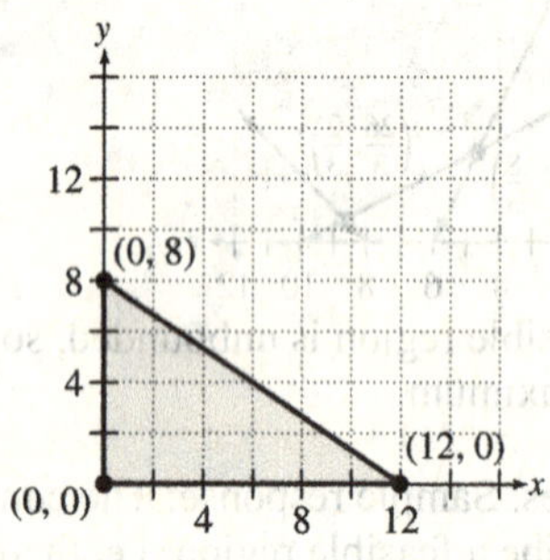

Corner Point	$z = 150x + 220y$
(0, 0)	$150(0)+220(0) = 0$
(12, 0)	$150(12) + 220(0) = 1800$ (maximum)
(0, 8)	$150(0) + 220(8) = 1760$

12 chain saws and no wood chippers should be assembled for maximum profit.

7. Let x = the number of pounds of deluxe coffee and let y = the number of pounds of regular coffee. The mixture of deluxe and regular coffee needs to be at least 50 pounds, so $x + y \geq 50$
The mixture must have at least 10 pounds of deluxe coffee, so $x \geq 10$.
The pounds of coffee cannot be negative, so $x \geq 0, y \geq 0$.
At \$6 per pound of deluxe coffee and \$5 per pound of regular coffee, the total cost, z, is $z = 6x + 5y$, which is the objective function to be minimized.
Minimize: $z = 6x + 5y$
subject to: $x + y \geq 50$
$x \geq 10$
$y \geq 0$

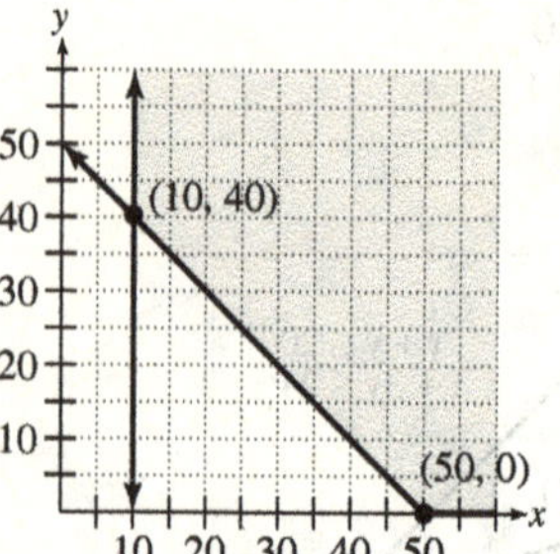

Corner Point	$z = 6x + 5y$
(50, 0)	$6(50) + 5(0) = 300$
(10, 40)	$6(10) + 5(40) = 260$ (minimum)

The mixture should contain 10 pounds of deluxe and 40 pounds of regular coffee to minimize cost.

9. Let x = the number of shock replacements and let y = the number of brake replacements.
The constraint imposed by his 48 hour week is $3x + 2y \leq 48$.
He sells at least two shock replacements and at least six brake replacements, so $x \geq 2$ and $y \geq 6$.
Thus, the constraints are
$3x + 2y \leq 48$
$x \geq 2$
$y \geq 6$

Graph the feasibility region

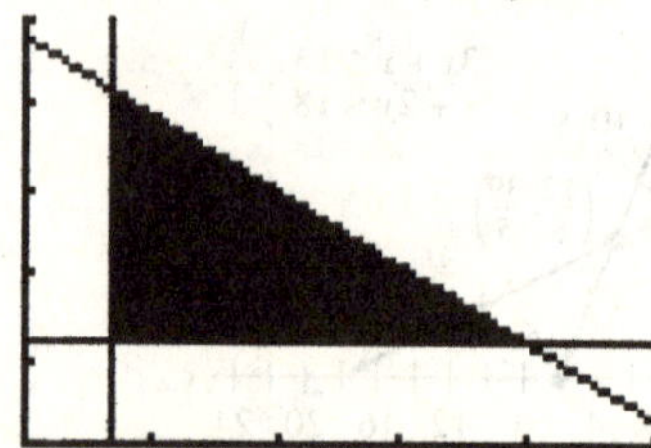

The points of intersection are summarized in the table below.
His labor cost, z, is $z = 500x + 300y$, which is the objective function to be maximized.

Corner Point	$z = 500x + 300y$
(2, 6)	$500(2) + 300(6) = 2800$
(2, 21)	$500(2) + 300(21) = 7300$
(12, 6)	$500(12) + 300(6) = 7800$ (maximum)

He should do 12 shock replacements and 6 brake replacements.

Copyright © 2019 Pearson Education, Inc.

11. From exercise 3, we let x = the number of radio spots and let y = the number of TV ads. Since each radio spot costs \$250 and each TV ad cost \$750, the constraint imposed by the cost is $250x + 750y \le 9500$. The constraints imposed by the number of ads are $x \ge 8$ and $y \ge 3$. Thus, the constraints are
$250x + 750y \le 9500$
$x \ge 8,\ y \ge 3$
The function that describes the number of people reached by the ads is $z = 600x + 2000y$. This is the function to be maximized.

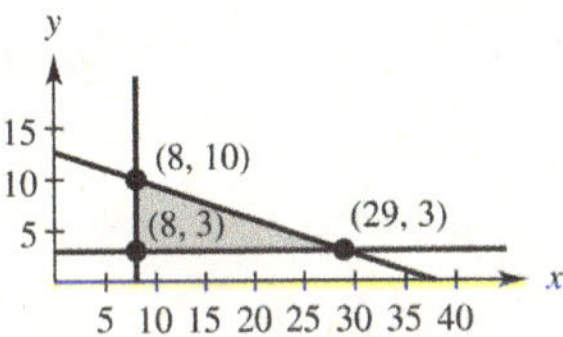

Corner Point	$z = 600x + 2000y$
(8, 3)	$600(8) + 2000(3) = 10{,}800$
(8, 10)	$600(8) + 2000(10) = 24{,}800$ maximum
(29, 3)	$600(29) + 2000(3) = 23{,}400$

The candidate should use 8 radio spots and 10 tv ads to maximize the number of people reached.

13. Let x = the number of brand X pills, let y = the number of brand Z pills. The constraint imposed by the amount of vitamin A is $8x + 2y \ge 16$. The constraint imposed by the amount of vitamin B-1 is $x + y \ge 5$. The constraint imposed by the amount of vitamin C is $2x + 7y \ge 20$. The cost function z is $z = 15x + 30y$. This is the function to be minimized subject to the constraints
$8x + 2y \ge 16$
$x + y \ge 5$
$2x + 7y \ge 20$
$x \ge 0,\ y \ge 0$

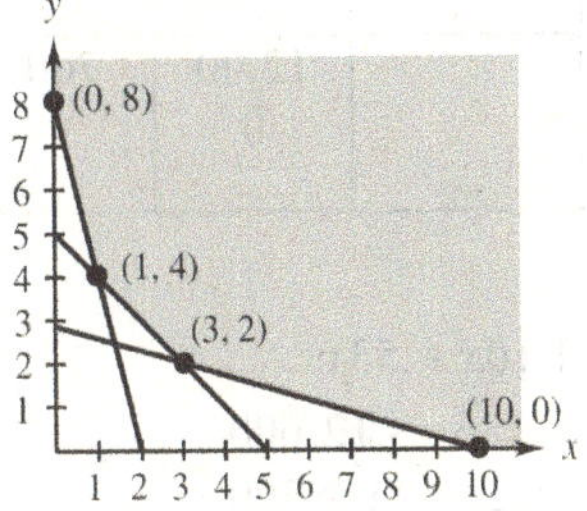

Corner Point	$z = .15x + .30y$
(0, 8)	$.15(0) + .30(8) = 2.40$
(1, 4)	$.15(1) + .30(4) = 1.35$
(3, 2)	$.15(3) + .30(2) = 1.05$ (minimum)
(10, 0)	$.15(10) + .30(0) = 1.50$

He can satisfy the requirements with 3 brand X pills and 2 brand Z pills for a minimum cost of \$1.05.

15. Let x = the number of Type 1 bolts and let y = the number of Type 2 bolts.
Maximize $z = .10x + .12y$
subject to: $.1x + .1y \le 240$
$.1x + .4y \le 720$
$.1x + .02y \le 160$
$x \ge 0$
$y \ge 0$

Graph the feasible region, and label the corner points. Use the corner points to find the maximum value of the objective function.

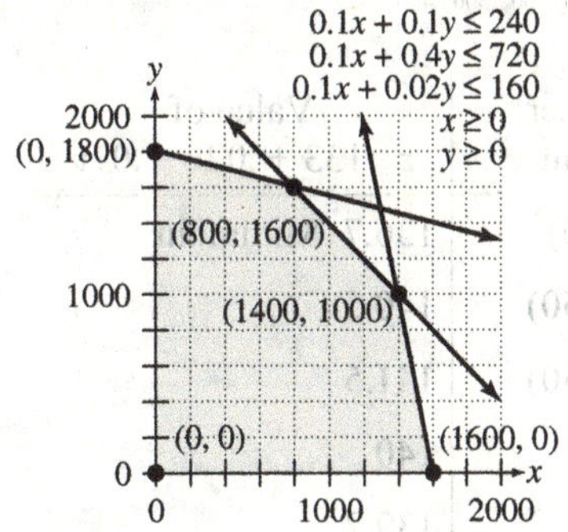

Corner Point	Value of $z = .10x + .12y$
(0, 0)	$.10(0) + .12(0) = 0$
(0, 1800)	$.10(0) + .12(1800) = 216$
(800, 1600)	$.10(800) + .12(1600) = 272$ Maximum
(1400, 1000)	$.10(1400) + .12(1000) = 260$
(1600, 0)	$.1(1600) + .12(0) = 160$

Manufacture 800 Type 1 bolts and 1600 Type 2 bolts for a maximum revenue of \$272/day.

Copyright © 2019 Pearson Education, Inc.

17. Let x = the number of cards from warehouse I to San Jose. Then $350 - x$ is the number of cards from warehouse II to San Jose. Let y = the number of cards from warehouse I to Memphis. Then $250 - y$ is the number of cards from warehouse II to Memphis. The constraints are represented by the inequalities

$$x \geq 0,\ y \geq 0$$
$$x \leq 350$$
$$y \leq 250$$
$$x + y \leq 500$$
$$(350 - x) + (250 - y) \leq 290 \Rightarrow x + y \geq 310.$$

Minimize
$$.25x + .23(350 - x) + .22y + .21(250 - y) = 133 + .02x + .01y.$$

The graph and corner points are shown below.

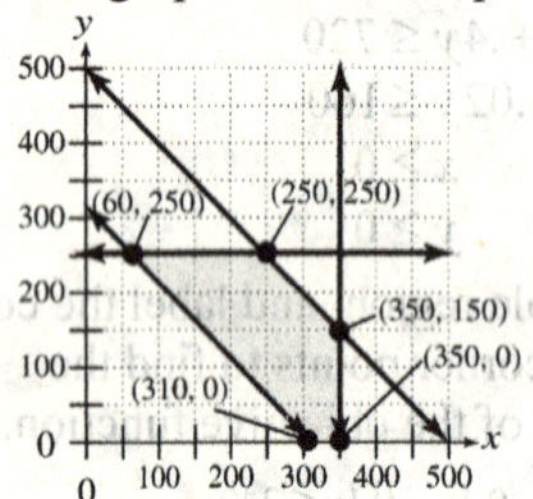

Corner Point	Value of $z = 133 + .02x + .01y$
(60, 250)	136.7 Minimum
(250, 250)	140.5
(350, 150)	141.5
(350, 0)	140
(310, 0)	139.2

The minimum cost is $136.70. From warehouse I ship 60 boxes to San Jose and 250 boxes to Memphis. From warehouse II ship 290 boxes to San Jose and none to Memphis.

19. Let x = the amount invested in bonds and let y = the amount invested in mutual funds. (Both are in millions of dollars.)
The amount of annual interest is $.04x + .06y$.
Maximize $z = .04x + .06y$
subject to:
$$x \geq 20$$
$$y \geq 6$$
$$300x + 100y \leq 8400$$
$$x + y \leq 50$$
$$x \geq 0,\ y \geq 0$$

Graph the feasible region, and label the corner points.

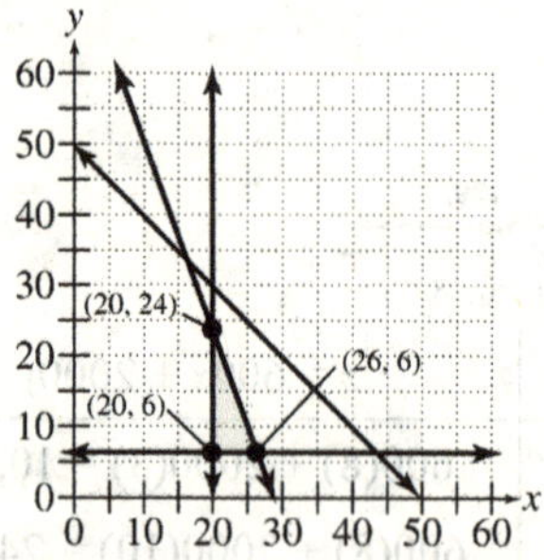

Corner Point	Value of $z = .04x + .06y$
(20, 24)	2.24 Maximum
(26, 6)	1.4
(20, 6)	1.16

He should invest $20 million in bonds and $24 million in mutual funds for maximum annual interest of $2.24 million.

21.

	Number of Shares	Cost per Share	Profit	Risk
GE	x	29	2	1.10
P&G	y	86	4	.53
Constraints		12,000	700	

Minimize $z = 1.10x + .53y$
subject to: $29x + 86y \leq 12{,}000$
$2x + 4y \geq 700$

Copyright © 2019 Pearson Education, Inc.

Graph the feasibility region.

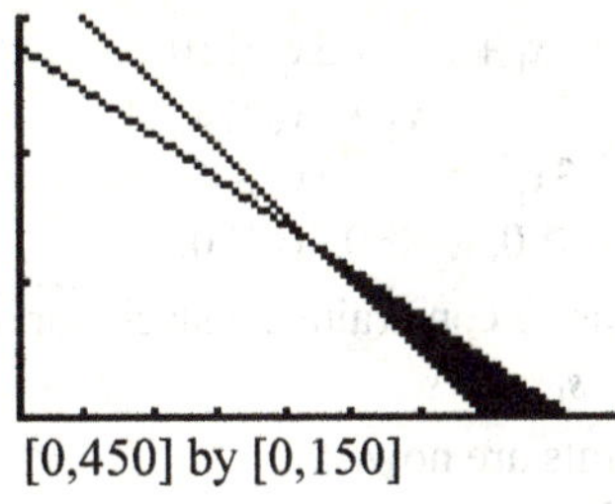

[0,450] by [0,150]

The points of intersection are summarized in the table below.

Corner Point	Value of $z = 1.10x + .53y$
(218,66)	274.78 Minimum
(414, 0)	455.4
(350, 0)	385

Tori should buy about 218 shares of General Electric stock and 66 shares of the Proctor and Gamble stock.

23.

	Number of Shares	Cost per Share	Profit	Risk
Visa Inc.	x	92	20	1.06
JPM Chase and Co.	y	87	17	1.21
Constraints		15,000	2500	

Minimize $z = 1.06x + 1.21y$

subject to: $92x + 87y \leq 15{,}000$

$20x + 17y \geq 2500$

Graph the feasibility region.

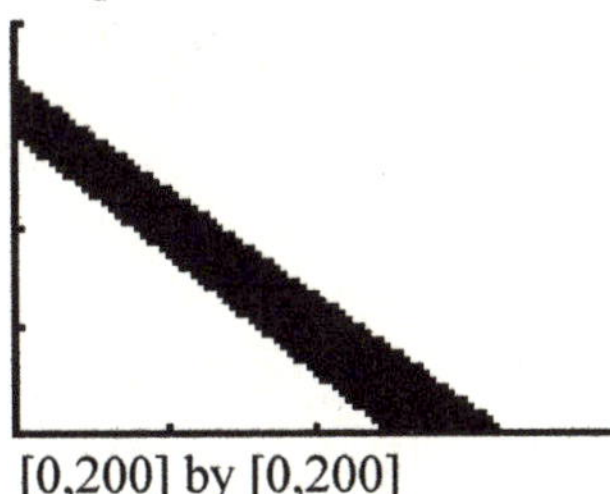

[0,200] by [0,200]

The points of intersection are summarized in the table below.

Corner Point	Value of $z = 1.06x + 1.21y$
(0,172)	208.12
(0,147)	177.87
(125, 0)	132.5 Minimum
(163, 0)	172.78

Derek should buy about 125 shares of Visa Inc. stock and none of the JP Morgan Chase and Co. stock.

25.

	Number of Shares	Cost per Share	Profit	Risk
VBMFX	x	11	.25	1.07
FFFDX	y	16	.92	.77
Constraints		3000	100	

Minimize $z = 1.07x + .77y$

subject to: $11x + 16y \leq 3000$

$.25x + .92y \geq 100$

Graph the feasibility region.

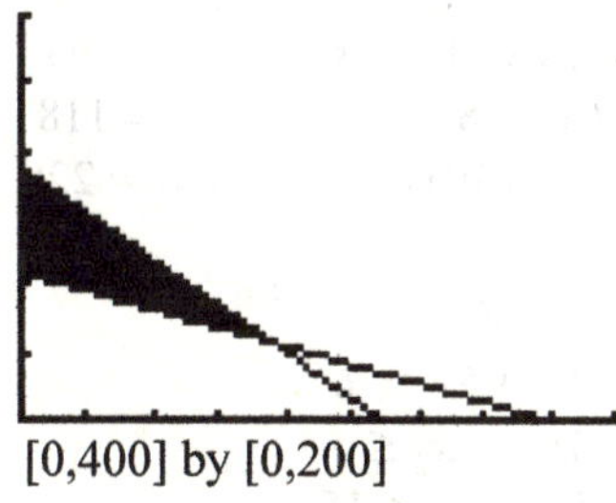

[0,400] by [0,200]

The points of intersection are summarized in the table below (Values have been rounded).

Corner Point	Value of $z = 1.07x + .77y$
(190,57)	247.19
(0, 109)	83.93 Minimum
(0, 188)	144.76

Jonquin should buy about 0 shares of the VBMFX fund and about 109 shares of the FFFDX fund.

Copyright © 2019 Pearson Education, Inc.

Section 7.4 The Simplex Method: Maximization

1. Maximize $z = 32x_1 + 9x_2$

subject to:
$$\begin{aligned} 4x_1 + 2x_2 &\le 20 \\ 5x_1 + x_2 &\le 50 \\ 2x_1 + 3x_2 &\le 25 \\ x_1 \ge 0,\ x_2 &\ge 0. \end{aligned}$$

a. There are 3 constraints, so 3 slack variables are needed.

b. Use s_1, s_2, and s_3 for the slack variables.

c.
$$\begin{aligned} 4x_1 + 2x_2 + s_1 \qquad\qquad &= 20 \\ 5x_1 + x_2 \qquad + s_2 \qquad &= 50 \\ 2x_1 + 3x_2 \qquad\qquad + s_3 &= 25 \end{aligned}$$

3. Maximize $z = 8x_1 + 3x_2 + x_3$

subject to:
$$\begin{aligned} 3x_1 - x_2 + 4x_3 &\le 95 \\ 7x_1 + 6x_2 + 8x_3 &\le 118 \\ 4x_1 + 5x_2 + 10x_3 &\le 220 \\ x_1 \ge 0,\ x_2 \ge 0,\ x_3 &\ge 0. \end{aligned}$$

a. There are 3 constraints, so 3 slack variables are needed.

b. Use s_1, s_2, and s_3 for the slack variables.

c.
$$\begin{aligned} 3x_1 - x_2 + 4x_3 + s_1 \qquad\qquad &= 95 \\ 7x_1 + 6x_2 + 8x_3 \qquad + s_2 \qquad &= 118 \\ 4x_1 + 5x_2 + 10x_3 \qquad\qquad + s_3 &= 220 \end{aligned}$$

5. Maximize $z = 5x_1 + x_2$

subject to:
$$\begin{aligned} 2x_1 + 5x_2 &\le 6 \\ 4x_1 + x_2 &\le 6 \\ 5x_1 + 3x_2 &\le 15 \\ x_1 \ge 0,\ x_2 &\ge 0. \end{aligned}$$

Since there are 3 constraints, 3 slack variables are needed: s_1, s_2, and s_3.

The constraints are now
$$\begin{aligned} 2x_1 + 5x_2 + s_1 \qquad\qquad &= 6 \\ 4x_1 + x_2 \qquad + s_2 \qquad &= 6 \\ 5x_1 + 3x_2 \qquad\qquad + s_3 &= 15. \end{aligned}$$

The initial simplex tableau is

$$\begin{array}{c} \begin{matrix} x_1 & x_2 & s_1 & s_2 & s_3 & z \end{matrix} \\ \left[\begin{array}{cccccc|c} 2 & 5 & 1 & 0 & 0 & 0 & 6 \\ 4 & 1 & 0 & 1 & 0 & 0 & 6 \\ 5 & 3 & 0 & 0 & 1 & 0 & 15 \\ \hline -5 & -1 & 0 & 0 & 0 & 1 & 0 \end{array}\right]. \end{array}$$

7. Maximize $z = x_1 + 5x_2 + 10x_3$

subject to:
$$\begin{aligned} x_1 + 2x_2 + 3x_3 &\le 10 \\ 2x_1 + x_2 + x_3 &\le 8 \\ 3x_1 \qquad + 4x_3 &\le 6 \\ x_1 \ge 0,\ x_2 \ge 0,\ x_3 &\ge 0. \end{aligned}$$

Since there are 3 constraints, 3 slack variables are needed.: s_1, s_2, s_3.

The constraints are now
$$\begin{aligned} x_1 + 2x_2 + 3x_3 + s_1 \qquad\qquad &= 10 \\ 2x_1 + x_2 + x_3 \qquad + s_2 \qquad &= 8 \\ 3x_1 \qquad + 4x_3 \qquad\qquad + s_3 &= 6 \end{aligned}$$

The initial tableau is

$$\begin{array}{c} \begin{matrix} x_1 & x_2 & x_3 & s_1 & s_2 & s_3 & z \end{matrix} \\ \left[\begin{array}{ccccccc|c} 1 & 2 & 3 & 1 & 0 & 0 & 0 & 10 \\ 2 & 1 & 1 & 0 & 1 & 0 & 0 & 8 \\ 3 & 0 & 4 & 0 & 0 & 1 & 0 & 6 \\ \hline -1 & -5 & -10 & 0 & 0 & 0 & 1 & 0 \end{array}\right]. \end{array}$$

9.
$$\begin{array}{c} \begin{matrix} x_1 & x_2 & x_3 & s_1 & s_2 & z \end{matrix} \\ \left[\begin{array}{cccccc|c} 2 & 2 & 0 & 3 & 1 & 0 & 15 \\ 3 & 4 & 1 & 6 & 0 & 0 & 20 \\ \hline -2 & -3 & 0 & 1 & 0 & 1 & 10 \end{array}\right] \end{array}$$

The most negative indicator is –3, so the pivot column is column 2. Since $\frac{20}{4} = 5$ is the smaller quotient, the pivot row is row two. The pivot is the 4 in row two, column two.

11.
$$\begin{array}{c} \begin{matrix} x_1 & x_2 & x_3 & s_1 & s_2 & s_3 & z \end{matrix} \\ \left[\begin{array}{ccccccc|c} 6 & 2 & 1 & 3 & 0 & 0 & 0 & 8 \\ 0 & 2 & 0 & 1 & 0 & 1 & 0 & 7 \\ 6 & 1 & 0 & 3 & 1 & 0 & 0 & 6 \\ \hline -3 & -2 & 0 & 2 & 0 & 0 & 1 & 12 \end{array}\right] \end{array}$$

The most negative indicator is –3, so the pivot column is column one. Since $\frac{6}{6} = 1$ is the smallest quotient, the pivot row is row three. Thus, the pivot is the 6 in row three, column one.

Copyright © 2019 Pearson Education, Inc.

13.

$$\begin{array}{c} \begin{matrix} x_1 & x_2 & x_3 & s_1 & s_2 & z \end{matrix} \\ \left[\begin{array}{cccccc|c} 1 & 2 & 4 & 1 & 0 & 0 & 56 \\ 2 & \boxed{2} & 1 & 0 & 1 & 0 & 40 \\ \hline -1 & -3 & -2 & 0 & 0 & 1 & 0 \end{array}\right] \end{array}$$

Start by multiplying each entry of row 2 by $\frac{1}{2}$ in order to change the pivot to 1.

$$\begin{array}{c} \begin{matrix} x_1 & x_2 & x_3 & s_1 & s_2 & z \end{matrix} \\ \left[\begin{array}{cccccc|c} 1 & 2 & 4 & 1 & 0 & 0 & 56 \\ 1 & 1 & \frac{1}{2} & 0 & \frac{1}{2} & 0 & 20 \\ \hline -1 & -3 & -2 & 0 & 0 & 1 & 0 \end{array}\right] \begin{array}{l} \\ \frac{1}{2}R_2 \\ \\ \end{array} \end{array}$$

Now use row operations to change the entry in row one column two and the indicator –3 to 0.

$$\begin{array}{c} \begin{matrix} x_1 & x_2 & x_3 & s_1 & s_2 & z \end{matrix} \\ \left[\begin{array}{cccccc|c} -1 & 0 & 3 & 1 & -1 & 0 & 16 \\ 1 & 1 & \frac{1}{2} & 0 & \frac{1}{2} & 0 & 20 \\ \hline 2 & 0 & -\frac{1}{2} & 0 & \frac{3}{2} & 1 & 60 \end{array}\right] \begin{array}{l} -2R_2 + R_1 \\ \\ 3R_2 + R_3 \end{array} \end{array}$$

15.

$$\begin{array}{c} \begin{matrix} x_1 & x_2 & x_3 & s_1 & s_2 & s_3 & z \end{matrix} \\ \left[\begin{array}{ccccccc|c} 1 & 1 & 1 & 1 & 0 & 0 & 0 & 60 \\ 3 & 1 & \boxed{2} & 0 & 1 & 0 & 0 & 100 \\ 1 & 2 & 3 & 0 & 0 & 1 & 0 & 200 \\ \hline -1 & -1 & -2 & 0 & 0 & 0 & 1 & 0 \end{array}\right] \end{array}$$

$$\begin{array}{c} \begin{matrix} x_1 & x_2 & x_3 & x_4 & x_5 & x_6 & z \end{matrix} \\ \left[\begin{array}{ccccccc|c} 1 & 1 & 1 & 1 & 0 & 0 & 0 & 60 \\ \frac{3}{2} & \frac{1}{2} & 1 & 0 & \frac{1}{2} & 0 & 0 & 50 \\ 1 & 2 & 3 & 0 & 0 & 1 & 0 & 200 \\ \hline -1 & -1 & -2 & 0 & 0 & 0 & 1 & 0 \end{array}\right] \begin{array}{l} \\ \frac{1}{2}R_2 \\ \\ \\ \end{array} \end{array}$$

$$\begin{array}{c} \begin{matrix} x_1 & x_2 & x_3 & s_1 & s_2 & s_3 & z \end{matrix} \\ \left[\begin{array}{ccccccc|c} -\frac{1}{2} & \frac{1}{2} & 0 & 1 & -\frac{1}{2} & 0 & 0 & 10 \\ \frac{3}{2} & \frac{1}{2} & 1 & 0 & \frac{1}{2} & 0 & 0 & 50 \\ -\frac{7}{2} & \frac{1}{2} & 0 & 0 & -\frac{3}{2} & 1 & 0 & 50 \\ \hline 2 & 0 & 0 & 0 & 1 & 0 & 1 & 100 \end{array}\right] \begin{array}{l} -1R_2 + R_1 \\ \\ -3R_2 + R_3 \\ 2R_2 + R_4 \end{array} \end{array}$$

17.

$$\begin{array}{c} \begin{matrix} x_1 & x_2 & x_3 & s_1 & s_2 & z \end{matrix} \\ \left[\begin{array}{cccccc|c} 3 & 2 & 0 & -3 & 1 & 0 & 29 \\ 4 & 0 & 1 & -2 & 0 & 0 & 16 \\ \hline -5 & 0 & 0 & -1 & 0 & 1 & 11 \end{array}\right] \end{array}$$

a. The basic variables are x_3, s_2, and z. x_1, x_2, and s_1 are nonbasic.

b. The basic feasible solution is $x_1 = 0$, $x_2 = 0$, $x_3 = 16$, $s_1 = 0$, $s_2 = 29$, $z = 11$

c. Because there are still negative indicators, this solution is not the maximum.

19.

$$\begin{array}{c} \begin{matrix} x_1 & x_2 & x_3 & s_1 & s_2 & s_3 & z \end{matrix} \\ \left[\begin{array}{ccccccc|c} 1 & 0 & 2 & \frac{1}{2} & 0 & \frac{1}{3} & 0 & 6 \\ 0 & 1 & -1 & 5 & 0 & -1 & 0 & 13 \\ 0 & 0 & 1 & \frac{3}{2} & 1 & -\frac{1}{3} & 0 & 21 \\ \hline 0 & 0 & 2 & \frac{1}{2} & 0 & 3 & 1 & 18 \end{array}\right] \end{array}$$

a. The basic variables are x_1, x_2, s_2, and z. x_3, s_1, and s_3 are nonbasic.

b. The basic feasible solution is $x_1 = 6$, $x_2 = 13$, $x_3 = 0$, $s_1 = 0$, $s_2 = 21$, $s_3 = 0$, $z = 18$.

c. Because there are no negative indicators, this solution is the maximum.

21. Maximize $z = x_1 + 3x_2$

subject to:

$$\begin{aligned} x_1 + x_2 &\le 10 \\ 5x_1 + 2x_2 &\le 20 \\ x_1 + 2x_2 &\le 36 \\ x_1 \ge 0,\ x_2 &\ge 0. \end{aligned}$$

Using slack variables, s_1, s_2, s_3. the constraints become:

$$\begin{aligned} x_1 + x_2 + s_1 \qquad\qquad &= 10 \\ 5x_1 + 2x_2 \qquad + s_2 \qquad &= 20 \\ x_1 + 2x_2 \qquad\qquad + s_3 &= 36. \end{aligned}$$

Copyright © 2019 Pearson Education, Inc.

The initial simplex tableau is

$$\begin{array}{c} \begin{matrix} x_1 & x_2 & s_1 & s_2 & s_3 & z \end{matrix} \\ \left[\begin{array}{cccccc|c} 1 & \boxed{1} & 1 & 0 & 0 & 0 & 10 \\ 5 & 2 & 0 & 1 & 0 & 0 & 20 \\ 1 & 2 & 0 & 0 & 1 & 0 & 36 \\ \hline -1 & -3 & 0 & 0 & 0 & 1 & 0 \end{array}\right]. \end{array}$$

Pivot on the 1 in row one, column two.

$$\begin{array}{c} \begin{matrix} x_1 & x_2 & s_1 & s_2 & s_3 & z \end{matrix} \\ \left[\begin{array}{cccccc|c} 1 & 1 & 1 & 0 & 0 & 0 & 10 \\ 3 & 0 & -2 & 1 & 0 & 0 & 0 \\ -1 & 0 & -2 & 0 & 1 & 0 & 16 \\ \hline 2 & 0 & 3 & 0 & 0 & 1 & 30 \end{array}\right] \begin{array}{l} \\ -2R_1+R_2 \\ -2R_1+R_3 \\ 3R_1+R_4 \end{array} \end{array}$$

Since there are no negative indicators, this matrix is the final tableau. The maximum is 30 when $x_1 = 0$, $x_2 = 10$, $s_1 = 0$, $s_2 = 0$, and $s_3 = 16$.

23. Maximize $z = 2x_1 + x_2$

subject to:
$$\begin{aligned} x_1 + 3x_2 &\le 12 \\ 2x_1 + x_2 &\le 10 \\ x_1 + x_2 &\le 4 \\ x_1 \ge 0, x_2 &\ge 0. \end{aligned}$$

Using slack variables s_1, s_2, s_3. the constraints become:

$$\begin{aligned} x_1 + 3x_2 + s_1 \quad\quad\quad &= 12 \\ 2x_1 + x_2 \quad + s_2 \quad &= 10 \\ x_1 + x_2 \quad\quad + s_3 &= 4. \end{aligned}$$

The initial simplex tableau is

$$\begin{array}{c} \begin{matrix} x_1 & x_2 & s_1 & s_2 & s_3 & z \end{matrix} \\ \left[\begin{array}{cccccc|c} 1 & 3 & 1 & 0 & 0 & 0 & 12 \\ 2 & 1 & 0 & 1 & 0 & 0 & 10 \\ \boxed{1} & 1 & 0 & 0 & 1 & 0 & 4 \\ \hline -2 & -1 & 0 & 0 & 0 & 1 & 0 \end{array}\right]. \\ \uparrow \end{array}$$

Pivot on the 1 in row three, column one.

$$\begin{array}{c} \begin{matrix} x_1 & x_2 & s_1 & s_2 & s_3 & z \end{matrix} \\ \left[\begin{array}{cccccc|c} 0 & 2 & 1 & 0 & -1 & 0 & 8 \\ 0 & -1 & 0 & 1 & -2 & 0 & 2 \\ 1 & 1 & 0 & 0 & 1 & 0 & 4 \\ \hline 0 & 1 & 0 & 0 & 2 & 1 & 8 \end{array}\right] \begin{array}{l} -1R_3+R_1 \\ -2R_3+R_2 \\ \\ 2R_3+R_4 \end{array} \end{array}$$

Since there are no negative indicators, this matrix is the final tableau.
The maximum is 8 when $x_1 = 4$, $x_2 = 0$, $s_1 = 8$, $s_2 = 2$, $s_3 = 0$.

25. Maximize $z = 5x_1 + 4x_2 + x_3$

subject to:
$$\begin{aligned} -2x_1 + x_2 + 2x_3 &\le 3 \\ x_1 - x_2 + x_3 &\le 1 \\ x_1 \ge 0, x_2 \ge 0, x_3 &\ge 0. \end{aligned}$$

Using the slack variables s_1 and s_2, the constraints become:

$$\begin{aligned} -2x_1 + x_2 + 2x_3 + s_1 \quad &= 3 \\ x_1 - x_2 + x_3 \quad + s_2 &= 1. \end{aligned}$$

The initial simplex tableau is

$$\begin{array}{c} \begin{matrix} x_1 & x_2 & x_3 & s_1 & s_2 & z \end{matrix} \\ \left[\begin{array}{cccccc|c} -2 & 1 & 2 & 1 & 0 & 0 & 3 \\ \boxed{1} & -1 & 1 & 0 & 1 & 0 & 1 \\ \hline -5 & -4 & -1 & 0 & 0 & 1 & 0 \end{array}\right]. \\ \uparrow \end{array}$$

Pivot on the 1 in row two, column one.

$$\begin{array}{c} \begin{matrix} x_1 & x_2 & x_3 & s_1 & s_2 & z \end{matrix} \\ \left[\begin{array}{cccccc|c} 0 & -1 & 4 & 1 & 2 & 0 & 5 \\ 1 & -1 & 1 & 0 & 1 & 0 & 1 \\ \hline 0 & -9 & 4 & 0 & 5 & 1 & 5 \end{array}\right] \begin{array}{l} 2R_2+R_1 \\ \\ 5R_2+R_3 \end{array} \end{array}$$

There is a negative indicator in column two, but all entries in that column are negative, so there is no place to continue pivoting. Therefore, there is no maximum.

27. Maximize $z = 2x_1 + x_2 + x_3$

subject to:
$$\begin{aligned} x_1 - 3x_2 + x_3 &\le 3 \\ x_1 - 2x_2 + 2x_3 &\le 12 \\ x_1 \ge 0, x_2 \ge 0, x_3 &\ge 0. \end{aligned}$$

Using slack variables s_1 and s_2, the constraints become:

$$\begin{aligned} x_1 - 3x_2 + x_3 + s_1 \quad &= 3 \\ x_1 - 2x_2 + 2x_3 \quad + s_2 &= 12. \end{aligned}$$

The initial simplex tableau is

$$\begin{array}{c} \begin{matrix} x_1 & x_2 & x_3 & s_1 & s_2 & z \end{matrix} \\ \left[\begin{array}{cccccc|c} \boxed{1} & -3 & 1 & 1 & 0 & 0 & 3 \\ 1 & -2 & 2 & 0 & 1 & 0 & 12 \\ \hline -2 & -1 & -1 & 0 & 0 & 1 & 0 \end{array}\right]. \end{array}$$

Copyright © 2019 Pearson Education, Inc.

Pivot on the 1 in row one, column one.

$$\begin{array}{cccccc|c|l} x_1 & x_2 & x_3 & s_1 & s_2 & z & & \\ 1 & -3 & 1 & 1 & 0 & 0 & 3 & \\ 0 & \boxed{1} & 1 & -1 & 1 & 0 & 9 & -R_1+R_2 \\ \hline 0 & -7 & 1 & 2 & 0 & 1 & 6 & 2R_1+R_3 \end{array}$$

$\uparrow$ (column two)

Pivot on the 1 in row two, column two.

$$\begin{array}{cccccc|c|l} x_1 & x_2 & x_3 & s_1 & s_2 & z & & \\ 1 & 0 & 4 & -2 & 3 & 0 & 30 & 3R_2+R_1 \\ 0 & 1 & 1 & -1 & 1 & 0 & 9 & \\ \hline 0 & 0 & 8 & -5 & 7 & 1 & 69 & 7R_2+R_3 \end{array}$$

The only negative indicator is in column four, which has all negative entries, so there is no place to continue pivoting. Therefore, there is no maximum.

29. Maximize $z = 2x_1 + 2x_2 - 4x_3$

Subject to: $3x_1 + 3x_2 - 6x_3 \le 51$
$5x_1 + 5x_2 + 10x_3 \le 99$
$x_1 \ge 0,\ x_2 \ge 0,\ x_3 \ge 0.$

Using slack variables s_1 and s_2, the constraints become:

$3x_1 + 3x_2 - 6x_3 + s_1 \qquad = 51$
$5x_1 + 5x_2 + 10x_3 \qquad + s_2 = 99.$

The initial simplex tableau is

$$\begin{array}{cccccc|c} x_1 & x_2 & x_3 & s_1 & s_2 & z & \\ \boxed{3} & 3 & -6 & 1 & 0 & 0 & 51 \\ 5 & 5 & 10 & 0 & 1 & 0 & 99 \\ \hline -2 & -2 & 4 & 0 & 0 & 1 & 0 \end{array}.$$

$\uparrow$ (column one)

Pivot on the 3 in row one, column one.

$$\begin{array}{cccccc|c|l} x_1 & x_2 & x_3 & s_1 & s_2 & z & & \\ 1 & 1 & -2 & \frac{1}{3} & 0 & 0 & 17 & \frac{1}{3}R_1 \\ 5 & 5 & 10 & 0 & 1 & 0 & 99 & \\ \hline -2 & -2 & 4 & 0 & 0 & 1 & 0 & \end{array}$$

$$\begin{array}{cccccc|c|l} x_1 & x_2 & x_3 & s_1 & s_2 & z & & \\ 1 & 1 & -2 & \frac{1}{3} & 0 & 0 & 17 & \\ 0 & 0 & 20 & -\frac{5}{3} & 1 & 0 & 14 & -5R_1+R_2 \\ \hline 0 & 0 & 0 & \frac{2}{3} & 0 & 1 & 34 & 2R_1+R_3 \end{array}$$

The maximum is 34 when
$x_1 = 17,\ x_2 = 0,\ x_3 = 0,\ s_1 = 0,\ s_2 = 14$
or when $x_1 = 0,\ x_2 = 17,\ x_3 = 0,\ s_1 = 0,$
and $s_2 = 14.$

Copyright © 2019 Pearson Education, Inc.

31. Maximize $z = 300x_1 + 200x_2 + 100x_3$

subject to:
$$\begin{aligned} x_1 + x_2 + x_3 &\le 100 \\ 2x_1 + 3x_2 + 4x_3 &\le 320 \\ 2x_1 + x_2 + x_3 &\le 160 \\ x_1 \ge 0, x_2 \ge 0, x_3 &\ge 0. \end{aligned}$$

Using slack variables s_1, s_2, s_3. the constraints become:

$$\begin{aligned} x_1 + x_2 + x_3 + s_1 &= 100 \\ 2x_1 + 3x_2 + 4x_3 + s_2 &= 320 \\ 2x_1 + x_2 + x_3 + s_3 &= 160. \end{aligned}$$

The initial simplex tableau is

$$\begin{array}{c} \begin{matrix} x_1 & x_2 & x_3 & s_1 & s_2 & s_3 & z \end{matrix} \\ \left[\begin{array}{ccccccc|c} 1 & 1 & 1 & 1 & 0 & 0 & 0 & 100 \\ 2 & 3 & 4 & 0 & 1 & 0 & 0 & 320 \\ \boxed{2} & 1 & 1 & 0 & 0 & 1 & 0 & 160 \\ \hline -300 & -200 & -100 & 0 & 0 & 0 & 1 & 0 \end{array}\right] \end{array}.$$

Pivot on the 2 in row three, column one.

$$\begin{array}{c} \begin{matrix} x_1 & x_2 & x_3 & s_1 & s_2 & s_3 & z \end{matrix} \\ \left[\begin{array}{ccccccc|c} 1 & 1 & 1 & 1 & 0 & 0 & 0 & 100 \\ 2 & 3 & 4 & 0 & 1 & 0 & 0 & 320 \\ \boxed{1} & \frac{1}{2} & \frac{1}{2} & 0 & 0 & \frac{1}{2} & 0 & 80 \\ \hline -300 & -200 & -100 & 0 & 0 & 0 & 1 & 0 \end{array}\right] \end{array} \begin{array}{l} \\ \\ \\ \frac{1}{2}R_3 \\ \\ \end{array}$$

$$\begin{array}{c} \begin{matrix} x_1 & x_2 & x_3 & s_1 & s_2 & s_3 & z \end{matrix} \\ \left[\begin{array}{ccccccc|c} 0 & \frac{1}{2} & \frac{1}{2} & 1 & 0 & -\frac{1}{2} & 0 & 20 \\ 0 & 2 & 3 & 0 & 1 & -1 & 0 & 160 \\ 1 & \frac{1}{2} & \frac{1}{2} & 0 & 0 & \frac{1}{2} & 0 & 80 \\ \hline 0 & -50 & 50 & 0 & 0 & 150 & 1 & 24,000 \end{array}\right] \end{array} \begin{array}{l} \\ -1R_3 + R_1 \\ -2R_3 + R_2 \\ \\ 300R_3 + R_4 \end{array}$$

Pivot on the $\frac{1}{2}$ in row one, column two.

$$\begin{array}{c} \begin{matrix} x_1 & x_2 & x_3 & s_1 & s_2 & s_3 & z \end{matrix} \\ \left[\begin{array}{ccccccc|c} 0 & 1 & 1 & 2 & 0 & -1 & 0 & 40 \\ 0 & 2 & 3 & 0 & 1 & -1 & 0 & 160 \\ 1 & \frac{1}{2} & \frac{1}{2} & 0 & 0 & \frac{1}{2} & 0 & 80 \\ \hline 0 & -50 & 50 & 0 & 0 & 150 & 1 & 24,000 \end{array}\right] \end{array} \begin{array}{l} \\ 2R_1 \\ \\ \\ \\ \end{array}$$

$$\begin{array}{c} \begin{matrix} x_1 & x_2 & x_3 & s_1 & s_2 & s_3 & z \end{matrix} \\ \left[\begin{array}{ccccccc|c} 0 & 1 & 1 & 2 & 0 & -1 & 0 & 40 \\ 0 & 0 & 1 & -4 & 1 & 1 & 0 & 80 \\ 1 & 0 & 0 & -1 & 0 & 1 & 0 & 60 \\ \hline 0 & 0 & 100 & 100 & 0 & 100 & 1 & 26,000 \end{array}\right] \end{array} \begin{array}{l} \\ \\ -2R_1 + R_2 \\ -\frac{1}{2}R_1 + R_3 \\ 50R_1 + R_4 \end{array}$$

The maximum value is 26,000 when $x_1 = 60$, $x_2 = 40$, $x_3 = 0$, $s_1 = 0$, $s_2 = 80$, and $s_3 = 0$.

Copyright © 2019 Pearson Education, Inc.

33. Maximize $z = 4x_1 - 3x_2 + 2x_3$

subject to:
$$2x_1 - x_2 + 8x_3 \le 40$$
$$4x_1 - 5x_2 + 6x_3 \le 60$$
$$2x_1 - 2x_2 + 6x_3 \le 24$$
$$x_1 \ge 0,\ x_2 \ge 0,\ x_3 \ge 0.$$

Note: The third constraint simplifies to $x_1 - x_2 + 3x_3 \le 12.$

Using slack variables s_1, s_2, s_3. the constraints become:
$$2x_1 - x_2 + 8x_3 + s_1 = 40$$
$$4x_1 - 5x_2 + 6x_3 + s_2 = 60$$
$$x_1 - x_2 + 3x_3 + s_3 = 12.$$

The initial simplex tableau is

x_1	x_2	x_3	s_1	s_2	s_3	z	
2	−1	8	1	0	0	0	40
4	−5	6	0	1	0	0	60
$\boxed{1}$	−1	3	0	0	1	0	12
−4	3	−2	0	0	0	1	0

Pivot on the 1 in row three, column one.

x_1	x_2	x_3	s_1	s_2	s_3	z		
0	$\boxed{1}$	2	1	0	−2	0	16	$-2R_3 + R_1$
0	−1	−6	0	1	−4	0	12	$-4R_3 + R_2$
1	−1	3	0	0	1	0	12	
0	−1	10	0	0	4	1	48	$4R_3 + R_4$

Pivot on the 1 in row one, column two.

x_1	x_2	x_3	s_1	s_2	s_3	z		
0	1	2	1	0	−2	0	16	
0	0	−4	1	1	−6	0	28	$R_1 + R_2$
1	0	5	1	0	−1	0	28	$R_1 + R_3$
0	0	12	1	0	2	1	64	$R_1 + R_4$

The maximum is 64 when $x_1 = 28,\ x_2 = 16,\ x_3 = 0,\ s_1 = 0,\ s_2 = 28,$ and $s_3 = 0.$

35. Maximize $z = x_1 + 2x_2 + x_3 + 5x_4$

subject to:
$$x_1 + 2x_2 + x_3 + x_4 \le 50$$
$$3x_1 + x_2 + 2x_3 + x_4 \le 100$$
$$x_1 \ge 0,\ x_2 \ge 0,\ x_3 \ge 0, x_4 \ge 0.$$

Using slack variables s_1 and s_2 the constraints become:
$$x_1 + 2x_2 + x_3 + x_4 + s_1 = 50$$
$$3x_1 + x_2 + 2x_3 + x_4 + s_2 = 100.$$

The initial simplex tableau is

x_1	x_2	x_3	x_4	s_1	s_2	z	
1	2	1	$\boxed{1}$	1	0	0	50
3	1	2	1	0	1	0	100
−1	−2	−1	−5	0	0	1	0

The pivot is the 1 in row one, column four.

x_1	x_2	x_3	x_4	s_1	s_2	z		
1	2	1	1	1	0	0	50	
2	−1	1	0	−1	1	0	50	$-1R_1 + R_2$
4	8	4	0	5	0	1	250	$5R_1 + R_3$

The maximum is 250 when $x_1 = 0, x_2 = 0,\ x_3 = 0,\ x_4 = 50,\ s_1 = 0,$ and $s_2 = 50.$

37.

x_1	x_2	x_3	s_1	s_2	z	
1	1	1	1	0	0	12
2	1	2	0	1	0	30
−2	−2	−1	0	0	1	0

a. Pivot on the 1 in row one, column one.

x_1	x_2	x_3	s_1	s_2	z		
1	1	1	1	0	0	12	
0	−1	0	−2	1	0	6	$-2R_1 + R_2$
0	0	1	2	0	1	24	$2R_1 + R_3$

The maximum is 24 when $x_1 = 12,\ x_2 = 0,\ x_3 = 0,\ s_1 = 0,\ s_2 = 6.$

b. Pivot on the 1 in row one, column two.

x_1	x_2	x_3	s_1	s_2	z		
1	1	1	1	0	0	12	
1	0	1	−1	1	0	18	$-1R_1 + R_2$
0	0	1	2	0	1	24	$2R_1 + R_3$

The maximum is 24 when $x_1 = 0,\ x_2 = 12,\ x_3 = 0,\ s_1 = 0,\ s_2 = 18.$

c. This problem has a unique maximum value of z, which is 24, but it occurs at two different basic feasible solutions.

Copyright © 2019 Pearson Education, Inc.

Section 7.5 Maximization Applications

1. Let x_1 = the number of WildSplash and let x_2 = the number of TastyTreat.

The problem is to maximize $z = 12x_1 + 10x_2$ subject to:

$$\begin{aligned} 2x_1 + x_2 &\le 90 \\ x_1 + 2x_2 &\le 80 \\ x_1 + x_2 &\le 50 \\ x_1 \ge 0, x_2 &\ge 0. \end{aligned}$$

There are three constraints to be changed into equalities, so introduce three slack variables, s_1, s_2, and s_3. The problem can now be restated as:

Find $x_1 \ge 0, x_2 \ge 0, s_1 \ge 0, s_2 \ge 0, s_3 \ge 0,$ such that

$$\begin{aligned} 2x_1 + x_2 + s_1 \quad &= 90 \\ x_1 + 2x_2 \quad + s_2 \quad &= 80 \\ x_1 + x_2 \quad + s_3 &= 50 \end{aligned}$$

and $z = 12x_1 + 10x_2$ is maximized.

The initial simplex tableau is

$$\begin{array}{c} \begin{matrix} x_1 & x_2 & s_1 & s_2 & s_3 \end{matrix} \\ \left[\begin{array}{rrrrr|r} 2 & 1 & 1 & 0 & 0 & 90 \\ 1 & 2 & 0 & 1 & 0 & 80 \\ 1 & 1 & 0 & 0 & 1 & 50 \\ \hline -12 & -10 & 0 & 0 & 0 & 0 \end{array}\right] \end{array}.$$

3. Let x_1 = the number of kg of P,
x_2 = the number of kg of Q,
x_3 = the number of kg of R, and
x_4 = the number of kg of S

The constraints are

$$\begin{aligned} .375x_3 + .625x_4 &\le 500 \\ .75x_2 + .50x_3 + .375x_4 &\le 600 \\ x_1 + .25x_2 + .125x_3 \quad &\le 300 \\ x_1 \ge 0, x_2 \ge 0, x_3 \ge 0, x_4 &\ge 0. \end{aligned}$$

(Notice the food contents are given in *percent* of nutrient per kilogram.) The objective function to maximize is $z = 90x_1 + 70x_2 + 60x_3 + 50x_4$.

The initial tableau is

$$\begin{array}{c} \begin{matrix} x_1 & x_2 & x_3 & x_4 & s_1 & s_2 & s_3 \end{matrix} \\ \left[\begin{array}{rrrrrrr|r} 0 & 0 & .375 & .625 & 1 & 0 & 0 & 500 \\ 0 & .75 & .5 & .375 & 0 & 1 & 0 & 600 \\ 1 & .25 & .125 & 0 & 0 & 0 & 1 & 300 \\ \hline -90 & -70 & -60 & -50 & 0 & 0 & 0 & 0 \end{array}\right] \end{array}$$

Copyright © 2019 Pearson Education, Inc.

5. a. The information is contained in the table.

	Aluminum	Steel	Profit
1 Speed	12	20	\$8
3 Speed	21	30	\$12
10 Speed	16	40	\$24
Amount Available	42,000	91,800	

Let x_1 = number of 1-speed bikes, let x_2 = number of 3-speed bikes, and let x_3 = number of 10-speed bikes.

The problem is to maximize
$z = 80x_1 + 120x_2 + 240x_3$
subject to:

$$12x_1 + 21x_2 + 16x_3 \le 42,000$$
$$20x_1 + 30x_2 + 40x_3 \le 91,800$$
$$x_1 \ge 0,\ x_2 \ge 0,\ x_3 \ge 0.$$

The initial simplex tableau is

$$\begin{array}{ccccc|c} x_1 & x_2 & x_3 & s_1 & s_2 & \\ 12 & 21 & 16 & 1 & 0 & 42,000 \\ 20 & 30 & 40 & 0 & 1 & 91,800 \\ \hline -80 & -120 & -240 & 0 & 0 & 0 \end{array}.$$

Pivot on the 40 in row two, column three.

$$\begin{array}{ccccc|cl} x_1 & x_2 & x_3 & s_1 & s_2 & & \\ 12 & 21 & 16 & 1 & 0 & 42,000 & \\ \frac{1}{2} & \frac{3}{4} & 1 & 0 & \frac{1}{40} & 2295 & \frac{1}{40}R_2 \\ \hline -80 & -120 & -240 & 0 & 0 & 0 & \end{array}$$

$$\begin{array}{ccccc|cl} x_1 & x_2 & x_3 & s_1 & s_2 & & \\ 4 & 9 & 0 & 1 & -\frac{2}{5} & 5280 & -16R_2 + R_1 \\ \frac{1}{2} & \frac{3}{4} & 1 & 0 & \frac{1}{40} & 2295 & \frac{1}{40}R_2 \\ \hline 40 & 60 & 0 & 0 & \frac{3}{5} & 550,800 & 24R_2 + R_3 \end{array}$$

From the final tableau, the maximum is 550,800 when $x_3 = 2295$, $s_1 = 5280$, and $x_1 = x_2 = s_2 = 0$. Thus, the manufacturer should make no 1-speed or 3-speed bicycles, and should make 2295 10-speed bicycles for a maximum profit of \$550,800.

b. In the optimal solution, $s_1 = 5280$ and $s_2 = 0$. $s_1 = 5280$ means 5280 units of aluminum should be left unused. $s_2 = 0$ means all of the steel is used.

7. a. Let x_1 = the number of minutes allotted to the sports, let x_2 = the number of minutes allotted to the news, and let x_3 = the number of minutes allotted to the weather.

Maximize $z = 40x_1 + 60x_2 + 50x_3$

Subject to: $x_1 + x_2 + x_3 = 27$

$$x_1 = 2x_3$$
$$x_1 + x_3 = 2x_2$$
$$x_1 \ge 0,\ x_2 \ge 0,\ x_3 \ge 0$$

Rewrite problem in standard maximum form.

Maximize $z = 40x_1 + 60x_2 + 50x_3$

$$x_1 + x_2 + x_3 = 27$$
$$-x_1 + 2x_3 = 0$$
$$-x_1 + 2x_2 - x_3 = 0$$
$$x_1 \ge 0,\ x_2 \ge 0,\ x_3 \ge 0$$

The initial simplex tableau is

$$\begin{array}{cccccc|c} x_1 & x_2 & x_3 & s_1 & s_2 & s_3 & \\ 1 & 1 & 1 & 1 & 0 & 0 & 27 \\ -1 & 0 & 2 & 0 & 1 & 0 & 0 \\ -1 & \boxed{2} & -1 & 0 & 0 & 1 & 0 \\ \hline -40 & -60 & -50 & 0 & 0 & 0 & 0 \end{array}$$

Pivot on the 2 in row three, column two. Multiply row three by $\frac{1}{2}$ and complete the pivot to get

$$\begin{array}{cccccc|cl} x_1 & x_2 & x_3 & s_1 & s_2 & s_3 & & \\ \frac{3}{2} & 0 & \frac{3}{2} & 1 & 0 & -\frac{1}{2} & 27 & -R_3 + R_1 \\ -1 & 0 & \boxed{2} & 0 & 1 & 0 & 0 & \\ -\frac{1}{2} & 1 & -\frac{1}{2} & 0 & 0 & \frac{1}{2} & 0 & \frac{1}{2}R_3 \\ \hline -70 & 0 & -80 & 0 & 0 & 30 & 0 & 60R_3 + R_4 \end{array}$$

Pivot on the 2 in row two, column three. Multiply row 2 by $\frac{1}{2}$ and complete the pivot to get

$$\begin{array}{cccccc|cl} x_1 & x_2 & x_3 & s_1 & s_2 & s_3 & & \\ \boxed{\frac{9}{4}} & 0 & 0 & 1 & -\frac{3}{4} & -\frac{1}{2} & 27 & -\frac{3}{2}R_2 + R_1 \\ -\frac{1}{2} & 0 & 1 & 0 & \frac{1}{2} & 0 & 0 & \frac{1}{2}R_2 \\ -\frac{3}{4} & 1 & 0 & 0 & \frac{1}{4} & \frac{1}{2} & 0 & \frac{1}{2}R_2 + R_3 \\ \hline -110 & 0 & 0 & 0 & 40 & 30 & 0 & 80R_2 + R_4 \end{array}$$

Pivot on the $\frac{9}{4}$ in row one, column one.

$$\begin{array}{cccccc|cl} x_1 & x_2 & x_3 & s_1 & s_2 & s_3 & & \\ 1 & 0 & 0 & \frac{4}{9} & -\frac{1}{3} & -\frac{2}{9} & 12 & \frac{4}{9}R_1 \\ 0 & 0 & 1 & \frac{2}{9} & \frac{1}{3} & -\frac{1}{9} & 6 & \frac{1}{2}R_1 + R_2 \\ 0 & 1 & 0 & \frac{1}{3} & 0 & \frac{1}{3} & 9 & \frac{3}{4}R_1 + R_3 \\ \hline 0 & 0 & 0 & \frac{440}{9} & \frac{10}{3} & \frac{50}{9} & 1320 & 110R_1 + R_4 \end{array}$$

Copyright © 2019 Pearson Education, Inc.

The sports should get 12 minutes of airtime, the news 9 minutes and the weather 6 minutes for a maximum of 1,320,000 viewers.

b. In the optimal solution, $s_1 = 0$, $s_2 = 0$ and $s_3 = 0$. $s_1 = 0$ means that all of the 27 minutes will be used by the new show, i.e. there is no slack. $s_2 = 0$ means that the sports has exactly twice as much time as the weather. $s_3 = 0$ means that the sports and weather have exactly twice as much time as the news.

9. a. Let x_1 = the number of Japanese Maples and

let x_2 = the number of tri-color beech trees.

We want to maximize the profit function, $z = 350x_1 + 500x_2$ subject to the constraints

$$5x_1 + 7x_2 \le 3600$$
$$x_1 + 2x_2 \le 900$$
$$4x_1 + 4x_2 \le 2600.$$

The third constraint simplifies to

$x_1 + x_2 \le 650$

The initial simplex tableau is

$$\begin{array}{ccccc|c} x_1 & x_2 & s_1 & s_2 & s_3 & \\ 5 & 7 & 1 & 0 & 0 & 3600 \\ 1 & \boxed{2} & 0 & 1 & 0 & 900 \\ 1 & 1 & 0 & 0 & 1 & 650 \\ \hline -350 & -500 & 0 & 0 & 0 & 0 \end{array}.$$

Pivot on the 2 in row two, column two. First divide each entry in row 2 by 2.

$$\begin{array}{ccccc|c} x_1 & x_2 & s_1 & s_2 & s_3 & \\ 5 & 7 & 1 & 0 & 0 & 3600 \\ \frac{1}{2} & 1 & 0 & \frac{1}{2} & 0 & 450 \\ 1 & 1 & 0 & 0 & 1 & 650 \\ \hline -350 & -500 & 0 & 0 & 0 & 0 \end{array}.$$

$$\begin{array}{ccccc|c|l} x_1 & x_2 & s_1 & s_2 & s_3 & & \\ \boxed{\frac{3}{2}} & 0 & 1 & -\frac{7}{2} & 0 & 450 & -7R_2 + R_1 \\ \frac{1}{2} & 1 & 0 & \frac{1}{2} & 0 & 450 & \\ \frac{1}{2} & 0 & 0 & -\frac{1}{2} & 1 & 200 & -R_2 + R_3 \\ \hline -100 & 0 & 0 & 250 & 0 & 225{,}000 & 500R_2 + R_4 \end{array}$$

Now pivot on $\frac{3}{2}$ in row 1 column 1. First multiply each entry in row 1 by $\frac{1}{3}$.

$$\begin{array}{ccccc|c} x_1 & x_2 & s_1 & s_2 & s_3 & \\ \frac{1}{2} & 0 & \frac{1}{3} & -\frac{7}{6} & 0 & 150 \\ \frac{1}{2} & 1 & 0 & \frac{1}{2} & 0 & 450 \\ \frac{1}{2} & 0 & 0 & -\frac{1}{2} & 1 & 200 \\ \hline -100 & 0 & 0 & 250 & 0 & 225{,}000 \end{array}$$

$$\begin{array}{ccccc|c|l} x_1 & x_2 & s_1 & s_2 & s_3 & & \\ \frac{1}{2} & 0 & \frac{1}{3} & -\frac{7}{6} & 0 & 150 & \\ 0 & 1 & -\frac{1}{3} & \frac{5}{3} & 0 & 300 & -R_1 + R_2 \\ 0 & 0 & -\frac{1}{3} & \frac{2}{3} & 1 & 50 & -R_1 + R_3 \\ \hline 0 & 0 & \frac{200}{3} & \frac{50}{3} & 0 & 255{,}000 & 200R_1 + R_4 \end{array}$$

$$\begin{array}{ccccc|c|l} x_1 & x_2 & s_1 & s_2 & s_3 & & \\ 1 & 0 & \frac{2}{3} & -\frac{7}{3} & 0 & 300 & 2R_1 \\ 0 & 1 & -\frac{1}{3} & \frac{5}{3} & 0 & 300 & \\ 0 & 0 & -\frac{1}{3} & \frac{2}{3} & 1 & 50 & \\ \hline 0 & 0 & \frac{200}{3} & \frac{50}{3} & 0 & 255{,}000 & \end{array}$$

From the final tableau, we have $x_1 = 300$ and $x_2 = 300$. Thus, the nursery should acquire 300 Japanese Maples and 300 tri-color beech trees, for a maximum profit of \$255,000

b. s_1 and s_2 both are zero.

$4(300) + 4(300) + s_3 = 2600 \Rightarrow s_3 = 200$

Thus, there are 200 unused hours in the delivery to the client.

Copyright © 2019 Pearson Education, Inc.

11. Let x_1, x_2, and x_3 = the number of newspapers, radio, and TV ads respectively.

Maximize $z = 2000x_1 + 1200x_2 + 10{,}000x_3$

subject to:

$400x_1 + 200x_2 + 1200x_3 \le 8000$

$x_1 \le 20,\ x_2 \le 30,\ x_3 \le 6$

$x_1 \ge 0,\ x_2 \ge 0,\ x_3 \ge 0.$

The initial simplex tableau is

$$\begin{array}{c} \\ \\ \end{array}\left[\begin{array}{ccccccc|c} x_1 & x_2 & x_3 & s_1 & s_2 & s_3 & s_4 & \\ 400 & 200 & 1200 & 1 & 0 & 0 & 0 & 8000 \\ 1 & 0 & 0 & 0 & 1 & 0 & 0 & 20 \\ 0 & 1 & 0 & 0 & 0 & 1 & 0 & 30 \\ 0 & 0 & \boxed{1} & 0 & 0 & 0 & 1 & 6 \\ \hline -2000 & -1200 & -10{,}000 & 0 & 0 & 0 & 0 & 0 \end{array}\right]$$

Pivot on the 1 in row four, column three.

$$\left[\begin{array}{ccccccc|c} x_1 & x_2 & x_3 & s_1 & s_2 & s_3 & s_4 & \\ \boxed{400} & 200 & 0 & 1 & 0 & 0 & -1200 & 800 \\ 1 & 0 & 0 & 0 & 1 & 0 & 0 & 20 \\ 0 & 1 & 0 & 0 & 0 & 1 & 0 & 30 \\ 0 & 0 & 1 & 0 & 0 & 0 & 1 & 6 \\ \hline -2000 & -1200 & 0 & 0 & 0 & 0 & 10{,}000 & 60{,}000 \end{array}\right]\begin{array}{l} \\ -1200R_4 + R_1 \\ \\ \\ \\ 10{,}000R_4 + R_5 \end{array}$$

Pivot on the 400 in row one, column one.

$$\left[\begin{array}{ccccccc|c} x_1 & x_2 & x_3 & s_1 & s_2 & s_3 & s_4 & \\ 1 & \frac{1}{2} & 0 & \frac{1}{400} & 0 & 0 & -3 & 2 \\ 1 & 0 & 0 & 0 & 1 & 0 & 0 & 20 \\ 0 & 1 & 0 & 0 & 0 & 1 & 0 & 30 \\ 0 & 0 & 1 & 0 & 0 & 0 & 1 & 6 \\ \hline -2000 & -1200 & 0 & 0 & 0 & 0 & 10{,}000 & 60{,}000 \end{array}\right]\begin{array}{l} \\ \frac{1}{400}R_1 \\ \\ \\ \\ \\ \end{array}$$

$$\left[\begin{array}{ccccccc|c} x_1 & x_2 & x_3 & s_1 & s_2 & s_3 & s_4 & \\ 1 & \boxed{\frac{1}{2}} & 0 & \frac{1}{400} & 0 & 0 & -3 & 2 \\ 0 & -\frac{1}{2} & 0 & -\frac{1}{400} & 1 & 0 & 3 & 18 \\ 0 & 1 & 0 & 0 & 0 & 1 & 0 & 30 \\ 0 & 0 & 1 & 0 & 0 & 0 & 1 & 6 \\ \hline 0 & -200 & 0 & 5 & 0 & 0 & 4000 & 64{,}000 \end{array}\right]\begin{array}{l} \\ \\ -1R_1 + R_2 \\ \\ \\ 2000R_1 + R_5 \end{array}$$

Pivot on the $\frac{1}{2}$ in row one, column two.

$$\left[\begin{array}{ccccccc|c} x_1 & x_2 & x_3 & s_1 & s_2 & s_3 & s_4 & \\ 2 & 1 & 0 & \frac{1}{200} & 0 & 0 & -6 & 4 \\ 0 & -\frac{1}{2} & 0 & -\frac{1}{400} & 1 & 0 & 3 & 18 \\ 0 & 1 & 0 & 0 & 0 & 1 & 0 & 30 \\ 0 & 0 & 1 & 0 & 0 & 0 & 1 & 6 \\ \hline 0 & -200 & 0 & 5 & 0 & 0 & 4000 & 64{,}000 \end{array}\right]\begin{array}{l} \\ 2R_1 \\ \\ \\ \\ \\ \end{array}$$

Copyright © 2019 Pearson Education, Inc.

$$\begin{array}{ccccccc|c|l} x_1 & x_2 & x_3 & s_1 & s_2 & s_3 & s_4 & & \\ 2 & 1 & 0 & \frac{1}{200} & 0 & 0 & -6 & 4 & \\ 1 & 0 & 0 & 0 & 1 & 0 & 0 & 20 & \frac{1}{2}R_1 + R_2 \\ -2 & 0 & 0 & -\frac{1}{200} & 0 & 1 & 6 & 26 & -1R_1 + R_3 \\ 0 & 0 & 1 & 0 & 0 & 0 & 1 & 6 & \\ \hline 400 & 0 & 0 & 6 & 0 & 0 & 2800 & 64,800 & 200R_1 + R_5 \end{array}$$

From the final tableau, we have $x_1 = 0$, $x_2 = 4$, and $x_3 = 6$.

No newspaper ads, 4 radio ads, and 6 TV ads will give a maximum exposure of 64,800 people.

13. a. Let x_1 = the number of fund-raising parties, let x_2 = the number of mailings, and let x_3 = the number of dinner parties.

Maximize $z = 200,000x_1 + 100,000x_2 + 600,000x_3$

Subject to:

$$x_1 + x_2 + x_3 \le 25$$
$$3000x_1 + 1000x_2 + 12,000x_3 \le 102,000$$
$$x_1 \ge 0,\ x_2 \ge 0,\ x_3 \ge 0$$

The initial simplex tableau is

$$\begin{array}{ccccc|c} x_1 & x_2 & x_3 & s_1 & s_2 & \\ 1 & 1 & 1 & 1 & 0 & 25 \\ 3000 & 1000 & \boxed{12,000} & 0 & 1 & 102,000 \\ \hline -200,000 & -100,000 & -600,000 & 0 & 0 & 0 \end{array}$$

Pivot on the 12,000, row two, column three.

$$\begin{array}{ccccc|c|l} x_1 & x_2 & x_3 & s_1 & s_2 & & \\ \frac{3}{4} & \boxed{\frac{11}{12}} & 0 & 1 & -\frac{1}{12,000} & \frac{33}{2} & -R_2 + R_1 \\ \frac{1}{4} & \frac{1}{12} & 1 & 0 & \frac{1}{12,000} & \frac{17}{2} & \frac{1}{12,000}R_2 \\ \hline -50,000 & -50,000 & 0 & 0 & 50 & 5,100,000 & 600,000R_2 + R_3 \end{array}$$

Pivot on $\frac{11}{12}$ in row one, column two.

$$\begin{array}{ccccc|c|l} x_1 & x_2 & x_3 & s_1 & s_2 & & \\ \boxed{\frac{9}{11}} & 1 & 0 & \frac{12}{11} & -\frac{1}{11,000} & 18 & \frac{12}{11}R_1 \\ \frac{2}{11} & 0 & 1 & -\frac{1}{11} & \frac{1}{11,000} & 7 & -\frac{1}{12}R_1 + R_2 \\ \hline -\frac{100,000}{11} & 0 & 0 & \frac{600,000}{11} & \frac{500}{11} & 6,000,000 & 50,000R_1 + R_3 \end{array}$$

Pivot on $\frac{9}{11}$ in row one, column one.

$$\begin{array}{ccccc|c|l} x_1 & x_2 & x_3 & s_1 & s_2 & & \\ 1 & \frac{11}{9} & 0 & \frac{4}{3} & -\frac{1}{9000} & 22 & \frac{11}{9}R_1 \\ 0 & -\frac{2}{9} & 1 & -\frac{1}{3} & \frac{1}{9000} & 3 & -\frac{2}{11}R_1 + R_2 \\ \hline 0 & \frac{100,000}{9} & 0 & \frac{200,000}{3} & \frac{400}{9} & 6,200,000 & \frac{100,000}{11}R_1 + R_3 \end{array}$$

The party should plan 22 fund raising parties and 3 dinner parties and no mailings to raise a maximum of \$6,200,000.

b. Answers will vary.

Copyright © 2019 Pearson Education, Inc.

15. Let x_1 = the number of hours running, x_2 = the number of hours biking, and x_3 = the number of hours walking. We are looking to maximize $z = 531x_1 + 472x_2 + 354x_3$ subject to the constraints

$$\begin{aligned} x_1 + x_2 + x_3 &\le 15 \\ x_1 &\le 3 \\ 2x_2 - x_3 &\le 0 \end{aligned}$$

$x_1 \ge 0, x_2 \ge 0, x_3 \ge 0.$

The initial simplex tableau is

$$\begin{array}{cccccc|c} x_1 & x_2 & x_3 & s_1 & s_2 & s_3 & \\ 1 & 1 & 1 & 1 & 0 & 0 & 15 \\ \boxed{1} & 0 & 0 & 0 & 1 & 0 & 3 \\ 0 & 2 & -1 & 0 & 0 & 1 & 0 \\ \hline -531 & -472 & -354 & 0 & 0 & 0 & 0 \end{array}$$

Pivot on the 1 in row 2, column 1.

$$\begin{array}{cccccc|cl} x_1 & x_2 & x_3 & s_1 & s_2 & s_3 & & \\ 0 & 1 & 1 & 1 & -1 & 0 & 12 & -R_2 + R_1 \\ 1 & 0 & 0 & 0 & 1 & 0 & 3 & \\ 0 & \boxed{2} & -1 & 0 & 0 & 1 & 0 & \\ \hline 0 & -472 & -354 & 0 & 531 & 0 & 1593 & 531R_2 + R_4 \end{array}$$

Pivot on the 2 in row 3, column 2.

$$\begin{array}{cccccc|cl} x_1 & x_2 & x_3 & s_1 & s_2 & s_3 & & \\ 0 & 0 & \boxed{3} & 2 & -2 & -1 & 24 & -R_3 + 2R_1 \\ 1 & 0 & 0 & 0 & 1 & 0 & 3 & \\ 0 & 2 & -1 & 0 & 0 & 1 & 0 & \\ \hline 0 & 0 & -590 & 0 & 531 & 236 & 1593 & 236R_2 + R_4 \end{array}$$

Finally, pivot on the 3 in row 1, column 3.

$$\begin{array}{cccccc|cl} x_1 & x_2 & x_3 & s_1 & s_2 & s_3 & & \\ 0 & 0 & 3 & 2 & -2 & -1 & 24 & \\ 1 & 0 & 0 & 0 & 1 & 0 & 3 & \\ 0 & 6 & 0 & 2 & -2 & 2 & 24 & R_1 + 3R_3 \\ \hline 0 & 0 & 0 & 1180 & 413 & 118 & 18{,}939 & 590R_1 + 3R_4 \end{array}$$

$$\begin{array}{cccccc|cl} x_1 & x_2 & x_3 & s_1 & s_2 & s_3 & & \\ 0 & 0 & 1 & \frac{2}{3} & -\frac{2}{3} & -\frac{1}{3} & 8 & \frac{1}{3}R_1 \\ 1 & 0 & 0 & 0 & 1 & 0 & 3 & \\ 0 & 1 & 0 & \frac{1}{3} & -\frac{1}{3} & \frac{1}{3} & 4 & \frac{1}{6}R_3 \\ \hline 0 & 0 & 0 & 1180 & 413 & 118 & 18{,}939 & \end{array}$$

Thus, $x_1 = 3$, $x_2 = 4$, and $x_3 = 8$, so

$z = 531(3) + 472(4) + 354(8) = 6313.$

Ramona should run 3 hours, bike 4 hours, and walk 8 hours for a maximum calorie expenditure of 6313 calories.

17. From Exercise 1, the initial simplex tableau is

$$\begin{array}{ccccc|c} x_1 & x_2 & s_1 & s_2 & s_3 & \\ \boxed{2} & 1 & 1 & 0 & 0 & 90 \\ 1 & 2 & 0 & 1 & 0 & 80 \\ 1 & 1 & 0 & 0 & 1 & 50 \\ \hline -12 & -10 & 0 & 0 & 0 & 0 \end{array}.$$

Pivot on the 2 in row one, column one.

$$\begin{array}{ccccc|cl} x_1 & x_2 & s_1 & s_2 & s_3 & & \\ 1 & \frac{1}{2} & \frac{1}{2} & 0 & 0 & 45 & \frac{1}{2}R_1 \\ 1 & 2 & 0 & 1 & 0 & 80 & \\ 1 & 1 & 0 & 0 & 1 & 50 & \\ \hline -12 & -10 & 0 & 0 & 0 & 0 & \end{array}$$

$$\begin{array}{ccccc|cl} x_1 & x_2 & s_1 & s_2 & s_3 & & \\ 1 & \frac{1}{2} & \frac{1}{2} & 0 & 0 & 45 & \\ 0 & \frac{3}{2} & -\frac{1}{2} & 1 & 0 & 35 & -1R_1 + R_2 \\ 0 & \boxed{\frac{1}{2}} & -\frac{1}{2} & 0 & 1 & 5 & -1R_1 + R_3 \\ \hline 0 & -4 & 6 & 0 & 0 & 540 & 12R_1 + R_4 \end{array}$$

Pivot on the $\frac{1}{2}$ in row three, column two.

$$\begin{array}{ccccc|cl} x_1 & x_2 & s_1 & s_2 & s_3 & & \\ 1 & \frac{1}{2} & \frac{1}{2} & 0 & 0 & 45 & \\ 0 & \frac{3}{2} & -\frac{1}{2} & 1 & 0 & 35 & \\ 0 & 1 & -1 & 0 & 2 & 10 & 2R_3 \\ \hline 0 & -4 & 6 & 0 & 0 & 540 & \end{array}$$

$$\begin{array}{ccccc|cl} x_1 & x_2 & s_1 & s_2 & s_3 & & \\ 1 & 0 & 1 & 0 & -1 & 40 & -\frac{1}{2}R_3 + R_1 \\ 0 & 0 & 1 & 1 & -3 & 20 & -\frac{3}{2}R_3 + R_2 \\ 0 & 1 & -1 & 0 & 2 & 10 & \\ \hline 0 & 0 & 2 & 0 & 8 & 580 & 4R_3 + R_4 \end{array}$$

The juice blender should make 40 units of WildSplash and 10 units of TastyTreat for a maximum gross income of \$580.

Copyright © 2019 Pearson Education, Inc.

Exercise 19 require the use of a computer.

19. The initial simplex tableau is

$$\begin{array}{ccccccc} x_1 & x_2 & x_3 & x_4 & s_1 & s_2 & s_3 & \\ \end{array}$$
$$\left[\begin{array}{ccccccc|c} 0 & 0 & .375 & .625 & 1 & 0 & 0 & 500 \\ 0 & .75 & .5 & .375 & 0 & 1 & 0 & 600 \\ 1 & .25 & .125 & 0 & 0 & 0 & 1 & 300 \\ \hline -90 & -70 & -60 & -50 & 0 & 0 & 0 & 0 \end{array}\right]$$

Using the SIMPLEX program on a TI-84 Plus, we get the maximum total growth value is 87,454.5 when 163.6 kg of food P, none of food Q, 1090.9 kg of food R, and 145.5 kg of food S are used.

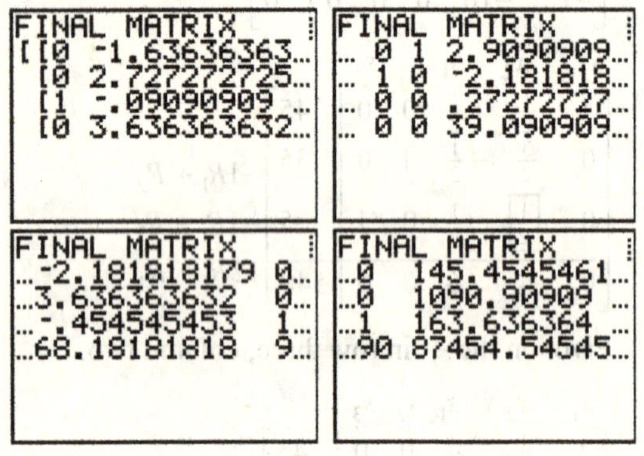

Section 7.6 The Simplex Method: Duality and Minimization

1. The transpose of

$$\begin{bmatrix} 3 & -4 & 5 \\ 1 & 10 & 7 \\ 0 & 3 & 6 \end{bmatrix} \text{ is } \begin{bmatrix} 3 & 1 & 0 \\ -4 & 10 & 3 \\ 5 & 7 & 6 \end{bmatrix}.$$

3. The transpose of

$$\begin{bmatrix} 3 & 0 & 14 & -5 & 3 \\ 4 & 17 & 8 & -6 & 1 \end{bmatrix} \text{ is } \begin{bmatrix} 3 & 4 \\ 0 & 17 \\ 14 & 8 \\ -5 & -6 \\ 3 & 1 \end{bmatrix}.$$

5. Minimize $w = 3y_1 + 5y_2$

subject to:
$$3y_1 + y_2 \ge 4$$
$$-y_1 + 2y_2 \ge 6$$
$$y_1 \ge 0,\ y_2 \ge 0.$$

The augmented matrix is

$$\left[\begin{array}{cc|c} 3 & 1 & 4 \\ -1 & 2 & 6 \\ \hline 3 & 5 & 0 \end{array}\right].$$

The transpose of this matrix is

$$\left[\begin{array}{cc|c} 3 & -1 & 3 \\ 1 & 2 & 5 \\ \hline 4 & 6 & 0 \end{array}\right].$$

The entries in this second matrix can be used to write the following dual maximization problem:

Maximize $z = 4x_1 + 6x_2$

subject to:
$$3x_1 - x_2 \le 3$$
$$x_1 + 2x_2 \le 5$$
$$x_1 \ge 0,\ x_2 \ge 0.$$

7. Minimize $w = 2y_1 + 8y_2$

subject to:
$$y_1 + 7y_2 \ge 18$$
$$4y_1 + y_2 \ge 15$$
$$5y_1 + 3y_2 \ge 20$$
$$y_1 \ge 0,\ y_2 \ge 0.$$

The augmented matrix is

$$\left[\begin{array}{cc|c} 1 & 7 & 18 \\ 4 & 1 & 15 \\ 5 & 3 & 20 \\ \hline 2 & 8 & 0 \end{array}\right].$$

The transpose of this matrix is

$$\left[\begin{array}{ccc|c} 1 & 4 & 5 & 2 \\ 7 & 1 & 3 & 8 \\ \hline 18 & 15 & 20 & 0 \end{array}\right].$$

The entries of this second matrix can be used to write the following dual maximization problem:

Maximize $z = 18x_1 + 15x_2 + 20x_3$

subject to:
$$x_1 + 4x_2 + 5x_3 \le 2$$
$$7x_1 + x_2 + 3x_3 \le 8$$
$$x_1 \ge 0,\ x_2 \ge 0,\ x_3 \ge 0.$$

Copyright © 2019 Pearson Education, Inc.

9. Minimize $w = 5y_1 + y_2 + 3y_3$

subject to: $7y_1 + 6y_2 + 8y_3 \geq 18$
$4y_1 + 5y_2 + 10y_3 \geq 20$
$y_1 \geq 0,\ y_2 \geq 0,\ y_3 \geq 0.$

The augmented matrix is

$$\left[\begin{array}{ccc|c} 7 & 6 & 8 & 18 \\ 4 & 5 & 10 & 20 \\ \hline 5 & 1 & 3 & 0 \end{array}\right].$$

The transpose of this matrix is

$$\left[\begin{array}{cc|c} 7 & 4 & 5 \\ 6 & 5 & 1 \\ 8 & 10 & 3 \\ \hline 18 & 20 & 0 \end{array}\right].$$

The entries of this second matrix can be used to write the following dual maximization problem.

Maximize $z = 18x_1 + 20x_2$

subject to: $7x_1 + 4x_2 \leq 5$
$6x_1 + 5x_2 \leq 1$
$8x_1 + 10x_2 \leq 3$
$x_1 \geq 0,\ x_2 \geq 0.$

11. Minimize $w = 8y_1 + 9y_2 + 3y_3$

subject to: $y_1 + y_2 + y_3 \geq 5$
$y_1 + y_2 \geq 4$
$2y_1 + y_2 + 3y_3 \geq 15$
$y_1 \geq 0,\ y_2 \geq 0,\ y_3 \geq 0.$

The augmented matrix is

$$\left[\begin{array}{ccc|c} 1 & 1 & 1 & 5 \\ 1 & 1 & 0 & 4 \\ 2 & 1 & 3 & 15 \\ \hline 8 & 9 & 3 & 0 \end{array}\right].$$

The transpose of this matrix is

$$\left[\begin{array}{ccc|c} 1 & 1 & 2 & 8 \\ 1 & 1 & 1 & 9 \\ 1 & 0 & 3 & 3 \\ \hline 5 & 4 & 15 & 0 \end{array}\right].$$

The entries of this second matrix can be used to write the following dual maximization problem:

Maximize $z = 5x_1 + 4x_2 + 15x_3$

subject to: $x_1 + x_2 + 2x_3 \leq 8$
$x_1 + x_2 + x_3 \leq 9$
$x_1 + 3x_3 \leq 3$
$x_1 \geq 0,\ x_2 \geq 0,\ x_3 \geq 0.$

13. From exercise 9, we are to minimize $w = 5y_1 + y_2 + 3y_3$ subject to the constraints

$7y_1 + 6y_2 + 8y_3 \geq 18$
$4y_1 + 5y_2 + 10y_3 \geq 20$
$y_1 \geq 0,\ y_2 \geq 0,\ y_3 \geq 0$

We write the dual maximization problem, maximize $z = 18x_1 + 20x_2$

subject to: $7x_1 + 4x_2 \leq 5$
$6x_1 + 5x_2 \leq 1$
$8x_1 + 10x_2 \leq 3$
$x_1 \geq 0,\ x_2 \geq 0.$

The initial simplex tableau is

$$\begin{array}{c} \begin{array}{cccccc} x_1 & x_2 & s_1 & s_2 & s_3 & \end{array} \\ \left[\begin{array}{ccccc|c} 7 & 4 & 1 & 0 & 0 & 5 \\ 6 & \boxed{5} & 0 & 1 & 0 & 1 \\ 8 & 10 & 0 & 0 & 1 & 3 \\ \hline -18 & -20 & 0 & 0 & 0 & 0 \end{array}\right] \end{array}.$$

Pivot on the 5 in row 2, column 2.

$$\begin{array}{c} \begin{array}{cccccc} x_1 & x_2 & s_1 & s_2 & s_3 & \end{array} \\ \left[\begin{array}{ccccc|c} 11 & 0 & 5 & -4 & 0 & 21 \\ 6 & 5 & 0 & 1 & 0 & 1 \\ -4 & 0 & 0 & -2 & 1 & 1 \\ \hline 6 & 0 & 0 & 4 & 0 & 4 \end{array}\right] \end{array} \begin{array}{l} -4R_2 + 5R_1 \\ \\ -2R_2 + R_3 \\ 4R_2 + R_4 \end{array}$$

From the final tableau, the minimum is 4 for $y_1 = 0,\ y_2 = 4,$ and $y_3 = 0.$

Copyright © 2019 Pearson Education, Inc.

15. From exercise 11, we are to minimize
$w = 8y_1 + 9y_2 + 3y_3$
subject to:
$$\begin{aligned} y_1 + y_2 + y_3 &\ge 5 \\ y_1 + y_2 &\ge 4 \\ 2y_1 + y_2 + 3y_3 &\ge 15 \\ y_1 \ge 0,\ y_2 \ge 0,\ y_3 &\ge 0. \end{aligned}$$
We write the dual maximization problem,
maximize $z = 5x_1 + 4x_2 + 15x_3$
subject to:
$$\begin{aligned} x_1 + x_2 + 2x_3 &\le 8 \\ x_1 + x_2 + x_3 &\le 9 \\ x_1 + 3x_3 &\le 3 \\ x_1 \ge 0,\ x_2 \ge 0,\ x_3 &\ge 0. \end{aligned}$$
The initial simplex tableau is
$$\begin{array}{cccccc|c} x_1 & x_2 & x_3 & s_1 & s_2 & s_3 & \\ 1 & 1 & 2 & 1 & 0 & 0 & 8 \\ 1 & 1 & 1 & 0 & 1 & 0 & 9 \\ 1 & 0 & \boxed{3} & 0 & 0 & 1 & 3 \\ \hline -5 & -4 & -15 & 0 & 0 & 0 & 0 \end{array}.$$
Pivot on the 3 in row 3, column 3.
$$\begin{array}{cccccc|c|l} x_1 & x_2 & x_3 & s_1 & s_2 & s_3 & & \\ 1 & 1 & 2 & 1 & 0 & 0 & 8 & \\ 1 & 1 & 1 & 0 & 1 & 0 & 9 & \\ \frac{1}{3} & 0 & 1 & 0 & 0 & \frac{1}{3} & 1 & \frac{1}{3}R_3 \\ \hline -5 & -4 & -15 & 0 & 0 & 0 & 0 & \end{array}$$
$$\begin{array}{cccccc|c|l} x_1 & x_2 & x_3 & s_1 & s_2 & s_3 & & \\ \frac{1}{3} & \boxed{1} & 0 & 1 & 0 & -\frac{2}{3} & 6 & -2R_3 + R_1 \\ \frac{2}{3} & 1 & 0 & 0 & 1 & -\frac{1}{3} & 8 & -R_3 + R_2 \\ \frac{1}{3} & 0 & 1 & 0 & 0 & \frac{1}{3} & 1 & \\ \hline 0 & -4 & 0 & 0 & 0 & 5 & 15 & 15R_3 + R_4 \end{array}$$
Now pivot on the 1 in row 1, column 2.
$$\begin{array}{cccccc|c|l} x_1 & x_2 & x_3 & s_1 & s_2 & s_3 & & \\ \frac{1}{3} & \boxed{1} & 0 & 1 & 0 & -\frac{2}{3} & 6 & \\ \frac{2}{3} & 1 & 0 & 0 & 1 & -\frac{1}{3} & 8 & \\ \frac{1}{3} & 0 & 1 & 0 & 0 & \frac{1}{3} & 1 & \\ \hline \frac{4}{3} & 0 & 0 & 4 & 0 & \frac{7}{3} & 39 & 4R_1 + R_4 \end{array}$$
From the final tableau, the minimum is 39 for $y_1 = 4$, $y_2 = 0$, and $y_3 = \frac{7}{3}$.

17. Minimize $w = 2y_1 + y_2 + 3y_3$
subject to:
$$\begin{aligned} y_1 + y_2 + y_3 &\ge 100 \\ 2y_1 + y_2 &\ge 50 \\ y_1 \ge 0,\ y_2 \ge 0,\ y_3 &\ge 0. \end{aligned}$$
The augmented matrix is
$$\left[\begin{array}{ccc|c} 1 & 1 & 1 & 100 \\ 2 & 1 & 0 & 50 \\ \hline 2 & 1 & 3 & 0 \end{array}\right].$$
The transpose is
$$\left[\begin{array}{cc|c} 1 & 2 & 2 \\ 1 & 1 & 1 \\ 1 & 0 & 3 \\ \hline 100 & 50 & 0 \end{array}\right].$$
The dual maximization problem is:
Maximize $z = 100x_1 + 50x_2$
subject to:
$$\begin{aligned} x_1 + 2x_2 &\le 2 \\ x_1 + x_2 &\le 1 \\ x_1 &\le 3 \\ x_1 \ge 0,\ x_2 &\ge 0. \end{aligned}$$
The initial simplex tableau is
$$\begin{array}{ccccc|c} x_1 & x_2 & s_1 & s_2 & s_3 & \\ 1 & 2 & 1 & 0 & 0 & 2 \\ \boxed{1} & 1 & 0 & 1 & 0 & 1 \\ 1 & 0 & 0 & 0 & 1 & 3 \\ \hline -100 & -50 & 0 & 0 & 0 & 0 \end{array}.$$
Pivot on the 1 in row two, column one.
$$\begin{array}{ccccc|c|l} x_1 & x_2 & s_1 & s_2 & s_3 & & \\ 0 & 1 & 1 & -1 & 0 & 1 & -1R_2 + R_1 \\ 1 & 1 & 0 & 1 & 0 & 1 & \\ 0 & -1 & 0 & -1 & 1 & 2 & -R_2 + R_3 \\ \hline 0 & 50 & 0 & 100 & 0 & 100 & 100R_2 + R_4 \end{array}$$
The solution to the minimization problem is found in the bottom row of the final matrix in the entries corresponding to the slack variables.
The minimum is 100 when
$y_1 = 0$, $y_2 = 100$, and $y_3 = 0$.

Copyright © 2019 Pearson Education, Inc.

19. Minimize $w = 3y_1 + y_2 + 4y_3$

subject to:
$$\begin{aligned} 2y_1 + y_2 + y_3 &\ge 6 \\ y_1 + 2y_2 + y_3 &\ge 8 \\ 2y_1 + y_2 + 2y_3 &\ge 12 \\ y_1 \ge 0,\ y_2 \ge 0,\ y_3 &\ge 0 \end{aligned}$$

The augmented matrix is
$$\left[\begin{array}{ccc|c} 2 & 1 & 1 & 6 \\ 1 & 2 & 1 & 8 \\ 2 & 1 & 2 & 12 \\ \hline 3 & 1 & 4 & 0 \end{array}\right].$$

The transpose is
$$\left[\begin{array}{ccc|c} 2 & 1 & 2 & 3 \\ 1 & 2 & 1 & 1 \\ 1 & 1 & 2 & 4 \\ \hline 6 & 8 & 12 & 0 \end{array}\right].$$

The dual maximization problem is

Maximize $z = 6x_1 + 8x_2 + 12x_3$

subject to:
$$\begin{aligned} 2x_1 + x_2 + 2x_3 &\le 3 \\ x_1 + 2x_2 + x_3 &\le 1 \\ x_1 + x_2 + 2x_3 &\le 4 \\ x_1 \ge 0,\ x_2 \ge 0,\ x_3 &\ge 0. \end{aligned}$$

The initial tableau is
$$\begin{array}{c} \begin{array}{cccccc} x_1 & x_2 & x_3 & s_1 & s_2 & s_3 \end{array} \\ \left[\begin{array}{cccccc|c} 2 & 1 & 2 & 1 & 0 & 0 & 3 \\ 1 & 2 & \boxed{1} & 0 & 1 & 0 & 1 \\ 1 & 1 & 2 & 0 & 0 & 1 & 4 \\ \hline -6 & -8 & -12 & 0 & 0 & 0 & 0 \end{array}\right]. \end{array}$$

Pivot on the 1 in row two, column three.
$$\begin{array}{c} \begin{array}{cccccc} x_1 & x_2 & x_3 & s_1 & s_2 & s_3 \end{array} \\ \left[\begin{array}{cccccc|c} 0 & -3 & 0 & 1 & -2 & 0 & 1 \\ 1 & 2 & 1 & 0 & 1 & 0 & 1 \\ -1 & -3 & 0 & 0 & -2 & 1 & 2 \\ \hline 6 & 16 & 0 & 0 & 12 & 0 & 12 \end{array}\right] \begin{array}{l} -2R_2 + R_1 \\ \\ -2R_2 + R_3 \\ 12R_2 + R_4 \end{array} \end{array}$$

The minimum is 12 when $y_1 = 0,\ y_2 = 12,\ y_3 = 0.$

21. Minimize $w = 6y_1 + 4y_2 + 2y_3$

subject to:
$$\begin{aligned} 2y_1 + 2y_2 + y_3 &\ge 2 \\ y_1 + 3y_2 + 2y_3 &\ge 3 \\ y_1 + y_2 + 2y_3 &\ge 4 \\ y_1 \ge 0,\ y_2 \ge 0,\ y_3 &\ge 0. \end{aligned}$$

The augmented matrix is
$$\left[\begin{array}{ccc|c} 2 & 2 & 1 & 2 \\ 1 & 3 & 2 & 3 \\ 1 & 1 & 2 & 4 \\ \hline 6 & 4 & 2 & 0 \end{array}\right].$$

The transpose is
$$\left[\begin{array}{ccc|c} 2 & 1 & 1 & 6 \\ 2 & 3 & 1 & 4 \\ 1 & 2 & 2 & 2 \\ \hline 2 & 3 & 4 & 0 \end{array}\right].$$

The dual maximization problem is

Maximize $z = 2x_1 + 3x_2 + 4x_3$

subject to:
$$\begin{aligned} 2x_1 + x_2 + x_3 &\le 6 \\ 2x_1 + 3x_2 + x_3 &\le 4 \\ x_1 + 2x_2 + 2x_3 &\le 2 \\ x_1 \ge 0,\ x_2 \ge 0,\ x_3 &\ge 0. \end{aligned}$$

The initial simplex tableau is
$$\begin{array}{c} \begin{array}{cccccc} x_1 & x_2 & x_3 & s_1 & s_2 & s_3 \end{array} \\ \left[\begin{array}{cccccc|c} 2 & 1 & 1 & 1 & 0 & 0 & 6 \\ 2 & 3 & 1 & 0 & 1 & 0 & 4 \\ 1 & 2 & \boxed{2} & 0 & 0 & 1 & 2 \\ \hline -2 & -3 & -4 & 0 & 0 & 0 & 0 \end{array}\right]. \end{array}$$

Pivot on the 2 in row three, column three.
$$\begin{array}{c} \begin{array}{cccccc} x_1 & x_2 & x_3 & s_1 & s_2 & s_3 \end{array} \\ \left[\begin{array}{cccccc|c} 2 & 1 & 1 & 1 & 0 & 0 & 6 \\ 2 & 3 & 1 & 0 & 1 & 0 & 4 \\ \frac{1}{2} & 1 & 1 & 0 & 0 & \frac{1}{2} & 1 \\ \hline -2 & -3 & -4 & 0 & 0 & 0 & 0 \end{array}\right] \begin{array}{l} \\ \\ \frac{1}{2}R_3 \\ \\ \end{array} \end{array}$$

$$\begin{array}{c} \begin{array}{cccccc} x_1 & x_2 & x_3 & s_1 & s_2 & s_3 \end{array} \\ \left[\begin{array}{cccccc|c} \frac{3}{2} & 0 & 0 & 1 & 0 & -\frac{1}{2} & 5 \\ \frac{3}{2} & 2 & 0 & 0 & 1 & -\frac{1}{2} & 3 \\ \frac{1}{2} & 1 & 1 & 0 & 0 & \frac{1}{2} & 1 \\ \hline 0 & 1 & 0 & 0 & 0 & 2 & 4 \end{array}\right] \begin{array}{l} -R_3 + R_1 \\ -R_3 + R_2 \\ \\ 4R_3 + R_4 \end{array} \end{array}$$

The minimum is 4 when $y_1 = 0,\ y_2 = 0,$ and $y_3 = 2.$

Copyright © 2019 Pearson Education, Inc.

23. Minimize $w = 20y_1 + 12y_2 + 40y_3$

subject to:
$$\begin{aligned} y_1 + y_2 + 5y_3 &\ge 20 \\ 2y_1 + y_2 + y_3 &\ge 30 \\ y_1 \ge 0,\ y_2 \ge 0,\ y_3 &\ge 0. \end{aligned}$$

The augmented matrix is

$$\left[\begin{array}{ccc|c} 1 & 1 & 5 & 20 \\ 2 & 1 & 1 & 30 \\ \hline 20 & 12 & 40 & 0 \end{array}\right].$$

The transpose is

$$\left[\begin{array}{cc|c} 1 & 2 & 20 \\ 1 & 1 & 12 \\ 5 & 1 & 40 \\ \hline 20 & 30 & 0 \end{array}\right].$$

The dual maximization problem is

Maximize $z = 20x_1 + 30x_2$

subject to:
$$\begin{aligned} x_1 + 2x_2 &\le 20 \\ x_1 + x_2 &\le 12 \\ 5x_1 + x_2 &\le 40 \\ x_1 \ge 0,\ x_2 &\ge 0. \end{aligned}$$

The initial simplex tableau is

$$\begin{array}{c} \begin{matrix} x_1 & x_2 & s_1 & s_2 & s_3 \end{matrix} \\ \left[\begin{array}{ccccc|c} 1 & \boxed{2} & 1 & 0 & 0 & 20 \\ 1 & 1 & 0 & 1 & 0 & 12 \\ 5 & 1 & 0 & 0 & 1 & 40 \\ \hline -20 & -30 & 0 & 0 & 0 & 0 \end{array}\right]. \end{array}$$

Pivot on the 2 in row one, column two.

$$\begin{array}{c} \begin{matrix} x_1 & x_2 & s_1 & s_2 & s_3 \end{matrix} \\ \left[\begin{array}{ccccc|c} \frac{1}{2} & 1 & \frac{1}{2} & 0 & 0 & 10 \\ 1 & 1 & 0 & 1 & 0 & 12 \\ 5 & 1 & 0 & 0 & 1 & 40 \\ \hline -20 & -30 & 0 & 0 & 0 & 0 \end{array}\right] \begin{matrix} \frac{1}{2}R_1 \\ \\ \\ \\ \end{matrix} \end{array}$$

$$\begin{array}{c} \begin{matrix} x_1 & x_2 & s_1 & s_2 & s_3 \end{matrix} \\ \left[\begin{array}{ccccc|c} \frac{1}{2} & 1 & \frac{1}{2} & 0 & 0 & 10 \\ \boxed{\frac{1}{2}} & 0 & -\frac{1}{2} & 1 & 0 & 2 \\ \frac{9}{2} & 0 & -\frac{1}{2} & 0 & 1 & 30 \\ \hline -5 & 0 & 15 & 0 & 0 & 300 \end{array}\right] \begin{matrix} \\ -R_1 + R_2 \\ -R_1 + R_3 \\ 30R_1 + R_4 \end{matrix} \end{array}$$

Pivot on the $\frac{1}{2}$ in row two, column one.

$$\begin{array}{c} \begin{matrix} x_1 & x_2 & s_1 & s_2 & s_3 \end{matrix} \\ \left[\begin{array}{ccccc|c} \frac{1}{2} & 1 & \frac{1}{2} & 0 & 0 & 10 \\ 1 & 0 & -1 & 2 & 0 & 4 \\ \frac{9}{2} & 0 & -\frac{1}{2} & 0 & 1 & 30 \\ \hline -5 & 0 & 15 & 0 & 0 & 300 \end{array}\right] \begin{matrix} \\ 2R_2 \\ \\ \\ \end{matrix} \end{array}$$

$$\begin{array}{c} \begin{matrix} x_1 & x_2 & s_1 & s_2 & s_3 \end{matrix} \\ \left[\begin{array}{ccccc|c} 0 & 1 & 1 & -1 & 0 & 8 \\ 1 & 0 & -1 & 2 & 0 & 4 \\ 0 & 0 & 4 & -9 & 1 & 12 \\ \hline 0 & 0 & 10 & 10 & 0 & 320 \end{array}\right] \begin{matrix} -\frac{1}{2}R_2 + R_1 \\ \\ -\frac{9}{2}R_2 + R_3 \\ 5R_2 + R_4 \end{matrix} \end{array}$$

The minimum is 320 when $y_1 = 10$, $y_2 = 10$, and $y_3 = 0$.

25. Minimize $w = 4y_1 + 2y_2 + y_3$

subject to:
$$\begin{aligned} y_1 + y_2 + y_3 &\ge 4 \\ 3y_1 + y_2 + 3y_3 &\ge 6 \\ y_1 + y_2 + 3y_3 &\ge 5 \\ y_1 \ge 0,\ y_2 \ge 0,\ y_3 &\ge 0. \end{aligned}$$

The augmented matrix is

$$\left[\begin{array}{ccc|c} 1 & 1 & 1 & 4 \\ 3 & 1 & 3 & 6 \\ 1 & 1 & 3 & 5 \\ \hline 4 & 2 & 1 & 0 \end{array}\right].$$

The transpose is

$$\left[\begin{array}{ccc|c} 1 & 3 & 1 & 4 \\ 1 & 1 & 1 & 2 \\ 1 & 3 & 3 & 1 \\ \hline 4 & 6 & 5 & 0 \end{array}\right].$$

The dual maximization problem is

Maximize $z = 4x_1 + 6x_2 + 5x_3$

subject to:
$$\begin{aligned} x_1 + 3x_2 + x_3 &\le 4 \\ x_1 + x_2 + x_3 &\le 2 \\ x_1 + 3x_2 + 3x_3 &\le 1 \\ x_1 \ge 0,\ x_2 > 0,\ x_3 &\ge 0. \end{aligned}$$

The initial simplex tableau is

$$\begin{array}{c} \begin{matrix} x_1 & x_2 & x_3 & s_1 & s_2 & s_3 \end{matrix} \\ \left[\begin{array}{cccccc|c} 1 & 3 & 1 & 1 & 0 & 0 & 4 \\ 1 & 1 & 1 & 0 & 1 & 0 & 2 \\ 1 & \boxed{3} & 3 & 0 & 0 & 1 & 1 \\ \hline -4 & -6 & -5 & 0 & 0 & 0 & 0 \end{array}\right]. \end{array}$$

Copyright © 2019 Pearson Education, Inc.

Pivot on the 3 in row three, column two.

$$\begin{array}{c} \\ \\ \\ \\ \end{array}\begin{array}{cccccc} x_1 & x_2 & x_3 & s_1 & s_2 & s_3 \end{array}$$

$$\left[\begin{array}{cccccc|c} 1 & 3 & 1 & 1 & 0 & 0 & 4 \\ 1 & 1 & 1 & 0 & 1 & 0 & 2 \\ \frac{1}{3} & 1 & 1 & 0 & 0 & \frac{1}{3} & \frac{1}{3} \\ \hline -4 & -6 & -5 & 0 & 0 & 0 & 0 \end{array}\right]\begin{array}{l} \\ \\ \frac{1}{3}R_3 \\ \\ \end{array}$$

$$\begin{array}{cccccc} x_1 & x_2 & x_3 & s_1 & s_2 & s_3 \end{array}$$

$$\left[\begin{array}{cccccc|c} 0 & 0 & -2 & 1 & 0 & -1 & 3 \\ \frac{2}{3} & 0 & 0 & 0 & 1 & -\frac{1}{3} & \frac{5}{3} \\ \boxed{\frac{1}{3}} & 1 & 1 & 0 & 0 & \frac{1}{3} & \frac{1}{3} \\ \hline -2 & 0 & 1 & 0 & 0 & 2 & 2 \end{array}\right]\begin{array}{l} -3R_3 + R_1 \\ -1R_3 + R_2 \\ \\ 6R_3 + R_4 \end{array}$$

Pivot on the $\frac{1}{3}$ in row three, column one.

$$\begin{array}{cccccc} x_1 & x_2 & x_3 & s_1 & s_2 & s_3 \end{array}$$

$$\left[\begin{array}{cccccc|c} 0 & 0 & -2 & 1 & 0 & -1 & 3 \\ \frac{2}{3} & 0 & 0 & 0 & 1 & -\frac{1}{3} & \frac{5}{3} \\ 1 & 3 & 3 & 0 & 0 & 1 & 1 \\ \hline -2 & 0 & 1 & 0 & 0 & 2 & 2 \end{array}\right]\begin{array}{l} \\ \\ 3R_3 \\ \\ \end{array}$$

$$\begin{array}{cccccc} x_1 & x_2 & x_3 & s_1 & s_2 & s_3 \end{array}$$

$$\left[\begin{array}{cccccc|c} 0 & 0 & -2 & 1 & 0 & -1 & 3 \\ 0 & -2 & -2 & 0 & 1 & -1 & 1 \\ 1 & 3 & 3 & 0 & 0 & 1 & 1 \\ \hline 0 & 6 & 7 & 0 & 0 & 4 & 4 \end{array}\right]\begin{array}{l} \\ -\frac{2}{3}R_3 + R_2 \\ \\ 2R_3 + R_4 \end{array}$$

The minimum is 4 when $y_1 = 0$, $y_2 = 0$, and $y_3 = 4$.

27. Let y_1 = the amount of product A and let y_2 = the amount of product B.
The given information can be expressed as the following minimization problem:
Minimize $w = 24y_1 + 40y_2$
subject to: $4y_1 + 2y_2 \geq 20$
$2y_1 + 5y_2 \geq 18$
$y_1 \geq 0,\ y_2 \geq 0.$
The augmented matrix is

$$\left[\begin{array}{cc|c} 4 & 2 & 20 \\ 2 & 5 & 18 \\ \hline 24 & 40 & 0 \end{array}\right].$$

The transpose is

$$\left[\begin{array}{cc|c} 4 & 2 & 24 \\ 2 & 5 & 40 \\ \hline 20 & 18 & 0 \end{array}\right].$$

The dual maximization problem is
Maximize $z = 20x_1 + 18x_2$
subject to: $4x_1 + 2x_2 \leq 24$
$2x_1 + 5x_2 \leq 40$
$x_1 \geq 0,\ x_2 \geq 0.$
The initial simplex tableau is

$$\begin{array}{cccc} x_1 & x_2 & s_1 & s_2 \end{array}$$

$$\left[\begin{array}{cccc|c} \boxed{4} & 2 & 1 & 0 & 24 \\ 2 & 5 & 0 & 1 & 40 \\ \hline -20 & -18 & 0 & 0 & 0 \end{array}\right].$$

Pivot on the 4 in row one, column one.

$$\begin{array}{ccccc} x_1 & x_2 & x_3 & x_4 & x_5 \end{array}$$

$$\left[\begin{array}{cccc|c} 1 & \frac{1}{2} & \frac{1}{4} & 0 & 6 \\ 2 & 5 & 0 & 1 & 40 \\ \hline -20 & -18 & 0 & 0 & 0 \end{array}\right]\begin{array}{l} \frac{1}{4}R_1 \\ \\ \\ \end{array}$$

$$\begin{array}{cccc} x_1 & x_2 & s_1 & s_2 \end{array}$$

$$\left[\begin{array}{cccc|c} 1 & \frac{1}{2} & \frac{1}{4} & 0 & 6 \\ 0 & \boxed{4} & -\frac{1}{2} & 1 & 28 \\ \hline 0 & -8 & 5 & 0 & 120 \end{array}\right]\begin{array}{l} \\ -2R_1 + R_2 \\ 20R_1 + R_3 \end{array}$$

Pivot on the 4 in row two, column two.

$$\begin{array}{ccccc} x_1 & x_2 & x_3 & x_4 & x_5 \end{array}$$

$$\left[\begin{array}{cccc|c} 1 & \frac{1}{2} & \frac{1}{4} & 0 & 6 \\ 0 & 1 & -\frac{1}{8} & \frac{1}{4} & 7 \\ \hline 0 & -8 & 5 & 0 & 120 \end{array}\right]\begin{array}{l} \\ \frac{1}{4}R_2 \\ \\ \end{array}$$

$$\begin{array}{cccc} x_1 & x_2 & x_3 & x_4 \end{array}$$

$$\left[\begin{array}{cccc|c} 1 & 0 & \frac{5}{16} & -\frac{1}{8} & \frac{5}{2} \\ 0 & 1 & -\frac{1}{8} & \frac{1}{4} & 7 \\ \hline 0 & 0 & 4 & 2 & 176 \end{array}\right]\begin{array}{l} -\frac{1}{2}R_2 + R_1 \\ \\ 8R_2 + R_3 \end{array}$$

Glenn should use 4 servings of product A and 2 servings of product B for a minimum cost of \$1.76.

Copyright © 2019 Pearson Education, Inc.

29. Let y_1 = the number of additional units of regular beer to produce, and let y_2 = the number of additional units of light beer to produce
The sale of 12 units of regular beer and 10 units of light beer already generates
$12 \cdot 100{,}000 + 10 \cdot 300{,}000 = \$4{,}200{,}000$ in revenue. Additional units need only to generate $7{,}000{,}000 - 4{,}200{,}000 = \$2{,}800{,}000$ in revenue.
Minimize $w = 36{,}000y_1 + 48{,}000y_2$
subject to: $100{,}000y_1 + 300{,}000y_2 \ge 2{,}800{,}000$
$y_1 + y_2 \ge 20$
$y_1 \ge 0,\ y_2 \ge 0$
The augmented matrix is

$$\left[\begin{array}{cc|c} 100{,}000 & 300{,}000 & 2{,}800{,}000 \\ 1 & 1 & 20 \\ \hline 36{,}000 & 48{,}000 & 0 \end{array}\right]$$

The transpose is

$$\left[\begin{array}{cc|c} 100{,}000 & 1 & 36{,}000 \\ 300{,}000 & 1 & 48{,}000 \\ \hline 2{,}800{,}000 & 20 & 0 \end{array}\right]$$

The dual maximization problem is
Maximize $z = 2{,}800{,}000x_1 + 20x_2$
subject to: $100{,}000x_1 + x_2 \le 36{,}000$
$300{,}000x_1 + x_2 \le 48{,}000$
$x_1 \ge 0,\ x_2 \ge 0$
The initial simplex tableau is

$$\begin{array}{c} \quad x_1 \qquad\quad x_2 \quad s_1 \quad s_2 \\ \left[\begin{array}{cccc|c} 100{,}000 & 1 & 1 & 0 & 36{,}000 \\ \boxed{300{,}000} & 1 & 0 & 1 & 48{,}000 \\ \hline -2{,}800{,}000 & -20 & 0 & 0 & 0 \end{array}\right] \end{array}$$

Pivot on the 300,000 in row two, column one.

$$\begin{array}{c} x_1 \quad x_2 \quad s_1 \quad s_2 \\ \left[\begin{array}{cccc|c} 0 & \boxed{\frac{2}{3}} & 1 & -\frac{1}{3} & 20{,}000 \\ 1 & \frac{1}{300{,}000} & 0 & \frac{1}{300{,}000} & \frac{16}{100} \\ \hline 0 & -\frac{32}{3} & 0 & \frac{28}{3} & 448{,}000 \end{array}\right] \begin{array}{l} -100{,}000R_2 + R_1 \\ \frac{1}{300{,}000}R_2 \\ 2{,}800{,}000R_2 + R_3 \end{array} \end{array}$$

Pivot on the $\frac{2}{3}$ in row one, column two.

$$\left[\begin{array}{cccc|c} 0 & 1 & \frac{3}{2} & -\frac{1}{2} & 30{,}000 \\ 1 & 0 & -\frac{1}{200{,}000} & \frac{1}{200{,}000} & \frac{3}{50} \\ \hline 0 & 0 & 16 & 4 & 768{,}000 \end{array}\right] \begin{array}{l} \frac{3}{2}R_1 \\ -\frac{1}{300{,}000}R_1 + R_2 \\ \frac{32}{3}R_1 + R_3 \end{array}$$

The total cost is the cost of the original 12 and 10 units of beer plus the cost of the additional units of beer.
$768{,}000 + 12 \cdot 36{,}000 + 10 \cdot 48{,}000 = \$1{,}680{,}000$
The brewery should make 16 additional units of regular beer and 4 additional units of light beer for a total of 28 units of regular beer and 14 units of light beer at a minimum cost of $1,680,000.

31. a. Let x = the number of bags of feed 1 and let y = the number of bags of feed 2.
Solve the system of equations:
$3x + 2y = 7$
$x + 3y = 7$
$2x + y = 4$
Use the elimination method (or matrix methods) to get $x = 1,\ y = 2$.
Use 1 bag of feed 1, 2 bags of feed 2.

b. Use the shadow values of $x_1 = \frac{1}{5}$ and $x_2 = \frac{7}{5}$. The cost of 7 units of *A* and 4 of *B* less the cost of 2 units of *A* (same amount of *B*) = \$7.00 – \$.40 = \$6.60.
To find the combination of bags costing \$6.60 and giving 5 units of *A* and 4 units of *B*, solve the system of equations:
$3x + 2y = 6.60$ (1)
$x + 3y = 5$ (2)
$2x + y = 4$ (3)
The solution is $x = 1.4$ and $y = 1.2$.
For a daily cost of \$6.60, use 1.4 bags of feed 1 and 1.2 bags of feed 2.

Copyright © 2019 Pearson Education, Inc.

33. Let x_1 = the number of toy bears and let x_2 = the number of monkeys.
Maximize $z = x_1 + 1.5x_2$
subject to: $x_1 + 2x_2 \le 200$
$4x_1 + 3x_2 \le 600$
$x_2 \le 90$
$x_1 \ge 0,\ x_2 \ge 0.$

a. The augmented matrix is

$$\left[\begin{array}{cc|c} 1 & 2 & 200 \\ 4 & 3 & 600 \\ 0 & 1 & 90 \\ \hline 1 & 1.5 & 0. \end{array}\right].$$

The transpose is

$$\left[\begin{array}{ccc|c} 1 & 4 & 0 & 1 \\ 2 & 3 & 1 & 1.5 \\ \hline 200 & 600 & 90 & 0 \end{array}\right].$$

The dual minimization problem is
Minimize $w = 200y_1 + 600y_2 + 90y_3$
subject to: $y_1 + 4y_2 \ge 1$
$2y_1 + 3y_2 + y_3 \ge 1.5$
$y_1 \ge 0,\ y_2 \ge 0,\ y_3 \ge 0.$

b. The solution to the dual minimization problem is found in the last row of the final simplex. tableau of the original problem. The minimum is 180 when $y_1 = .6$, $y_2 = .1$, and $y_3 = 0$.

c. The shadow cost for felt is in row four, column three of the final matrix, \$.60/unit. So, an increase of 10 squares of felt would increase the profit by \$6. The profit would be \$186.

Section 7.7 The Simplex Method: Nonstandard Problems

1. Maximize $z = -5x_1 + 4x_2 - 2x_3$
subject to $-2x_2 + 5x_3 \ge 8$
$4x_1 - x_2 + 3x_3 \le 12$
$x_1 \ge 0,\ x_2 \ge 0,\ x_3 \ge 0.$

a. Insert a surplus variable in the first constraint and slack variables in the last constraint. The problem then becomes:
Maximize $z = -5x_1 + 4x_2 - 2x_3$
subject to:
$-2x_2 + 5x_3 - s_1 = 8$
$4x_1 - x_2 + 3x_3 + s_2 = 12$
$x_1 \ge 0,\ x_2 \ge 0,\ x_3 \ge 0,\ s_1 \ge 0,\ s_2 \ge 0.$

b. The initial simplex tableau is

$$\begin{array}{c} \begin{matrix} x_1 & x_2 & x_3 & s_1 & s_2 \end{matrix} \\ \left[\begin{array}{ccccc|c} 0 & -2 & 5 & -1 & 0 & 8 \\ 4 & -1 & 3 & 0 & 1 & 12 \\ \hline 5 & -4 & 2 & 0 & 0 & 0 \end{array}\right] \end{array}.$$

3. Maximize $z = 2x_1 - 3x_2 + 4x_3$
subject to: $x_1 + x_2 + x_3 \le 100$
$x_1 + x_2 + x_3 \ge 75$
$x_1 + x_2 \ge 27$
$x_1 \ge 0,\ x_2 \ge 0,\ x_3 \ge 0.$

a. Insert a slack variable in the first constraint and surplus variables in the last two constraints. The problem then becomes:
Maximize $z = 2x_1 - 3x_2 + 4x_3$
subject to:
$x_1 + x_2 + x_3 + s_1 = 100$
$x_1 + x_2 + x_3 - s_2 = 75$
$x_1 + x_2 - s_3 = 27$
$x_1 \ge 0,\ x_2 \ge 0,\ x_3 \ge 0,\ s_1 \ge 0,\ s_2 \ge 0,\ s_3 \ge 0.$

b. The initial simplex tableau is

$$\begin{array}{c} \begin{matrix} x_1 & x_2 & x_3 & s_1 & s_2 & s_3 \end{matrix} \\ \left[\begin{array}{cccccc|c} 1 & 1 & 1 & 1 & 0 & 0 & 100 \\ 1 & 1 & 1 & 0 & -1 & 0 & 75 \\ 1 & 1 & 0 & 0 & 0 & -1 & 27 \\ \hline -2 & 3 & -4 & 0 & 0 & 0 & 0 \end{array}\right] \end{array}.$$

Copyright © 2019 Pearson Education, Inc.

5. Minimize $w = 2y_1 + 5y_2 - 3y_3$
subject to: $y_1 + 2y_2 + 3y_3 \ge 115$
$2y_1 + y_2 + y_3 \le 200$
$y_1 + y_3 \ge 50$
$y_1 \ge 0,\ y_2 \ge 0,\ y_3 \ge 0.$

Rewrite the objective function to get a maximization problem.
Maximize $z = -2y_1 - 5y_2 + 3y_3$
subject to: $y_1 + 2y_2 + 3y_3 \ge 115$
$2y_1 + y_2 + y_3 \le 200$
$y_1 + y_3 \ge 50$
$y_1 \ge 0,\ y_2 \ge 0,\ y_3 \ge 0.$

Insert surplus variables in the first and third constraints and a slack variable in the second to get the initial simplex tableau.

y_1	y_2	y_3	s_1	s_2	s_3	
1	2	3	−1	0	0	115
2	1	1	0	1	0	200
1	0	1	0	0	−1	50
2	5	−3	0	0	0	0

7. Minimize $w = 10y_1 + 8y_2 + 15y_3$
subject to: $y_1 + y_2 + y_3 \ge 12$
$5y_1 + 4y_2 + 9y_3 \ge 48$
$y_1 \ge 0,\ y_2 \ge 0,\ y_3 \ge 0.$

Rewrite the objective function to get a maximization problem.
Maximize $z = -10y_1 - 8y_2 - 15y_3$
subject to: $y_1 + y_2 + y_3 \ge 12$
$5y_1 + 4y_2 + 9y_3 \ge 48$
$y_1 \ge 0,\ y_2 \ge 0,\ y_3 \ge 0.$

Insert surplus variables in both constraints to get the initial simplex tableau.

y_1	y_2	y_3	s_1	s_2	
1	1	1	−1	0	12
5	4	9	0	−1	48
10	8	15	0	0	0

9. Maximize $z = 12x_1 + 10x_2$
subject to: $x_1 + 2x_2 \ge 24$
$x_1 + x_2 \le 40$
$x_1 \ge 0,\ x_2 \ge 0.$

The initial simplex tableau is

x_1	x_2	s_1	s_2	
[1]	2	−1	0	24
1	1	0	1	40
−12	−10	0	0	0

For Stage I pivoting, pivot on the 1 in row one, column one.

x_1	x_2	s_1	s_2		
1	2	−1	0	24	
0	−1	[1]	1	16	$-R_1 + R_2$
0	14	−12	0	288	$12R_2 + R_3$

For Stage II pivoting, pivot on the 1 in row two, column three.

x_1	x_2	s_1	s_2		
1	1	0	1	40	$R_2 + R_1$
0	−1	1	1	16	
0	2	0	12	480	$12R_1 + R_3$

The maximum is 480 when $x_1 = 40$ and $x_2 = 0$.

Copyright © 2019 Pearson Education, Inc.

11. Find $x_1 \geq 0$, $x_2 \geq 0$, and $x_3 \geq 0$ such that

$$\begin{aligned} x_1 + x_2 + 2x_3 &\leq 38 \\ 2x_1 + x_2 + x_3 &\geq 24 \end{aligned}$$

and $z = 3x_1 + 2x_2 + 2x_3$ is maximized.

The initial simplex tableau is

$$\begin{array}{c} \\ \\ \\ \end{array}\begin{array}{ccccc|c} x_1 & x_2 & x_3 & s_1 & s_2 & \\ 1 & 1 & 2 & 1 & 0 & 38 \\ \boxed{2} & 1 & 1 & 0 & -1 & 24 \\ \hline -3 & -2 & -2 & 0 & 0 & 0 \end{array}$$

For Stage I pivoting, pivot on the 2 in row two, column one.

$$\begin{array}{ccccc|c|l} x_1 & x_2 & x_3 & s_1 & s_2 & & \\ 1 & 1 & 2 & 1 & 0 & 38 & \\ 1 & \frac{1}{2} & \frac{1}{2} & 0 & -\frac{1}{2} & 12 & \frac{1}{2}R_2 \\ \hline -3 & -2 & -2 & 0 & 0 & 0 & \end{array}$$

$$\begin{array}{ccccc|c|l} x_1 & x_2 & x_3 & s_1 & s_2 & & \\ 0 & \frac{1}{2} & \frac{3}{2} & 1 & \boxed{\frac{1}{2}} & 26 & -1R_2 + R_1 \\ 1 & \frac{1}{2} & \frac{1}{2} & 0 & -\frac{1}{2} & 12 & \\ \hline 0 & -\frac{1}{2} & -\frac{1}{2} & 0 & -\frac{3}{2} & 36 & 3R_2 + R_3 \end{array}$$

For Stage II pivoting, pivot on the $\frac{1}{2}$ in row one, column five.

$$\begin{array}{ccccc|c|l} x_1 & x_2 & x_3 & s_1 & s_2 & & \\ 0 & 1 & 3 & 2 & 1 & 52 & 2R_1 \\ 1 & \frac{1}{2} & \frac{1}{2} & 0 & -\frac{1}{2} & 12 & \\ \hline 0 & -\frac{1}{2} & -\frac{1}{2} & 0 & -\frac{3}{2} & 36 & \end{array}$$

$$\begin{array}{ccccc|c|l} x_1 & x_2 & x_3 & s_1 & s_2 & & \\ 0 & 1 & 3 & 2 & 1 & 52 & \\ 1 & 1 & 2 & 1 & 0 & 38 & \frac{1}{2}R_1 + R_2 \\ \hline 0 & 1 & 4 & 3 & 0 & 114 & \frac{3}{2}R_1 + R_3 \end{array}$$

The maximum is 114 when $x_1 = 38$, $x_2 = 0$, and $x_3 = 0$.

13. Find $x_1 \geq 0$ and $x_2 \geq 0$ such that

$$\begin{aligned} x_1 + 2x_2 &\leq 18 \\ x_1 + 3x_2 &\geq 12 \\ 2x_1 + 2x_2 &\leq 30 \end{aligned}$$

and $z = 5x_1 + 10x_2$ is maximized.

The initial simplex tableau is

$$\begin{array}{ccccc|c} x_1 & x_2 & s_1 & s_2 & s_3 & \\ 1 & 2 & 1 & 0 & 0 & 18 \\ \boxed{1} & 3 & 0 & -1 & 0 & 12 \\ 2 & 2 & 0 & 0 & 1 & 30 \\ \hline -5 & -10 & 0 & 0 & 0 & 0 \end{array}.$$

For Stage I pivoting, we can pivot on the 1 in row two, column one or the 3 in row two, column two. Choosing the 1 in row two, column one as the pivot, we proceed as follows.

$$\begin{array}{ccccc|c|l} x_1 & x_2 & s_1 & s_2 & s_3 & & \\ 0 & -1 & 1 & 1 & 0 & 6 & -R_2 + R_1 \\ 1 & 3 & 0 & -1 & 0 & 12 & \\ 0 & -4 & 0 & \boxed{2} & 1 & 6 & -2R_2 + R_3 \\ \hline 0 & 5 & 0 & -5 & 0 & 60 & 5R_2 + R_4 \end{array}$$

For Stage II, pivot on the 2 in row three, column four.

$$\begin{array}{ccccc|c|l} x_1 & x_2 & s_1 & s_2 & s_3 & & \\ 0 & -1 & 1 & 1 & 0 & 6 & \\ 1 & 3 & 0 & -1 & 0 & 12 & \\ 0 & -2 & 0 & 1 & \frac{1}{2} & 3 & \frac{1}{2}R_3 \\ \hline 0 & 5 & 0 & -5 & 0 & 60 & \end{array}$$

$$\begin{array}{ccccc|c|l} x_1 & x_2 & s_1 & s_2 & s_3 & & \\ 0 & \boxed{1} & 1 & 0 & -\frac{1}{2} & 3 & -R_3 + R_1 \\ 1 & 1 & 0 & 0 & \frac{1}{2} & 15 & R_3 + R_2 \\ 0 & -2 & 0 & 1 & \frac{1}{2} & 3 & \\ \hline 0 & -5 & 0 & 0 & \frac{5}{2} & 75 & 5R_3 + R_4 \end{array}$$

Now pivot on the 1 in row one, column two.

$$\begin{array}{ccccc|c|l} x_1 & x_2 & s_1 & s_2 & s_3 & & \\ 0 & 1 & 1 & 0 & -\frac{1}{2} & 3 & \\ 1 & 0 & -1 & 0 & 1 & 12 & -R_1 + R_2 \\ 0 & 0 & 2 & 1 & -\frac{1}{2} & 9 & 2R_1 + R_3 \\ \hline 0 & 0 & 5 & 0 & 0 & 90 & 5R_1 + R_4 \end{array}$$

The above tableau gives the solution:
The maximum is 90 when $x_1 = 12$ and $x_2 = 3$.

Copyright © 2019 Pearson Education, Inc.

However, in Stage I, we could also choose the 3 in row two, column two as the pivot and proceed as follows.

$$\begin{array}{ccccc|c}
x_1 & x_2 & s_1 & s_2 & s_3 & \\
1 & 2 & 1 & 0 & 0 & 18 \\
\frac{1}{3} & \boxed{1} & 0 & -\frac{1}{3} & 0 & 4 \\
2 & 2 & 0 & 0 & 1 & 30 \\ \hline
-5 & -10 & 0 & 0 & 0 & 0
\end{array} \quad \begin{array}{l} \\ \\ \frac{1}{3}R_2 \\ \\ \\ \end{array}$$

$$\begin{array}{ccccc|c}
x_1 & x_2 & s_1 & s_2 & s_3 & \\
\frac{1}{3} & 0 & 1 & \boxed{\frac{2}{3}} & 0 & 10 \\
\frac{1}{3} & 1 & 0 & -\frac{1}{3} & 0 & 4 \\
\frac{4}{3} & 0 & 0 & \frac{2}{3} & 1 & 22 \\ \hline
-\frac{5}{3} & 0 & 0 & -\frac{10}{3} & 0 & 40
\end{array} \quad \begin{array}{l} \\ -2R_2 + R_1 \\ \\ -2R_2 + R_3 \\ 10R_2 + R_4 \end{array}$$

For Stage II, pivot on the $\frac{2}{3}$ in row one, column four.

$$\begin{array}{ccccc|c}
x_1 & x_2 & s_1 & s_2 & s_3 & \\
\frac{1}{2} & 0 & \frac{3}{2} & 1 & 0 & 15 \\
\frac{1}{3} & 1 & 0 & -\frac{1}{3} & 0 & 4 \\
\frac{4}{3} & 0 & 0 & \frac{2}{3} & 1 & 22 \\ \hline
-\frac{5}{3} & 0 & 0 & -\frac{10}{3} & 0 & 40
\end{array} \quad \begin{array}{l} \\ \frac{3}{2}R_1 \\ \\ \\ \\ \end{array}$$

$$\begin{array}{ccccc|c}
x_1 & x_2 & s_1 & s_2 & s_3 & \\
\frac{1}{2} & 0 & \frac{3}{2} & 1 & 0 & 15 \\
\frac{1}{2} & 1 & \frac{1}{2} & 0 & 0 & 9 \\
1 & 0 & -1 & 0 & 1 & 12 \\ \hline
0 & 0 & 5 & 0 & 0 & 90
\end{array} \quad \begin{array}{l} \\ \\ \frac{1}{3}R_1 + R_2 \\ -\frac{2}{3}R_1 + R_3 \\ \frac{10}{3}R_1 + R_4 \end{array}$$

This tableau gives the solution:
The maximum is 90 when $x_1 = 0$ and $x_2 = 9$.
Thus, the maximum is 90 when $x_1 = 12$ and $x_2 = 3$ or when $x_1 = 0$ and $x_2 = 9$.

15. Minimize $w = 3y_1 + 2y_2$
subject to: $2y_1 + 3y_2 \ge 60$
$y_1 + 4y_2 \ge 40$
$y_1 \ge 0,\ y_2 \ge 0.$

The initial simplex tableau is

$$\begin{array}{cccc|c}
y_1 & y_2 & s_1 & s_2 & \\
\boxed{2} & 3 & -1 & 0 & 60 \\
1 & 4 & 0 & -1 & 40 \\ \hline
3 & 2 & 0 & 0 & 0
\end{array}.$$

For Stage I, pivot on the 2 in row one, column one.

$$\begin{array}{cccc|c}
y_1 & y_2 & s_1 & s_2 & \\
1 & \frac{3}{2} & -\frac{1}{2} & 0 & 30 \\
0 & \boxed{\frac{5}{2}} & \frac{1}{2} & -1 & 10 \\ \hline
0 & -\frac{5}{2} & \frac{3}{2} & 0 & -90
\end{array} \quad \begin{array}{l} \\ \frac{1}{2}R_1 \\ -R_1 + R_2 \\ -3R_1 + R_3 \end{array}$$

To continue Stage I, pivot on the $\frac{5}{2}$ in row two, column two.

$$\begin{array}{cccc|c}
y_1 & y_2 & s_1 & s_2 & \\
1 & 0 & -\frac{4}{5} & \boxed{\frac{3}{5}} & 24 \\
0 & 1 & \frac{1}{5} & -\frac{2}{5} & 4 \\ \hline
0 & 0 & 2 & -1 & -80
\end{array} \quad \begin{array}{l} \\ -\frac{3}{5}R_2 + R_1 \\ \frac{2}{5}R_2 \\ \frac{5}{2}R_2 + R_3 \end{array}$$

For Stage II, pivot on the $\frac{3}{5}$ in row one, column four.

$$\begin{array}{cccc|c}
y_1 & y_2 & s_1 & s_2 & \\
\frac{5}{3} & 0 & -\frac{4}{3} & 1 & 40 \\
\frac{2}{3} & 1 & -\frac{1}{3} & 0 & 20 \\ \hline
\frac{5}{3} & 0 & \frac{2}{3} & 0 & -40
\end{array} \quad \begin{array}{l} \\ \frac{5}{3}R_1 \\ \frac{2}{5}R_1 + R_2 \\ R_1 + R_3 \end{array}$$

The minimum is 40 when $y_1 = 0$ and $y_2 = 20$.

Copyright © 2019 Pearson Education, Inc.

17. Maximize $z = 3x_1 + 2x_2$

subject to:
$$\begin{aligned} x_1 + x_2 &= 50 \\ 4x_1 + 2x_2 &\ge 120 \\ 5x_1 + 2x_2 &\le 200 \\ x_1 \ge 0,\ x_2 &\ge 0 \end{aligned}$$

The initial simplex tableau is

$$\begin{array}{cccccc|c} x_1 & x_2 & s_1 & s_2 & s_3 & s_4 & \\ \hline 1 & 1 & -1 & 0 & 0 & 0 & 50 \\ 1 & 1 & 0 & 1 & 0 & 0 & 50 \\ \boxed{4} & 2 & 0 & 0 & -1 & 0 & 120 \\ 5 & 2 & 0 & 0 & 0 & 1 & 200 \\ \hline -3 & -2 & 0 & 0 & 0 & 0 & 0 \end{array}$$

For Stage I, pivot on the 4 in row three, column one.

$$\begin{array}{cccccc|c|l} x_1 & x_2 & s_1 & s_2 & s_3 & s_4 & & \\ \hline 0 & \boxed{\frac{1}{2}} & -1 & 0 & \frac{1}{4} & 0 & 20 & -R_3 + R_1 \\ 0 & \frac{1}{2} & 0 & 1 & \frac{1}{4} & 0 & 20 & -R_3 + R_2 \\ 1 & \frac{1}{2} & 0 & 0 & -\frac{1}{4} & 0 & 30 & \frac{1}{4}R_3 \\ 0 & -\frac{1}{2} & 0 & 0 & \frac{5}{4} & 1 & 50 & -5R_3 + R_4 \\ \hline 0 & -\frac{1}{2} & 0 & 0 & -\frac{3}{4} & 0 & 90 & 3R_3 + R_5 \end{array}$$

To continue Stage I, pivot on the $\frac{1}{2}$ in row one column two.

$$\begin{array}{cccccc|c|l} x_1 & x_2 & s_1 & s_2 & s_3 & s_4 & & \\ \hline 0 & 1 & -2 & 0 & \frac{1}{2} & 0 & 40 & 2R_1 \\ 0 & 0 & \boxed{1} & 1 & 0 & 0 & 0 & -\frac{1}{2}R_1 + R_2 \\ 1 & 0 & 1 & 0 & -\frac{1}{2} & 0 & 10 & -\frac{1}{2}R_1 + R_3 \\ 0 & 0 & -1 & 0 & \frac{3}{2} & 1 & 70 & \frac{1}{2}R_1 + R_4 \\ \hline 0 & 0 & -1 & 0 & -\frac{1}{2} & 0 & 110 & \frac{1}{2}R_1 + R_5 \end{array}$$

For Stage II, pivot on the 1 in row two, column three.

$$\begin{array}{cccccc|c|l} x_1 & x_2 & s_1 & s_2 & s_3 & s_4 & & \\ \hline 0 & 1 & 0 & 2 & \frac{1}{2} & 0 & 40 & 2R_2 + R_1 \\ 0 & 0 & 1 & 1 & 0 & 0 & 0 & \\ 1 & 0 & 0 & -1 & -\frac{1}{2} & 0 & 10 & -R_2 + R_3 \\ 0 & 0 & 0 & 1 & \boxed{\frac{3}{2}} & 1 & 70 & R_2 + R_4 \\ \hline 0 & 0 & 0 & 1 & -\frac{1}{2} & 0 & 110 & R_2 + R_5 \end{array}$$

Pivot on the $\frac{3}{2}$ in row four, column five.

$$\begin{array}{cccccc|c|l} x_1 & x_2 & s_1 & s_2 & s_3 & s_4 & & \\ \hline 0 & 1 & 0 & \frac{5}{3} & 0 & -\frac{1}{3} & \frac{50}{3} & -\frac{1}{2}R_4 + R_1 \\ 0 & 0 & 1 & 1 & 0 & 0 & 0 & \\ 1 & 0 & 0 & -\frac{2}{3} & 0 & \frac{1}{3} & \frac{100}{3} & \frac{1}{2}R_4 + R_3 \\ 0 & 0 & 0 & \frac{2}{3} & 1 & \frac{2}{3} & \frac{140}{3} & \frac{2}{3}R_4 \\ \hline 0 & 0 & 0 & \frac{4}{3} & 0 & \frac{1}{3} & \frac{400}{3} & \frac{1}{2}R_4 + R_5 \end{array}$$

The maximum is $133\frac{1}{3}$ when $x_1 = 33\frac{1}{3}$ and $x_2 = 16\frac{2}{3}$.

19. Minimize $w = 32y_1 + 40y_2$

Maximize $z = -w = -32y_1 - 40y_2$.

subject to:
$$\begin{aligned} 20y_1 + 10y_2 &= 200 \\ 25y_1 + 40y_2 &\le 500 \\ 18y_1 + 24y_2 &\ge 300 \\ y_1 \ge 0,\ y_2 &\ge 0 \end{aligned}$$

The initial simplex tableau is

$$\begin{array}{cccccc|c} y_1 & y_2 & s_1 & s_2 & s_3 & s_4 & \\ \hline \boxed{20} & 10 & -1 & 0 & 0 & 0 & 200 \\ 20 & 10 & 0 & 1 & 0 & 0 & 200 \\ 25 & 40 & 0 & 0 & 1 & 0 & 500 \\ 18 & 24 & 0 & 0 & 0 & -1 & 300 \\ \hline 32 & 40 & 0 & 0 & 0 & 0 & 0 \end{array}$$

For Stage I, pivot on the 20 in row one, column one.

$$\begin{array}{cccccc|c|l} y_1 & y_2 & s_1 & s_2 & s_3 & s_4 & & \\ \hline 1 & \frac{1}{2} & -\frac{1}{20} & 0 & 0 & 0 & 10 & \frac{1}{20}R_1 \\ 0 & 0 & 1 & 1 & 0 & 0 & 0 & -20R_1 + R_2 \\ 0 & \frac{55}{2} & \frac{5}{4} & 0 & 1 & 0 & 250 & -25R_1 + R_3 \\ 0 & \boxed{15} & \frac{9}{10} & 0 & 0 & -1 & 120 & -18R_1 + R_4 \\ \hline 0 & 24 & \frac{8}{5} & 0 & 0 & 0 & -320 & -32R_1 + R_5 \end{array}$$

Pivot on the 15 in row four, column two.

$$\begin{array}{cccccc|c|l} y_1 & y_2 & s_1 & s_2 & s_3 & s_4 & & \\ \hline 1 & 0 & -\frac{4}{50} & 0 & 0 & \frac{1}{30} & 6 & -\frac{1}{2}R_4 + R_1 \\ 0 & 0 & 1 & 1 & 0 & 0 & 0 & \\ 0 & 0 & -\frac{2}{5} & 0 & 1 & \frac{11}{6} & 30 & -\frac{55}{2}R_4 + R_3 \\ 0 & 1 & \frac{3}{50} & 0 & 0 & -\frac{1}{15} & 8 & \frac{1}{15}R_4 \\ \hline 0 & 0 & \frac{4}{25} & 0 & 0 & \frac{8}{5} & -512 & -24R_4 + R_5 \end{array}$$

The program is optimal after Stage I.

The minimum is 512 when $y_1 = 6$ and $y_2 = 8$.

Copyright © 2019 Pearson Education, Inc.

21. Maximize $z = -5x_1 + 4x_2 - 2x_3$
subject to
$$-2x_2 + 5x_3 \ge 8$$
$$4x_1 - x_2 + 3x_3 \le 12$$
$$x_1 \ge 0,\ x_2 \ge 0,\ x_3 \ge 0.$$

Insert a surplus variable in the first constraint and slack variables in the last two constraints. The problem then becomes:
Maximize $z = -5x_1 + 4x_2 - 2x_3$
subject to:
$$-2x_2 + 5x_3 - s_1 = 8$$
$$4x_1 - x_2 + 3x_3 + s_2 = 12$$
$$x_1 \ge 0,\ x_2 \ge 0,\ x_3 \ge 0,\ s_1 \ge 0,\ s_2 \ge 0.$$

The initial simplex tableau is

$$\begin{array}{ccccc} x_1 & x_2 & x_3 & s_1 & s_2 \end{array}$$
$$\left[\begin{array}{ccccc|c} 0 & -2 & \boxed{5} & -1 & 0 & 8 \\ 4 & -1 & 3 & 0 & 1 & 12 \\ \hline 5 & -4 & 2 & 0 & 0 & 0 \end{array}\right].$$

For Stage I, pivot on the 5 in row 1, column 3.

$$\begin{array}{ccccc} x_1 & x_2 & x_3 & s_1 & s_2 \end{array}$$
$$\left[\begin{array}{ccccc|c} 0 & -\frac{2}{5} & 1 & -\frac{1}{5} & 0 & \frac{8}{5} \\ 4 & -1 & 3 & 0 & 1 & 12 \\ \hline 5 & -4 & 2 & 0 & 0 & 0 \end{array}\right] \begin{array}{l} \frac{1}{5}R_1 \\ \\ \\ \end{array}$$

$$\begin{array}{ccccc} x_1 & x_2 & x_3 & s_1 & s_2 \end{array}$$
$$\left[\begin{array}{ccccc|c} 0 & -\frac{2}{5} & 1 & -\frac{1}{5} & 0 & \frac{8}{5} \\ 4 & \boxed{\frac{1}{5}} & 0 & \frac{3}{5} & 1 & \frac{36}{5} \\ \hline 5 & -\frac{16}{5} & 0 & \frac{2}{5} & 0 & -\frac{16}{5} \end{array}\right] \begin{array}{l} \\ -3R_1 + R_2 \\ -2R_1 + R_3 \end{array}$$

This completes Stage I because. the solution given in the usual way from the matrix has nonnegative values for all variables. Since there are negative indicators in the objective row, we continue with Stage II. Now pivot on the $\frac{1}{5}$ in row 1, column 2.

$$\begin{array}{ccccc} x_1 & x_2 & x_3 & s_1 & s_2 \end{array}$$
$$\left[\begin{array}{ccccc|c} 8 & 0 & 1 & 1 & 2 & 16 \\ 4 & \frac{1}{5} & 0 & \frac{3}{5} & 1 & \frac{36}{5} \\ \hline 69 & 0 & 0 & 10 & 16 & 112 \end{array}\right] \begin{array}{l} R_1 + 2R_2 \\ \\ R_3 + 16R_2 \end{array}$$

$$\begin{array}{ccccc} x_1 & x_2 & x_3 & s_1 & s_2 \end{array}$$
$$\left[\begin{array}{ccccc|c} 8 & 0 & 1 & 1 & 2 & 16 \\ 20 & 1 & 0 & 3 & 5 & 36 \\ \hline 69 & 0 & 0 & 10 & 16 & 112 \end{array}\right] \begin{array}{l} \\ 5R_2 \\ \\ \end{array}$$

The maximum is 112 when
$x_1 = 0,\ x_2 = 36,\ x_3 = 16.$

23. Maximize $z = 2x_1 - 3x_2 + 4x_3$
subject to:
$$x_1 + x_2 + x_3 \le 100$$
$$x_1 + x_2 + x_3 \ge 75$$
$$x_1 + x_2 \ge 27$$
$$x_1 \ge 0,\ x_2 \ge 0,\ x_3 \ge 0.$$

The initial simplex tableau is

$$\begin{array}{cccccc} x_1 & x_2 & x_3 & s_1 & s_2 & s_3 \end{array}$$
$$\left[\begin{array}{cccccc|c} 1 & 1 & 1 & 1 & 0 & 0 & 100 \\ 1 & 1 & \boxed{1} & 0 & -1 & 0 & 75 \\ 1 & 1 & 0 & 0 & 0 & -1 & 27 \\ \hline -2 & 3 & -4 & 0 & 0 & 0 & 0 \end{array}\right].$$

For Stage I, pivot on the 1 in row 2, column 3.

$$\begin{array}{cccccc} x_1 & x_2 & x_3 & s_1 & s_2 & s_3 \end{array}$$
$$\left[\begin{array}{cccccc|c} 0 & 0 & 0 & 1 & 1 & 0 & 25 \\ 1 & 1 & 1 & 0 & -1 & 0 & 75 \\ \boxed{1} & 1 & 0 & 0 & 0 & -1 & 27 \\ \hline 2 & 7 & 0 & 0 & -4 & 0 & 300 \end{array}\right] \begin{array}{l} -R_2 + R_1 \\ \\ \\ 4R_2 + R_4 \end{array}$$

Now pivot on the 1 in row 3 column 1.

$$\begin{array}{cccccc} x_1 & x_2 & x_3 & s_1 & s_2 & s_3 \end{array}$$
$$\left[\begin{array}{cccccc|c} 0 & 0 & 0 & 1 & 1 & 0 & 25 \\ 0 & 0 & 1 & 0 & -1 & 1 & 48 \\ 1 & 1 & 0 & 0 & 0 & -1 & 27 \\ \hline 0 & 5 & 0 & 0 & -4 & 2 & 246 \end{array}\right] \begin{array}{l} \\ -R_3 + R_2 \\ \\ -2R_3 + R_4 \end{array}$$

This completes Stage I because the solution given in the usual way from the matrix has nonnegative values for all variables. Since there is a negative indicator in the objective row, we continue with Stage II.

Pivot on the 1 in row 1 column 5.

$$\begin{array}{cccccc} x_1 & x_2 & x_3 & s_1 & s_2 & s_3 \end{array}$$
$$\left[\begin{array}{cccccc|c} 0 & 0 & 0 & 1 & \boxed{1} & 0 & 25 \\ 0 & 0 & 1 & 1 & 0 & 1 & 73 \\ 1 & 1 & 0 & 0 & 0 & -1 & 27 \\ \hline 0 & 5 & 0 & 4 & 0 & 2 & 346 \end{array}\right] \begin{array}{l} \\ R_1 + R_2 \\ \\ 4R_1 + R_4 \end{array}$$

The maximum is 346 when
$x_1 = 27,\ x_2 = 0,\ x_3 = 73.$

Copyright © 2019 Pearson Education, Inc.

25. Minimize $w = 2y_1 + 5y_2 - 3y_3$

subject to: $y_1 + 2y_2 + 3y_3 \geq 115$
$2y_1 + y_2 + y_3 \leq 200$
$y_1 + y_3 \geq 50$
$y_1 \geq 0,\ y_2 \geq 0,\ y_3 \geq 0.$

The initial simplex tableau is

$$\begin{array}{c} \quad y_1 \;\; y_2 \;\; y_3 \;\; s_1 \;\; s_2 \;\; s_3 \\ \left[\begin{array}{cccccc|c} 1 & 2 & 3 & -1 & 0 & 0 & 115 \\ 2 & 1 & 1 & 0 & 1 & 0 & 200 \\ \boxed{1} & 0 & 1 & 0 & 0 & -1 & 50 \\ \hline 2 & 5 & -3 & 0 & 0 & 0 & 0 \end{array}\right] \end{array}.$$

Stage I of the two-stage program produces the matrix

$$\begin{array}{c} \quad y_1 \;\; y_2 \;\; y_3 \;\; s_1 \;\; s_2 \;\; s_3 \\ \left[\begin{array}{cccccc|c} 0 & 1 & 1 & -\frac{1}{2} & 0 & \frac{1}{2} & \frac{65}{2} \\ 0 & 0 & -2 & \frac{1}{2} & 1 & \frac{3}{2} & \frac{135}{2} \\ 1 & 0 & 1 & 0 & 0 & -1 & 50 \\ \hline 0 & 0 & -10 & \frac{5}{2} & 0 & -\frac{1}{2} & -\frac{525}{2} \end{array}\right] \end{array}.$$

This gives the feasible solution $y_1 = 50$, $y_2 = \frac{65}{2}$, $y_3 = 0$, $s_1 = 0$, $s_2 = \frac{135}{2}$, and $s_3 = 0$.

Stage II produces the final matrix

$$\begin{array}{c} \quad y_1 \;\; y_2 \;\; y_3 \;\; s_1 \;\; s_2 \;\; s_3 \\ \left[\begin{array}{cccccc|c} 2 & 1 & 1 & 0 & 1 & 0 & 200 \\ 1 & 1 & 0 & 0 & 1 & 1 & 150 \\ 5 & 1 & 0 & 1 & 3 & 0 & 485 \\ \hline 8 & 8 & 0 & 0 & 3 & 0 & 600 \end{array}\right] \end{array}.$$

The minimum is –600 when $y_1 = 0$, $y_2 = 0$, and $y_3 = 200$.

27. Minimize $w = 10y_1 + 8y_2 + 15y_3$

subject to: $y_1 + y_2 + y_3 \geq 12$
$5y_1 + 4y_2 + 9y_3 \geq 48$
$y_1 \geq 0,\ y_2 \geq 0,\ y_3 \geq 0.$

Insert surplus variables in both constraints to get the initial simplex tableau.

$$\begin{array}{c} \quad y_1 \;\; y_2 \;\; y_3 \;\; s_1 \;\; s_2 \\ \left[\begin{array}{ccccc|c} 1 & 1 & 1 & -1 & 0 & 12 \\ 5 & 4 & 9 & 0 & -1 & 48 \\ \hline 10 & 8 & 15 & 0 & 0 & 0 \end{array}\right] \end{array}$$

Stage I of the two-stage program produces the matrix

$$\begin{array}{c} \quad y_1 \;\; y_2 \;\; y_3 \;\; s_1 \;\; s_2 \\ \left[\begin{array}{ccccc|c} 0 & 1 & -4 & -5 & 1 & 12 \\ 1 & 0 & 5 & 4 & -1 & 0 \\ \hline 0 & 0 & -3 & 0 & 2 & -96 \end{array}\right] \end{array}.$$

This gives the feasible solution $y_1 = 0$, $y_2 = 12$, $y_3 = 0$, $s_1 = 0$, and $s_2 = 0$.

Stage II produces the final matrix

$$\begin{array}{c} \quad y_1 \;\; y_2 \;\; y_3 \;\; s_1 \;\; s_2 \\ \left[\begin{array}{ccccc|c} \frac{4}{5} & 1 & 0 & -\frac{9}{5} & \frac{1}{5} & 12 \\ \frac{1}{5} & 0 & 1 & \frac{4}{5} & -\frac{1}{5} & 0 \\ \hline \frac{3}{5} & 0 & 0 & \frac{12}{5} & \frac{7}{5} & -96 \end{array}\right] \end{array}.$$

The minimum is 96 when $y_1 = 0$, $y_2 = 12$, and $y_3 = 0$.

Copyright © 2019 Pearson Education, Inc.

29. Let y_1 = the amount of ingredient I per barrel of gasoline, let y_2 = the amount of ingredient II per barrel of gasoline, let y_3 = the amount of ingredient III per barrel of gasoline.

Minimize $w = .30y_1 + .09y_2 + .27y_3$

Maximize $z = -w = -.30y_1 - .09y_2 - .27y_3$

subject to:

$$\begin{aligned} y_1 + y_2 + y_3 &\geq 10 \\ y_1 + y_2 + y_3 &\leq 15 \\ y_1 - \tfrac{1}{4}y_2 \quad &\geq 0 \quad \left(\text{Since } y_1 \geq \tfrac{1}{4}y_2\right) \\ -y_1 \quad + y_3 &\geq 0 \quad (\text{Since } y_3 \geq y_1) \\ y_1 \geq 0,\ y_2 \geq 0,\ y_3 &\geq 0 \end{aligned}$$

The initial simplex tableau is

$$\begin{array}{c} \begin{matrix} y_1 & y_2 & y_3 & s_1 & s_2 & s_3 & s_4 \end{matrix} \\ \left[\begin{array}{ccccccc|c} 1 & 1 & 1 & -1 & 0 & 0 & 0 & 10 \\ 1 & 1 & 1 & 0 & 1 & 0 & 0 & 15 \\ 1 & -\frac{1}{4} & 0 & 0 & 0 & -1 & 0 & 0 \\ -1 & 0 & 1 & 0 & 0 & 0 & -1 & 0 \\ \hline .30 & .09 & .27 & 0 & 0 & 0 & 0 & 0 \end{array}\right]. \end{array}$$

31. Let y_1 = the number of computers from W_1 to D_1; let y_2 = the number of computers from W_2 to D_1; let y_3 = the number of computers from W_1 to D_2; let y_4 = the number of computers from W_2 to D_2.

Minimize $w = 14y_1 + 12y_2 + 22y_3 + 10y_4$

Maximize $z = -w = -14y_1 - 12y_2 - 22y_3 - 10y_4$

subject to:

$$\begin{aligned} y_1 + y_2 \quad &= 32 \\ y_3 + y_4 &= 20 \\ y_1 + \quad y_3 \quad &\leq 25 \\ y_2 + \quad y_4 &\leq 30 \\ y_1 \geq 0,\ y_2 \geq 0, y_3 \geq 0,\ y_4 &\geq 0. \end{aligned}$$

The initial simplex tableau is

$$\begin{array}{c} \begin{matrix} y_1 & y_2 & y_3 & y_4 & s_1 & s_2 & s_3 & s_4 & s_5 & s_6 \end{matrix} \\ \left[\begin{array}{cccccccccc|c} 1 & 1 & 0 & 0 & -1 & 0 & 0 & 0 & 0 & 0 & 32 \\ 1 & 1 & 0 & 0 & 0 & 1 & 0 & 0 & 0 & 0 & 32 \\ 0 & 0 & 1 & 1 & 0 & 0 & -1 & 0 & 0 & 0 & 20 \\ 0 & 0 & 1 & 1 & 0 & 0 & 0 & 1 & 0 & 0 & 20 \\ 1 & 0 & 1 & 0 & 0 & 0 & 0 & 0 & 1 & 0 & 25 \\ 0 & 1 & 0 & 1 & 0 & 0 & 0 & 0 & 0 & 1 & 30 \\ \hline 14 & 12 & 22 & 10 & 0 & 0 & 0 & 0 & 0 & 0 & 0 \end{array}\right]. \end{array}$$

Copyright © 2019 Pearson Education, Inc.

33. Let x_1 = the number of barrels of oil supplied by S_1 to D_1.
Let x_2 = the number of barrels of oil supplied by S_2 to D_1.
Let x_3 = the number of barrels of oil supplied by S_1 to D_2.
Let x_4 = the number of barrels of oil supplied by S_2 to D_2.
Minimize $w = 30x_1 + 25x_2 + 20x_3 + 22x_4$
subject to:

$$\begin{aligned} x_1 + x_2 &\geq 3000 \\ x_3 + x_4 &\geq 5000 \\ x_1 + x_3 &\leq 5000 \\ x_2 + x_4 &\leq 5000 \\ 2x_1 + 5x_2 + 6x_3 + 4x_4 &\leq 40{,}000 \end{aligned}$$

The initial simplex tableau is

x_1	x_2	x_3	x_4	s_1	s_2	s_3	s_4	s_5	
$\boxed{1}$	1	0	0	−1	0	0	0	0	3000
0	0	1	1	0	−1	0	0	0	5000
1	0	1	0	0	0	1	0	0	5000
0	1	0	1	0	0	0	1	0	5000
2	5	6	4	0	0	0	0	1	40,000
30	25	20	22	0	0	0	0	0	0

For Stage I, pivot on the 1 in row one, column one.

x_1	x_2	x_3	x_4	s_1	s_2	s_3	s_4	s_5		
1	1	0	0	−1	0	0	0	0	3000	
0	0	1	1	0	−1	0	0	0	5000	
0	−1	$\boxed{1}$	0	1	0	1	0	0	2000	$-R_1 + R_3$
0	1	0	1	0	0	0	1	0	5000	
0	3	6	4	2	0	0	0	1	34,000	$-2R_1 + R_5$
0	−5	20	22	30	0	0	0	0	−90,000	$-30R_1 + R_6$

To continue Stage I, pivot on the 1 in row three, column three.

x_1	x_2	x_3	x_4	s_1	s_2	s_3	s_4	s_5		
1	1	0	0	−1	0	0	0	0	3000	
0	1	0	$\boxed{1}$	−1	−1	−1	0	0	3000	$-R_3 + R_2$
0	−1	1	0	1	0	1	0	0	2000	
0	1	0	1	0	0	0	1	0	5000	
0	9	0	4	−4	0	−6	0	1	22,000	$-6R_3 + R_5$
0	15	0	22	10	0	−20	0	0	−130,000	$-20R_3 + R_6$

Copyright © 2019 Pearson Education, Inc.

To continue Stage I, pivot on the 1 in row two, column four.

$$\begin{array}{ccccccccc|r l}
x_1 & x_2 & x_3 & x_4 & s_1 & s_2 & s_3 & s_4 & s_5 & & \\
\hline
1 & 1 & 0 & 0 & -1 & 0 & 0 & 0 & 0 & 3000 & \\
0 & 1 & 0 & 1 & -1 & -1 & -1 & 0 & 0 & 3000 & \\
0 & -1 & 1 & 0 & 1 & 0 & 1 & 0 & 0 & 2000 & \\
0 & 0 & 0 & 0 & 1 & 1 & 1 & 1 & 0 & 2000 & -R_2 + R_4 \\
0 & \boxed{5} & 0 & 0 & 0 & 4 & -2 & 0 & 1 & 10,000 & -4R_2 + R_5 \\
\hline
0 & -7 & 0 & 0 & 32 & 22 & 2 & 0 & 0 & -196,000 & -22R_2 + R_6
\end{array}$$

We have completed Stage I. For Stage II, pivot on the 5 in row five, column two.

$$\begin{array}{ccccccccc|r l}
x_1 & x_2 & x_3 & x_4 & s_1 & s_2 & s_3 & s_4 & s_5 & & \\
\hline
1 & 1 & 0 & 0 & -1 & 0 & 0 & 0 & 0 & 3000 & \\
0 & 1 & 0 & 1 & -1 & -1 & -1 & 0 & 0 & 3000 & \\
0 & -1 & 1 & 0 & 1 & 0 & 1 & 0 & 0 & 2000 & \\
0 & 0 & 0 & 0 & 1 & 1 & 1 & 1 & 0 & 2000 & \\
0 & 1 & 0 & 0 & 0 & \frac{4}{5} & -\frac{2}{5} & 0 & \frac{1}{5} & 2000 & \frac{1}{5}R_5 \\
\hline
0 & -7 & 0 & 0 & 32 & 22 & 2 & 0 & 0 & -196,000 &
\end{array}$$

$$\begin{array}{ccccccccc|r l}
x_1 & x_2 & x_3 & x_4 & s_1 & s_2 & s_3 & s_4 & s_5 & & \\
\hline
1 & 0 & 0 & 0 & -1 & -\frac{4}{5} & \frac{2}{5} & 0 & -\frac{1}{5} & 1000 & -R_5 + R_1 \\
0 & 0 & 0 & 1 & -1 & -\frac{9}{5} & -\frac{3}{5} & 0 & -\frac{1}{5} & 1000 & -R_5 + R_2 \\
0 & 0 & 1 & 0 & 1 & \frac{4}{5} & \frac{3}{5} & 0 & \frac{1}{5} & 4000 & R_5 + R_3 \\
0 & 0 & 0 & 0 & 1 & 1 & \boxed{1} & 1 & 0 & 2000 & \\
0 & 1 & 0 & 0 & 0 & \frac{4}{5} & -\frac{2}{5} & 0 & \frac{1}{5} & 2000 & \\
\hline
0 & 0 & 0 & 0 & 32 & \frac{138}{5} & -\frac{4}{5} & 0 & \frac{7}{5} & -182,000 & 7R_5 + R_6
\end{array}$$

Finally, pivot on the 1 in row four, column seven.

$$\begin{array}{ccccccccc|r l}
x_1 & x_2 & x_3 & x_4 & s_1 & s_2 & s_3 & s_4 & s_5 & & \\
\hline
1 & 0 & 0 & 0 & -\frac{7}{5} & -\frac{6}{5} & 0 & -\frac{2}{5} & -\frac{1}{5} & 200 & -\frac{2}{5}R_4 + R_1 \\
0 & 0 & 0 & 1 & -\frac{2}{5} & -\frac{6}{5} & 0 & \frac{3}{5} & -\frac{1}{5} & 2200 & \frac{3}{5}R_4 + R_2 \\
0 & 0 & 1 & 0 & \frac{2}{5} & \frac{1}{5} & 0 & -\frac{3}{5} & \frac{1}{5} & 2800 & -\frac{3}{5}R_4 + R_3 \\
0 & 0 & 0 & 0 & 1 & 1 & 1 & 1 & 0 & 2000 & \\
0 & 1 & 0 & 0 & \frac{2}{5} & \frac{6}{5} & 0 & \frac{2}{5} & \frac{1}{5} & 2800 & \frac{2}{5}R_4 + R_5 \\
\hline
0 & 0 & 0 & 0 & \frac{164}{5} & \frac{142}{5} & 0 & \frac{4}{5} & \frac{7}{5} & -180,400 & \frac{4}{5}R_4 + R_6
\end{array}$$

This gives $x_1 = 200$, $x_2 = 2800$, $x_3 = 2800$, and $x_4 = 2200$. This means the company should ship 200 barrels from supplier S_1 to distributor D_1, 2800 barrels from supplier S_2 to distributor D_1, 2800 barrels from supplier S_1 to distributor D_2 and 2200 barrels from supplier S_2 to distributor D_2 at a minimum cost of \$180,400.

Copyright © 2019 Pearson Education, Inc.

35. Let y_1 = the amount of bluegrass seed, y_2 = the amount of rye seed, and y_3 = the amount of Bermuda seed. The problem is to minimize $w = .12y_1 + .15y_2 + .05y_3$ subject to

$$y_1 \geq .2(y_1 + y_2 + y_3)$$
$$y_3 \leq \frac{2}{3}y_2$$
$$y_1 + y_2 + y_3 \geq 5000$$

Or, maximize: $z = -w = -.12y_1 - .15y_2 - .05y_3$ subject to

$$.8y_1 - .2y_2 - .2y_3 \geq 0$$
$$2y_2 - 3y_3 \geq 0$$
$$y_1 + y_2 + y_3 \geq 5000$$

Adding surplus variables s_1, s_2, and s_3, the initial tableau is

$$\begin{array}{c} \begin{matrix} y_1 & y_2 & y_3 & s_1 & s_2 & s_3 & \end{matrix} \\ \left[\begin{array}{cccccc|c} \boxed{.8} & -.2 & -.2 & -1 & 0 & 0 & 0 \\ 0 & 2 & -3 & 0 & -1 & 0 & 0 \\ 1 & 1 & 1 & 0 & 0 & -1 & 5000 \\ \hline .12 & .15 & .05 & 0 & 0 & 0 & 0 \end{array}\right] \end{array}$$

Since y_4 through y_6 are negative, this does not have a feasible solution. For Stage I, use the .8 in row one, column one.

$$\begin{array}{c} \begin{matrix} y_1 & y_2 & y_3 & s_1 & s_2 & s_3 & \end{matrix} \\ \left[\begin{array}{cccccc|c} 1 & -.25 & -.25 & -1.25 & 0 & 0 & 0 \\ 0 & \boxed{2} & -3 & 0 & -1 & 0 & 0 \\ 0 & 1.25 & 1.25 & 1.25 & 0 & -1 & 5000 \\ \hline 0 & .18 & .08 & .15 & 0 & 0 & 0 \end{array}\right] \begin{array}{l} 1.25R_1 \\ \\ -R_1 + R_3 \\ -.12R_1 + R_4 \end{array} \end{array}$$

Continue Stage 1 by pivoting on the 2 in row two, column two.

$$\begin{array}{c} \begin{matrix} y_1 & y_2 & y_3 & s_1 & s_2 & s_3 & \end{matrix} \\ \left[\begin{array}{cccccc|c} 1 & 0 & -.625 & -1.25 & -.125 & 0 & 0 \\ 0 & 1 & -1.5 & 0 & -.5 & 0 & 0 \\ 0 & 0 & \boxed{3.125} & 1.25 & .625 & -1 & 5000 \\ \hline 0 & 0 & .35 & .15 & .09 & 0 & 0 \end{array}\right] \begin{array}{l} .25R_2 + R_1 \\ .5R_2 \\ -1.25R_1 + R_3 \\ -.18R_1 + R_4 \end{array} \end{array}$$

To finish Stage 1, pivot on the 3.125 in row three, column three.

$$\begin{array}{c} \begin{matrix} y_1 & y_2 & y_3 & s_1 & s_2 & s_3 & \end{matrix} \\ \left[\begin{array}{cccccc|c} 1 & 0 & 0 & -1 & 0 & -.2 & 1000 \\ 0 & 1 & 0 & .6 & -.2 & -.48 & 2400 \\ 0 & 0 & 1 & .4 & .2 & -.32 & 1600 \\ \hline 0 & 0 & 0 & .01 & .02 & .112 & -560 \end{array}\right] \begin{array}{l} .625R_3 + R_1 \\ 1.5R_3 + R_2 \\ .32R_3 \\ -.35R_3 + R_4 \end{array} \end{array}$$

Stage II pivoting is not necessary. The solution is $y_1 = 1000$, $y_2 = 2400$, $y_3 = 1600$ and $w = 560$. In other words, Topgrade Turf should use 1000 lbs.of bluegrass seed, 2400 lbs. of rye seed, and 1600 lbs. of Bermuda seed for a minimum cost of \$560.

Copyright © 2019 Pearson Education, Inc.

37. Let x_1 = amount allotted to commercial loans (in millions) and x_2 = amount allotted to home loans (in millions)

Maximize $z = .10x_1 + .12x_2$

subject to:
$$\begin{aligned} x_1 + x_2 &\le 25 \\ 2x_1 + 3x_2 &\le 72 \\ -4x_1 + x_2 &\ge 0 \\ x_1 + x_2 &\ge 10 \\ x_1 \ge 0,\ x_2 &\ge 0 \end{aligned}$$

The initial simplex tableau is

$$\begin{array}{cccccc|c} x_1 & x_2 & s_1 & s_2 & s_3 & s_4 & \\ 1 & 1 & 1 & 0 & 0 & 0 & 25 \\ 2 & 3 & 0 & 1 & 0 & 0 & 72 \\ -4 & \boxed{1} & 0 & 0 & -1 & 0 & 0 \\ 1 & 1 & 0 & 0 & 0 & -1 & 10 \\ \hline -.10 & -.12 & 0 & 0 & 0 & 0 & 0 \end{array}$$

For Stage I, pivot on the 1 in row three, column two.

$$\begin{array}{cccccc|c|l} x_1 & x_2 & s_1 & s_2 & s_3 & s_4 & & \\ 5 & 0 & 1 & 0 & 1 & 0 & 25 & -R_3 + R_1 \\ 14 & 0 & 0 & 1 & 3 & 0 & 72 & -3R_3 + R_2 \\ -4 & 1 & 0 & 0 & -1 & 0 & 0 & \\ \boxed{5} & 0 & 0 & 0 & 1 & -1 & 10 & -R_3 + R_4 \\ \hline -\frac{29}{50} & 0 & 0 & 0 & -\frac{3}{25} & 0 & 0 & \frac{3}{25}R_3 + R_5 \end{array}$$

To continue Stage I, pivot on the 5 in row four, column one.

$$\begin{array}{cccccc|c|l} x_1 & x_2 & s_1 & s_2 & s_3 & s_4 & & \\ 0 & 0 & 1 & 0 & 0 & \boxed{1} & 15 & -5R_4 + R_1 \\ 0 & 0 & 0 & 1 & \frac{1}{5} & \frac{14}{5} & 44 & -14R_4 + R_2 \\ 0 & 1 & 0 & 0 & -\frac{1}{5} & -\frac{4}{5} & 8 & 4R_4 + R_3 \\ 1 & 0 & 0 & 0 & \frac{1}{5} & -\frac{1}{5} & 2 & \frac{1}{5}R_4 \\ \hline 0 & 0 & 0 & 0 & -\frac{1}{250} & -\frac{29}{250} & \frac{29}{25} & \frac{29}{50}R_4 + R_5 \end{array}$$

For Stage II, pivot on the 1 in row one, column six.

$$\begin{array}{cccccc|c|l} x_1 & x_2 & s_1 & s_2 & s_3 & s_4 & & \\ 0 & 0 & 1 & 0 & 0 & 1 & 15 & \\ 0 & 0 & -\frac{14}{5} & 1 & \boxed{\frac{1}{5}} & 0 & 2 & -\frac{14}{5}R_1 + R_2 \\ 0 & 1 & \frac{4}{5} & 0 & -\frac{1}{5} & 0 & 20 & \frac{4}{5}R_1 + +R_3 \\ 1 & 0 & \frac{1}{5} & 0 & \frac{1}{5} & 0 & 5 & \frac{1}{5}R_1 + R_4 \\ \hline 0 & 0 & \frac{29}{250} & 0 & -\frac{1}{250} & 0 & \frac{29}{10} & \frac{29}{250}R_1 + R_5 \end{array}$$

To continue stage II, pivot on the $\frac{1}{5}$ in row two, column five.

$$\begin{array}{cccccc|c|l} x_1 & x_2 & s_1 & s_2 & s_3 & s_4 & & \\ 0 & 0 & 1 & 0 & 0 & 1 & 15 & \\ 0 & 0 & -14 & 5 & 1 & 0 & 10 & 5R_2 \\ 0 & 1 & -2 & 1 & 0 & 0 & 22 & \frac{1}{5}R_2 + R_3 \\ 1 & 0 & 3 & -1 & 0 & 0 & 3 & -\frac{1}{5}R_2 + R_4 \\ \hline 0 & 0 & \frac{3}{50} & \frac{1}{50} & 0 & 0 & \frac{147}{50} & \frac{1}{250}R_2 + R_5 \end{array}$$

This gives $x_1 = 3$, $x_2 = 22$, and the maximum is $\frac{147}{50} = 2.94$. Allot $3,000,000 for commercial loans and $22,000,000 for home loans for a maximum interest income of $2,940,000.

39. Let y_1 = the amount of regular beer and y_2 = the amount of light beer.

Minimize $36{,}000y_1 + 48{,}000y_2$

subject to:
$$\begin{aligned} y_1 &\ge 12 \\ y_2 &\ge 10 \\ -2y_1 + y_2 &\le 0 \quad (\text{Since } y_2 \le 2y_1) \\ y_1 + y_2 &\ge 42. \end{aligned}$$

(Since the company already produces at least $12 + 10 = 22$ units and can produce at least 20 additional units: $y_1 + y_2 \ge 22 + 20 = 42$.)

The initial simplex tableau is

$$\begin{array}{cccccc|c} y_1 & y_2 & s_1 & s_2 & s_3 & s_4 & \\ \boxed{1} & 0 & -1 & 0 & 0 & 0 & 12 \\ 0 & 1 & 0 & -1 & 0 & 0 & 10 \\ -2 & 1 & 0 & 0 & 1 & 0 & 0 \\ 1 & 1 & 0 & 0 & 0 & -1 & 42 \\ \hline 36 & 48 & 0 & 0 & 0 & 0 & 0 \end{array}.$$

For Stage I, pivot on the 1 in row one, column one.

$$\begin{array}{cccccc|c|l} y_1 & y_2 & s_1 & s_2 & s_3 & s_4 & & \\ 1 & 0 & -1 & 0 & 0 & 0 & 12 & \\ 0 & \boxed{1} & 0 & -1 & 0 & 0 & 10 & \\ 0 & 1 & -2 & 0 & 1 & 0 & 24 & 2R_1 + R_3 \\ 0 & 1 & 1 & 0 & 0 & -1 & 30 & -R_1 + R_4 \\ \hline 0 & 48 & 36 & 0 & 0 & 0 & -432 & -36R_1 + R_5 \end{array}$$

Copyright © 2019 Pearson Education, Inc.

To continue Stage I, pivot on the 1 in row two, column two.

$$\begin{array}{c} \\ \\ \\ \\ \\ \end{array}\begin{array}{cccccc|c} y_1 & y_2 & s_1 & s_2 & s_3 & s_4 & \\ 1 & 0 & -1 & 0 & 0 & 0 & 12 \\ 0 & 1 & 0 & -1 & 0 & 0 & 10 \\ 0 & 0 & -2 & 1 & 1 & 0 & 14 \\ 0 & 0 & 1 & \boxed{1} & 0 & -1 & 20 \\ \hline 0 & 0 & 36 & 48 & 0 & 0 & -912 \end{array}\begin{array}{l} \\ \\ \\ -R_2 + R_3 \\ -R_2 + R_4 \\ -48R_2 + R_5 \end{array}$$

Now pivot on the 1 in row four, column four.

$$\begin{array}{cccccc|c} y_1 & y_2 & s_1 & s_2 & s_3 & s_4 & \\ 1 & 0 & -1 & 0 & 0 & 0 & 12 \\ 0 & 1 & 1 & 0 & 0 & -1 & 30 \\ 0 & 0 & \boxed{-3} & 0 & 1 & 1 & -6 \\ 0 & 0 & 1 & 1 & 0 & -1 & 20 \\ \hline 0 & 0 & -12 & 0 & 0 & 48 & -1872 \end{array}\begin{array}{l} \\ \\ R_4 + R_2 \\ -R_4 + R_3 \\ \\ -48R_4 + R_5 \end{array}$$

For Stage II, pivot on the –3 in row three, column three.

$$\begin{array}{cccccc|c} y_1 & y_2 & s_1 & s_2 & s_3 & s_4 & \\ 1 & 0 & 0 & 0 & -\frac{1}{3} & -\frac{1}{3} & 14 \\ 0 & 1 & 0 & 0 & \frac{1}{3} & -\frac{2}{3} & 28 \\ 0 & 0 & 1 & 0 & -\frac{1}{3} & -\frac{1}{3} & 2 \\ 0 & 0 & 0 & 1 & \boxed{\frac{1}{3}} & -\frac{2}{3} & 18 \\ \hline 0 & 0 & 0 & 0 & -4 & 44 & -1848 \end{array}\begin{array}{l} \\ R_3 + R_1 \\ -R_3 + R_2 \\ -\frac{1}{3}R_3 \\ -R_3 + R_4 \\ 12R_3 + R_5 \end{array}$$

Finally, pivot on the $\frac{1}{3}$ in row four, column five.

$$\begin{array}{cccccc|c} y_1 & y_2 & s_1 & s_2 & s_3 & s_4 & \\ 1 & 0 & 0 & 1 & 0 & -1 & 32 \\ 0 & 1 & 0 & -1 & 0 & 0 & 10 \\ 0 & 0 & 1 & 1 & 0 & -1 & 20 \\ 0 & 0 & 0 & 3 & 1 & -2 & 54 \\ \hline 0 & 0 & 0 & 12 & 0 & 36 & -1632 \end{array}\begin{array}{l} \\ \frac{1}{3}R_4 + R_1 \\ -\frac{1}{3}R_4 + R_2 \\ \frac{1}{3}R_4 + R_3 \\ 3R_4 \\ 4R_4 + R_5 \end{array}$$

Make 32 units of regular beer and 10 units of light beer for a minimum cost of $1,632,000.

41. From Exercise 29, we know the initial simplex tableau is

$$\begin{array}{ccccccc|c} y_1 & y_2 & y_3 & s_1 & s_2 & s_3 & s_4 & \\ 1 & 1 & 1 & -1 & 0 & 0 & 0 & 10 \\ 1 & 1 & 1 & 0 & 1 & 0 & 0 & 15 \\ 1 & -\frac{1}{4} & 0 & 0 & 0 & -1 & 0 & 0 \\ -1 & 0 & 1 & 0 & 0 & 0 & -1 & 0 \\ \hline .30 & .09 & .27 & 0 & 0 & 0 & 0 & 0 \end{array}.$$

Using the LINPROG program on a TI-84 Plus, we find the optimal solution is

$$y_1 = 1\frac{2}{3},\ y_2 = 6\frac{2}{3},\ \text{and}\ y_3 = 1\frac{2}{3}.$$

For each barrel of gasoline, the mixture should contain $1\frac{2}{3}$ ounces of ingredient I, $6\frac{2}{3}$ ounces of ingredient II, and $1\frac{2}{3}$ ounces of ingredient III at a minimum cost of $1.55 per barrel.

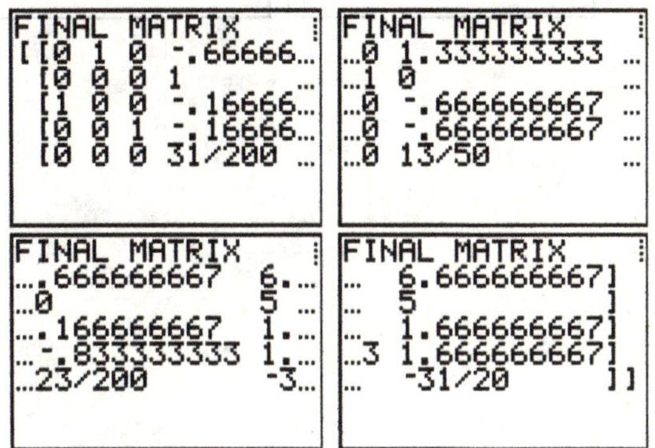

Copyright © 2019 Pearson Education, Inc.

43. From Exercise 31, we know the initial simplex tableau is

$$\begin{array}{cccccccccc|c}
y_1 & y_2 & y_3 & y_4 & s_1 & s_2 & s_3 & s_4 & s_5 & s_6 & \\
1 & 1 & 0 & 0 & -1 & 0 & 0 & 0 & 0 & 0 & 32 \\
1 & 1 & 0 & 0 & 0 & 1 & 0 & 0 & 0 & 0 & 32 \\
0 & 0 & 1 & 1 & 0 & 0 & -1 & 0 & 0 & 0 & 20 \\
0 & 0 & 1 & 1 & 0 & 0 & 0 & 1 & 0 & 0 & 20 \\
1 & 0 & 1 & 0 & 0 & 0 & 0 & 0 & 1 & 0 & 25 \\
0 & 1 & 0 & 1 & 0 & 0 & 0 & 0 & 0 & 1 & 30 \\
\hline
14 & 12 & 22 & 10 & 0 & 0 & 0 & 0 & 0 & 0 & 0
\end{array}.$$

Using the LINPROG program on a TI-84 Plus, we find the optimal solution is $y_1 = 22$, $y_2 = 0$, $y_3 = 10$, and $y_4 = 20$. The manufacturer should send 22 computers from W_1 to D_1, 0 from W_1 to D_2, 10 from W_2 to D_1, and 20 computers from W_2 to D_2 for a minimum cost of \$628.

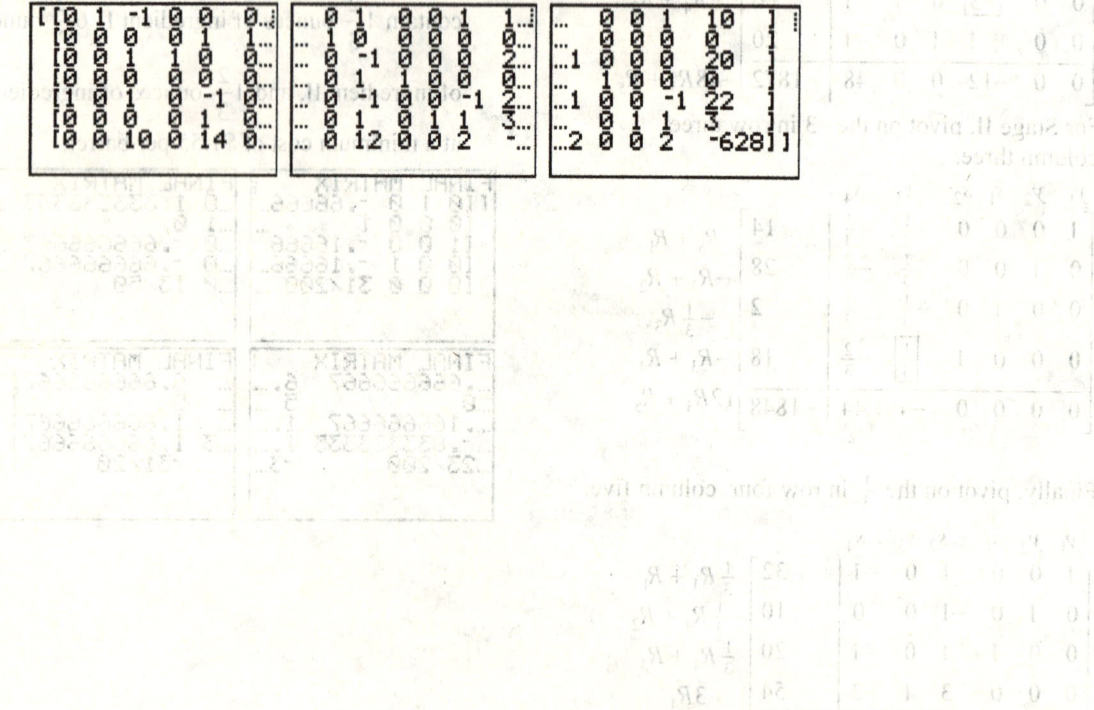

Copyright © 2019 Pearson Education, Inc.

Chapter 7 Review Exercises

1. $y \le 3x + 2$

   Graph the solid boundary line $y = 3x + 2$ which goes through the points (0, 2) and (2, 8). The inequality is $\le$, so shade the half-plane below the line.

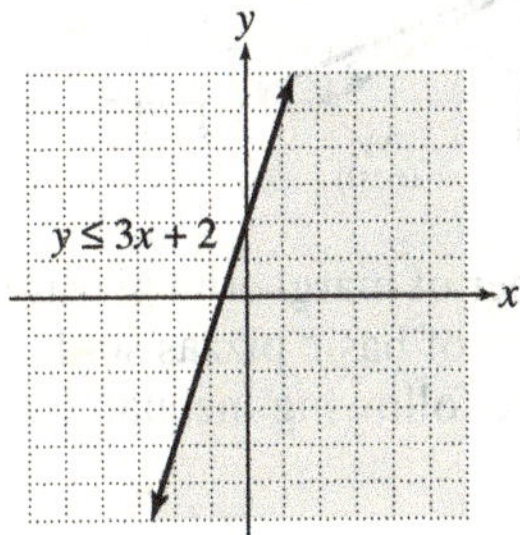

2. $2x - y \ge 6$

   Graph the solid boundary line $2x - y = 6$, which goes through the points (0, –6) and (3, 0). Testing the origin gives the statement $2(0) - 0 \ge 6$, which is false, so shade the half-plane not containing the origin, which is the region below the line.

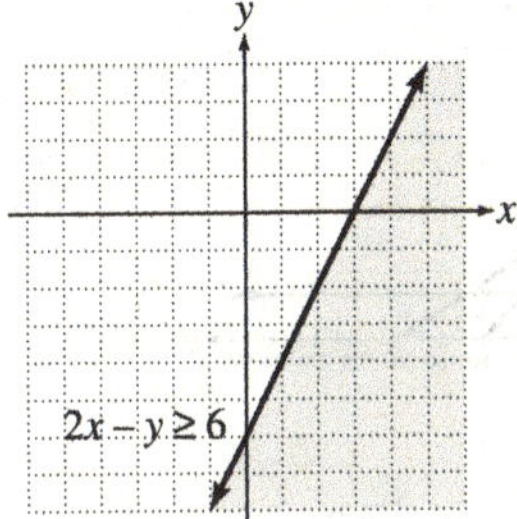

3. $3x + 4y \ge 12$

   Graph the solid boundary line $3x + 4y = 12$, which goes through (0, 3) and (4, 0). Testing the origin gives $3(0) + 4(0) \ge 12$, which is false, so shade the half-plane not containing the origin, which is the region above the line.

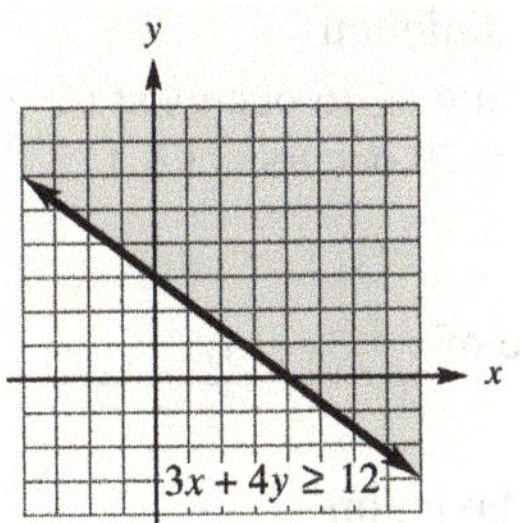

4. $y \le 4$

   Graph the solid boundary line $y = 4$, which is the horizontal line crossing the y-axis at (0, 4). Shade the half-plane below the line.

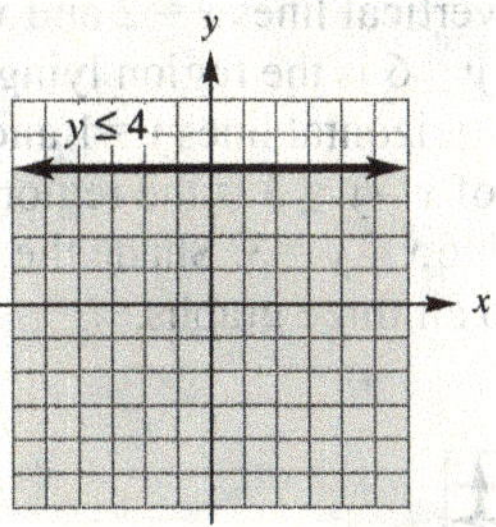

5. $x + y \le 6$

   $2x - y \ge 3$

   $x + y \le 6$ is the region on or below the line $x + y = 6$; $2x - y \ge 3$ is the region on or below the line $2x - y = 3$. The system of inequalities must meet both conditions so we shade the overlap of the two half-planes.

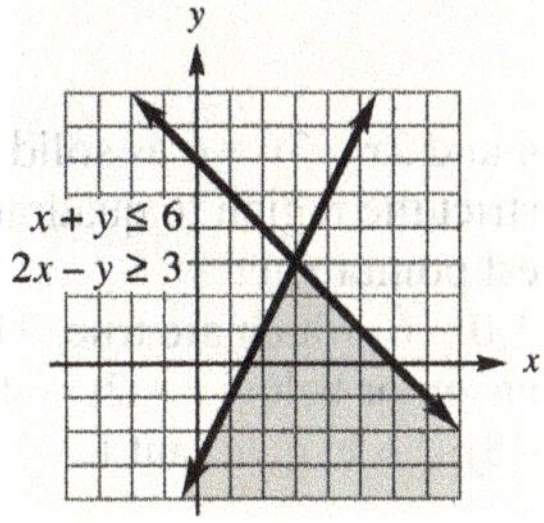

6. $4x + y \ge 8$

   $2x - 3y \le 6$

   Graph $4x + y = 8$ as a solid line using (2, 0) and (0, 8), and $2x - 3y = 6$ as a solid line using (3, 0) and (0, –2). The test point (0, 0) gives $0 + 0 \ge 8$, which is false and $0 - 0 \le 6$, which is true. Shade all points above $4x + y = 8$ and above $2x - 3y = 6$. The solution region is to the right of $4x + y = 8$ and above $2x - 3y = 6$.

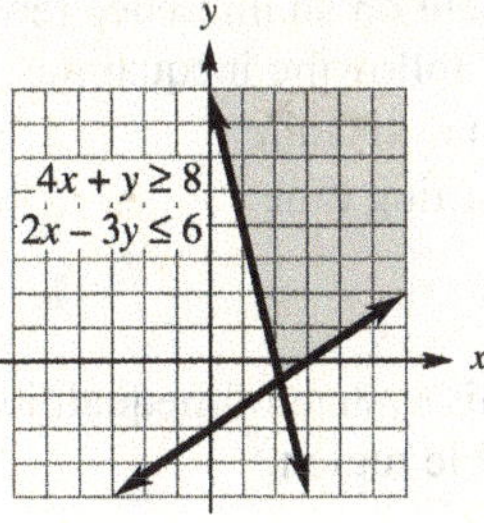

Copyright © 2019 Pearson Education, Inc.

7. $2 \le x \le 5$
$1 \le y \le 6$
$x - y \le 3$
The graph of $2 \le x \le 5$ is the region lying on or between the two vertical lines $x = 2$ and $x = 5$. The graph of $1 \le y \le 6$ is the region lying on or between the two horizontal lines $y = 1$ and $y = 6$. The graph of $x - y \le 3$ is the region lying on or below the line $x - y = 3$. Shade the region that is common to all three graphs.

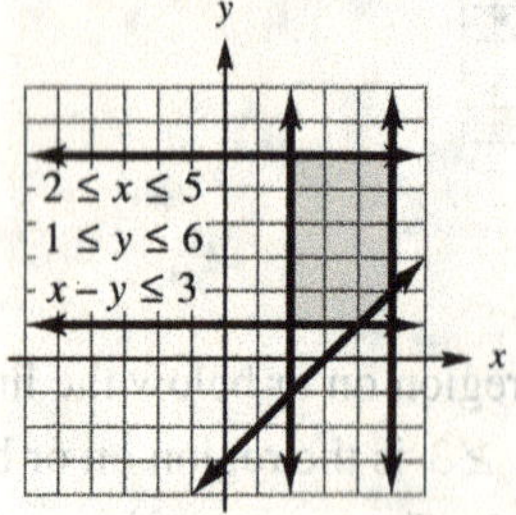

8. $x + 2y \le 4$
$2x - 3y \le 6$
$x \ge 0$
$y \ge 0$
Graph $x + 2y = 4$ and $2x - 3y = 6$ as solid lines. $x \ge 0$ and $y \ge 0$ restrict the region to quadrant I. Use (0, 0) as a test point to get $0 + 0 \le 4$ and $0 + 0 \le 6$, which are true. The region is all points on or below $x + 2y = 4$, and on or above $2x - 3y = 6$ in quadrant I.

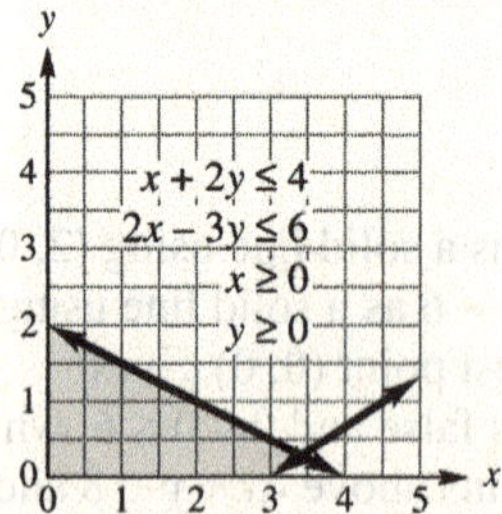

9. Let x = the time spent on a summary report and let y = the time spent on an inference report. Then we have the following inequalities:
$2x + 4y \le 15$ Analysis time
$1.5x + 2y \le 9$ Writing time
$x \ge 0$
$y \ge 0.$
The solution of this system of inequalities is the graph of the feasible region.

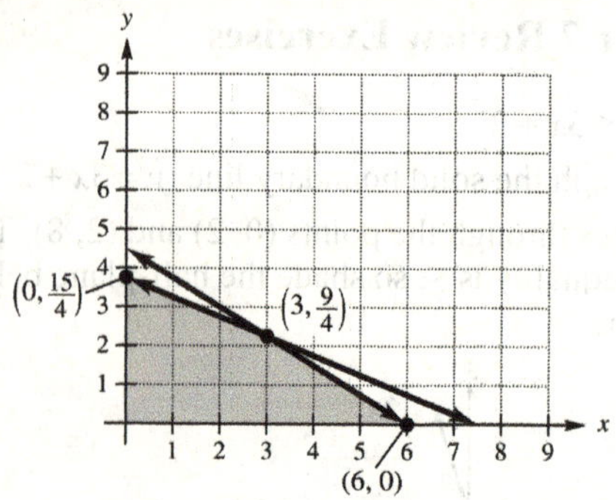

10. Let x = the number of margherita pizzas sold and let y = the number of basic pizzas sold
Then we have the following inequalities:
$5x + 4y \le 1000$
$2x + y \le 320$
$x \ge 60$
$y \ge 40$
The solution of this system is the graph of the feasible region.

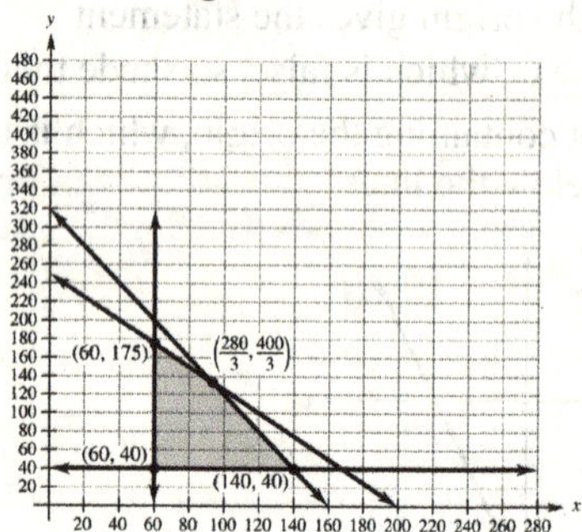

11.

Corner Point	Value of $z = 3x + 4y$
(1, 6)	27
(6 ,7)	46 Maximum
(7, 3)	33
$\left(1, 2\frac{1}{2}\right)$	13
(2, 1)	10 Minimum

The maximum value of 46 occurs at (6, 7) the minimum value of 10 occurs at (2, 1).

12.

Corner Point	Value of $z = 3x + 4y$
(0, 8)	32
(8, 8)	56 Maximum
(5, 2)	23
(2, 0)	6 Minimum

The maximum value is 56 at (8, 8); the minimum value is 6 at (2, 0).

Copyright © 2019 Pearson Education, Inc.

13. Maximize $z = 6x + 2y$
subject to $2x + 7y \le 14$
$2x + 3y \le 10$
$x \ge 0,\ y \ge 0.$
Graph the feasible region.

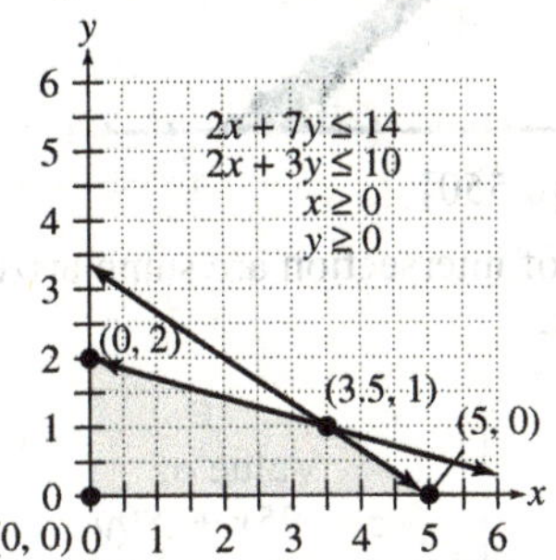

Corner Point	Value of $z = 6x + 2y$
(0, 0)	0
(0, 2)	4
(3.5, 1)	23
(5, 0)	30 Maximum

The maximum is 30 when $x = 5$, $y = 0$.

14. Find $x \ge 0$ and $y \ge 0$ such that
$8x + 9y \ge 72$
$6x + 8y \ge 72$
and $w = 2x + 10y$ is minimized.
Graph the feasible region.

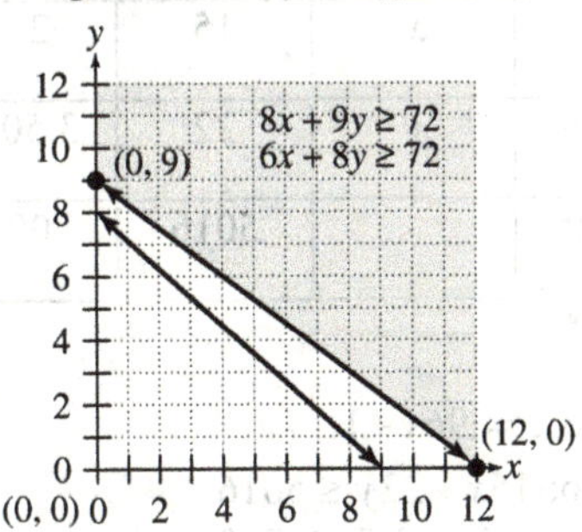

Corner Point	Value of $z = 2x + 10y$
(0, 9)	90
(12, 0)	24 Minimum

The minimum is 24 when $x = 12$, $y = 0$.

15. Find $x \ge 0$ and $y \ge 0$ such that
$x + y \le 50$
$2x + y \ge 20$
$x + 2y \ge 30$
and $w = 5x + 2y$ is minimized.
Graph the feasible region.

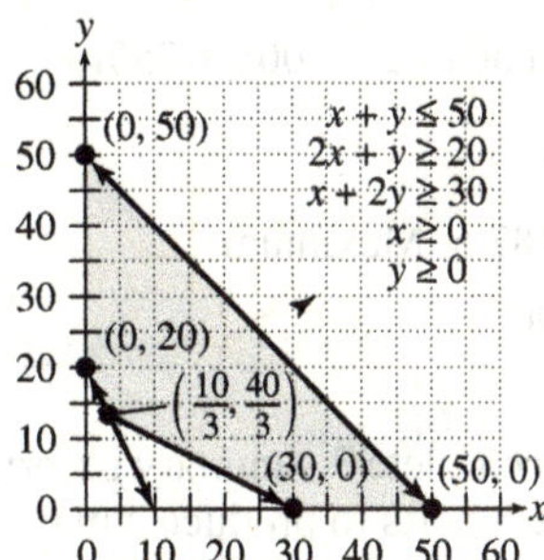

Corner Point	Value of $w = 5x + 2y$
(0, 20)	40 Minimum
(0, 50)	100
(50, 0)	250
(30, 0)	150
$\left(\frac{10}{3}, \frac{40}{3}\right)$	$\frac{130}{3} = 43\frac{1}{3}$

The minimum is 40 when $x = 0$, $y = 20$.

16. Maximize $z = 5x - 2y$
subject to: $3x + 2y \le 12$
$5x + y \ge 5$
$x \ge 0,\ y \ge 0.$
Graph the feasible region.

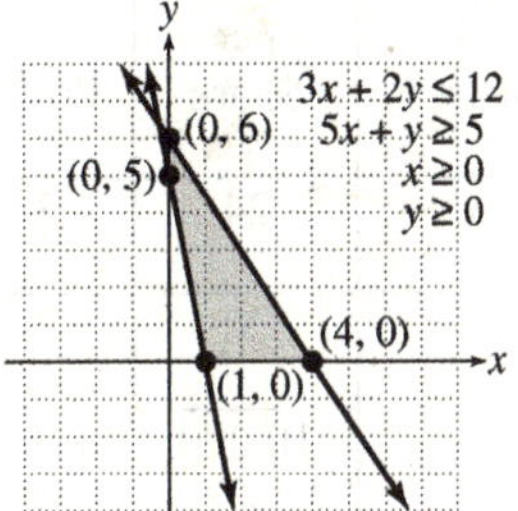

Corner Point	Value of $z = 5x - 2y$
(0, 5)	−10
(0, 6)	−12
(4, 0)	20 Maximum
(1, 0)	5

The maximum is 20 when $x = 4$, $y = 0$.

Copyright © 2019 Pearson Education, Inc.

17. From the graph for Exercise 9, the corner points are (0, 3.75), (3, 2.25), (6, 0), and (0, 0). Since x was the number of summary reports and y the number of inference reports, the revenue function is $z = 500x + 750y$. Evaluate this objective function at each corner point.

Corner Point	Value of $z = 500x + 750y$
(0, 3.75)	2812.5
(3, 2.25)	3187.5 Maximum
(6,0)	3000
(0, 0)	0

Linda should complete 3 summary reports and 2.25 inference reports to produce a maximum profit of $3187.50.

18. The profit function is $z = 20x + 15y$. Refer to the graph in Exercise 10, and evaluate at the corner points.

Corner Point	Value of $z = 20x + 15y$
(60, 40)	1800
(60, 175)	3825
(93, 133)	3855
(140, 40)	3400

A maximum profit of $3855 is obtained by making 133 basic pizzas and 93 margherita pizzas.

19.

	Number of Shares	Cost per Share	Profit	Risk
Blackrock	x	24	3	.95
Columbia	y	20	2	.89
Constraints		10,000	1100	

Minimize $z = .95x + .89y$

subject to: $24x + 20y \le 10{,}000$

$3x + 2y \ge 1100$

Graph the feasibility region.

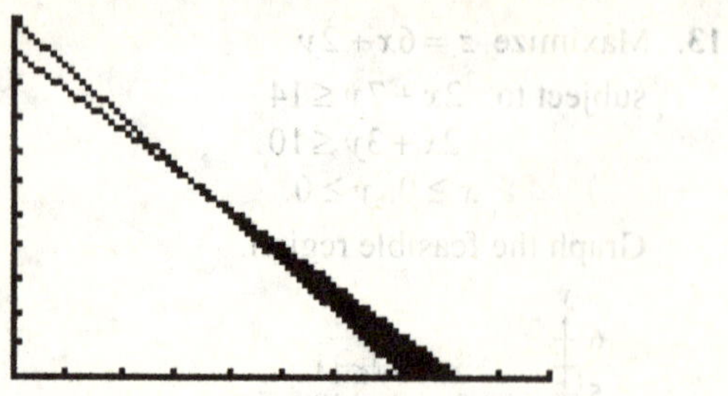

[0,500] by [0,550]

The points of intersection are summarized in the table below.

Corner Point	Value of $z = .95x + .89y$
(22, 474)	442.76
(417,0)	396.15
(367, 0)	348.65 Minimum

Donald should buy about 367 shares of Blackrock Equity Dividend Fund and none of the Columbia Dividend Income Fund. The risk has a value of about 348.65 for Donald.

20.

	Number of Shares	Cost per Share	Profit	Risk
Alger	x	15	2	.96
JP Morgan	y	22	3.50	1.14
Constraints		5016	700	

Minimize $z = .96x + 1.14y$

subject to: $15x + 22y \le 5016$

$2x + 3.5y \ge 700$

Graph the feasibility region

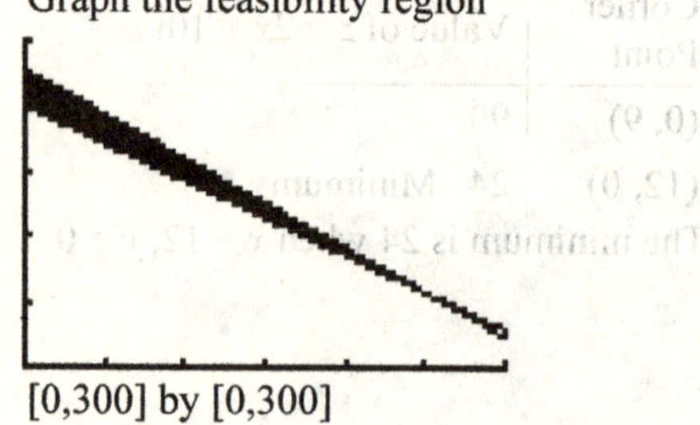

[0,300] by [0,300]

The points of intersection are summarized in the table below.

Copyright © 2019 Pearson Education, Inc.

Corner Point	Value of $z = .96x + 1.14y$
(0, 200)	228 Minimum
(254, 55)	306.54
(0, 228)	259.92

Ming-Na should buy 200 shares of JP Morgan U.S. Dynamic Plus Fund and none of the Alger Spectra Z Fund. The risk has a value of 228 for Ming-Na.

21. Maximize $z = 5x_1 + 6x_2 + 3x_3$

subject to:
$$\begin{aligned} x_1 + x_2 + x_3 &\le 100 \\ 2x_1 + 3x_2 &\le 500 \\ x_1 + 2x_3 &\le 350 \\ x_1 \ge 0,\ x_2 \ge 0,\ x_3 &\ge 0. \end{aligned}$$

a.
$$\begin{aligned} x_1 + x_2 + x_3 + s_1 &= 100 \\ 2x_1 + 3x_2 + s_2 &= 500 \\ x_1 + 2x_3 + s_3 &= 350 \end{aligned}$$

b.
$$\begin{array}{cccccc} x_1 & x_2 & x_3 & s_1 & s_2 & s_3 \end{array}$$
$$\left[\begin{array}{cccccc|c} 1 & 1 & 1 & 1 & 0 & 0 & 100 \\ 2 & 3 & 0 & 0 & 1 & 0 & 500 \\ 1 & 0 & 2 & 0 & 0 & 1 & 350 \\ \hline -5 & -6 & -3 & 0 & 0 & 0 & 0 \end{array}\right]$$

22. Maximize $z = 2x_1 + 9x_2$

subject to:
$$\begin{aligned} 3x_1 + 5x_2 &\le 47 \\ x_1 + x_2 &\le 25 \\ 5x_1 + 2x_2 &\le 35 \\ 2x_1 + x_2 &\le 30 \\ x_1 \ge 0,\ x_2 &\ge 0. \end{aligned}$$

a.
$$\begin{aligned} 3x_1 + 5x_2 + s_1 &= 47 \\ x_1 + x_2 + s_2 &= 25 \\ 5x_1 + 2x_2 + s_3 &= 35 \\ 2x_1 + x_2 + s_4 &= 30 \end{aligned}$$

b.
$$\begin{array}{cccccc} x_1 & x_2 & s_1 & s_2 & s_3 & s_4 \end{array}$$
$$\left[\begin{array}{cccccc|c} 3 & 5 & 1 & 0 & 0 & 0 & 47 \\ 1 & 1 & 0 & 1 & 0 & 0 & 25 \\ 5 & 2 & 0 & 0 & 1 & 0 & 35 \\ 2 & 1 & 0 & 0 & 0 & 1 & 30 \\ \hline -2 & -9 & 0 & 0 & 0 & 0 & 0 \end{array}\right]$$

23. Maximize $z = x_1 + 8x_2 + 2x_3$

subject to:
$$\begin{aligned} x_1 + x_2 + x_3 &\le 90 \\ 2x_1 + 5x_2 + x_3 &\le 120 \\ x_1 + 3x_2 &\le 80 \\ x_1 \ge 0,\ x_2 \ge 0,\ x_3 &\ge 0. \end{aligned}$$

a.
$$\begin{aligned} x_1 + x_2 + x_3 + s_1 &= 90 \\ 2x_1 + 5x_2 + x_3 + s_2 &= 120 \\ x_1 + 3x_2 + s_3 &= 80 \end{aligned}$$

b.
$$\begin{array}{cccccc} x_1 & x_2 & x_3 & s_1 & s_2 & s_3 \end{array}$$
$$\left[\begin{array}{cccccc|c} 1 & 1 & 1 & 1 & 0 & 0 & 90 \\ 2 & 5 & 1 & 0 & 1 & 0 & 120 \\ 1 & 3 & 0 & 0 & 0 & 1 & 80 \\ \hline -1 & -8 & -2 & 0 & 0 & 0 & 0 \end{array}\right]$$

24. Maximize $z = 15x_1 + 12x_2$

subject to:
$$\begin{aligned} 2x_1 + 5x_2 &\le 50 \\ x_1 + 3x_2 &\le 25 \\ 4x_1 + x_2 &\le 18 \\ x_1 + x_2 &\le 12 \\ x_1 \ge 0,\ x_2 &\ge 0. \end{aligned}$$

a.
$$\begin{aligned} 2x_1 + 5x_2 + s_1 &= 50 \\ x_1 + 3x_2 + s_2 &= 25 \\ 4x_1 + x_2 + s_3 &= 18 \\ x_1 + x_2 + s_4 &= 12 \end{aligned}$$

b.
$$\begin{array}{cccccc} x_1 & x_2 & s_1 & s_2 & s_3 & s_4 \end{array}$$
$$\left[\begin{array}{cccccc|c} 2 & 5 & 1 & 0 & 0 & 0 & 50 \\ 1 & 3 & 0 & 1 & 0 & 0 & 25 \\ 4 & 1 & 0 & 0 & 1 & 0 & 18 \\ 1 & 1 & 0 & 0 & 0 & 1 & 12 \\ \hline -15 & -12 & 0 & 0 & 0 & 0 & 0 \end{array}\right]$$

Copyright © 2019 Pearson Education, Inc.

25.

$$\begin{array}{c} \begin{array}{ccccc|c} x_1 & x_2 & x_3 & s_1 & s_2 & \end{array} \\ \left[\begin{array}{ccccc|c} 1 & 2 & 3 & 1 & 0 & 28 \\ \boxed{2} & 4 & 8 & 0 & 1 & 32 \\ \hline -5 & -2 & -3 & 0 & 0 & 0 \end{array}\right] \end{array}$$

Pivot on the 2 in row two, column one.

$$\left[\begin{array}{ccccc|c} x_1 & x_2 & x_3 & s_1 & s_2 & \\ 1 & 2 & 3 & 1 & 0 & 28 \\ 1 & 2 & 4 & 0 & \frac{1}{2} & 16 \\ \hline -5 & -2 & -3 & 0 & 0 & 0 \end{array}\right] \begin{array}{l} \\ \\ \frac{1}{2}R_2 \\ \\ \end{array}$$

$$\left[\begin{array}{ccccc|c} x_1 & x_2 & x_3 & s_1 & s_2 & \\ 0 & 0 & -1 & 1 & -\frac{1}{2} & 12 \\ 1 & 2 & 4 & 0 & \frac{1}{2} & 16 \\ \hline 0 & 8 & 17 & 0 & \frac{5}{2} & 80 \end{array}\right] \begin{array}{l} \\ -R_2 + R_1 \\ \\ 5R_2 + R_3 \end{array}$$

The maximum is 80 when
$x_1 = 16,\ x_2 = 0,\ x_3 = 0,\ s_1 = 12,\ s_2 = 0.$

26.

$$\left[\begin{array}{cccc|c} x_1 & x_2 & s_1 & s_2 & \\ 2 & 1 & 1 & 0 & 10 \\ 9 & \boxed{3} & 0 & 1 & 15 \\ \hline -2 & -3 & 0 & 0 & 0 \end{array}\right]$$

Pivot on the 3 in row two, column two.

$$\left[\begin{array}{cccc|c} x_1 & x_2 & s_1 & s_2 & \\ 2 & 1 & 1 & 0 & 10 \\ 3 & 1 & 0 & \frac{1}{3} & 5 \\ \hline -2 & -3 & 0 & 0 & 0 \end{array}\right] \begin{array}{l} \\ \\ \frac{1}{3}R_2 \\ \\ \end{array}$$

$$\left[\begin{array}{cccc|c} x_1 & x_2 & s_1 & s_2 & \\ -1 & 0 & 1 & -\frac{1}{3} & 5 \\ 3 & 1 & 0 & \frac{1}{3} & 5 \\ \hline 7 & 0 & 0 & 1 & 15 \end{array}\right] \begin{array}{l} \\ -R_2 + R_1 \\ \\ 3R_2 + R_3 \end{array}$$

The maximum is 15 when $x_1 = 0,\ x_2 = 5,\ s_1 = 5,$ and $s_2 = 0.$

27.

$$\left[\begin{array}{cccccc|c} x_1 & x_2 & x_3 & s_1 & s_2 & s_3 & \\ 1 & 2 & 2 & 1 & 0 & 0 & 50 \\ \boxed{4} & 24 & 0 & 0 & 1 & 0 & 20 \\ 1 & 0 & 2 & 0 & 0 & 1 & 15 \\ \hline -5 & -3 & -2 & 0 & 0 & 0 & 0 \end{array}\right]$$

Pivot on the 4 in row two, column one.

$$\left[\begin{array}{cccccc|c} x_1 & x_2 & x_3 & s_1 & s_2 & s_3 & \\ 1 & 2 & 2 & 1 & 0 & 0 & 50 \\ 1 & 6 & 0 & 0 & \frac{1}{4} & 0 & 5 \\ 1 & 0 & 2 & 0 & 0 & 1 & 15 \\ \hline -5 & -3 & -2 & 0 & 0 & 0 & 0 \end{array}\right] \begin{array}{l} \\ \\ \frac{1}{4}R_2 \\ \\ \\ \end{array}$$

$$\left[\begin{array}{cccccc|c} x_1 & x_2 & x_3 & s_1 & s_2 & s_3 & \\ 0 & -4 & 2 & 1 & -\frac{1}{4} & 0 & 45 \\ 1 & 6 & 0 & 0 & \frac{1}{4} & 0 & 5 \\ 0 & -6 & \boxed{2} & 0 & -\frac{1}{4} & 1 & 10 \\ \hline 0 & 27 & -2 & 0 & \frac{5}{4} & 0 & 25 \end{array}\right] \begin{array}{l} \\ -R_2 + R_1 \\ \\ -R_2 + R_3 \\ 5R_2 + R_4 \end{array}$$

Pivot on the 2 in row three, column three.

$$\left[\begin{array}{cccccc|c} x_1 & x_2 & x_3 & s_1 & s_2 & s_3 & \\ 0 & -4 & 2 & 1 & -\frac{1}{4} & 0 & 45 \\ 1 & 6 & 0 & 0 & \frac{1}{4} & 0 & 5 \\ 0 & -3 & 1 & 0 & -\frac{1}{8} & \frac{1}{2} & 5 \\ \hline 0 & 27 & -2 & 0 & \frac{5}{4} & 0 & 25 \end{array}\right] \begin{array}{l} \\ \\ \\ \frac{1}{2}R_3 \\ \\ \end{array}$$

$$\left[\begin{array}{cccccc|c} x_1 & x_2 & x_3 & s_1 & s_2 & s_3 & \\ 0 & 2 & 0 & 1 & 0 & -1 & 35 \\ 1 & 6 & 0 & 0 & \frac{1}{4} & 0 & 5 \\ 0 & -3 & 1 & 0 & -\frac{1}{8} & \frac{1}{2} & 5 \\ \hline 0 & 21 & 0 & 0 & 1 & 1 & 35 \end{array}\right] \begin{array}{l} \\ -2R_3 + R_1 \\ \\ \\ 2R_3 + R_4 \end{array}$$

The maximum is 35 when
$x_1 = 5,\ x_2 = 0,\ x_3 = 5,\ s_1 = 35,\ s_2 = 0,$ and $s_3 = 0.$

Copyright © 2019 Pearson Education, Inc.

28.
$$\begin{array}{ccccc|c} x_1 & x_2 & s_1 & s_2 & s_3 & \\ \end{array}$$
$$\left[\begin{array}{ccccc|c} 1 & -2 & 1 & 0 & 0 & 38 \\ 1 & -1 & 0 & 1 & 0 & 12 \\ 2 & \underline{1} & 0 & 0 & 1 & 30 \\ \hline -1 & -2 & 0 & 0 & 0 & 0 \end{array}\right]$$

Pivot on the 1 in row three, column two.

$$\begin{array}{c} \\ \end{array}\begin{array}{ccccc} x_1 & x_2 & s_1 & s_2 & s_3 \end{array}$$
$$\left[\begin{array}{ccccc|c} 5 & 0 & 1 & 0 & 2 & 98 \\ 3 & 0 & 0 & 1 & 1 & 42 \\ 2 & 1 & 0 & 0 & 1 & 30 \\ \hline 3 & 0 & 0 & 0 & 2 & 60 \end{array}\right]\begin{array}{l} 2R_3 + R_1 \\ R_3 + R_2 \\ \\ 2R_3 + R_4 \end{array}$$

The maximum is 60 when
$x_1 = 0, x_2 = 30,\ s_1 = 98,\ s_2 = 42,$ and $s_3 = 0.$

29. From exercise 21, we have

$$\begin{array}{cccccc} x_1 & x_2 & x_3 & s_1 & s_2 & s_3 \end{array}$$
$$\left[\begin{array}{cccccc|c} 1 & 1 & 1 & 1 & 0 & 0 & 100 \\ 2 & 3 & 0 & 0 & 1 & 0 & 500 \\ 1 & 0 & 2 & 0 & 0 & 1 & 350 \\ \hline -5 & -6 & -3 & 0 & 0 & 0 & 0 \end{array}\right].$$

Pivot on the 1 in row 1 column 2.

$$\begin{array}{cccccc} x_1 & x_2 & x_3 & s_1 & s_2 & s_3 \end{array}$$
$$\left[\begin{array}{cccccc|c} 1 & \boxed{1} & 1 & 1 & 0 & 0 & 100 \\ -1 & 0 & -3 & -3 & 1 & 0 & 200 \\ 1 & 0 & 2 & 0 & 0 & 1 & 350 \\ \hline 1 & 0 & 3 & 6 & 0 & 0 & 600 \end{array}\right]\begin{array}{l} \\ -3R_1 + R_2 \\ \\ 6R_1 + R_4 \end{array}$$

The maximum is 600 when $x_1 = 0$, $x_2 = 100$, and $x_3 = 0$.

30. From exercise 20, we have

$$\begin{array}{cccccc} x_1 & x_2 & s_1 & s_2 & s_3 & s_4 \end{array}$$
$$\left[\begin{array}{cccccc|c} 3 & 5 & 1 & 0 & 0 & 0 & 47 \\ 1 & 1 & 0 & 1 & 0 & 0 & 25 \\ 5 & 2 & 0 & 0 & 1 & 0 & 35 \\ 2 & 1 & 0 & 0 & 0 & 1 & 30 \\ \hline -2 & -9 & 0 & 0 & 0 & 0 & 0 \end{array}\right].$$

Pivot on the 5 in row 1 column 2. First divide row 1 by 5, then continue the process.

$$\begin{array}{cccccc} x_1 & x_2 & s_1 & s_2 & s_3 & s_4 \end{array}$$
$$\left[\begin{array}{cccccc|c} \frac{3}{5} & 1 & \frac{1}{5} & 0 & 0 & 0 & \frac{47}{5} \\ 1 & 1 & 0 & 1 & 0 & 0 & 25 \\ 5 & 2 & 0 & 0 & 1 & 0 & 35 \\ 2 & 1 & 0 & 0 & 0 & 1 & 30 \\ \hline -2 & -9 & 0 & 0 & 0 & 0 & 0 \end{array}\right]\begin{array}{l} \frac{1}{5}R_1 \\ \\ \\ \\ \\ \end{array}$$

$$\begin{array}{cccccc} x_1 & x_2 & s_1 & s_2 & s_3 & s_4 \end{array}$$
$$\left[\begin{array}{cccccc|c} \frac{3}{5} & 1 & \frac{1}{5} & 0 & 0 & 0 & \frac{47}{5} \\ \frac{2}{5} & 0 & -\frac{1}{5} & 1 & 0 & 0 & \frac{78}{5} \\ \frac{19}{5} & 0 & -\frac{2}{5} & 0 & 1 & 0 & \frac{81}{5} \\ \frac{7}{5} & 0 & -\frac{1}{5} & 0 & 0 & 1 & \frac{103}{5} \\ \hline \frac{17}{5} & 0 & \frac{9}{5} & 0 & 0 & 0 & \frac{423}{5} \end{array}\right]\begin{array}{l} \\ -R_1 + R_2 \\ -2R_1 + R_3 \\ -R_1 + R_4 \\ 9R_1 + R_5 \end{array}$$

The maximum is $\frac{423}{5}$ when $x_1 = 0$ and $x_2 = \frac{47}{5}$.

31. From exercise 21, we have

$$\begin{array}{cccccc} x_1 & x_2 & x_3 & s_1 & s_2 & s_3 \end{array}$$
$$\left[\begin{array}{cccccc|c} 1 & 1 & 1 & 1 & 0 & 0 & 90 \\ 2 & \boxed{5} & 1 & 0 & 1 & 0 & 120 \\ 1 & 3 & 0 & 0 & 0 & 1 & 80 \\ \hline -1 & -8 & -2 & 0 & 0 & 0 & 0 \end{array}\right].$$

Pivot on the 5 in row 2 column 2. First divide row 2 by 5, then continue the process.

$$\begin{array}{cccccc} x_1 & x_2 & x_3 & s_1 & s_2 & s_3 \end{array}$$
$$\left[\begin{array}{cccccc|c} 1 & 1 & 1 & 1 & 0 & 0 & 90 \\ \frac{2}{5} & 1 & \frac{1}{5} & 0 & \frac{1}{5} & 0 & 24 \\ 1 & 3 & 0 & 0 & 0 & 1 & 80 \\ \hline -1 & -8 & -2 & 0 & 0 & 0 & 0 \end{array}\right]\begin{array}{l} \\ \frac{1}{5}R_1 \\ \\ \\ \end{array}$$

$$\begin{array}{cccccc} x_1 & x_2 & x_3 & s_1 & s_2 & s_3 \end{array}$$
$$\left[\begin{array}{cccccc|c} \frac{3}{5} & 0 & \boxed{\frac{4}{5}} & 1 & -\frac{1}{5} & 0 & 66 \\ \frac{2}{5} & 1 & \frac{1}{5} & 0 & \frac{1}{5} & 0 & 24 \\ -\frac{1}{5} & 0 & -\frac{3}{5} & 0 & -\frac{3}{5} & 1 & 8 \\ \hline \frac{11}{5} & 0 & -\frac{2}{5} & 0 & \frac{8}{5} & 0 & 192 \end{array}\right]\begin{array}{l} -R_2 + R_1 \\ \\ -3R_2 + R_3 \\ \\ \end{array}$$

Now pivot on the $\frac{4}{5}$ in row 1 column 3. First multiply row 1 by $\frac{5}{4}$, then continue the process.

$$\begin{array}{cccccc} x_1 & x_2 & x_3 & s_1 & s_2 & s_3 \end{array}$$
$$\left[\begin{array}{cccccc|c} \frac{3}{4} & 0 & 1 & \frac{5}{4} & -\frac{1}{4} & 0 & \frac{165}{2} \\ \frac{2}{5} & 1 & \frac{1}{5} & 0 & \frac{1}{5} & 0 & 24 \\ -\frac{1}{5} & 0 & -\frac{3}{5} & 0 & -\frac{3}{5} & 1 & 8 \\ \hline \frac{11}{5} & 0 & -\frac{2}{5} & 0 & \frac{8}{5} & 0 & 192 \end{array}\right]\begin{array}{l} \frac{5}{4}R_1 \\ \\ \\ \\ \end{array}$$

Copyright © 2019 Pearson Education, Inc.

$$\begin{array}{c} \begin{matrix} x_1 & x_2 & x_3 & s_1 & s_2 & s_3 & \end{matrix} \\ \left[\begin{array}{cccccc|c} \frac{3}{4} & 0 & 1 & \frac{5}{4} & -\frac{1}{4} & 0 & \frac{165}{2} \\ \frac{1}{4} & 1 & 0 & -\frac{1}{4} & \frac{1}{4} & 0 & \frac{15}{2} \\ \frac{1}{4} & 0 & 0 & \frac{3}{4} & -\frac{3}{4} & 1 & \frac{115}{2} \\ \hline \frac{5}{2} & 0 & 0 & \frac{1}{2} & \frac{3}{2} & 0 & 225 \end{array}\right] \end{array} \begin{array}{l} \\ -\frac{1}{5}R_1 + R_2 \\ \frac{3}{5}R_1 + R_3 \\ \frac{2}{5}R_1 + R_4 \end{array}$$

The maximum is 225 when $x_1 = 0$, $x_2 = \frac{15}{2}$, and $x_3 = \frac{165}{2}$.

32. From exercise 22, we have

$$\begin{array}{c} \begin{matrix} x_1 & x_2 & s_1 & s_2 & s_3 & s_4 & \end{matrix} \\ \left[\begin{array}{cccccc|c} 2 & 5 & 1 & 0 & 0 & 0 & 50 \\ 1 & 3 & 0 & 1 & 0 & 0 & 25 \\ \boxed{4} & 1 & 0 & 0 & 1 & 0 & 18 \\ 1 & 1 & 0 & 0 & 0 & 1 & 12 \\ \hline -15 & -12 & 0 & 0 & 0 & 0 & 0 \end{array}\right] \end{array}.$$

Pivot on the 4 in row 3 column 1. First divide row 3 by 4, then continue the process.

$$\begin{array}{c} \begin{matrix} x_1 & x_2 & s_1 & s_2 & s_3 & s_4 & \end{matrix} \\ \left[\begin{array}{cccccc|c} 2 & 5 & 1 & 0 & 0 & 0 & 50 \\ 1 & 3 & 0 & 1 & 0 & 0 & 25 \\ 1 & \frac{1}{4} & 0 & 0 & \frac{1}{4} & 0 & \frac{9}{2} \\ 1 & 1 & 0 & 0 & 0 & 1 & 12 \\ \hline -15 & -12 & 0 & 0 & 0 & 0 & 0 \end{array}\right] \end{array} \begin{array}{l} \\ \\ \\ \frac{1}{4}R_3 \\ \\ \\ \end{array}$$

$$\begin{array}{c} \begin{matrix} x_1 & x_2 & s_1 & s_2 & s_3 & s_4 & \end{matrix} \\ \left[\begin{array}{cccccc|c} 0 & \frac{9}{2} & 1 & 0 & -\frac{1}{2} & 0 & 41 \\ 0 & \boxed{\frac{11}{4}} & 0 & 1 & -\frac{1}{4} & 0 & \frac{41}{2} \\ 1 & \frac{1}{4} & 0 & 0 & \frac{1}{4} & 0 & \frac{9}{2} \\ 0 & \frac{3}{4} & 0 & 0 & -\frac{1}{4} & 1 & \frac{15}{2} \\ \hline 0 & -\frac{33}{4} & 0 & 0 & \frac{15}{4} & 0 & \frac{135}{2} \end{array}\right] \end{array} \begin{array}{l} -2R_3 + R_1 \\ -R_3 + R_2 \\ \\ -R_3 + R_4 \\ 15R_3 + R_5 \end{array}$$

Now pivot on the $\frac{11}{4}$ in row 2 column 2. First multiply row 2 by $\frac{4}{11}$, then continue the process.

$$\begin{array}{c} \begin{matrix} x_1 & x_2 & s_1 & s_2 & s_3 & s_4 & \end{matrix} \\ \left[\begin{array}{cccccc|c} 0 & \frac{9}{2} & 1 & 0 & -\frac{1}{2} & 0 & 41 \\ 0 & 1 & 0 & \frac{4}{11} & -\frac{1}{11} & 0 & \frac{82}{11} \\ 1 & \frac{1}{4} & 0 & 0 & \frac{1}{4} & 0 & \frac{9}{2} \\ 0 & \frac{3}{4} & 0 & 0 & -\frac{1}{4} & 1 & \frac{15}{2} \\ \hline 0 & -\frac{33}{4} & 0 & 0 & \frac{15}{4} & 0 & \frac{135}{2} \end{array}\right] \end{array} \begin{array}{l} \\ \frac{11}{4}R_2 \\ \\ \\ \\ \end{array}$$

$$\begin{array}{c} \begin{matrix} x_1 & x_2 & s_1 & s_2 & s_3 & s_4 & \end{matrix} \\ \left[\begin{array}{cccccc|c} 0 & 0 & 1 & -\frac{18}{11} & -\frac{1}{11} & 0 & \frac{82}{11} \\ 0 & 1 & 0 & \frac{4}{11} & -\frac{1}{11} & 0 & \frac{82}{11} \\ 1 & 0 & 0 & -\frac{1}{11} & \frac{3}{11} & 0 & \frac{29}{11} \\ 0 & 0 & 0 & -\frac{3}{11} & -\frac{2}{11} & 1 & \frac{21}{11} \\ \hline 0 & 0 & 0 & 3 & 3 & 0 & 129 \end{array}\right] \end{array} \begin{array}{l} -\frac{9}{2}R_2 + R_1 \\ \\ -\frac{1}{4}R_2 + R_3 \\ -\frac{3}{4}R_2 + R_4 \\ \frac{33}{4}R_2 + R_5 \end{array}$$

The maximum is 129 when $x_1 = \frac{29}{11}$ and $x_2 = \frac{82}{11}$.

33. **a.** Let x_1 = the number of item *A*, let x_2 = the number of item *B*, and x_3 = the number of item *C*.

b. The objective function is $z = 4x_1 + 3x_2 + 3x_3$.

c. The constraints are

$$\begin{aligned} 2x_1 + 3x_2 + 6x_3 &\le 1200 \\ x_1 + 2x_2 + 2x_3 &\le 800 \\ 2x_1 + 2x_2 + 4x_3 &\le 500 \\ x_1 \ge 0,\ x_2 \ge 0,\ x_3 &\ge 0. \end{aligned}$$

34. **a.** Let x_1 = the amount invested in oil leases, x_2 = the amount in bonds, and x_3 = the amount invested in stock.

b. We want to maximize the objective function $z = .15x_1 + .09x_2 + .05x_3$.

c.

$$\begin{aligned} x_1 + x_2 + x_3 &\le 50{,}000 \\ x_1 + x_2 \qquad &\le 15{,}000 \\ x_1 \qquad + x_3 &\le 25{,}000 \\ x_1 \ge 0,\ x_2 \ge 0,\ x_3 &\ge 0 \end{aligned}$$

35. **a.** Let x_1 = the number of gallons of Fruity wine, and x_2 = the number of gallons of Crystal wine to be made.

b. The objective function is $z = 12x_1 + 15x_2$.

c. The ingredients available are the limitations. The constraints are

$$\begin{aligned} 2x_1 + x_2 &\le 110 \\ 2x_1 + 3x_2 &\le 125 \\ 2x_1 + x_2 &\le 90 \\ x_1 \ge 0,\ x_2 &\ge 0. \end{aligned}$$

Copyright © 2019 Pearson Education, Inc.

36. **a.** Let x_1 = the number of 5-gallon bags, x_2 = the number of 10-gallon bags, and x_3 = the number of 20-gallon bags.

b. The objective function is $z = x_1 + .9x_2 + .95x_3$.

c.

$$\begin{aligned} x_1 + 1.1x_2 + 1.5x_3 &\le 8 \\ x_1 + 1.2x_2 + 1.3x_3 &\le 8 \\ 2x_1 + 3x_2 + 4x_3 &\le 8 \\ x_1 \ge 0,\ x_2 \ge 0,\ x_3 &\ge 0 \end{aligned}$$

37. It is necessary to use the simplex method when there are more than two variables.

38. Problem with constraints involving "≤" can be solved with the use of slack variables, constraints involving "≥" can be solved with the use of surplus variables, while constraints involving "=" require both slack and surplus variables.

39. Any standard minimization problem can be solved using the method of duals.

40. **a.** Maximize $z = 6x_1 + 7x_2 + 5x_3$

subject to:
$$\begin{aligned} 4x_1 + 2x_2 + 3x_3 &\le 9 \\ 5x_1 + 4x_2 + x_3 &\le 10 \\ x_1 \ge 0,\ x_2 \ge 0,\ x_3 &\ge 0 \end{aligned}$$

b. The first constraint would become $4x_1 + 2x_2 + 3x_3 \ge 9$

c. $z = 22.7$ when $x_1 = 0$, $x_2 = 2.1$, $x_3 = 1.6$

d. Minimize $w = 9y_1 + 10y_2$

subject to:
$$\begin{aligned} 4y_1 + 5y_2 &\ge 6 \\ 2y_1 + 4y_2 &\ge 7 \\ 3y_1 + y_2 &\ge 5 \\ y_1 \ge 0, y_2 &\ge 0 \end{aligned}$$

e. The minimum value is $w = 22.7$, when $y_1 = 1.3$, and $y_2 = 1.1$.

41. Using the method of duals,

$$\left[\begin{array}{cccccc|c} 1 & 0 & 0 & 3 & 1 & 2 & 12 \\ 0 & 0 & 1 & 4 & 5 & 3 & 5 \\ 0 & 1 & 0 & -2 & 7 & -6 & 8 \\ \hline 0 & 0 & 0 & 5 & 7 & 3 & 172 \end{array}\right]$$

indicates a minimum value of 172 at (5, 7, 3, 0, 0, 0).

42. Using the method of duals,

$$\left[\begin{array}{cccccc|c} 0 & 0 & 1 & 6 & 3 & 1 & 2 \\ 1 & 0 & 0 & 4 & -2 & 2 & 8 \\ 0 & 1 & 0 & 10 & 7 & 0 & 12 \\ \hline 0 & 0 & 0 & 9 & 5 & 8 & 62 \end{array}\right]$$

indicates a minimum of 62 at (9, 5, 8, 0, 0, 0).

43. Using the method of duals,

$$\left[\begin{array}{cccc|c} 1 & 0 & 7 & -1 & 100 \\ 0 & 1 & 1 & 3 & 27 \\ \hline 0 & 0 & 7 & 2 & 640 \end{array}\right]$$

indicates a minimum of 640 at (7, 2, 0, 0).

44. Minimize $w = 5y_1 + 2y_2$

subject to
$$\begin{aligned} 2y_1 + 3y_2 &\ge 6 \\ 2y_1 + y_2 &\ge 7 \\ y_1 \ge 0,\ y_2 &\ge 0 \end{aligned}$$

Rewrite the objective function by forming the augmented matrix of the system and then form its transpose.

$$\left[\begin{array}{cc|c} 2 & 3 & 6 \\ 2 & 1 & 7 \\ \hline 5 & 2 & 0 \end{array}\right] \to \left[\begin{array}{cc|c} 2 & 2 & 5 \\ 3 & 1 & 2 \\ \hline 6 & 7 & 0 \end{array}\right].$$

This gives the problem

Maximize $z = 6x_1 + 7x_2$

subject to
$$\begin{aligned} 2x_1 + 2x_2 &\le 5 \\ 3x_1 + x_2 &\le 2 \\ x_1 \ge 0,\ x_2 &\ge 0 \end{aligned}$$

The initial simplex tableau is

$$\begin{array}{c} \begin{array}{cccc} x_1 & x_2 & s_1 & s_2 \end{array} \\ \left[\begin{array}{cccc|c} 2 & 2 & 1 & 0 & 5 \\ 3 & \boxed{1} & 0 & 1 & 2 \\ \hline -6 & -7 & 0 & 0 & 0 \end{array}\right] \end{array}.$$

Pivot on the 1 in row 2 column 2.

$$\begin{array}{c} \begin{array}{cccc} x_1 & x_2 & s_1 & s_2 \end{array} \\ \left[\begin{array}{cccc|c} -4 & 0 & 1 & -2 & 1 \\ 3 & 1 & 0 & 1 & 2 \\ \hline 15 & 0 & 0 & 7 & 14 \end{array}\right] \begin{array}{l} -2R_2 + R_1 \\ \\ 7R_2 + R_3 \end{array} \\ \begin{array}{ccc} \uparrow & \uparrow & \uparrow \\ y_1 & y_2 & w \end{array} \end{array}$$

The minimum is 14 when $y_1 = 0$ and $y_2 = 7$.

Copyright © 2019 Pearson Education, Inc.

45. Minimize $w = 18y_1 + 10y_2$

subject to: $y_1 + y_2 \ge 17$
$5y_1 + 8y_2 \ge 42$
$y_1 \ge 0,\ y_2 \ge 0.$

Rewrite the objective function by forming the augmented matrix of the system and then form its transpose.

$$\left[\begin{array}{cc|c} 1 & 1 & 17 \\ 5 & 8 & 42 \\ \hline 18 & 10 & 0 \end{array}\right] \to \left[\begin{array}{cc|c} 1 & 5 & 18 \\ 1 & 8 & 10 \\ \hline 17 & 42 & 0 \end{array}\right].$$

This gives the problem
Maximize $z = 17x_1 + 42x_2$
subject to $x_1 + 5x_2 \le 18$.
$x_1 + 8x_2 \le 10$
$x_1 \ge 0,\ x_2 \ge 0$

The initial simplex tableau is

$$\begin{array}{c} \begin{array}{cccc} x_1 & x_2 & s_1 & s_2 \end{array} \\ \left[\begin{array}{cccc|c} 1 & 5 & 1 & 0 & 18 \\ 1 & 8 & 0 & 1 & 10 \\ \hline -17 & -42 & 0 & 0 & 0 \end{array}\right] \end{array}.$$

Pivot on the 8 in row 2 column 2.

$$\begin{array}{c} \begin{array}{cccc} x_1 & x_2 & s_1 & s_2 \end{array} \\ \left[\begin{array}{cccc|c} 1 & 5 & 1 & 0 & 18 \\ \frac{1}{8} & 1 & 0 & \frac{1}{8} & \frac{5}{4} \\ \hline -17 & -42 & 0 & 0 & 0 \end{array}\right] \begin{array}{l} \\ \frac{1}{8}R_2 \\ \\ \end{array} \end{array}$$

$$\begin{array}{c} \begin{array}{cccc} x_1 & x_2 & s_1 & s_2 \end{array} \\ \left[\begin{array}{cccc|c} \frac{3}{8} & 0 & 1 & -\frac{5}{8} & \frac{47}{4} \\ \frac{1}{8} & 1 & 0 & \frac{1}{8} & \frac{5}{4} \\ \hline -\frac{47}{4} & 0 & 0 & \frac{21}{4} & \frac{105}{2} \end{array}\right] \begin{array}{l} -5R_2 + R_1 \\ \\ 42R_2 + R_3 \end{array} \end{array}$$

Now pivot on the $\frac{1}{8}$ in row 2 column 1.

$$\begin{array}{c} \begin{array}{cccc} x_1 & x_2 & s_1 & s_2 \end{array} \\ \left[\begin{array}{cccc|c} 0 & -3 & 1 & -1 & 8 \\ \frac{1}{8} & 1 & 0 & \frac{1}{8} & \frac{5}{4} \\ \hline 0 & 94 & 0 & 17 & 170 \end{array}\right] \begin{array}{l} -3R_2 + R_1 \\ \\ 94R_2 + R_3 \end{array} \\ \begin{array}{ccc} \uparrow & \uparrow & \uparrow \\ y_1 & y_2 & w \end{array} \end{array}$$

The minimum is 170 when $y_1 = 0$ and $y_2 = 17$.

46. Minimize $w = 4y_1 + 5y_2$

subject to $10y_1 + 5y_2 \ge 100$
$20y_1 + 10y_2 \ge 150$
$y_1 \ge 0,\ y_2 \ge 0$

Rewrite the objective function by forming the augmented matrix of the system and then form its transpose.

$$\left[\begin{array}{cc|c} 10 & 5 & 100 \\ 20 & 10 & 150 \\ \hline 4 & 5 & 0 \end{array}\right] \to \left[\begin{array}{cc|c} 10 & 20 & 4 \\ 5 & 10 & 5 \\ \hline 100 & 150 & 0 \end{array}\right].$$

This gives the problem
Maximize $z = 100x_1 + 150x_2$
subject to $10x_1 + 20x_2 \le 4$.
$5x_1 + 10x_2 \le 5$
$x_1 \ge 0,\ x_2 \ge 0$

The initial simplex tableau is

$$\begin{array}{c} \begin{array}{cccc} x_1 & x_2 & s_1 & s_2 \end{array} \\ \left[\begin{array}{cccc|c} 10 & 20 & 1 & 0 & 4 \\ 5 & 10 & 0 & 1 & 5 \\ \hline -100 & -150 & 0 & 0 & 0 \end{array}\right] \end{array}.$$

Pivot on the 20 in row 1 column 2.

$$\begin{array}{c} \begin{array}{cccc} x_1 & x_2 & s_1 & s_2 \end{array} \\ \left[\begin{array}{cccc|c} \frac{1}{2} & 1 & \frac{1}{20} & 0 & \frac{1}{5} \\ 5 & 10 & 0 & 1 & 5 \\ \hline -100 & -150 & 0 & 0 & 0 \end{array}\right] \begin{array}{l} \frac{1}{20}R_1 \\ \\ \\ \end{array} \end{array}$$

$$\begin{array}{c} \begin{array}{cccc} x_1 & x_2 & s_1 & s_2 \end{array} \\ \left[\begin{array}{cccc|c} \frac{1}{2} & 1 & \frac{1}{20} & 0 & \frac{1}{5} \\ 0 & 0 & -\frac{1}{2} & 1 & 3 \\ \hline -25 & 0 & \frac{15}{2} & 0 & 30 \end{array}\right] \begin{array}{l} \\ -10R_1 + R_2 \\ 150R_1 + R_3 \end{array} \end{array}$$

Now pivot on the $\frac{1}{2}$ in row 1 column 1.

$$\begin{array}{c} \begin{array}{cccc} x_1 & x_2 & s_1 & s_2 \end{array} \\ \left[\begin{array}{cccc|c} 1 & 2 & \frac{1}{10} & 0 & \frac{2}{5} \\ 0 & 0 & -\frac{1}{2} & 1 & 3 \\ \hline -25 & 0 & \frac{15}{2} & 0 & 30 \end{array}\right] \begin{array}{l} 2R_1 \\ \\ \\ \end{array} \end{array}$$

$$\begin{array}{c} \begin{array}{cccc} x_1 & x_2 & s_1 & s_2 \end{array} \\ \left[\begin{array}{cccc|c} 1 & 2 & \frac{1}{10} & 0 & \frac{2}{5} \\ 0 & 0 & -\frac{1}{2} & 1 & 3 \\ \hline 0 & 50 & 10 & 0 & 40 \end{array}\right] \begin{array}{l} 2R_1 \\ \\ 25R_1 + R_3 \end{array} \\ \begin{array}{ccc} \uparrow & \uparrow & \uparrow \\ y_1 & y_2 & w \end{array} \end{array}$$

The minimum is 40 when $y_1 = 10$ and $y_2 = 0$.

Copyright © 2019 Pearson Education, Inc.

47. Maximize $z = 20x_1 + 30x_2$
subject to $5x_1 + 10x_2 \le 120$
$10x_1 + 15x_2 \ge 200$
$x_1 \ge 0,\ x_2 \ge 0$

The initial simplex tableau is

$$\begin{array}{c} \begin{matrix} x_1 & x_2 & s_1 & s_2 \end{matrix} \\ \left[\begin{array}{cccc|c} 5 & 10 & 1 & 0 & 120 \\ 10 & 15 & 0 & -1 & 200 \\ \hline -20 & -30 & 0 & 0 & 0 \end{array}\right]. \end{array}$$

48. Minimize $w = 4y_1 + 2y_2$
subject to $y_1 + 3y_2 \ge 6$
$2y_1 + 8y_2 \le 21$
$y_1 \ge 0,\ y_2 \ge 0$

The initial simplex tableau is

$$\begin{array}{c} \begin{matrix} y_1 & y_2 & s_1 & s_2 \end{matrix} \\ \left[\begin{array}{cccc|c} 1 & 3 & -1 & 0 & 6 \\ 2 & 8 & 0 & 1 & 21 \\ \hline 4 & 2 & 0 & 0 & 0 \end{array}\right]. \end{array}$$

49. Minimize $w = 12y_1 + 20y_2 - 8y_3$
subject to: $y_1 + y_2 + 2y_3 \ge 48$
$y_1 + y_2 \le 12$
$y_3 \ge 10$
$3y_1 + y_3 \ge 30$
$y_1 \ge 0,\ y_2 \ge 0, y_3 \ge 0.$

The initial simplex tableau is

$$\begin{array}{c} \begin{matrix} y_1 & y_2 & y_3 & s_1 & s_2 & s_3 & s_4 \end{matrix} \\ \left[\begin{array}{ccccccc|c} 1 & 1 & 2 & -1 & 0 & 0 & 0 & 48 \\ 1 & 1 & 0 & 0 & 1 & 0 & 0 & 12 \\ 0 & 0 & 1 & 0 & 0 & -1 & 0 & 10 \\ 3 & 0 & 1 & 0 & 0 & 0 & -1 & 30 \\ \hline 12 & 20 & -8 & 0 & 0 & 0 & 0 & 0 \end{array}\right]. \end{array}$$

50. Maximize $w = 6x_1 - 3x_2 + 4x_3$
subject to: $2x_1 + x_2 + x_3 \le 112$
$x_1 + x_2 + x_3 \ge 80$
$x_1 + x_2 \le 45$
$x_1 \ge 0,\ x_2 \ge 0,\ x_3 \ge 0.$

The initial simplex tableau is

$$\begin{array}{c} \begin{matrix} x_1 & x_2 & x_3 & s_1 & s_2 & s_3 \end{matrix} \\ \left[\begin{array}{cccccc|c} 2 & 1 & 1 & 1 & 0 & 0 & 112 \\ 1 & 1 & 1 & 0 & -1 & 0 & 80 \\ 1 & 1 & 0 & 0 & 0 & 1 & 45 \\ \hline -6 & 3 & -4 & 0 & 0 & 0 & 0 \end{array}\right]. \end{array}$$

51. If $w = -z$

$$\left[\begin{array}{cccccc|c} 0 & 1 & 0 & 2 & 5 & 0 & 17 \\ 0 & 0 & 1 & 3 & 1 & 1 & 25 \\ 1 & 0 & 0 & 4 & 2 & \frac{1}{2} & 8 \\ \hline 0 & 0 & 0 & 2 & 5 & 0 & -427 \end{array}\right]$$

indicates a minimum value of 427 at (8, 17, 25, 0 ,0, 0).

52.

$$\left[\begin{array}{ccccccc|c} 0 & 0 & 2 & 1 & 0 & 6 & 6 & 92 \\ 1 & 0 & 3 & 0 & 0 & 0 & 2 & 47 \\ 0 & 1 & 0 & 0 & 0 & 1 & 0 & 68 \\ 0 & 0 & 4 & 0 & 1 & 0 & 3 & 35 \\ \hline 0 & 0 & 5 & 0 & 0 & 2 & 9 & -1957 \end{array}\right]$$

The minimum value is 1957 at (47, 68, 0, 92, 35, 0, 0).

53. Maximize $z = 20x_1 + 30x_2$
subject to $5x_1 + 10x_2 \le 120$
$10x_1 + 15x_2 \ge 200$
$x_1 \ge 0,\ x_2 \ge 0$

The initial simplex tableau is

$$\begin{array}{c} \begin{matrix} x_1 & x_2 & s_1 & s_2 \end{matrix} \\ \left[\begin{array}{cccc|c} 5 & 10 & 1 & 0 & 120 \\ \boxed{10} & 15 & 0 & -1 & 200 \\ \hline -20 & -30 & 0 & 0 & 0 \end{array}\right]. \end{array}$$

Pivot on the 10 in row 2 column 1.

$$\begin{array}{c} \begin{matrix} x_1 & x_2 & s_1 & s_2 \end{matrix} \\ \left[\begin{array}{cccc|c} 0 & 5 & 2 & 1 & 40 \\ 10 & 15 & 0 & -1 & 200 \\ \hline 0 & 0 & 0 & -2 & 400 \end{array}\right] \begin{matrix} -R_2 + 2R_1 \\ \\ 2R_2 + R_3 \end{matrix} \end{array}$$

This completes Stage I because the solution given in the usual way from the matrix has nonnegative values for all variables. Since there is a negative indicator in the objective row, we continue with Stage II. Pivot on the 1 in row 1 column 4.

$$\begin{array}{c} \begin{matrix} x_1 & x_2 & s_1 & s_2 \end{matrix} \\ \left[\begin{array}{cccc|c} 0 & 5 & 2 & \boxed{1} & 40 \\ 10 & 20 & 2 & 0 & 240 \\ \hline 0 & 10 & 4 & 0 & 480 \end{array}\right] \begin{matrix} \\ \\ 2R_1 + R_3 \end{matrix} \end{array}$$

$$\begin{array}{c} \begin{matrix} x_1 & x_2 & s_1 & s_2 \end{matrix} \\ \left[\begin{array}{cccc|c} 0 & 5 & 2 & 1 & 40 \\ 1 & 2 & \frac{1}{5} & 0 & 24 \\ \hline 0 & 10 & 4 & 0 & 480 \end{array}\right] \begin{matrix} \\ \frac{1}{10}R_2 \\ \\ \end{matrix} \end{array}$$

The maximum is 480 when $x_1 = 24$ and $x_2 = 0$.

Copyright © 2019 Pearson Education, Inc.

54. Minimize $w = 4y_1 + 2y_2$
subject to
$$y_1 + 3y_2 \ge 6$$
$$2y_1 + 8y_2 \le 21$$
$$y_1 \ge 0,\ y_2 \ge 0$$
The initial simplex tableau is

$$\begin{array}{c} \begin{matrix} y_1 & y_2 & s_1 & s_2 & \end{matrix} \\ \left[\begin{array}{cccc|c} \boxed{1} & 3 & -1 & 0 & 6 \\ 2 & 8 & 0 & 1 & 21 \\ \hline 4 & 2 & 0 & 0 & 0 \end{array}\right] \end{array}.$$

Pivot on the 1 in row 1 column 1.

$$\begin{array}{c} \begin{matrix} y_1 & y_2 & s_1 & s_2 & \end{matrix} \\ \left[\begin{array}{cccc|c} 1 & 3 & -1 & 0 & 6 \\ 0 & 2 & 2 & 1 & 9 \\ \hline 0 & -10 & 4 & 0 & -24 \end{array}\right] \end{array} \begin{array}{l} \\ \\ -2R_1 + R_2 \\ -4R_1 + R_3 \end{array}$$

This completes Stage I because the solution given in the usual way from the matrix has nonnegative values for all variables. Since there is a negative indicator in the objective row, we continue with Stage II. Pivot on the 3 in row 1 column 2.

$$\begin{array}{c} \begin{matrix} y_1 & y_2 & s_1 & s_2 & \end{matrix} \\ \left[\begin{array}{cccc|c} \frac{1}{3} & 1 & -\frac{1}{3} & 0 & 2 \\ 0 & 2 & 2 & 1 & 9 \\ \hline 0 & -10 & 4 & 0 & -24 \end{array}\right] \end{array} \begin{array}{l} \\ \frac{1}{3}R_1 \\ \\ \\ \end{array}$$

$$\begin{array}{c} \begin{matrix} y_1 & y_2 & s_1 & s_2 & \end{matrix} \\ \left[\begin{array}{cccc|c} \frac{1}{3} & 1 & -\frac{1}{3} & 0 & 2 \\ -\frac{2}{3} & 0 & \frac{8}{3} & 1 & 5 \\ \hline \frac{10}{3} & 0 & \frac{2}{3} & 0 & -4 \end{array}\right] \end{array} \begin{array}{l} \\ \\ -2R_1 + R_2 \\ 10R_1 + R_3 \end{array}$$

The minimum is 4 when $y_1 = 0$ and $y_2 = 2$.

55. Minimize $w = 4y_1 - 8y_2$
subject to:
$$y_1 + y_2 \le 50$$
$$2y_1 - 4y_2 \ge 20$$
$$y_1 - y_2 \le 22$$
$$y_1 \ge 0,\ y_2 \ge 0.$$
The initial simplex tableau is

$$\begin{array}{c} \begin{matrix} y_1 & y_2 & s_1 & s_2 & s_3 & \end{matrix} \\ \left[\begin{array}{ccccc|c} 1 & 1 & 1 & 0 & 0 & 50 \\ \boxed{2} & -4 & 0 & -1 & 0 & 20 \\ 1 & -1 & 0 & 0 & 1 & 22 \\ \hline 4 & -8 & 0 & 0 & 0 & 0 \end{array}\right] \end{array}.$$

For Stage I, pivot on the 2 in row two, column one.

$$\begin{array}{c} \begin{matrix} y_1 & y_2 & s_1 & s_2 & s_3 & \end{matrix} \\ \left[\begin{array}{ccccc|c} 1 & 1 & 1 & 0 & 0 & 50 \\ 1 & -2 & 0 & -\frac{1}{2} & 0 & 10 \\ 1 & -1 & 0 & 0 & 1 & 22 \\ \hline 4 & -8 & 0 & 0 & 0 & 0 \end{array}\right] \end{array} \begin{array}{l} \\ \\ \frac{1}{2}R_2 \\ \\ \\ \end{array}$$

$$\begin{array}{c} \begin{matrix} y_1 & y_2 & s_1 & s_2 & s_3 & \end{matrix} \\ \left[\begin{array}{ccccc|c} 0 & 3 & 1 & \frac{1}{2} & 0 & 40 \\ 1 & -2 & 0 & -\frac{1}{2} & 0 & 10 \\ 0 & 1 & 0 & \frac{1}{2} & 1 & 12 \\ \hline 0 & 0 & 0 & 2 & 0 & -40 \end{array}\right] \end{array} \begin{array}{l} \\ -R_2 + R_1 \\ \\ -R_2 + R_3 \\ -4R_2 + R_4 \end{array}$$

The minimum is 40 when $y_1 = 10$ and $y_2 = 0$.

56. Maximize $z = 2x_1 + 4x_2$
subject to:
$$3x_1 + 2x_2 \le 12$$
$$5x_1 + x_2 \ge 5$$
$$x_1 \ge 0,\ x_2 \ge 0.$$
The initial simplex tableau is

$$\begin{array}{c} \begin{matrix} x_1 & x_2 & s_1 & s_2 & \end{matrix} \\ \left[\begin{array}{cccc|c} 3 & 2 & 1 & 0 & 12 \\ 5 & \boxed{1} & 0 & -1 & 5 \\ \hline -2 & -4 & 0 & 0 & 0 \end{array}\right] \end{array}.$$

For Stage I, pivot on the 1 in row 2 column 1.

$$\begin{array}{c} \begin{matrix} x_1 & x_2 & s_1 & s_2 & \end{matrix} \\ \left[\begin{array}{cccc|c} -7 & 0 & 1 & \boxed{2} & 2 \\ 5 & 1 & 0 & -1 & 5 \\ \hline 18 & 0 & 0 & -4 & 20 \end{array}\right] \end{array} \begin{array}{l} \\ -2R_2 + R_1 \\ \\ 4R_2 + R_3 \end{array}$$

For Stage II, pivot on the 2 in row 1 column 4.

$$\begin{array}{c} \begin{matrix} x_1 & x_2 & s_1 & s_2 & \end{matrix} \\ \left[\begin{array}{cccc|c} -\frac{7}{2} & 0 & \frac{1}{2} & 1 & 1 \\ \frac{3}{2} & 1 & \frac{1}{2} & 0 & 6 \\ \hline 4 & 0 & 2 & 0 & 24 \end{array}\right] \end{array} \begin{array}{l} \\ \frac{1}{2}R_1 \\ R_1 + R_2 \\ 4R_1 + R_3 \end{array}$$

The maximum is 24 when $x_1 = 0$ and $x_2 = 6$.

Copyright © 2019 Pearson Education, Inc.

57. From Exercise 33, the initial simplex tableau is

$$\begin{array}{c} \\ \\ \\ \\ \\ \end{array}\begin{array}{cccccc|c} x_1 & x_2 & x_3 & s_1 & s_2 & s_3 & \\ \hline 2 & 3 & 6 & 1 & 0 & 0 & 1200 \\ 1 & 2 & 2 & 0 & 1 & 0 & 800 \\ \boxed{2} & 2 & 4 & 0 & 0 & 1 & 500 \\ \hline -4 & -3 & -3 & 0 & 0 & 0 & 0 \end{array}.$$

Pivot on the 2 in row three, column one.

$$\begin{array}{cccccc|cl} x_1 & x_2 & x_3 & s_1 & s_2 & s_3 & & \\ \hline 2 & 3 & 6 & 1 & 0 & 0 & 1200 & \\ 1 & 2 & 2 & 0 & 1 & 0 & 800 & \\ 1 & 1 & 2 & 0 & 0 & \frac{1}{2} & 250 & \frac{1}{2}R_3 \\ \hline -4 & -3 & -3 & 0 & 0 & 0 & 0 & \end{array}$$

$$\begin{array}{cccccc|cl} x_1 & x_2 & x_3 & s_1 & s_2 & s_3 & & \\ \hline 0 & 1 & 2 & 1 & 0 & -1 & 700 & -2R_3 + R_1 \\ 0 & 1 & 0 & 0 & 1 & -\frac{1}{2} & 550 & -1R_3 + R_2 \\ 1 & 1 & 2 & 0 & 0 & \frac{1}{2} & 250 & \\ \hline 0 & 1 & 5 & 0 & 0 & 2 & 1000 & 4R_3 + R_4 \end{array}$$

Roberta should get 250 units of item A, none of item B, and none of item C for a maximum profit of $1000.

58. From Exercise 34, we have the initial simplex tableau (after multiplying the fourth row by 100 to clear the decimals):

$$\begin{array}{cccccc|c} x_1 & x_2 & x_3 & s_1 & s_2 & s_3 & \\ \hline 1 & 1 & 1 & 1 & 0 & 0 & 50{,}000 \\ \boxed{1} & 1 & 0 & 0 & 1 & 0 & 15{,}000 \\ 1 & 0 & 1 & 0 & 0 & 1 & 25{,}000 \\ \hline -15 & -9 & -5 & 0 & 0 & 0 & 0 \end{array}.$$

$$\begin{array}{cccccc|cl} x_1 & x_2 & x_3 & s_1 & s_2 & s_3 & & \\ \hline 0 & 0 & 1 & 1 & -1 & 0 & 35{,}000 & -1R_2 + R_1 \\ 1 & 1 & 0 & 0 & 1 & 0 & 15{,}000 & \\ 0 & -1 & \boxed{1} & 0 & -1 & 1 & 10{,}000 & -1R_2 + R_3 \\ \hline 0 & 6 & -5 & 0 & 15 & 0 & 225{,}000 & 15R_2 + R_4 \end{array}$$

$$\begin{array}{cccccc|cl} x_1 & x_2 & x_3 & s_1 & s_2 & s_3 & & \\ \hline 0 & 1 & 0 & 1 & 0 & -1 & 25{,}000 & -1R_3 + R_1 \\ 1 & 1 & 0 & 0 & 1 & 0 & 15{,}000 & \\ 0 & -1 & 1 & 0 & -1 & 1 & 10{,}000 & \\ \hline 0 & 1 & 0 & 0 & 10 & 5 & 275{,}000 & 5R_3 + R_4 \end{array}$$

Since $x_1 = 15{,}000$ and $x_3 = 10{,}000$, he should invest $15,000 in oil leases and $10,000 in stock for a maximum return of .01(275,000) = $2750.

59. From Exercise 35, we have the initial simplex tableau:

$$\begin{array}{ccccc|c} x_1 & x_2 & s_1 & s_2 & s_3 & \\ \hline 2 & 1 & 1 & 0 & 0 & 110 \\ 2 & \boxed{3} & 0 & 1 & 0 & 125 \\ 2 & 1 & 0 & 0 & 1 & 90 \\ \hline -12 & -15 & 0 & 0 & 0 & 0 \end{array}$$

$$\begin{array}{ccccc|cl} x_1 & x_2 & s_1 & s_2 & s_3 & & \\ \hline 2 & 1 & 1 & 0 & 0 & 110 & \\ \frac{2}{3} & \boxed{1} & 0 & \frac{1}{3} & 0 & \frac{125}{3} & \frac{1}{3}R_2 \\ 2 & 1 & 0 & 0 & 1 & 90 & \\ \hline -12 & -15 & 0 & 0 & 0 & 0 & \end{array}$$

$$\begin{array}{ccccc|cl} x_1 & x_2 & s_1 & s_2 & s_3 & & \\ \hline \frac{4}{3} & 0 & 1 & -\frac{1}{3} & 0 & \frac{205}{3} & -R_2 + R_1 \\ \frac{2}{3} & 1 & 0 & \frac{1}{3} & 0 & \frac{125}{3} & \\ \boxed{\frac{4}{3}} & 0 & 0 & -\frac{1}{3} & 1 & \frac{145}{3} & -R_2 + R_3 \\ \hline -2 & 0 & 0 & 5 & 0 & 625 & 15R_2 + R_4 \end{array}$$

$$\begin{array}{ccccc|cl} x_1 & x_2 & s_1 & s_2 & s_3 & & \\ \hline \frac{4}{3} & 0 & 1 & -\frac{1}{3} & 0 & \frac{205}{3} & \\ \frac{2}{3} & 1 & 0 & \frac{1}{3} & 0 & \frac{125}{3} & \\ \boxed{1} & 0 & 0 & -\frac{1}{4} & \frac{3}{4} & \frac{145}{4} & \frac{4}{3}R_3 \\ \hline -2 & 0 & 0 & 5 & 0 & 625 & \end{array}$$

$$\begin{array}{ccccc|cl} x_1 & x_2 & s_1 & s_2 & s_3 & & \\ \hline 0 & 0 & 1 & 0 & -1 & 20 & -\frac{4}{3}R_3 + R_1 \\ 0 & 1 & 0 & \frac{1}{2} & -\frac{1}{2} & \frac{35}{2} & -\frac{2}{3}R_3 + R_2 \\ 1 & 0 & 0 & -\frac{1}{4} & \frac{3}{4} & \frac{145}{4} & \\ \hline 0 & 0 & 0 & \frac{9}{2} & \frac{3}{2} & \frac{1395}{2} & 2R_3 + R_4 \end{array}$$

The winery should make 17.5 gallons of Crystal wine and 36.25 gallons of Fruity wine for a maximum profit of $697.50.

Copyright © 2019 Pearson Education, Inc.

60. From Exercise 36, we have the initial tableau:

$$\begin{array}{cccccc} x_1 & x_2 & x_3 & s_1 & s_2 & s_3 \end{array}$$
$$\left[\begin{array}{cccccc|c} 1 & 1.1 & 1.5 & 1 & 0 & 0 & 8 \\ 1 & 1.2 & 1.3 & 0 & 1 & 0 & 8 \\ \boxed{2} & 3 & 4 & 0 & 0 & 1 & 8 \\ \hline -1 & -.9 & -.95 & 0 & 0 & 0 & 0 \end{array}\right].$$

$$\begin{array}{cccccc} x_1 & x_2 & x_3 & s_1 & s_2 & s_3 \end{array}$$
$$\left[\begin{array}{cccccc|c} 1 & 1.1 & 1.5 & 1 & 0 & 0 & 8 \\ 1 & 1.2 & 1.3 & 0 & 1 & 0 & 8 \\ \boxed{1} & 1.5 & 2 & 0 & 0 & .5 & 4 \\ \hline -1 & -.9 & -.95 & 0 & 0 & 0 & 0 \end{array}\right]\begin{array}{l} \\ \\ \frac{1}{2}R_3 \\ \\ \end{array}$$

$$\begin{array}{cccccc} x_1 & x_2 & x_3 & s_1 & s_2 & s_3 \end{array}$$
$$\left[\begin{array}{cccccc|c} 0 & -.4 & -.5 & 1 & 0 & -.5 & 4 \\ 0 & -.3 & -.7 & 0 & 1 & -.5 & 4 \\ 1 & 1.5 & 2 & 0 & 0 & .5 & 4 \\ \hline 0 & .6 & 1.05 & 0 & 0 & .5 & 4 \end{array}\right]\begin{array}{l} -1R_3 + R_1 \\ -1R_3 + R_2 \\ \\ R_3 + R_4 \end{array}$$

The final tableau gives $x_1 = 4$, $x_2 = 0$, $x_3 = 0$, and $z = 4$. Therefore, 4 units of 5-gallon bags (and none of the others) should be made for a maximum profit of $4 per unit.

61. Let y_1 = the number of cases of corn
y_2 = the number of cases of beans
y_3 = the number of cases of carrots

Minimize $w = 10y_1 + 15y_2 + 25y_3$

Maximize $z = -w = -10y_1 - 15y_2 - 25y_3$

subject to:
$$\begin{aligned} y_1 + y_2 + y_3 &\geq 1000 \\ y_1 - 2y_2 &\geq 0 \\ y_3 &\geq 340 \\ y_1 \geq 0,\ y_2 \geq 0,\ y_3 &\geq 0 \end{aligned}$$

The initial simplex tableau is

$$\begin{array}{cccccc} y_1 & y_2 & y_3 & s_1 & s_2 & s_3 \end{array}$$
$$\left[\begin{array}{cccccc|c} 1 & 1 & 1 & -1 & 0 & 0 & 1000 \\ \boxed{1} & -2 & 0 & 0 & -1 & 0 & 0 \\ 0 & 0 & 1 & 0 & 0 & -1 & 340 \\ \hline 10 & 15 & 25 & 0 & 0 & 0 & 0 \end{array}\right]$$

For Stage I, pivot on the 1 in row two, column one.

$$\begin{array}{cccccc} y_1 & y_2 & y_3 & s_1 & s_2 & s_3 \end{array}$$
$$\left[\begin{array}{cccccc|c} 0 & \boxed{3} & 1 & -1 & 1 & 0 & 1000 \\ 1 & -2 & 0 & 0 & -1 & 0 & 0 \\ 0 & 0 & 1 & 0 & 0 & -1 & 340 \\ \hline 0 & 35 & 25 & 0 & 10 & 0 & 0 \end{array}\right]\begin{array}{l} -R_2 + R_1 \\ \\ \\ -10R_2 + R_4 \end{array}$$

Continuing Stage I, pivot on the 3 in row one, column two.

$$\begin{array}{cccccc} y_1 & y_2 & y_3 & s_1 & s_2 & s_3 \end{array}$$
$$\left[\begin{array}{cccccc|c} 0 & 1 & \frac{1}{3} & -\frac{1}{3} & \frac{1}{3} & 0 & \frac{1000}{3} \\ 1 & 0 & \frac{2}{3} & -\frac{2}{3} & -\frac{1}{3} & 0 & \frac{2000}{3} \\ 0 & 0 & \boxed{1} & 0 & 0 & -1 & 340 \\ \hline 0 & 0 & \frac{40}{3} & \frac{35}{3} & -\frac{5}{3} & 0 & -\frac{35,000}{3} \end{array}\right]\begin{array}{l} \frac{1}{3}R_1 \\ 2R_1 + R_2 \\ \\ -35R_1 + R_4 \end{array}$$

Pivot on the 1 in row three, column three.

$$\begin{array}{cccccc} y_1 & y_2 & y_3 & s_1 & s_2 & s_3 \end{array}$$
$$\left[\begin{array}{cccccc|c} 0 & 1 & 0 & -\frac{1}{3} & \boxed{\frac{1}{3}} & \frac{1}{3} & 220 \\ 1 & 0 & 0 & -\frac{2}{3} & -\frac{1}{3} & \frac{2}{3} & 440 \\ 0 & 0 & 1 & 0 & 0 & -1 & 340 \\ \hline 0 & 0 & 0 & \frac{35}{3} & -\frac{5}{3} & \frac{40}{3} & -16,200 \end{array}\right]\begin{array}{l} -\frac{1}{3}R_3 + R_1 \\ -\frac{2}{3}R_3 + R_2 \\ \\ -\frac{40}{3}R_3 + R_4 \end{array}$$

For Stage II, pivot on the $\frac{1}{3}$ in row one, column five.

$$\begin{array}{cccccc} y_1 & y_2 & y_3 & s_1 & s_2 & s_3 \end{array}$$
$$\left[\begin{array}{cccccc|c} 0 & 3 & 0 & -1 & 1 & 1 & 660 \\ 1 & 1 & 0 & -1 & 0 & 1 & 660 \\ 0 & 0 & 1 & 0 & 0 & -1 & 340 \\ \hline 0 & 5 & 0 & 10 & 0 & 15 & -15,100 \end{array}\right]\begin{array}{l} 3R_1 \\ \frac{1}{3}R_1 + R_2 \\ \\ \frac{5}{3}R_1 + R_4 \end{array}$$

The Cauchy Canners should produce 660 cases of corn, no cases of beans, and 340 cases of carrots for a minimum cost of $15,100.

Copyright © 2019 Pearson Education, Inc.

62. First put the data into a table.

	Lumber	Concrete	Advertising	Total Spent
Atlantic	1000	3000	2000	\$3000
Pacific	2000	3000	3000	\$4000
Minimum Use	8000	18000	15000	

Let y_1 = the number of Atlantic boats and let y_2 = the number of Pacific boats.
The problem is to minimize $w = 3000y_1 + 4000y_2$
subject to:

$$\begin{aligned} 1000y_1 + 2000y_2 &\geq 8000 \\ 3000y_1 + 3000y_2 &\geq 18,000 \\ 2000y_1 + 3000y_2 &\geq 15,000 \\ y_1 \geq 0,\ y_2 &\geq 0. \end{aligned}$$

The matrix for this problem is

$$\left[\begin{array}{cc|c} 1000 & 2000 & 8000 \\ 3000 & 3000 & 18,000 \\ 2000 & 3000 & 15,000 \\ \hline 3000 & 4000 & 0 \end{array}\right].$$

The transpose of this matrix is

$$\left[\begin{array}{ccc|c} 1000 & 3000 & 2000 & 3000 \\ 2000 & 3000 & 3000 & 4000 \\ \hline 8000 & 18,000 & 15,000 & 0 \end{array}\right].$$

The dual problem is as follows.
Maximize $z = 8000x_1 + 18,000x_2 + 15,000x_3$
subject to:

$$\begin{aligned} 1000x_1 + 3000x_2 + 2000x_3 &\leq 3000 \\ 2000x_1 + 3000x_2 + 3000x_3 &\leq 4000 \\ x_1 \geq 0,\ x_2 \geq 0,\ x_3 &\geq 0. \end{aligned}$$

The initial simplex tableau is

$$\begin{array}{c} \begin{array}{cccccc} x_1 & x_2 & x_3 & s_1 & s_2 & \end{array} \\ \left[\begin{array}{ccccc|c} 1000 & \boxed{3000} & 2000 & 1 & 0 & 3000 \\ 2000 & 3000 & 3000 & 0 & 1 & 4000 \\ \hline -8000 & -18,000 & -15,000 & 0 & 0 & 0 \end{array}\right]. \end{array}$$

Pivot on the 3000 in row one, column two.

$$\begin{array}{c} \begin{array}{cccccc} x_1 & x_2 & x_3 & s_1 & s_2 & \end{array} \\ \left[\begin{array}{ccccc|c} \frac{1}{3} & \boxed{1} & \frac{2}{3} & \frac{1}{3000} & 0 & 1 \\ 2000 & 3000 & 3000 & 0 & 1 & 4000 \\ \hline -8000 & -18,000 & -15,000 & 0 & 0 & 0 \end{array}\right] \begin{array}{l} \frac{1}{3000}R_1 \\ \\ \\ \end{array} \end{array}$$

$$\begin{array}{c} \begin{array}{cccccc} x_1 & x_2 & x_3 & s_1 & s_2 & \end{array} \\ \left[\begin{array}{ccccc|c} \frac{1}{3} & 1 & \frac{2}{3} & \frac{1}{3000} & 0 & 1 \\ 1000 & 0 & \boxed{1000} & -1 & 1 & 1000 \\ \hline -2000 & 0 & -3000 & 6 & 0 & 18,000 \end{array}\right] \begin{array}{l} \\ -3000R_1 + R_2 \\ 18,000R_1 + R_3 \end{array} \end{array}$$

Copyright © 2019 Pearson Education, Inc.

Pivot on the 1000 in row two, column three.

$$\begin{array}{ccccc} x_1 & x_2 & x_3 & x_4 & x_5 \end{array}$$
$$\left[\begin{array}{ccccc|c} \frac{1}{3} & 1 & \frac{2}{3} & \frac{1}{3000} & 0 & 1 \\ 1 & 0 & 1 & -\frac{1}{1000} & \frac{1}{1000} & 1 \\ \hline -2000 & 0 & -3000 & 6 & 0 & 18,000 \end{array}\right] \begin{array}{l} \\ \frac{1}{1000}R_2 \\ \\ \end{array}$$

$$\begin{array}{ccccc} x_1 & x_2 & x_3 & x_4 & x_5 \end{array}$$
$$\left[\begin{array}{ccccc|c} -\frac{1}{3} & 1 & 0 & \frac{1}{1000} & -\frac{1}{1500} & \frac{1}{3} \\ 1 & 0 & 1 & -\frac{1}{1000} & \frac{1}{1000} & 1 \\ \hline 1000 & 0 & 0 & 3 & 3 & 21,000. \end{array}\right] \begin{array}{l} -\frac{2}{3}R_2 + R_1 \\ \\ 3000R_2 + R_3 \end{array}$$

The contractor should build 3 Atlantic and 3 Pacific models for a minimum cost of $21,000.

63. Let y_1 = kg of whole tomatoes and let y_2 = kg of tomato sauce.

Minimize $w = 4y_1 + 3.25y_2$

subject to:
$$\begin{aligned} y_1 + y_2 &\le 3,000,000 \\ y_1 &\ge 800,000 \\ y_2 &\ge 80,000 \\ \frac{6}{60}y_1 + \frac{3}{60}y_2 &\ge 110,000 \\ y_1 \ge 0,\ y_2 &\ge 0 \end{aligned}$$

The initial simplex tableau is

$$\begin{array}{cccccc} y_1 & y_2 & s_1 & s_2 & s_3 & s_4 \end{array}$$
$$\left[\begin{array}{cccccc|c} 1 & 1 & 1 & 0 & 0 & 0 & 3,000,000 \\ \boxed{1} & 0 & 0 & -1 & 0 & 0 & 800,000 \\ 0 & 1 & 0 & 0 & -1 & 0 & 80,000 \\ .1 & .05 & 0 & 0 & 0 & -1 & 110,000 \\ \hline 4 & 3.25 & 0 & 0 & 0 & 0 & 0 \end{array}\right]$$

For Stage I, pivot on the 1 in row two, column one.

$$\begin{array}{cccccc} y_1 & y_2 & s_1 & s_2 & s_3 & s_4 \end{array}$$
$$\left[\begin{array}{cccccc|c} 0 & 1 & 1 & 1 & 0 & 0 & 2,200,000 \\ 1 & 0 & 0 & -1 & 0 & 0 & 800,000 \\ 0 & \boxed{1} & 0 & 0 & -1 & 0 & 80,000 \\ 0 & .05 & 0 & .1 & 0 & -1 & 30,000 \\ \hline 0 & 3.25 & 0 & 4 & 0 & 0 & -3,200,000 \end{array}\right] \begin{array}{l} -R_2 + R_1 \\ \\ \\ -.1R_2 + R_4 \\ -4R_2 + R_5 \end{array}$$

Pivot on the 1 in row three, column two.

$$\begin{array}{cccccc} y_1 & y_2 & s_1 & s_2 & s_3 & s_4 \end{array}$$
$$\left[\begin{array}{cccccc|c} 0 & 0 & 1 & 1 & 1 & 0 & 2,120,000 \\ 1 & 0 & 0 & -1 & 0 & 0 & 800,000 \\ 0 & 1 & 0 & 0 & -1 & 0 & 80,000 \\ 0 & 0 & 0 & \boxed{.1} & .05 & -1 & 26,000 \\ \hline 0 & 0 & 0 & 4 & 3.25 & 0 & -3,460,000 \end{array}\right] \begin{array}{l} -R_3 + R_1 \\ \\ \\ -.05R_3 + R_4 \\ -3.25R_3 + R_5 \end{array}$$

Copyright © 2019 Pearson Education, Inc.

Pivot on the .1 in row four, column four.

$$\begin{array}{cccccc|c|l} y_1 & y_2 & s_1 & s_2 & s_3 & s_4 & & \\ 0 & 0 & 1 & 0 & .5 & 10 & 1{,}860{,}000 & -R_4 + R_1 \\ 1 & 0 & 0 & 0 & .5 & -10 & 1{,}060{,}000 & R_4 + R_2 \\ 0 & 1 & 0 & 0 & -1 & 0 & 80{,}000 & \\ 0 & 0 & 0 & 1 & .5 & -10 & 260{,}000 & 10R_4 \\ \hline 0 & 0 & 0 & 0 & 1.25 & 40 & -4{,}500{,}000 & -4R_4 + R_5 \end{array}$$

This program is optimal after Stage I.
1,060,000 kg of tomatoes should be used for canned whole tomatoes and 80,000 kg of tomatoes should be used for sauce at a minimum cost of $4,500,000.

64. Let y_1 = the number of runs of type I and let y_2 = the number of runs of type II.

Minimize $w = 15{,}000y_1 + 6000y_2$

Maximize $z = -w = -15{,}000y_1 - 6000y_2$

subject to:
$$\begin{aligned} 3000y_1 + 3000y_2 &= 18{,}000 \\ 2000y_1 + 1000y_2 &= 7000 \\ 2000y_1 + 3000y_2 &\geq 14{,}000 \\ y_1 \geq 0,\ y_2 &\geq 0 \end{aligned}$$

The initial simplex tableau is

$$\begin{array}{ccccccc|c} y_1 & y_2 & s_1 & s_2 & s_3 & s_4 & s_5 & \\ 3000 & 3000 & -1 & 0 & 0 & 0 & 0 & 18{,}000 \\ 3000 & 3000 & 0 & 1 & 0 & 0 & 0 & 18{,}000 \\ \boxed{2000} & 1000 & 0 & 0 & -1 & 0 & 0 & 7000 \\ 2000 & 1000 & 0 & 0 & 0 & 1 & 0 & 7000 \\ 2000 & 3000 & 0 & 0 & 0 & 0 & -1 & 14{,}000 \\ \hline 15{,}000 & 6000 & 0 & 0 & 0 & 0 & 0 & 0 \end{array}$$

For Stage I, pivot on the 2000 in row three, column one.

$$\begin{array}{ccccccc|c|l} y_1 & y_2 & s_1 & s_2 & s_3 & s_4 & s_5 & & \\ 0 & 1500 & -1 & 0 & 1.5 & 0 & 0 & 7500 & -3000R_3 + R_1 \\ 0 & 1500 & 0 & 1 & 1.5 & 0 & 0 & 7500 & -3000R_3 + R_2 \\ 1 & .5 & 0 & 0 & -.0005 & 0 & 0 & 3.5 & .0005R_3 \\ 0 & 0 & 0 & 0 & 1 & 1 & 0 & 0 & -2000R_3 + R_4 \\ 0 & \boxed{2000} & 0 & 0 & 1 & 0 & -1 & 7000 & -2000R_3 + R_5 \\ \hline 0 & -1500 & 0 & 0 & 7.5 & 0 & 0 & -52{,}500 & -15{,}000R_3 + R_6 \end{array}$$

Pivot on the 2000 in row five, column two.

$$\begin{array}{ccccccc|c|l} y_1 & y_2 & s_1 & s_2 & s_3 & s_4 & s_5 & & \\ 0 & 0 & -1 & 0 & .75 & 0 & \boxed{.75} & 2250 & -1500R_5 + R_1 \\ 0 & 0 & 0 & 1 & .75 & 0 & .75 & 2250 & -1500R_5 + R_2 \\ 1 & 0 & 0 & 0 & -.00075 & 0 & .00025 & 1.75 & -.5R_5 + R_3 \\ 0 & 0 & 0 & 0 & 1 & 1 & 0 & 0 & \\ 0 & 1 & 0 & 0 & .0005 & 0 & -.0005 & 3.5 & .0005R_5 \\ \hline 0 & 0 & 0 & 0 & 8.25 & 0 & -.75 & -47{,}250 & 1500R_5 + R_6 \end{array}$$

Copyright © 2019 Pearson Education, Inc.

Pivot on the .75 in row one column seven.

$$\begin{array}{ccccccc|c|l} y_1 & y_2 & s_1 & s_2 & s_3 & s_4 & s_5 & & \\ 0 & 0 & -\frac{4}{3} & 0 & 1 & 0 & 1 & 3000 & \frac{4}{3}R_1 \\ 0 & 0 & \boxed{1} & 1 & 0 & 0 & 0 & 0 & -.75R_1 + R_2 \\ 1 & 0 & \frac{1}{3000} & 0 & -.001 & 0 & 0 & 1 & -.00025R_1 + R_3 \\ 0 & 0 & 0 & 0 & 1 & 1 & 0 & 0 & \\ 0 & 1 & -\frac{1}{1500} & 0 & .001 & 0 & 0 & 5 & .0005R_1 + R_5 \\ \hline 0 & 0 & -1 & 0 & 9 & 0 & 0 & -45{,}000 & .75R_1 + R_6 \end{array}$$

For Stage II, pivot on the 1 in row two, column three.

$$\begin{array}{ccccccc|c|l} y_1 & y_2 & y_3 & y_4 & y_5 & y_6 & y_7 & & \\ 0 & 0 & 0 & \frac{4}{3} & 1 & 0 & 1 & 3000 & \frac{4}{3}R_2 + R_1 \\ 0 & 0 & 1 & 1 & 0 & 0 & 0 & 0 & \\ 1 & 0 & 0 & -\frac{1}{3000} & -.001 & 0 & 0 & 1 & -\frac{1}{3000}R_2 + R_3 \\ 0 & 0 & 0 & 0 & 1 & 1 & 0 & 0 & \\ 0 & 1 & 0 & \frac{1}{1500} & .001 & 0 & 0 & 5 & \frac{1}{1500}R_2 + R_5 \\ \hline 0 & 0 & 0 & 1 & 9 & 0 & 0 & -45{,}000 & R_2 + R_6 \end{array}$$

The company should produce 1 run of type I and 5 runs of type II for a minimum cost of $45,000.

Case 7 Cooking with Linear Programming

1. Let x_1 = the number of 100 gram units of feta cheese, x_2 = the number of 100 gram units of lettuce, x_3 = the number of 100 gram units of salad dressing, and x_4 = the number of 100 gram units of tomato.

Maximize $z = 4.09x_1 + 2.37x_2 + 2.5x_3 + 4.64x_4$

subject to:

$$\begin{aligned} 263x_1 + 14x_2 + 448.8x_3 + 21x_4 &< 260 \\ 492.5x_1 + 36x_2 + 5x_4 &> 210 \\ 10.33x_1 + 1.62x_2 + .85x_4 &> 6 \\ x_1 + x_2 + x_3 + x_4 &< 4 \\ x_3 &\geq .3125 \end{aligned}$$

$x_1 \geq 0,\ x_2 \geq 0,\ x_3 \geq 0,\ x_4 \geq 0$

Using a computer with linear programming software, the optimal solution is:

$x_1 = .243037,\ x_2 = 2.35749,\ x_3 = .3125,$

$x_4 = 1.08698$

Converting to kitchen units gives approximately $\frac{1}{6}$ cup feta cheese $4\frac{1}{4}$ cups of lettuce, $\frac{1}{8}$ cup of salad dressing and $\frac{7}{8}$ of a tomato for a salad with a maximum of about 12.41 g of carbohydrates.

2. Let y_1 = the number of 100 gram units of beef, y_2 = the number of 100 gram units of oil, y_3 = the number of 100 gram units of onion, and y_4 = the number of 100 gram units of soy sauce.

Minimize $215y_1 + 884y_2 + 38y_3 + 60y_4$

subject to:

$$\begin{aligned} y_2 + 8.63y_3 + 5.57y_4 &< 10 \\ 26y_1 + 1.16y_3 + 10.51y_4 &> 50 \\ 6.4y_3 &> 3.5 \\ y_2 &\geq .045 \end{aligned}$$

$y_1 \geq 0,\ y_2 \geq 0,\ y_3 \geq 0,\ y_4 \geq 0$

Using a computer with linear programming software, the optimal solution is

$y_1 = 1.51873,\ y_2 = .045,\ y_3 = .546875,$

$y_4 = .939941$

Converting to kitchen units gives approximately $5\frac{1}{3}$ ounces of beef, $\frac{1}{3}$ tablespoon of oil, $\frac{1}{2}$ an onion, and $5\frac{1}{4}$ tablespoons of soy sauce for a stir-fry with a minimum 443.48 calories.

Copyright © 2019 Pearson Education, Inc.

Chapter 8 Sets and Probability

Section 8.1 Sets

1. $3 \in \{2, 5, 7, 9, 10\}$
This statement is false. The number 3 is not an element of the set.

3. $9 \notin \{2, 1, 5, 8\}$
The statement is true. 9 is not an element of the set.

5. $\{2, 5, 8, 9\} = \{2, 5, 9, 8\}$
The statement is true. The sets contain exactly the same elements, so they are equal. The ordering of the elements in a set is unimportant.

7. {all whole numbers greater than 7 and less than 10} = {8, 9}
The statement is true. 8 and 9 are the only such numbers.

9. $\{x \mid x \text{ is an odd integer}, 6 \le x \le 18\} = \{7, 9, 11, 15, 17\}$
The statement is false. The number 13 should be included.

11. Answers vary.
Possible answer:
{0} is a set containing 1 element, namely zero.
$\{\varnothing\}$ is a set containing 1 element, namely, a mathematical symbol representing the empty set.

13. Since every element of A is also an element of U, $A \subseteq U$.

15. $A \not\subseteq E$ since A contains elements that do not belong to E, namely –3, 0, 3.

17. $\varnothing \subset A$ since the empty set is a subset of every set.

19. $D \subset B$, since every element of D is also an element of B.

21. $\{A, B, C\}$ contains 3 elements. Therefore, it has $2^3 = 8$ subsets.

23. $\{x \mid x \text{ is an integer strictly between 0 and 8}\} = \{1, 2, 3, 4, 5, 6, 7\}$. This set contains 7 elements. Therefore it has $2^7 = 128$ subsets.

25. $\left\{x \middle| x \text{ is an integer less than or equal to 0 or greater than or equal to 8}\right\}$

27. Answers vary.

29. $\{8, 11, 15\} \cap \{8, 11, 19, 20\} = \{8, 11\}$

31. $\{6, 12, 14, 16\} \cap \{6, 14, 19\} = \{6, 14\}$
{6, 14} is the set of all elements belonging to both of the sets, so it is the intersection of those sets.

33. $\{3, 5, 9, 10\} \cup \varnothing = \{3, 5, 9, 10\}$.

35. $\{1, 2, 4\} \cup \{1, 2\} = \{1, 2, 4\}$
The answer set {1, 2, 4} consists of all elements belonging to the first set, to the second set, or to both sets, and therefore it is the union of the first two sets.

37. $X \cap Y = \{b, 1, 3\}$ since only these elements are contained in both.

39. $X' = \{d, e, f, 4, 5, 6\}$ since these are the elements that are in U but not in X.

41. $X' \cap Y'$
$= \{d, e, f, 4, 5, 6\} \cap \{a, c, e, 2, 4, 6\}$
$= \{e, 4, 6\}$

43. $X \cup (Y \cap Z)$
$Y \cap Z = \{b, d, f, 1, 3, 5\} \cap \{b, d, 2, 3, 5\}$
$= \{b, d, 3, 5\}$
$X \cup (Y \cap Z) = \{a, b, c, 1, 2, 3\} \cup \{b, d, 3, 5\}$
$= \{a, b, c, d, 1, 2, 3, 5\}$

45. M' is the set of all students in this school not taking this course.

47. $N \cap P$ is the set of all students in this school taking both accounting and philosophy.

49. A pair of sets is disjoint if the two sets have no elements in common. The pairs of these sets that are disjoint are C and D, A and E, C and E, and D and E.

51. A' is the set of all stocks with a high price less than or equal to \$50. A' = {Ford}

Copyright © 2019 Pearson Education, Inc.

53. $A' \cap B'$ is the set of all stocks with a high price less than or equal to \$50 and a last price either less than or equal to \$25 or greater than or equal to \$80. $A' \cap B'$ = {Ford}.

55. $M \cap E$ is the set of all male employed clients.

57. $M' \cup S'$ is the set of all female or married clients.

59. {T}

61. {C}

63. F = {Direct TV, Comcast, Charter, Dish}

65. H = {Charter, Dish, Verizon FIOS, Altice}

67. $F \cup G$ = {Direct TV, Comcast, Charter, Dish} $\cup$ {Charter, Dish} = {Direct TV, Comcast, Charter, Dish}

69. I' = {Dish, Verizon FIOS, Altice}

71. $J = \varnothing$

73. {Corn, Rice}

75. $\varnothing$

77. {Wheat, Corn, Rice}

Section 8.2 Applications of Venn Diagrams

1. $A \cap B'$
Shade the region inside B that is outside A.

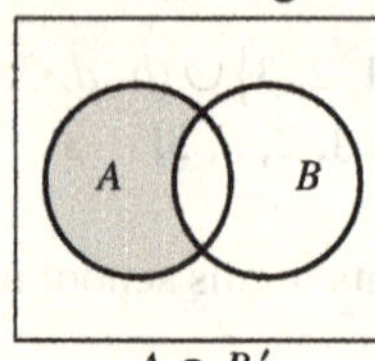

A ∩ B′

3. $B' \cup A'$
Shade the region outside B and the region outside of region A.

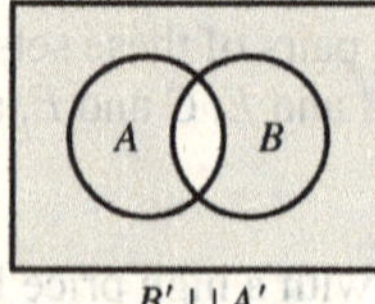

B′ ∪ A′

5. $B' \cup (A \cap B')$
First shade the common region that is in A and outside B. Then shade the rest of the region outside B.

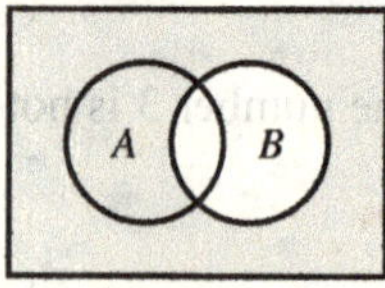

B′ ∪ (A ∩ B′)

7. $U' = \varnothing$
There are no elements outside the universal set. Therefore, there is no region to be shaded.

9. Three sets divide the universal set into at most <u>8</u> regions.

11. $(A \cap C') \cup B$
First find $A \cap C'$, the region in A *and* not in C.

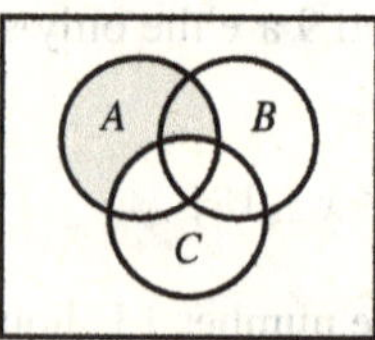

A ∩ C′

For the union, we want the region in $(A \cap C')$ *or* in B, or both.

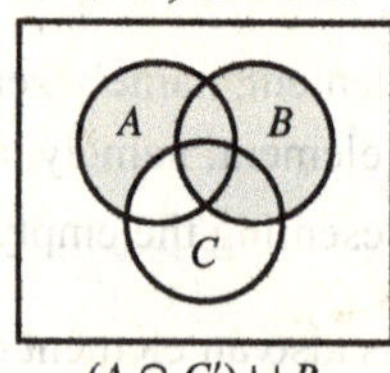

(A ∩ C′) ∪ B

13. $A' \cap (B \cap C)$
First find $B \cap C$, the region in B *and* in C.

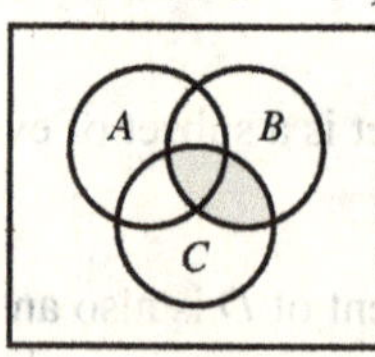

B ∩ C

Now find A', the region not in A.

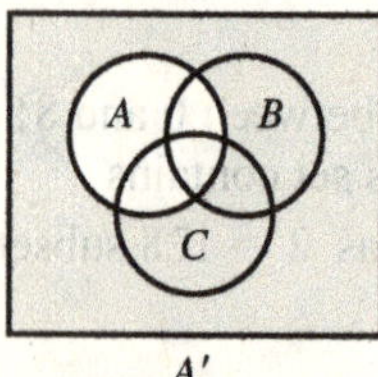

A′

For the intersection, we want the region in A' *and* in $(B \cap C)$.

Copyright © 2019 Pearson Education, Inc.

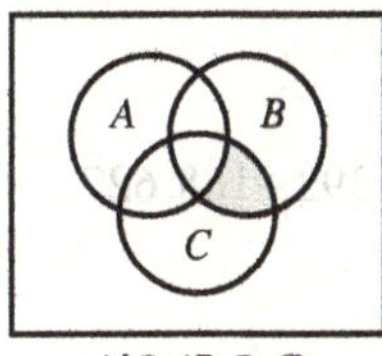

$A' \cap (B \cap C)$

15. $(A \cap B') \cup C$
First find $A \cap B'$, the region in *A and* not in *B*.

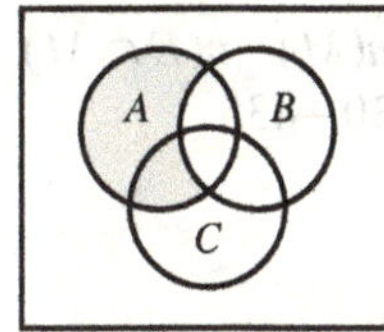

$A \cap B'$

For the union, we want the region in $(A \cap B')$ *or* in *C*, or both.

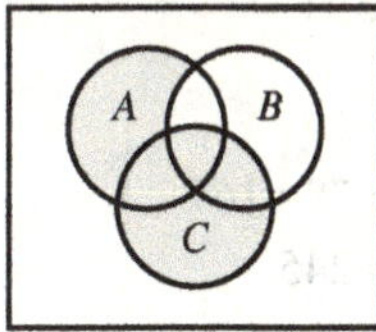

$(A \cap B') \cup C$

17. Number of movies that earned over \$100 million and that were not rated PG-13:
$100 - 28 - 38 - 15 = 13$

19. There are 26 applicants with sales experience and a college degree, of which 11 also have a real estate license. Therefore, 15 have only sales experience and a college degree. There are 16 applicants with sales experience and a real estate license, of which 11 also have a college degree. Therefore, there are 5 applicants with only sales experience and a real estate license. There are $66 - 15 - 11 - 5 = 35$ applicants with only sales experience.
Using similar reasoning, we find that there are 10 applicants with only a college degree, 4 applicants with a only a college degree and a real estate license, and 3 applicants with only a real estate license. This is illustrated in the Venn diagram below.

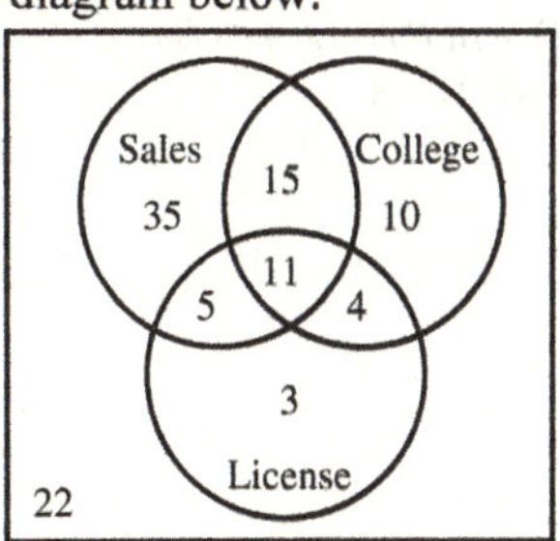

a. There were $35 + 15 + 11 + 5 + 10 + 4 + 3 + 22 = 105$ applicants.

b. $22 + 3 + 4 + 10 = 39$ applicants did not have sales experience.

c. 15 applicants had sales experience and a college degree, but not a real estate license.

d. Three applicants had only a real estate license.

21. **a.** The total number of people surveyed is the sum of all those who caught at least one fish minus the sum of those who caught at least two fish minus the number of those who caught all three fish.
$(124 + 133 + 146) - (75 + 67 + 79 - 45)$
$= 227$

b. The total number of people who caught at least one walleye or at least one smallmouth bass equals the sum of the number of those who caught at least one walleye and the number of those who caught at least one smallmouth bass minus the number of those who caught at least one of each fish.
$124 + 133 - 75 = 182$

c. 45 people caught all three fish, while 75 people caught at least one walleye and at least one smallmouth bass, so 30 people caught at least one walleye and at least one smallmouth bass only. 67 people caught at least one walleye and at least one yellow perch, so $67 - 45 = 22$ people caught at least one walleye and at least one yellow perch only. Therefore, $124 - (30 + 45 + 22) = 27$ people caught at least one walleye only. The information in the exercise is summarized in the Venn diagram below.

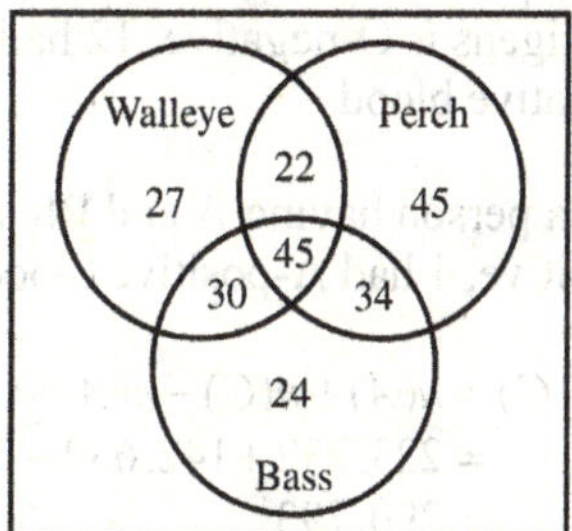

Copyright © 2019 Pearson Education, Inc.

23. In the diagram, the regions are labeled (a) to (g). First, the number for each region is found by starting with innermost region (d). 15 had all three; then proceed to work outward as follows:
Region (a):
Since 25 had A, 25 – (2 + 15 + 1) = 7.
Region (b):
Since 17 had A and B, 17 – 15 = 2.
Region (c):
Since 27 had B, 27 – (2 + 15 + 7) = 3.
Region (d):
15 had all three.
Region (e):
Since 22 had B and Rh, 22 – 15 = 7.
Region (f):
Since 30 had Rh, 30 – (1 + 15 + 7) = 7.
Region (g):
12 had none.
Region with no label:
Since 16 had A and Rh, 16 – 15 = 1.

a. 7 + 2 + 3 + 15 + 7 + 7 + 12 + 1 = 54 patients were represented.

b. 7 + 3 + 7 = 17 patients had exactly one antigen.

c. 2 + 1 + 7 = 10 patients had exactly two antigens.

d. Since a person having only the Rh antigen has type O-positive blood, 7 had type O-positive blood.

e. Since a person having A, B, and Rh antigens is AB-positive, 15 had AB-positive blood.

f. Since a person having only the B antigen is B-negative, 3 had B-negative blood.

g. Since a person having neither A, B, nor Rh antigens is O-negative, 12 had O-negative blood.

h. Since a person having A and Rh antigens is A-positive, 1 had A-positive blood.

25. a. $n(A \cup C) = n(A) + n(C) - n(A \cap C)$
$= 223,232 + 142,632 - 75,071$
$= 290,793$

b. $n(M \cup W) = n(M) + n(W) - n(M \cap W)$
$= 66,775 + 82,392 - 0$
$= 149,167$

c. $n(R \cap B) = 81,878$

d. $n[C'] = 66,775 + 82,392 + 168,627$
$= 317,394$

27. a. $n(E) = 945$

b. $n(D \cap F) = 55$

c. $n(E \cup M) = n(E) + n(M) - n(E \cap M)$
$= 945 + 830 - 435$
$= 1,340$

d. $n[E'] = 835 + 93$
$= 928$

29. a. $n(CS \cup VS) = n(CS) + n(VS)$
$= 164 + 343$
$= 507$

b. $n((CS \cup VS) \cap F) = 78 + 167$
$= 245$

c. The number of females who are completely satisfied or very satisfied with their job.

31. a. $n(CS \cup VS \cup FS) = n(CS) + n(VS) + n(FS)$
$= 164 + 343 + 293$
$= 800$

b. $n((CS \cup VS \cup FS) \cap M) = 86 + 176 + 149$
$= 411$

c. The number of males who are completely, very, or fairly satisfied with their job.

33. Answers will vary.

35. $n(A \cup B) = n(A) + n(B) - n(A \cap B)$
$30 = 12 + 27 - n(A \cap B)$
$30 = 39 - n(A \cap B)$
$n(A \cap B) = 9$

37. $n(A \cup B) = n(A) + n(B) - n(A \cap B)$
$35 = 13 + n(B) - 5$
$n(B) = 27$

Copyright © 2019 Pearson Education, Inc.

39. $n(A) = 28,\ n(B) = 12,\ n(A \cup B) = 30,$
$n(A') = 19$

This gives

$n(A \cup B) = n(A) + n(B) - n(A \cap B) \Rightarrow$
$30 = 28 + 12 - n(A \cap B)$
$n(A \cap B) = 28 + 12 - 30 = 10$

$n(A) = 28$ and $n(A \cap B) = 10$, so
$n(A \cap B') = 18$
$n(B) = 12$ and $n(A \cap B) = 10$, so
$n(B \cap A') = 2$
Since $n(A')$=19, 2 of which are accounted for in $B \cap A'$, 17 remain in $A' \cap B'$.

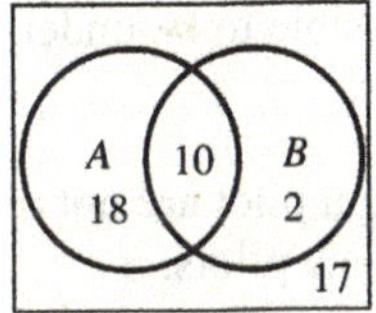

41. $n(A') = 28,\ n(B) = 25,\ n(A' \cup B') = 45,$
$n(A \cap B) = 12$

$n(B) = 25$ and $n(A \cap B) = 12$, so
$n(B \cap A') = 13$
Since $n(A')$=28, of which 13 are accounted for, 15 are in $A' \cap B'$.

$$n(A' \cup B') = n(A') + n(B') - n(A' \cap B')$$
$$45 = 28 + n(B') - 15$$
$$45 = 13 + n(B')$$
$$32 = n(B')$$

15 are in $A' \cap B'$, so the rest are in $A \cap B'$, and $n(A \cap B')$=17.

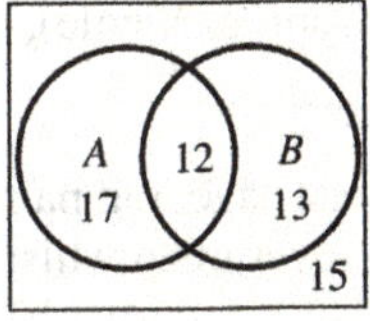

43.
$$n(A) = 54$$
$$n(A \cap B) = 22$$
$$n(A \cup B) = 85$$
$$n(A \cap B \cap C) = 4$$
$$n(A \cap C) = 15$$
$$n(B \cap C) = 16$$
$$n(C) = 44$$
$$n(B') = 63$$

Start with $A \cap B \cap C$. Now $n(A \cap C) = 15$, of which 4 are in $A \cap B \cap C$, so $n(A \cap B' \cap C) = 11$. $n(B \cap C) = 16$, of which 4 are in $A \cap B \cap C$, so $n(B \cap C \cap A') = 12$.
$n(C) = 44$, so 17 are in $C \cap A' \cap B'$.
$n(A \cap B) = 22$, so 18 are in $A \cap B \cap C'$.
$n(A) = 54$, so $54 - 11 - 18 - 4 = 21$ are in $A \cap B' \cap C'$.

$$n(A \cup B) = n(A) + n(B) - n(A \cap B)$$
$$85 = 54 + n(B) - 22$$
$$53 = n(B)$$

This leaves 19 in $B \cap A' \cap C'$. $n(B') = 63$, of which $21 + 11 + 17 = 49$ are accounted for, leaving 14 in $A' \cap B' \cap C'$.

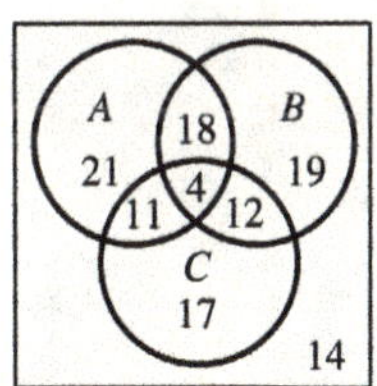

45. $(A \cap B)'$ is the complement of the intersection of A and B; hence it contains all elements not in $A \cap B$.

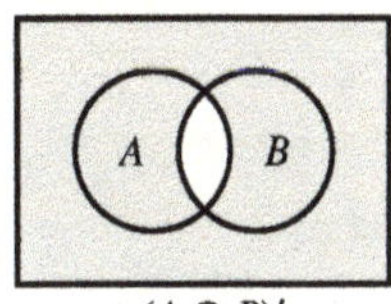

$(A \cap B)'$

$A' \cup B'$ is the union of the complements of A and B; hence it contains any element that is either not in A or not in B.

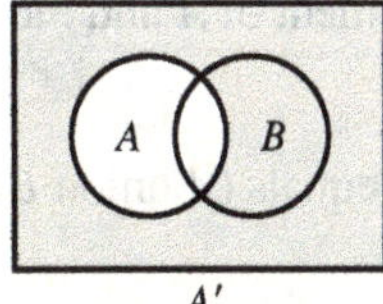

A'

B'

Note that $(A \cap B)' = A' \cup B'$, as claimed.

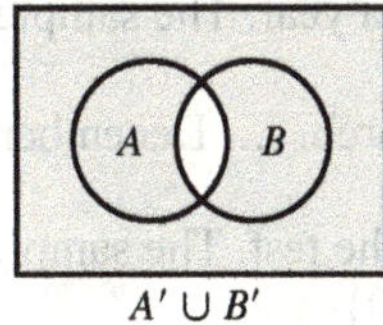

$A' \cup B'$

Copyright © 2019 Pearson Education, Inc.

47. $A\cup(B\cap C)$ contains all points in A and the points where B and C overlap.

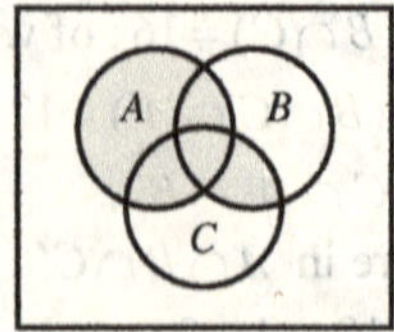

$A \cup (B \cap C)$

$(A\cup B)\cap(A\cup C)$ contains the points where $A\cup B$ and $A\cup C$ overlap.

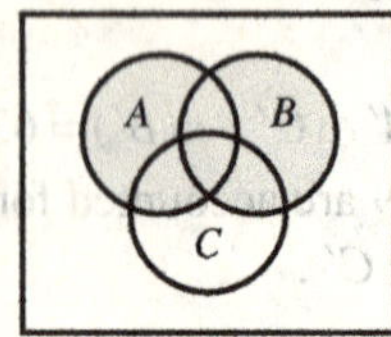

$A \cup B$

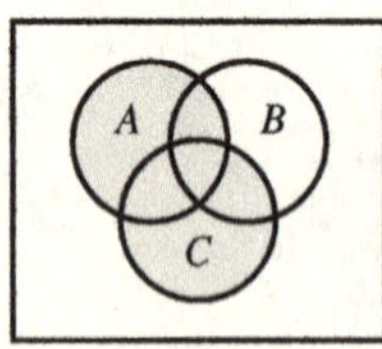

$A \cup C$

Note that $A\cup(B\cap C)=(A\cup B)\cap(A\cup C)$, as claimed.

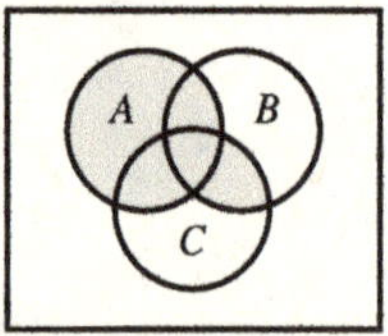

$(A \cup B) \cap (A \cup C)$

49. The complement of A and (intersection) B equals the union of the complement of A and the complement of B.

51. A union (B intersect C) equals (A union B) intersect (A union C).

Section 8.3 Introduction to Probability

1. Answers vary. Possible answer: A coin or die is fair if the probability of any result is the same, that is, all results are equally likely.

3. There are 12 months in a year. The sample space is the set {January, February, March, ..., December}.

5. There are 80 points on the test. The sample space is the set {0, 1, 2, ..., 80}.

7. There are only 2 choices the management can make. The sample space is the set {go ahead, cancel}.

9. There are four possibilities, so the sample space is the set {Q1, Q2, Q3, Q4}.

11. Answers vary. Possible answer: Disjoint events are events that cannot occur at the same time.

13. Wearing a hat and wearing glasses are not disjoint, because it is possible to wear both at the same time.

15. Being a doctor and being under 5 years old are disjoint, since it is impossible to be under 5 years old and be a doctor.

17. Being a female and being a pilot are not disjoint, since there are many female pilots.

19. S = {(Connie, Kate), (Connie, Lindsey), (Connie, Jackie), (Connie, Taisa), (Connie, Nicole), (Kate, Lindsey), (Kate, Jackie), (Kate, Taisa), (Kate, Nicole), (Lindsey, Jackie), (Lindsey, Taisa), (Lindsey, Nicole), (Jackie, Taisa), (Jackie, Nicole), (Taisa, Nicole)}

a. Taisa is selected.
{(Connie, Taisa),(Kate, Taisa), (Lindsey, Taisa), (Jackie, Taisa), (Taisa, Nicole)}

b. The two names selced have the same number of letters.
{(Connie, Jackie), (Connie, Nicole), (Jackie, Nicole)}

21. The sample space is {(forest sage, rag painting), (forest sage, colorwash), (evergreen whisper, rag painting), (evergreen whisper, colorwash), (opaque emerald, rag painting), (opaque emerald, colorwash)}

a. There are 3 combinations with colorwash: {(colorwash, forest sage), (colorwash, evergreen whisper), (colorwash, opaque emerald)}.

$$P_C = \frac{n(C)}{n(S)} = \frac{3}{6} = \frac{1}{2}.$$

Copyright © 2019 Pearson Education, Inc.

b. There are 4 combinations with opaque emerald or rag painting: {(opaque emerald, rag painting), (opaque emerald, colorwash), (forest sage, rag painting), (evergreen whisper, rag painting)}.

$$P_{O\cup R} = \frac{n(O \cup R)}{n(S)} = \frac{4}{6} = \frac{2}{3}$$

23. There are 2 possibilities for size and 3 possibilities for color. The sample space is {(10-foot, beige), (10-foot, forest green) (10-foot, rust), (12-foot, beige), (12-foot, forest green), (12-foot, rust)}

a. Doug buys a 12-foot forest green umbrella. There is 1 possibility: {(12-foot, forest green)}.

$$P_{12'\cap G} = \frac{n(12' \cap G)}{n(S)} = \frac{1}{6}$$

b. Doug buys a 10-foot umbrella. There are 3 possibilities: {(10-foot, beige), (10-foot, forest green), (10-foot, rust)}

$$P_{10'} = \frac{n(10')}{n(S)} = \frac{3}{6} = \frac{1}{2}$$

c. Doug buys a rust-colored umbrella. There are 2 possibilities: {(10-foot, rust), (12-foot, rust)}

$$P_R = \frac{n(R)}{n(S)} = \frac{2}{6} = \frac{1}{3}$$

25. Use the sample space $S = \{1, 2, 3, 4, 5, 6\}$. Let E be the event "getting a number less than 4." So, $E = \{1, 2, 3\}$. Since S contains six elements,

$$P(E) = \frac{3}{6} = \frac{1}{2}.$$

27. Use the sample space $S = \{1, 2, 3, 4, 5, 6\}$. Let E be the event "getting a 2 or a 5." So, $E = \{2, 5\}$. Since S contains 6 elements,

$$P(E) = \frac{2}{6} = \frac{1}{3}.$$

29. {male beagle, male boxer, male collie, male Labrador, female beagle, female boxer, female collie, female Labrador}

31. There are four possibilities, {male beagle, male boxer, male collie, male Labrador}.

$$P(E) = \frac{4}{8} = \frac{1}{2}.$$

33. There is one possibility, {female Labrador}.

$$P(E) = \frac{1}{8}.$$

35. There are six possibilities, {male beagle, male boxer, male collie, female beagle, female boxer, female collie}.

$$P(E) = \frac{6}{8} = \frac{3}{4}.$$

For exercise 37, the total number of fatalities in 2015 is 4836.

37. $P(E) = \frac{800}{4836} \approx .1654$

39. The table lists religious service attendance as number of respondents. The total number of respondents is 2837.

a. Several times a year: $P(E) = \frac{312}{2837} \approx .1088$

b. 2–3 times a month: $P(E) = \frac{250}{2837} \approx .0872$

c. nearly every week or more frequently:

$$P(E) = \frac{126 + 481 + 210}{2837} = \frac{817}{2837} \approx .2850$$

For exercises 41 – 44, the total number of stocks as of March 7, 2017 is 505.

41. $P(E) = \frac{35}{505} \approx .0693$

42. $P(E) = \frac{65}{505} \approx .1287$

43. $P(E) = \frac{30 + 28}{505} = \frac{58}{505} \approx .1149$

44. $P(E) = \frac{60 + 68}{505} = \frac{128}{505} \approx .2535$

For exercise 45, the total number of stocks as of March 7, 2017 is 505.

45. $P(E) = \frac{159}{505} \approx .3149$

47. This experiment is possible, since all probabilities are non-negative and $.09 + .32 + .21 + .25 + .13 = 1$

Copyright © 2019 Pearson Education, Inc.

49. This experiment is not possible, since the sum of the probabilities is $\frac{39}{40}$, which is less than 1.

51. This experiment is not possible, since a probability cannot be negative.

Section 8.4 Basic Concepts of Probability

1. Let G be the event of the marble landing in a green slot, and let B be the event of the marble landing in a black slot. There are 2 slots that are green and 18 slots that are black, so

$$P(G \cup B) = P(G) + P(B) - P(G \cap B) = \frac{2}{38} + \frac{18}{38} - \frac{0}{38} = \frac{20}{38} = \frac{10}{19}.$$

3. Let O be the event of the marble landing in an odd slot, and let B be the event of the marble landing in a black slot. There are 18 slots that are odd, 18 slots that are black, and 8 slots that are odd and black, so

$$P(O \cup B) = P(O) + P(B) - P(O \cap B) = \frac{18}{38} + \frac{18}{38} - \frac{8}{38} = \frac{28}{38} = \frac{14}{19}.$$

5. Let E be the event of the marble landing in a slot numbered 0, 00, 1, 2, or 3. There are 5 such slots, so $P(E) = \frac{5}{38}$.

7. Let E be the event of the marble landing in a slot numbered 25–36. There are 12 such slots, so

$$P(E) = \frac{12}{38} = \frac{6}{19}.$$

For Exercises 8–13, count outcomes by referring to Figure 8.22 in the text.

9. **a.** $P(\text{sum is } 8) = \frac{5}{36}$

b. $P(\text{sum is } 9) = \frac{4}{36} = \frac{1}{9}$

c. $P(\text{sum is } 10) = \frac{3}{36} = \frac{1}{12}$

d. $P(\text{sum is } 13) = \frac{0}{36} = 0$

11. **a.** $P(\text{not more than } 5) = \frac{10}{36} = \frac{5}{18}$

b. $P(\text{not less than } 8) = \frac{15}{36} = \frac{5}{12}$

c. $P(\text{between 3 and 7 (exclusive)}) = \frac{12}{36} = \frac{1}{3}$

13. The shoes come in two shades of beige (light and dark) and black, so $P(\text{shoes are black}) = \frac{1}{3}$.

15. **a.** $P(\text{a sister or an aunt}) = \frac{2}{10} + \frac{3}{10} = \frac{5}{10} = \frac{1}{2}$

b. $P(\text{a sister or a cousin}) = \frac{2}{10} + \frac{2}{10} = \frac{4}{10} = \frac{2}{5}$

c. $P(\text{a sister or her mother}) = \frac{2}{10} + \frac{1}{10} = \frac{3}{10}$

17.

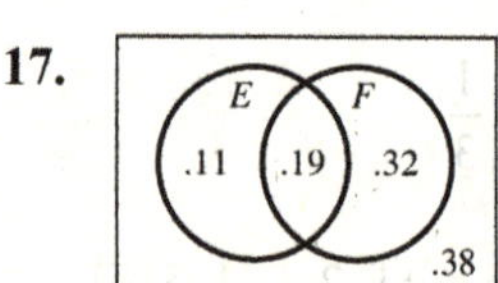

a. $P(E \cup F) = P(E) + P(F) - P(E \cap F) = .30 + .51 - .19 = .62$

b. $P(E' \cap F) = P(F) - P(E \cap F) = .51 - .19 = .32$

c. $P(E \cap F') = P(E) - P(E \cap F) = .30 - .19 = .11$

d. $P(E' \cup F') = P(E') + P(F') - P(E' \cap F') = .70 + .49 - .38 = .81$

Copyright © 2019 Pearson Education, Inc.

19.

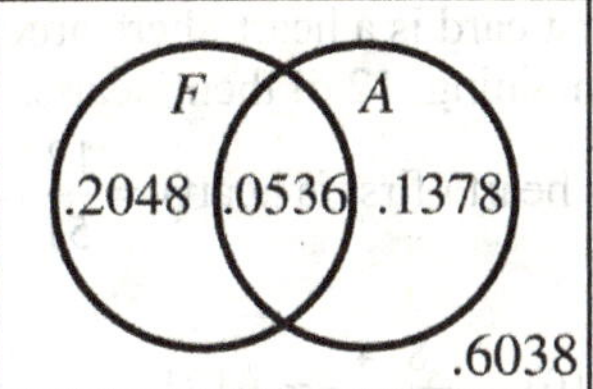

a. $P(F \cup A) = P(F) + P(A) - P(F \cap A)$
$= .2584 + .1914 - .0536$
$= .3962$

b. $P(F' \cup A) = P(F') + P(A) - P(F' \cap A)$
$= .7416 + .1914 - .1378$
$= .7952$

c. $P(F \cap A') = P(F' \cup A)' = 1 - P(F' \cup A)$
$= 1 - .7952 = .2048$

d. $P(F' \cup A') = P(F') + P(A') - P(F' \cap A')$
$= (1 - .2584) + (1 - .1914) - .6032$
$= .9464$

21.

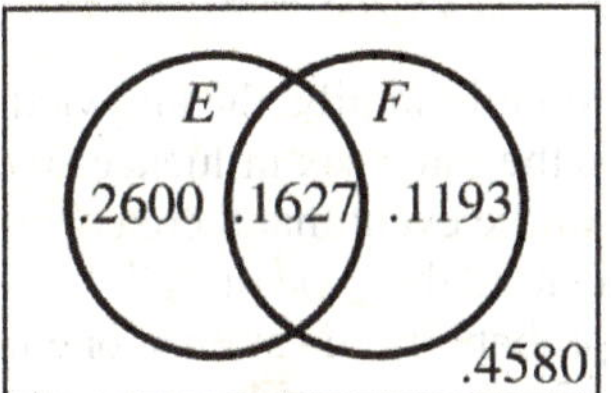

a. $P(E \cap F') = P(E) - P(E \cap F)$
$= .4227 - .1627 = .2600$

b. $P(E' \cap F') = 1 - [P(E) + P(F) - P(E \cap F)]$
$= 1 - [.4227 + .2820 - .1627]$
$= .4580$

c. $P(E' \cup F') = 1 - P(E \cap F)$
$= 1 - .1627$
$= .8373$

23. When rolling a die, there are 6 equally likely outcomes. The sample space is $S = \{1, 2, 3, 4, 5, 6\}$. Let E be the event "2 is rolled." Then $P(E) = \frac{1}{6}$ and $P(E') = \frac{5}{6}$. The odds in favor of rolling a 5 are $\frac{P(E)}{P(E')} = \frac{\frac{1}{6}}{\frac{5}{6}} = \frac{1}{5}$, written 1 to 5.

25. Let E be the event " 2, 3, 5, or 6 is rolled." Then $P(E) = \frac{4}{6} = \frac{2}{3}$ and $P(E') = \frac{1}{3}$. The odds in favor of rolling 2, 3, 5, or 6 are $\frac{P(E)}{P(E')} = \frac{\frac{2}{3}}{\frac{1}{3}} = 2$, written 2 to 1.

27. There are 3 yellow, 4 white, and 8 blue marbles.

a. Yellow: There are 3 ways to win and 12 ways to lose. The odds are 3 to 12 or 1 to 4.

b. Blue: There are 8 ways to win and 7 ways to lose; the odds are 8 to 7.

c. White: There are 4 ways to win and 11 ways to lose; the odds are 4 to 11.

29. Let E be the event "rolling a 7 or 11." Then $P(E) = \frac{8}{36}$ and $P(E') = \frac{28}{36}$. The odds in favor of rolling a 7 or 11 are $\frac{\frac{8}{36}}{\frac{28}{36}} = \frac{8}{28} = \frac{2}{7}$, written 2 to 7.

31. 19:81

33. 1:19

35. $\frac{31}{100}$

37. $\frac{23}{100}$

For exercises 39 – 46, the total number of stocks as of March 7, 2017 is 505.

39. $P(E) = \frac{452}{505} \approx .8950$

41. $P(E) = \frac{30}{505} \approx .0594$

43. $P(E) = \frac{207 + 452 - 189}{505} = \frac{470}{505} \approx .9307$

45. $P(E) = \frac{189 + 135}{505} = \frac{324}{505} \approx .6416$

Copyright © 2019 Pearson Education, Inc.

For exercises 47 – 56, the total number of workers from 2016 is 1582.

47. $P(E)=\frac{1274}{1582}\approx .8053$

49. $P(E)=\frac{401}{1582}\approx .2535$

51. $P(E)=\frac{41+24}{1582}=\frac{65}{1582}\approx .0411$

53. $P(E)=\frac{1274+(732+401)-(691+377)}{1582}$
$=\frac{1339}{1582}\approx .8464$

55. Using a graphing calculator,

a. $P(\text{the sum is 9 or more})\approx .2778$

b. $P(\text{the sum is less than 7})\approx .4167$

The probabilities compare very well to the results in Exercise 10.

57. Using a graphing calculator,

a. $P(\text{the sum is 5 or less})\approx .0463$

b. $P(\text{neither a 1 nor a 6 is rolled})\approx .2963$

59. Since red is dominant, red = {*RR*, *RW*, *WR*} out of a total of 4.

$P(\text{red})=\frac{3}{4}$

Section 8.5 Conditional Probability and Independent Events

1. Roll a fair die.

$$P(3\mid\text{odd})=\frac{P(3\cap\text{odd})}{P(\text{odd})}=\frac{\frac{1}{6}}{\frac{1}{2}}=\frac{1}{3}$$

3. $P(\text{odd}\mid 3)=\frac{P(\text{odd}\cap 3)}{P(3)}=\frac{\frac{1}{6}}{\frac{1}{6}}=1$

5. There are 6 doubles, 1 of which has a sum of 6.

$P(\text{sum of 6}\mid\text{double})=\frac{1}{6}$

7. Since the first card is a heart, there are 51 cards remaining, 12 of them hearts.

$P(\text{second is heart}\mid\text{first is heart})=\frac{12}{51}=\frac{4}{17}$

9. $P(\text{jack and 10})=\frac{8\cdot 4}{52\cdot 51}\approx .0121$

There are 8 possibilities for the first card (4 jacks and 4 tens), but for the second card there are only 4 (the 4 tens if a jack was picked or the 4 jacks if a 10 was picked).

11. Answers vary.

13. Answers vary.

15. No. Information about the event that it snows tomorrow does affect the probability that an instructor will be late for class. So, the events are dependent.

17. Yes. Knowledge of the event that it rains in the Amazon jungle gives no information about the occurrence or nonoccurrence of the event that an instructor in New York will write a difficult exam. So, the events are independent.

19. No, for a two-child family, the knowledge that each child is the same sex influences the probability of the event that there is at most one male. Eliminating the "one of each" possibility lowers the probability of "at most one male" from .75 to .5. However, the events are independent for a three-child family because the first event does not influence the probability of the second event. The probability of at most one male is .5 whether male-female mixes are allowed or not.

21. $P(E\cap C)=P(E\mid C)P(C)$
$=(.6930)(.1135)$
$\approx .0787$

23. Let *C* represent "construction" and *E* represent "1 – 4 employees" We are seeking

$P(C\mid E)=\frac{P(C\cap E)}{P(E)}.$

$P(C\mid E)=\frac{.0787}{.6214}\approx .1266.$

The probability of a firm being construction, given that it has 1 – 4 employees is .1226.

Copyright © 2019 Pearson Education, Inc.

25. $P(\text{FT} \mid > 50 \text{ hours}) = \dfrac{P(FT \cap > 50 \text{ hours})}{P(> \text{hours})}$
$= \dfrac{377}{401} \approx .9401$

27. $P(\text{0-19 hours} \mid FT) = \dfrac{P(0-19 \text{ hours} \cap FT)}{P(\text{FT})}$
$= \dfrac{26}{1274} \approx .0204$

29. $P(<30 \text{ hours} \mid PT) = \dfrac{P(<30 \text{ hours} \cap PT)}{P(\text{PT})}$
$= \dfrac{156}{308} \approx .5065$

31. $P(\text{gd con in banks} \mid \text{ha con in corp})$
$= \dfrac{P(\text{gd con in banks} \cap \text{ha con in corp})}{P(\text{ha con in corp})}$
$= \dfrac{25}{339} \approx .0737$

33. $P(\text{gd con in corp} \mid \text{some con in banks})$
$= \dfrac{P(\text{gd con in corp} \cap \text{some con in banks})}{P(\text{some con in banks})}$
$= \dfrac{172}{1044} \approx .1648$

35. $P(M) = .527$ (directly from the chart)

37. $P(M \cap C) = P(M \text{ and } C) = .042$

39. $P(M \mid C) = \dfrac{P(M \cap C)}{P(C)} = \dfrac{.042}{.049} = \dfrac{6}{7} \approx .857$

41. $P(C \mid M) = \dfrac{P(M \cap C)}{P(M)} = \dfrac{.042}{.527} \approx .0797$
$P(C) = .049$
Since $P(C \mid M) \neq P(C)$, they are dependent.

43. $P(C) = .0800$; $P(D) = .0050$; $P(C \cap D) = .0004$;
$P(C \mid D) = \dfrac{P(C \cap D)}{P(D)} = \dfrac{.0004}{.0050} = .08$; Yes,
$P(C) = P(C \mid D)$, therefore the two events are independent.

45. Complete the probability tree.

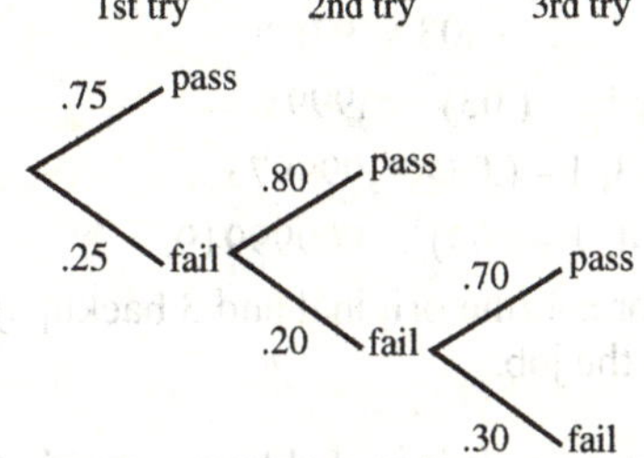

P(fails 1st and 2nd test)
$= P(\text{fails 1st}) \cdot P(\text{fails 2nd} \mid \text{fails 1st})$
$= (.25)(.20) = .05$

47. $P(\text{On time} \mid \text{American}) = \dfrac{62{,}642}{78{,}114} \approx .8019$

49. $P(\text{On time} \mid \text{SW})$
$= \dfrac{91{,}664}{115{,}144} \approx .7961$

51. $P(\text{United} \mid \text{On time}) = \dfrac{39{,}262}{262{,}289} \approx .1497$

53. $P(\text{Not Delta} \mid \text{Delay}) = \dfrac{48{,}187}{58{,}536} \approx .8232$

55. Based on exercises 47 – 50, Delta has the highest probability of an on-time flight.

57. $P(\text{at least 1 hit}) = 1 - P(\text{no hit})$
$P(\text{no hit}) = P(\text{no hits in the first four})$
$= (1 - .32)(1 - .16)(1 - .08)(1 - .04)$
$= .5045$
$P(\text{at least 1 hit}) = 1 - .5045 = .4955$

59. **a.** The probability of all 3 flights will be on time $(.87)^3 \approx .6585$.

b. The assumption is not very realistic. If a flight from Seattle to Chicago is canceled because of bad weather in Seattle, and that same aircraft is used for a flight from Chicago to Detroit, the canceling of one flight could affect the other, that is, they are dependent.

Copyright © 2019 Pearson Education, Inc.

61. a. The probability of success with one component is

$1 - .03 = .97;$

with 2, $1 - (.03)^2 = .9991;$

with 3, $1 - (.03)^3 = .999973;$

with 4, $1 - (.03)^4 = .99999919.$

Therefore, 4 (the original and 3 backups) will do the job.

b. Answers vary. It is probably reasonable to assume independence here so long as the cause of failure is some internal defect and the components are from different manufacture lots.

63. Let A be the event "hardly any confidence in banks" and B be the event "hardly any confidence in major corporations." Since $P(A) \cdot P(B)$ is not equal to $P(A \cap B)$, A and B are not independent. Rather, they are dependent events.

Section 8.6 Bayes' Formula

For Exercise 1:

$P(M) = .4$: M — $P(N \mid M) = .3$: N .12; $P(N' \mid M) = .7$: N' .28

$P(M') = .6$: M' — $P(N \mid M') = .4$: N .24; $P(N' \mid M') = .6$: N' .36

1. $P(M \mid N) = \dfrac{.12}{.12 + .24} = \dfrac{.12}{.36} = \dfrac{1}{3}$

For Exercises 3–6:

$P(R_1) = .05$: $P(Q \mid R_1) = .4$: Q .02; $P(Q' \mid R_1) = .6$: Q' .03

$P(R_2) = .6$: $P(Q \mid R_2) = .3$: Q .18; $P(Q' \mid R_2) = .7$: Q' .42

$P(R_3) = .35$: $P(Q \mid R_3) = .6$: Q .21; $P(Q' \mid R_3) = .4$: Q' .14

3. $P(R_1 \mid Q) = \dfrac{.02}{.02 + .18 + .21} = \dfrac{.02}{.41} = \dfrac{2}{41} \approx .0488$

5. $P(R_3 \mid Q) = \dfrac{.21}{.02 + .18 + .21} = \dfrac{.21}{.41} \approx .5122$

For Exercise 7:

$P(\text{jar 1}) = 1/2$: jar 1 — $P(\text{white} \mid \text{jar 1}) = 1/3$ — white 1/6

$P(\text{jar 2}) = 1/3$: jar 2 — $P(\text{white} \mid \text{jar 2}) = 2/3$ — white 2/9

$P(\text{jar 3}) = 1/6$: jar 3 — $P(\text{white} \mid \text{jar 3}) = 1/2$ — white 1/12

7. $P(\text{jar 2} \mid \text{white})$

$= \dfrac{\frac{2}{9}}{\frac{1}{6} + \frac{2}{9} + \frac{1}{12}} = \dfrac{8}{17} \approx .4706$

9. $P(IT \mid \text{yield} \geq 2.00\% \text{under } 60)$

$= \dfrac{P(\geq 2.00\%)P(\text{IT} \mid \geq 2.00\%)}{P(\geq 2.00\%)P(\text{IT} \mid \geq 2.00\%) + P(< 2.00\%)P(\text{IT} \mid < 2.00\%)}$

$= \dfrac{(.1347)(.2941)}{(.1347)(.2941) + (.8653)(.4462)}$

$= \dfrac{.03961527}{.42571213} \approx .0931$

11. $P(\text{some college or assos} \mid \geq \$75\text{K})$

$= \dfrac{P(\text{some col.})P(\geq \$75\text{K} \mid \text{some col})}{[P(\text{HS})P(\geq \$75\text{K} \mid \text{HS}) + P(\text{some col.})P(\geq \$75\text{K} \mid \text{some col}) + P(\text{Bach.})P(\geq \$75\text{K} \mid \text{Bach})]}$

$= \dfrac{(.279)(.135)}{(.333)(.074) + (.279)(.135) + (.389)(.370)}$

$= \dfrac{.037665}{.206237} \approx .1826$

13. $P(\text{HS} \mid <\$75\text{K})$

$= \dfrac{P(\text{HS})P(<\$75\text{K} \mid \text{HS})}{[P(\text{HS})P(<\$75\text{K} \mid \text{HS}) + P(\text{some col.})P(<\$75\text{K} \mid \text{some col}) + P(\text{Bach.})P(<\$75\text{K} \mid \text{Bach})]}$

$= \dfrac{(.333)(.926)}{(.333)(.926) + (.279)(.865) + (.389)(.630)}$

$= \dfrac{.308358}{.794763} \approx .3880$

Copyright © 2019 Pearson Education, Inc.

15. $P(\text{Amer} \mid \text{On time})$

$$= \frac{P(\text{Amer})P(\text{OT}|\text{Amer})}{[P(\text{Delta})P(\text{OT}|\text{Delta}) + P(\text{Other})P(\text{OT}|\text{Other}) + P(\text{Amer})P(\text{OT}|\text{Amer})]}$$

$$= \frac{(.1599)(.8019)}{(.1618)(.8691)+(.6783)(.7820)+(.1599)(.8019)}$$

$$= \frac{.12822381}{.79927479} \approx .1604$$

17. $P(\text{Other} \mid \text{Delay})$

$$= \frac{P(\text{Other})P(\text{Delay}|\text{Other})}{[P(\text{Delta})P(\text{Delay}|\text{Delta}) + P(\text{Other})P(\text{Delay}|\text{Other}) + P(\text{Amer})P(\text{Delay}|\text{Amer})]}$$

$$= \frac{(.6783)(.2180)}{(.1618)(.1309)+(.6783)(.2180)+(.1599)(.1981)}$$

$$= \frac{.1478694}{.20072521} \approx .7367$$

For Exercise 19:

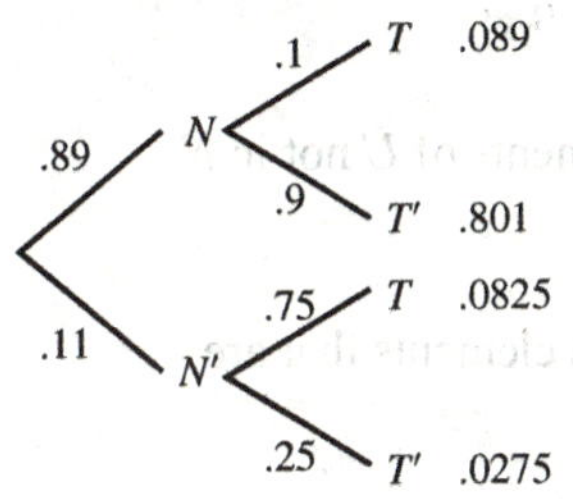

19. $P(N \mid T) = \dfrac{.089}{.089+.0825} = \dfrac{.089}{.1715} \approx .519$

21. $P(\text{GM} \mid \text{car})$

$$= \frac{P(\text{GM})P(\text{car}|\text{GM})}{[P(\text{GM})P(\text{car}|\text{GM}) + P(\text{Ford})P(\text{car}|\text{Ford}) + P(\text{Chrys.})P(\text{car}|\text{Chrys})]}$$

$$= \frac{(.3873)(.2765)}{(.3873)(.2765)+(.3385)(.2336)+(.2742)(.1253)}$$

$$= \frac{.10708845}{.22051931} \approx .4856$$

23. $P(\text{Ford} \mid \text{truck})$

$$= \frac{P(\text{Ford})P(\text{truck}|\text{Ford})}{[P(\text{GM})P(\text{truck}|\text{GM}) + P(\text{Ford})P(\text{truck}|\text{Ford}) + P(\text{Chrys.})P(\text{truck}|\text{Chrys})]}$$

$$= \frac{(.3385)(.7664)}{(.3873)(.7235)+(.3385)(.7664)+(.2742)(.8747)}$$

$$= \frac{.2594264}{.77948069} \approx .3328$$

25. $P(\text{adv.} \mid \text{NYSE})$

$$= \frac{P(\text{NYSE})P(\text{adv.}|\text{NYSE})}{P(\text{NYSE})P(\text{adv}|\text{NYSE}) + P(\text{NASDAQ})P(\text{adv}|\text{NASDAQ})}$$

$$= \frac{(.5085)(.7240)}{(.5085)(.7240)+(.4915)(.6024)}$$

$$= \frac{.368154}{.6642336} \approx .5543$$

27. $P(\text{unch.} \mid \text{NYSE})$

$$= \frac{P(\text{NYSE})P(\text{unch}|\text{NYSE})}{P(\text{NYSE})P(\text{unch}|\text{NYSE}) + P(\text{NASDAQ})P(\text{unch}|\text{NASDAQ})}$$

$$= \frac{(.5085)(.0306)}{(.5085)(.0306)+(.4915)(.0540)}$$

$$= \frac{.0155601}{.0421011} \approx .3696$$

29. $P(\text{adv.} \mid \text{NASDAQ})$

$$= \frac{P(\text{NASDAQ})P(\text{adv}|\text{NASDAQ})}{P(\text{NYSE})P(\text{adv}|\text{NYSE}) + P(\text{NASDAQ})P(\text{adv}|\text{NASDAQ})}$$

$$= \frac{(.4915)(.6024)}{(.5085)(.7240)+(.4915)(.6024)}$$

$$= \frac{.2960796}{.6642336} \approx .4457$$

Copyright © 2019 Pearson Education, Inc.

31. $P(\text{male} \mid \text{VH})$

$$= \frac{P(\text{male})P(\text{VH}|\text{male})}{P(\text{male})P(\text{VH}|\text{male}) + P(\text{female})P(\text{VH}|\text{female})}$$

$$= \frac{(.492)(.287)}{(.492)(.287)+(.508)(.277)}$$

$$= \frac{.141204}{.28192} \approx .5009$$

33. $P(\text{female} \mid \text{NTH})$

$$= \frac{P(\text{female})P(\text{NTH}|\text{female})}{P(\text{male})P(\text{NTH}|\text{male}) + P(\text{female})P(\text{NTH}|\text{female})}$$

$$= \frac{(.508)(.165)}{(.492)(.149)+(.508)(.165)}$$

$$= \frac{.08382}{.157128} \approx .5335$$

35. $P(\text{male} \mid \text{PH})$

$$= \frac{P(\text{male})P(\text{PH}|\text{male})}{P(\text{male})P(\text{PH}|\text{male}) + P(\text{female})P(\text{PH}|\text{female})}$$

$$= \frac{(.492)(.564)}{(.492)(.564)+(.508)(.558)}$$

$$= \frac{.277488}{.560952} \approx .4947$$

Chapter 8 Review Exercises

1. $9 \in \{8, 4, -3, -9, 6\}$
Because 9 is not an element of the given set, the statement is false.

2. $4 \in \{3, 9, 7\}$
Because 4 is not an element of the given set, the statement is false.

3. $2 \notin \{0, 1, 2, 3, 4\}$
Because 2 is an element of the given set, the statement is false.

4. $0 \notin \{0, 1, 2, 3, 4\}$
Because 0 is an element of the given set, the statement is false.

5. $\{3, 4, 5\} \subseteq \{2, 3, 4, 5, 6\}$
The statement is true because every member of the first set is in the second set.

6. $(1, 2, 5, 8) \subseteq \{1, 2, 5, 10, 11\}$
This statement is false because 8 is an element of the first set but not of the second.

7. $\{1,5,9\} \subset \{1,5,6,9,10\}$
This statement is true because every member of the first set is a member of the second set and the second set contains at least one element not in the first set.

8. $0 \subseteq \varnothing$
This statement is false because the empty set has no subsets except itself. Also, 0 is an element, not a set, so it cannot be a subset.

9. $\{x \mid x$ is a national holiday$\}$ = {New Year's Day, Martin Luther King's Birthday, Presidents' Day, Memorial Day, Independence Day, Labor Day, Columbus Day, Veterans' Day, Thanksgiving, Christmas}

10. $\{x \mid x \text{ is an integer}, -3 \le x < 1\} = \{-3, -2, -1, 0\}$

11. {all counting numbers less than 5} = {1, 2, 3, 4}

12. $\{x \mid x$ is a leap year between 1999 and 2018$\}$ = {2000, 2004, 2008, 2012, 2016}

13. M' contains all the elements of U not in M.
$M' = \{B_1, B_2, B_3, B_6, B_{12}\}$

14. N' contains all the elements of U not in N.
$N' = \{B_3, B_6, B_{12}, D\}$

15. $M \cap N$ contains all the elements that are common to M and N.
$M \cap N = \{A, C, E\}$

16. $M \cup N$ contains all the elements in either M or N or both.
$M \cup N = \{A, B_1, B_2, C, D, E\}$

17. $M \cup N'$ contains all the elements either in M or not in N, or both.
$M \cup N' = \{A, B_3, B_6, B_{12}, C, D, E\}$

18. $M' \cap N$ contains all the elements in N and not in M.
$M' \cap N = \{B_1, B_2\}$

19. $A \cap C$ is the set of all female students older than 22.

20. $B \cap D$ is the set of all finance majors with a GPA > 3.5.

21. $A \cup D$ is the set of females or students with a GPA > 3.5.

Copyright © 2019 Pearson Education, Inc.

22. $A' \cap D$ is the set of male students with a GPA > 3.5.

23. $B' \cap C'$ is the set of non-finance majors who are 22 or younger

24. $B \cup A'$
Shade the region inside B as well as all of the region outside of A.

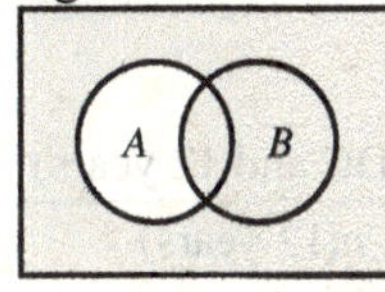

$B \cup A'$

25. $A' \cap B$
Shade all the region inside B that is also outside of A.

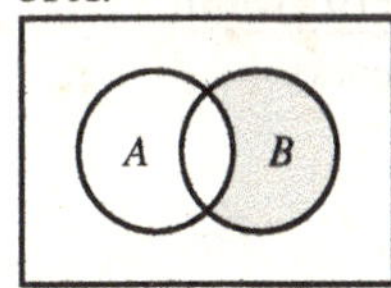

26. $A' \cap (B' \cap C)$
First choose the regions that are inside C and outside of B. From those regions, then shade the region outside of A.

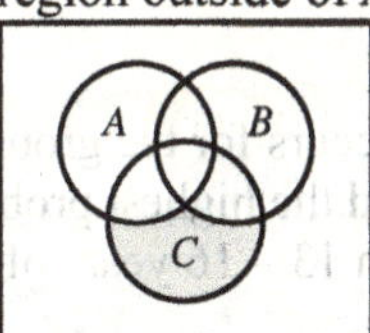

For Exercises 27–30, use the following Venn diagram. Let A be action movies, P be movies rated PG-13, and M be movies made after 2005. Since 20 movies were action, rated PG-13, and made after 2005, and 23 were rated PG-13 and made after 2005, 3 movies were rated PG-13 and made after 2005 only. Since 23 movies were action and made after 2005, then 3 were action and made after 2005 only. 24 movies were action and PG-13, so 24 – 20 = 4 was action and PG-13 only. 28 movies were action, so 28 – (20 + 3 + 4) = 1 were action only. Similar reasoning shows that there were 3 movies that were PG-13 only and 10 movies that were made after 2005 only.

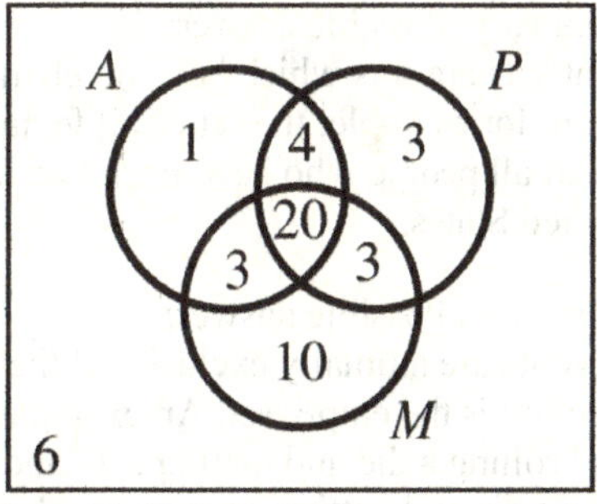

27. $1 + 3 = 4$
4 movies were action movies and were not rated PG–13.

28. $50 - 10 - 6 = 34$
34 movies were action or rated PG-13.

29. $50 - 1 - 6 = 43$
43 movies were made after 2005 or rated PG-13.

30. $50 - 1 - 4 - 3 - 3 - 20 - 3 - 6 = 10$
10 movies were neither action or rated PG-13, and made before 2005.

31. The sample space for rolling a die is {1, 2, 3, 4, 5, 6}.

32. The sample space for choosing a color and then a number is {(red, 10), (red, 20), (red, 30), (blue, 10), (blue, 20), (blue, 30), (green, 10), (green, 20), (green, 30)}.

33. {(rock, rock), (rock, pop), (rock, alternative), (pop, rock), (pop, pop), (pop, alternative), (alternative, rock), (alternative, pop), (alternative, alternative)}

34. No, the outcomes in the sample space are not equally likely because the probabilities for each type of music are not equal.

35. {(Dell, Epson), (Dell, HP), (Apple, Epson), (Apple, HP), (HP, Epson), (HP, HP)}

36. No, the outcomes in the sample space are not equally likely because the probabilities for each type of computer and printer are not equal.

37. "A customer buys neither" is written $E' \cap F'$. (This event can also be written as $(E \cup F)'$.)

38. "A customer buys at least one" is written $E \cup F$.

39. Answers vary. Possible answer:
This answer must be incorrect because any probability must be between 0 and 1 inclusive.

Copyright © 2019 Pearson Education, Inc.

40. Answers vary. Possible answer: Disjoint sets are sets which have no elements in common, for example, the set of all females and the set of all people who have been President of the United States.

41. Answers vary. Possible answer: Two events are mutually exclusive if their intersection is the empty set. An example is the events "rolling a die and getting a 3" and "rolling a die and getting an even number."

42. Answers vary. Possible answer: Disjoint sets and mutually exclusive events both have no elements in common, that is, their intersection is the empty set. Mutually exclusive events are in fact disjoint subsets of the sample space.

43. P(consumer discretionary or consumer staples) $= .1746 + .0411 = .2157$

44. P(information tech or industrials) $= .1594 + .1906 = .3500$

45. P(not financials) $= 1 - .1616 = .8384$

46. P(Not information Technology) $= 1 - .1594 = .8406$

47. P(Not Consumer Discretionary Or Staples) $= 1 - .1746 - .0411 = .7843$

48. P(Not Materials Or Industrials) $= 1 - .0989 - .1906 = .7105$

49. $P(\leq 12 \text{ years}) = \dfrac{248+566}{1913} = \dfrac{814}{1913} \approx .4255$

50. $P(\text{GDC or SC}) = \dfrac{339+1237}{1913} = \dfrac{1516}{1913} \approx .8238$

51. $P(> 16 \text{ years and SC}) = \dfrac{197}{1913} \approx .1030$

52. $P(< 12 \text{ years and HAC}) = \dfrac{57}{1913} \approx .0298$

53. $P(13-16 \text{ years or SC}) = \dfrac{808+1237-502}{1913} = \dfrac{1543}{1913} \approx .8066$

54. $P(\text{more than 16 years or HAC}) = \dfrac{291+337-45}{1913} = \dfrac{583}{1913} \approx .3048$

55. $P(\text{GDC}|{<}12 \text{ years}) = \dfrac{n(\text{GDC and} < 12 \text{ years})}{n(< 12 \text{ years})} = \dfrac{42}{248} \approx .1694$

56. $P(\text{GDC}|12 \text{ years}) = \dfrac{n(\text{GDC and 12 years})}{n(12 \text{ years})} = \dfrac{95}{566} \approx .1678$

57. $P(\text{GDC}|13\text{-}16 \text{ years}) = \dfrac{n(\text{GDC and 13-16 years})}{n(13\text{-}16 \text{ years})} = \dfrac{153}{808} \approx .1894$

58. $P(\text{GDC}|{>}\ 16 \text{ years}) = \dfrac{n(\text{GDC and} > 16 \text{ years})}{n(> 16 \text{ years})} = \dfrac{49}{291} \approx .1684$

59. The lowest probability occurs for the group with 12 years of education and the highest probability occurs for the group with 13 – 16 years of education.

60. No; all four probabilities are very close in value.

61.

		2nd Parent	
		N_2	T_2
1st	N_1	N_1N_2	N_1T_2
Parent	T_1	T_1N_2	T_1T_2

62. $P(\text{child has disease}) = P(T_1T_2) = \dfrac{1}{4}$

63. There are 4 possible combinations, but only 2 have a normal cell combined with a trait cell $(N_1T_2,\ T_1N_2)$.

$P(\text{child is carrier}) = \dfrac{2}{4} = \dfrac{1}{2}$

Copyright © 2019 Pearson Education, Inc.

64. $P(\text{child is neither carrier nor has disease})$
$= P(N_1N_2) = \frac{1}{4}$

65. There are 36 possibilities, with 5 having a sum of 8:
(4, 4), (3, 5), (5, 3), (2, 6) and (6, 2).
$P(8) = \frac{5}{36} \approx .1389$

66. $P(\text{no more than 4})$
$= P(4) + P(3) + P(2)$
$= \frac{3}{36} + \frac{2}{36} + \frac{1}{36} = \frac{6}{36} \approx .1667$

67. $P(\text{at least 9})$
$= P(9) + P(10) + P(11) + P(12)$
$= \frac{4}{36} + \frac{3}{36} + \frac{2}{36} + \frac{1}{36} = \frac{10}{36} \approx .2778$

68. $P(\text{odd and greater than 8})$
$= P(9) + P(11)$
$= \frac{4}{36} + \frac{2}{36} = \frac{6}{36} \approx .1667$

69. A roll less than 4 means 3 or 2. There are 2 ways to get 3 and 1 way to get 2. Hence,
$P(2 \mid \text{less than } 4) = \frac{1}{3} \approx .3333$.

70. $P(7 \mid \text{at least one is a } 4) = \frac{2}{11} \approx .1818$, since there are 11 possibilities with at least one of the dice being a 4 {(4, 1), (1, 4), (4, 2), (2, 4), (4, 3), (3, 4), (4, 4), (5, 4), (4, 5), (6, 4), (4, 6)} with only (4, 3) and (3, 4) having a sum of 7.

For Exercises 71–74, draw a Venn diagram and use the given information to fill in the probabilities for each of the regions.

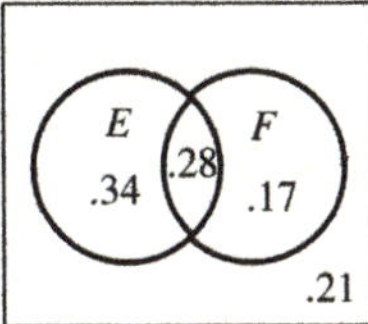

71. $P(E \cup F) = .34 + .28 + .17 = .79$

72. $P(E \cap F') = .34$

73. $P(E' \cup F) = .28 + .17 + .21 = .66$

74. $P(E' \cap F') = 1 - (.34 + .28 + .17) = .21$

75. $P(E' \mid F) = 1 - P(E \mid F) = 1 - .3 = .7$

76. $P(E \mid F') = \frac{.08}{.6} = \frac{2}{15} \approx .1333$

77. Answers vary.

78. Answers vary.

79. $\frac{419}{772} \approx .5427$

80. $\frac{1635}{2983} \approx .5481$

81. $\frac{.5452}{.4548} = \frac{5452}{4548} = \frac{1363}{1137}$

82. $\frac{.4879}{.5121} = \frac{4879}{5121}$

Use the following probability tree for Exercises 83–86.

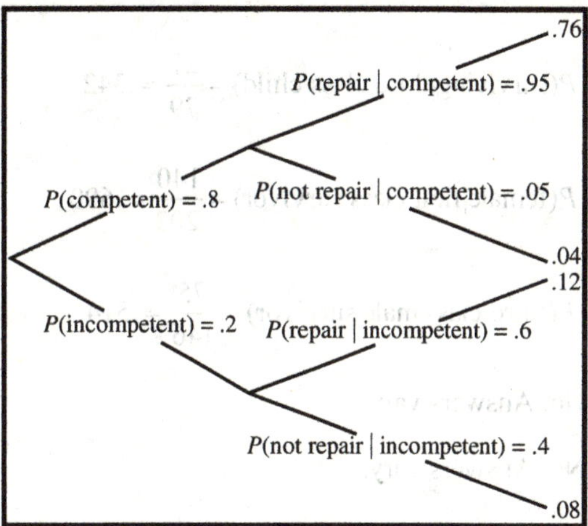

83. $P(\text{competent} \mid \text{repaired})$
$= \frac{.76}{.76 + .12} = \frac{.76}{.88} = \frac{19}{22} \approx .8636$

84. $P(\text{incompetent} \mid \text{repaired})$
$= \frac{.12}{.76 + .12} = \frac{12}{88} = \frac{3}{22} \approx .1364$

85. $P(\text{competent} \mid \text{not repaired})$
$= \frac{.04}{.04 + .08} = \frac{.04}{.12} = \frac{1}{3} \approx .3333$

86. $P(\text{incompetent} \mid \text{not repaired})$
$= \frac{.08}{.04 + .08} = \frac{8}{12} = \frac{2}{3} \approx .6667$

Copyright © 2019 Pearson Education, Inc.

87. Let D mean defective.

$P(D) = .17(.04)+.39(.02)+.35(.07)+.09(.03) = .0418$

$$P(4 \mid D) = \frac{P(4 \cap D)}{P(D)} = \frac{.09(.03)}{.0418} = \frac{.0027}{.0418} \approx .0646$$

88. $$P(2 \mid D) = \frac{P(2 \cap D)}{P(D)} = \frac{.39(.02)}{.0418} = \frac{.0078}{.0418} \approx .1866$$

89. $P(\text{second class}) = \frac{357}{1316} \approx .271$

90. $P(\text{surviving}) = \frac{499}{1316} \approx .379$

91. $P(\text{surviving} \mid \text{first class}) = \frac{203}{325} \approx .625$

92. $P(\text{surviving} \mid \text{third class child}) = \frac{27}{79} \approx .342$

93. $P(\text{female} \mid \text{first class survivor}) = \frac{140}{203} \approx .690$

94. $P(\text{third class} \mid \text{male survivor}) = \frac{75}{146} \approx .514$

95. No. Answers vary.

96. No. Answers vary.

Case 8 Medical Diagnosis

1. $P(T' \mid D') = .999 \Rightarrow P(T \mid D') = 1 - .999 = .001$

2. $P(D) = .005 \Rightarrow P(D') = 1 - .005 = .995$

$$P(D \mid T) = \frac{P(D)P(T \mid D)}{P(D)P(T \mid D) + P(D')P(T \mid D')} = \frac{.005(.99)}{.005(.99) + .995(.001)} = \frac{.00495}{.005945} \approx .833$$

3. $$P(D \mid T) = \frac{P(D)P(T \mid D)}{P(D)P(T \mid D) + P(D')P(T \mid D')} = \frac{.0005(.99)}{.0005(.99) + .9995(.001)} = \frac{.000495}{.0014945} \approx .331$$

4. $P(D) = .0145 \Rightarrow P(D') = 1 - .005 = .9855$

$$P(D \mid T) = \frac{P(D)P(T \mid D)}{P(D)P(T \mid D) + P(D')P(T \mid D')} = \frac{.0145(.8)}{.0145(.8) + .9855(.1)} = \frac{.0116}{.11015} \approx .1053$$

5. $$P(D \mid T) = \frac{P(D)P(T \mid D)}{P(D)P(T \mid D) + P(D')P(T \mid D')} = \frac{.0145(.9)}{.0145(.9) + .9855(.05)} = \frac{.01305}{.062325} \approx .2094$$

Copyright © 2019 Pearson Education, Inc.

Chapter 9 Counting, Probability Distributions, and Further Topics in Probability

Section 9.1 Probability Distributions and Expected Value

1. The number of possible samples is 16.

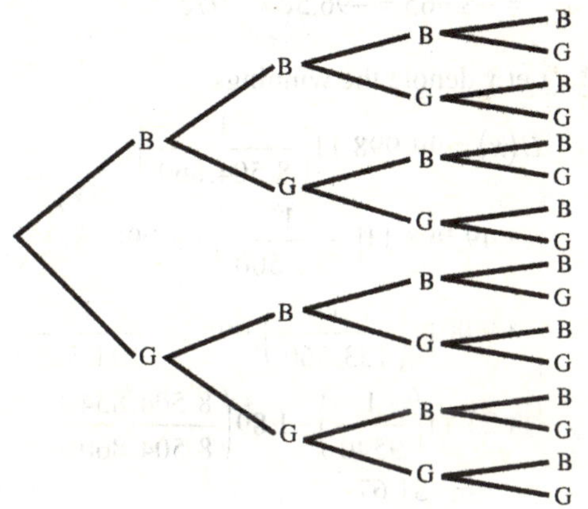

Number of Boys	$P(x)$
0	$\frac{1}{16} \approx .063$
1	$\frac{4}{16} = .25$
2	$\frac{6}{16} = .375$
3	$\frac{4}{16} = .25$
4	$\frac{1}{16} \approx .063$

3. Let x be the number of queens drawn. Then x can take on values 0, 1, 2, or 3.

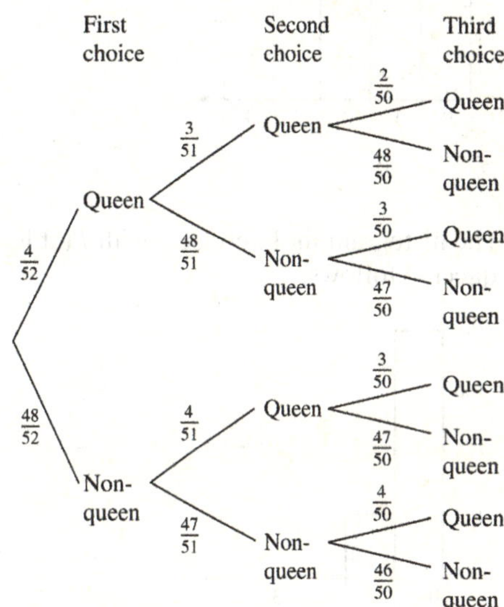

Number of queens	$P(x)$
0	$\frac{48}{52}\cdot\frac{47}{51}\cdot\frac{46}{50} \approx .7826$
1	$\frac{4}{52}\cdot\frac{48}{51}\cdot\frac{47}{50}+\frac{48}{52}\cdot\frac{4}{51}\cdot\frac{47}{50}+\frac{48}{52}\cdot\frac{47}{51}\cdot\frac{4}{50} \approx .2042$
2	$\frac{4}{52}\cdot\frac{3}{51}\cdot\frac{48}{50}+\frac{4}{52}\cdot\frac{48}{51}\cdot\frac{3}{50}+\frac{48}{52}\cdot\frac{4}{51}\cdot\frac{3}{50} \approx .0130$
3	$\frac{4}{52}\cdot\frac{3}{51}\cdot\frac{2}{50} \approx .0002$

Copyright © 2019 Pearson Education, Inc.

5. 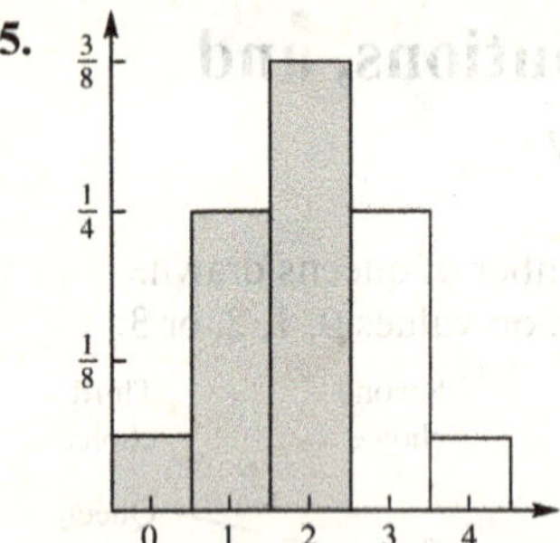

7. The histogram for Exercise 3 with P(at least one queens) follows.

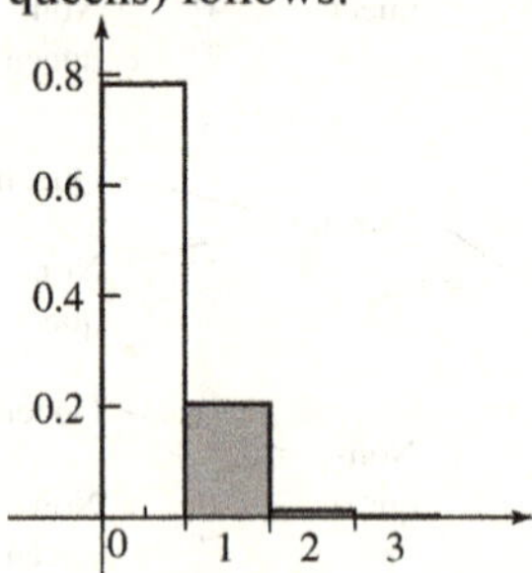

9. expected value $= 1(.1) + 3(.5) + 5(.2) + 7(.2) = 4$

11. expected value
$= 0(.21) + 2(.24) + 4(.21) + 8(.17) + 16(.17) = 5.4$

13. $E(x) = 1(.2) + 2(.3) + 3(.1) + 4(.4) = 2.7$

15. $E(x) = 1(.3) + 2(.25) + 3(.2) + 4(.15) + 5(.1) = 2.5$

17. $E(x) = 1\left(\frac{18}{38}\right) - 1\left(\frac{20}{38}\right) = -\frac{2}{38} = -\frac{1}{19} \approx -0.0526$

19. You have one chance in a thousand of winning \$500 on a \$1 bet for a net return of \$499. In the 999 other outcomes you lose your dollar

$$E(x) = 499\left(\frac{1}{1000}\right) + (-1)\left(\frac{999}{1000}\right) = -\frac{500}{1000} = -\$.50 \text{ or } -50¢$$

21. Let x denote the winnings.

$$E(x) = 49,999\left(\frac{1}{2,000,000}\right) + 9,999\left(\frac{2}{2,000,000}\right) - 1\left(\frac{1,999,997}{2,000,000}\right) = -\$.965 = -96.5¢ \approx -97¢$$

23. Let x denote the winnings.

$$E(x) = 99,998.11\left(\frac{1}{8,504,860}\right) + 49,998.11\left(\frac{1}{302,500}\right) + 9,998.11\left(\frac{1}{282,735}\right) + 998.11\left(\frac{1}{153,560}\right) + 98.11\left(\frac{1}{104,560}\right) + 23.11\left(\frac{1}{9540}\right) - 1.89\left(\frac{8,504,854}{8,504,860}\right) \approx -\$1.67$$

25. $E(x) = 0(.2643) + 1(.4173) + 2(.2470) + 3(.0650) + 4(.0064) = 1.1319$

27. The distribution is not valid since no probability can be less than 0.

29. The sum of the probabilities total 1.0, so this is a valid distribution.

31. Let x denote the missing probability. Then,
$.01 + .09 + .25 + .45 + .05 + x = 1.0 \Rightarrow .85 + x = 1.0 \Rightarrow x = 1.0 - .85 = .15$

33. Let x denote the missing probability. Then
$.20 + x + .25 + .30 = 1.0 \Rightarrow x + .75 = 1.0 \Rightarrow x = 1.0 - .75 = .25$

35. Let x and y denote the missing probabilities. Then
$.10 + .10 + .20 + .25 + .05 + x + y = 1.0 \Rightarrow .70 + x + y = 1.0 \Rightarrow x + y = 1.0 - .7 = .3$
Answers may vary. One possible answer would be .15 and .15.

37. $E(x) = .0007[100(15,000) + 250(10,000) + 500(5,000)] = .0007(6,500,000) = \$4550$

39. $E(x) = 0(.5697) + 1(.3526) + 2(.0727) + 3(.005) = .513$

Copyright © 2019 Pearson Education, Inc.

41. **a.** Let x denote the cost of using each antibiotic. For Drug A,
$E(x) = .7(160) + .3(350) = \$217.$
For Drug B,
$E(x) = .90(210) + .10(400) = \$229.$

b. Drug A, since the total expected cost is less.

43. **a.** $E(x) = 2{,}000{,}000(.5) + 1{,}000{,}000(.5) = 1{,}500{,}000$ pounds

b. $E(x) = 2{,}000{,}000(.64) + 1{,}000{,}000(.36) = 1{,}640{,}000$ pounds

Section 9.2 The Multiplication Principle, Permutations and Combinations

1. $_4P_2 = \frac{4!}{(4-2)!} = \frac{4!}{2!} = 4 \cdot 3 = 12$

3. $_8C_5 = \frac{8!}{5!(8-5)!} = \frac{8!}{3!5!} = \frac{8 \cdot 7 \cdot 6}{3 \cdot 2 \cdot 1} = 56$

5. $_8P_1 = \frac{8!}{(8-1)!} = \frac{8!}{7!} = 8$

7. $4! = 4 \cdot 3 \cdot 2 \cdot 1 = 24$

9. $_9C_6 = \frac{9!}{6!(9-6)!} = \frac{9!}{6!3!} = \frac{9 \cdot 8 \cdot 7}{3 \cdot 2 \cdot 1} = 84$

11. $_{13}P_3 = \frac{13!}{(13-3)!} = \frac{13!}{10!} = 13 \cdot 12 \cdot 11 = 1716$

13. $_{25}P_5 = 6{,}375{,}600$

15. $_{14}P_5 = 240{,}240$

17. $_{18}C_5 = \frac{18!}{5!13!} = \frac{18 \cdot 17 \cdot 16 \cdot 15 \cdot 14}{5 \cdot 4 \cdot 3 \cdot 2 \cdot 1} = 8568$

19. $_{28}C_{14} = 40{,}116{,}600$

21. If $0! = 0$, $_4P_4 = \frac{4!}{(4-4)!} = \frac{24}{0} =$ undefined

23. Since there are 264 kinds of glass tile, 31 colors of grout, 12 kinds of mortar trowels, and 17 kinds of mortar, by the multiplication principle there would be $264 \cdot 31 \cdot 12 \cdot 17 = 1{,}669{,}536$ different tile projects.

25. $6 \cdot 8 \cdot 4 \cdot 3 = 576$
There are 576 varieties of autos available.

27. Yes; Since a social security number has 9 digits with no restrictions, there are $10^9 = 1{,}000{,}000{,}000$ (1 billion) different social security numbers. This is enough for every one of the people in the United States to have a social security number.

29. Since a zip code has nine digits with no restrictions, there are 10^9 or 1,000,000,000 different 9-digit zip codes.

31. There are $12 \cdot 10 = 120$ different ways Sherri can buy 1 Pantene shampoo and 1 Pantene conditioner.

33. **a.** There are 8 possibilities for the first digit, 2 possibilities for the second digit, and 10 possibilities for the last digit. The total number of possible area codes is
$8 \cdot 2 \cdot 10 = 160$
There are 8 possibilities for the first digit and 10 possibilities for each of the next six digits. The total number of phone numbers is
$8 \cdot 10 \cdot 10 \cdot 10 \cdot 10 \cdot 10 \cdot 10 = 8 \cdot 10^6 = 8{,}000{,}000$

b. Some numbers, like 911, 800, and 900, are reserved for special purposes.

35. In this new plan, there would be 8 possibilities for the first digit, 2 possibilities for the second digit, and 10 possibilities for each of the last 2 digits. The total number of area codes is
$8 \cdot 2 \cdot 10 \cdot 10 = 1600$.

37. Answers vary. Possible answer:
A permutation of a elements ($a \geq 1$) from a set of b elements is any arrangement, without repetition, of the a elements.

39. Since order makes a difference, the number of arrangements is
$_{12}P_5 = \frac{12!}{(10-5)!} = \frac{12!}{5!} = 95{,}040.$

Copyright © 2019 Pearson Education, Inc.

41. Order is important, and there are no repeats, so there are ${}_{24}P_3 = \frac{24!}{(24-3)!} = 24 \cdot 23 \cdot 22 = 12{,}144$ ways to name a fraternity using 3 Greek letters.

43. Four people are being selected. Since each receives a different job, order is important. The total number of different officer selections is

$${}_{32}P_4 = \frac{32!}{(32-4)!} = \frac{32!}{28!} = 32 \cdot 31 \cdot 30 \cdot 29$$
$$= 863{,}040$$

45. ${}_{17}P_2 = \frac{17!}{(17-2)!} = \frac{17!}{15!} = 17 \cdot 16 = 272$

47. This is a combinations problem.

a. ${}_{10}C_4 = \frac{10!}{4!6!} = 210$

b. ${}_{10}C_6 = \frac{10!}{6!4!} = 210$

49. Order is not important, so use a combination.

$${}_{11}C_4 = \frac{11!}{(11-4)!4!} = 330$$

There are 330 ways to select a group of 4 co-captains from a group of 11.

51. Answers vary. Possible answer: Combinations are not ordered, while permutations are ordered.

53. a. With repetition permitted, the tree diagram shows 9 different pairs.

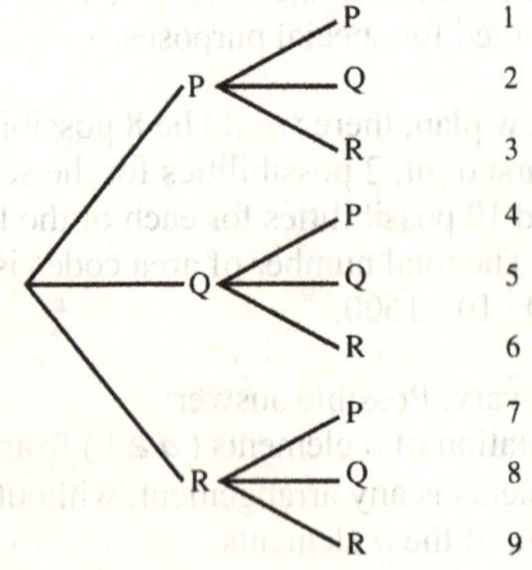

b. If repetition is not permitted, one branch is missing from each of the clusters of second branches, for a total of 6 different pairs.

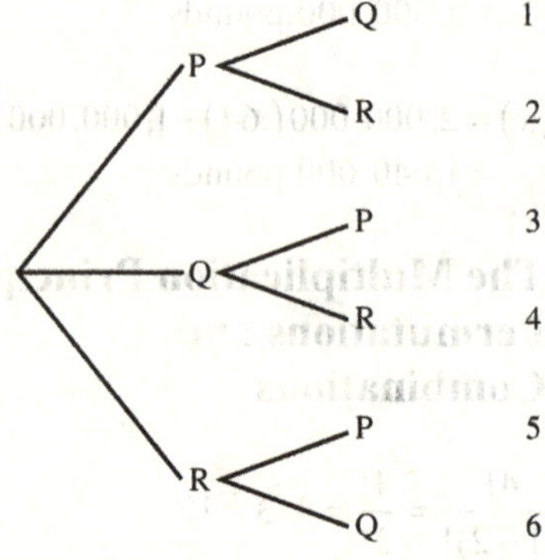

c. Find the number of combinations of 3 elements taken 2 at a time.

$${}_3C_2 = \frac{3!}{2!1!} = 3$$

No repetitions are allowed, so the answer cannot equal that for part a. However, since order does not matter, our answer is only half of the answer for part b. For example *PQ* and *QP* are distinct in b. but not in c. Thus, the answer differs from both a. and b.

55. These are combination problems since order is not important.

a. ${}_{21}C_5 = 20{,}349$ delegations

b. ${}_{11}C_5 = 462$ delegations

c. ${}_{10}C_3 \cdot {}_{11}C_2 = 120 \cdot 55 = 6600$ delegations

d. At least one means one or two or three or four or five.

$${}_{10}C_1 \cdot {}_{11}C_4 + {}_{10}C_2 \cdot {}_{11}C_3 + {}_{10}C_3 \cdot {}_{11}C_2 + {}_{10}C_4 \cdot {}_{11}C_1 + {}_{10}C_5$$
$$= 3300 + 7425 + 6600 + 2310 + 252$$
$$= 19{,}887$$

19,887 delegations would have at least one Democrat.

Copyright © 2019 Pearson Education, Inc.

57. a. The number of different double-scoops will be ${}_{25}C_2 = 300$.

b. The number of different triple-scoops will be ${}_{25}P_3 = 13{,}800$ since order matters.

59. a. Since order is not important, this is a combination problem.
${}_{99}C_6 = 1{,}120{,}529{,}256$

b. Since order is important, this is a permutation problem.
${}_{99}P_6 = 806{,}781{,}064{,}320$

61. Using the multiplication principle, we have $3 \cdot 3 \cdot 3 \cdot 3 = 81$

63. It is not possible since $26^3 = 17{,}576$ (the number of different 3-initial names) and the biologist needs 52,000 names. Actually, 4-initial names would do the biologist's job since $26^4 = 456{,}976$.

65. Since order is not important, these are combination problems.

a. ${}_5C_3 = 10$

b. Since you are taking 3 Diet Coke, and only 1 Diet Coke exists, this situation is impossible. The answer is 0.

c. ${}_3C_3 = 1$

d. ${}_5C_2 \cdot {}_1C_1 = 10 \cdot 1 = 10$

e. ${}_5C_2 \cdot {}_3C_1 = 10 \cdot 3 = 30$

f. ${}_3C_2 \cdot {}_5C_1 = 3 \cdot 5 = 15$

g. Again, since you are picking 2 Diet Coke and only 1 Diet Coke exists, this situation is impossible. The answer is 0.

67. Since order is not important, we use combinations. ${}_{35}C_{20} = 3{,}247{,}943{,}160$
There are 3,247,943,160 ways to select 20 volunteers from the group of 35.

69. a. martini
There are 7 total letters, $1m$, $1a$, $1r$, $1t$, $2i$'s, and $1n$.

$$\frac{7!}{1!1!1!1!2!1!} = 2520$$

b. nunnery
There are 7 letters, $3n$'s, $1u$, $1e$, $1r$, and $1y$.

$$\frac{7!}{3!1!1!1!1!} = 840$$

c. grinding
There are 8 letters, $2g$'s, $1r$, $2i$'s, $2n$'s, and $1d$.

$$\frac{8!}{2!1!2!2!1!} = 5040$$

71. a. The key word is "distinguishable", so use permutations. Total of 12 dinners:
${}_{12}P_{12} = 479{,}001{,}600$ ways

b. Since dinners of the same company are considered identical, the problem is to find the number of different arrangements of the three colors. ${}_3P_3 = 6$

c. $\dfrac{12!}{4!3!5!} = 27{,}720$

Section 9.3 Applications of Counting

1. There are ${}_{24}C_3 = 2024$ ways to select 3 tiles from the batch of 24. There are ${}_5C_1 = 5$ ways to select one defective tile from the batch, so there are ${}_{19}C_2 = 171$ ways to choose 2 good tiles from the 19 good tiles in the batch. Thus,

$$P(1 \text{ defective}) = \frac{{}_5C_1 \cdot {}_{19}C_2}{{}_{24}C_3} = \frac{5 \cdot 171}{2024} \approx .4224.$$

3. There are 8 computers, 5 non-defective, and a sample of size 1 is chosen.

$$P(\text{no defectives}) = \frac{{}_5C_1}{{}_8C_1} = \frac{5}{8}$$

5. There are 8 computers, 5 non-defective, and a sample of size 3 is chosen.

$$P(\text{no defectives}) = \frac{{}_5C_3}{{}_8C_3} = \frac{10}{56} = \frac{5}{28}$$

Copyright © 2019 Pearson Education, Inc.

7. There are 42 prizes, 10 of which are \$100 prizes; 3 are drawn.

$$P(\text{all are \$100 prizes}) = \frac{{}_{10}C_3}{{}_{242}C_3} = \frac{120}{2,332,880} \approx .00005$$

9. There are 42 prizes, 20 of which are \$25 prizes and 200 "dummy" tickets; 3 are drawn.

$$P(\text{two \$25 prizes}) = \frac{{}_{20}C_2 \cdot {}_{200}C_1}{{}_{242}C_3} = \frac{190 \cdot 1 \cdot 200}{2,332,880} \approx .0163$$

11. There are 200 dummy tickets, and all 3 must come from that group.
P(no winning ticket)

$$= \frac{{}_{200}C_3}{{}_{242}C_3} = \frac{1,313,400}{2,332,880} \approx .5630$$

13. Since order does not make a difference, the number of 2-card hands is ${}_{52}C_2 = 1326$.

15. There are 48 non-deuces (2's) in a deck of cards.

$$P(\text{no deuces}) = \frac{{}_{48}C_2}{{}_{52}C_2} = \frac{1128}{1326} \approx .8507$$

17. From Exercise 13, we know there are 1326 different 2-card hands. To see how many hands have the same suit, there are ${}_{13}C_2 = 78$ ways to get a 2-card hand of one particular suit. Because there are 4 different suits, there are 4 · 78 = 312 hands of the same suit. Therefore, there are 1326 – 312 = 1014 hands of different suits.

$$P(\text{different suits}) = \frac{1014}{1326} \approx .7647$$

19. P(no more than 1 diamond)
= P(no diamonds) + P(1 diamond)
Because there are 39 non-diamonds in a deck,

$$P(\text{no more than 1 diamond}) = \frac{741}{1326} + \frac{39 \cdot 13}{1326} = \frac{1248}{1326} \approx .9412.$$

Alternate solution:

$$P(\text{at least 1 black card}) = 1 - P(\text{2 diamonds}) = 1 - \frac{{}_{13}C_2}{{}_{52}C_2} = \frac{78}{1326} \approx .9412$$

21. Answer varies. Possible answer:
The advantage of using this rule is that it is many times easier to calculate the probability of the complement of an event than of the given event. For example in dealing a hand of 5 cards, if E is the event to get at least one heart, it is much easier to calculate $P(E)'$, the probability of getting no hearts and subtracting it from 1.

23. There are 22 rocks in all, so there are ${}_{22}C_5 = 26,334$ ways to select 5 rocks. There are ${}_7C_3 = 35$ ways to select 3 sedimentary rocks and ${}_{15}C_2 = 105$ ways to select the remaining 2 rocks. Therefore, the probability of selecting 5 rocks of which 3 are sedimentary is

$$\frac{{}_7C_3 \cdot {}_{15}C_2}{{}_{22}C_5} = \frac{35 \cdot 105}{26,334} \approx .1396.$$

25. a. We must pick all 6 of our numbers from the 99 total numbers.

$$P(\text{all 6}) = \frac{{}_6C_6}{{}_{99}C_6} \approx 8.9 \times 10^{-10}$$

b. $P(\text{all 6}) = \frac{1}{{}_{99}P_6} \approx 1.2 \times 10^{-12}$

27. The probability of two individuals independently selecting the winning numbers is

$$\left(\frac{1}{292,201,338}\right)^2 \approx 1.17 \times 10^{-17}.$$

29. a. $P(\text{7 men and 1 woman}) = \frac{{}_{51}C_7 \cdot {}_9C_1}{{}_{60}C_8} = \frac{115,775,100 \cdot 9}{2,558,620,845} \approx .4072$

b. $P(\text{5 men and 3 women}) = \frac{{}_{51}C_5 \cdot {}_9C_3}{{}_{60}C_8} = \frac{2,349,060 \cdot 84}{2,558,620,845} \approx .0771$

c. P(at least 1 woman)
= 1 – P(no women)
= 1 – P(all men)
= 1 – .2489 = .7511

Copyright © 2019 Pearson Education, Inc.

31. There are ${}_{20}C_7 = 77,520$ ways to select 7 volunteers from the pool of 20.

a. $P(\text{all 20-39}) = \frac{{}_{10}C_7}{{}_{20}C_7} = \frac{120}{77,520} \approx .0015$

b. $P(\text{5 20-39, 2 60 or older}) = \frac{{}_{10}C_5 \cdot {}_2C_2}{{}_{20}C_7} = \frac{252 \cdot 1}{77,520} \approx .0033$

c. $P(\text{3 40-59, 4 other}) = \frac{{}_8C_3 \cdot {}_{12}C_4}{{}_{20}C_7} = \frac{56 \cdot 495}{77,520} \approx .3576$

33. The probability that at least 2 of the 100 U.S. Senators have the same birthday is $1 - \frac{{}_{365}P_{100}}{(365)^{100}} \approx 1.$

35. There are ${}_{20}C_4 = 4845$ ways to pick the correct 4 numbers out of the 20 the state picks. There are ${}_{80}C_4 = 1,581,580$ ways to pick 4 numbers out of 80. The probability of winning \$55 is $\frac{{}_{20}C_4}{{}_{80}C_4} = \frac{4845}{1581580} \approx .0031.$

37. There are 60 losing numbers from which all 4 must be picked. There are 80 numbers from which 4 are picked. The probability of losing is $\frac{{}_{20}C_0 \cdot {}_{60}C_4}{{}_{80}C_4} = \frac{1 \cdot 487,635}{1,581,580} \approx .3083$

Section 9.4 Binomial Probability

1. $n = 10, p = .44, x = 6, 1 - p = .56$
$P(\text{exactly } 6) = {}_{10}C_6(.44)^6(.56)^4 \approx .1499$

3. $n = 10, p = .44, x = 0, 1 - p = 1 - .7 = .56$
$P(\text{none}) = {}_{10}C_0(.44)^0(.56)^{10} \approx .0030$

5. $P(\text{at least } 1) = 1 - P(\text{none})$
From Exercise 3, we have $P(\text{none}) \approx .0030$, so
$P(\text{at least } 1) = 1 - .0030 \approx .9970.$

For exercises 7–12, we have $n = 8,\ p = .39,$ and $1 - p = .61$.

7. $x = 2$; $P(\text{exactly } 2) = {}_8C_2(.39)^2(.61)^8 \approx .2194$.

9. $x = 0$; $P(\text{none}) = {}_8C_0(.39)^0(.61)^{10} \approx .0192$

11. $P(\text{at least } 1) = 1 - P(\text{none})$. From Exercise 9, we have $P(\text{none}) \approx .0192$, so
$P(\text{at least } 1) \approx 1 - .0192 \approx .9808.$

For exercises 13–16, we have $n = 5,\ p = \frac{1}{2},$ and $1 - p = \frac{1}{2}$.

13. $x = 5$; $P(\text{all heads}) = {}_5C_5\left(\frac{1}{2}\right)^5\left(\frac{1}{2}\right)^0 = \frac{1}{32}$

15. "No more than 3 heads" means 0 heads, 1 head, 2 heads, or 3 heads.

$P(\text{0 heads}) = {}_5C_0\left(\frac{1}{2}\right)^0\left(\frac{1}{2}\right)^5 = \frac{1}{32}$

$P(\text{1 head}) = {}_5C_1\left(\frac{1}{2}\right)^1\left(\frac{1}{2}\right)^4 = \frac{5}{32}$

$P(\text{2 heads}) = {}_5C_2\left(\frac{1}{2}\right)^2\left(\frac{1}{2}\right)^3 = \frac{10}{32}$

$P(\text{3 heads}) = {}_5C_3\left(\frac{1}{2}\right)^3\left(\frac{1}{2}\right)^2 = \frac{10}{32}$

$P(\text{no more than 3 heads}) = \frac{1}{32} + \frac{5}{32} + \frac{10}{32} + \frac{10}{32} = \frac{26}{32} = \frac{13}{16}$

17. Answer varies. Possible answer:
A problem involves a binomial experiment if the experiment is repeated several times, there are only two possible outcomes, and the repeated trials are independent.

For exercises 19–22, we have $n = 15, p = .272$, and $1 - p = .728$.

19. $P(3) = {}_{15}C_3(.272)^3(.728)^{12} \approx .2029$

21. $P(\text{at most } 2) = P(0) + P(1) + P(2)$
$= {}_{15}C_0(.272)^0(.728)^{15} + {}_{15}C_1(.272)^1(.728)^{14} + {}_{15}C_2(.272)^2(.728)^{13}$
$\approx .0086 + .04792 + .12532 \approx .1818$

Copyright © 2019 Pearson Education, Inc.

For exercises 23–28, we have $n = 9, p = .31$, and $1 - p = .69$.

23. $P(0) = {}_9C_0(.31)^0(.69)^9 \approx .0355$

25. $P(5 \text{ or more})$
$= P(5)+P(6)+P(7)+P(8)+P(9)$
$= {}_9C_5(.31)^5(.69)^4 + {}_9C_6(.31)^6(.69)^3 + {}_9C_7(.31)^7(.69)^2 + {}_9C_8(.31)^8(.69)^1 + {}_9C_9(.31)^9(.69)^0$
$\approx .08177+.02449+.00472+.00053+.00003$
$\approx .1115$

27. Since 31% of adults aged 18 – 29 have money invested in the stock market, we would expect that $.31 \cdot 500 = 155$ adults aged 18 – 29 are invested in the stock market.

29. $n=100,\ p=.027$, and $1-p=.973$
$P(\text{exactly 2 sets of twins})$
$= {}_{100}C_2(.027)^2(.973)^{98} \approx .2468$

31. $n=9,\ p=.11,\ 1-p=.89$
$P(2) = {}_9C_2(.11)^2(.89)^7 \approx .1927$

33. $P(\text{none}) = {}_9C_0(.11)^0(.89)^9 \approx .3504$

35. Since 11% of Americans are left-handed, $.11 \times 35 = 3.85$ would be expected to be left-handed from the sample of 35.

For exercises 36–40, we have $n = 16, p = .14$, and $1 - p = .86$.

37. $P(\text{at most } 3) = P(0)+P(1)+P(2)+P(3)$
$= {}_{16}C_0(.14)^0(.86)^{16} + {}_{16}C_1(.14)^1(.86)^{15} + {}_{16}C_2(.14)^2(.86)^{14} + {}_{16}C_3(.14)^3(.86)^{13}$
$\approx .08953+.23320+.28472+.21630 \approx .8237$

39. We would expect that $.14 \cdot 300 = 42$ US residents would have a great deal of confidence in banks and financial institutions.

For exercises 41–44, we have $n = 100, p = .44$, and $1 - p = .56$.

41. In order to find the probability, we will use the TI – 83 distribution mode. Enter the distribution mode by hitting the 2nd key followed by the VARS key (distribution mode on the secondary level). Since we are looking for the probability of 35 or less, use the binomcdf function. Enter the number of trial, probability of success and stopping number, all separated by commas to get the answer. Therefore the P(at most 35) = binomcdf (100, .44, 35) $\approx .0424$.

43. We would expect that $.44 \cdot 100 = 44$ would be made by one of the Big 3 automakers.

Section 9.5 Markov Chains

1. $\begin{bmatrix} \frac{1}{4} & \frac{3}{4} \end{bmatrix}$ could be a probability vector because it is a matrix with only one row containing only nonnegative entries whose sum is $\frac{1}{4}+\frac{3}{4}=1$.

3. $[0 \quad 1]$ could be a probability vector because it has only one row, the entries are nonnegative, and 0 + 1 = 1.

5. $[.3 \quad -.1 \quad .6]$ cannot be a probability vector because it has a negative entry, $-.1$.

7. $\begin{bmatrix} .7 & .2 \\ .5 & .5 \end{bmatrix}$ cannot be a transition matrix because the sum of the entries in the first row is $.7 + .2 = .9 \neq 1$.

Copyright © 2019 Pearson Education, Inc.

9. $\begin{bmatrix} \frac{4}{9} & \frac{1}{3} \\ \frac{1}{5} & \frac{7}{10} \end{bmatrix}$

This cannot be a transition matrix because the sum of the entries in row 1 is $\frac{4}{9}+\frac{1}{3}=\frac{7}{9}\neq 1$ and the sum of the entries in row 2 is $\frac{1}{5}+\frac{7}{10}=\frac{9}{10}\neq 1.$

11. $\begin{bmatrix} \frac{1}{2} & \frac{1}{4} & 1 \\ \frac{2}{3} & 0 & \frac{1}{3} \\ \frac{1}{3} & 1 & 0 \end{bmatrix}$

This could not be a transition matrix because the sum of the entries in row 1 is $\frac{1}{2}+\frac{1}{4}+1=\frac{7}{4}\neq 1,$ and the sum of the entries in row 3 is $\frac{1}{3}+1=\frac{4}{3}\neq 1.$

13. This is not a transition diagram because the sum of the probabilities for changing from state A to states A, B, and C is $\frac{1}{3}+\frac{1}{2}+1\neq 1.$

15. This is a transition diagram. The information given in this diagram can also be given by the following matrix.

$$\begin{array}{c} \\ A \\ B \\ C \end{array}\begin{array}{c} \begin{array}{ccc} A & B & C \end{array} \\ \begin{bmatrix} .6 & .20 & .20 \\ .9 & .02 & .08 \\ .4 & .0 & .6 \end{bmatrix} \end{array}$$

17. Let $A=\begin{bmatrix} .2 & .8 \\ .9 & .1 \end{bmatrix}$

A is a regular transition matrix since A^2 contains all positive entries.

19. Let $A=\begin{bmatrix} .25 & .40 & .30 & .05 \\ .18 & .23 & .59 & 0 \\ 0 & .15 & .36 & .49 \\ .28 & .32 & .24 & .16 \end{bmatrix}$

$$A^2=\begin{bmatrix} .1485 & .253 & .431 & .1675 \\ .0864 & .2134 & .4021 & .2981 \\ .1642 & .2453 & .3357 & .2548 \\ .1724 & .2728 & .3976 & .1572 \end{bmatrix}$$

A is a regular transition matrix since A^2 contains all positive entries.

21. $\begin{bmatrix} v_1 & v_2 \end{bmatrix}\begin{bmatrix} .3 & .7 \\ .4 & .6 \end{bmatrix}=[.3v_1+.4v_2 \quad .7v_1+.6v_2]$

We obtain the two equations
$v_1=.3v_1+.4v_2,\ v_2=.7v_1+.6v_2.$
Simplify these equations.
$0=-.7v_1+.4v_2,\ 0=.7v_1-.4v_2.$
These equations are dependent. Since **V** is a probability vector, $v_1+v_2=1\Rightarrow v_1=1-v_2.$
Substituting $1-v_2$ for v_1 in the first equation gives
$0=-.7(1-v_2)+.4v_2\Rightarrow 0=-.7+.7v_2+.4v_2\Rightarrow$
$0=-.7+1.1v_2\Rightarrow v_2=\frac{7}{11}.$

Then, $v_1=1-v_2=1-\frac{7}{11}=\frac{4}{11}.$

The equilibrium vector is $\begin{bmatrix} \frac{4}{11} & \frac{7}{11} \end{bmatrix}.$

Copyright © 2019 Pearson Education, Inc.

23. $[v_1 \quad v_2]\begin{bmatrix} \frac{5}{8} & \frac{3}{8} \\ \frac{7}{9} & \frac{2}{9} \end{bmatrix} = \begin{bmatrix} \frac{5}{8}v_1 + \frac{7}{9}v_2 & \frac{3}{8}v_1 + \frac{2}{9}v_2 \end{bmatrix}$

We obtain the two equations

$\frac{5}{8}v_1 + \frac{7}{9}v_2 = v_1, \quad \frac{3}{8}v_1 + \frac{2}{9}v_2 = v_2.$

Simplify these equations.

$-\frac{3}{8}v_1 + \frac{7}{9}v_2 = 0, \quad \frac{3}{8}v_1 - \frac{7}{9}v_2 = 0$

$v_1 + v_2 = 1 \Rightarrow v_1 = 1 - v_2.$

Substituting $1 - v_2$ for v_1 in the first equation gives

$-\frac{3}{8}(1 - v_2) + \frac{7}{9}v_2 = 0 \Rightarrow$

$-\frac{3}{8} + \frac{3}{8}v_2 + \frac{7}{9}v_2 = 0 \Rightarrow \frac{83}{72}v_2 = \frac{3}{8} \Rightarrow$

$v_2 = \frac{3}{8} \cdot \frac{72}{83} = \frac{27}{83} \approx .325$

Then, $v_1 = 1 - v_2 = 1 - .325 = .675.$

The equilibrium vector is

$[.675 \quad .325].$

25. Let **V** be the probability vector $[v_1 \quad v_2 \quad v_3].$

$[v_1 \quad v_2 \quad v_3]\begin{bmatrix} .25 & .35 & .4 \\ .1 & .3 & .6 \\ .55 & .4 & .05 \end{bmatrix} = [v_1 \quad v_2 \quad v_3]$

We obtain the three equations

$.25v_1 + .1v_2 + .55v_3 = v_1$

$.35v_1 + .3v_2 + .4v_3 = v_2$

$.4v_1 + .6v_2 + .05v_3 = v_3.$

Simplify these equations to get the dependent system

$-.75v_1 + .1v_2 + .55v_3 = 0$

$.35v_1 - .7v_2 + .4v_3 = 0$

$.4v_1 + .6v_2 - .95v_3 = 0.$

Since **V** is a probability vector,

$v_1 + v_2 + v_3 = 1.$

Solving this system we obtain

$v_1 = \frac{170}{563}, v_2 = \frac{197}{563},$ and $v_3 = \frac{196}{563}.$

Thus, the equilibrium vector is

$\begin{bmatrix} \frac{170}{563} & \frac{197}{563} & \frac{196}{563} \end{bmatrix}$

or $[.302 \quad .350 \quad .348]$ in decimal notation.

27. Let **V** be the probability vector $[v_1 \quad v_2 \quad v_3].$

$[v_1 \quad v_2 \quad v_3]\begin{bmatrix} .15 & .15 & .70 \\ .42 & .38 & .20 \\ .16 & .28 & .56 \end{bmatrix} = [v_1 \quad v_2 \quad v_3]$

We obtain the three equations

$.15v_1 + .42v_2 + .16v_3 = v_1$

$.15v_1 + .38v_2 + .28v_3 = v_2$

$.70v_1 + .20v_2 + .56v_3 = v_3.$

Simplify these equations to get the dependent system

$-.85v_1 + .42v_2 + .16v_3 = 0$

$.15v_1 - .62v_2 + .28v_3 = 0$

$.70v_1 + .20v_2 - .44v_3 = 0.$

Since **V** is a probability vector,

$v_1 + v_2 + v_3 = 1.$

Solving the above system of 4 equations using the Gauss-Jordan method, we obtain

$v_1 = \frac{542}{2357}, v_2 = \frac{655}{2357},$ and $v_3 = \frac{1160}{2357}.$

Thus, the equilibrium vector is

$\begin{bmatrix} \frac{542}{2357} & \frac{655}{2357} & \frac{1160}{2357} \end{bmatrix}$ or $[.230 \quad .278 \quad .492]$

in decimal notation.

Copyright © 2019 Pearson Education, Inc.

29. The following solution assumes the use of a TI-83 graphing calculator. Similar results can be obtained from other graphing calculators.
Store the given transition matrix as matrix [A].

$$A = \begin{bmatrix} .3 & .2 & .3 & .1 & .1 \\ .4 & .2 & .1 & .2 & .1 \\ .1 & .3 & .2 & .2 & .2 \\ .2 & .1 & .3 & .2 & .2 \\ .1 & .1 & .4 & .2 & .2 \end{bmatrix}; \quad A^2 = \begin{bmatrix} .23 & .21 & .24 & .17 & .15 \\ .26 & .18 & .26 & .16 & .14 \\ .23 & .18 & .24 & .19 & .16 \\ .19 & .19 & .27 & .18 & .17 \\ .17 & .2 & .26 & .19 & .18 \end{bmatrix}$$

$$A^3 = \begin{bmatrix} .226 & .192 & .249 & .177 & .156 \\ .222 & .196 & .252 & .174 & .156 \\ .219 & .189 & .256 & .177 & .159 \\ .213 & .192 & .252 & .181 & .162 \\ .213 & .189 & .252 & .183 & .163 \end{bmatrix}$$

$$A^4 = \begin{bmatrix} .2205 & .1916 & .2523 & .1774 & .1582 \\ .2206 & .1922 & .2512 & .1778 & .1582 \\ .2182 & .1920 & .2525 & .1781 & .1592 \\ .2183 & .1909 & .2526 & .1787 & .1595 \\ .2176 & .1906 & .2533 & .1787 & .1598 \end{bmatrix}$$

$$A^5 = \begin{bmatrix} .21932 & .19167 & .25227 & .17795 & .15879 \\ .21956 & .19152 & .25226 & .17794 & .15872 \\ .21905 & .19152 & .25227 & .17818 & .15898 \\ .21880 & .19144 & .25251 & .17817 & .15908 \\ .21857 & .19148 & .25253 & .17824 & .15918 \end{bmatrix}$$

The entry in row 2, column 4 of A^5 is .17794, which gives the probability that state 2 changes to state 4 after 5 repetitions.

Copyright © 2019 Pearson Education, Inc.

31. a. The transition matrix is

$$\begin{array}{c} \\ \text{Shot} \\ \text{No Shot} \end{array}\begin{array}{c} \text{Shot}\ \ \text{No Shot} \\ \begin{bmatrix} .85 & .15 \\ .40 & .60 \end{bmatrix} \end{array}$$

b. $$\begin{bmatrix} .26 & .74 \end{bmatrix}\begin{bmatrix} .85 & .15 \\ .40 & .60 \end{bmatrix} = \begin{bmatrix} .517 & .483 \end{bmatrix}$$

In year two, .517 of the patients will receive a flu shot and .483 will not receive a flu shot.

c.

$$\begin{bmatrix} v_1 & v_2 \end{bmatrix}\begin{bmatrix} .85 & .15 \\ .40 & .60 \end{bmatrix} = \begin{bmatrix} .85v_1 + .40v_2 & .15v_1 + .60v_2 \end{bmatrix}$$

We obtain the equations
$.85v_1 + .40v_2 = v_1$
$.15v_1 + .60v_2 = v_2$
which simplify to
$-.15v_1 + .40v_2 = 0$
$.15v_1 - .40v_2 = 0$

$v_1 + v_2 = 1 \Rightarrow v_1 = 1 - v_2.$
Substituting $1 - v_2$ for v_1 in the second equation gives
$.15(1 - v_2) - .40v_2 = 0 \Rightarrow .15 - .55v_2 = 0 \Rightarrow$
$v_2 = \frac{.15}{.55} = \frac{3}{11}$ and $v_1 = 1 - v_2 = \frac{8}{11}.$

The long-range trend is that $\frac{8}{11}$ of the patients will receive flu shots and $\frac{3}{11}$ will not receive flu shots. This is represented as $\begin{bmatrix} \frac{8}{11} & \frac{3}{11} \end{bmatrix}.$

33. The transition matrix would be

$$\begin{array}{c} \\ \text{Works} \\ \text{Doesn't Work} \end{array}\begin{array}{c} \text{Works}\ \ \text{Doesn't Work} \\ \begin{bmatrix} .91 & .09 \\ .68 & .32 \end{bmatrix} \end{array}.$$

$$\begin{bmatrix} v_1 & v_2 \end{bmatrix}\begin{bmatrix} .91 & .09 \\ .68 & .32 \end{bmatrix} = \begin{bmatrix} .91v_1 + .68v_2 & .09v_1 + .32v_2 \end{bmatrix}$$

The first equation is
$.91v_1 + .68v_2 = v_1$
$-.09v_1 + .68v_2 = 0.$
The second equation is
$.09v_1 + .32v_2 = v_2$
$.09v_1 - .68v_2 = 0.$
Since $v_1 + v_2 = 1,$
$v_1 = 1 - v_2.$
Substitute $1 - v_2$ for v_1 in the first equation.
$-.09(1 - v_2) + .68v_2 = 0 \Rightarrow$
$-.09 + .09v_2 + .68v_2 = 0 \Rightarrow .77v_2 = .09 \Rightarrow$
$v_2 = \frac{9}{77} \approx .117$

Then $v_1 = 1 - v_2 = 1 - \frac{9}{77} = \frac{68}{77} \approx .883.$

In the long run, the assembly line will run $\frac{68}{77}$ of the time.

35. a. The transition matrix is

$$\begin{array}{c} \\ \text{Red} \\ \text{White} \end{array}\begin{array}{c} \text{Red}\ \ \text{White} \\ \begin{bmatrix} .75 & .25 \\ .50 & .50 \end{bmatrix} \end{array}.$$

b. The probability vector is $\begin{bmatrix} .75 & .25 \end{bmatrix}.$

c. The square of the transition matrix is
$$\begin{bmatrix} .75 & .25 \\ .50 & .50 \end{bmatrix}\begin{bmatrix} .75 & .25 \\ .50 & .50 \end{bmatrix} = \begin{bmatrix} .6875 & .3125 \\ .625 & .375 \end{bmatrix}.$$
The transition matrix to the fourth power is
$$\begin{bmatrix} .6875 & .3125 \\ .625 & .375 \end{bmatrix}\begin{bmatrix} .6875 & .3125 \\ .625 & .375 \end{bmatrix} = \begin{bmatrix} .6680 & .3320 \\ .6641 & .3359. \end{bmatrix}.$$

$$\begin{bmatrix} .75 & .25 \end{bmatrix}\begin{bmatrix} .6680 & .3320 \\ .6641 & .3359 \end{bmatrix} = \begin{bmatrix} .667 & .333 \end{bmatrix}$$

Copyright © 2019 Pearson Education, Inc.

d.

$$\begin{bmatrix} v_1 & v_2 \end{bmatrix}\begin{bmatrix} .75 & .25 \\ .50 & .50 \end{bmatrix} = \begin{bmatrix} .75v_1 + .50v_2 & .25v_1 + .50v_2 \end{bmatrix}$$

The first equation is

$v_1 = .75v_1 + .50v_2$
$0 = -.25v_1 + .50v_2.$

The second equation is

$v_2 = .25v_1 + .50v_2$
$0 = .25v_1 - .50v_2.$

Substitute $1 - v_2$ for v_1 in the first equation.

$0 = -.25(1 - v_2) + .50v_2 \Rightarrow$
$0 = -.25 + .25v_2 + .50v_2 \Rightarrow .25 = .75v_2 \Rightarrow$
$\frac{1}{3} = v_2$ and $v_1 = 1 - v_2 = 1 - \frac{1}{3} = \frac{2}{3}.$

The long-range distribution of colors is $\begin{bmatrix} \frac{2}{3} & \frac{1}{3} \end{bmatrix}$ or $\frac{2}{3}$ red and $\frac{1}{3}$ white.

37. a.

$$\begin{bmatrix} .582 & .410 & .008 \end{bmatrix}\begin{bmatrix} .90 & .10 & 0 \\ .09 & .909 & .001 \\ 0 & .34 & .66 \end{bmatrix} \approx \begin{bmatrix} .5607 & .43361 & .005691 \end{bmatrix}$$

About 56.07% will own a home, 43.361% will rent, and .5691% will be homeless.

b.

$$\begin{bmatrix} v_1 & v_2 & v_3 \end{bmatrix}\begin{bmatrix} .9 & .10 & 0 \\ .09 & .909 & .001 \\ 0 & .34 & .66 \end{bmatrix} = \begin{bmatrix} v_1 & v_2 & v_3 \end{bmatrix}$$

$$\begin{aligned} .90v_1 + .09v_2 + 0v_3 &= v_1 \\ .10v_1 + .909v_2 + .34v_3 &= v_2 \\ 0v_1 + .001v_2 + .66v_3 &= v_3. \end{aligned}$$

Therefore, we have the system

$$\begin{aligned} v_1 + v_2 + v_3 &= 1 \\ .90v_1 + .09v_2 + 0v_3 &= v_1 \\ .10v_1 + .909v_2 + .34v_3 &= v_2 \\ 0v_1 + .001v_2 + .66v_3 &= v_3. \end{aligned}$$

Solve this system by the Gauss-Jordan method to obtain

$v_1 \approx .473,\ v_2 = .526,\ v_3 = .002.$ Therefore, 47.3% will be homeowners, 52.6% will be renters, and there will be 0.2% homeless in the long-range.

39. a. The transition matrix is

$$P = \begin{matrix} G_0 \\ G_1 \\ G_2 \end{matrix}\overset{\begin{matrix} G_0 & G_1 & G_2 \end{matrix}}{\begin{bmatrix} .85 & .10 & .05 \\ 0 & .80 & .20 \\ 0 & 0 & 1 \end{bmatrix}}.$$

We have 50,000 new policy holders, all in G_0. The probability vector for these people is $\overset{\begin{matrix} G_0 & G_1 & G_2 \end{matrix}}{\begin{bmatrix} 1 & 0 & 0 \end{bmatrix}}.$

b. After 1 yr, the distribution of people in each group is

$$\begin{bmatrix} 1 & 0 & 0 \end{bmatrix}\begin{bmatrix} .85 & .10 & .05 \\ 0 & .80 & .20 \\ 0 & 0 & 1 \end{bmatrix} = \begin{bmatrix} .85 & .10 & .05 \end{bmatrix}.$$

There are

$.85(50,000) = 42,500$ people in G_0,
$.1(50,000) = 5000$ people in G_1, and
$.05(50,000) = 2500$ people in G_2.

c.

$$P^2 = \begin{bmatrix} .85 & .10 & .05 \\ 0 & .80 & .20 \\ 0 & 0 & 1 \end{bmatrix}\begin{bmatrix} .85 & .10 & .05 \\ 0 & .80 & .20 \\ 0 & 0 & 1 \end{bmatrix} = \begin{bmatrix} .7225 & .165 & .1125 \\ 0 & .64 & .36 \\ 0 & 0 & 1 \end{bmatrix}$$

After 2 yr, the distribution of people in each group is

$$\begin{bmatrix} 1 & 0 & 0 \end{bmatrix}\begin{bmatrix} .7225 & .165 & .1125 \\ 0 & .64 & .36 \\ 0 & 0 & 1 \end{bmatrix} = \begin{bmatrix} .7225 & .165 & .1125 \end{bmatrix}.$$

There are

$.7225(50,000) = 36,125$ people in G_0,
$.165(50,000) = 8250$ people in G_1, and
$.1125(50,000) = 5625$ people in G_2.

Copyright © 2019 Pearson Education, Inc.

d.

$$P^3 = \begin{bmatrix} .85 & .10 & .05 \\ 0 & .80 & .20 \\ 0 & 0 & 1 \end{bmatrix} \begin{bmatrix} .7225 & .165 & .1125 \\ 0 & .64 & .36 \\ 0 & 0 & 1 \end{bmatrix} = \begin{bmatrix} .61413 & .20425 & .18163 \\ 0 & .512 & .488 \\ 0 & 0 & 1 \end{bmatrix}$$

After 3 yr, the distribution is

$$\begin{bmatrix} 1 & 0 & 0 \end{bmatrix} \begin{bmatrix} .61413 & .20425 & .18163 \\ 0 & .512 & .488 \\ 0 & 0 & 1 \end{bmatrix} = \begin{bmatrix} .61413 & .20425 & .18163 \end{bmatrix}.$$

There are
$.61413(50,000) = 30,706$ people in G_0,
$.20425(50,000) = 10,213$ people in G_1, and
$.18163(50,000)=9081$ people in G_2.

e.

$$P^4 = \begin{bmatrix} .85 & .10 & .05 \\ 0 & .80 & .20 \\ 0 & 0 & 1 \end{bmatrix} \begin{bmatrix} .6141 & .2043 & .1816 \\ 0 & .512 & .488 \\ 0 & 0 & 1 \end{bmatrix} = \begin{bmatrix} .52201 & .22481 & .25318 \\ 0 & .4096 & .5904 \\ 0 & 0 & 1 \end{bmatrix}$$

After 4 yr, the distribution is

$$\begin{bmatrix} 1 & 0 & 0 \end{bmatrix} \begin{bmatrix} .52201 & .22481 & .25318 \\ 0 & .4096 & .5904 \\ 0 & 0 & 1 \end{bmatrix} = \begin{bmatrix} .52201 & .22481 & .25318 \end{bmatrix}.$$

There are
$.52201(50,000) \approx 26,100$ people in G_0,
$.22481(50,000) \approx 11,241$ people in G_1,
and
$.25318(50,000) \approx 12,659$ people in G_2.

f. From the transition matrix we obtain

$$\begin{aligned} .85v_1 &= v_1 \\ .10v_1 + .80v_2 &= v_2 \\ .05v_1 + .20v_2 + v_3 &= v_3 \\ v_1 + v_2 + v_3 &= 1. \end{aligned}$$

Solving, we obtain
$v_1 = 0, v_2 = 0,$ and $v_3 = 1.$
The equilibrium vector is $\begin{bmatrix} 0 & 0 & 1 \end{bmatrix}$ which means that, in the long run, everybody has two or more accidents.

41. $P = \begin{bmatrix} .91 & .07 & .02 \\ .03 & .87 & .10 \\ .14 & .04 & .82 \end{bmatrix}$

The initial probability vector is
$\begin{bmatrix} .351 & .324 & .325 \end{bmatrix}$.

a. The share held by each type after 1 yr is given by

$$\begin{bmatrix} .351 & .324 & .325 \end{bmatrix} \begin{bmatrix} .91 & .07 & .02 \\ .03 & .87 & .10 \\ .14 & .04 & .82 \end{bmatrix} = \begin{bmatrix} .375 & .319 & .306 \end{bmatrix} = \begin{bmatrix} 37.5\% & 31.9\% & 30.6\% \end{bmatrix}$$

b.

$$P^2 = \begin{bmatrix} .91 & .07 & .02 \\ .03 & .87 & .10 \\ .14 & .04 & .82 \end{bmatrix} \begin{bmatrix} .91 & .07 & .02 \\ .03 & .87 & .10 \\ .14 & .04 & .82 \end{bmatrix} = \begin{bmatrix} .8330 & .1254 & .0416 \\ .0674 & .7630 & .1696 \\ .2434 & .0774 & .6792 \end{bmatrix}$$

The share held by each after 2 years is

$$\begin{bmatrix} .351 & .324 & .325 \end{bmatrix} \begin{bmatrix} .8330 & .1254 & .0416 \\ .0674 & .7630 & .1696 \\ .2434 & .0774 & .6792 \end{bmatrix} = \begin{bmatrix} .393 & .316 & .290 \end{bmatrix} = \begin{bmatrix} 39.3\% & 31.6\% & 29.0\% \end{bmatrix}.$$

Copyright © 2019 Pearson Education, Inc.

c.

$$P^3 = \begin{bmatrix} .91 & .07 & .02 \\ .03 & .87 & .10 \\ .14 & .04 & .82 \end{bmatrix}\begin{bmatrix} .8330 & .1254 & .0416 \\ .0674 & .7630 & .1696 \\ .2434 & .0774 & .6792 \end{bmatrix}$$
$$= \begin{bmatrix} .767616 & .169072 & .063312 \\ .107968 & .675312 & .216720 \\ .318904 & .111544 & .569552 \end{bmatrix}$$

The share held by each after 3 years is

$$[.351 \quad .324 \quad .325]\begin{bmatrix} .767616 & .169072 & .063312 \\ .107968 & .675312 & .216720 \\ .318904 & .111544 & .569552 \end{bmatrix}$$
$$= [.408 \quad .314 \quad .278]$$
$$= [40.8\% \quad 31.4\% \quad 27.8\%].$$

d. From the transition matrix we obtain
$.91v_1 + .03v_2 + .14v_3 = v_1$
$.07v_1 + .87v_2 + .04v_3 = v_2$
$.02v_1 + .10v_2 + .82v_3 = v_3$
and $v_1 + v_2 + v_3 = 1.$
Solving, we obtain
$v_1 = .4575, v_2 = .3160,$ and $v_3 = .2264.$

The equilibrium vector is
$[.4575 \quad .3160 \quad .2264]$
which means that, in the long run, 45.75% of the market is held by Verizon, 31.60% by AT&T, and 22.64% by all other carriers.

43. a.

$$[.142 \quad .097 \quad .761]\begin{bmatrix} .95 & .02 & .03 \\ .04 & .92 & .04 \\ .09 & .07 & .84 \end{bmatrix}$$
$$\approx [.2073 \quad .1454 \quad .6474].$$

The probability that a vehicle was purchased from Toyota in the next year was about .2073.

b. From the transition matrix we obtain
$.95v_1 + .04v_2 + .09v_3 = v_1$
$.02v_1 + .92v_2 + .07v_3 = v_2$
$.03v_1 + .04v_2 + .84v_3 = v_3$
and $v_1 + v_2 + v_3 = 1.$
Solve using the Gauss-Jordan method to obtain $v_1 = .541, v_2 = .286,$ and $v_3 = .173.$
Thus, the long-term probability that a vehicle is purchased from Honda is about .2865.

45. Store the given transition matrix as matrix [A].

a.

$$A^5 = \begin{bmatrix} .01024 & .03285125 & .10948625 & .8474225 \\ 0 & .0184528125 & .0892315625 & .892315625 \\ 0 & 0 & 1 & 0 \\ 0 & 0 & 0 & 1 \end{bmatrix}$$

The probability of changing from state 1 to state 4 after 5 repetitions is the entry in the first row, fourth column of A^5,.8474225. Approximately 85% complete the program after it has been offered five times.

b. Store the probability vector $[.5 \quad .5 \quad 0 \quad 0]$ as matrix [B]. Determine BA^4.
After four times the percents are 1.28%, 5.13%, 9.62%, and 83.97%.

Section 9.6 Decision Making

1. a. An optimist should choose the coast; $150,000 is the largest profit.

b. A pessimist should choose the highway; the worst case of $30,000 is better than –$40,000 if the coast is chosen.

c. If the possibility of heavy opposition is .8, the probability of light opposition is .2. Find his expected profit for each strategy.
Highway:
70,000(.2) + 30,000(.8) = $38,000
Coast:
150,000(.2) + –(40,000)(.8) = –$2000
He should choose the highway for an expected profit of $38,000

d. If the probability of heavy opposition is .4, the probability of light opposition is .6. Find his expected profit for each strategy.
Highway:
70,000(.6) + 30,000(.4) = $54,000
Coast:
150,000(.6) + (–40,000)(.4) = $74,000
He should choose the coast.

Copyright © 2019 Pearson Education, Inc.

3. Note that the costs given in the payoff matrix are given in hundreds of dollars.

a. An optimist should make no upgrade; minimum cost is $2800.

b. A pessimist should make the upgrade; a worst case of $13,000 is better than a possible cost of $45,000 if no upgrade is made.

c. Find the expected cost of each strategy.
Make upgrade:
.7(130) + .2(130) + .1(130) = 130,
or $13,000
Make no upgrade:
.7(28) + .2(180) + .1(450) = 100.6
or $10,600
He should not upgrade. The expected cost to the company if this strategy is chosen is $10,060.

5. a.

	Fails	Doesn't Fail
Overhaul	−$8600	−$2600
Don't Overhaul	−$6000	$0

b. Find the expected cost under each strategy.
Overhaul:
.1(−8600) + .9(−2600) = −$3200
Don't Overhaul:
.3(−6000) + .7(0) = −$1800
To minimize his expected costs, the business should not overhaul the machine before shipping.

7. a.

	No Rain	Rain
Tent	$2500	$1500
No tent	$3000	$0

b. Find her expected revenue for each scenario.
No rain:
.6(2500) + .4(1500) = $2100
Rain:
.6(1500) + .4(0) = $900
She should rent the tent since the expected revenue is $2100.

9. Find the expected utility under each strategy.
Jobs:
(.35)(40) + (.65)(−10) = 7.5
Environment:
(.35)(−12) + (.65)(30) = 15.3
She should emphasize the environment. The expected utility of this strategy is 15.3.

Chapter 9 Review Exercises

1. $E(x) = 0(.22) + 1(.54) + 2(.16) + 3(.08) = 1.1$

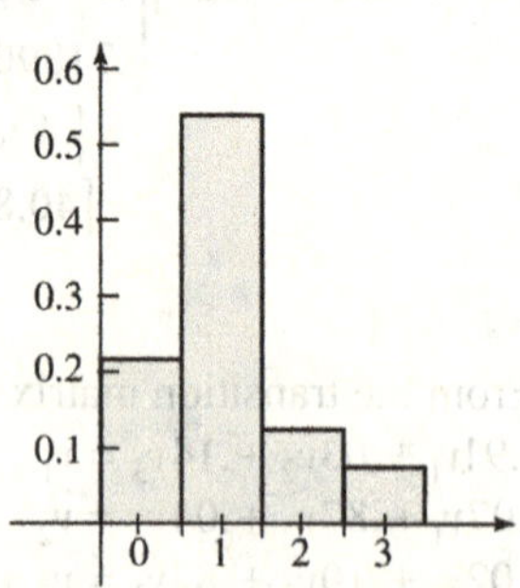

2. $E(x) = -3(.15) - 2(.20) - 1(.25) + 0(.18) + 1(.12) + 2(.06) + 3(.04) = -.74$

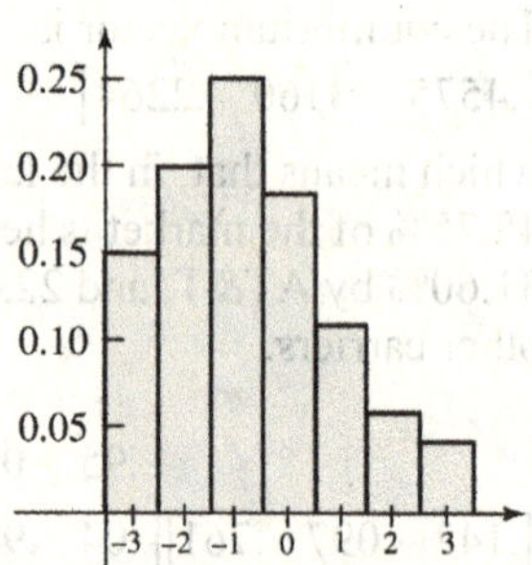

3. $E(x) = -10(.333) + 0(.333) + 10(.333) = 0$

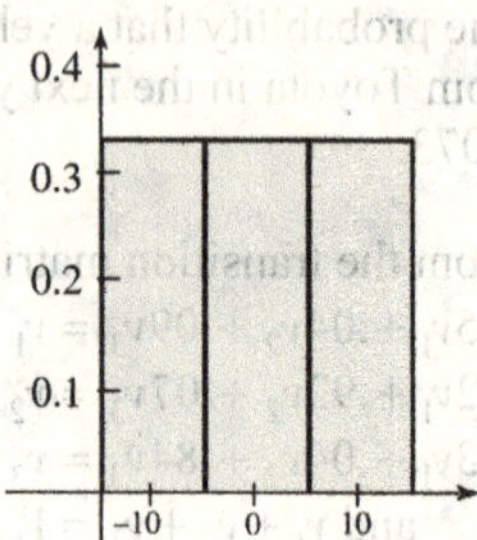

Copyright © 2019 Pearson Education, Inc.

4. $E(x) = 0(.35) + 2(.15) + 4(.2) + 6(.3) = 2.9$

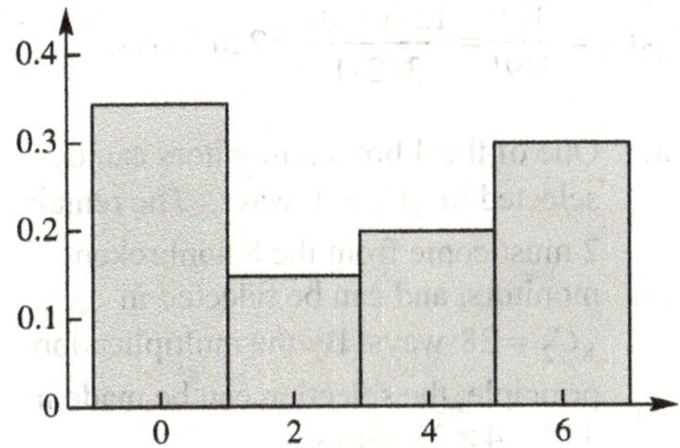

5.

$$E(x) = 0(.079) + 1(.321) + 2(.381) + 3(.151) + 4(.049) + 5(.013) + 6(.006) = 1.833$$

6. $E(x) = 0(.0216) + 1(.0939) + 2(.2510) + 3(.4136) + 4(.1740) + 5(.0390) + 6(.0042) + 7(.0026) + 8(.0001) = 2.7719$

7. a. In selecting 2 bouquets from a group of 10, there are ${}_{10}C_2 = 45$ ways.

$$P(0\text{ roses}) = \frac{{}_3C_0 \cdot {}_7C_2}{{}_{10}C_2} = \frac{21}{45} = \frac{7}{15}$$

$$P(1\text{ rose}) = \frac{{}_3C_1 \cdot {}_7C_1}{{}_{10}C_2} = \frac{21}{45} = \frac{7}{15}$$

$$P(2\text{ roses}) = \frac{{}_3C_2 \cdot {}_7C_0}{{}_{10}C_2} = \frac{3}{45} = \frac{1}{15}$$

x	0	1	2
$P(x)$	$\frac{7}{15}$	$\frac{7}{15}$	$\frac{1}{15}$

b. $E(x) = 0\left(\frac{7}{15}\right) + 1\left(\frac{7}{15}\right) + 2\left(\frac{1}{15}\right) = .6$

8. a. Three of the 10 members did not do their homework, and 3 members of the 10 are selected. Three members of the 10 can be selected in ${}_{10}C_3 = 120$ ways. Let $P(n)$ represent the probability of selecting n students who did not do homework.

$$P(0) = \frac{{}_3C_0 \cdot {}_7C_3}{{}_{10}C_3} = \frac{35}{120} = .292$$

$$P(1) = \frac{{}_3C_1 \cdot {}_7C_2}{{}_{10}C_3} = \frac{63}{120} = .525$$

$$P(2) = \frac{{}_3C_2 \cdot {}_7C_1}{{}_{10}C_3} = \frac{21}{120} = .175$$

$$P(3) = \frac{{}_3C_3 \cdot {}_7C_0}{{}_{10}C_3} = \frac{1}{120} = .008$$

x	0	1	2	3
$P(x)$	.292	.525	.175	.008

b.

$$E(x) = 0(.292) + 1(.525) + 2(.175) + 3(.008) = .899$$

9. Probability of getting 3 hearts

$$= \frac{{}_{13}C_3}{{}_{52}C_3} = \frac{286}{22{,}100} \approx .0129$$

Probability of not getting 3 hearts $= 1 - .0129 = .9871.$

Let x = the amount to pay for the game.

If you win, you get $100 - x$.

If you lose, you get $-x$. The expected value must be 0.

$(100 - x)(.0129) + (-x)(.9871) = 0 \Rightarrow$

$1.29 - .0129x - .9871x = 0 \Rightarrow 1.29 = x$

You must pay \$1.29.

Copyright © 2019 Pearson Education, Inc.

10. The probability of winning if you bet "under" is $\frac{15}{36}$ and has a net return of \$2. The probability of 7 is $\frac{6}{36}$ and has a net return of \$4. Betting "over" has a probability of $\frac{15}{36}$ with a net return of \$2.

$$P(\text{sum} < 7) = P(2) + P(3) + P(4) + P(5) + P(6) = \frac{1}{36} + \frac{2}{36} + \frac{3}{36} + \frac{4}{36} + \frac{5}{36} = \frac{15}{36}$$

$$P(\text{sum} = 7) = \frac{6}{36}$$

$$P(\text{sum} > 7) = P(8) + P(9) + P(10) + P(11) + P(12) = \frac{5}{36} + \frac{4}{36} + \frac{3}{36} + \frac{2}{36} + \frac{1}{36} = \frac{15}{36}$$

The expected return for each type of bet is as follows:

Under:

$$E(x) = 2\left(\frac{15}{36}\right) - 2\left(\frac{21}{36}\right) = -\$.33$$

Exactly 7:

$$E(x) = 4\left(\frac{6}{36}\right) - 2\left(\frac{30}{36}\right) = -\$1.00$$

Over:

$$E(x) = 2\left(\frac{15}{36}\right) - 2\left(\frac{21}{36}\right) = -\$.33$$

11. $E(x) = 0(.8018) + 1(.1838) + 2(.0140) + 3(.0004) \approx .213$

12. $E(x) = 0(.069) + 1(.244) + 2(.345) + 3(.244) + 4(.086) + 5(.012) = 2.07$

13. Since order is important, this is a permutation. Eight taxis can line up in ${}_8P_8 = 8! = 40{,}320$ ways.

14. Since order is important, this is a permutation. If there are 8 finalists, there are

$${}_8P_3 = \frac{8!}{(8-3)!} = 8 \cdot 7 \cdot 6 = 336 \text{ variations.}$$

15. Since order is not important, this is a combination. Three monitors can be selected from 12 monitors in

$${}_{12}C_3 = \frac{12!}{3!9!} = \frac{12 \cdot 11 \cdot 10}{3 \cdot 2 \cdot 1} = 220 \text{ ways.}$$

16. **a.** One of the 4 broken monitors can be selected in ${}_4C_1 = 4$ ways. The remaining 2 must come from the 8 nonbroken monitors, and can be selected in ${}_8C_2 = 28$ ways. By the multiplication principle, the selection can be made in $112 = 4 \times 28$ ways.

b. All 3 monitors must come from the non-broken group of 8. This can be accomplished in ${}_8C_3 = 56$ ways.

c. At least one broken monitor can be accomplished by selecting 1, 2, or 3 defective monitors. If 1 monitor is broken, 2 must be non-broken. If 2 are broken, 1 must be non-broken. If 3 are broken, then 0 must be non-broken. The number of ways to select
1 broken: ${}_4C_1 \cdot {}_8C_2 = 4 \times 28 = 112$
2 broken: ${}_4C_2 \cdot {}_8C_1 = 6 \times 8 = 48$
3 broken: ${}_4C_3 \cdot {}_8C_0 = 4 \times 1 = 4$
is then $112 + 48 + 4 = 164$.

17. Since order is important, this is a permutation. There are 30 choices for the first seat, 29 for the second, 28 for the third, 27 for the fourth, 26 for the fifth, and 25 for the sixth, or ${}_{30}P_6 = 427{,}518{,}000$ ways.

18. The first seat will be occupied by the given student. That leaves 29 choices for the second seat, 28 choices for the third seat, 27 choices for the fourth seat, 26 choices for the fifth seat, and 25 for the sixth seat, or ${}_{29}P_5 = 14{,}250{,}600$ possible ways.

19. **a.** Since there are 15 students in each major, there are ${}_{15}P_3 = 2730$ ways to arrange the students within each major. There are also ${}_2P_2 = 2$ ways to arrange the two different groups. Thus, the total number of arrangements is $2 \times 2730 \times 2730 = 14{,}905{,}800$.

Copyright © 2019 Pearson Education, Inc.

b. Assume the odd seats are occupied by science majors, then because order is important, there are $_{15}P_3 = 2730$ possible arrangements for that group. Then the even seats would be occupied by the business majors with $_{15}P_3 = 2730$ possible arrangements By the multiplication principle, there would be $2730 \times 2730 = 7{,}452{,}900$ possible arrangements. However, if the odd seats were occupied by the business majors and the even by the science majors, there would also be $2730 \times 2730 = 7{,}452{,}900$ possible arrangements. So, there must be $7{,}452{,}900 + 7{,}452{,}900 = 14{,}905{,}800$ possibilities.

20. Answers vary.

21. Answers vary.

22. There are 26 black cards in the deck.

$$P(\text{both black}) = \frac{_{26}C_2}{_{52}C_2} = \frac{325}{1326} \approx .2451$$

23. There are 13 hearts in a deck.

$$P(\text{both hearts}) = \frac{_{13}C_2}{_{52}C_2} = \frac{78}{1326} \approx .0588$$

24. To get exactly one face card, you must have one non-face card. There are 12 face cards and 40 non-face cards in a deck.

$$P(\text{exactly one face card}) = \frac{_{12}C_1 \cdot {_{40}C_1}}{_{52}C_2} = \frac{480}{1326} \approx .3620$$

25. There are 4 aces in a deck of cards.

$$P(\text{at most one ace}) = \frac{_{4}C_0 \cdot {_{48}C_2}}{_{52}C_2} + \frac{_{4}C_1 \cdot {_{48}C_1}}{_{52}C_2} = \frac{1128}{1326} + \frac{192}{1326} = \frac{1320}{1326} \approx .9955$$

26. There are 12 possible selections, 6 of them being ice cream, and 6 not ice cream.

$$P(3 \text{ ice cream}) = \frac{_{6}C_3 \cdot {_{6}C_0}}{_{12}C_3} = \frac{20 \cdot 1}{220} \approx .0909$$

27. There are 12 possible selections, 4 custard and 8 non-custard.

$$P(4 \text{ custard}) = \frac{_{4}C_3 \cdot {_{8}C_0}}{_{12}C_3} = \frac{4 \cdot 1}{220} \approx .0182$$

28. There are 12 possible selections, 2 frozen yogurt and 10 non-yogurt.

$$\begin{aligned} P(\text{at least 1 yogurt}) &= 1 - P(\text{no yogurt}) \\ &= 1 - \frac{_{2}C_0 \cdot {_{10}C_3}}{_{12}C_3} \\ &= 1 - \frac{1 \cdot 120}{220} = \frac{100}{220} \\ &\approx .4545 \end{aligned}$$

29. There are 12 possible selections, 4 custard, 6 ice cream, and 2 frozen yogurt.

$$P(1 \text{ custard}, 1 \text{ ice cream}, 1 \text{ yogurt}) = \frac{_{4}C_1 \cdot {_{6}C_1} \cdot {_{2}C_1}}{_{12}C_3} = \frac{4 \cdot 6 \cdot 2}{220} = \frac{48}{220} \approx .2182$$

30. There are 12 possible selections, 6 ice cream, and 6 not ice cream.

$$\begin{aligned} P(\text{at most 1 ice cream}) &= P(0) + P(1) \\ &= \frac{_{6}C_0 \cdot {_{6}C_3}}{_{12}C_3} + \frac{_{6}C_1 \cdot {_{6}C_2}}{_{12}C_3} \\ &= \frac{20}{220} + \frac{90}{220} = \frac{110}{220} = .5 \end{aligned}$$

31. a. The number of subsets of size 0 is $_nC_0$ or 1.
The number of subsets of size 1 is $_nC_1$ or n.
The number of subsets of size 2 is $_nC_2$ or $\frac{n(n-1)}{2}$.
The number of subsets of size n is $_nC_n$ or 1.

b. The total number of subsets of a set with n elements is $_nC_0 + {_nC_1} + {_nC_2} + \text{L} + {_nC_n}$.

32. a. Answers vary. Possible Answer: Each element is either in or out of the subset, so there are 2^n ways to make these choices.

Copyright © 2019 Pearson Education, Inc.

b. Let $n = 4$.

$${}_4C_0 + {}_4C_1 + {}_4C_2 + {}_4C_3 + {}_4C_4 = 1+4+6+4+1 = 16$$

Since $2^4 = 16$, the equation from part a holds.

Let $n = 5$.

$${}_5C_0 + {}_5C_1 + {}_5C_2 + {}_5C_3 + {}_5C_4 + {}_5C_5 = 1+5+10+10+5+1 = 32$$

Since $2^5 = 32$, the equation from part c holds.

33. a. $n = 7,\ p = .507,\ 1 - p = .493$

x	$P(x)$	
0	.0071	$= {}_7C_0(.507)^0(.493)^7$
1	.051	$= {}_7C_1(.507)^1(.493)^6$
2	.1572	$= {}_7C_2(.507)^2(.493)^5$
3	.2695	$= {}_7C_3(.507)^3(.493)^4$
4	.2771	$= {}_7C_4(.507)^4(.493)^3$
5	.171	$= {}_7C_5(.507)^5(.493)^2$
6	.0586	$= {}_7C_6(.507)^6(.493)^1$
7	.0086	$= {}_7C_7(.507)^7(.493)^0$

b. $E(x) = 0(.0071) + 1(.051) + 2(.1572) + 3(.2695) + 4(.2771) + 5(.171) + 6(.0586) + 7(.0086) \approx 3.549$

34. a. $n = 6,\ p = .457,\ 1 - p = .543$

x	$P(x)$	
0	.0256	$= {}_6C_0(.457)^0(.543)^6$
1	.1294	$= {}_6C_1(.457)^1(.543)^5$
2	.2723	$= {}_6C_2(.457)^2(.543)^4$
3	.3056	$= {}_6C_3(.457)^3(.543)^3$
4	.1929	$= {}_6C_4(.457)^4(.543)^2$
5	.0649	$= {}_6C_5(.457)^5(.543)^1$
6	.0091	$= {}_6C_6(.457)^6(.543)^0$

b. $E(x) = 0(.0256) + 1(.1294) + 2(.2723) + 3(.3056) + 4(.1929) + 5(.0649) + 6(.0091) \approx 2.742$

35. a. $n = 4,\ p = .885,\ 1 - p = .115$

$$P(0) = {}_4C_0(.885)^0(.115)^4 \approx .0002$$

b.
$$\begin{aligned} P(\text{at least } 2) &= P(2) + P(3) + P(4) \\ &= 1 - P(0) - P(1) \\ &= 1 - {}_4C_0(.885)^0(.115)^4 - {}_4C_1(.885)^1(.115)^3 \\ &= 1 - .0002 - .0054 \\ &\approx .9944 \end{aligned}$$

c. $P(4) = {}_4C_4(.885)^4(.115)^0 \approx .6134$

36. a. $n = 5,\ p = .77,\ 1 - p = .23$

$$P(4) = {}_5C_4(.77)^4(.23)^1 \approx .4043$$

b.
$$\begin{aligned} P(\text{at most } 4) &= P(0) + P(1) + P(2) + P(3) + P(4) \\ &= 1 - P(5) \\ &= 1 - {}_5C_5(.77)^5(.23)^0 \\ &\approx 1 - .2707 = .7293 \end{aligned}$$

c.
$$\begin{aligned} P(\text{at least } 1) &= P(1) + P(2) + P(3) + P(4) + P(5) \\ &= 1 - P(5) \\ &= 1 - {}_5C_0(.77)^0(.23)^5 \\ &\approx 1 - .0006 = .9994 \end{aligned}$$

37. a. $n = 5,\ p = .652,\ 1 - p = .348$

x	$P(x)$	
0	.0051	$= {}_5C_0(.652)^0(.348)^5$
1	.0478	$= {}_5C_1(.652)^1(.348)^4$
2	.1792	$= {}_5C_2(.652)^2(.348)^3$
3	.3357	$= {}_5C_3(.652)^3(.348)^2$
4	.3144	$= {}_5C_4(.652)^4(.348)^1$
5	.1178	$= {}_5C_5(.652)^5(.348)^0$

b.
$$\begin{aligned} P(\text{at least } 2) &= P(2) + P(3) + P(4) + P(5) \\ &= .1792 + .3357 + .3144 + .1178 \\ &\approx .9471 \end{aligned}$$

Copyright © 2019 Pearson Education, Inc.

c. $E(x) = 0(.0051) + 1(.0478) + 2(.1792) + 3(.3357) + 4(.3144) + 5(.1178) \approx 3.26$

d. Answers Vary

38. a. $n = 5,\ p = .225,\ 1 - p = .775$

x	$P(x)$	
0	.2796	$= {}_5C_0(.225)^0(.775)^5$
1	.4058	$= {}_5C_1(.225)^1(.775)^4$
2	.2357	$= {}_5C_2(.225)^2(.775)^3$
3	.0684	$= {}_5C_3(.225)^3(.775)^2$
4	.0099	$= {}_5C_4(.225)^4(.775)^1$
5	.0006	$= {}_5C_5(.225)^5(.775)^0$

b. $P(\text{at most } 3) = P(0) + P(1) + P(2) + P(3) = .2796 + .4058 + .2357 + .0684 \approx .9895$

c. $E(x) = 0(.2796) + 1(.4058) + 2(.2357) + 3(.0684) + 4(.0099) + 5(.0006) \approx 1.125$

d. Answers Vary

39. $\begin{bmatrix} 0 & 1 \\ .77 & .23 \end{bmatrix}$

This is a regular transition matrix because

$$\begin{bmatrix} 0 & 1 \\ .77 & .23 \end{bmatrix}\begin{bmatrix} 0 & 1 \\ .77 & .23 \end{bmatrix} = \begin{bmatrix} .77 & .23 \\ .1771 & .8229 \end{bmatrix},$$

in which all entries are positive.

40. $\begin{bmatrix} -.2 & .4 \\ .3 & .7 \end{bmatrix}$

This is not a regular transition matrix because there is a negative entry in the first row and first column. In fact, this makes it not even a transition matrix.

41. $\begin{bmatrix} .21 & .15 & .64 \\ .50 & .12 & .38 \\ 1 & 0 & 0 \end{bmatrix}$

This is a regular transition matrix because

$$\begin{bmatrix} .21 & .15 & .64 \\ .50 & .12 & .38 \\ 1 & 0 & 0 \end{bmatrix}\begin{bmatrix} .21 & .15 & .64 \\ .50 & .12 & .38 \\ 1 & 0 & 0 \end{bmatrix} = \begin{bmatrix} .7591 & .0495 & .1914 \\ .545 & .0894 & .3656 \\ .21 & .15 & .64 \end{bmatrix},$$

in which all entries are positive.

42. $\begin{bmatrix} .22 & 0 & .78 \\ .40 & .33 & .27 \\ 0 & .61 & .39 \end{bmatrix}$

This is a regular transition matrix because

$$\begin{bmatrix} .22 & 0 & .78 \\ .40 & .33 & .27 \\ 0 & .61 & .39 \end{bmatrix}\begin{bmatrix} .22 & 0 & .78 \\ .40 & .33 & .27 \\ 0 & .61 & .39 \end{bmatrix} = \begin{bmatrix} .0484 & .4758 & .4758 \\ .220 & .2736 & .5064 \\ .244 & .4392 & .3168 \end{bmatrix},$$

in which all entries are positive.

43. $P = \begin{bmatrix} .35 & .15 & .50 \\ .30 & .35 & .35 \\ .15 & .30 & .55 \end{bmatrix}$

$I = [.2\ \ .4\ \ .4]$

a. The distribution after one month is

$$[.2\ \ .4\ \ .4]\begin{bmatrix} .35 & .15 & .50 \\ .30 & .35 & .35 \\ .15 & .30 & .55 \end{bmatrix} = [.25\ \ .29\ \ .46].$$

b. $I = [.2\ \ .4\ \ .4]$

$$A = \begin{bmatrix} .35 & .15 & .50 \\ .30 & .35 & .35 \\ .15 & .30 & .55 \end{bmatrix}$$

$$A^2 = \begin{bmatrix} .2425 & .2550 & .5025 \\ .2625 & .2725 & .4650 \\ .2250 & .2925 & .4825 \end{bmatrix}$$

The distribution after 2 months is given by

$IA^2 = [.2435\ \ .277\ \ .4795].$

Copyright © 2019 Pearson Education, Inc.

c. To find the long-range distribution, we use the system

$$v_1 + v_2 + v_3 = 1$$
$$.35v_1 + .3v_2 + .15v_3 = v_1$$
$$.15v_1 + .35v_2 + .3v_3 = v_2$$
$$.5v_1 + .35v_2 + .55v_3 = v_3.$$

Simplify these equations to obtain the system

$$v_1 + v_2 + v_3 = 1$$
$$-.65v_1 + .3v_2 + .15v_3 = 0$$
$$.15v_1 - .65v_2 + .3v_3 = 0$$
$$.5v_1 + .35v_2 - .45v_3 = 0.$$

Solve this system by the Gauss-Jordan method to obtain

$$v_1 = \frac{75}{313}, v_2 = \frac{87}{313}, \text{ and } v_3 \frac{151}{313}.$$

The long-range distribution is $[.240 \quad .278 \quad .482]$.

44. a.

$$I = [.15 \quad .60 \quad .25]; \quad A = \begin{bmatrix} .80 & .14 & .06 \\ .04 & .85 & .11 \\ .03 & .13 & .84 \end{bmatrix}$$

The distribution after one month is

$$[.15 \quad .60 \quad .25] \begin{bmatrix} .80 & .14 & .06 \\ .04 & .85 & .11 \\ .03 & .13 & .84 \end{bmatrix} = [.1515 \quad .5635 \quad .2850]$$

b.

$$A^2 = \begin{bmatrix} .80 & .14 & .06 \\ .04 & .85 & .11 \\ .03 & .13 & .84 \end{bmatrix} \begin{bmatrix} .80 & .14 & .06 \\ .04 & .85 & .11 \\ .03 & .13 & .84 \end{bmatrix} = \begin{bmatrix} .6474 & .2388 & .1138 \\ .0693 & .7424 & .1883 \\ .0544 & .2239 & .7217 \end{bmatrix}$$

$$A^3 = \begin{bmatrix} .80 & .14 & .06 \\ .04 & .85 & .11 \\ .03 & .13 & .84 \end{bmatrix} \begin{bmatrix} .6474 & .2388 & .1138 \\ .0693 & .7424 & .1883 \\ .0544 & .2239 & .7217 \end{bmatrix} = \begin{bmatrix} .530886 & .308410 & .160704 \\ .090785 & .665221 & .243994 \\ .074127 & .291752 & .634121 \end{bmatrix}$$

The distribution after 3 years is given by $IA^3 = [.1526 \quad .5183 \quad .3290]$

c. To find the long range distribution, we use the system

$$v_1 + v_2 + v_3 = 1$$
$$.80v_1 + .04v_2 + .03v_3 = v_1$$
$$.14v_1 + .85v_2 + .13v_3 = v_3$$
$$.06v_1 + .11v_2 + .84v_3 = v_3$$

This system yields

$$v_1 + v_2 + v_3 = 1$$
$$-.20v_1 + .04v_2 + .03v_3 = 0$$
$$.14v_1 - .15v_2 + .13v_3 = 0$$
$$.06v_1 + .11v_2 - .16v_3 = 0$$

Solving this system yields $v_1 = \frac{97}{643}$, $v_2 = \frac{302}{643}$ and $v_3 = \frac{244}{643}$. Thus the long range distribution is $[.1509 \quad .4697 \quad .3795]$.

Copyright © 2019 Pearson Education, Inc.

45. a. Since the candidate is an optimist, look for the biggest value in the matrix, which is 5000. Hence, she should oppose it.

b. A pessimistic candidate wants to find the best of the worst things that can happen. If she favors, the worst is –4000. If she waffles the worst is –500. If she opposes, then worst is 0. Since the best of these is 0, she should oppose.

c. Since there is a 40% chance the opponent favors the plant and a 35% that he will waffle, the chance he will oppose is $1 - .4 - .35 = .25$
Expected gain if she favors:
$(0)(.4) + (-1000)(.35) + (-4000)(.25) = -1350$
Expected gain if she waffles:
$(1000)(.4)+0(.35)+(-500)(.25) = 275$
Expected gain if she opposes:
$(5000)(.4) + (2000)(.35) + (0)(.25) = 2700$
She should oppose and get 2700 additional votes.

d. Now the opponent has 0 probability of favoring, .7 of waffling and .3 of opposing.
Expected gain if she favors:
$(0)(0) + (.7)(-1000) + (.3)(-4000) = -1900$
Expected gain if she waffles:
$(0)(1000) + (.7)(0) + (.3)(-500) = -150$
Expected gain if she opposes:
$(0)(5000) + (.7)(2000) + (.3)(0) = 1400$
She should oppose and get 1400 additional votes.

46. a. Since the department chair is an optimist, look for the biggest value in the matrix, which is 100. Hence, she should use active learning.

b. A pessimistic wants to find the best of the worst things that can happen. If she lectures, the worst is –80. If she uses active learning, the worst is –30. Since the best of these is –30, she should use active learning.

c. Since there is a 75% chance the class will prefer the lecture format, there is a 25% chance the class will support the active learning format.
Expected gain for lecture format:
$50(.75) - 80(.25) = 17.5$ points.
Expected gain for active learning format:
$-30(.75) + 100(.25) = 2.5$ points. She should use the lecture format where the expected gain is 17.5.

d. Since there is a 60% chance the class will prefer the active learning format, there is a 40% chance the class will support the lecture format.
Expected gain for lecture format:
$50(.4) - 80(.6) = -28$ points.
Expected gain for active learning format:
$-30(.4) + 100(.6) = 48$ points. She should use the active learning format where the expected gain is 48.

Case 9 Quick Draw® from the New York State Lottery

1. Using the probabilities determined in the case study, we have

x	Net winnings	$P(x)$
0	$2(0) - 1 = -\$1$	.16660
1	$2(0) - 1 = -\$1$	.36349
2	$2(0) - 1 = -\$1$	.30832
3	$2(1) - 1 = \$1$	.12982
4	$2(6) - 1 = \$11$	.02854
5	$2(55) - 1 = \$109$	.00310
6	$2(1000) - 1 = \$1999$	.00013

$$E(W) = -1(.16660) - 1(.36349) - 1(.30832) + 1(.12982) + 11(.02854) + 109(.00310) + 1999(.00013) \approx .2031$$

Copyright © 2019 Pearson Education, Inc.

2. It is not in the state's interest to offer such a promotion because the players win about 20 cents on average.

3.

x	$P(x)$
0	$\frac{{}_{20}C_0 \cdot {}_{60}C_4}{{}_{80}C_4} \approx .30832$
1	$\frac{{}_{20}C_1 \cdot {}_{60}C_3}{{}_{80}C_4} \approx .43273$
2	$\frac{{}_{20}C_2 \cdot {}_{60}C_2}{{}_{80}C_4} \approx .21264$
3	$\frac{{}_{20}C_3 \cdot {}_{60}C_1}{{}_{80}C_4} \approx .04325$
4	$\frac{{}_{20}C_4 \cdot {}_{60}C_0}{{}_{80}C_4} \approx .00306$

For exercises 4 and 5, carry all decimal places in order to compute the expected winnings.

4.

x	Net winnings	$P(x)$
0	–$1	.30832
1	–$1	.43273
2	$0	.21264
3	$4	.04325
4	$54	.00306

$$E(W) = -1(.30832) - 1(.43273) + 0(.21264) + 4(.04325) + 54(.00306) \approx -.40281$$

5.

x	Net winnings	$P(x)$
0	2(0) – 1 = –$1	.30832
1	2(0) – 1 = –$1	.43273
2	2(1) – 1 = $1	.21264
3	2(5) – 1 = $9	.04325
4	2(55) – 1 = $109	.00306

$$E(W) = -1(.30832) - 1(.43273) + 1(.21264) + 9(.04325) + 109(.00306) \approx .19438$$

Copyright © 2019 Pearson Education, Inc.

Chapter 10 Introduction to Statistics

Section 10.1 Frequency Distributions

1. a–b. Since 0–39.99 is to be the first interval and there are 40 numbers between 0 and 39.99 inclusive, we will let all six intervals be of size 40. The other five intervals are 40.00–79.99, 80.00–119.99, 160.00–199.99, and 200.00–239.99. Keeping a tally of how many data values lie in each interval leads to the following frequency distribution.

Interval	Freque
0–39.99	5
40.00–79.99	9
80.00–119.99	1
120.00–159.99	2
160.00–199.99	2
200.00–239.99	1

c. Draw the histogram. It consists of 6 bars of equal width, having heights as determined by the frequency of each interval.

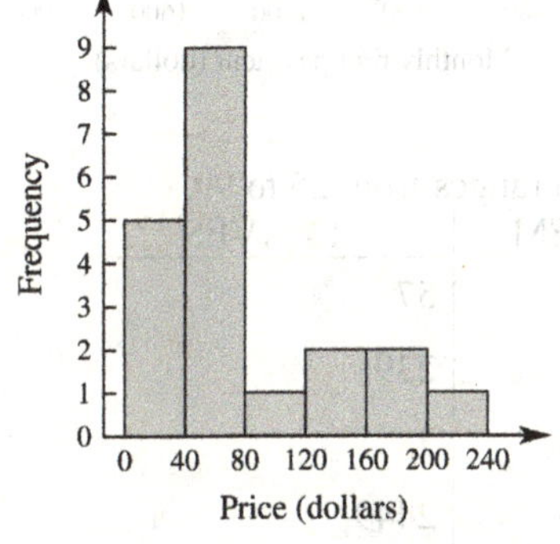

3. a–b. Since 0– 14.99 is to be the first interval, we let all the intervals be of size 15. The largest data value is 79.11, so the last interval that will be needed is 75.00–89.99. The frequency distribution is as follows.

Interval	Frequency
0–14.99	6
15.00–29.99	11
30.00–44.99	2
45.00–59.99	0
60.00–74.99	0
75.00–89.99	1

c. Draw the histogram. It consists of 6 bars of equal width and having heights as determined by the frequency of each interval.

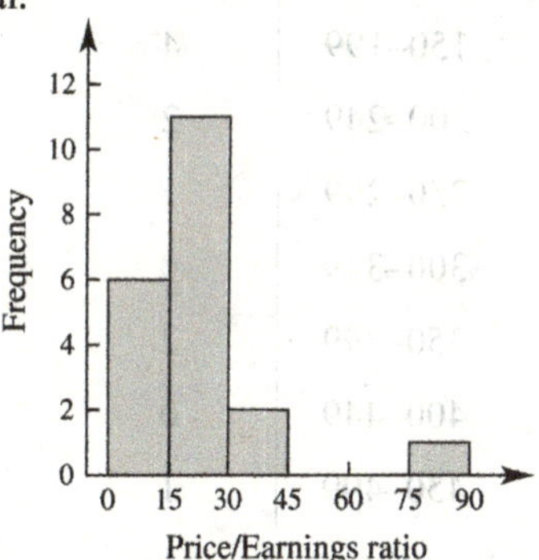

5. The data ranges from a low of 3 to a high of 221, so we will use intervals of size 25 starting with 0–24. The frequency distribution is as follows.

Interval	Frequency
0–24	3
25–49	4
50–74	3
75–99	2
100–124	2
125–149	3
150–174	1
175–199	1
200–224	1

Copyright © 2019 Pearson Education, Inc.

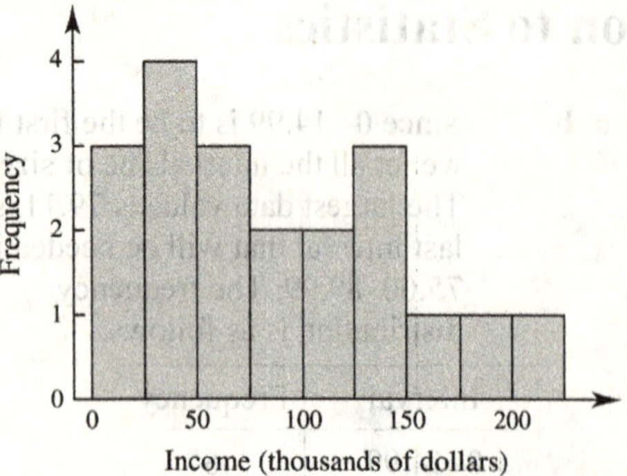

7. The data ranges from a low of 30 to a high of 460, so we will use intervals of size 50 starting with 0–49. The frequency distribution is as follows.

Interval	Frequency
0–49	1
50–99	3
100–149	6
150–199	4
200–249	2
250–299	1
300–349	0
350–399	2
400–449	0
450–499	1

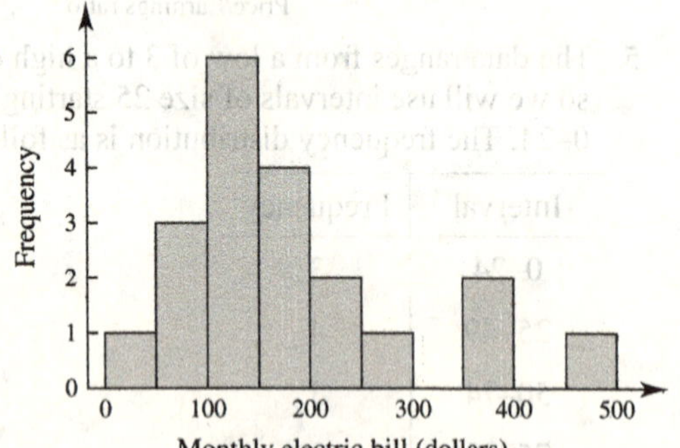

9. The data ranges from a low of 300 to a high of 1900, so we will use intervals of size 200 starting with 200–399. The frequency distribution is as follows.

Interval	Frequency
200–399	1
400–599	1
600–799	4
800–999	6
1000–1199	0
1200–1399	3
1400–1599	4
1600–1799	0
1800–1999	1

11. The data ranges from 25 to 99.

STEM	LEAVES
2	57
3	239
4	3
5	2449
6	4667
7	2245
8	3
9	9

Units: 9|9 = $99

Copyright © 2019 Pearson Education, Inc.

13. The data ranges from 54 to 69.

STEM	LEAVES
5	4
5	
5	
6	0111
6	22223
6	444455
6	66
6	89

Units: 6|8 = 68 inches

15. The data ranges from 5 to 79.

STEM	LEAVES
0	566
1	134577
2	00234679
3	14
4	
5	
6	
7	9

Units: 7|9 = 79

17. The data ranges from 82 to 93.

STEM	LEAVES
8	23
8	445
8	6667
8	8999
9	00000011
9	2223

Units: 9|2 = 92%

19. The data ranges from 21 to 41.

STEM	LEAVES
2	11333
2	557778899
3	0022234
3	888
4	1

Units: 4|1 = 41%

21. uniform

23. left skewed

25. **a.** The distribution is normal.

b. 2 states had their percentage between 6% and 6.9%.

c. 9 states had their percentage above 14%.

27. **a.** The distribution is right skewed.

b. 26 states had percentages in the thirties.

c. 7 states had percentages above 35%.

29. **a.** The distribution is left skewed.

b. 12 students earned 90% or better.

c. 8 students earned less than 60%.

Section 10.2 Measures of Center

1. $\Sigma x = 21,900 + 22,850 + 24,930 + 29,710 + 28,340 + 40,000 = 167,730$

The mean of the 6 numbers is

$$\bar{x} = \frac{167,730}{6} = 27,955.$$

3. $\Sigma x = 3.5 + 4.2 + 5.8 + 6.3 + 7.1 + 2.8 + 3.7 + 4.2 + 4.2 + 5.7 = 47.5$

$$\bar{x} = \frac{47.5}{10} = 4.75 \approx 4.8$$

5. $\Sigma x = 9.2 + 10.4 + 13.5 + 8.7 + 9.7 = 51.5$

$$\bar{x} = \frac{51.5}{5} = 10.3$$

Copyright © 2019 Pearson Education, Inc.

7.

Value	Frequency	Value × Frequency	
19	3	19·3	57
20	5	20·5	100
21	25	21·25	525
22	8	22·8	176
23	2	23·2	46
24	1	24·1	24
28	1	28·1	28
	Total: 45	Total:	956

The mean is $\overline{x} = \frac{956}{45} \approx 21.2$.

9.

x	f	xf
9	5	45
11	10	110
15	12	180
17	9	153
20	6	120
28	1	28
Totals	43	636

$\overline{x} = \frac{636}{43} \approx 14.8$

11. First arrange the numbers in numerical order, from smallest to largest.
$28458, $29679, $33679, $38400, $39720
There are 5 numbers; the median is the middle term, in this case $33,679.

13. First arrange the numbers in numerical order, from smallest to largest.
94.1, 96.8, 97.4, 98.6, 98.4, 98.7, 99.2, 99.9
There are 8 numbers; the median is the mean of the 2 middle numbers, which is
$\frac{98.4+98.6}{2} = \frac{197}{2} = 98.5$.

15. 1, 2, 2, 1, 2, 2, 1, 1, 2, 2, 3, 4, 2, 3, 4, 2, 3, 2, 3,
The mode is the number that occurs most often.
The mode is 2.

17. 62, 65, 71, 74, 71, 76, 71, 63, 59, 65, 65, 64, 72, 71, 77, 63, 65
The mode is the number that occurs most often. There are two modes, 65 and 71, since they both appear four times.

19. 3.2, 2.7, 1.9, 3.7, 3.9
There is no one number that occurs more times than any other number; so, there is no mode.

21. Answers vary. Possible answer: The mode has advantages of being easily found and not being influenced by data that are very large or very small compared to the rest of the data. It is often used in samples where the data to be "averaged" are not numerical.

Copyright © 2019 Pearson Education, Inc.

23.

Interval	Midpoint, x	f	xf
10–14.99	12.5	1	12.5
15-19.99	17.5	0	0
20–24.99	22.5	1	22.5
25–29.99	27.5	18	495
30–34.99	32.5	22	715
35–39.99	37.5	4	150
40–44.99	42.5	1	42.5
45–49.99	47.5	2	95
	Totals:	49	1532.5

Mean $\bar{x} = \frac{1532.5}{49} = 31.28$

The modal class is 30–34.99.

25. a. Mean $\bar{x} = \frac{242+202+200+200+178+176+172+172+168+164}{10} = \frac{1,874}{10} = \$187.4$ million.

b. The middle data when the data is in ascending order are 176 and 178. The median value is $\frac{176+178}{2} = \$177$ million.

27. a. Mean $\bar{x} = \frac{486+425+408+368+364+363+341+330+325+230+224+210}{12}$

$= \frac{4074}{12} = \$339.5$ million

b. The two middle receipts are \$341 million and \$363 million. The median revenue is $\frac{341+363}{2} = \$352$ million

29. a. Mean $\bar{x} = \frac{5748+5710+5531+5636+5903+5995+5997+6121+6282+6440}{10}$

$= \frac{59,363}{10} = \$5,936.3$ million

The two middle data are 5903 and 5995.

Median $= \frac{5903+5995}{2} = \$5,949$ million

b. The revenue in 2011 was closest to the mean.

31. The average monthly low temperatures in ascending order: 30, 32, 33, 40, 42, 48, 49, 57, 62, 65, 68, 69

$\bar{x} = \frac{\Sigma x}{n} = \frac{595}{12} \approx 49.6°\text{F}$

Median $= \frac{48+49}{2} = 48.5°\text{F}$

33. The distribution is right skewed, so the median is the better measure of center. There are 20 values, so the median is the average of the 10th and 11th values. These values fall in the 50–74 and the 75–99 ranges, so the median is the average of the upper limit 74 and lower limit 75, \$74.5 thousand.

Copyright © 2019 Pearson Education, Inc.

35. **a.** The distribution is normal.

b. There are 25 values, so the median is the 13th value, so the median value is 29%.

37. The final average of Nan will be:

$$\bar{x} = (.15)(73) + (.15)(78) + (.12)(91) + (.12)(93) + (.12)(83) + (.34)(88) = 84.61$$

39. The final average of Trey will be:

$$\bar{x} = (.05)(100) + (.25)(95) + (.20)(81) + (.20)(89) + (.30)(x) = 62.75 + .3x$$

Since Trey wants a final average of 90, the score he will need on the final exam will be:

$$90 = 62.75 + .3x$$
$$27.75 = .3x$$
$$x = 90.84$$

Section 10.3 Measures of Variation

Note: Answers may vary slightly throughout this section due to the number of decimal places carried throughout the computations.

1. Answers vary. Possible answer: The standard deviation of a sample of numbers is the square root of the variance of the sample.

3. Range $= 79.8 - 9.7 = 70.1$

$$\bar{x} = \frac{9.7 + 79.8 + 17.7 + 13.5 + 23.5 + 17.1 + 22.9}{7} = \frac{184.2}{7} = 26.31$$

x	$x - \bar{x}$	$(x - \bar{x})^2$
9.7	–16.6	275.56
79.8	53.5	2862.25
17.7	–8.6	73.96
13.5	–12.8	163.84
23.5	–2.8	7.84
17.1	–9.2	84.64
22.9	–3.4	11.56
Total		3479.65

$$s = \sqrt{\frac{3479.65}{7-1}} = \sqrt{579.9417} \approx 24.1$$

5. Range $= 9.8 - .5 = 9.3$

$$\bar{x} = \frac{.9 + 9.8 + 1.1 + .5 + 3.2 + 2.3 + 4}{7} = \frac{21.8}{7} \approx 3.1$$

x	$x - \bar{x}$	$(x - \bar{x})^2$
.9	–2.2	4.84
9.8	6.7	44.89
1.1	–2.0	4
.5	–2.6	6.76
3.2	.1	.01
2.3	–.8	.64
4.0	.9	.81
	Total:	61.95

$$s = \sqrt{\frac{61.95}{7-1}} = \sqrt{10.325} \approx 3.2$$

7. Range $= 12.6 - 1.8 = 10.8$

$$\bar{x} = \frac{1.8 + 12.6 + 5.3 + 2.2 + 2.6 + 3.6 + 6.9}{7} = \frac{35}{7} = 5$$

x	$x - \bar{x}$	$(x - \bar{x})^2$
1.8	–3.2	10.24
12.6	7.6	57.76
5.3	.3	.09
2.2	–2.8	7.84
2.6	–2.4	5.76
3.6	–1.4	1.96
6.9	1.9	3.61
	Total:	87.26

$$s = \sqrt{\frac{87.26}{7-1}} = \sqrt{14.5433} \approx 3.8$$

Copyright © 2019 Pearson Education, Inc.

9. Range = 8.9 – 1.1 = 7.8

$$\bar{x} = \frac{1.1+8.9+1.4+1.2+1.9+1.4+2}{7} = \frac{17.9}{7} \approx 2.6$$

x	$x-\bar{x}$	$(x-\bar{x})^2$
1.1	–1.5	2.25
8.9	6.3	39.69
1.4	–1.2	1.44
1.2	–1.4	1.96
1.9	–.7	.49
1.4	–1.2	1.44
2.0	–.6	.36
	Total:	47.63

$$s = \sqrt{\frac{47.63}{7-1}} = \sqrt{7.9383} \approx 2.8$$

11. Expand the table to include columns for the midpoint x of each interval, and for fx, x^2, and fx^2.

Interval	f	x	fx	x^2	fx^2
0–24	4	12	48	144	576
25–49	3	37	111	1369	4107
50–74	6	62	372	3844	23,064
75–99	3	87	261	7569	22,707
100–124	5	112	560	12,544	62,720
125–129	9	137	1233	18,769	168,921
Totals	30		2585		282,095

The mean of the grouped data is

$$\bar{x} = \frac{\Sigma fx}{n} = \frac{2585}{30} \approx 86.2$$

The standard deviation for the grouped data is

$$s = \sqrt{\frac{\Sigma fx^2 - n\bar{x}^2}{n-1}} = \sqrt{\frac{282{,}095 - 30(86.2)^2}{30-1}} \approx \sqrt{2040.8} \approx 45.2$$

13. $$\bar{x} = \frac{25+27+64+66+32+72+99+83+74+54}{10} = \frac{596}{10} = 59.6$$

x	$x-\bar{x}$	$(x-\bar{x})^2$
25	–34.6	1197.16
27	–32.6	1062.76
64	4.4	19.36
66	6.4	40.96
32	–27.6	761.76
72	12.4	153.76
99	39.4	1552.36
83	23.4	547.56
74	14.4	207.36
54	–5.6	31.36
	Total:	5574.4

$$s = \sqrt{\frac{5574.4}{10-1}} = \sqrt{619.3778} \approx 24.9$$

15. Use Chebyshev's theorem with $k = 2$.

$$1 - \frac{1}{2^2} = 1 - \frac{1}{4} = \frac{3}{4}$$

So, at least $\frac{3}{4}$ of the numbers lie within 2 standard deviations of the mean.

17. Use Chebyshev's theorem with $k = 1.5$.

$1 - \frac{1}{(1.5)^2} = 1 - \frac{4}{9} = \frac{5}{9}$, so at least $\frac{5}{9}$ of the numbers lie within 1.5 standard deviations of the mean.

19. Between 26 and 74, $\bar{x} = 50$ and $s = 6$; we have $26 = 50 - 4 \cdot 6$ and $74 = 50 + 4 \cdot 6$, so $k = 4$.

At least $1 - \frac{1}{k^2} = \frac{15}{16} = 93.75\%$ of the numbers lie between 26 and 74.

21. Less than 32 or more than 68
From exercise 18, 88.9% of the data lie between 32 and 68. So, no more than 100% – 88.9% = 11.1% of the data are less then 32 or more than 68.

Copyright © 2019 Pearson Education, Inc.

23. Columbia movie studios

$$\bar{x} = \frac{911+966+1262+1445+1792+1274+1283}{7}$$

$$= \frac{8933}{7} \approx 1276.14286$$

The mean revenue for Columbia studios is about $1,276.14286 million.

x	$x-\bar{x}$	$(x-\bar{x})^2$
911	−365.1	133,298.01
966	−310.1	96,162.01
1262	−14.1	198.81
1445	168.9	28,527.21
1792	515.9	266,152.81
1274	−2.1	4.41
1283	6.9	47.61
	Total:	524,390.87

$$s = \sqrt{\frac{524,390.87}{7-1}} \approx 295.6$$

The standard deviation is about $295.6 million.

25. Warner Brothers studios

$$\bar{x} = \frac{1902+1604+1563+1864+1665+1826+1924}{7}$$

$$= \frac{12,348}{7} = 1764$$

The mean revenue for Warner Brothers studios is $1764 million.

x	$x-\bar{x}$	$(x-\bar{x})^2$
1902	138	19,044
1604	−160	25,600
1563	−201	40,401
1864	100	10,000
1665	−99	9,801
1826	62	3,844
1924	160	25,600
	Total:	134,290

$$s = \sqrt{\frac{134,290}{7-1}} \approx 149.6$$

The standard deviation is about $149.6 million.

27. Columbia studios have more variation because they have a higher standard deviation.

29. Let x be the midpoint of the interval.

Interval	f	x	fx	x^2	fx^2
0–29	12	14.5	174	210.25	2523
30–59	3	44.5	133.5	1980.25	5940.75
60–89	10	74.5	745	5550.25	55,502.5
90–119	2	104.5	209	10,920.25	21,840.5
120–149	0	134.5	0	18,090.25	0
150–179	1	164.5	164.5	27,060.25	27,060.25
180–209	2	194.5	389	37,830.25	75,660.5
Totals	30		1815		188,527.5

$$\bar{x} = \frac{\Sigma(fx)}{30} = \frac{1815}{30} = 60.5$$

$$s = \sqrt{\frac{188,527.5-30(60.5)^2}{30-1}}$$

$$\approx \sqrt{2714.483} \approx 52.1$$

31.

Interval	f	x	xf	x^2	$x^2 f$
50–59	2	54.5	109	2970.25	5940.5
60–69	4	64.5	258	4160.25	16,641
70–79	7	74.5	521.5	5550.25	38,851.75
80–89	9	84.5	760.5	7140.25	64,262.25
90–99	8	94.5	756	8930.25	71,442
Totals:	30		2405		197,137.5

$$\bar{x} = \frac{2405}{30} \approx 80.17$$

$$s = \sqrt{\frac{197,137.5-30(80.17)^2}{30-1}} \approx 12.2$$

For individualized instruction, the mean is 80.17 and the standard deviation is 12.2.

Copyright © 2019 Pearson Education, Inc.

33. Answers vary. Possible answer: You would expect the mean of the traditional instruction to be smaller, since 13 of the 34 students lie in the first two intervals while only 6 of 30 students lie in the same intervals for individualized instruction. As far as standard deviation is concerned, there is a small difference between the two, 1.1. This gives the impression that the data is pretty much spread out in nearly the same fashion in both types of instruction.

35. Arrange the data in ascending order:
80, 82, 93, 96, 99, 100, 100, 104, 105, 107
Minimum = 80; maximum = 107

$$\text{Median} = Q_2 = \frac{99+100}{2} = 99.5$$

$n = 10$, so $.25 \cdot 10 = 2.5$, which rounds up to 3. Count to the third data point to obtain $Q_1 = 93$. $.75 \cdot 10 = 7.5$, which rounds up to 8, so count to the eighth data point to obtain $Q_3 = 104$.

37.

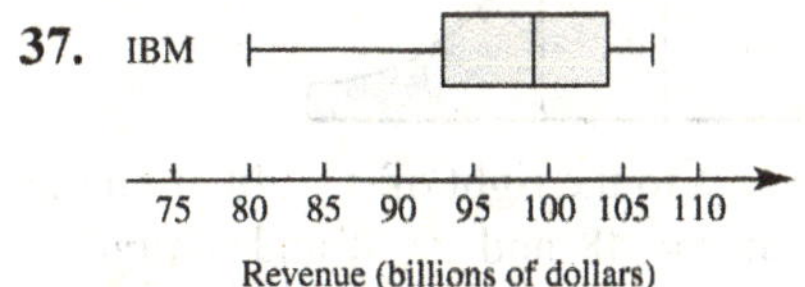

39.

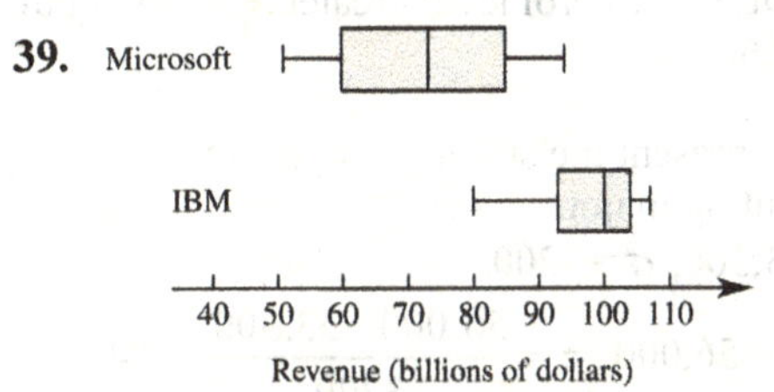

The distributions are not similar. IBM generally has higher revenue and less variability.

41. Arrange the expenditures on education data in ascending order:
9.7, 10.4, 12.0, 13.5, 17.1, 17.4, 17.7, 22.9, 23.5, 25.1, 50.4, 79.8
Minimum = 9.7; maximum = 79.8

$$\text{Median} = Q_2 = \frac{17.4+17.7}{2} = 17.55.$$

$n = 12$, so $.25 \cdot 12 = 3$. Count to the third data point to obtain $Q_1 = 12.0$. $.75 \cdot 12 = 9$, so count to the ninth data point to obtain $Q_3 = 23.5$.

43.

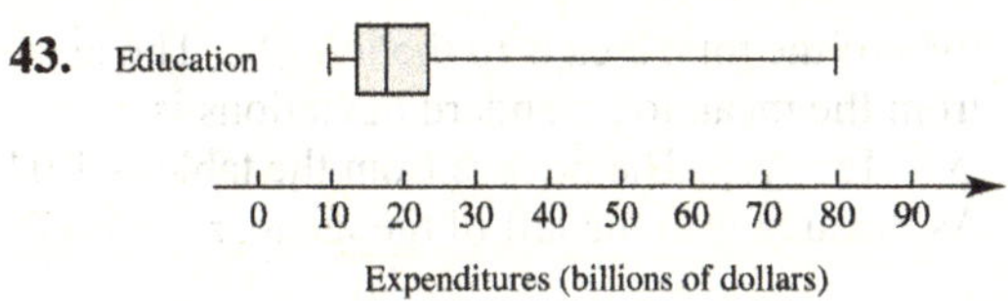

45. Answers vary, but estimates are:
Hospital : $Q_1 \approx .9, Q_2 \approx 1.7, Q_3 \approx 3.2$;
Highway: $Q_1 \approx 2.2, Q_2 \approx 3.1, Q_3 \approx 6.6$.

47. Health, because the box is wider; that is, the distance from Q_1 to Q_3 is greater.

Section 10.4 Normal Distributions

1. The peak in a normal curve occurs directly above the mean.

3. Answers vary. Possible answer:
If a normal distribution has mean μ and standard deviation σ, then the *z*-score for the number *x* is

$$z = \frac{x-\mu}{\sigma}.$$

5. By looking up 1.75 in Table 2, we get .4599 = 45.99%. Thus, the percent of the total area between the mean and 1.75 standard deviations from the mean is 45.99%.

7. –.43 indicates that we are below the mean. Looking up .43 in Table 2, we get .1664 = 16.64%. Thus, the percent of the total area between the mean and .43 standard deviations from the mean is 16.64%.

9. The entry corresponding to $z = 1.41$ is .4207, and the entry for $z = 2.83$ is .4977, so the area between $z = 1.41$ and $z = 2.83$ is $.4977 - .4207 = .0770$ or 7.7%.

11. For $z = 2.48$, the entry is .4934, and for $z = .05$, the entry is .0199. The area between $z = -2.48$ and $z = -.05$ is $.4934 - .0199 = .4735$ or 47.35%.

13. To find the area between $z = -3.05$ and $z = 1.36$, add the area between $z = -3.05$ and $z = 0$ to the area between $z = 1.36$ and $z = 0$. The area is $.4989 + .4131 = .9120 = 91.20\%$.

15. 5% of the total area is to the right of *z*. The mean divides the area in half or .5. The area from the mean to *z* standard deviations is $.5 - .05 = .45$. Using Table 2 backwards, we get the *z*-score corresponding to the area .45 as 1.64 or 1.65 (approximately).

Copyright © 2019 Pearson Education, Inc.

17. 15% of the total area is to the left of z. The area from the mean to z standard deviations is $.5 - .15 = .35$. The z-score from the table is 1.04. As the area is to the left of the mean, $z = -1.04$.

19. To find $P(x \le \mu)$ and $P(x \ge \mu)$ consider the normal curve. The curve is symmetric about the vertical line through μ, so $P(x \le \mu)$ represents half the area under the curve. Since the area under the curve is 1, $P(x \le \mu) = .5$. Similarly, $P(x \ge \mu) = .5$.

21. Using Chebyshev's theorem with $k = 3$:

$$1 - \frac{1}{3^2} = 1 - \frac{1}{9} = \frac{8}{9} \approx .889$$

Using the normal distribution, the area between $z = -3$ and $z = 3$ is $.4987 + .4987 = .997$

The probability a number will lie within 3 standard deviations of the mean is greater as indicated by the normal distribution at .997 than Chebyshev's theorem shows at .889.

23. Let x represent the height of a female.

$\mu = 63.7, \sigma = 2.8$

Find the z-score for $x = 60$.

$$z = \frac{x - \mu}{\sigma} = \frac{60 - 63.7}{2.8} \approx -1.32$$

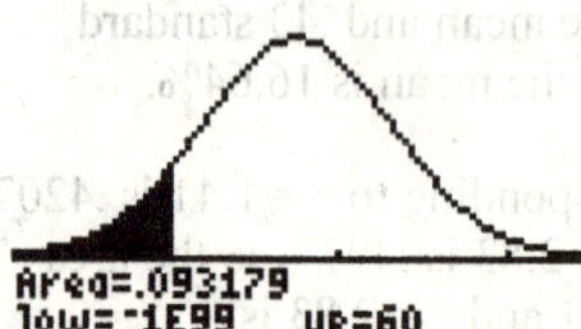

To find the area to the left of $z = -1.32$, find the area between $z = 0$ and $z = -1.32$ and subtract that answer from .5. The area is $.5 - .4066 = .0934$. Thus, the probability that a female is shorter than 60 inches is .0934.

25. Let x represent the height of a female.

$\mu = 63.7, \sigma = 2.8$

For $x = 60$, $z = \frac{60 - 63.7}{2.8} = \frac{-3.7}{2.8} \approx -1.32$,

and for $x = 72$, $z = \frac{72 - 63.7}{2.8} \approx 2.96$.

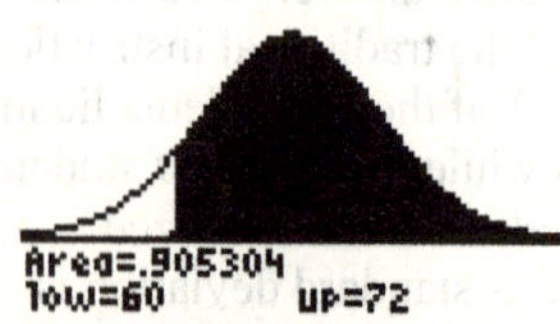

To find the area between $z = -1.32$ and $z = 2.96$, find the area between $z = 0$ and $z = -1.32$ and add it to the area between $z = 0$ and $z = 2.96$. The area is $.4066 + .4985 = .9051$. Thus, the probability that a female will be between 60 and 72 inches is .9051.

27. Let x represent HDL cholesterol level.

$\mu = 52.6, \sigma = 15.5$

For $x = 60$, $z = \frac{60 - 52.6}{15.5} \approx .48$

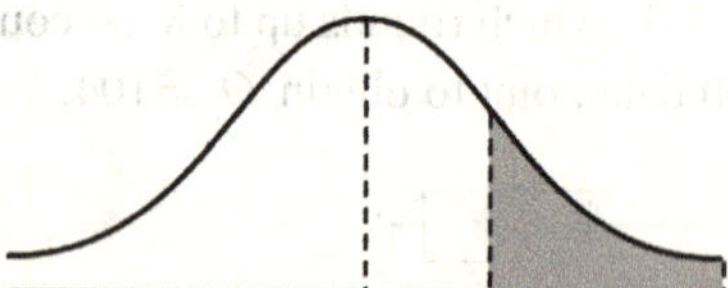

To find the area to the right of $z = .48$, find the area between $z = .48$ and $z = 0$ and subtract that from 0.5. The area is $0.5 - .1844 = .3156$. Thus, the probability that an individual will have an HDL cholesterol level greater than 60 mg/dL is .3156.

29. Let x represent the starting salaries for accounting majors.

$\mu = 53{,}300, \sigma = 3200$

For $x = 56{,}000$, $z = \frac{56{,}000 - 53{,}300}{3200} \approx .84$.

To find the area to the right of $z = .84$, find the area between $z = 0$ and $z = .84$ and subtract it from .5. The area is $.5 - .2996 = .2004$. Thus, the probability that an individual will have a starting salary above \$56,000 is .2004.

For exercises 31–37, $\mu = 552$ and $\sigma = 121$. Refer to Table 2 in the back of the text.

Copyright © 2019 Pearson Education, Inc.

31. For $x = 540$, $z = \frac{540-552}{121} \approx -.10$.

For $x = 700$, $z = \frac{700-552}{121} \approx 1.22$.

The area between $z = 0$ and $z = -.10$ is .0398, while the area between $z = 0$ and $z = 1.22$ is .3888. Therefore, the area between $z = -.10$ and $z = 1.22$ is $.3888 + .0398 = .4286$, which is the probability that a GMAT test taker earns a score in the range 540–700.

33. From exercise 31, we know that $z = 1.22$ for $x = 700$ and $P = .3888$. From exercise 32, we know that $z = -2.08$ for $x = 300$ and $P = .4812$. Therefore, the probability that a GMAT test taker earns a score in the range 300–700 is $.3888 + .4812 = .8700$.

35. For $x = 750$, $z = \frac{750-552}{121} \approx 1.64$.

The area between $z = 0$ and $z = 1.64$ is .4495. The area to the right of $z = 0$ is .5, so the area to the right of $z = 1.64$ is $.5 - .4495 = .0505$. Thus, the probability that a GMAT test taker earns a score greater than 750 is .0505.

37. For $x = 300$, $z = \frac{300-552}{121} \approx -2.08$.

For $x = 400$, $z = \frac{400-552}{121} \approx -1.26$.

The area between $z = 0$ and $z = -2.08$ is .4812, while the area between $z = 0$ and $z = -1.26$ is .3962. Therefore, the area between $z = -2.08$ and $z = -1.26$ is $.4812 - .3962 = .0850$, which is the probability that a GMAT test taker earns a score in the range 300–400.

39. If 85% of the total area is to the left of z, then $z > 0$ and 35% of the area is between z and the mean. Look for a value of z in Table 2 with an area of .35. The closest match is $z = 1.04$. Let x stand for the speed.

$\frac{x-40}{5} = 1.04 \Rightarrow x - 40 = 5.2 \Rightarrow x = 45.2$

The 85th percentile speed is 45.2 mph.

41.

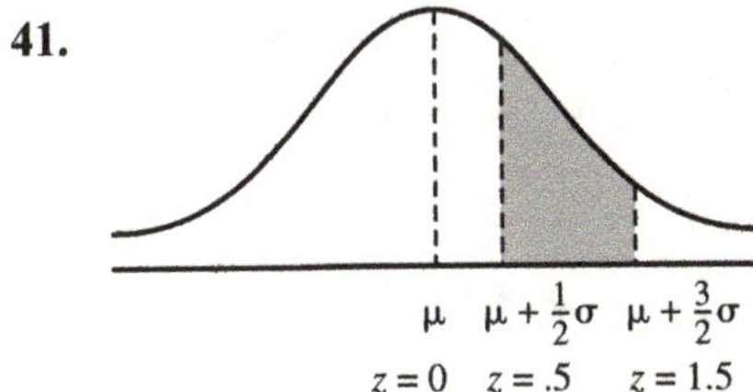

To find the area between $z = .5$ and $z = 1.5$, first find the area between $z = 1.5$ and $z = 0$. Then subtract the area between $z = 0$ and $z = .5$. The area is $.4332 - .1915 = .2417 = 24.17\%$, which means 24.17% of the students receive a B.

43. Answers vary. Possible answer: This system would be more fair in a large freshman class in psychology than in a graduate seminar of five students since the large class is more apt to have grades ranging the entire spectrum. The graduate students might have all grades in the 90's which would mean that the graduate student with the lowest score in the 90's would get an F.

45. $\mu = 550$ units; $\sigma = 45$ units
The recommended daily allowance is
$\mu + 2.5\sigma = 550 + (2.5)(46) = 665$ units

47. $\mu = 155$ units; $\sigma = 14$ units
The recommended daily allowance is
$\mu + 2.5\sigma = 155 + (2.5)(14) = 190$ units.

49. To find the area to the right of $z = 1$, find the area between $z = 0$ and $z = 1$, and subtract that from .5. The area is $.5 - .3413 = .1587$, which means 15.87% of the students had scores more than 1 standard deviation above the mean.

51. $p = .38$ and $n = 200$
$\mu = np = 200(.38) = 76$
$\sigma = \sqrt{200(.38)(.62)} \approx 6.9$

53. For between 70 and 80 employees, the area is between 69.5 and 80.5. For $x = 69.5$,

$z = \frac{69.5-76}{6.9} \approx -.94$

which gives an area of .3263. For $x = 80.5$,

$z = \frac{80.5-76}{6.9} \approx .65$

which gives an area of .2423. The probability is $.3263 + .2423 = .5686$.

55. For at least 88 employees, the area is greater than 87.5. For $x = 87.5$,

$z = \frac{87.5-76}{6.9} \approx 1.67$

which gives an area of .4525. The probability is $.5 - .4525 = .0475$.

57. $p = .44$ and $n = 75$
$\mu = np = 75(.44) = 33$
$\sigma = \sqrt{75(.44)(.56)} \approx 4.3$

Copyright © 2019 Pearson Education, Inc.

59. For between 25 and 35 females, the area is between 24.5 and 35.5. For $x = 24.5$,

$$z = \frac{24.5 - 33}{4.3} \approx -1.98$$

which gives an area of .4758. For $x = 35.5$,

$$z = \frac{35.5 - 33}{4.3} \approx .58$$

which gives an area of .2194. The probability is .4758 + .2194 = .6952.

61. For 27 or less female, the area is less than 27.5. For $x = 27.5$,

$$z = \frac{27.5 - 33}{4.3} \approx -1.28$$

which gives an area of .3997. The probability is .5 − .3997 = .1003.

Chapter 10 Review Exercises

1. a.

Interval	Frequenc
20–39.99	2
40–59.99	5
60–79.99	5
80–99.99	0
100–119.99	5
120–139.99	1
140–159.99	1
160–179.99	0
180–199.99	1

b.

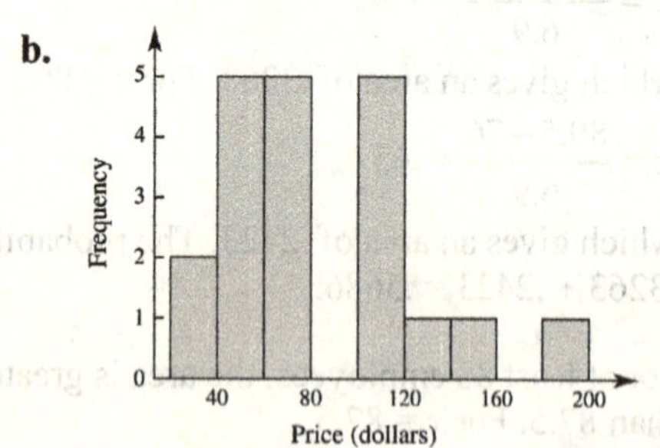

2. a.

Interval	Frequency
0–999	3
1000–1999	6
2000–2999	6
3000–3999	0
4000–4999	2
5000–5999	2
6000–6999	0
7000–7999	1

b.

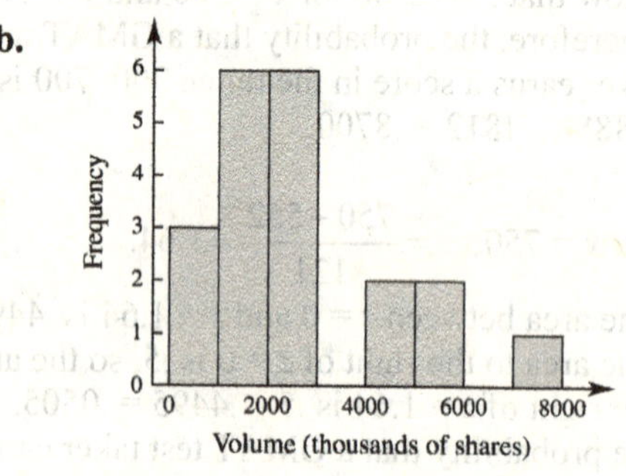

3.

STEM	LEAVES
0	8
1	
1	
1	4555
1	666
1	8
2	00011
2	22
2	5
2	6
2	88

Units: 2|8 = 28

Copyright © 2019 Pearson Education, Inc.

4.

STEM	LEAVES
0	78
1	138
2	156789
3	677
4	45
5	
6	125
7	4

Units: 7|4 = 7.4

5. $\Sigma x = 123 + 45 + 180 + 66 + 57 + 35 + 62 + 108 + 100 + 107 = 883$
The mean of the 10 numbers is

$\overline{x} = \frac{\Sigma x}{10} = \frac{883}{10} = \$88.3$.

The median is $\frac{66+100}{2} = \$83$

6. $\Sigma x = 241 + 1400 + 2900 + 2600 + 2500 + 7300 + 1800 + 411 + 2900 + 2300 = 24{,}352$

$\overline{x} = \frac{\Sigma x}{n} = \frac{24{,}352}{10} = 2435.2$

The median is $\frac{2300+2500}{2} = 2400$

7. Arrange the data in ascending order:
35, 45, 57, 62, 66, 100, 107, 108, 123, 180
Minimum = 35; maximum = 180

Median $= Q_2 = \frac{66+100}{2} = 83.$

$n = 10$, so $.25 \cdot 10 = 2.5$. Count to the third data point to obtain $Q_1 = 57$. $.75 \cdot 10 = 7.5$, so count to the eighth data point to obtain $Q_3 = 108$.

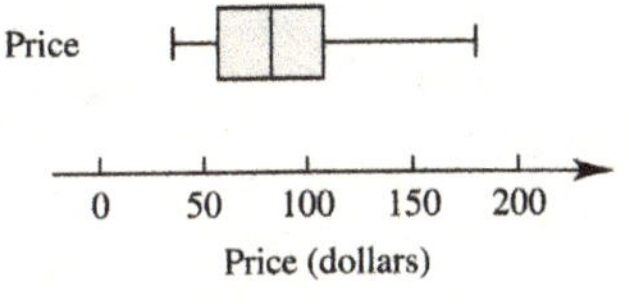

8. Arrange the data in ascending order:
241, 411, 1400, 1800, 2300, 2500, 2600, 2900, 2900, 7300
Minimum = 241; maximum = 7300

Median $= Q_2 = \frac{2300+2500}{2} = 2400.$

$n = 10$, so $.25 \cdot 10 = 2.5$. Count to the third data point to obtain $Q_1 = 1400$. $.75 \cdot 10 = 7.5$, so count to the eighth data point to obtain $Q_3 = 2900$.

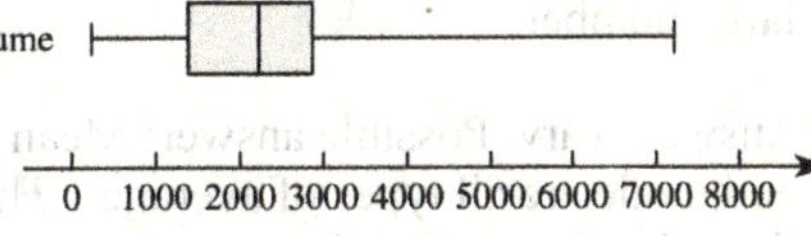

9.

Interval	Midpoint, x	Frequency, f	Product, xf
2–2.99	2.5	2	5
3–3.99	3.5	5	17.5
4–4.99	4.5	2	9
5–5.99	5.5	5	27.5
6–6.99	6.5	1	6.5
7–7.99	7.5	3	22.5
8–8.99	8.5	2	17
	Total:	20	105

The mean of this collection of grouped data is $\overline{x} = \frac{105}{20} = 5.25$. The median is the average of the 10th and 11th term. Both of these terms are in the 5 – 5.99 interval, so the median would be 5.5.

10.

Interval	Midpoint, x	Freq, f	Product xf
0–4.99	2.5	1	2.5
5–9.99	7.5	1	7.5
10–14.99	12.5	5	62.5
15–19.99	17.5	8	140
20–24.99	22.5	3	67.5
25–29.99	27.5	2	55
	Total:	20	335

Copyright © 2019 Pearson Education, Inc.

Use the formula for the mean of a grouped frequency distribution.

$\overline{x} = \frac{\Sigma xf}{n} = \frac{335}{20} = 16.75$ The median is the average of the 10th and 11th term. Both of these terms are in the 15 – 19.99 interval, so the median would be 17.5.

11. Answers vary. Possible answer: Grouped frequency distribution would be used when the data becomes too varied, when the range is a large number.

12. Answers vary. Possible answer: Mean, median, and mode are all types of averages. The mode is the value that occurs the most. The median is the middle value when the data are ranked from highest to lowest. The mean is the sum of the data divided by the number of data items.

13. There are 10 numbers here; the median is the mean of the 2 middle numbers after the data is placed in order, which is $\frac{40+45}{2} = 42.5$ hours..
The mode = 40 and 50, which both occur 2 times.

14. There are 10 numbers here; the median is the mean of the 2 middle numbers after the data is placed in order, which is $\frac{27+28}{2} = 27.5\%$..
The mode = 26, 28, and 36, which all occur 2 times.

15. The modal class is the interval with the greatest frequency. For the distribution of Exercise 9, the modal class is 3–3.99 and 5–5.99.

16. The modal class for the distribution of Exercise 10 is the interval 15–19.99, since it contains more data values than any of the other intervals.

17. right skewed

18. uniform

19. normal

20. left skewed

21. Answers vary. Possible answer: The range of a distribution is the difference between the largest and smallest data values.

22. Answers vary. Possible answer: The standard deviation is the square root of the variance. The standard deviation measures how spread out the data are from the mean.

23. The range is the difference between the largest and smallest numbers. For this distribution, the range is 70 – 16 = 54.
To find the standard deviation, the first step is to find the mean.

$$\overline{x} = \frac{30+45+16+20+70+40+48+40+50+50}{10}$$

$$= \frac{409}{10} = 40.9$$

Now complete the following chart.

x	x^2
30	900
45	2025
16	256
20	400
70	4900
40	1600
48	2304
40	1600
50	2500
50	2500
Total: 18,985	

$$s = \sqrt{\frac{\Sigma x^2 - n\overline{x}^2}{n-1}} = \sqrt{\frac{18{,}985 - 10(40.9)^2}{9}}$$

$$= \sqrt{250.767} \approx 15.8$$

Copyright © 2019 Pearson Education, Inc.

24. The range is 36 – 20 = 16, the difference of the highest and lowest numbers in the distribution.

The mean is $\overline{x} = \frac{\Sigma x}{n} = \frac{279}{10} = 27.9$.

Construct a table with the values of x, $x - \overline{x}$, and $(x - \overline{x})^2$.

x	$x - \overline{x}$	$(x - \overline{x})^2$
22	–5.9	34.81
28	.1	.01
26	–1.9	3.61
20	–7.9	62.41
30	2.1	4.41
36	8.1	65.61
36	8.1	65.61
28	.1	.01
26	–1.9	3.61
27	–.9	.81
Totals: 279		240.9

The standard deviation is

$$s = \sqrt{\frac{240.9}{10-1}} \approx \sqrt{26.767} \approx 5.2.$$

25. Recall that when working with grouped data, x represents the midpoint of each interval. Complete the following table, which extends the table from Exercise 9.

Interval	f	x	xf	x^2	fx^2
2–2.99	2	2.5	5	6.25	12.5
3–3.99	5	3.5	17.5	12.25	61.25
4–4.99	2	4.5	9	20.25	40.5
5–5.99	5	5.5	27.5	30.25	151.25
6–6.99	1	6.5	6.5	42.25	42.25
7–7.99	3	7.5	22.5	56.25	168.75
8– 8.99	2	8.5	17	72.25	144.5
Totals:	20		105		621

Use the formulas for grouped frequency distributions to find the mean and then the standard deviation. (The mean was also calculated in Exercise 7.)

$$\overline{x} = \frac{\Sigma xf}{n} = \frac{105}{20} = 5.25$$

$$s = \sqrt{\frac{\Sigma fx^2 - n\overline{x}^2}{n-1}} = \sqrt{\frac{621 - 20(5.25)^2}{19}} \approx 1.9$$

Copyright © 2019 Pearson Education, Inc.

26. Start with the frequency distribution that was the answer to Exercise 10, and expand the table to include columns for the midpoint x of each interval, and for xf, x^2, and fx^2.

Interval	f	x	xf	x^2	fx^2
0–4.99	1	2.5	2.5	6.25	6.25
5–9.99	1	7.5	7.5	56.25	56.25
10–14.99	5	12.5	62.5	156.25	781.25
15–19.99	8	17.5	140	306.25	2450
20–24.99	3	22.5	67.5	506.25	1518.75
25–29.99	2	27.5	55	756.25	1512.5
Totals:	20		335		6325

The mean of the grouped data is

$$\bar{x} = \frac{\Sigma xf}{n} = \frac{335}{20} = 16.75.$$

The standard deviation for the grouped data is

$$s = \sqrt{\frac{\Sigma fx^2 - n\bar{x}^2}{n-1}} = \sqrt{\frac{6325 - 20(16.75)^2}{20-1}} \approx \sqrt{37.565} \approx 6.1$$

27. To find the standard deviation, the first step is to find the mean.

$$\bar{x} = \frac{27.7 + 27.9 + 28.1 + 27.1 + 25.8}{5}$$

$$= \frac{136.6}{5} = 27.32$$

Now complete the following chart.

X	x^2
27.7	767.29
27.9	778.41
28.1	789.61
27.1	734.41
25.8	665.64
Total:	3735.56

$$s = \sqrt{\frac{\Sigma x^2 - n\bar{x}^2}{n-1}} = \sqrt{\frac{3735.36 - 5(27.32)^2}{4}}$$

$$= \sqrt{.862} \approx .93$$

28. To find the standard deviation, the first step is to find the mean.

$$\bar{x} = \frac{19.3 + 19.0 + 19.0 + 19.2 + 18.7}{5}$$

$$= \frac{95.2}{5} = 19.04$$

Now complete the following chart.

x	x^2
19.3	372.49
19.0	361
19.0	361
19.2	368.64
18.7	349.69
Total:	1812.82

$$s = \sqrt{\frac{\Sigma x^2 - n\bar{x}^2}{n-1}} = \sqrt{\frac{1812.82 - 5(19.04)^2}{4}}$$

$$= \sqrt{.053} \approx .23$$

Copyright © 2019 Pearson Education, Inc.

29. Answers vary. Possible answer: A normal distribution is a continuous distribution with the following properties:

- The highest frequency is at the mean;
- The graph is symmetric about a vertical line through the mean; and
- The total area under the curve, above the x-axis, is 1.

30. Answers vary. Possible answer: A distribution in which the peak is not at the center, or mean, is called skewed.

31. Between $z = 0$ and $z = 1.35$
By Table 2, the area between $z = 0$ and $z = 1.35$ is .4115.

32. To the left of $z = .38$
The area between $z = 0$ and $z = .38$ is .1480, so the area to the left of $z = .38$ is $.5 + .1480 = .6480$.

33. Between $z = -1.88$ and $z = 2.41$
The area between $z = -1.88$ and $z = 0$ is .4700.
The area between $z = 0$ and $z = 2.41$ is .4920.
The total area between $z = -1.88$ and $z = 2.41$ is $.4700 + .4920 = .9620$.

34. Between $z = 1.53$ and $z = 2.82$
The area between $z = 2.82$ and $z = 1.53$ is $.4976 - .4370 = .0606$.

35. Since 8% of the area is to the right of the z-score, 42% or .4200 is between $z = 0$ and the appropriate z-score. Use Table 2 to find an appropriate z-score whose area is .4200. The closest approximation is $z = 1.41$.

36. Answers vary. Possible answer: The normal distribution is not a good approximation of a binomial distribution that has a value of p close to 0 or 1 because the histogram of such a binomial distribution is skewed and therefore not close to the shape of a normal distribution.

37. For $x = 45$, $z = \dfrac{45 - 38.6}{2.3} \approx 2.78$.
The area between $z = 0$ and $z = 2.78$ is .4973.
The area to the left of $z = 0$ is .5, so the area to the left of $z = 2.78$ is $.5 + .4973 = .9973$. Thus, the probability that a female has an upper arm circumference less than 45 cm is .9973.

38. For $x = 35$, $z = \dfrac{35 - 38.6}{2.3} \approx -1.57$.

For $x = 45$, $z = \dfrac{45 - 38.6}{2.3} \approx 2.78$.
The area between $z = 0$ and $z = -1.57$ is .4571.
The area between $z = 0$ and $z = 2.78$ is .4820, so the area between $z = -1.57$ and $z = 2.78$ is $.4571 + .4820 = .9391$. Thus, the probability that a female has an upper arm circumference between 35 and 45 cm is .9391.

39. For $x = 38$, $z = \dfrac{38 - 41.5}{2.5} \approx -1.4$.
The area between $z = 0$ and $z = -1.4$ is .4192.
The area to the right of $z = 0$ is .5, so the area to the right of $z = -1.4$ is $.5 + .4192 = .9192$. Thus, the probability that a male has an upper arm circumference more than 38 cm is .9192.

40. For $x = 38$, $z = \dfrac{38 - 41.5}{2.5} \approx -1.4$.

For $x = 46$, $z = \dfrac{46 - 41.5}{2.5} \approx 1.8$.
The area between $z = 0$ and $z = -1.4$ is .4187.
The area between $z = 0$ and $z = 1.8$ is .4646, so the area between $z = -1.4$ and $z = 1.8$ is $.4187 + .4646 = .8833$. Thus, the probability that a male has an upper arm circumference between 38 and 46 cm is .8833.

41. a. May 2016

x	x^2
240	57,600
235	55,225
192	36,864
219	47,961
147	21,609
133	17,689
71	5,041
28	784
63	3,969
50	2,500
32	1,024
29	841
1439	251,107

Copyright © 2019 Pearson Education, Inc.

$\bar{x} = \frac{1439}{12} \approx 119.9167$

$s = \sqrt{\frac{\Sigma x^2 - n\bar{x}^2}{n-1}} = \sqrt{\frac{251{,}107 - 12(119.9167)^2}{12-1}}$

≈ 84.5

b. Nissan is closest to the mean sales.

42. **a.** May 2017

x	x^2
237	56,169
240	57,600
189	35,721
218	47,524
148	21,904
137	18,769
60	3,600
26	676
59	3,481
56	3,136
30	900
30	900
1430	250,308

$\bar{x} = \frac{1430}{12} \approx 119.1667$

$s = \sqrt{\frac{\Sigma x^2 - n\bar{x}^2}{n-1}} = \sqrt{\frac{250{,}380 - 12(119.1667)^2}{12-1}}$

≈ 85.3

b. Nissan is closest to the mean sales.

43. May 2016:
Arrange the data in ascending order:
28, 29, 32, 50, 63, 71, 133, 147, 192, 219, 235, 240
Minimum: 28, maximum: 240

$Q_2 = \frac{71+133}{2} = \frac{204}{2} = 102$

$n = 12$, so Q_1 = the third data entry, 32, and
Q_3 = the ninth data entry, 192.

44. May 2017:
Arrange the data in ascending order:
26, 30, 30, 56, 59, 60, 137, 148, 189, 218, 237, 240
Minimum: 26, maximum: 240

$Q_2 = \frac{60+137}{2} = \frac{197}{2} = 98.5$

$n = 12$, so Q_1 = the third data entry, 30, and
Q_3 = the ninth data entry, 189.

45.

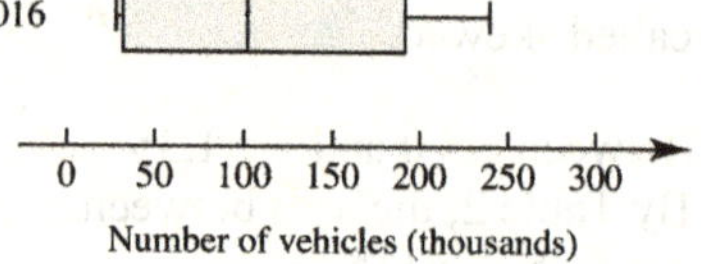

46.

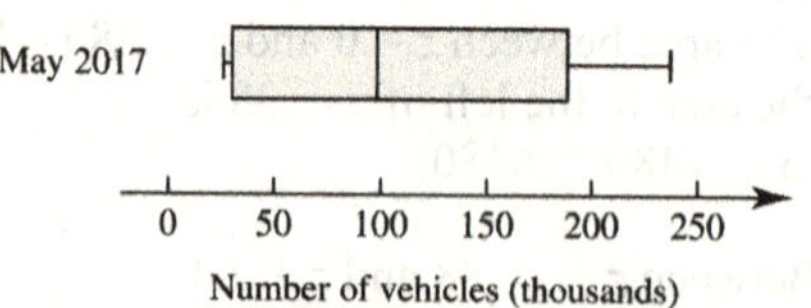

47. $n = 500, p = .51, 1 - p = .49$

$\mu = np = 500(.51) = 255$

$\sigma = \sqrt{np(1-p)} = \sqrt{255(.49)} \approx 11.2$

For more than 260, let $x = 259.5$

$z = \frac{259.5 - 255}{11.2} \approx .40$

The area is .1879, so the probability is $.5 - .1879 = .3121$.

48. $n = 500, p = .51, 1 - p = .49$

$\mu = np = 500(.51) = 255$

$\sigma = \sqrt{np(1-p)} = \sqrt{255(.49)} \approx 11.2$

For between 265 and 280, let $x = 264.5$ and $x = 280.5$

$z = \frac{264.5 - 255}{11.2} \approx .85$

The area is .3023.

$z = \frac{280.5 - 255}{11.2} \approx 2.28$

The area is .4887, so the probability is $.4887 - .3023 = .1864$.

Copyright © 2019 Pearson Education, Inc.

49. $n = 500, p = .51, 1-p = .49$
$\mu = np = 500(.51) = 255$
$\sigma = \sqrt{np(1-p)} = \sqrt{255(.49)} \approx 11.2$
For between 240 and 280, let $x = 239.5$ and $x = 280.5$
$z = \frac{239.5-255}{11.2} \approx -1.38$
The area is .4162.
$z = \frac{280.5-255}{11.2} \approx 2.28$
The area is .4887, so the probability is $.4887 + .4162 = .9049$.

50. $n = 500, p = .51, 1-p = .49$
$\mu = np = 500(.51) = 255$
$\sigma = \sqrt{np(1-p)} = \sqrt{255(.49)} \approx 11.2$
For less than 240, let $x = 240.5$
$z = \frac{240.5-255}{11.2} \approx -1.29$
The area is .4162, so the probability is $.5 - .4162 = .0838$.

51. $n = 1000, p = .15, 1-p = .85$
$\mu = np = 1000(.15) = 150$
$\sigma = \sqrt{np(1-p)} = \sqrt{150(.85)} \approx 11.3$
For more than 130, let $x = 129.5$
$z = \frac{129.5-150}{11.3} \approx -1.81$
The area is .4582, so the probability is $.5 + .4582 = .9582$.

52. $n = 1000, p = .15, 1-p = .85$
$\mu = np = 1500(.15) = 150$
$\sigma = \sqrt{np(1-p)} = \sqrt{150(.85)} \approx 11.3$
For between 120 and 130, let $x = 119.5$ and $x = 130.5$
$z = \frac{119.5-150}{11.3} \approx -2.70$
The area is .4961.
$z = \frac{130.5-150}{11.3} \approx -1.73$
The area is .4578, so the probability is $.4961 - .4578 = .0383$.

53. $n = 1000, p = .15, 1-p = .85$
$\mu = np = 1500(.15) = 150$
$\sigma = \sqrt{np(1-p)} = \sqrt{150(.85)} \approx 11.3$
For between 120 and 165, let $x = 119.5$ and $x = 165.5$
$z = \frac{119.5-150}{11.3} \approx -2.70$
The area is .4961.
$z = \frac{165.5-150}{11.3} \approx 1.37$
The area is .4151, so the probability is $.4961 + .4151 = .9112$.

54. $n = 1000, p = .15, 1-p = .85$
$\mu = np = 1000(.15) = 150$
$\sigma = \sqrt{np(1-p)} = \sqrt{150(.85)} \approx 11.3$
For 165 or fewer, let $x = 165.5$
$z = \frac{165.5-150}{11.3} \approx 1.37$
The area is .4147, so the probability is $.5 + .4151 = .9147$.

Case 10 Standard Deviation as a Measure of Risk

1. The mean monthly return for the Vanguard fund is $\overline{x} = \frac{\Sigma x}{n} = \frac{37.14}{36} \approx 1.032$. The annual return is $12\overline{x} = 12(1.032) \approx 12.38$.

2. The mean monthly return for the S&P 500 fund is $\overline{x} = \frac{\Sigma x}{n} = \frac{30.69}{36} \approx .8525$. The annual return is $12\overline{x} = 12(.8525) \approx 10.23$.

3. The mean monthly return for the Large Growth fund is $\overline{x} = \frac{\Sigma x}{n} = \frac{29.03}{36} \approx .8064$. The annual return is $12\overline{x} = 12(.8064) \approx 9.68$.

4. The Vanguard fund has the highest annual return.

5. Through the use of a computer or calculator, the standard deviation for the Vanguard fund will be $s \approx 3.2555$. The annual return is $s\sqrt{12} = (3.2555)\sqrt{12} \approx 11.28$.

Copyright © 2019 Pearson Education, Inc.

6. Through the use of a computer or calculator, the standard deviation for the S&P 500 will be $s \approx 2.9947$. The annual return is $s\sqrt{12} = (2.9947)\sqrt{12} \approx 10.37$.

7. Through the use of a computer or calculator, the standard deviation for the Large Growth fund will be $s \approx 3.2307$. The annual return is $s\sqrt{12} = (3.2307)\sqrt{12} \approx 11.19$.

8. The Vanguard fund has the highest level of risk.

9. The bounds at the 68% level are found by $\overline{x} \pm s$. Therefore the bounds are:
$(12.38 - 11.28, 12.38 + 11.28)$
$(1.1, 23.66)$

10. The bounds at the 95% level are found by $\overline{x} \pm 2s$. Therefore the bounds are:
$(12.38 - 2 \cdot 11.28, 12.38 + 2 \cdot 11.28)$
$(-10.18, 34.94)$

Copyright © 2019 Pearson Education, Inc.

Chapter 11 Differential Calculus

Section 11.1 Limits

1. a. By reading the graph, as x gets closer to 3 from the left or the right, $f(x)$ approaches 8, so $\lim_{x \to 3} f(x) = 8.$

   b. As x gets closer to –1 from the left or right,

3. a. By reading the graph, as x gets closer to 0 from the left or the right, $F(x)$ approaches 4, so $\lim_{x \to 0} F(x) = 4.$

   b. By reading the graph, as x gets closer to 1 from the left or the right, $F(x)$ approaches 2, so $\lim_{x \to 1} F(x) = 2.$

5. a. By reading the graph, as x gets closer to –3 from the left $f(x)$ approaches –1. As x gets closer to –3 from the right, $f(x)$ approaches 2. Since these two values of the limit are not equal, $\lim_{x \to -3} f(x)$ does not exist.

   b. By reading the graph, as x gets closer to 2 from the left or right, $f(x)$ approaches 2 so $\lim_{x \to 2} f(x) = 2.$

7. a. By reading the graph, as x gets closer to –2 from the left or the right, $h(x)$ becomes infinitely large. Since $h(x)$ does not approach any fixed number as x approaches –2 from either side, $\lim_{x \to -2} h(x)$ does not exist.

   b. By reading the graph, as x gets closer to 0 from the left or the right, $h(x)$ approaches 2, so $\lim_{x \to 0} h(x) = 2.$

9. $\lim_{x \to 1} F(x)$ in Exercise 3(b) exists because, as x gets closer to 1 from the left or the right, $F(x)$ gets closer to 2, a single number. On the other hand, $\lim_{x \to -3} f(x)$ in Exercise 5(a) does not exist because, as x approaches –3 from the left, $f(x)$ gets closer to –1, but, as x approaches –3 from the right, $f(x)$ approaches another number, 2.

11. $\lim_{x \to 1} \frac{\ln x}{x - 1}$

x	$\frac{\ln x}{x-1}$
.9	1.0536
.99	1.0050
.999	1.0005
1	
1.001	.9995
1.01	.9950
1.1	.9531

As x gets closer to 1 from the left or right, the value of $\frac{\ln x}{x-1}$ gets closer to 1, so $\lim_{x \to 1} \frac{\ln x}{x-1} = 1.$ Verify this by examining the graph of the function.

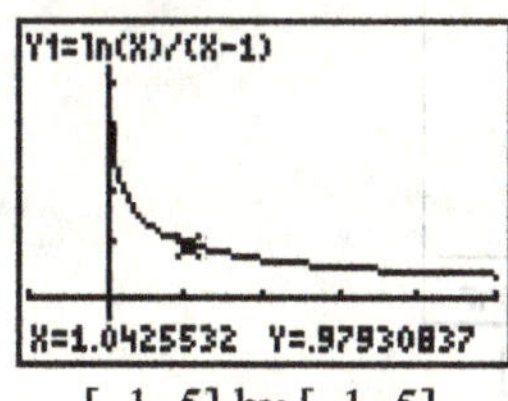

[–1, 5] by [–1, 5]

13. $\lim_{x \to 0} (x \cdot \ln|x|)$

x	$x \cdot \ln\|x\|$
–.1	.2303
–.01	.0461
–.001	.0069
0	
.001	–.0069
.01	–.0461
.1	–.2303

As x gets closer to 0 from the left or right, the value of $x \cdot \ln|x|$ gets closer to 0, so $\lim_{x \to 0} (x \cdot \ln|x|) = 0.$ Verify this by examining the graph of the function.

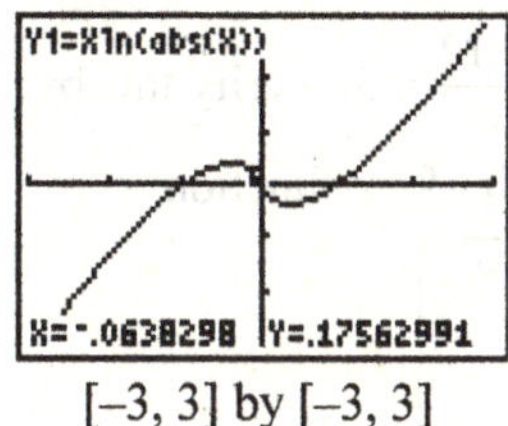

[–3, 3] by [–3, 3]

Copyright © 2019 Pearson Education, Inc.

15. $\lim_{x\to 0} \frac{e^{3x}-1}{x}$

x	$\frac{e^{3x}-1}{x}$
–.1	2.5918
–.01	2.9554
–.001	2.9955
0	
.001	3.0045
.01	3.0455
.1	3.4986

As x gets closer to 0 from the left or right, the value of $\frac{e^{3x}-1}{x}$ gets closer to 3, so $\lim_{x\to 0} \frac{e^{3x}-1}{x} = 3$. Verify this by examining the graph of the function.

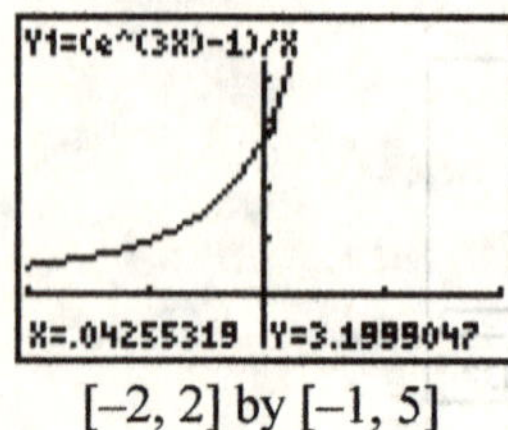

[–2, 2] by [–1, 5]

17. $\lim_{x\to 3} \frac{x^3-3x^2-4x+12}{x-3}$

x	$\frac{x^3-3x^2-4x+12}{x-3}$
2.9	4.4100
2.99	4.9401
2.999	4.9940
3	
3.001	5.0060
3.01	5.0601
3.1	5.6100

As x gets closer to 3 from the left or right, the value of $\frac{x^3-3x^2-4x+12}{x-3}$ gets closer to 5, so $\lim_{x\to 3} \frac{x^3-3x^2-4x+12}{x-3} = 5$. Verify this by examining the graph of the function.

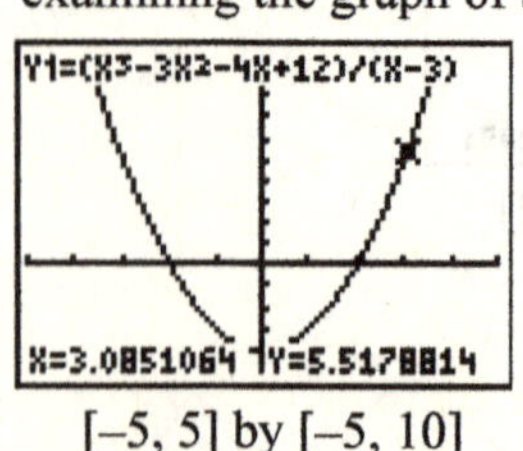

[–5, 5] by [–5, 10]

19. $\lim_{x\to -2} \frac{x^4+2x^3-x^2+3x+1}{x+2}$

x	$\frac{x^4+2x^3-x^2+3x+1}{x+2}$
–2.1	87.839
–2.01	898.8894
–2.001	8998.9889
–2	
–1.999	–9000.989
–1.99	–900.8906
–1.9	–89.959

As x gets closer to –2 from the left, the value of $\frac{x^4+2x^3-x^2+3x+1}{x+2}$ becomes infinitely large, and from the right, the value becomes infinitely small, so $\lim_{x\to -2} \frac{x^4+2x^3-x^2+3x+1}{x+2}$ does not exist. Verify this by examining the graph of the function.

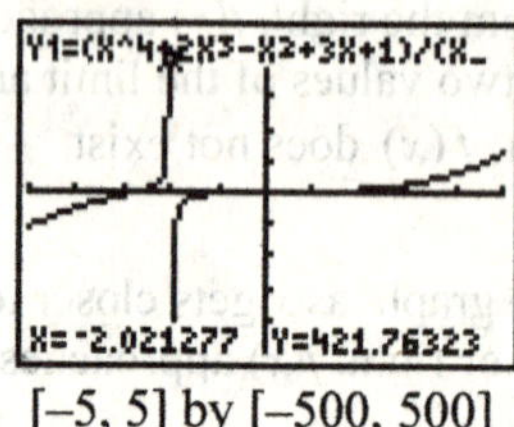

[–5, 5] by [–500, 500]

21. $\lim_{x\to 4}[f(x)-g(x)]$

Use property 2.

$$\lim_{x\to 4}[f(x)-g(x)] = \lim_{x\to 4} f(x) - \lim_{x\to 4} g(x) = 25-10=15$$

23. $\lim_{x\to 4} \frac{f(x)}{g(x)}$

Use property 4.

$$\lim_{x\to 4} \frac{f(x)}{g(x)} = \frac{\lim_{x\to 4} f(x)}{\lim_{x\to 4} g(x)} = \frac{25}{10} = \frac{5}{2}$$

25. $\lim_{x\to 4} \sqrt{f(x)} = \lim_{x\to 4}[f(x)]^{1/2}$

Use property 5.

$$\lim_{x\to 4} \sqrt{f(x)} = \left[\lim_{x\to 4} f(x)\right]^{1/2} = 25^{1/2} = 5$$

Copyright © 2019 Pearson Education, Inc.

27. $\lim\limits_{x\to 4} \dfrac{f(x)+g(x)}{2g(x)}$

$= \dfrac{\lim\limits_{x\to 4}[f(x)+g(x)]}{\lim\limits_{x\to 4}[2g(x)]}$ Property 4

$= \dfrac{\lim\limits_{x\to 4} f(x) + \lim\limits_{x\to 4} g(x)}{2\lim\limits_{x\to 4} g(x)}$ Property 1 and L.C.F.

$= \dfrac{25+10}{2(10)} = \dfrac{35}{20} = \dfrac{7}{4}$

29. a. $f(x)\begin{cases} 3-x & \text{if } x<-2 \\ x+2 & \text{if } -2\le x<2 \\ 1 & \text{if } x\ge 2 \end{cases}$

The graph of $f(x) = 3 - x$ if $x < -2$ is the ray through (–4, 7) and (–2, 5), with excluded (open) endpoint (–2, 5). The graph of $f(x) = x + 2$ if $-2 \le x < 2$ is the segment with endpoint (–2, 0) and open endpoint (2, 4). The graph of $f(x)$ if $x \ge 2$ is the horizontal ray to the right with endpoint (2, 1).

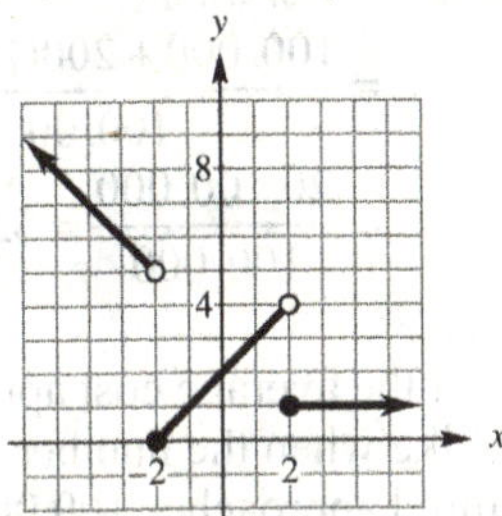

b. $\lim\limits_{x\to -2} f(x)$

As x gets closer to –2 from the left, $f(x) = 3 - x$ gets closer to $3 - (-2) = 5$ because $f(x) = 3 - x$ if $x < -2$. As x gets closer to –2 from the right, $f(x + 2)$ gets closer to 0 because $f(x) = x + 2$ if $-2 \le x < 2$. $f(x)$ does not get closer to a single real number. Therefore $\lim\limits_{x\to -2} f(x)$ does not exist.

c. $\lim\limits_{x\to 1} f(x)$

$f(x)$ is the polynomial function $f(x) = x + 2$ if $-2 \le x < 2$, so $\lim\limits_{x\to 1} f(x) = \lim\limits_{x\to 1} f(1) = 1 + 2 = 3.$

d. $\lim\limits_{x\to 2} f(x)$

As x gets closer to 2 from the left, $f(x) = x + 2$ gets closer to $2 + 2 = 4$ because $f(x) = x + 2$ if $-2 \le x < 2$. As x gets closer to 2 from the right, $f(x) = 1$ gets closer to 1 because $f(x) = 1$ if $x \ge 2$. $f(x)$ does not get closer to a single real number. Therefore, $\lim\limits_{x\to 2} f(x)$ does not exist.

31. $\lim\limits_{x\to 2}\left(3x^3 - 4x^2 - 5x + 2\right)$

$= 3(2)^3 - 4(2)^2 - 5(2) + 2 = 0$

33. $\lim\limits_{x\to 3} \dfrac{4x+7}{10x+1} = \dfrac{\lim\limits_{x\to 3}(4x+7)}{\lim\limits_{x\to 3}(10x+1)} = \dfrac{4(3)+7}{10(3)+1} = \dfrac{19}{31}$

35. $\lim\limits_{x\to 5} \dfrac{x^2-25}{x-5} = \lim\limits_{x\to 5} \dfrac{(x-5)(x+5)}{x-5}$

$= \lim\limits_{x\to 5}(x+5) = 5+5 = 10$

37. $\lim\limits_{x\to 4} \dfrac{x^2-x-12}{x-4} = \lim\limits_{x\to 4} \dfrac{(x-4)(x+3)}{x-4}$

$= \lim\limits_{x\to 4}(x+3) = 4+3 = 7$

39. $\lim\limits_{x\to 2} \dfrac{x^2-5x+6}{x^2-6x+8} = \lim\limits_{x\to 2} \dfrac{(x-2)(x-3)}{(x-2)(x-4)}$

$= \lim\limits_{x\to 2} \dfrac{x-3}{x-4} = \dfrac{\lim\limits_{x\to 2}(x-3)}{\lim\limits_{x\to 2}(x-4)}$

$= \dfrac{2-3}{2-4} = \dfrac{1}{2}$

41. $\lim\limits_{x\to -4} \dfrac{(x+4)^2(x-5)}{(x-4)(x+4)^2} = \lim\limits_{x\to -4} \dfrac{x-5}{x-4}$

$= \dfrac{-9}{-8} = \dfrac{9}{8}$

43. $\lim\limits_{x\to 3} \sqrt{x^2-3} = \sqrt{\lim\limits_{x\to 3}(x^2-3)} = \sqrt{3^2-3} = \sqrt{6}$

45. $\lim\limits_{x\to 4} \dfrac{-6}{(x-4)^2} = \dfrac{\lim\limits_{x\to 4}(-6)}{\lim\limits_{x\to 4}(x-4)^2}$

$= \dfrac{-6}{0}$ Undefined

The limit does not exist.

Copyright © 2019 Pearson Education, Inc.

47. $\lim\limits_{x\to 0}\dfrac{\frac{1}{x+3}-\frac{1}{3}}{x}=\lim\limits_{x\to 0}\left(\dfrac{1}{x+3}-\dfrac{1}{3}\right)\left(\dfrac{1}{x}\right)$

$=\lim\limits_{x\to 0}\dfrac{3-x-3}{3(x+3)(x)}=\lim\limits_{x\to 0}\dfrac{-x}{3(x+3)x}$

$=\lim\limits_{x\to 0}\dfrac{-1}{3(x+3)}=-\dfrac{1}{9}$

49. $\lim\limits_{x\to 25}\dfrac{\sqrt{x}-5}{x-25}=\lim\limits_{x\to 25}\dfrac{\sqrt{x}-5}{x-25}\cdot\dfrac{\sqrt{x}+5}{\sqrt{x}+5}$

$=\lim\limits_{x\to 25}\dfrac{x-25}{(x-25)(\sqrt{x}+5)}$

$=\lim\limits_{x\to 25}\dfrac{1}{\sqrt{x}+5}=\dfrac{1}{\sqrt{25}+5}$

$=\dfrac{1}{5+5}=\dfrac{1}{10}$

51. $\lim\limits_{x\to 5}\dfrac{\sqrt{x}-\sqrt{5}}{x-5}=\lim\limits_{x\to 5}\dfrac{\sqrt{x}-\sqrt{5}}{x-5}\cdot\dfrac{\sqrt{x}+\sqrt{5}}{\sqrt{x}+\sqrt{5}}$

$=\lim\limits_{x\to 5}\dfrac{x-5}{(x-5)(\sqrt{x}+\sqrt{5})}$

$=\lim\limits_{x\to 5}\dfrac{1}{\sqrt{x}+\sqrt{5}}$

$=\dfrac{1}{2\sqrt{5}}$ or $\dfrac{\sqrt{5}}{10}$

53. $P(s)=\dfrac{105s}{s+7}$

a. $P(1)=\dfrac{105(1)}{1+7}=\dfrac{105}{8}=13.125$

b. $P(13)=\dfrac{105(13)}{13+7}=\dfrac{1365}{20}=68.25$

c. $\lim\limits_{s\to 13}P(s)=\lim\limits_{s\to 13}\dfrac{105s}{s+7}=\dfrac{105(13)}{13+7}=68.25$

55. $c(x)=100{,}000+200x$

$\overline{c}(x)=\dfrac{c(x)}{x}$

a. $\overline{c}(1000)=\dfrac{100{,}000+200(1000)}{1000}$

$=\dfrac{300{,}000}{1000}=\$300.00$

When manufacturing 1000 bikes, the average cost is $300 per bike.

b. $\overline{c}(10{,}000)=\dfrac{100{,}000+200(10{,}000)}{10{,}000}$

$=\dfrac{2{,}100{,}000}{10{,}000}=\$210.00$

When manufacturing 10,000 bikes, the average cost is $210 per bike.

c. $\lim\limits_{x\to 100{,}000}\overline{c}(x)=\lim\limits_{x\to 100{,}000}\dfrac{100{,}000+200x}{x}$

$=\dfrac{100{,}000+200(100{,}000)}{100{,}000}$

$=\dfrac{20{,}100{,}000}{100{,}000}=\$201.00$

The limit of the average cost approaches $201 per bike when the number of bikes manufactured approaches 100,000.

57. a. $\lim\limits_{t\to 2015}L(t)-M(t)$

$=\lim\limits_{t\to 2015}L(t)-\lim\limits_{t\to 2015}M(t)$

$=L(2015)-M(2015)$

$\approx 2.5-2.0\approx .5\%$

b. $\lim\limits_{t\to 2012}L(t)-M(t)$

$=\lim\limits_{t\to 2012}L(t)-\lim\limits_{t\to 2012}M(t)$

$=L(2012)-M(2012)$

$\approx 3.0-2.0\approx 1.0\%$

c. $\lim\limits_{t\to 2009}M(t)-S(t)$

$=\lim\limits_{t\to 2009}M(t)-\lim\limits_{t\to 2009}S(t)$

$=L(2009)-M(2009)$

$\approx 2.5-.5\approx 2.0\%$

Copyright © 2019 Pearson Education, Inc.

Section 11.2 One-Sided Limits and Limits Involving Infinity

1. As x approaches 2 from the left, $f(x)$ approaches 3. Thus, $\lim_{x\to 2^-} f(x) = 3.$

3. As x approaches –3 from the left, $f(x)$ approaches –2. Thus, $\lim_{x\to -3^-} f(x) = -2.$

5. As x approaches –1 from the left, $f(x)$ approaches 6. Thus, $\lim_{x\to -1^-} f(x) = 6.$

7. As x approaches –4 from the left, $f(x)$ approaches 0. Thus, $\lim_{x\to -4^-} f(x) = 0.$

9. a. As x approaches –2 from the left, $f(x)$ approaches 0. Thus $\lim_{x\to -2^-} f(x) = 0$.

   b. As x approaches 0 from the right, $f(x)$ approaches 1. Thus, $\lim_{x\to 0^+} f(x) = 1$.

   c. As x approaches 3 from the left, $f(x)$ approaches –1. Thus, $\lim_{x\to 3^-} f(x) = -1$.

   d. As x approaches 3 from the right, $f(x)$ approaches –1. Thus, $\lim_{x\to 3^+} f(x) = -1$.

11. a. As x approaches –2 from the left, $f(x)$, approaches 1. Thus $\lim_{x\to -2^-} f(x) = 1$.

    b. As x approaches 0 from the right, $f(x)$ approaches 0. Thus, $\lim_{x\to 0^+} f(x) = 0$.

    c. As x approaches 3 from the left, $f(x)$ approaches positive infinity. Thus, $\lim_{x\to 3^-} f(x) = \infty$.

    d. As x approaches 3 from the right, $f(x)$ approaches negative infinity. Thus, $\lim_{x\to 3^+} f(x) = -\infty$.

13. $\lim_{x\to 2^+} \sqrt{x^2-4} = \lim_{x\to 2^+} \left(x^2-4\right)^{1/2}$

$$= \left[\lim_{x\to 2^+} \left(x^2-4\right)\right]^{1/2} = 0^{1/2} = 0$$

15. $\lim_{x\to 0^+} \sqrt{x} + x + 1$

$$= \lim_{x\to 0^+} \sqrt{x} + \lim_{x\to 0^+} x + \lim_{x\to 0^+} 1$$
$$= 0 + 0 + 1 = 1$$

17. $\lim_{x\to -2^+} \left(x^3 - x^2 - x + 1\right)$

$$= \lim_{x\to -2^+} x^3 - \lim_{x\to -2^+} x^2 - \lim_{x\to -2^+} x + \lim_{x\to -2^+} 1$$
$$= -8 - 4 + 2 + 1 = -9$$

19. $\lim_{x\to -5^+} \dfrac{\sqrt{x+5}+5}{x^2-5}$

$$= \lim_{x\to -5^+} \left[\frac{\sqrt{x+5}+5}{x^2-5} \cdot \frac{\sqrt{x+5}-5}{\sqrt{x+5}-5}\right]$$
$$= \lim_{x\to -5^+} \left[\frac{x-20}{x^2\sqrt{x+5} - 5\sqrt{x+5} - 5x^2 + 25}\right]$$
$$= \lim_{x\to -5^+} \left[\frac{\frac{x-20}{x^2}}{\sqrt{x+5} - \frac{5\sqrt{x+5}}{x^2} - 5 + \frac{25}{x^2}}\right]$$
$$= \frac{-1}{-5+1} = \frac{1}{4}$$

21.

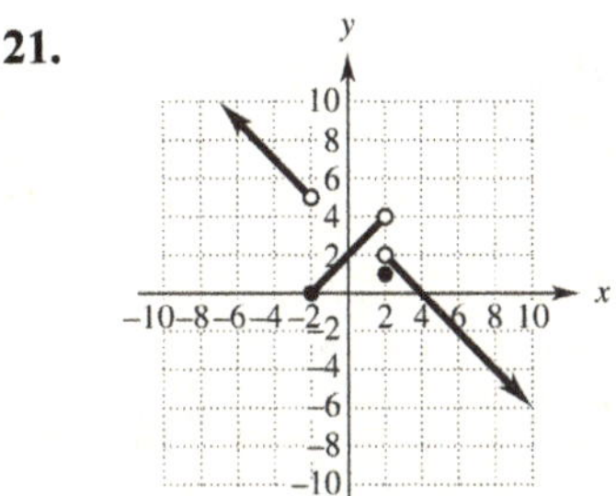

a. As x approaches –2 from the left, $f(x)$ approaches $3-(-2)=5$. Thus, $\lim_{x\to -2^-} f(x) = 5$.

b. As x approaches –2 from the right, $f(x)$ approaches $-2+2=0$. Thus, $\lim_{x\to -2^+} f(x) = 0$.

c. Since $f(x)$ does not approach the same value as x approaches –2 from the left and right, $\lim_{x\to -2} f(x)$ does not exist.

Copyright © 2019 Pearson Education, Inc.

23. a. As x approaches 2 from the left and the right, $f(x)$ approaches positive infinity. Thus, $\lim_{x\to 2} f(x) = \infty$.

b. As x approaches –2 from the right, $f(x)$ approaches positive infinity. Thus, $\lim_{x\to -2^+} f(x) = \infty$.

c. As x approaches –2 from the left, $f(x)$ approaches negative infinity. Thus, $\lim_{x\to -2^-} f(x) = -\infty$.

25. $\lim_{x\to 4} \frac{-6}{(x-4)^2} = -\infty$

27. $\lim_{x\to -3^+} \frac{2-x}{x+3} = \infty$

29. $\lim_{x\to\infty} f(x) = \infty$ and $\lim_{x\to -\infty} f(x) = 0$

31. $\lim_{x\to\infty} f(x) = -10$ and $\lim_{x\to -\infty} f(x) = -10$

33. $\lim_{x\to\infty} f(x) = \infty$ and $\lim_{x\to -\infty} f(x) = -\infty$

35. $\lim_{x\to\infty} \left[\sqrt{x^2+1} - (x+1)\right] = -1$

37. $\lim_{x\to -\infty} \frac{x^{2/3} - x^{4/3}}{x^3} = 0$

39. $\lim_{x\to -\infty} e^{1/x} = 1$

41. $\lim_{x\to\infty} \frac{\ln x}{x} = 0$

43. $$\lim_{x\to\infty} \frac{10x^2 - 5x + 8}{3x^2 + 8x - 29} = \lim_{x\to\infty} \frac{10 - \frac{5}{x} + \frac{8}{x^2}}{3 + \frac{8}{x} - \frac{29}{x^2}} = \frac{10 - 0 + 0}{3 + 0 - 0} = \frac{10}{3}$$

45. $$\lim_{x\to -\infty} \frac{5x + 9x^2 - 17}{21x + 5x^3 + 12x^2} = \lim_{x\to -\infty} \frac{\frac{5}{x^2} + \frac{9}{x} - \frac{17}{x^3}}{\frac{21}{x^2} + 5 + \frac{12}{x}} = \frac{0 + 0 - 0}{0 + 5 + 0} = 0$$

47. $$\lim_{x\to -\infty} \frac{8x^5 + 7x^4 - 10x^3}{25 - 4x^5} = \lim_{x\to -\infty} \frac{8 + \frac{7}{x} - \frac{10}{x^2}}{\frac{25}{x^5} - 4} = \frac{8 + 0 - 0}{0 - 4} = -2$$

49. $$\lim_{x\to -\infty} \frac{(4x-1)(x+2)}{5x^2 - 3x + 1} = \lim_{x\to -\infty} \frac{4x^2 + 7x - 2}{5x^2 - 3x + 1} = \lim_{x\to -\infty} \frac{4 + \frac{7}{x} - \frac{2}{x^2}}{5 - \frac{3}{x} + \frac{1}{x^2}} = \frac{4 + 0 - 0}{5 - 0 + 0} = \frac{4}{5}$$

51. $\lim_{x\to\infty} \left(\frac{1}{4x^2}\right) = 0$

53. $$\lim_{x\to -\infty} \left(\frac{18x}{x-9} - \frac{3x}{x+4}\right) = \lim_{x\to -\infty} \left(\frac{18x}{x-9}\right) - \lim_{x\to -\infty} \left(\frac{3x}{x+4}\right) = \lim_{x\to -\infty} \frac{18}{1 - \frac{9}{x}} - \lim_{x\to -\infty} \frac{3}{1 + \frac{4}{x}} = \frac{18}{1-0} - \frac{3}{1+0} = 15$$

55. $\lim_{x\to\infty} \frac{x}{|x|}$

Since x approaches positive infinity, $|x| = x$.

Thus, $\lim_{x\to\infty} \frac{x}{|x|} = \lim_{x\to\infty} \frac{x}{x} = \lim_{x\to\infty} 1 = 1$.

Copyright © 2019 Pearson Education, Inc.

57. $\lim\limits_{x\to-\infty} \frac{x}{|x|+1}$

Since x approaches negative infinity, $|x| = -x$.

Thus, $\lim\limits_{x\to-\infty} \frac{x}{|x|+1} = \lim\limits_{x\to-\infty} \frac{x}{-x+1}$

$= \lim\limits_{x\to\infty} \frac{1}{-1+\frac{1}{x}} = \frac{1}{-1} = -1$

59. a. As x approaches 4 from the left, $c(x)$ approaches 20. Thus, $\lim\limits_{x\to4^-} c(x) = \$20$.

b. As x approaches 4 from the right, $c(x)$ approaches 20. Thus, $\lim\limits_{x\to4^+} c(x) = \$20$.

c. Since the left and right limits are the same value, $\lim\limits_{x\to4} c(x) = \$20$.

d. As x approaches 10 from the left, $c(x)$ approaches 50. Thus, $\lim\limits_{x\to10^-} c(x) = \$50$.

e. As x approaches 10 from the right, $c(x)$ approaches 60. Thus, $\lim\limits_{x\to10^+} c(x) = \$60$.

f. Since the left and right limits are different values, $\lim\limits_{x\to10} c(x)$ does not exist.

g. \$8; The slope of the line after 10,000 is the rate charged.

h. \$10; The "jump" in the graph is the amount of the surcharge.

61. a. As x approaches 8 from the left, $p(x)$ approaches 3. Thus, $\lim\limits_{x\to8^-} p(x) = 3$ thousand or \$3000.

b. As x approaches 8 from the right, $p(x)$ approaches 1. Thus, $\lim\limits_{x\to8^+} p(x) = 1$ thousand or \$1000.

c. Since the left and right limits are different values, $\lim\limits_{x\to8} p(x)$ does not exist.

d. 3; The x value when the graph crosses the x axis.

e. 12; The point where the second "piece" surpasses the first "piece".

63. For this problem, use the function defined as follows: $c(t) = \begin{cases} 3 & x \le 1 \\ 3+2(t-1) & 1 < x \le 7 \\ 15 & x > 7 \end{cases}$

a. $c(1.5) = 3 + 2(1) = \$5$.

b. $c(2) = 3 + 2(1) = \$5$.

c. $c(2.5) = 3 + 2(2) = \$7$.

d. $c(3) = 3 + 2(2) = \$7$.

e. $c(8) = \$15$.

f. Since the left and right limits are the same value as t approaches 1.5, $\lim\limits_{x\to1.5} c(t) = \$5$.

g. $\lim\limits_{x\to2^-} c(t) = 3 + 2(1) = \$5$.

h. $\lim\limits_{x\to2^+} c(t) = 3 + 2(2) = \$7$.

i. Since the left and right limits are different values, $\lim\limits_{x\to2} c(t)$ does not exist.

j. Since the left and right limits are the same value as t approaches 8, $\lim\limits_{x\to8} c(t) = \$15$.

65. a. $\overline{c}(100) = \frac{c(100)}{100} = \frac{6000+5(100)}{100} = \$65$.

b. $\overline{c}(1000) = \frac{c(1000)}{1000} = \frac{6000+5(1000)}{1000} = \$11$.

c. $\lim\limits_{x\to\infty} \overline{c}(x) = \lim\limits_{x\to\infty} \frac{6000+5x}{x}$

$= \lim\limits_{x\to\infty} \frac{\frac{6000}{x}+\frac{5x}{x}}{\frac{x}{x}}$

$= \frac{0+5}{1} = 5$

Copyright © 2019 Pearson Education, Inc.

Section 11.3 Rates of Change

1. The average rate of change for $f(x) = x^2 + 2x$ between $x = 0$ and $x = 6$ is $\frac{f(6) - f(0)}{6 - 0}$.

$f(6) = 6^2 + 2(6) = 36 + 12 = 48$

$f(0) = 0^2 + 2(0) = 0$

The average rate of change is $\frac{48 - 0}{6 - 0} = \frac{48}{6} = 8$.

3. $f(x) = 2x^3 - 4x^2 + 6$ between $x = -1$ and $x = 2$.

The average rate of change is

$$\frac{f(2) - f(-1)}{2 - (-1)} = \frac{6 - (-0)}{3} = \frac{6}{3} = 2.$$

5. $f(x) = \sqrt{x}$ between $x = 1$ and $x = 9$.

The average rate of change is

$$\frac{f(9) - f(1)}{9 - 1} = \frac{3 - 1}{9 - 1} = \frac{2}{8} = \frac{1}{4}.$$

7. $f(x) = \frac{1}{x - 1}$ between $x = -2$ and $x = 0$

The average rate of change is

$$\frac{f(0) - f(-2)}{0 - (-2)} = \frac{-1 - \left(-\frac{1}{3}\right)}{2} = \frac{-1 + \frac{1}{3}}{2} = \frac{-\frac{2}{3}}{2} = -\frac{1}{3}$$

9. Let $S(x)$ be the function representing sales in thousands of dollars of x thousand catalogs.

a. $S(20) = 40$

$S(10) = 30$

$$\text{Average rate of change} = \frac{S(20) - S(10)}{20 - 10} = \frac{40 - 30}{10} = 1$$

As catalog distribution changes from 10,000 to 20,000, sales will have an average increase of \$1000 for each additional 1000 catalogs distributed.

b. $S(30) = 46$

$S(20) = 40$

$$\text{Average rate of change} = \frac{S(30) - S(20)}{30 - 20} = \frac{46 - 40}{10} = \frac{6}{10} = \frac{3}{5}$$

As catalog distribution changes from 20,000 to 30,000, sales will have an average increase of $\left(\frac{3}{5}\right)(1000)$ or \$600 for each additional 1000 catalogs distributed.

c. $S(40) = 50$

$S(30) = 46$

$$\text{Average rate of change} = \frac{S(40) - S(30)}{40 - 30} = \frac{50 - 46}{10} = \frac{4}{10} = \frac{2}{5}$$

As catalog distribution changes from 30,000 to 40,000, sales will have an average increase of $\left(\frac{2}{5}\right)(1000)$ or \$400 for each additional 1000 catalogs distributed.

d. As more catalogs are distributed, sales increase at a smaller and smaller rate.

e. It might be that the market for items in the catalog is becoming saturated.

11. a. Average rate of change in revenue from 2007 to 2009 is

$$\frac{850 - 1450}{2} = -300$$

Average rate of change in revenue from 2011 to 2012is

$$\frac{1350 - 1250}{1} = 100$$

Average rate of change in revenue from 2013 to 2016 is

$$\frac{1950 - 1500}{3} = 150$$

The greatest average change occurred from 2013 to 2016.

Copyright © 2019 Pearson Education, Inc.

b. Average rate of change in revenue from 2010 to 2013 is

$$\frac{1500-1050}{3}=150$$

Average rate of change in revenue from 2013 to 2015 is

$$\frac{2000-1500}{2}=250$$

The greater average change occurred from 2013 to 2015.

c. Average rate of change in revenue from 2008 to 2009 is

$$\frac{850-1400}{1}=-550$$

Average rate of change in revenue from 2015 to 2016 is

$$\frac{1950-2000}{1}=-50$$

The greater average change occurred from 2015 to 2016.

13. a. Average rate of change in employees from \$5 billion to \$108 billion is

$$\frac{60-10}{103}\approx .485$$

The average rate of change in employees from \$5 billion to \$108 billion was about .485 thousand or 485 employees per \$1 billion.

b. Average rate of change in employees from \$108 billion to \$234 billion is

$$\frac{110-60}{126}\approx .397$$

The average rate of change in employees from \$108 billion to \$234 billion was about .397 thousand or 397 employees per \$1 billion.

c. Average rate of change in employees from \$5 billion to \$234 billion is

$$\frac{110-10}{229}\approx .437$$

The average rate of change in employees from \$5 billion to \$234 billion was about .437 thousand or 437 employees per \$1 billion.

15.a. Average velocity $=\frac{s(2)-s(0)}{2-0}=\frac{10-0}{2}$
$=5$ ft/sec

b. Average velocity $=\frac{s(4)-s(2)}{4-2}=\frac{14-10}{2}$
$=2$ ft/sec

c. Average velocity $=\frac{s(6)-s(4)}{6-4}=\frac{20-14}{2}$
$=3$ ft/sec

d. **i.** Let $t=4$ and $h=2$

instantaneous velocity $\approx\frac{s(4+2)-s(4)}{2}$
$=\frac{20-14}{2}=3$ ft/sec

ii. instantaneous velocity $\approx\frac{2+3}{2}$
$=2.5$ ft/sec

17. The average rate of change of y as x changes from a to b is found by $\frac{y(b)-y(a)}{b-a}$.
The instantaneous rate of change of y as $x=a$ is found by $\lim\limits_{h\to 0}\frac{y(a+h)-y(a)}{h}$.

19. $s(t)=3t^2$

Time Interval	Average Speed
a. $t=10$ to $t=10.1$	$\frac{s(10.1)-s(10)}{10.1-10}=\frac{306.03-300}{.1}$ $=60.3$
b. $t=10$ to $t=10.01$	$\frac{s(10.01)-s(10)}{10.01-10}=60.03$
c. $t=10$ to $t=5.001$	$\frac{s(10.001)-s(10)}{10.001-10}=60.003$

Copyright © 2019 Pearson Education, Inc.

21. $s(t) = 3t^2$.

For $t = 10$, the instantaneous velocity is

$$\lim_{h\to 0} \frac{s(10+h)-s(10)}{10+h-10}$$
$$= \lim_{h\to 0} \frac{3(10+h)^2-(300)}{h}$$
$$= \lim_{h\to 0} \frac{300+60h+3h^2-300}{h}$$
$$= \lim_{h\to 0} \frac{60h+3h^2}{h} = \lim_{h\to 0} \frac{h(60+3h)}{h}$$
$$= \lim_{h\to 0}(60+3h) = 60 \text{ ft/sec.}$$

This chart from exercise 19 suggests that the instantaneous velocity at $t = 10$ is 60 ft/sec which is the same answer.

23. $s(t) = t^2 + 4t + 3$

For $t = 1$, the instantaneous velocity is

$$\lim_{h\to 0} \frac{s(1+h)-s(1)}{h}.$$

We have

$$s(1+h) = (1+h)^2 + 4(1+h) + 3$$
$$= 1 + 2h + h^2 + 4 + 4h + 3$$
$$= h^2 + 6h + 8,$$

and $s(1) = 1^2 + 4(1) + 3 = 8$.

Thus,

$$s(1+h) - s(1) = (h^2 + 6h + 8) - 8$$
$$= h^2 + 6h,$$

and the instantaneous velocity at $t = 1$ is

$$\lim_{h\to 0} \frac{h^2+6h}{h} = \lim_{h\to 0} \frac{h(h+6)}{h} = \lim_{h\to 0}(h+6)$$
$$= 6 \text{ ft/sec.}$$

25. $f(x) = x^2 - 3x$

a. $$f(a+h) = (a+h)^2 - 3(a+h)$$
$$= a^2 + 2ah + h^2 - 3a - 3h$$

b. $$\frac{f(a+h)-f(a)}{h}$$
$$= \frac{a^2+2ah+h^2-3a-3h-(a^2-3a)}{h}$$
$$= \frac{2ah+h^2-3h}{h} = \frac{h(2a+h-3)}{h}$$
$$= 2a + h - 3$$

c. The instantaneous rate of change is

$$\lim_{h\to 0} \frac{f(a+h)-f(a)}{h} = \lim_{h\to 0} 2a + h - 3$$
$$= 2a - 3$$

So the instantaneous rate of change when $a = 5$ is $2(5) - 3 = 7$.

27. $f(x) = x^2 - x - 1$

a. $$f(a+h) = (a+h)^2 - (a+h) - 1$$
$$= a^2 + 2ah + h^2 - a - h - 1$$

b. $$\frac{f(a+h)-f(a)}{h}$$
$$= \frac{a^2+2ah+h^2-a-h-1-(a^2-a-1)}{h}$$
$$= \frac{2ah+h^2-h}{h} = \frac{h(2a+h-1)}{h}$$
$$= 2a + h - 1$$

c. The instantaneous rate of change is

$$\lim_{h\to 0} \frac{f(a+h)-f(a)}{h} = \lim_{h\to 0} 2a + h - 1$$
$$= 2a - 1,$$

so the instantaneous rate of change when $a = 5$ is $2(5) - 1 = 9$.

29. $f(x) = x^3$

a. $f(a+h) = (a+h)^3 = a^3 + 3a^2h + 3ah^2 + h^3$

b. $$\frac{f(a+h)-f(a)}{h} = \frac{a^3+3a^2h+3ah^2+h^3-a^3}{h} = \frac{h(3a^2+3ah+h^2)}{h} = 3a^2 + 3ah + h^2$$

Copyright © 2019 Pearson Education, Inc.

c. The instantaneous rate of change is $\lim_{h\to 0}\frac{f(a+h)-f(a)}{h}=\lim_{h\to 0}3a^2+3ah+h^2=3a^2$,

the instantaneous rate of change when $a=5$ is $3(5)^2=75$.

31. $R=10x-.002x^2$

$R(1000)=10(1000)-.002(1000)^2=8000$

$R(1001)=10(1001)-.002(1001)^2=8005.998$

a. Average rate of change $=\frac{R(1001)-R(1000)}{1001-1000}=\frac{8005.998-8000}{1}=5.998$

Since the revenue is given in thousands of dollars, the average rate of change is \$5998 per unit.

b. $$\text{Marginal revenue}=\lim_{h\to 0}\frac{R(1000+h)-R(1000)}{h}=\lim_{h\to 0}\frac{10(1000+h)-.002(1000+h)^2-8000}{h}$$

$$=\lim_{h\to 0}\frac{10{,}000+10h-.002(1{,}000{,}000+2000h+h^2)-8000}{h}$$

$$=\lim_{h\to 0}\frac{6h-.002h^2}{h}=\lim_{h\to 0}\frac{h(6-.002h)}{h}=\lim_{h\to 0}6-.002h=6$$

The marginal revenue is \$6000 per unit.

33. $p(t)=t^2+t$ for $0\le t\le 5$

a. Average rate of change $=\frac{p(4)-p(1)}{4-1}=\frac{(4^2+4)-(1^2+1)}{3}=\frac{18}{3}=6\%$ per day

b. $$\text{Instantaneous rate of change}=\lim_{h\to 0}\frac{p(t+h)-p(t)}{h}=\lim_{h\to 0}\frac{(t+h)^2+t+h-t^2-t}{h}$$

$$=\lim_{h\to 0}\frac{t^2+2th+h^2+t+h-t^2-t}{h}=\lim_{h\to 0}\frac{2th+h^2+h}{h}$$

$$=\lim_{h\to 0}2t+h+1=2t+1$$

When $t=3$, the instantaneous rate of change is $2(3)+1=7\%$ per day.

35. a. Let $t=20$.

$$\lim_{h\to 0}\frac{f(20+h)-f(20)}{h}=\lim_{h\to 0}\frac{\left(21e^{-.03(20+h)}\right)-\left(21e^{-.03(20)}\right)}{h}$$

$$=\lim_{h\to 0}\frac{21e^{-.6-.03h}-21e^{-.6}}{h}$$

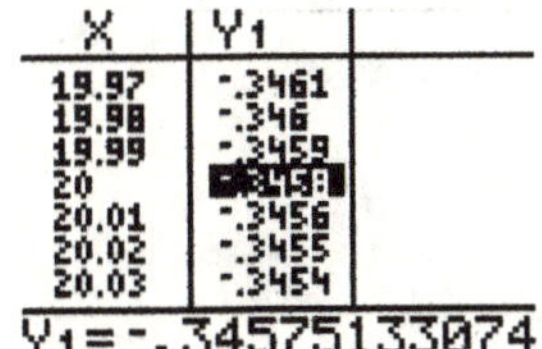

X	Y1
19.97	-.3461
19.98	-.346
19.99	-.3459
20	-.3458
20.01	-.3456
20.02	-.3455
20.03	-.3454

Y1=-.34575133074

When a woman is 20 years old, the rate of change of the unemployment will be approximately −.35% per year.

Copyright © 2019 Pearson Education, Inc.

b. Let $t = 50$.

$$\lim_{h\to 0}\frac{f(50+h)-f(50)}{h} = \lim_{h\to 0}\frac{\left(21e^{-.03(50+h)}\right)-\left(21e^{-.03(50)}\right)}{h}$$
$$= \lim_{h\to 0}\frac{21e^{-1.5-.03h}-21e^{-1.5}}{h}$$

X	Y1
49.97	-.1407
49.98	-.1407
49.99	-.1406
50	[illegible]
50.01	-.1405
50.02	-.1405
50.03	-.1404

Y1=-.14057200089

When a woman is 50 years old, the rate of change of the unemployment will be approximately – .14% per year r.

37. Let $F(t)$ represent the thermal effect of food in kJ/hr, where t is the hours elapsed since eating a meal.

a. $F(t) = -10.28 + 175.9te^{-t/1.3}$

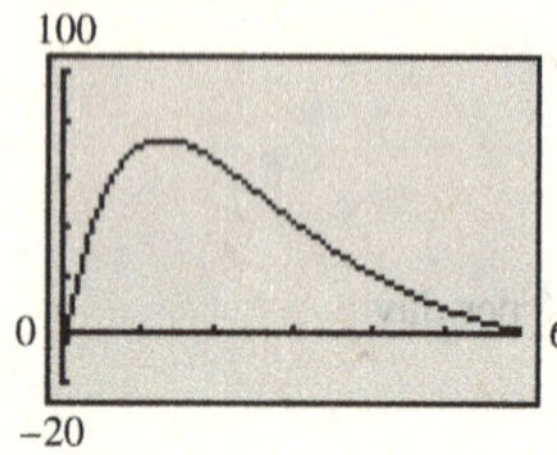

b. Average rate of change of the thermal effect for $F(t) = -10.28 + 175.9te^{-t/1.3}$ between 0 and 1 hour is $\dfrac{F(1)-F(0)}{1-0}$.

$F(1) = -10.28 + 175.9(1)e^{-1/1.3} = 71.23$
$F(0) = -10.28 + 175.9(0)e^{0/1.3} = -10.28$

The average rate of change is $\dfrac{71.23-(-10.28)}{1-0} = \dfrac{81.51}{1} = 81.51$ kJ/hr.

c. $$\lim_{h\to 0}\frac{F(1+h)-F(1)}{h} = \lim_{h\to h}\frac{-10.28+175.9(1+h)e^{-(1+h)/1.3}-\left(-10.28+175.9(1)e^{-1/1.3}\right)}{h}$$

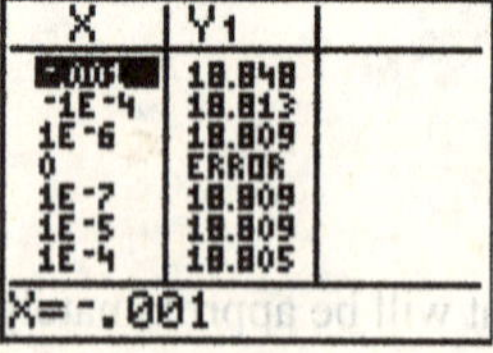

X	Y1
-.001	18.848
-1E-4	18.813
1E-6	18.809
0	ERROR
1E-7	18.809
1E-5	18.809
1E-4	18.805

X=-.001

At exactly 1 hour after eating, the rate of change for the thermal effect is 18.81 kJ/hr.

Copyright © 2019 Pearson Education, Inc.

d. The function stops increasing and begins to decrease at approximately $t = 1.3$ hours .

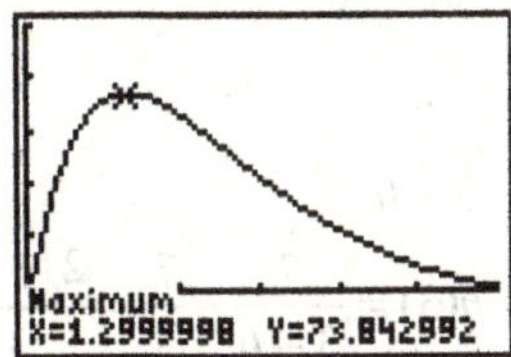

Section 11.4 Tangent Lines and Derivatives

1. $f(x) = 8x + 6$

We condense the steps as follows

$$\frac{f(x+h)-f(x)}{h} = \frac{[8(x+h)+6]-(8x+6)}{h} = \frac{(8x+8h+6)-(8x+6)}{h} = \frac{8h}{h} = 8$$

$f'(x) = \lim\limits_{h\to 0} 8 = 8$

$f'(2) = 8;\ f'(0) = 8;\ f(-3) = 8$

3. $f(x) = -4x^2 + 11x$

Step 1

$$f(x+h) = -4(x+h)^2 + 11(x+h) = -4(x^2 + 2xh + h^2) + 11x + 11h = -4x^2 - 8xh - 4h^2 + 11x + 11h$$

Step 2

$$f(x+h) - f(x) = (-4x^2 - 8xh - 4h^2 + 11x + 11h) - (-4x^2 + 11x) = -8xh - 4h^2 + 11h$$

Step 3

$$\frac{f(x+h)-f(x)}{h} = \frac{-8xh - 4h^2 + 11h}{h} = \frac{h(-8x - 4h + 11)}{h} = -8x - 4h + 11$$

Step 4

$$f'(x) = \lim_{h\to 0} \frac{f(x+h)-f(x)}{h} = \lim_{h\to 0}(-8x - 4h + 11) = -8x + 11$$

$f'(2) = -8(2) + 11 = -5$

$f'(0) = -8(0) + 11 = 11$

$f'(-3) = -8(-3) + 11 = 35$

5. $f(x) = -\dfrac{2}{x}$

$$\frac{f(x+h)-f(x)}{h} = \frac{\frac{-2}{x+h} - \left(\frac{-2}{x}\right)}{h} = \frac{\frac{-2x+2(x+h)}{(x+h)x}}{h} = \frac{2h}{h(x+h)(x)} = \frac{2}{(x+h)(x)}$$

$f'(x) = \lim\limits_{h\to 0} \dfrac{2}{(x+h)(x)} = \dfrac{2}{x^2}$

$f'(2) = \dfrac{2}{2^2} = \dfrac{1}{2}$

$f'(0) = \dfrac{2}{0^2}$ Undefined

The derivative does not exist at $x = 0$.

$f'(-3) = \dfrac{2}{(-3)^2} = \dfrac{2}{9}$

7. $f(x) = 2\sqrt{x}$

$$\frac{f(x+h)-f(x)}{h} = \frac{2\sqrt{x+h} - 2\sqrt{x}}{h} \cdot \frac{2\sqrt{x+h} + 2\sqrt{x}}{2\sqrt{x+h} + 2\sqrt{x}} = \frac{4(x+h) - 4x}{h(2\sqrt{x+h} + 2\sqrt{x})} = \frac{4h}{h(2\sqrt{x+h} + 2\sqrt{x})} = \frac{4}{2\sqrt{x+h} + 2\sqrt{x}}$$

$f'(x) = \lim\limits_{h\to 0} \dfrac{4}{2\sqrt{x+h} + 2\sqrt{x}} = \dfrac{1}{\sqrt{x}}$

$f'(2) = \dfrac{1}{\sqrt{2}}$

$f'(0)$ is undefined.

$f'(-3)$ is undefined.

Copyright © 2019 Pearson Education, Inc.

9. $f'(x) = -20x + 3$

$f'(0) = -20(0) + 3 = 3$

The slope of the tangent line at $(0, f(0)) = (0, -7)$ is 3.

$y - y_1 = m(x - x_1) \Rightarrow$
$y - (-7) = 3(x - 0) \Rightarrow$
$y = 3x - 7$

11. $f'(x) = -\frac{1}{x^2}$

$f'(-2) = -\frac{1}{(-2)^2} = -\frac{1}{4}$

The slope of the tangent line at $(-2, f(-2)) = \left(-2, -\frac{1}{2}\right)$ is $-\frac{1}{4}$.

$y - y_1 = m(x - x_1) \Rightarrow y + \frac{1}{2} = -\frac{1}{4}(x + 2) \Rightarrow$

$y = -\frac{1}{4}x - 1$

13. $g'(x) = \frac{1}{2\sqrt{x}}$

$g'(9) = \frac{1}{2\sqrt{9}} = \frac{1}{6}$

The slope of the tangent line at $(9, f(9)) = (9, 3)$ is $\frac{1}{6}$.

$y - y_1 = m(x - x_1) \Rightarrow y - 3 = \frac{1}{6}(x - 9) \Rightarrow$

$y = \frac{1}{6}x + \frac{3}{2}$

15. $f(x) = x^2 + 3$ at $x = 2$.

a. $f(2 + h) = (2 + h)^2 + 3 = 7 + 4h + h^2$

$f(2) = 2^2 + 3 = 7$

$f(2 + h) - f(2) = 7 + 4h + h^2 - 7 = 4h + h^2$

$\frac{f(2 + h) - f(2)}{h} = \frac{4h + h^2}{h} = 4 + h$

Letting $h \to 0$, slope of tangent = 4

b. The slope of the tangent line at $(2, f(2)) = (2, 7)$ is 4.

$y - y_1 = m(x - x_1) \Rightarrow y - 7 = 4(x - 2) \Rightarrow$
$y = 4x - 1$

17. $h(x) = \frac{7}{x}$ at $x = 3$.

a. $h(3 + h) = \frac{7}{3 + h}$

$h(3 + h) - h(3) = \frac{7}{3 + h} - \frac{7}{3} = \frac{21 - (21 + 7h)}{3(3 + h)}$

$= \frac{-7h}{3(3 + h)}$

$\frac{h(3 + h) - h(3)}{h} = \frac{\frac{-7h}{3(3+h)}}{h} = \frac{-7}{3(3 + h)}$

Letting $h \to 0$,

slope of tangent $= \frac{-7}{3(3)} = -\frac{7}{9}$

b. The slope of the tangent line at $(3, h(3)) = \left(3, \frac{7}{3}\right)$ is $-\frac{7}{9}$.

$y - \frac{7}{3} = -\frac{7}{9}(x - 3) \Rightarrow y = -\frac{7}{9}x + \frac{14}{3}$

19. $g(x) = 5\sqrt{x}$ at $x = 9$.

a. $g(9 + h) = 5\sqrt{9 + h}$

$g(9) = 5\sqrt{9} = 15$

$g(9 + h) - g(9) = 5\sqrt{9 + h} - 15$

$= 5\left(\sqrt{9 + h} - 3\right)$

$\frac{g(9 + h) - g(9)}{h}$

$= \frac{5\left(\sqrt{9 + h} - 3\right)}{h}$

$= 5 \cdot \frac{\sqrt{9 + h} - 3}{h} \cdot \frac{\left(\sqrt{9 + h} + 3\right)}{\left(\sqrt{9 + h} + 3\right)}$

$= \frac{5\left[(9 + h) - 9\right]}{h\left(\sqrt{9 + h} + 3\right)} = \frac{5h}{h\left(\sqrt{9 + h} + 3\right)}$

$= \frac{5}{\sqrt{9 + h} + 3}$

Letting $h \to 0$,

slope of tangent $= \frac{5}{\sqrt{9} + 3} = \frac{5}{6}$

Copyright © 2019 Pearson Education, Inc.

b. The slope of the tangent line at $(9, g(9)) = (9, 15)$ is $\frac{5}{6}$.

$$y-15=\frac{5}{6}(x-9) \Rightarrow y=\frac{5}{6}x+\frac{15}{2}$$

21. a. From the graph, $f(x)$ is largest at $x = x_5$.

b. $f(x)$ is smallest at $x = x_4$.

c. $f'(x)$ is smallest at $x = x_3$ because x_3 is the only labeled value at which the tangent line has a negative slope.

d. $f'(x)$ is closest to 0 at $x = x_2$ because the tangent line is closer to being horizontal (where the slope is 0) at x_2 than at any other labeled value.

23. If $g'(x) > 0$ for $x < 0$, the slope of the tangent line must be positive, so $g(x)$ must increase for $x < 0$. If $g'(x) < 0$ for $x > 0$, the slope of the tangent line must be negative, so $g(x)$ must decrease for $x > 0$. One of the many possible graphs follows.

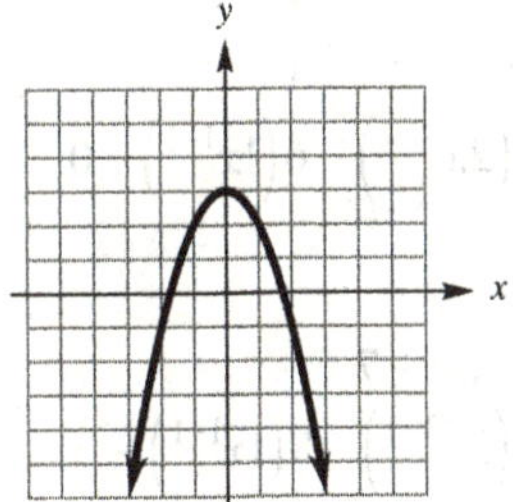

25. a. At $x = .5$, the slope of $T(x)$ is 1000, so the oven temperature is increasing at 1000° per hour.

b. At $x = 2$, the slope of $T(x)$ is 0, so the oven temperature is not changing.

c. At $x = 3.5$, the slope of $T(x)$ is -1000, so the oven temperature is decreasing at 1000° per hour.

27. Since velocity is the rate of change of distance from a starting point, we must decide which graph represents the function (distance over time) and which represents the derivative of that function (velocity over time). For $t > 0$, the graph of (a) is positive. If (a) is the derivative, the graph of (b) would have no high or low points when (a) would be 0, which is not the case.
Furthermore, assuming (b) is the derivative of (a), we see that (b) is positive when (a) is increasing ($0 < t < 2$ and $t > 4$) and (b) is negative when (a) is decreasing ($2 < t < 4$). Therefore, graph (a) is distance and graph (b) is velocity.

29. For $x = \pm 6$, the graph of $f(x)$ has sharp points. Therefore, there is no derivative for $x = 6$ or $x = -6$.

31. The derivative does not exist at the following x-values.

x	Reason
−5	Function is not defined.
−3	Function has a sharp point.
0	Function is not defined.
2	Function has a sharp point.
4	Function has a vertical tangent

33. $s(7) = 0$ means that seven seconds after the stone is dropped, the stone hits the ocean. $s'(7) = -224$ means its velocity at the instant the stone hits the water is −224 ft/s.

35. a. $P(6) = 25$ means six weeks after entering the race, the candidate has the support of 25% of likely voters.

b. $P'(6) = 3$ means six weeks after entering the race, support for the candidate increased by 3 percentage point in the next week.

37. a. $P(5) = 1610.46$ means if the interest rate on your mortgage is 5%, your monthly payment will be $1610.46.

b. $P'(5) = 183.37$ means if the interest rate you can get on your mortgage increases from 5% to 6%, your monthly payment will increase by approximately $183.37.

Copyright © 2019 Pearson Education, Inc.

39. a. $A(5) = 223{,}537.94$ means if the interest rate on your mortgage is 5%, the amount you can afford to borrow is approximately $223,537.94.

b. Negative; if you have to pay a higher interest rate, the amount you can afford to borrow is smaller.

c. The units for $A'(r)$ are dollars/percent.

d. $A'(5) = -25{,}449.21$ means if the interest rate you can get on your mortgage increases from 5% to 6%, the amount you can afford to borrow will decrease by $25,449.21.

41. At $t = 2$, the slope of the tangent line is 1000. Thus, the population is increasing at a rate of 1000 shellfish per time unit.
At $t = 10$, the slope of the tangent line is 570. Thus, the population is increasing more slowly at 570 shellfish per time unit.
At $t = 13$, the slope of the tangent line is 250; the population is increasing at a much slower rate of 250 shellfish per time unit.

43. $f(t) = 42\left(e^{.15t}\right)$

Set $Y_1 = f(t)$.

a.

```
nDeriv(Y1,X,5)
        13.33710016
nDeriv(Y1,X,10)
        28.23464125
nDeriv(Y1,X,15)
        59.77273599
```

$f'(5) \approx \$13.34$ billion per year
$f'(10) \approx \$28.24$ billion per year
$f'(15) \approx \$59.77$ billion per year

b. The rate of spending on research and development is increasing.

45. $g(x) = 183x^{.06}$

Set $Y_1 = g(x)$.

```
nDeriv(Y1,X,18)
        .725514645
```

$g'(18) \approx .726$

In 2018, the number of licensed drivers will increase by approximately .726 million or 726,000 drivers in the next year.

47. a. $f(x) = .5x^5 - 2x^3 + x^2 - 3x + 2;\ -3 \le x \le 3$

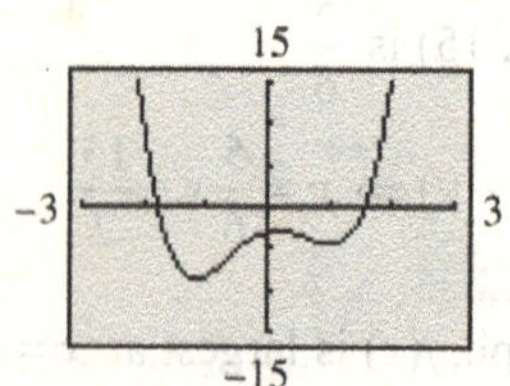

b. $g(x) = 2.5x^4 - 6x^2 + 2x - 3$

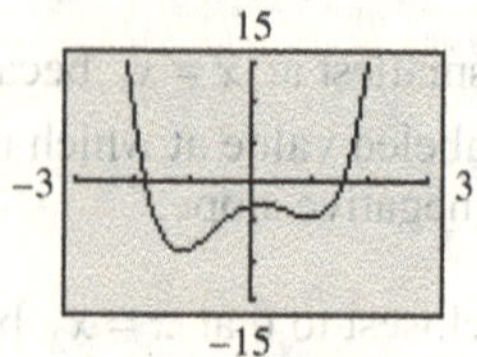

c. The graphs appear identical, which suggests that $f'(x) = g(x)$.

Section 11.5 Techniques for Finding Derivatives

1. $f(x) = 4x^2 - 3x + 5$

$$f'(x) = 4\left(2x^{2-1}\right) - 3\left(1x^{1-1}\right) + 0$$
$$= 8x - 3$$

3. $y = 2x^3 + 3x^2 - 6x + 2$

$$y' = 2\left(3x^{3-1}\right) + 3\left(2x^{2-1}\right) - 6\left(1x^{1-1}\right) + 0$$
$$= 6x^2 + 6x - 6$$

5. $g(x) = x^4 + 3x^3 - 8x - 7$

$$g'(x) = 4x^{4-1} + 3\left(3x^{3-1}\right) - 8\left(1x^{1-1}\right) - 0$$
$$= 4x^3 + 9x^2 - 8$$

7. $f(x) = 6x^{1.5} - 4x^{.5}$

$$f'(x) = 6\left(1.5x^{1.5-1}\right) - 4\left(.5x^{-.5}\right)$$
$$= 9x^{.5} - 2x^{-.5} = 9x^{.5} - \frac{2}{x^{.5}}$$

9. $y = -15x^{3/2} + 2x^{1.9} + x$

$$y' = -15\left(\frac{3}{2}x^{3/2-1}\right) + 2\left(1.9x^{1.9-1}\right) + 1x^{1-1}$$
$$= -22.5x^{1/2} + 3.8x^{.9} + 1$$

Copyright © 2019 Pearson Education, Inc.

11. $y = 24t^{3/2} + 4t^{1/2}$

$$y' = 24\left(\frac{3}{2}t^{1/2}\right) + 4\left(\frac{1}{2}t^{-1/2}\right) = 36t^{1/2} + 2t^{-1/2} = 36t^{1/2} + \frac{2}{t^{1/2}}$$

13. $y = 8\sqrt{x} + 6x^{3/4} = 8x^{1/2} + 6x^{3/4}$

$$y' = 8\left(\frac{1}{2}x^{-1/2}\right) + 6\left(\frac{3}{4}x^{-1/4}\right) = 4x^{-1/2} + \frac{9}{2}x^{-1/4} = \frac{4}{x^{1/2}} + \frac{9}{2x^{1/4}}$$

15. $g(x) = 6x^{-5} - 2x^{-1}$

$$g'(x) = 6\left(-5x^{-6}\right) - 2(-1)x^{-2} = -30x^{-6} + 2x^{-2} = -\frac{30}{x^6} + \frac{2}{x^2}$$

17. $y = 10x^{-2} + 6x^{-4} + 3x$

$$y' = 10(-2x^{-3}) + 6(-4x^{-5}) + 3 = -20x^{-3} - 24x^{-5} + 3 = -\frac{20}{x^3} - \frac{24}{x^5} + 3$$

19. $f(t) = \frac{-4}{t} + \frac{8}{t^2} = -4t^{-1} + 8t^{-2}$

$$f'(t) = 4t^{-2} - 16t^{-3} = \frac{4}{t^2} - \frac{16}{t^3}$$

21. $f(t) = -\frac{1}{4t} + \frac{1}{8t^2} = -\frac{1}{4}t^{-1} + \frac{1}{8}t^{-2}$

$$f'(t) = \frac{1}{4}t^{-2} - \frac{2}{8}t^{-3} = \frac{1}{4t^2} - \frac{1}{4t^3}$$

23. $g(x) = 5x^{-1/2} - 7x^{1/2} + 101x$

$$g'(x) = 5\left(-\frac{1}{2}x^{-3/2}\right) - 7\left(\frac{1}{2}x^{-1/2}\right) + 101 = -\frac{5}{2}x^{-3/2} - \frac{7}{2}x^{-1/2} + 101 = -\frac{5}{2x^{3/2}} - \frac{7}{2x^{1/2}} + 101$$

25. $y = 7x^{-3/2} + 8x^{-1/2} + x^3 - 9$

$$y' = 7\left(-\frac{3}{2}x^{-5/2}\right) + 8\left(-\frac{1}{2}x^{-3/2}\right) + 3x^2 + 0 = -\frac{21}{2}x^{-5/2} - 4x^{-3/2} + 3x^2 = -\frac{21}{2x^{5/2}} - \frac{4}{x^{3/2}} + 3x^2$$

27. $y = \frac{19}{\sqrt[4]{x}} = 19x^{-1/4}$

$$y' = 19\left(-\frac{1}{4}x^{-5/4}\right) = -\frac{19}{4}x^{-5/4} = -\frac{19}{4x^{5/4}}$$

29. $y = \frac{-8t}{\sqrt[3]{t^2}} = -8t\left(t^{-2/3}\right) = -8t^{1/3}$

$$y' = -8\left(\frac{1}{3}t^{-2/3}\right) = -\frac{8}{3}t^{-2/3} = -\frac{8}{3t^{2/3}}$$

31. $y = 8x^{-5} - 9x^{-4} + 9x^4$

$$\frac{dy}{dx} = 8\left(-5x^{-6}\right) - 9\left(-4x^{-5}\right) + 9\left(4x^{4-1}\right) = -40x^{-6} + 36x^{-5} + 36x^3 = -\frac{40}{x^6} + \frac{36}{x^5} + 36x^3$$

33. $D_x\left(9x^{-1/2} + \frac{2}{x^{3/2}}\right)$

$$= D_x\left(9x^{-1/2} + 2x^{-3/2}\right) = 9\left(-\frac{1}{2}x^{-3/2}\right) + 2\left(-\frac{3}{2}x^{-5/2}\right) = -\frac{9}{2}x^{-3/2} - 3x^{-5/2} = -\frac{9}{2x^{3/2}} - \frac{3}{x^{5/2}}$$

35. $f(x) = 6x^2 - 2x$

$f'(x) = 12x - 2$

$f'(-2) = 12(-2) - 2 = -24 - 2 = -26$

37. $f(t) = 2\sqrt{t} - \frac{3}{\sqrt{t}} = 2t^{1/2} - 3t^{-1/2}$

$$f'(t) = 2\left(\frac{1}{2}t^{-1/2}\right) - 3\left(-\frac{1}{2}t^{-3/2}\right) = t^{-1/2} + \frac{3}{2}t^{-3/2} = \frac{1}{t^{1/2}} + \frac{3}{2t^{3/2}}$$

$$f'(4) = \frac{1}{4^{1/2}} + \frac{3}{2(4^{3/2})} = \frac{1}{2} + \frac{3}{16} = \frac{11}{16}$$

39. $y = \frac{12 - 7x + 6x^3}{x^4} = \frac{12}{x^4} - \frac{7x}{x^4} + \frac{6x^3}{x^4} = 12x^{-4} - 7x^{-3} + 6x^{-1}$

$$y' = -48x^{-5} + 21x^{-4} - 6x^{-2} = -\frac{48}{x^5} + \frac{21}{x^4} - \frac{6}{x^2}$$

Copyright © 2019 Pearson Education, Inc.

41. $f(x) = x^4 - 2x^2 + 1; x = 1$
$f'(x) = 4x^3 - 4x$
The slope of the tangent line at $x = 1$ is
$f'(1) = 4(1^3) - 4(1) = 4 - 4 = 0.$
The tangent line at $x = 1$ is the horizontal line through $f(1) = 1^4 - 2(1^2) + 1 = 0.$
The equation of the tangent line is $y = 0$.

43. $y = 4x^{1/2} + 2x^{3/2} + 1; x = 4.$

$$y' = 4\left(\frac{1}{2}x^{-1/2}\right) + 2\left(\frac{3}{2}x^{1/2}\right)$$
$$= \frac{2}{x^{1/2}} + 3x^{1/2}$$

The slope of the tangent line at $x = 4$ is

$$y'(4) = \frac{2}{4^{1/2}} + 3(4^{1/2}) = 7$$

Because $y(4) = 4(4^{1/2}) + 2(4^{3/2}) + 1 = 25,$ the tangent line $x = 4$ goes through (4, 25). Using the point-slope form, the equation of the tangent line at $x = 4$ is
$y - 25 = 7(x - 4) \Rightarrow y = 7x - 28 + 25 \Rightarrow$
$y = 7x - 3.$

45. $C(x) = 100 + 12x + .1x^2 + .001x^3$
Marginal cost $= C'(x)$
$= 0 + 12 + .1(2x) + .001(3x^2)$
$= 12 + .2x + .003x^2$

a. $C'(0) = 12 + .2(0) + .003(0)^2 = 12$
The marginal cost is \$12.

b. $C'(10) = 12 + .2(10) + .003(10)^2$
$= 12 + 2 + .3 = 14.3$
The marginal cost is \$14.30.

c. $C'(30) = 12 + .2(30) + .003(30)^2$
$= 12 + 6 + 2.7 = 20.7$
The marginal cost is \$20.70.

d. $C'(50) = 12 + .2(50) + .003(50)^2$
$= 12 + 10 + 7.5 = 29.5$
The marginal cost is \$29.50.

47. $P(x) = .03x^2 - 4x - 3x^{.8} + -5000$
Marginal profit
$= P'(x) = .03(2x) - 4 + 3(.8x^{-.2})$
$= .06x - 4 + 2.4x^{-.2}$

a. $P'(100) = .06(100) - 4 + 2.4(100)^{-.2} = \$2.96$

b. $P'(1000) = .06(1000) - 4 + 2.4(1000)^{-.2}$
$= \$56.60$

c. $P'(5000) = .06(5000) - 4 + 2.4(5000)^{-.2}$
$= \$296.44$

d. $P'(10{,}000) = .06(10{,}000) - 4 + 2.4(10{,}000)^{-.2}$
$= \$596.38$

49. $R(x) = .0608x^2 + 1.4284x + 2.3418$

a. $R'(x) = .1216x + 1.4284$

b. $R'(5) = .1216(5) + 1.4284 = 2.0364$
The marginal revenue for 5 hundred million bushels is \$2.0364 hundred million.

c. $R'(7) = .1216(7) + 1.4284 = 2.2796$
The marginal revenue for 7 hundred million bushels is \$2.2796 hundred million.

51. $S(t) = 10{,}000 - 10{,}000t^{-.2} + 100t^{.1}$
$S'(t) = -10{,}000(-.2t^{-1.2}) + 100(.1t^{-.9})$

$$= \frac{2000}{t^{1.2}} + \frac{2010}{t^{.9}}$$

a. $S'(1) = \dfrac{2000}{(1)^{1.2}} + \dfrac{10}{1^{.9}} = \dfrac{100}{1} = 2010$

b. $S'(10) = \dfrac{2000}{(10)^{1.2}} + \dfrac{10}{10^{.9}} = 127.45$

Copyright © 2019 Pearson Education, Inc.

53. $R(x) = 20x - \dfrac{x^2}{500}$

a. The marginal revenue is the rate of change of the revenue, or $R'(x)$, so find $R'(x)$.

$$\frac{R(x+h) - R(x)}{h} = \frac{20(x+h) - \frac{(x+h)^2}{500} - 20x + \frac{x^2}{500}}{h} = \frac{20h - \frac{2xh+h^2}{500}}{h} = 20 - \frac{2x+h}{500}$$

$$R'(x) = \lim_{h\to 0}\frac{R(x+h) - R(x)}{h} = \lim_{h\to 0}\left(20 - \frac{2x+h}{500}\right) = 20 - \frac{2x}{500}$$

At $x = 1000$, $R'(x) = 20 - \dfrac{2(1000)}{500} = 16$

The marginal revenue is \$16/table.

b. The actual revenue from the sale of the 1001st item is

$$R(1001) - R(1000) = 20(1001) - \frac{1001^2}{500} - \left[20(1000) - \frac{1000^2}{500}\right] = 18{,}015.998 - 18{,}000 = 15.998 \text{ or } 16.$$

The actual revenue is \$15.998 or \$16.

c. The marginal revenue found in part (a) approximates the actual revenue from the sale of the 1001st item found in part (b).

55. $R(x) = x(24 - x) = 24x - x^2$
$R'(x) = 24 - 2x$

a. Let $x = 5$
$R'(5) = 24 - 2(5)$
$= 14$
The marginal revenue when 5 handcrafted violins are produced is \$14,000 per violin.

b. Let $x = 10$
$R'(10) = 24 - 2(10)$
$= 4$
The marginal revenue when 10 handcrafted violins are produced is \$4,000 per violin.

c. Let $x = 12$
$R'(12) = 24 - 2(12)$
$= 0$
The marginal revenue when 12 handcrafted violins are produced is 0.

57. Demand is $x = 5000 - 100p \Rightarrow p = \dfrac{5000 - x}{100}$

$$R(x) = x\left(\frac{5000 - x}{100}\right) = \frac{5000x - x^2}{100}$$

$$R'(x) = \frac{5000 - 2x}{100}$$

a. $R'(1000) = \dfrac{5000 - 2000}{100} = 30$

b. $R'(2500) = \dfrac{5000 - 2(2500)}{100} = 0$

c. $R'(3000) = \dfrac{5000 - 2(3000)}{100} = -10$

59. $C(x) = 3x^{3/4}$
$C'(x) = \dfrac{9}{4}x^{-1/4}$

a. Let $x = 100$
$C(100) = 3(100)^{3/4} \approx 94.87$
$C'(100) = \dfrac{9}{4}(100)^{-1/4} \approx .71$
These values mean for such an individual, the cost of \$100,000 in life insurance is \$94.87 and adding an additional 1,000 in coverage would increase the cost by approximately \$0.71.

Copyright © 2019 Pearson Education, Inc.

b. Let $x = 1000$

$C(1000) = 3(1000)^{3/4} \approx 533.48$

$C'(1000) = \frac{9}{4}(1000)^{-1/4} \approx .40$

These values mean for such an individual, the cost of $1,000,000 in life insurance is $533.48 and adding an additional 1,000 in coverage would increase the cost by approximately $0.40.

61. $P(x) = 26x^{-.03}$

$P'(x) = -.78x^{-1.03}$

a. Let $x = 14$

$P(14) = 26(14)^{-.03} \approx 24.02$

$P'(14) = -.78(14)^{-1.03} \approx -.05$

These values mean that in 2014, the percentage of the population that is 0–17 years is estimated to be 24.02% and to decrease by approximately .05% in the next year.

b. Let $x = 24$

$P(24) = 26(24)^{-.03} \approx 23.64$

$P'(24) = -.78(24)^{-1.03} \approx -.03$

These values mean that in 2024, the percentage of the population that is 0–17 years is estimated to be 23.64% and to decrease by approximately .03% in the next year.

63. $C(x) = 2x^2 - 47x + 415$

$C'(x) = 4x - 47$

a. Let $x = 10$

$C(10) = 2(10)^2 - 47(10) + 415$
$= 145$

$C'(10) = 4(10) - 47$
$= -7$

In 2010, the National Home Price Index was 145 and was decreasing at approximately 7 index points per year.

b. Let $x = 1$

$C(16) = 2(16)^2 - 47(16) + 415$
$= 175$

$C'(16) = 4(16) - 47$
$= 17$

In 2016, the National Home Price Index was 175 and was increasing at approximately 17 index points per year.

65. a. $f(x) = .27x^2 + 3.52x + 51.78$

$f'(x) = .54x + 3.52$

b. Let $x = 20$

$f'(20) = .54(20) + 3.52 \approx 14$

In 2020, the rate of change in the number of Americans over 100 years old will be approximately 14 persons/year.

c. Increasing because the value found in part b is positive.

67. $V = \pi r^2 h$

Since $h = 80$ mm, $V = \pi r^2(80) = 80\pi r^2$.

a. $\frac{dV}{dr} = 80\pi(2r) = 160\pi r$

b. When $r = 4$, $\frac{dV}{dr} = 160\pi(4) = 640\pi$.

The volume of blood is increasing at a rate of 640π mm^3/mm of change in the radius.

c. When $r = 6$, $\frac{dV}{dr} = 160\pi(6) = 960\pi$.

The volume of blood is increasing at a rate of 960π mm^3/mm of change in the radius.

d. When $r = 8$, $\frac{dV}{dr} = 160\pi(8) = 1280\pi$.

The volume of blood is increasing at a rate of 1280π mm^3/mm of change in the radius.

69. a. $f(h) = 70,300h^{-2}$

$f'(h) = -2(70,300)h^{-3}$
$= -\frac{140,600}{h^3}$

b. Negative because taller individuals of the same weight would have lower BMI.

c. $f'(60) = -\frac{140,600}{(60)^3}$
$= -\frac{140,600}{216,000} \approx -.65$

Copyright © 2019 Pearson Education, Inc.

71. $s(t) = 8t^2 + 3t + 1$

a. $v(t) = 8(2t) + 3 = 16t + 3$

b.
$$v(0) = 16(0) + 3 = 3$$
$$v(5) = 16(5) + 3 = 83$$
$$v(10) = 16(10) + 3 = 163$$

73. $s(t) = 2t^3 + 6t^2$

a. $v(t) = 2(3t^2) + 6(2t) = 6t^2 + 12t$

b.
$$v(0) = 6(0^2) + 12(0) = 0$$
$$v(5) = 6(5^2) + 12(5) = 210$$
$$v(10) = 6(10^2) + 12(10) = 720$$

75. $s(t) = -16t^2 + 144$

velocity $= s'(t) = -32t$

a. $s'(1) = -32$ ft/sec
$s'(2) = -32 \cdot 2 = -64$ ft/sec

b. The rock will hit the ground when $s(t) = 0$.

$$-16t^2 + 144 = 0 \Rightarrow t = \sqrt{\frac{144}{16}} = 3 \text{ sec}$$

c. The velocity at impact is the velocity at 3 sec.
$v = -32 \cdot 3 = -96$ ft/sec

77. Proof that if $y = x^n, y' = n \cdot x^{n-1}$:

a. $y = (x+h)^n$
By the binomial theorem,
$$y = x^n + n \cdot x^{n-1}h + \frac{n(n-1)}{2!}x^{n-2}h^2 + \ldots + h^n$$

b.
$$\frac{(x+h)^n - x^n}{h} = n \cdot x^{n-1} + \frac{n(n-1)}{2!}x^{n-2}h + \ldots + h^{n-1}$$

c.
$$\lim_{h \to 0} \frac{(x+h)^n - x^n}{h} = n \cdot x^{n-1} = y'$$

79. $g(x) = 6 - 4x + 3x^2 - x^3$

$g'(x) = -4 + 6x - 3x^2$

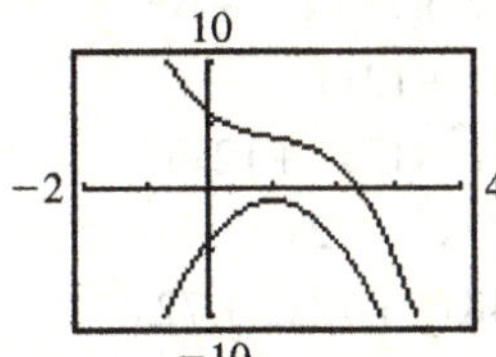

a. $g'(x) > 0$ for no value of x.

b. $g'(x) = 0$ for no value of x.

c. $g'(x) < 0$ on $(-\infty, \infty)$.

d. The derivative is always negative, so the graph of $g(x)$ is always decreasing.

Section 11.6 Derivatives of Products and Quotients

1. $y = (x^2 - 2)(3x + 2)$
$$\begin{aligned} y' &= (x^2-2)D_x(3x+2) \\ &\quad + (3x+2)D_x(x^2-2) \\ &= (x^2-2)(3) + (3x+2)(2x) \\ &= 3x^2 - 6 + 6x^2 + 4x \\ &= 9x^2 + 4x - 6 \end{aligned}$$

3. $y = (6x^3 + 2)(5x - 3)$
$$\begin{aligned} y' &= (6x^3+2)D_x(5x-3) \\ &\quad + (5x-3)D_x(6x^3+2) \\ &= (6x^3+2)(5) + (5x-3)(18x^2) \\ &= 30x^3 + 10 + 90x^3 - 54x^2 \\ &= 120x^3 - 54x^2 + 10 \end{aligned}$$

5. $y = (3x^4 - x + 6)(-x^5 + 4)$
$$\begin{aligned} y' &= (3x^4 - x + 6)[D_x(-x^5+4)] \\ &\quad + (-x^5+4)[D_x(3x^4 - x + 6)] \\ &= (3x^4 - x + 6)(-5x^4) \\ &\quad + (-x^5+4)(12x^3 - 1) \\ &= -15x^8 + 5x^5 - 30x^4 - 12x^8 + x^5 + 48x^3 - 4 \\ &= -27x^8 + 6x^5 - 30x^4 + 48x^3 - 4 \end{aligned}$$

7. $y = (3x-4)^2$
$$\begin{aligned} y' &= (3x-4)D_x(3x-4) + (3x-4)D_x(3x-4) \\ &= (3x-4)(3) + (3x-4)(3) \\ &= 9x - 12 + 9x - 12 = 18x - 24 \end{aligned}$$

Copyright © 2019 Pearson Education, Inc.

9. $y = (4x^3 + 6x^2)^2$

$$\begin{aligned} y' &= (4x^3 + 6x^2)[D_x(4x^3 + 6x^2)] \\ &\quad + (4x^3 + 6x^2)[D_x(4x^3 + 6x^2)] \\ &= (4x^3 + 6x^2)(12x^2 + 12x) \\ &\quad + (4x^3 + 6x^2)(12x^2 + 12x) \\ &= 48x^5 + 48x^4 + 72x^4 + 72x^3 \\ &\quad + 48x^5 + 48x^4 + 72x^4 + 72x^3 \\ &= 96x^5 + 240x^4 + 144x^3 \end{aligned}$$

11. $y = \dfrac{3x-5}{x-3}$

$$\begin{aligned} y' &= \frac{(x-3)D_x(3x-5) - (3x-5)D_x(x-3)}{(x-3)^2} \\ &= \frac{(x-3)(3) - (3x-5)(1)}{(x-3)^2} \\ &= \frac{3x-9-3x+5}{(x-3)^2} = \frac{-4}{(x-3)^2} \end{aligned}$$

13. $f(x) = \dfrac{7x-3}{x^2-8}$

$$\begin{aligned} f'(x) &= \frac{(x^2-8)[D_x(7x-3)] - (7x-3)[D_x(x^2-8)]}{(x^2-8)^2} \\ &= \frac{(x^2-8)(7) - (7x-3)(2x)}{(x^2-8)^2} \\ &= \frac{7x^2-56-14x^2+6x}{(x^2-8)^2} = \frac{-7x^2+6x-56}{(x^2-8)^2} \end{aligned}$$

15. $f(t) = \dfrac{t^2-4t}{t+3}$

$D_t(t^2-4t) = 2t-4$ and $D_t(t+3) = 1$

$$\begin{aligned} f'(t) &= \frac{(t+3)(2t-4) - (t^2-4t)(1)}{(t+3)^2} \\ &= \frac{2t^2-4t+6t-12-t^2+4t}{(t+3)^2} \\ &= \frac{t^2+6t-12}{(t+3)^2} \end{aligned}$$

17. $g(x) = \dfrac{3x^2+x}{2x^3-2}$

$D_x(3x^2+x) = 6x+1$ and $D_x(2x^3-2) = 6x^2$

$$\begin{aligned} g'(x) &= \frac{(2x^3-2)(6x+1) - (3x^2+x)(6x^2)}{(2x^3-2)^2} \\ &= \frac{12x^4+2x^3-12x-2-18x^4-6x^3}{(2x^3-2)^2} \\ &= \frac{-6x^4-4x^3-12x-2}{(2x^3-2)^2} \end{aligned}$$

19. $y = \dfrac{x^2-4x+2}{x+3}$

$D_x(x^2-4x+2) = 2x-4$ and $D_x(x+3) = 1$

$$\begin{aligned} y' &= \frac{(x+3)(2x-4)}{(x+3)^2} - \frac{(x^2-4x+2)(1)}{(x+3)^2} \\ &= \frac{2x^2-4x+6x-12-x^2+4x-2}{(x+3)^2} \\ &= \frac{x^2+6x-14}{(x+3)^2} \end{aligned}$$

21. $y = \dfrac{9x-8}{\sqrt{x}} = \dfrac{9x-8}{x^{1/2}}$

$D_x(9x-8) = 9$ and $D_x(x^{1/2}) = \dfrac{1}{2}x^{-1/2}$

$$\begin{aligned} y' &= \frac{x^{1/2}(9) - (9x-8)\left(\frac{1}{2}x^{-1/2}\right)}{(x^{1/2})^2} \\ &= \frac{9x^{1/2} - \frac{9}{2}x^{1/2} + 4x^{-1/2}}{x} \\ &= \frac{\frac{9x^{1/2}}{2} + 4x^{-1/2}}{x} \\ &= \frac{\frac{9x^{1/2}}{2} + \frac{4}{x^{1/2}}}{x} = \frac{\frac{9x+8}{2x^{1/2}}}{x} = \frac{9x+8}{2x^{3/2}} = \frac{9x+8}{2x\sqrt{x}} \end{aligned}$$

23. $y = \dfrac{59}{4x+11}$

$D_x(59) = 0$ and $D_x(4x+11) = 4$

$$y' = \frac{(4x+11)(0) - 59(4)}{(4x+11)^2} = \frac{-236}{(4x+11)^2}$$

Copyright © 2019 Pearson Education, Inc.

25. $f(p)=\dfrac{(6p-7)(11p-1)}{4p+3}$

$D_p\left[(6p-7)(11p-1)\right]=(6p-7)(11)+(11p-1)(6)=66p-77+66p-6=132p-83$

$D_p(4p+3)=4$

$$f'(p)=\frac{(4p+3)D_p\left[(6p-7)(11p-1)\right]-(6p-7)(11p-1)D_p(4p+3)}{(4p+3)^2}$$

$$=\frac{(4p+3)(132p-83)-(6p-7)(11p-1)4}{(4p+3)^2}=\frac{\left(528p^2+64p-249\right)-\left(264p^2-332p+28\right)}{(4p+3)^2}$$

$$=\frac{264p^2+396p-277}{(4p+3)^2}$$

27. $g(x)=\dfrac{x^3-8}{(3x+9)(2x-1)}$

$D_x\left(x^3-8\right)=3x^2$

$D_x\left[(3x+9)(2x-1)\right]=(3x+9)(2)+(2x-1)(3)=6x+18+6x-3=12x+15$

$$g'(x)=\frac{\left[(3x+9)(2x-1)\right]D_x\left(x^3-8\right)-\left(x^3-8\right)D_x\left[(3x+9)(2x-1)\right]}{\left[(3x+9)(2x-1)\right]^2}$$

$$=\frac{\left(6x^2+15x-9\right)\left(3x^2\right)-\left(x^3-8\right)(12x+15)}{(3x+9)^2(2x-1)^2}$$

$$=\frac{\left(18x^4+45x^3-27x^2\right)-\left(12x^4+15x^3-96x-120\right)}{(3x+9)^2(2x-1)^2}$$

$$=\frac{6x^4+30x^3-27x^2+96x+120}{(3x+9)^2(2x-1)^2}$$

29. In the first step, the numerator should be $(x^2-1)(2)-(2x+5)(2x)$.

31. $f(x)=(x^5+1)(4x-3)$

$f'(x)=(x^5+1)[D_x(4x-3)]-(4x-3)[D_x(x^5+1)]=$

$=(x^5+1)(4)+(4x-3)(5x^4)$

$=4x^5+4+20x^5-15x^4$

$=24x^5-15x^4+4$

$f'(1)=24(1)^5-15(1)^4+4$

$=24-15+4=13$

The slope is equal to $f'(1)=13$.

$(y-2)=13(x-1)\Rightarrow y=13x-11$

Copyright © 2019 Pearson Education, Inc.

33. $f(x)=\dfrac{x}{x-2}$

$$f'(x)=\frac{(x-2)[D_x(x)]-(x)[D_x(x-2)]}{(x-2)^2}=\frac{(x-2)(1)-(x)(1)}{(x-2)^2}=\frac{x-2-x}{(x-2)^2}=\frac{-2}{(x-2)^2}$$

$$f'(3)=\frac{-2}{(3-2)^2}=-2$$

The slope is equal to $f'(3)=-2$.

$(y-3)=-2(x-3)\Rightarrow y=-2x+9$

35. a. $\bar{C}(x)=\dfrac{C(x)}{x}=\dfrac{200+75x}{x}$

b. $\bar{C}(10)=\dfrac{200+75(10)}{10}=95$

$\bar{C}(50)=\dfrac{200+75(50)}{50}=79$

The first average cost means that, at a sales level of 1000 hotdogs, the average cost is $95 per 100 hotdogs (or $.95/hotdog). Similarly, at a sales level of 5000 hotdogs, the average cost is $79 per 100 hotdogs.

c. $\bar{C}(x)=\dfrac{200+75x}{x}$

Marginal average cost function:

$$\bar{C}'(x)=\frac{d}{dx}\left(\frac{200+75x}{x}\right)$$
$$=\frac{(x)(75)-(200+75x)(1)}{x^2}$$
$$=\frac{75x-200-75x}{x^2}$$
$$=\frac{-200}{x^2}$$

d. $\bar{C}'(10)=\dfrac{-200}{(10)^2}=-2$

$\bar{C}'(50)=\dfrac{-200}{(50)^2}=-.08$

The first marginal average cost means that, at a sales level of 1000 hotdogs, if an additional 100 hotdogs are sold, the average cost decreases by $2 per 100 hotdogs. Similarly, at a sales level of 5000 hotdogs, if an additional 100 hotdogs are sold, the average cost decreases by $.08 per 100 hotdogs.

Copyright © 2019 Pearson Education, Inc.

37. a. $C'(x) = \dfrac{(x+100)(10)-(10x+1)(1)}{(x+100)^2}$

$= \dfrac{10x+1000-10x-1}{(x+100)^2}$

$= \dfrac{999}{(x+100)^2}$

$C'(20) = \dfrac{999}{(20+100)^2} \approx .0694$ hundred

$C'(40) = \dfrac{999}{(40+100)^2} \approx .0510$ hundred

The marginal cost for 20 baskets is \$6.94/basket and for 40 baskets is \$5.10/basket.

b. $\overline{C}(x) = \dfrac{10x+1}{x(x+100)} = \dfrac{10x+1}{x^2+100x}$

$\overline{C}(20) = \dfrac{10(20)+1}{(20)^2+100(20)} \approx .0838$ hundred

$\overline{C}(40) = \dfrac{10(40)+1}{(40)^2+100(40)} \approx .0716$ hundred

The average cost for 20 baskets is \$8.38/basket and for 40 baskets is \$7.16/basket

c.

$\overline{C}'(x) = \dfrac{(x^2+100x)(10)-(10x+1)(2x+100)}{(x^2+100x)^2}$

$= \dfrac{10x^2+1000x-20x^2-1002x-100}{(x^2+100x)^2}$

$= -\dfrac{10x^2+2x+100}{(x^2+100x)^2}$

$\overline{C}'(20) = -\dfrac{10(20)^2+2(20)+100}{((20)^2+100(20))^2}$

$\approx -.0007$ hundred

$\overline{C}'(40) = -\dfrac{10(40)^2+2(40)+100}{((40)^2+100(40))^2}$

$\approx -.0005$ hundred

The marginal average cost for 20 baskets is –\$.07/basket and for 40 baskets is –\$.05/basket.

39. $T(x) = \dfrac{10x}{x^2+5} + 98.6$

a.

$\dfrac{dT}{dx} = \dfrac{(x^2+5)[D_x(10x)]-(10x)[D_x(x^2+5)]}{(x^2+5)^2}$

$= \dfrac{(x^2+5)(10)-(10x)(2x)}{(x^2+5)^2}$

$= \dfrac{10x^2+50-20x^2}{(x^2+5)^2} = \dfrac{-10x^2+50}{(x^2+5)^2}$

b. $T'(0) = \dfrac{-10(0)^2+50}{((0)^2+5)^2} = 2$

The temperature is increasing at 2 degrees per hour.

c. $T'(1) = \dfrac{-10(1)^2+50}{((1)^2+5)^2} \approx 1.1111$

The temperature is increasing at about 1.1111 degrees per hour.

d. $T'(3) = \dfrac{-10(3)^2+50}{((3)^2+5)^2} \approx -.2041$

The temperature is decreasing at about .2041 degrees per hour.

41. $R(w) = 30\dfrac{w-4}{w-1.5}$

a. $R(5) = 30\left(\dfrac{5-4}{5-1.5}\right) \approx 8.57$ minutes

b. $R(7) = 30\left(\dfrac{7-4}{7-1.5}\right) \approx 16.36$ minutes

c. $R'(x) = 30\left(\dfrac{(w-1.5)(1)-(w-4)(1)}{(w-1.5)^2}\right)$

$= 30\left(\dfrac{w-1.5-w+4}{(w-1.5)^2}\right)$

$= \dfrac{75}{(w-1.5)^2}$

$R'(5) = \dfrac{75}{(5-1.5)^2} \approx 6.12$ min/(kcal/min)

$R'(7) = \dfrac{75}{(7-1.5)^2} \approx 2.48$ min/(kcal/min)

Copyright © 2019 Pearson Education, Inc.

43. $f(x)=\dfrac{x^2}{2-2x}$

The rate of change of the waiting time is

$$f'(x)=\frac{2(1-x)(2x)-x^2(-2)}{[2(1-x)]^2}$$
$$=\frac{(2-2x)(2x)+2x^2}{(2-2x)^2}=\frac{4x-4x^2+2x^2}{(2-2x)^2}$$
$$=\frac{-2x^2+4x}{(2-2x)^2}$$

a. $f'(.1)=\dfrac{-2(.1)^2+4(.1)}{[2-2(.1)]^2}=\dfrac{.38}{3.24}\approx .1173$

The rate of change of the number of vehicles waiting when the traffic intensity is .1 is .1173.

b. $f'(.6)=\dfrac{-2(.6)^2+4(.6)}{[2-2(.6)]^2}=\dfrac{1.68}{.64}=2.625$

The rate of change of the number of vehicles waiting when the traffic intensity is .6 is 2.625.

45. $M(x)=\dfrac{80x^2}{x^2+75}$

a. $M'(x)=\dfrac{(x^2+75)(160x)-(80x^2)(2x)}{(x^2+75)^2}$
$$=\frac{160x^3+12{,}000x-160x^3}{(x^2+75)^2}$$
$$=\frac{12{,}000x}{(x^2+75)^2}$$

b. $M'(1)=\dfrac{12{,}000(1)}{((1)^2+75)^2}$
≈ 2 codes/shift

$M'(5)=\dfrac{12{,}000(5)}{((5)^2+75)^2}$
≈ 6 codes/shift

$M'(20)=\dfrac{12{,}000(20)}{((20)^2+75)^2}$
≈ 1 code/shift

c. Answers vary; Possible answer: The rate of memorization increases for the first few shifts and then decreases as time goes on (since there are a finite number of codes to memorize).

Section 11.7 The Chain Rule

In Exercises 1–4, $f(x)=2x^2+3x$ and $g(x)=4x-1$.

1. $f[g(5)]$

Find $g(5)$ first.

$g(5)=4(5)-1=19$

$f[g(5)]=f(19)$
$=2(19^2)+3(19)=779$

3. $g[f(5)]$

Find $f(5)$ first.

$f(5)=2(5^2)+3(5)=65$

$g[f(5)]=g(65)$
$=4(65)-1=259$

5. $f(x)=8x+12;\ g(x)=2x+3$

$f[g(x)]=8(2x+3)+12=16x+36$

$g[f(x)]=2(8x+12)+3=16x+27$

7. $f(x)=3x^2-8x+1;\ g(x)=x-4$

$f[g(x)]=3(x-4)^2-8(x-4)+1$
$=3x^2-24x+48-8x+32+1$
$=3x^2-32x+81$

$g[f(x)]=(3x^2-8x+1)-4$
$=3x^2-8x-3$

9. $f(x)=\dfrac{1}{x};\ g(x)=x^2$

$f[g(x)]=\dfrac{1}{x^2}$

$g[f(x)]=\left(\dfrac{1}{x}\right)^2=\dfrac{1}{x^2}$

11. $f(x)=\sqrt{2x+5};\ g(x)=3x-8$

$f[g(x)]=\sqrt{2(3x-8)+5}$
$=\sqrt{6x-11}$

$g[f(x)]=3\sqrt{2x+5}-8$

13. $y=(4x+3)^5$

If $f(x)=x^5$ and $g(x)=4x+3$, then

$y=f[g(x)]=(4x+3)^5$.

Copyright © 2019 Pearson Education, Inc.

15. $y=\sqrt{6+3x^2}$

If $f(x)=\sqrt{x}$ and $g(x)=6+3x^2$, then

$y=f[g(x)]=\sqrt{6+3x^2}$.

17. $y=\dfrac{1}{(4x-1)^2}$

If $f(x)=x^{-2}$ and $g(x)=4x-1$, then

$y=f[g(x)]=\dfrac{1}{(4x-1)^2}$.

19. $y=(x^{1/2}-3)^3+(x^{1/2}-3)+5$

If $f(x)=x^3+x+5$ and

$g(x)=x^{1/2}-3$, then

$y=f[g(x)]=(x^{1/2}-3)^3+(x^{1/2}-3)+5$.

21. $y=(7x-12)^5$

$$y'=5(7x-12)^{5-1}\left[\frac{d}{dx}(7x-12)\right]$$
$$=5(7x-12)^4(7)=35(7x-12)^4$$

23. $y=7(4x+3)^4$

$$y'=7\left[\frac{d}{dx}\left((4x+3)^4\right)\right]+\frac{d}{dx}(7)\cdot(4x+3)^4$$
$$=7\left[4(4x+3)^3\right]\left[\frac{d}{dx}(4x+3)\right]+0$$
$$=28(4x+3)^3(4)=112(4x+3)^3$$

25. $y=-4\left(10x^2+8\right)^5$

$$y'=-4\left[\frac{d}{dx}\left(\left(10x^2+8\right)^5\right)\right]$$
$$+\frac{d}{dx}(-4)\cdot\left(10x^2+8\right)^5$$
$$=-4\left[5\left(10x^2+8\right)^4\right]\left[\frac{d}{dx}\left(10x^2+8\right)\right]+0$$
$$=-20\left(10x^2+8\right)^4(20x)=-400x\left(10x^2+8\right)^4$$

27. $y=(4x-9)^{3/2}$

$$y'=\frac{3}{2}\left[(4x-9)^{1/2}\right]\left[\frac{d}{dx}(4x-9)\right]$$
$$=\frac{3}{2}(4x-9)^{1/2}(4)=6(4x-9)^{1/2}$$

29. $y=-7(4x^2+9x)^{3/2}$

$$y'=-7\left(\frac{3}{2}\right)(4x^2+9x)^{1/2}(8x+9)$$
$$=-\frac{21}{2}(8x+9)(4x^2+9x)^{1/2}$$

31. $y=8\sqrt{4x+7}=8(4x+7)^{1/2}$

$$y'=8\left(\frac{1}{2}\right)(4x+7)^{-1/2}(4)$$
$$=16(4x+7)^{-1/2}=\frac{16}{\sqrt{4x+7}}$$

33. $y=-2\sqrt{x^2+4x}=-2(x^2+4x)^{1/2}$

$$y'=-2\left(\frac{1}{2}\right)(x^2+4x)^{-1/2}(2x+4)$$
$$=-(2x+4)(x^2+4x)^{-1/2}=\frac{-2x-4}{\sqrt{x^2+4x}}$$

35. $y=(x+1)(x-3)^2$

$$y'=(x+1)D_x(x-3)^2+(x-3)^2D_x(x+1)$$
$$=(x+1)(2)(x-3)(1)+(x-3)^2(1)$$
$$=2(x+1)(x-3)+(x-3)^2$$

37. $y=5(x+3)^2(2x-1)^5$

$$y'=5(x+3)^2(5)(2x-1)^4(2)$$
$$+(5)(2x-1)^5(2)(x+3)$$
$$=50(x+3)^2(2x-1)^4+10(2x-1)^5(x+3)$$
$$=10(x+3)(2x-1)^4\cdot[5(x+3)+(2x-1)]$$
$$=10(x+3)(2x-1)^4(7x+14)$$
$$=70(x+3)(2x-1)^4(x+2)$$

39. $y=(3x+1)^3\sqrt{x}=(3x+1)^3(x^{1/2})$

$$y'=(3x+1)^3\left(\frac{1}{2}\right)(x^{-1/2})$$
$$+(x^{1/2})(3)(3x+1)^2(3)$$
$$=\frac{1}{2}(x^{-1/2})(3x+1)^3+9(x^{1/2})(3x+1)^2$$
$$=(x^{-1/2})(3x+1)^2\left[\frac{1}{2}(3x+1)+9x\right]$$
$$=(x^{-1/2})(3x+1)^2\left(\frac{3x+1+18x}{2}\right)$$
$$=(x^{-1/2})(3x+1)^2\left(\frac{21x+1}{2}\right)$$
$$=\frac{(3x+1)^2(21x+1)}{2\sqrt{x}}$$

Copyright © 2019 Pearson Education, Inc.

41. $y = \dfrac{1}{(x-2)^3} = (x-2)^{-3}$

$y' = -3(x-2)^{-4} = \dfrac{-3}{(x-2)^4}$

43. $y = \dfrac{(2x-7)^2}{4x-1}$

$y' = \dfrac{(4x-1)(2)(2x-7)(2) - (2x-7)^2(4)}{(4x-1)^2}$

$= \dfrac{(16x-4)(2x-7) - (4)(2x-7)^2}{(4x-1)^2}$

$= \dfrac{(2x-7)[(16x-4) - 4(2x-7)]}{(4x-1)^2}$

$= \dfrac{(2x-7)(16x-4-8x+28)}{(4x-1)^2}$

$= \dfrac{(2x-7)(8x+24)}{(4x-1)^2}$

45. $y = \dfrac{x^2+5x}{(3x+1)^3}$

$y' = \dfrac{(3x+1)^3(2x+5) - (x^2+5x)(3)(3x+1)^2(3)}{(3x+1)^6}$

$= \dfrac{(3x+1)^3(2x+5) - (9x^2+45x)(3x+1)^2}{(3x+1)^6}$

$= \dfrac{(3x+1)^2\left[(3x+1)(2x+5) - (9x^2+45x)\right]}{(3x+1)^6}$

$= \dfrac{(3x+1)^2(6x^2+17x+5-9x^2-45x)}{(3x+1)^6}$

$= \dfrac{-3x^2-28x+5}{(3x+1)^4}$

47. $y = (x^{1/2}+2)(2x+3)$

$y' = (x^{1/2}+2)(2) + (2x+3)\left(\dfrac{1}{2}x^{-1/2}\right)$

$= 2\sqrt{x} + 4 + \dfrac{2x+3}{2\sqrt{x}}$

$= \dfrac{4x+8\sqrt{x}+2x+3}{2\sqrt{x}} = \dfrac{6x+8\sqrt{x}+3}{2\sqrt{x}}$

49. $f(x) = (5x-1)^3$

$f'(x) = 3(5x-1)^2[D_x(5x-1)]$

$= 3(5x-1)^2(5)$

$= 15(5x-1)^2$

$f'(0) = 15(5(0)-1)^2$

$= 15$

The slope is equal to $f'(0) = 15$.

$(y+1) = 15(x-0) \Rightarrow y = 15x-1$

51. $f(x) = \sqrt{20x+5} = (20x+5)^{1/2}$

$f'(x) = \frac{1}{2}(20x+5)^{-1/2}[D_x(20x+5)]$

$= \frac{1}{2}(20x+5)^{-1/2}(20)$

$= \dfrac{10}{\sqrt{20x+5}}$

$f'(1) = \dfrac{10}{\sqrt{20(1)+5}} = \dfrac{10}{\sqrt{25}} = \dfrac{10}{5} = 2$

The slope is equal to $f'(1) = 2$.

$(y-5) = 2(x-1) \Rightarrow y = 2x+3$

53. $D(p) = \dfrac{-p^2}{100} + 500;\ p(c) = 2c-10$

The demand in terms of the cost is

$D(c) = D[p(c)] = \dfrac{-(2c-10)^2}{100} + 500$

$= \dfrac{-4(c-5)^2}{100} + 500$

$= \dfrac{-c^2+10c-25}{25} + 500.$

55. **a.** Let $x = 57$.

$M(57) = -(10(57)-470)^2 + 55{,}000$

$= -(570-100)^2 + 55{,}000$

$= 45{,}000$

The average mortgage debt for 57-year-old borrowers is $45,000.

b. $M'(x) = -2(10x-470)[D_x(10x-470)]$.

$= -2(10x-470)(10)$

$= -200x + 9{,}400$

c. $M'(57) = -200(57) + 9{,}400$.

$= -11{,}400 + 9{,}400$

$= -2000$

The mortgage debt is decreasing by $2000 per year when the borrower is 57-years-old.

Copyright © 2019 Pearson Education, Inc.

57. $C(x) = 600 + \sqrt{200 + 25x^2 - x}$
$= 600 + \left(200 + 25x^2 - x\right)^{1/2}$

The marginal cost function is

$$C'(x) = \left(\frac{1}{2}\right)\left(200 + 25x^2 - x\right)^{-1/2}(50x - 1)$$
$$= \frac{50x - 1}{2\sqrt{200 + 25x^2 - x}}.$$

59. $R(x) = 15\sqrt{500x - 2x^2}, \quad 0 \le x \le 200$
$= 15\left(500x - 2x^2\right)^{1/2}$

a. The marginal revenue function is

$$R'(x) = 15\left(\frac{1}{2}\right)\left(500x - 2x^2\right)^{-1/2} \cdot (500 - 4x)$$
$$= \frac{15(500 - 4x)}{2\sqrt{500x - 2x^2}} = \frac{15(250 - 2x)}{\sqrt{500x - 2x^2}}$$

b. $R'(40) = \dfrac{15(250 - 2 \cdot 40)}{\sqrt{500 \cdot 40 - 2 \cdot 40^2}} \approx \$19.67$

$R'(80) = \dfrac{15(250 - 2 \cdot 80)}{\sqrt{500 \cdot 80 - 2 \cdot 80^2}} \approx \$8.19$

$R'(120) = \dfrac{15(250 - 2 \cdot 120)}{\sqrt{500 \cdot 120 - 2 \cdot 120^2}} \approx \$0.85$

$R'(160) = \dfrac{15(250 - 2 \cdot 160)}{\sqrt{500 \cdot 160 - 2 \cdot 160^2}} \approx -\$6.19$

c. Answers vary; a possible solution could be: As the number of items sold increases, the rate of change of revenue per item sold decreases.

61. $V(t) = 500(2t + 1)^{-2}$
$(0 \le t \le 4)$

a. $V(2) = 500(2(2) + 1)^{-2} = 20$

When the tablet computer is 2 years old, it's value will be \$20.

b. The marginal profit function is

$$V'(t) = -1000(2t + 1)^{-3}(2)$$
$$= \frac{-2000}{(2t + 1)^3}$$

$$V'(2) = \frac{-2000}{(2(2) + 1)^3} = -16$$

The value of the tablet computer will decrease by \$16 over the next year.

63. a. $R(x) = x\left(\sqrt{900 - 2x}\right)$
$= x(900 - 2x)^{1/2}$

b. $R'(x) = x\frac{1}{2}(900 - 2x)^{-1/2}(-2)$
$+ (900 - 2x)^{1/2}(1)$

$$= \frac{-x}{\sqrt{900 - 2x}} + \sqrt{900 - 2x}$$
$$= \frac{-3x + 900}{\sqrt{900 - 2x}}$$

c. $R'(250) = \dfrac{-3(250) + 900}{\sqrt{900 - 2(250)}} = 7.5$

When 250 tickets are sold, you will have a revenue of \$7.50/ticket.

65. a. $A = 12{,}000\left(1 + \dfrac{r}{1200}\right)^{96}$

$$A' = 12{,}000(96)\left(1 + \frac{r}{1200}\right)^{95}\left(\frac{1}{1200}\right)$$
$$= 960\left(1 + \frac{r}{1200}\right)^{95}$$

b. $A'(4) = 960\left(1 + \dfrac{4}{1200}\right)^{95}$
$\approx \$1316.95$ per percentage point

When the interest rate is 4%, increasing the interest rate by 1 percentage point would increase the balance by approximately \$1316.95.

67. a. $A = 24{,}000\left(1 + \dfrac{r}{1200}\right)^{24}$

$$\frac{dA}{dr} = 576{,}000\left(1 + \frac{r}{1200}\right)^{23}\left(\frac{1}{1200}\right)$$
$$= 480\left(1 + \frac{r}{1200}\right)^{23}$$

b. $\dfrac{dA}{dr} = 480\left(1 + \dfrac{2}{1200}\right)^{23} \approx 498.74$

When the interest rate is 2%, increasing the interest rate by 1 percentage point would increase the balance by approximately \$498.74.

Copyright © 2019 Pearson Education, Inc.

69. $P = \frac{200-x}{4} = 50 - \frac{x}{4}, \frac{dp}{dx} = -\frac{1}{4},$
$\frac{dx}{dn} = 6, \quad n = 10, \quad x = 60$

$$\frac{dR}{dn} = \left(p + x \cdot \frac{dp}{dx}\right)\frac{dx}{dn}$$
$$= \left(\frac{200-60}{4} - \frac{60}{4}\right)6$$
$$= \$120 \text{ per worker}$$

71. $C(x) = (6x+5)^{1/2}$

a. $\bar{C}(x) = \frac{\sqrt{6x+5}}{x} = x^{-1}(6x+5)^{1/2}$

b. $\bar{C}'(x) = x^{-1}\frac{1}{2}(6x+5)^{-1/2}(6) + (6x+5)^{1/2}(-x^{-2})$
$$= \frac{3}{x\sqrt{6x+5}} - \frac{\sqrt{6x+5}}{x^2}$$
$$= \frac{-3x-5}{x^2\sqrt{6x+5}}$$

73. $C(t) = \frac{1}{2}(2t+1)^{-1/2}$

The rate of change of C with respect to time is
$$C'(t) = \frac{1}{2}\left(-\frac{1}{2}\right)(2t+1)^{-3/2}(2)$$
$$= -\frac{1}{2}(2t+1)^{-3/2}.$$

a. $C'(0) = -\frac{1}{2}[2(0)+1]^{-3/2}$
$$= -\frac{1}{2} = -.5$$

b. $C'(4) = -\frac{1}{2}[2(4)+1]^{-3/2} = -\frac{1}{2}(9)^{-3/2}$
$$= -\frac{1}{2}\left(\frac{1}{27}\right) = -\frac{1}{54} \approx -.02$$

c. $C'(6) = -\frac{1}{2}[2(6)+1]^{-3/2}$
$$= -\frac{1}{2}(13)^{-3/2} \approx -.010667$$

d. $C'(7.5) = -\frac{1}{2}[2(7.5)+1]^{-3/2} = -\frac{1}{2}(16)^{-3/2}$
$$= -\frac{1}{2}\left(\frac{1}{64}\right) = -\frac{1}{128} \approx -.008$$

75. $F(t) = 60 - \frac{150}{\sqrt{8+t^2}} = 60 - 150(8+t^2)^{-1/2}$

a. $F'(t) = -150\left(-\frac{1}{2}\right)(8+t^2)^{-3/2}(2t)$
$$= 150t(8+t^2)^{-3/2} = \frac{150t}{(8+t^2)^{3/2}}$$

b. The rate at which the cashier's speed is increasing after 5 hours is:
$$F'(5) = \frac{150(5)}{(8+5^2)^{3/2}}$$
$$\approx 3.96 \text{ items/min per hr}$$

c. The rate at which the cashier's speed is increasing after 40 hours is:
$$F'(40) = \frac{150(40)}{(8+40^2)^{3/2}}$$
$$\approx .09 \text{ items/min per hr}$$

d. The cashiers speed is increasing more rapidly after 5 hours.

Section 11.8 Derivatives of Exponential and Logarithmic Functions

1. $y = e^{5x}$

Let $g(x) = 5x$. Then $y = e^{g(x)}$, so $y' = g'(x) \cdot e^{g(x)}$. Since $g'(x) = 5$, $y' = 5e^{5x}$.

3. $f(x) = 5e^{2x}$
$f'(x) = 5(2e^{2x}) = 10e^{2x}$

5. $g(x) = -4e^{-7x}$
$g'(x) = -4(-7)e^{-7x} = 28e^{-7x}$

7. $y = e^{x^2}$
$y' = 2xe^{x^2}$

9. $f(x) = e^{x^3/3}$
$f'(x) = \frac{1}{3}\left(3x^2\right)e^{x^3/3} = x^2e^{x^3/3}$

11. $y = -3e^{3x^2+5}$
$y' = (-3)(6x)e^{3x^2+5} = -18xe^{3x^2+5}$

Copyright © 2019 Pearson Education, Inc.

13. $y = \ln\left(-10x^2 + 7x\right)$

$$y' = \frac{D_x(-10x^2 + 7x)}{-10x^2 + 7x} = \frac{-20x + 7}{-10x^2 + 7x}$$

15. $y = \ln\sqrt{5x+1} = \ln(5x+1)^{1/2}$

Use a property of logarithms.

$$y = \frac{1}{2}\ln(5x+1)$$

$$y' = \frac{1}{2}\cdot\frac{D_x(5x+1)}{5x+1} = \frac{1}{2}\cdot\frac{5}{5x+1} = \frac{5}{2(5x+1)}$$

17. $f(x) = \ln\left[(5x+9)\left(x^2+1\right)\right]$

$$f'(x) = \frac{(5x+9)(2x) + \left(x^2+1\right)(5)}{\left(x^2+1\right)(5x+9)}$$

$$= \frac{10x^2 + 18x + 5x^2 + 5}{(5x+9)\left(x^2+1\right)} = \frac{15x^2 + 18x + 5}{\left(x^2+1\right)(5x+9)}$$

19. $y = \ln\left(\frac{4-x}{3x+8}\right)$

Use a property of logarithms.

$$y = \ln(4-x) - \ln(3x+8)$$

$$y' = \frac{-1}{4-x} - \frac{3}{3x+8} = \frac{-(3x+8) - 3(4-x)}{(4-x)(3x+8)}$$

$$= \frac{-3x - 8 - 12 + 3x}{(4-x)(3x+8)} = \frac{-20}{(4-x)(3x+8)}$$

21. $y = x^4e^{-3x}$

$$y' = 4x^3e^{-3x} + x^4(-3)e^{-3x}$$

$$= 4x^3e^{-3x} - 3x^4e^{-3x} = \left(4x^3 - 3x^4\right)e^{-3x}$$

23. $y = \left(6x^2 - 5x\right)e^{-3x}$

$$y' = \left(6x^2 - 5x\right)(-3)e^{-3x} + e^{-3x}(12x - 5)$$

$$= e^{-3x}\left(-18x^2 + 15x + 12x - 5\right)$$

$$= e^{-3x}\left(-18x^2 + 27x - 5\right)$$

25. $y = \ln\left(8x^3 - 3x\right)^{1/2} = \frac{1}{2}\ln\left(8x^3 - 3x\right)$

$$= \frac{1}{2}\ln\left[x\left(8x^2 - 3\right)\right] = \frac{1}{2}\left[\ln x + \ln\left(8x^2 - 3\right)\right]$$

$$y' = \frac{1}{2}\left[\frac{1}{x} + \frac{16x}{8x^2 - 3}\right] = \frac{1}{2}\left[\frac{8x^2 - 3 + 16x^2}{x\left(8x^2 - 3\right)}\right]$$

$$= \frac{24x^2 - 3}{2x\left(8x^2 - 3\right)} = \frac{3\left(8x^2 - 1\right)}{2x\left(8x^2 - 3\right)}$$

27. $y = x\ln\left(9 - x^4\right)$

$$y' = x\cdot\frac{-4x^3}{9 - x^4} + \ln\left(9 - x^4\right)\cdot 1$$

$$= \frac{-4x^4}{9 - x^4} + \ln\left(9 - x^4\right)$$

29. $y = x^4\ln|x|$

$$y' = x^4\cdot\frac{1}{x} + \ln|x|(4x^3) = x^3(1 + 4\ln|x|)$$

31. $y = \frac{-5\ln|x|}{5 - 2x}$

$$y' = \frac{(5 - 2x)\left(\frac{-5}{x}\right) - (-5\ln|x|)(-2)}{(5 - 2x)^2}$$

$$= \frac{\frac{-25}{x} + 10 - 10\ln|x|}{(5 - 2x)^2}$$

33. $y = \sqrt{\ln(x-3)} = [\ln(x-3)]^{1/2}$

$$y' = \frac{1}{2}[\ln(x-3)]^{-1/2}\cdot D_x[\ln(x-3)]$$

$$= \frac{1}{2}[\ln(x-3)]^{-1/2}\frac{1}{(x-3)}$$

$$= \frac{1}{2(x-3)\sqrt{\ln(x-3)}}$$

35. $y = \frac{e^x - 1}{\ln|x|}$

$$y' = \frac{\ln|x|\cdot e^x - (e^x - 1)\frac{1}{x}}{(\ln|x|)^2}$$

$$= \frac{xe^x\ln|x| - e^x + 1}{x(\ln|x|)^2}$$

Copyright © 2019 Pearson Education, Inc.

37. $y = \dfrac{e^x - e^{-x}}{x}$

$$y' = \frac{x\left[e^x - (-1)e^{-x}\right] - \left(e^x - e^{-x}\right)(1)}{x^2}$$
$$= \frac{xe^x + xe^{-x} - e^x + e^{-x}}{x^2}$$
$$\text{or } \frac{e^x(x-1) + e^{-x}(x+1)}{x^2}$$

39. $f(x) = e^{3x+2}\ln(4x-5)$

$$f'(x) = e^{3x+2}\left(\frac{4}{4x-5}\right) + [\ln(4x-5)](e^{3x+2})(3)$$
$$= \frac{4e^{3x+2}}{4x-5} + 3e^{3x+2}\ln(4x-5)$$

41. $y = \dfrac{700}{7-10e^{.4x}} = 700(7-10e^{.4x})^{-1}$

$$y' = -700(7-10e^{.4x})^{-2}[-10e^{.4x}(.4)]$$
$$= \frac{2800e^{.4x}}{(7-10^{.4x})^2}$$

43. $y = \dfrac{500}{12+5e^{-.5x}}$

$$= 500(12+5e^{-.5x})^{-1}$$
$$y' = -500\left(12+5e^{-.5x}\right)^{-2}\left[(5)e^{-.5x}(-.5)\right]$$
$$= -500\left(12+5e^{-.5x}\right)^{-2}\left(-2.5e^{-.5x}\right)$$
$$= 1250e^{-.5x}\left(12+5e^{-.5x}\right)^{-2}$$
$$= \frac{1250e^{-.5x}}{(12+5e^{-.5x})^2}$$

45. $y = \ln(\ln|x|)$

$$y' = \frac{1}{\ln|x|}\cdot\frac{1}{x} = \frac{1}{x\ln|x|}$$

47. $y = 8^x$

$$y = \left(e^{\ln 8}\right)^x = e^{(\ln 8)x}$$
$$y' = (\ln 8)e^{(\ln 8)x} = (\ln 8)8^x$$

49. $y = 15^{2x}$

$$y = \left(e^{\ln 15}\right)^{2x} = e^{(\ln 15)(2x)}$$
$$y' = 2(\ln 15)e^{(\ln 15)(2x)}$$
$$= 2(\ln 15)\cdot 15^{2x}$$

51. $g(x) = \log 6x = \dfrac{\ln 6x}{\ln 10}$

$$= \frac{1}{\ln 10}\cdot\ln 6x$$
$$g'(x) = \frac{1}{\ln 10}\cdot\frac{d}{dx}(\ln 6x)$$
$$= \frac{1}{\ln 10}\cdot\left(\frac{1}{6x}\right)\cdot(6) = \frac{1}{(\ln 10)x}$$

53. $f(x) = \dfrac{e^x}{2x+1}$

$$f'(x) = \frac{(2x+1)(e^x) - (e^x)(2)}{(2x+1)^2} = \frac{e^x(2x-1)}{(2x+1)^2}$$
$$f'(0) = \frac{e^0(2(0)-1)}{(2(0)+1)^2} = \frac{-1}{1} = -1$$

The slope of the tangent line to the graph of f is $f'(0) = -1$.

$(y-1) = -1(x-0) \Rightarrow y = -x+1$

55. $f(x) = \ln(1+x^2)$

$$f'(x) = \frac{1}{1+x^2}(2x) = \frac{2x}{1+x^2}$$
$$f'(0) = \frac{2(0)}{1+(0)^2} = 0$$

The slope of the tangent line to graph of f is $f'(0) = 0$.

$(y-0) = 0(x-0) \Rightarrow y = 0$

57. $f(x) = e^{2x}$

$$f'(x) = 2e^{2x}$$
$$f'[\ln(5)] = 2e^{2\ln(5)} = 2e^{\ln(5)^2}$$
$$= 2(5)^2 = 2(25) = 50$$

59. a. $P(t) = 1000e^{.165t}$

$$P'(t) = 1000(.165)e^{.165t}$$
$$= 165e^{.165t}$$

b. The year 2000 corresponds to $t = 0$.

$P'(0) = 165e^{.165(0)} = 165$

In 2000, the Chinese gross national income is increasing by \$165 billion per year.

c. The year 2016 corresponds to $t = 16$.

$P'(16) = 165e^{.165(16)} \approx 2{,}312$

In 2016, the Chinese gross national income will be increasing by about \$2,312 billion per year.

Copyright © 2019 Pearson Education, Inc.

61. **a.** $P(12) = 76e^{.13(12)} \approx 362$
In 2012, the population of breeding adult ferrets in the wild will be about 362.

b. $P(x) = 76e^{.13x}$
$P'(x) = (76)(.13)e^{.13x} = 9.88e^{.13x}$

$P'(12) = 9.88e^{.13(12)} \approx 47$
In 2012, the population of breeding adult ferrets in the wild was increasing by about 47 ferrets per year.

c. In 2012, there were about 362 ferrets, and the population was increasing at a rate of 47 ferrets per year.

63. $M(t) = 23e^{-.062t}$
$M'(t) = 2.3(-.062)e^{-.062t} = -1.426e^{-.062t}$

a. $M(22) = 23e^{-.062(22)} \approx 5.9$
In 2022, the preschool mortality rate will be about 5.9 deaths per 1000 people.

b. $M'(22) = -1.426e^{-.062(22)} \approx -.4$
In 2022, the preschool mortality rate will be decreasing by about .4 deaths per 1000 people.

c. According to part b of this exercise, the mortality rate will be decreasing in the year 2022.

65. $P(t) = 98.6 + te^{-.2t}$
$P'(t) = t \cdot (-.2)e^{-.2t} + e^{-.2t}(1)$
$= e^{-.2t}(1 - .2t)$

a. For $t = 1$.
$P'(1) = e^{-.2(1)}(1 - .2(1)) \approx .7$
After 1 hour, the patient's temperature is increasing by .7 degrees Fahrenheit per hour.

b. For $t = 12$.
$P'(12) = e^{-.2(12)}(1 - .2(12)) \approx -.1$
After 12 hour, the patient's temperature is decreasing by .1 degrees Fahrenheit per hour.

67. **a.** $f(t) = \dfrac{100}{1 + 25e^{-.4(t)}}$
$$f'(t) = \frac{(1 + 25e^{-.4t})(0) - (100)(-10e^{-.4t})}{(1 + 25e^{-.4t})^2} = \frac{1{,}000e^{-.4t}}{(1 + 25e^{-.4t})^2}$$

b. $f'(17) = \dfrac{1{,}000e^{-.4(17)}}{(1 + 25e^{-.4(17)})^2} \approx 1.1$
In 2017, the number of cell phone subscriptions is increasing by about 1.1 per 100 people per year.

c. In 2017, cell phone subscriptions per 100 people in middle income countries were increasing at about 1.1 subscriptions per 100 people per year.

69. $G(t) = \dfrac{52{,}000}{1 + 12e^{-.52(t)}}$
$$G'(t) = \frac{(1 + 12e^{-.52t})(0) - (52{,}000)(-6.24e^{-.52t})}{(1 + 12e^{-.52t})^2} = \frac{324{,}480e^{-.52t}}{(1 + 12e^{-.52t})^2}$$

a. $G(1) = \dfrac{52{,}000}{1 + 12e^{-.52(1)}} \approx 6393$
$G'(1) = \dfrac{324{,}480e^{-.52(1)}}{(1 + 12e^{-.52(1)})^2} \approx 2916$
After 1 year, the clam population is about 6393, and it is increasing at about 2916 per year.

b. $G(4) = \dfrac{52{,}000}{1 + 12e^{-.52(4)}} \approx 20{,}807$
$G'(4) = \dfrac{324{,}480e^{-.52(4)}}{(1 + 12e^{-.52(4)})^2} \approx 6490$
After 4 years, the clam population is about 20,807 and it is increasing at about 6490 per year.

Copyright © 2019 Pearson Education, Inc.

c. $G(10) = \dfrac{52,000}{1+12e^{-.52(10)}} \approx 48,771$

$G'(10) = \dfrac{324,480e^{-.52(10)}}{\left(1+12e^{-.52(10)}\right)^2} \approx 1575$

After 10 years, the clam population is about 48,771 and it is increasing at about 1575 per year.

d. The rate of growth increases for a while and then gradually decreases to almost 0.

71. a. $f(33) = 66\ln(33) + 135 \approx 366$

In 2033, the population of the United States is projected to be about 366 million.

b. $f(x) = 66\ln(x) + 135$

$f'(x) = (66)\left(\dfrac{1}{x}\right) + 0 = \dfrac{66}{x}$

c. $f'(33) = \dfrac{66}{33} = 2$

In 2033, the population of the United States is projected to be increasing by 2 million people per year.

73. Let $p = 100 + \dfrac{50}{\ln x}$.

a. The revenue function is given by $R(x) = xp = x\left(100 + \dfrac{50}{\ln x}\right)$. The marginal revenue is

$$R'(x) = x\frac{d}{dx}\left(100 + \frac{50}{\ln x}\right) + \left(100 + \frac{50}{\ln x}\right)\frac{d}{dx}x$$
$$= x\left(-\frac{50}{x(\ln x)^2}\right) + \left(100 + \frac{50}{\ln x}\right)$$
$$= \frac{-50}{(\ln x)^2} + \frac{50}{\ln x} + 100$$
$$= \frac{50(\ln x - 1)}{(\ln x)^2} + 100$$

b. The revenue from the next thousand items when $x = 8$ is

$$R'(8) = \frac{50(\ln 8 - 1)}{(\ln 8)^2} + 100 = \$112.48$$

75. $f(A) = 429\ln\left(\dfrac{A}{250}\right)$

$f'(A) = 429\left(\dfrac{250}{A}\right)\left(\dfrac{1}{250}\right)$

$= \dfrac{429}{A}$

a. $f(300) = 429\ln\left(\dfrac{300}{250}\right) \approx 78.2$

It takes 78.2 months to reach a balance of $300.

b. Positive; the more money you want to make, the longer time it will take.

c. $f'(300) = \dfrac{429}{300} \approx 1.4$

If you want the balance to increase by another $1, it will take approximately 1.4 more months.

Section 11.9 Continuity and Differentiability

1. $\lim\limits_{x\to 0} f(x)$ does not exist and $f(x)$ is not defined at $x = 0$. The open circle at $x = 2$ means $f(2)$ does not exist. Thus, the function is discontinuous at $x = 0$ and $x = 2$.

3. The limit of $h(x)$ as x approaches 1 is –2.5. The value of $h(x)$ at $x = 1$ is 2 (solid circle). Since the limit of $h(x)$ and the value of $h(x)$ are not equal, $h(x)$ is discontinuous at $x = 1$.

5. $\lim\limits_{x\to -3} y$ does not exist because y approaches 2 from the left and 3 from the right at $x = -3$, so y is not continuous at $x = -3$. Likewise, $\lim\limits_{x\to 0} y$ and $\lim\limits_{x\to 3} y$ do not exist because y approaches different values from the left and right at both $x = 0$ and $x = 3$. Therefore, y is discontinuous at $x = -3$, $x = 0$, and $x = 3$.

7. $\lim\limits_{x\to 0} y$ and $\lim\limits_{x\to 6} y$ do not exist because y approaches different values from the left and from the right at both $x = 0$ and $x = 6$. Thus, y is discontinuous at $x = 0$ and $x = 6$.

Copyright © 2019 Pearson Education, Inc.

In Exercises 9–20, the value of the limit of a function is found by the methods established earlier in the chapter. Details are omitted here.

9. $f(x)=\dfrac{4}{x-2}; x=0, x=2$

$f(0)=\dfrac{4}{-2}=-2$

$f(x)$ is continuous at $x=0$ because $f(0)$ is defined, $\lim\limits_{x\to 0} f(x)$ exists, and $\lim\limits_{x\to 0} f(x)=f(0)$.

$f(2)=\dfrac{4}{2-2}=\dfrac{4}{0}$, so $f(x)$ is not defined at $x=2$.

Hence, $f(x)$ is not continuous at $x=2$.

11. $h(x)=\dfrac{1}{x(x-3)}; x=0; x=3; x=5$

$h(0)=\dfrac{1}{0(0-3)}=\dfrac{1}{0}$

so $h(x)$ is not continuous at $x=0$.

$h(3)=\dfrac{1}{3(3-3)}=\dfrac{1}{0}$

so $h(x)$ is not continuous at $x=3$.

$h(5)=\dfrac{1}{5(5-3)}=\dfrac{1}{10}$

$\lim\limits_{x\to 5}\dfrac{1}{x(x-3)}=\dfrac{1}{10}$, so $h(x)$ is continuous at $x=5$.

13. $g(x)=\dfrac{x+2}{x^2-x-2}=\dfrac{x+2}{(x-2)(x+1)};$
$x=1, x=2, x=-2$

$g(1)=\dfrac{1+2}{(1-2)(1+1)}=-\dfrac{3}{2}$

$\lim\limits_{x\to 1}\dfrac{x+2}{x^2-x-2}=-\dfrac{3}{2}$, so $g(x)$ is continuous at $x=1$.

$g(2)=\dfrac{2+2}{(2-2)(2+1)}=\dfrac{4}{0(3)}=\dfrac{4}{0}$

$g(x)$ is not defined at $x=2$, so $g(x)$ is not continuous at $x=2$.

$g(-2)=\dfrac{-2+2}{(-2-2)(-2+1)}=\dfrac{0}{-4(-1)}$
$=0$

$\lim\limits_{x\to -2}\dfrac{x+2}{x^2-x-2}=0$, so $g(x)$ is continuous at $x=-2$.

15. $g(x)=\dfrac{x^2-4}{x-2}; x=0, x=2, x=-2$

$g(0)=\dfrac{0-4}{0-2}=2$

$\lim\limits_{x\to 0} g(x)=2$, so $g(x)$ is continuous at $x=0$.

$g(2)=\dfrac{4-4}{2-2}=\dfrac{0}{0}$

so $g(x)$ is not continuous at $x=2$.

$g(-2)=\dfrac{4-4}{-4}=0$

$\lim\limits_{x\to -2} g(x)=0$, so $g(x)$ is continuous at $x=-2$.

17. $f(x)=\begin{cases} x-2 & \text{if } x\le 3 \\ 2-x & \text{if } x>3 \end{cases}; x=2, x=3$

$f(2)=2-2=0$

$\lim\limits_{x\to 2} f(x)=0$

so $f(x)$ is continuous at $x=2$.

$f(3)=3-2=1$

$\lim\limits_{x\to 3} f(x)$ does not exist because the limit approaches different values when approached from the left and right so, $f(x)$ is not continuous at $x=3$.

19. $g(x)=x^2-5x+12$

$g(x)$ is continuous everywhere.

21. $h(x)=\dfrac{|x+4|}{x+4}$

$h(x)$ is not defined at $x=-4$, so $h(x)$ is not continuous at $x=-4$.

23. $k(x)=e^{\sqrt{x-2}}$

Since $\sqrt{x-2}$ is not defined for $x-2<0$, or $x<2$, $k(x)$ is not defined for $x<2$. Thus, $k(x)$ is not continuous for $x<2$.

Copyright © 2019 Pearson Education, Inc.

25. a.

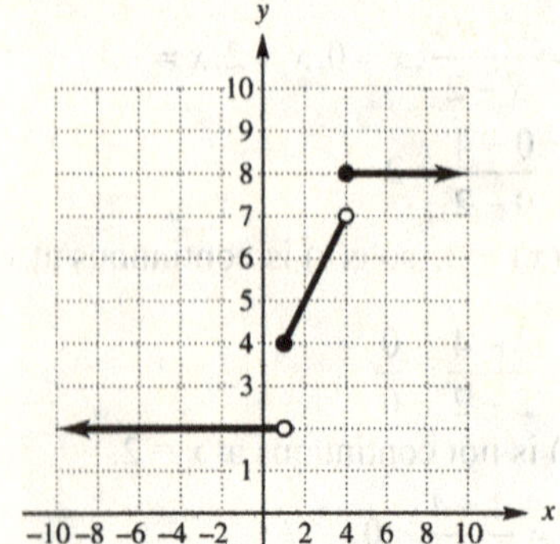

b. Since $\lim\limits_{x\to 1} f(x)$ and $\lim\limits_{x\to 4} f(x)$ do not exist, $f(x)$ is discontinuous at $x = 1$ and $x = 4$.

27. a.

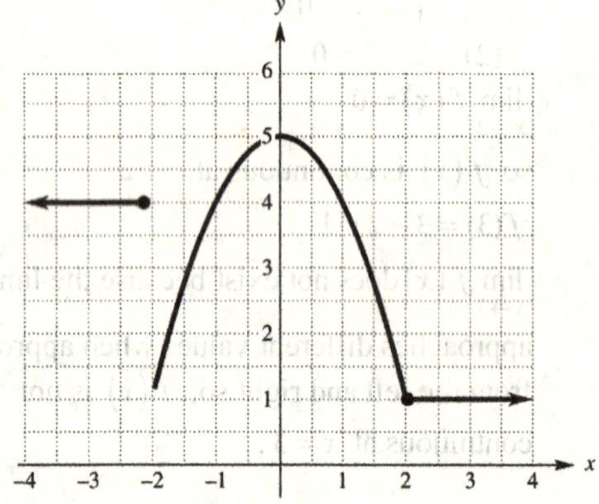

b. Since $\lim\limits_{x\to -2} f(x)$ does not exist, $f(x)$ is discontinuous at $x = -2$.

29. a. $\lim\limits_{x\to 2010^-} T(x) = 7.25\%$

b. $\lim\limits_{x\to 2010^+} T(x) = 7.25\%$

c. $\lim\limits_{x\to 2010} T(x) = 7.25\%$

d. $\lim\limits_{x\to 2017^-} T(x) = 6.5\%$

e. $\lim\limits_{x\to 2017^+} T(x) = 6.25\%$

f. $\lim\limits_{x\to 2017} T(x)$ does not exists since $T(x)$ does not have the same value as x approaches from the left and right.

g. The graph of $T(x)$ is discontinuous for years 2002, 2004, 2009, 2011, 2013 and 2017.

h. The graph of $T(x)$ is not differentiable for years 2002, 2004, 2009, 2011, 2013 and 2017.

31. The function is discontinuous only at $t = m$, where $\lim\limits_{t\to m} f(t) = 45\%$ from the left and $\lim\limits_{t\to m} f(t) = 70\%$ from the right. The function is differentiable everywhere except at $t = m$ and at the endpoints.

33.

a.

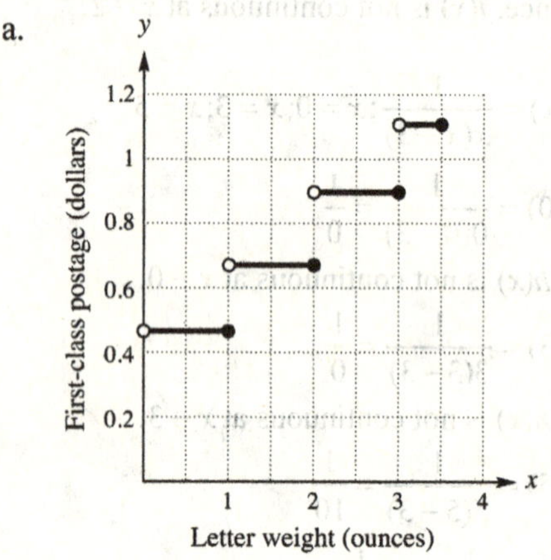

b. $\lim\limits_{x\to 1^-} C(x) = \$.47$

c. $\lim\limits_{x\to 1^+} C(x) = \$.68$

d. $\lim\limits_{x\to 1} C(x)$ does not exists since $C(x)$ does not have the same value as x approaches from the left and right.

e. $\lim\limits_{x\to 2.5^-} C(x) = \$.89$

f. $\lim\limits_{x\to 2.5^+} C(x) = \$.89$

g. $\lim\limits_{x\to 2.5} C(x) = \$.89$ since both the left and right limits approach the same value.

h. $C(x)$ is discontinuous at 1, 2, and 3 ounces.

Copyright © 2019 Pearson Education, Inc.

35. a. $C(1) = 88$
The cost for a 1 mile tow is $88.

b. $\lim_{x\to1^-} C(x) = \$88$ and $\lim_{x\to1^+} C(x) = \$88$
Therefore, $\lim_{x\to1} C(x) = \$88$.

c. $C(4) = 88 + 4.75(2) = 97.50$
The cost for a 4 mile tow is $97.50.

d. $\lim_{x\to4^-} C(x) = \$92.75$ and
$\lim_{x\to4^+} P(x) = \$97.50$ Therefore, $\lim_{x\to4} P(x)$ does not exist.

e. $C(x)$ is discontinuous at $x = 2$, 3, and 4.

37. $(-5, 7)$ since x must be greater than –5 and less than 7.

39. $(-\infty, -3]$ since x must be less than or equal to –3.

41. From the graph in Exercise 6, the function is continuous on (–6, 0), (4, 8). Because the function is not continuous as x approaches 0 from the right, the function is not continuous on [0, 3].

43. a.

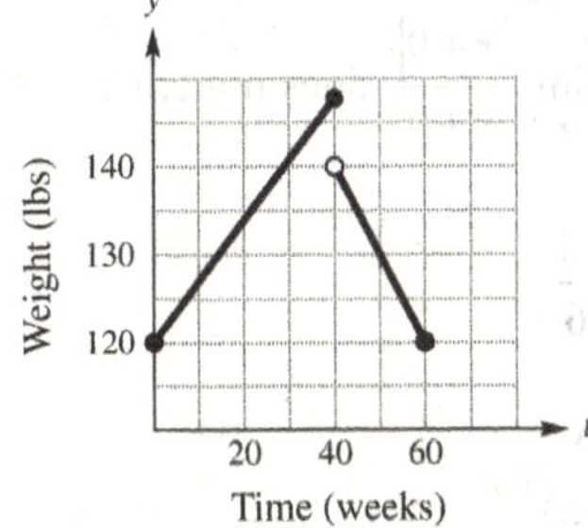

b. The function is not continuous at $t = 40$ since the function does not get close to the same value as t approaches 40 from the left and right.

Chapter 11 Review Exercises

1. The graph of $f(x)$ approaches 4 from the left and from the right of $x = -3$. Therefore, $\lim_{x\to-3} f(x) = 4$.

2. $\lim_{x\to-1} g(x)$ does not exist, since the graph shows that $g(x)$ approaches 2 as x approaches –1 from the right but $g(x)$approaches –2 as x approaches –1 from the left.

3. $\lim_{x\to0} \dfrac{\sqrt{2-x}-\sqrt{2}}{x}$

X	Y1	
-.01	-.3531	
-.001	-.3535	
-1E-5	-.3536	
0	ERROR	
1E-4	-.3536	
.001	-.3536	
.01	-.354	
X=0		

As x gets closer to 0 from the left or right, $\dfrac{\sqrt{2-x}-\sqrt{2}}{x}$ gets closer to –.3536, so $\lim_{x\to0} \dfrac{\sqrt{2-x}-\sqrt{2}}{x} = -.35$.

4. $\lim_{x\to-1} \dfrac{10^x - .1}{x+1}$

X	Y1	
-1.1	.20567	
-1.01	.22763	
-1.001	.22999	
-1	ERROR	
-.9999	.23029	
-.999	.23052	
-.99	.23293	
X=-1		

As x approaches –1 from the left or right, $\dfrac{10^x - .1}{x+1}$ approaches .23, so $\lim_{x\to-1} \dfrac{10^x - .1}{x+1} = .23026$.

5. $\dfrac{5x-4}{x-2}$ is continuous except at 2, where the expression is not defined. The function is defined for $x = 5$, so
$$\lim_{x\to5} \frac{5x-4}{x-2} = \frac{5(5)-4}{5-2} = \frac{21}{3} = 7..$$

6. $\dfrac{7-3x}{x-4}$ is continuous except at 4, where the expression is not defined. The function is defined for $x = -1$, so
$$\lim_{x\to-1} \frac{7-3x}{x-4} = \frac{7-3(-1)}{-1-4} = \frac{10}{-5} = -2.$$

7. $\lim_{x\to3} \dfrac{x^2-9}{x+3}$
$$\lim_{x\to3} \frac{x^2-9}{x+3} = \frac{3^2-9}{3+3} = \frac{0}{9} = 0$$

Copyright © 2019 Pearson Education, Inc.

8. $\lim_{x\to 7}\frac{49-x^2}{x+7}$

$\lim_{x\to 7}\frac{49-x^2}{x+7}=\frac{49-(7)^2}{7+7}=\frac{0}{14}=0.$

9. Although the function is discontinuous at $x = 4$, this limit exists.

$$\lim_{x\to 4}\frac{x^2-16}{x-4}=\lim_{x\to 4}\frac{(x+4)(x-4)}{x-4}=\lim_{x\to 4}(x+4)=4+4=8$$

10. $$\lim_{x\to -2}\frac{x^2-x-6}{x+2}=\lim_{x\to -2}\frac{(x-3)(x+2)}{x+2}=\lim_{x\to -2}(x-3)=-2-3=-5$$

11. $$\lim_{x\to 9}\frac{\sqrt{x}-3}{x-9}=\lim_{x\to 9}\frac{(\sqrt{x}-3)(\sqrt{x}+3)}{(x-9)(\sqrt{x}+3)}=\lim_{x\to 9}\frac{x-9}{(x-9)(\sqrt{x}+3)}=\lim_{x\to 9}\frac{1}{\sqrt{x}+3}=\frac{1}{\sqrt{9}+3}=\frac{1}{6}$$

12. $$\lim_{x\to 16}\frac{\sqrt{x}-4}{x-16}=\lim_{x\to 16}\frac{(\sqrt{x}-4)(\sqrt{x}+4)}{(x-16)(\sqrt{x}+4)}=\lim_{x\to 16}\frac{x-16}{(x-16)(\sqrt{x}+4)}=\lim_{x\to 16}\frac{1}{\sqrt{x}+4}=\frac{1}{\sqrt{16}+4}=\frac{1}{8}$$

13. $$\lim_{x\to 4^+}\sqrt{x^2-x-12}=\lim_{x\to 4^+}\sqrt{(x+3)(x-4)}=\left(\lim_{x\to 4^+}\sqrt{x+3}\right)\left(\lim_{x\to 4^+}\sqrt{x-4}\right)=\sqrt{7}\cdot 0=0$$

14. $$\lim_{x\to 3^-}\left(5x+\sqrt{9-x^2}\right)=\lim_{x\to 3^-}5x+\lim_{x\to 3^-}\sqrt{9-x^2}=15+0=15$$

15. $\lim_{x\to 7}\frac{|x-7|}{x-7}$

As x approaches 7 from the right, $|x-7|=(x-7)$, so

$$\lim_{x\to 7^+}\frac{|x-7|}{x-7}=\lim_{x\to 7^+}\frac{(x-7)}{x-7}=\lim_{x\to 7^+}1=1$$

As x approaches 7 from the left, $|x-7|=-(x-7)$, so

$$\lim_{x\to 7^+}\frac{|x-7|}{x-7}=\lim_{x\to 7^+}\frac{-(x-7)}{x-7}=\lim_{x\to 7^+}-1=-1$$

Since the right and left limits are different values, $\lim_{x\to 7}\frac{|x-7|}{x-7}$ does not exist.

16. $\lim_{x\to -9}\frac{|x+9|}{x+9}$

As x approaches –9 from the right, $|x+9|=(x+9)$, so

$$\lim_{x\to -9^+}\frac{|x+9|}{x+9}=\lim_{x\to -9^+}\frac{(x+9)}{x+9}=\lim_{x\to -9^+}1=1$$

As x approaches –9 from the left, $|x-7|=-(x-7)$, so

$$\lim_{x\to -9^+}\frac{|x+9|}{x+9}=\lim_{x\to -9^+}\frac{-(x+9)}{x+9}=\lim_{x\to -9^+}-1=-1$$

Since the right and left limits are different values, $\lim_{x\to -9}\frac{|x+9|}{x+9}$ does not exist

17. $$\lim_{x\to 6^-}\frac{x^2+1}{x-6}=-\infty$$

18. $$\lim_{x\to -5}\frac{8x}{(x+5)^2}=-\infty$$

19. $$\lim_{x\to\infty}\frac{9x+1}{3x-4x^2}=\lim_{x\to\infty}\frac{\frac{9}{x}+\frac{1}{x^2}}{\frac{3}{x}-4}=\frac{0}{-4}=0$$

20. $$\lim_{x\to\infty}\frac{5x^2-8x+3}{5-7x^3}=\lim_{x\to\infty}\frac{\frac{5}{x}-\frac{8}{x^2}+\frac{3}{x^3}}{\frac{5}{x^3}-7}=\frac{0}{-7}=0$$

Copyright © 2019 Pearson Education, Inc.

21. $\lim_{x\to\infty} \frac{6x^2 - x + 50}{3 - 2x^2} = \lim_{x\to\infty} \frac{6 - \frac{1}{x} + \frac{50}{x^2}}{\frac{3}{x^2} - 2}$

$= \frac{6}{-2} = -3$

22. $\lim_{x\to\infty} \frac{12x^3 + 3x^2 - 7x + 1}{6 - 3x - 4x^3} = \lim_{x\to\infty} \frac{12 + \frac{3}{x} - \frac{7}{x^2} + \frac{1}{x^3}}{\frac{6}{x^3} - \frac{3}{x^2} - 4}$

$= \frac{12}{-4} = -3$

23. $x = 0$ to $x = 4$

From the graph, $f(0) = 0$ and $f(4) = 1$.

Average rate of change

$= \frac{f(4) - f(0)}{4 - 0} = \frac{1 - 0}{4 - 0} = \frac{1}{4}$

24. $x = 2$ to $x = 8$

From the graph, $f(8) = 4$ and $f(2) = 4$.

Average rate of change

$= \frac{f(8) - f(2)}{8 - 2} = \frac{4 - 4}{8 - 2} = 0$

25. $f(x) = 3x^2 - 5$, from $x = 1$ to $x = 5$

$f(1) = 3(1^2) - 5 = -2$

$f(5) = 3(5^2) - 5 = 70$

Average rate of change

$= \frac{f(5) - f(1)}{5 - 1} = \frac{70 - (-2)}{4} = 18$

26. $g(x) = -x^3 + 2x^2 + 1$,

from $x = -2$ to $x = 3$

$g(-2) = -(-2)^3 + 2(-2)^2 + 1 = 17$

$g(3) = -3^3 + 2(3)^2 + 1 = -8$

Average rate of change

$= \frac{g(3) - g(-2)}{3 - (-2)} = \frac{-8 - 17}{5} = \frac{-25}{5} = -5$

27. $h(x) = \frac{6 - x}{2x + 3}$, from $x = 0$ to $x = 6$

$h(0) = \frac{6 - 0}{2(0) + 3} = \frac{6}{3} = 2$

$h(6) = \frac{6 - 6}{2(6) + 3} = \frac{0}{15} = 0$

Average rate of change

$= \frac{h(6) - h(0)}{6 - 0} = \frac{0 - 2}{6} = -\frac{1}{3}$

28. $f(x) = e^{2x} + 5\ln x$

from $x = 1$ to $x = 7$

$f(1) = e^2 + 5\ln 1 \approx 7.389056099$

$f(7) = e^{14} + 5\ln 7 \approx 1{,}202{,}614$

Average rate of change

$= \frac{f(7) - f(1)}{7 - 1} = \frac{1{,}202{,}606.625}{6}$

$\approx 200{,}434.4374$

29. $y = 2x + 3 = f(x)$

$f(x + h) = 2(x + h) + 3 = 2x + 2h + 3$

$f(x + h) - f(x) = (2x + 2h + 3) - (2x + 3) = 2h$

$\frac{f(x + h) - f(x)}{h} = \frac{2h}{h} = 2$

$f'(x) = y' = \lim_{h\to 0} \frac{f(x + h) - f(x)}{h} = \lim_{h\to 0} 2 = 2$

30. $y = 4 - 3x$

$y' = \lim_{h\to 0} \frac{[4 - 3(x + h)] - (4 - 3x)}{h}$

$= \lim_{h\to 0} \frac{4 - 3x - 3h - 4 + 3x}{h}$

$= \lim_{h\to 0} \frac{-3h}{h}$

$= \lim_{h\to 0} (-3) = -3$

31. $y = 2x^2 - x - 1$

$y' = \lim_{h\to 0} \frac{[2(x + h)^2 - (x + h) - 1] - (2x^2 - x - 1)}{h}$

$= \lim_{h\to 0} \frac{2x^2 + 4xh + 2h^2 - x - h - 1 - 2x^2 + x + 1}{h}$

$= \lim_{h\to 0} \frac{4xh + 2h^2 - h}{h} = \lim_{h\to 0} (4x + 2h - 1)$

$= 4x - 1$

Copyright © 2019 Pearson Education, Inc.

32. $y = x^2 + 2x$

$$y' = \lim_{h\to 0}\frac{[(x+h)^2 + 2(x+h)] - (x^2+2x)}{h}$$
$$= \lim_{h\to 0}\frac{x^2 + 2xh + h^2 + 2x + 2h - x^2 - 2x}{h}$$
$$= \lim_{h\to 0}\frac{2xh + h^2 + 2h}{h} = \lim_{h\to 0}(2x + h + 2)$$
$$= 2x + 2$$

33. $y = x^2 - 6x$; at $x = 2$

Let $f(x) = y$.

$f(2) = 2^2 - 6(2) = -8$

$f'(x) = 2x - 6$

The slope of the tangent line at (2, –8) is

$f'(2) = 2(2) - 6 = -2.$

Use point-slope form with $x_1 = 2$, $y_1 = -8$ and $m = -2$ to find equation of tangent line.

$y - y_1 = m(x - x_1)$

$y - (-8) = -2(x - 2) \Rightarrow y + 8 = -2x + 4 \Rightarrow$

$y = -2x - 4$

34. $y = 8 - x^2$; at $x = 1$

Let $f(x) = y$.

$f(1) = 8 - 1^2 = 7$

$f'(x) = -2x$

The slope of the tangent line at (1, 7) is

$f'(1) = -2(1) = -2$

The equation is

$y - y_1 = m(x - x_1)$

$y - 7 = -2(x - 1) \Rightarrow y - 7 = -2x + 2 \Rightarrow$

$y = -2x + 9.$

35. $y = \dfrac{-2}{x+5}$; at $x = -2$

Let $f(x) = y$.

$f(-2) = \dfrac{-2}{-2+5} = -\dfrac{2}{3}$

$f(x) = -2(x+5)^{-1}$

$f'(x) = -2[-1(x+5)^{-2}(1)] = \dfrac{2}{(x+5)^2}$

slope $= f'(-2) = \dfrac{2}{(-2+5)^2} = \dfrac{2}{(3)^2} = \dfrac{2}{9}$

The equation is

$y - y_1 = m(x - x_1)$

$y + \dfrac{2}{3} = \dfrac{2}{9}(x + 2) \Rightarrow 9y + 6 = 2(x + 2) \Rightarrow$

$9y + 6 = 2x + 4 \Rightarrow y = \dfrac{2}{9}x - \dfrac{2}{9}.$

36. $y = \sqrt{6x - 2}$; at $x = 3$

Let $f(x) = y$.

$f(3) = \sqrt{6(3) - 2} = 4$

$f(x) = (6x - 2)^{1/2}$

$f'(x) = \dfrac{1}{2}(6x-2)^{-1/2}(6) = 3(6x-2)^{-1/2}$

slope $= y'(3) = 3(6\cdot 3 - 2)^{-1/2}$

$= 3(16)^{-1/2} = \dfrac{3}{16^{1/2}} = \dfrac{3}{4}$

The equation is

$y - 4 = \dfrac{3}{4}(x - 3) \Rightarrow 4y - 16 = 3x - 9 \Rightarrow$

$4y = 3x + 7 \Rightarrow y = \dfrac{3}{4}x + \dfrac{7}{4}$

37. $y = 7x^8 - 8x + 3$

$y' = 7(8x^7) - 8 = 56x^7 - 8$

38. $y = 2 + 9x - 7x^6$

$y' = 9 - 7(6x^5) = 9 - 42x^5$

39. $y = e + ex + e^x + x^e$

$y' = e + e^x(1) + ex^{e-1}$

$= e + e^x + ex^{e-1}$

40. $y = 1 + \pi - \pi^2 + \sqrt{\pi} - \frac{1}{\pi}$

$y' = 0$

41. $f(x) = 8x^{1/2} + 7x^{-3}$

$f'(x) = 8(\frac{1}{2}x^{-1/2}) + 7(-3x^{-4})$

$= 4x^{-1/2} - 21x^{-4}$

42. $f(x) = 9x^{1/3} - 5x^{-7}$

$f'(x) = 9(\frac{1}{3}x^{-2/3}) - 5(-7x^{-8})$

$= 3x^{-2/3} + 35x^{-8}$

43. $f(x) = 6\sqrt{x} - \dfrac{3}{x} = 6x^{1/2} - 3x^{-1}$

$f'(x) = 6(\frac{1}{2}x^{-1/2}) - 3(-x^{-2})$

$= 3x^{-1/2} + 3x^{-2} = \dfrac{3}{\sqrt{x}} + \dfrac{3}{x^2}$

Copyright © 2019 Pearson Education, Inc.

44. $f(x)=4\sqrt{x}-\dfrac{3}{x^2}=4x^{1/2}-3x^{-2}$

$f'(x)=4(\tfrac{1}{2}x^{-1/2})-3(-2x^{-3})$

$=2x^{-1/2}+6x^{-3}=\dfrac{2}{\sqrt{x}}+\dfrac{6}{x^3}$

45. $y=(4t^2+9)(t^3-5)$

Use the product rule.

$y'=(4t^2+9)(3t^2)+(t^3-5)(8t)$

$=12t^4+27t^2+8t^4-40t$

$=20t^4+27t^2-40t$

46. $y=(-5t+3)(t^3-4t)$

Use the product rule.

$y'=(-5t+3)(3t^2-4)+(t^3-4t)(-5)$

$=-15t^3+9t^2+20t-12-5t^3+20t$

$=-20t^3+9t^2+40t-12$

47. $y=8x^{3/4}(5x+1)$

Use the product rule.

$y'=8x^{3/4}(5)+(5x+1)\left(\dfrac{3}{4}\right)8x^{-1/4}$

$=40x^{3/4}+6(5x+1)x^{-1/4}$

48. $y=40x^{-1/4}(x^2+1)$

Use the product rule.

$y'=40x^{-1/4}(2x)+(x^2+1)(40)\left(-\dfrac{1}{4}\right)x^{-5/4}$

$=80x^{3/4}-10x^{-5/4}(x^2+1)$

49. $f(x)=\dfrac{4x}{x^2-8}$

Use the quotient rule.

$f'(x)=\dfrac{(x^2-8)(4)-4x(2x)}{(x^2-8)^2}$

$=\dfrac{4x^2-32-8x^2}{(x^2-8)^2}=\dfrac{-4(x^2+8)}{(x^2-8)^2}$

50. $g(x)=\dfrac{-5x^2}{3x+1}$

Use the quotient rule.

$g'(x)=\dfrac{(3x+1)(-10x)-(-5x^2)(3)}{(3x+1)^2}$

$=\dfrac{-30x^2-10x+15x^2}{(3x+1)^2}=\dfrac{-5x(3x+2)}{(3x+1)^2}$

51. $y=\dfrac{\sqrt{x}-6}{3x+7}=\dfrac{x^{1/2}-6}{3x+7}$

Use the quotient rule.

$y'=\dfrac{(3x+7)\left(\frac{1}{2}x^{-1/2}\right)-(x^{1/2}-6)(3)}{(3x+7)^2}$

$=\dfrac{\frac{3}{2}x^{1/2}+\frac{7}{2}x^{-1/2}-3x^{1/2}+18}{(3x+7)^2}$

$=\dfrac{-\frac{3}{2}x^{1/2}+\frac{7}{2}x^{-1/2}+18}{(3x+7)^2}\cdot\dfrac{2}{2}$

$=\dfrac{-3x^{1/2}+7x^{-1/2}+36}{2(3x+7)^2}\cdot\dfrac{x^{1/2}}{x^{1/2}}$

$=\dfrac{-3x+7+36x^{1/2}}{2x^{1/2}(3x+7)^2}=\dfrac{-3x+36\sqrt{x}+7}{2\sqrt{x}(3x+7)^2}$

52. $y=\dfrac{\sqrt{x}+9}{x-4}=\dfrac{x^{1/2}+9}{x-4}$

Use the quotient rule.

$y'=\dfrac{(x-4)\left(\dfrac{1}{2}x^{-1/2}\right)-(x^{1/2}+9)(1)}{(x-4)^2}$

$=\dfrac{\frac{1}{2}x^{1/2}-2x^{-1/2}-x^{1/2}-9}{(x-4)^2}$

$=\dfrac{-\frac{1}{2}x^{1/2}-2x^{-1/2}-9}{(x-4)^2}\cdot\dfrac{2x^{1/2}}{2x^{1/2}}$

$=\dfrac{-x-4-18x^{1/2}}{2x^{1/2}(x-4)^2}=\dfrac{-x-18\sqrt{x}-4}{2\sqrt{x}(x-4)^2}$

53. $y=\dfrac{x^2-x+1}{x-1}$

Use the quotient rule.

$y'=\dfrac{(x-1)(2x-1)-(x^2-x+1)(1)}{(x-1)^2}$

$=\dfrac{2x^2-3x+1-x^2+x-1}{(x-1)^2}=\dfrac{x^2-2x}{(x-1)^2}$

Copyright © 2019 Pearson Education, Inc.

54. $y = \dfrac{2x^3 - 5x^2}{x+2}$

Use the quotient rule.

$$y' = \frac{(x+2)(6x^2 - 10x) - (2x^3 - 5x^2)(1)}{(x+2)^2}$$
$$= \frac{6x^3 - 10x^2 + 12x^2 - 20x - 2x^3 + 5x^2}{(x+2)^2}$$
$$= \frac{4x^3 + 7x^2 - 20x}{(x+2)^2}$$

55. $f(x) = (4x-2)^4$

Use the chain rule.

$f'(x) = 4(4x-2)^3(4) = 16(4x-2)^3$

56. $k(x) = (5x-1)^6$

Use the chain rule.

$k'(x) = 6(5x-1)^5(5) = 30(5x-1)^5$

57. $y = \sqrt{2t-5} = (2t-5)^{1/2}$

Use the chain rule.

$$y' = \frac{1}{2}(2t-5)^{-1/2}(2)$$
$$= (2t-5)^{-1/2} = \frac{1}{(2t-5)^{1/2}} = \frac{1}{\sqrt{2t-5}}$$

58. $y = -3\sqrt{8t-1} = -3(8t-1)^{1/2}$

$$y' = -3\left[\frac{1}{2}(8t-1)^{-1/2}\right](8)$$
$$= -12(8t-1)^{-1/2}$$
$$= \frac{-12}{(8t-1)^{1/2}} = \frac{-12}{\sqrt{8t-1}}$$

59. $y = 2x(3x-4)^3$

Use the product rule and the chain rule.

$$y' = 2x(3)(3x-4)^2(3) + (3x-4)^3(2)$$
$$= 18x(3x-4)^2 + 2(3x-4)^3$$

60. $y = 5x^2(2x+3)^5$

$$y' = 5x^2(5)(2x+3)^4(2) + (2x+3)^5(10x)$$
$$= 50x^2(2x+3)^4 + 10x(2x+3)^5$$

61. $f(u) = \dfrac{3u^2 - 4u}{(2u+3)^3}$

Use the quotient rule and the chain rule.

$f'(u)$
$$= \frac{(2u+3)^3(6u-4) - (3u^2-4u)(3)(2u+3)^2(2)}{(2u+3)^6}$$
$$= \frac{(2u+3)^3(6u-4) - 6(3u^2-4u)(2u+3)^2}{(2u+3)^6}$$
$$= \frac{(2u+3)(6u-4) - 6(3u^2-4u)}{(2u+3)^4}$$

62. $g(t) = \dfrac{t^3 + t - 2}{(2t-1)^5}$

$g'(t)$
$$= \frac{(2t-1)^5(3t^2+1) - (t^3+t-2)(5)(2t-1)^4(2)}{(2t-1)^{10}}$$
$$= \frac{(2t-1)^4\left[(2t-1)(3t^2+1) - 10(t^3+t-2)\right]}{(2t-1)^{10}}$$
$$= \frac{(2t-1)(3t^2+1) - 10(t^3+t-2)}{(2t-1)^6}$$

63. $y = -6e^{2x}$

$y' = -6(2)e^{2x} = -12e^{2x}$

64. $y = 8e^{.5x}$

$y' = 8(.5e^{.5x}) = 4e^{.5x}$

65. $y = e^{-2x^3}$

$y' = -6x^2e^{-2x^3}$

66. $y = -5e^{x^2}$

$y' = -5(2xe^{x^2}) = -10xe^{x^2}$

67. $y = 5x \cdot e^{2x}$

$y' = 5x(2)e^{2x} + e^{2x}(5) = 10xe^{2x} + 5e^{2x}$

68. $y = -7x^2 \cdot e^{-3x}$

$$y' = -7x^2(-3e^{-3x}) + e^{-3x}(-14x)$$
$$= 21x^2e^{-3x} - 14xe^{-3x}$$

69. $y = \ln(x^2 + 4x - 1)$

$$y' = \frac{D_x(x^2+4x+1)}{x^2+4x-1} = \frac{2x+4}{x^2+4x-1}$$

Copyright © 2019 Pearson Education, Inc.

70. $y = \ln(4x^3 + 2x)$

$$y' = \frac{12x^2 + 2}{4x^3 + 2x} = \frac{2(6x^2 + 1)}{2(2x^3 + x)} = \frac{6x^2 + 1}{2x^3 + x}$$

71. $y = \dfrac{\ln 6x}{x^2 - 1}$

$$y' = \frac{(x^2 - 1)\left(\frac{6}{6x}\right) - (\ln 6x)(2x)}{(x^2 - 1)^2}$$

$$= \frac{x - \frac{1}{x} - 2x(\ln 6x)}{(x^2 - 1)^2}$$

72. $y = \dfrac{\ln(3x + 5)}{x^2 + 5x}$

$$y' = \frac{(x^2 + 5x)\left(\frac{3}{3x+5}\right) - [\ln(3x + 5)](2x + 5)}{(x^2 + 5x)^2}$$

$$= \frac{3}{(x^2 + 5x)(3x + 5)} - \frac{(2x + 5)\ln(3x + 5)}{(x^2 + 5x)^2}$$

73. The graph is continuous throughout, because the function is defined for every x and the limit of the function is equal to the value of the function for every x. Thus, there is no point of discontinuity for this function.

74. The graph is discontinuous at $x = -4$ and $x = 2$, since the limit of the function fails to exist at each point.

75. $f(x) = \dfrac{x - 6}{x + 5}; x = 6, x = -5, x = 0$

Since $f(x)$ is a rational function, it will not be continuous at any x for which $x + 5 = 0$. Then, if $x = -5$, $f(x)$ will not be continuous, but it is continuous at all other points, including $x = 6$ and $x = 0$.

76. $f(x) = \dfrac{x^2 - 9}{x + 3}; x = 3, x = -3, x = 0$

$f(-3)$ is not defined, but $f(x)$ will be continuous for all other values of x. Thus, $f(x)$ is continuous at $x = 3$ and $x = 0$ and discontinuous at $x = -3$.

77. $h(x) = \dfrac{2 - 3x}{2 - x - x^2};\ x = -2,\ x = \dfrac{2}{3},\ x = 1$

$h(-2) = \dfrac{2 - 3(-2)}{2 - (-2) - (-2)^2} = \dfrac{8}{0}$, so $h(x)$ is not continuous at $x = -2$.

$$h\left(\frac{2}{3}\right) = \frac{2 - 3\left(\frac{2}{3}\right)}{2 - \frac{2}{3} - \left(\frac{2}{3}\right)^2} = \frac{0}{\frac{8}{9}} = 0$$

$\lim\limits_{x \to 2/3} h(x) = 0$, so $h(x)$ is continuous at $x = \dfrac{2}{3}$.

$h(1) = \dfrac{2 - 3 \cdot 1}{2 - 1 - 1^2} = \dfrac{-1}{0}$, so $h(x)$ is not continuous at $x = 1$.

78. $f(x) = \dfrac{x^2 - 4}{x^2 - x - 6}; x = 2, x = 3, x = 4$

$f(2) = \dfrac{2^2 - 4}{2^2 - 2 - 6} = \dfrac{0}{-4} = 0,$

$\lim\limits_{x \to 2} f(x) = 0$, so $f(x)$ is continuous at $x = 2$.

$f(3) = \dfrac{3^2 - 4}{3^2 - 3 - 6} = \dfrac{5}{0}$, so $f(x)$ is not continuous at $x = 3$.

$f(4) = \dfrac{4^2 - 4}{4^2 - 4 - 6} = \dfrac{12}{6} = 2$

$\lim\limits_{x \to 4} f(x) = 2$, so $f(x)$ is continuous at $x = 4$.

79. **a.** $\lim\limits_{x \to 50} S(x) = \$5$, since the left and right limits exist and have the same value.

b. $\lim\limits_{x \to 100} S(x)$ does not exist because $\lim\limits_{x \to 100^-} S(x) = 10$ and $\lim\limits_{x \to 100^+} S(x) = 0$.

80. **a.** $\lim\limits_{x \to 2} T(x) = \$350$, since the left and right limits exist and have the same value.

b. $\lim\limits_{x \to 3} T(x)$ does not exist because $\lim\limits_{x \to 3^-} T(x) = 310$ and $\lim\limits_{x \to 3^+} T(x) = 0$.

c. $\lim\limits_{x \to 4} T(x) = \$0$, since the left and right limits exist and have the same value.

d. $\lim\limits_{x \to 5} T(x)$ does not exist because $\lim\limits_{x \to 5^-} T(x) = 0$ and $\lim\limits_{x \to 5^+} T(x) = 110$.

Copyright © 2019 Pearson Education, Inc.

81. **a.** $C(2)=5+65(2)=\$135.$

b. $\lim_{x\to 2^-} C(x)=5+65(2)=\$135.$

c. $\lim_{x\to 2^+} C(x)=5+65(3)=\$200.$

d. $\lim_{x\to 2} C(x)$ does not exist because the left and right limits are different values.

82. $$\lim_{x\to\infty}\bar{C}(x)=\lim_{x\to\infty}\frac{450+115x}{x}=\lim_{x\to\infty}\left(\frac{450}{x}+115\right)=0+115=115 \text{ per hundred}$$

The average cost for an infinite number of brochures is \$115 per hundred or \$1.15 per brochure.

83. The average rate of change is
$$\frac{73.55-56.64}{2016-2014}=\frac{16.91}{2}=\$8.46 \text{ per year.}$$

84. The average rate of change is
$$\frac{73.61-50.81}{2015-2012}=\frac{22.8}{3}\approx\$7.60 \text{ per year.}$$

85. **a.** The cost of producing 100 paddleboats is \$80,000.

b. When producing 100 paddleboats, the cost of producing one additional paddleboat is approximately \$300.

86. **a.** If you are standing in the lake 20 feet from the bank, the water will have a depth of 5 feet.

b. If you are standing in the lake 20 feet from the bank and you walk another foot into the lake, the depth of the water will increase by approximately .3 feet.

87. $$R(t)=-.1t^3+3.25t^2-3.5t+134$$
$$R'(t)=-.3t^2+6.5t-3.5$$

88. $$P(t)=-.024t^3+.7t^2-6.2t+26$$
$$P'(t)=-.072t^2+1.4t-6.2$$

89. **a.** $$C(x)=\frac{250x+45}{2x+40}$$
$$C'(x)=\frac{(2x+40)(250)-(250x+45)(2)}{(2x+40)^2}=\frac{500x+10,000-500x-90}{(2x+40)^2}=\frac{9910}{(2x+40)^2}$$

b. 1000 sandwiches is 10 hundreds, so $x=10$.
$$C'(10)=\frac{9910}{(2(10)+40)^2}\approx 2.7528 \text{ per sandwich}$$
To make an additional 100 sandwiches, the cost will be approximately \$275.28.

90. **a.** $$\bar{C}(x)=\frac{250x+45}{x(2x+40)}=\frac{250x+45}{2x^2+40x}$$

b. $$C'(x)=\frac{(2x^2+40x)(250)-(250x+45)(4x+40)}{(2x^2+40x)^2}=\frac{500x^2+10,000x-1000x^2-10,180x-1800}{(2x^2+40x)^2}=-\frac{500x^2+180x+1800}{(2x^2+40x)^2}$$

c. 1000 sandwiches is 10 hundreds, so $x=10$.
$$C'(10)=\frac{-500(10)^2-180(10)-1800}{(2(10)^2+40(10))^2}\approx -.15 \text{ per sandwich}$$
To make an additional 100 sandwiches, the average cost will be decreasing by approximately \$.15 per sandwich.

91. **a.** $$R(x)=x\sqrt{25-.1x}=x(25-.1x)^{1/2}$$

b. $$R'(x)=x\cdot\frac{1}{2}(25-.1x)^{-1/2}(-.1)+(25-.1x)^{1/2}(1)=\frac{-.05x}{\sqrt{25-.1x}}+\sqrt{25-.1x}=\frac{25-.15x}{\sqrt{25-.1x}}$$

c. $$R(90)=90\sqrt{25-.1(90)}=\$360$$

d. $$R'(90)=\frac{25-.15(90)}{\sqrt{25-.1(90)}}=\frac{11.5}{4}\approx\$2.88$$

Copyright © 2019 Pearson Education, Inc.

92. **a.** $P(x) = R(x) - C(x)$
$= x\sqrt{25 - .1x} - 120 - .75x$

b. Using the derivative from problem 93b:

$$P'(x) = \frac{25 - .15x}{\sqrt{25 - .1x}} - .75$$

c. $P'(120) = \dfrac{25 - .15(120)}{\sqrt{25 - .1(120)}} - .75 \approx \$1.19$

93. $T(x) = 15\ln(x + 100) + 7.7 \quad T'(x) = \dfrac{15}{x + 100}$

a. Positive because life expectancy has improved over time.

b. For 2018, $x = 18$
$T(x) = 15\ln(18 + 100) + 7.7 \approx 79.3$
The life expectancy for a person born in the year 2018 will be approximately 79.3 years.

c. $T'(18) = \dfrac{15}{18 + 100} \approx .1$
If another person is born one year later, their life expectancy will be approximately .1 year longer.

94. $R(x) = -37\ln(x) + 172 \quad R'(x) = -\dfrac{37}{x}$

a. Negative because remaining life expectancy decreases as a person gets older.

b. For a person 69 years old, $x = 69$
$R(x) = -37\ln(69) + 172 \approx 15.3$ The remaining life expectancy for a 69 year old person approximately 15.3 years.

c. $R'(69) = -\dfrac{37}{69} \approx -.5$
Life expectancy would be approximately .5 years shorter for a 70 year-old person.

95. **a.** The function is discontinuous at the value $x =$ \$100.

b. The function is not differentiable at the values $x = \$50$ and $x = \$100$.

96. **a.** $C(400) = \$50$

b. $\lim_{x \to 400} C(x) = \$50$ because the left and right limits have the same value.

c. The function is continuous at $x = 400$ because all the conditions for continuity are met.

d. $C(460) = 50 + 10(.5) = \$55.$

e. $\lim_{x \to 460} C(x)$ does not exist because the left and right limits approach different values.

f. The function is not continuous at $x = 460$ because it fails the existence of a limit portion of continuity a shown in part e of the problem.

Case 11 Price Elasticity of Demand

1. $q = -3.003p + 675.23$

$$\frac{dq}{dp} = -3.003$$

$$E = -\frac{p}{q} \cdot \frac{dq}{dp} = \frac{3.003p}{-3.003p + 675.23}$$

For $p = 70$,

$$E = \frac{3.003(70)}{-3.003(70) + 675.23} \approx .452$$

2. $q = -1000p + 70{,}000$

$$\frac{dq}{dp} = -1000$$

$$E = -\frac{p}{q} \cdot \frac{dq}{dp} = \frac{1000p}{-1000p + 70{,}000}$$

a. For $p = 30$,

$$E = \frac{1000(30)}{-1000(30) + 70{,}000} = .75$$

b. For $p = 40$,

$$E = \frac{1000(40)}{-1000(40) + 70{,}000} \approx 1.33$$

Since the demand is elastic at $p = \$40$, a smaller price increase would be better. Therefore, no, raising the price is not a good idea.

3. $q = -2481.52p + 472{,}191.2$

$$\frac{dq}{dp} = -2481.52$$

$$E = -\frac{p}{q} \cdot \frac{dq}{dp} = \frac{2481.52p}{-2481.52p + 472{,}191.2}$$

Copyright © 2019 Pearson Education, Inc.

a. For $p = 100$,

$$E = \frac{2481.52(100)}{-2481.52(100) + 472{,}191.2} \approx 1.1$$

For $p = 75$,

$$E = \frac{2481.52(75)}{-2481.52(75) + 472{,}191.2} \approx .65$$

b. Let $1 = \frac{2481.52p}{-2481.52p + 472{,}191.2}$

$$-2481.52p + 472{,}191.2 = 2481.52p$$
$$472{,}191.2 = 4963.04p$$
$$p \approx 95.14$$

The demand has unit elasticity at $p \approx \$95.14$.

4. $q = 100 - 4p^2$

$$\frac{dq}{dp} = -8p$$

$$E = -\frac{p}{q} \cdot \frac{dq}{dp} = \frac{8p^2}{100 - 4p^2}$$

For $p = 4.00$,

$$E = \frac{8(4)^2}{100 - 4(4)^2} \approx 3.6$$

Choice (c).

5. $E = 0$ means that $-\frac{p}{q} \cdot \frac{dq}{dp} = 0.$

Because the price $p \neq 0$, $\frac{dq}{dp} = 0.$ Changes in price produce no change at all in demand, i.e., the demand is constant.

Copyright © 2019 Pearson Education, Inc.

Chapter 12 Applications of the Derivative

Section 12.1 Derivatives and Graphs

1. The function is increasing on $(1, \infty)$, and decreasing on $(-\infty, 1)$. The lowest point on the graph has coordinates $(1, -4)$, so the local minimum of -4 occurs at $x = 1$.

3. The function is increasing on $(-\infty, -2)$ and decreasing on $(-2, \infty)$. The highest point on the graph has coordinates $(-2, 3)$, so a local maximum of 3 occurs at $x = -2$.

5. The function is increasing on $(-\infty, -4)$ and $(-2, \infty)$, and decreasing on $(-4, -2)$. A high point occurs at $(-4, 3)$ and a low point occurs at $(-2, 1)$, so a local maximum of 3 occurs at $x = -4$ and a local minimum of 1 occurs at $x = -2$.

7. The function is increasing on $(-7, -4)$ and $(-2, \infty)$ and decreasing on $(-\infty, -7)$ and $(-4, -2)$. Low points occur at $(-7, -2)$ and $(-2, -2)$, and a high point occurs at $(-4, 3)$. Thus a local minimum of -2 occurs at $x = -7$ and $x = -2$, and a local maximum of 3 occurs at $x = -4$.

9. $f(x) = \frac{1}{3}x^3 - 25x + 7$
$f'(x) = x^2 - 25$
$x^2 - 25 = 0$
$(x-5)(x+5) = 0$
$x = -5$ or $x = 5$
Test $f'(x)$ at $x = -6, x = 0, x = 6$.
$f'(-6) = (-6)^2 - 25 = 11 > 0$
$f'(0) = (0)^2 - 25 = -25 < 0$
$f'(6) = (6)^2 - 25 = 11 > 0$
$f'(x)$ is positive on $(-\infty, -5)$, so $f(x)$ is increasing.
$f'(x)$ is negative on $(-5, 5)$, so $f(x)$ is decreasing.
$f'(x)$ is positive on $(5, \infty)$, so $f(x)$ is increasing.

11. $f(x) = 2x^3 + 3x^2 - 12x + 5$
$f'(x) = 6x^2 + 6x - 12$
$6x^2 + 6x - 12 = 0$
$6(x^2 + x - 2) = 0$
$6(x+2)(x-1) = 0$
$x = -2$ or $x = 1$
Test $f'(x)$ at $x = -3, x = 0, x = 3$
$f'(-3) = 6(-3)^2 + 6(-3) - 12 = 24 > 0$
$f'(0) = 6(0)^2 + 6(0) - 12 = -12 < 0$
$f'(3) = 6(3)^2 + 6(3) - 12 = 60 > 0$
$f'(x)$ is positive on $(-\infty, -2)$ and $(1, \infty)$, so $f(x)$ is increasing.
$f'(x)$ is negative on $(-2, 1)$, so $f(x)$ is decreasing.

13. $f(x) = 2x^3 - 5x^2 - 4x + 2$
$f'(x) = 6x^2 - 10x - 4$
$6x^2 - 10x - 4 = 0$
$2(3x^2 - 5x - 2) = 0$
$2(3x+1)(x-2) = 0$
$x = -\frac{1}{3}$ or $x = 2$
Test $f'(x)$ at $x = -2, x = 0, x = 3$.
$f'(-2) = 6(-2)^2 - 10(-2) - 4 = 40 > 0$
$f'(0) = 6(0)^2 - 10(0) - 4 = -4 < 0$
$f'(3) = 6(3)^2 - 10(3) - 4 = 20 > 0$
$f'(x)$ is positive on $\left(-\infty, -\frac{1}{3}\right)$, so $f(x)$ is increasing.
$f'(x)$ is negative on $\left(-\frac{1}{3}, 2\right)$, so $f(x)$ is decreasing.
$f'(x)$ is positive on $(2, \infty)$, so $f(x)$ is increasing.

Copyright © 2019 Pearson Education, Inc.

15. $f(x)=\dfrac{x+1}{x+3}$

$f'(x)=\dfrac{(x+3)(1)-(x+1)(1)}{(x+3)^2}=\dfrac{2}{(x+3)^2}$

so $f'(x)=\dfrac{2}{(x+3)^2}=0$ has no solution.

$f'(x)$ does not exist if $(x+3)^2=0\Rightarrow x=-3$

Test $f'(x)$ at $x=-4$ and $x=0$.

$f'(-4)=\dfrac{2}{(-4+3)^2}=2$

$f'(0)=\dfrac{2}{(0+3)^2}=\dfrac{2}{9}$

$f'(x)$ is positive on $(-\infty,-3)$ and $(-3,\infty)$, so $f(x)$ is increasing on those intervals.

17. $f(x)=\sqrt{x^2+9}=\left(x^2+9\right)^{1/2}$

$f'(x)=\dfrac{1}{2}\left(x^2+9\right)^{-1/2}(2x)=\dfrac{x}{\sqrt{x^2+9}}$

$\dfrac{x}{\sqrt{x^2+9}}=0$ when $x=0$.

Test $f'(x)$ at $x=3$, $x=-3$.

$f'(3)=\dfrac{3}{\sqrt{3^2+9}}=\dfrac{3}{\sqrt{18}}>0$

$f'(-3)=\dfrac{-3}{\sqrt{(-3)^2+9}}=\dfrac{-3}{\sqrt{18}}<0$

$f'(x)$ is positive on $(0,\infty)$, so $f(x)$ is increasing.

$f'(x)$ is negative on $(-\infty,0)$, so $f(x)$ is decreasing.

19. $f(x)=x^3+3x^2-3$

$f'(x)=3x^2+6x=3x(x+2)$

Find the critical numbers by solving the equation $f'(x)=0$

$3x(x+2)=0$

$3x(x+2)=0\Rightarrow x=0,\ -2$

Use the first derivative test by testing $f'(x)$ at $x=-3$, $x=1$, and $x=3$.

$f'(-3)=3(-3)^2+6(-3)=9>0$

$f'(-1)=3(-1)^2+6(-1)=-3<0$

$f'(3)=3(3)^2+6(3)=45>0$

$f'(x)$ is positive on $(-\infty,-2)$ and negative on $(-2,0)$, so a local maximum occurs at $x=-2$.

$f'(x)$ is negative on $(-2,0)$ and positive on $(0,\infty)$, so a local minimum occurs at $x=0$.

Values of extrema:

$f(0)=0^3+3(0)^2-3=-3$

$f(-2)=(-2)^3+3(-2)^2-3=1$

A local maximum of 1 occurs at $x=-2$ and a local minimum of -3 occurs at $x=0$.

21. $f(x)=x^3+6x^2+9x+2$

$f'(x)=3x^2+12x+9$

Find critical numbers by solving $f'(x)=0$.

$f'(x)=3x^2+12x+9=3\left(x^2+4x+3\right)$
$=3(x+3)(x+1)=0\Rightarrow x=-3,\ -1$

Test $f'(x)$ at $x=-4$, $x=-2$, and $x=0$.

$f'(-4)=3(-4)^2+12(-4)+9=9>0$

$f'(-2)=3(-2)^2+12(-2)+9=-3<0$

$f'(0)=3(0)^2+12(0)+9=9>0$

$f'(x)$ is positive on $(-\infty,-3)$ and negative on $(-3,-1)$, so a local maximum occurs at $x=-3$.

$f'(x)$ is negative on $(-3,-1)$ and positive on $(-1,\infty)$, so a local minimum occurs at $x=-1$.

$f(-3)=(-3)^3+6(-3)^2+9(-3)+2=2$

$f(-1)=(-1)^3+6(-1)^2+9(-1)+2=-2$

Thus, f has a local maximum of 2 at $x=-3$ and a local minimum of -2 at $x=-1$.

23. $f(x)=\dfrac{4}{3}x^3-\dfrac{21}{2}x^2+5x+3$

$f'(x)=4x^2-21x+5$

Find critical numbers by solving $f'(x)=0$.

$4x^2-21x+5=0\Rightarrow(4x-1)(x-5)=0\Rightarrow$

$x=\dfrac{1}{4},\ 5$

Test $f'(x)$ at 0, 1, and 6.

$f'(0)=5>0$

$f'(1)=-12<0$

$f'(6)=23>0$

Copyright © 2019 Pearson Education, Inc.

$f'(x)$ is positive on $(-\infty, \frac{1}{4})$ and positive on $(5, \infty)$. $f'(x)$ is negative on $\left(\frac{1}{4}, 5\right)$.

By the first derivative test, $f(x)$ has a local maximum at $x = \frac{1}{4}$, and a local minimum at $x = 5$.

$f(5) = -\frac{407}{6}$; $f\left(-\frac{1}{4}\right) = \frac{347}{96}$

Thus, f has a local maximum of $\frac{347}{96}$ at $x = \frac{1}{4}$ and a local minimum of $-\frac{407}{6}$ at $x = 5$.

25. $f(x) = \frac{2}{3}x^3 - x^2 - 12x + 2$

$f'(x) = 2x^2 - 2x - 12$

Find critical numbers by solving $f'(x) = 0$.

$2(x^2 - x - 6) = 0 \Rightarrow 2(x+2)(x-3) = 0 \Rightarrow$
$x = -2, 3$

Test $f'(x)$ at $x = -3$, $x = 0$, and $x = 4$.

$f'(-3) = 2(-3)^2 - 2(-3) - 12 = 12 > 0$

$f'(0) = 2(0)^2 - 2(0) - 12 = -12 < 0$

$f'(4) = 2(4)^2 - 2(4) - 12 = 12 > 0$

$f'(x)$ is positive on $(-\infty, -2)$ and negative on $(-2, 3)$, so $f(x)$ has a local maximum at $x = -2$. $f'(x)$ is negative on $(-2, 3)$ and positive on $(3, \infty)$, so $f(x)$ has a local minimum at $x = 3$.

$f(-2) = \frac{2}{3}(-2)^3 - (-2)^2 - 12(-2) + 2 = \frac{50}{3}$

$f(3) = \frac{2}{3}(3)^3 - 3^2 - 12(3) + 2 = -25$

Thus, f has a local maximum of $\frac{50}{3}$ at $x = -2$ and a local minimum of -25 at $x = 3$.

27. $f(x) = \frac{3}{2}x^4 + 4x^3 - 24x^2 + 20$

$f'(x) = 6x^3 + 12x^2 - 48x$

Find critical numbers by solving $f'(x) = 0$.

$6x^3 + 12x^2 - 48x = 0 \Rightarrow 6x(x^2 + 2x - 8) = 0 \Rightarrow$
$6x(x+4)(x-2) = 0 \Rightarrow x = 0, -4, 2$

Test $f'(x)$ at $x = -5$, $x = -1$, $x = 1$, and $x = 3$.

$f'(-5) = 6(-5)^3 + 12(-5)^2 - 48(-5) = -210 < 0$

$f'(-1) = 6(-1)^3 + 12(-1)^2 - 48(-1) = 54 > 0$

$f'(1) = 6(1)^3 + 12(1)^2 - 48(1) = -30 < 0$

$f'(3) = 6(3)^3 + 12(3)^2 - 48(3) = 126 > 0$

$f'(x)$ is negative on $(-\infty, -4)$ and positive on $(-4, 0)$, so $f(x)$ has a local minimum at $x = -2$. $f'(x)$ is positive on $(-4, 0)$ and negative on $(0, 2)$, so $f(x)$ has a local maximum at $x = 0$. $f'(x)$ is negative on $(0, 2)$ and positive on $(2, \infty)$, so $f(x)$ has a local minimum at $x = 2$. Thus, f has a local maximum at $x = 0$ and local minima at $x = -4, 2$.

29. $f(x) = 5 - x^{2/3}$

$f'(x) = -\frac{2}{3}x^{-1/3}$

Find critical numbers by solving $f'(x) = 0$.

$-\frac{2}{3}x^{-1/3} = 0 \Rightarrow x = 0$

Test $f'(x)$ at $x = -1$ and $x = 1$.

$f'(-1) = -\frac{2}{3}(-1)^{-1/3} = \frac{2}{3} > 0$

$f'(1) = -\frac{2}{3}(1)^{-1/3} = -\frac{2}{3} < 0$

$f'(x)$ is positive on $(-\infty, 0)$ and negative on $(0, \infty)$, so a local maximum occurs at $x = 0$.

$f(0) = 5 - 0^{2/3} = 5$

A local maximum of 5 occurs at $x = 0$, and there is no local minimum.

Copyright © 2019 Pearson Education, Inc.

31. $f(x)=\dfrac{x}{x^2+49}$

$f'(x)=\dfrac{(x^2+49)(1)-(x)(2x)}{(x^2+49)^2}$

$=\dfrac{-x^2+49}{(x^2+49)^2}$

Find critical numbers by solving $f'(x)=0$.

$\dfrac{-x^2+49}{(x^2+49)^2}=0\Rightarrow -x^2+49=0\Rightarrow$
$-(x-7)(x+7)=0\Rightarrow x=-7, 7$

Test $f'(x)$ at $x=-8$, $x=0$, and $x=8$.

$f'(-8)=-\dfrac{15}{113}<0$

$f'(0)=\dfrac{1}{49}>0$

$f'(8)=-\dfrac{15}{113}<$

$f'(x)$ is negative on $(-\infty,-7)$ and positive on $(-7, 7)$, so a local minimum occurs at $x=-7$. $f'(x)$ is positive on $(-7, 7)$ and negative on $(7,\infty)$, so a local maximum occurs at $x=7$.

$f(-7)=-\dfrac{1}{14}$; $f(7)=\dfrac{1}{14}$

A local minimum of $-\dfrac{1}{14}$ occurs at $x=-7$, and a local maximum of $\dfrac{1}{14}$ occurs at $x=7$.

33. $f(x)=\dfrac{x^3}{x^3+1}$

$f'(x)=\dfrac{\left(x^3+1\right)3x^2-x^3(3x^2)}{\left(x^3+1\right)^2}=\dfrac{3x^2}{\left(x^3+1\right)^2}$

$f'(x)=0$ when $x=0$.

Note that both $f(x)$ and $f'(x)$ do not exist at $x=-1$, so -1 is not a critical number.

Test $f'(x)$ at $x=-2$ and $x=1$.

$f'(-2)=\dfrac{12}{49}>0$; $f'(1)=\dfrac{3}{4}>0$

$f'(x)$ is positive on $(-\infty,0)$ and $(0,\infty)$.

f has no local extrema.

35. $f(x)=xe^{-x}$

$f'(x)=(1)e^{-x}+x\left(-e^{-x}\right)=e^{-x}-xe^{-x}$

$=e^{-x}(1-x)$

Find critical numbers by solving $f'(x)=0$.

$e^{-x}(1-x)=0$

$e^{-x}=0\Rightarrow$ No solution or

$1-x=0\Rightarrow x=1$

The critical number is $x=1$.

$f'(0)=e^{-0}(1-0)=1>0$

$f'(2)=e^{-2}(1-2)=-e^{-2}<0$

Values of extremum: $f(1)=1e^{-1}=\dfrac{1}{e}$

A local maximum of $\dfrac{1}{e}$ occurs at $x=1$.

37. $f(x)=x\cdot\ln|x|$

$f'(x)=(1)\ln|x|+x\left(\dfrac{1}{x}\right)=\ln|x|+1$

Find critical numbers by solving $f'(x)=0$.

$\ln|x|+1=0\Rightarrow \ln|x|=-1$

If $x>0$, $x=e^{-1}=\dfrac{1}{e}\approx .3679$

Test $f'(x)$ at $x=.2$ and $x=1$.

$f'(.2)=\ln .2+1\approx -.6094<0$

$f'(1)=\ln 1+1=1>0$

$f'(x)$ is negative on $\left(-\infty,\dfrac{1}{e}\right)$ and positive on $\left(\dfrac{1}{e},\infty\right)$.

$f\left(\dfrac{1}{e}\right)=\dfrac{1}{e}\ln\left(\dfrac{1}{e}\right)=\dfrac{1}{e}(-1)=-\dfrac{1}{e}$

f has a local minimum of $-\dfrac{1}{e}$ at $x=\dfrac{1}{e}$. By symmetry and absolute value, there is a local maximum of $\dfrac{1}{e}$ at $x=-\dfrac{1}{e}$.

Copyright © 2019 Pearson Education, Inc.

39. $f(x) = xe^{3x} - 2$

$f'(x) = (1)e^{3x} + (3)xe^{3x} = e^{3x}(1+3x)$

Find critical numbers by solving $f'(x) = 0$.

$e^{3x}(1+3x) = 0$

$e^{3x} = 0 \Rightarrow$ No solution or

$1 + 3x = 0 \Rightarrow x = -\frac{1}{3}$

The critical number is $x = -\frac{1}{3}$.

$f'(-1) = e^{-3}(1-3) = -2e^{-3} < 0$

$f'(0) = e^{3(0)}(1+3(0)) = 1 > 0$

Values of extremum:

$f\left(-\frac{1}{3}\right) = \frac{-1}{3}e^{3(-1/3)} - 2 = \frac{-1}{3e} - 2 \approx -2.1226$

A local minimum of -2.1226 occurs at $x = -\frac{1}{3}$.

Solutions 41–44 presuppose the use of a TI-83 graphics calculator. Similar results can be obtained from other graphics calculators.

41. Graph the function

$f(x) = .2x^4 - x^3 - 12x^2 + 99x - 5$

by entering it as y_1.

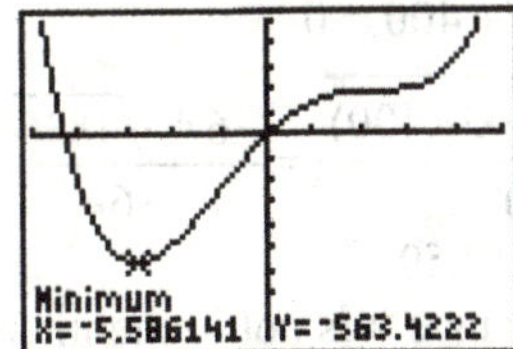

[−10, 10] by [−900, 500]

Under the CALC menu use, “minimum” and “maximum” to find a local minimum of –563.4222 at $x = -5.5861$.

43. Graph the function

$f(x) = .01x^5 + .2x^4 - x^3 - 6x^2 + 5x + 40$

by entering it as y_1.

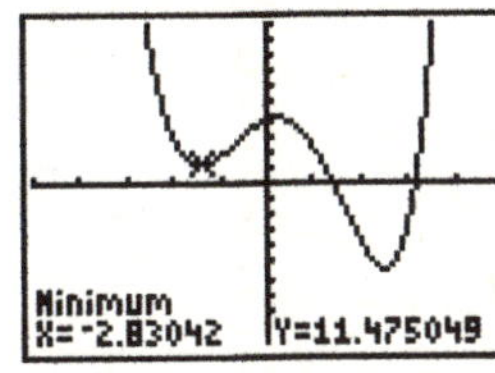

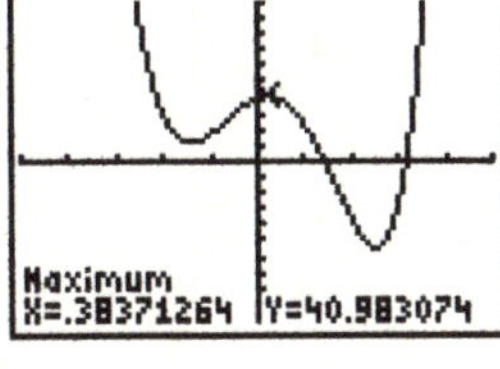

[−10, 10] by [−100, 100]

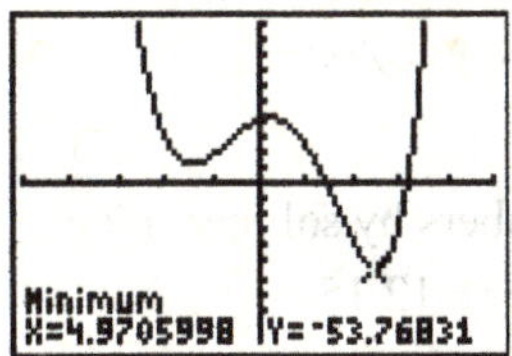

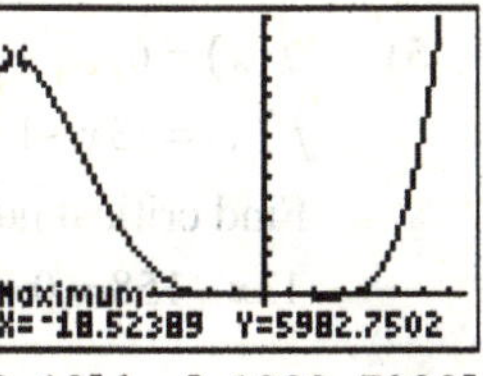

[−20, 10] by [−1000, 7000]

Under the CALC menu, use “minimum” and “maximum” to find local maxima of 5982.7502 at $x = -18.5239$ and 40.9831 at $x = .3837$, and local minima of 11.4750 at $x = -2.8304$, and -53.7683 at $x = 4.9706$. Note that two different viewing windows were used.

45. a. The power function is increasing on the interval (2500, 5000).

b. The power function is decreasing on the interval (5000, 6000).

c. The torque function is increasing on the interval (3500, 5000).

d. The torque function is decreasing on the intervals (2500, 3500) and (5000, 6000).

47. a. The average number of calls is increasing on (5, 16) and (18, 19).

b. The average number of calls is decreasing on (0, 5), (16, 18), and (19, 23).

c. Local maxima occur at $t = 16$ and $t = 19$, which correspond to 4:00 pm and 7:00 pm. Local minima occur at $t = 5$ and $t = 18$, which correspond to 5:00 am and 6:00 pm.

49. a. $f(x) = -29x^2 + 812x - 4559$

$f'(x) = -58x + 812$

Find critical numbers by solving $f'(x) = 0$.

$-58x + 812 = 0 \Rightarrow x = 14$, therefore 2014.

b. Test $f'(x)$ at $x = 0$ and $x = 15$.

$f'(0) = -58(0) + 812 = 812 > 0$

$f'(15) = -58(15) + 812 = -58 < 0$

$f'(x)$ is positive on (0, 14) and negative on $(14, \infty)$, so $f(x)$ has a local maximum at $x = 14$. The value at this point is

$f(14) = -29(14)^2 + 812(14) - 4559$
$= 1125$ million

Thus, there is a maximum profit of \$1,125 million which occurred in the year 2014.

Copyright © 2019 Pearson Education, Inc.

51. $f(x) = 6.5x^2 - 158x + 975$
$f'(x) = 13x - 158$
Find critical numbers by solving $f'(x) = 0$.
$13x - 158 = 0 \Rightarrow x \approx 12.15$

Thus, there was a local minimum in the year 2012.

53. a. $h(t) = -16t^2 + 32t + 48,$
$h'(t) = -32t + 32$
$-32t + 32 = 0 \Rightarrow t = 1$ second

b. Maximum height is at $t = 1$
$h(1) = -16(1)^2 + 32(1) + 48$
$= 64$ ft

c. The ball hits the ground when
$h(t) = -16t^2 + 32t + 48 = 0$
$$t = \frac{-32 \pm \sqrt{(32)^2 - 4(-16)(48)}}{2(-16)}$$
$$= \frac{-32 \pm \sqrt{4096}}{-32} \Rightarrow t = -1, 3$$
Since $t \geq 0$, $t = 3$ seconds.

d. The velocity of the ball when it hits the ground is $h'(3) = -32(3) + 32 = -64$ ft/sec.

55. $B(x) = x^3 - 7x^2 - 160x + 1800$, $0 \leq x \leq 20$

a. $B'(x) = 3x^2 - 14x - 160$
$3x^2 - 14x - 160 = 0 \Rightarrow$
$(3x + 16)(x - 10) = 0 \Rightarrow x = -\frac{16}{3}, 10$

$-\frac{16}{3}$ is not in the given interval.
$B'(9) < 0$ and $B'(11) > 0$, so $B(10)$ is a minimum. The oxygen concentration of 10 will lead to a minimum microbe concentration.

b. $B(10) = 10^3 - 7(10)^2 - 160(10) + 1800 = 500$
The minimum microbe concentration is 500.

57. $C(x) = \frac{7}{3}x^3 - 57x^2 + 360x + 377$
$C'(x) = 7x^2 - 114x + 360$
Find critical numbers by solving $f'(x) = 0$.
$C'(x) = 7x^2 - 114x + 360$
$= (7x - 30)(x - 12) = 0 \Rightarrow x = \frac{30}{7}, 12$

Test $f'(x)$ at $x = 0$, $x = 5$, and $x = 13$.
$C'(0) = 7(0)^2 - 114(0) + 360 = 360 > 0$
$C'(5) = 7(5)^2 - 114(5) + 360 = -35 < 0$
$C'(13) = 7(13)^2 - 114(13) + 360 = 61 > 0$
$C(12) = \frac{7}{3}(12)^3 - 57(12)^2 + 360(12) + 377 = 521$
$C(\frac{30}{7}) = \frac{7}{3}(\frac{30}{7})^3 - 57(\frac{30}{7})^2 + 360(\frac{30}{7}) + 377$
≈ 1056.6

a. Based on the above information, the local minimum occurs at (12, 521).

b. Based on the above information, the local maximum occurs at about (4.3, 1056.6).

59. $S(x) = -x^3 + 3x^2 + 400x + 5000$, $6 \leq x \leq 20$
$S'(x) = -3x^2 + 6x + 400$
$S'(x) = -3x^2 + 6x + 400 = 0$
$$x = \frac{-6 \pm \sqrt{6^2 - 4(-3)(400)}}{2(-3)} = \frac{-6 \pm \sqrt{4836}}{-6} \Rightarrow$$
$x = -10.59$ or $x = 12.59$
Reject $x = -10.59$ since it is not in the interval being considered.
$S'(10) = 160 > 0$
$S'(13) = -29 < 0$
So $S(12.59)$ is a maximum. Thus, 12.6° is the temperature that produces the maximum number of salmon.

Copyright © 2019 Pearson Education, Inc.

61. $A(x) = .004x^3 - .05x^2 + .16x + .05,\ 0 \le x \le 6$

$A'(x) = .012x^2 - .1x + .16$

Set the derivative equal to zero and solve for x.

$.012x^2 - .1x + .16 = 0$

$x \approx 2.16$ or $x \approx 6.17$

since $0 \le x \le 6$, $x \approx 2.16$ is the only critical number. Test $A'(x)$ for $x = 0$, and $x = 3$.

$A'(0) = .012(0)^2 - .1(0) + .16 = .16 > 0$

$A'(3) = .012(3)^2 - .1(3) + .16 = -.032 < 0$

a. $A'(x)$ is positive on [0, 2.16), so $A(x)$, the alcohol concentration, is increasing.

b. $A'(x)$ is negative on (2.16, 6], so $A(x)$ is decreasing.

63. $f(x) = \frac{1}{\sqrt{2\pi}} e^{-x^2/2}$

$f'(x) = \frac{-x}{\sqrt{2\pi}} e^{-x^2/2}$

$\frac{-x}{\sqrt{2\pi}} e^{-x^2/2} = 0$ when $x = 0$.

Test $f'(x)$ at $x = -1, x = 1$.

$f'(-1) = \frac{-(-1)}{\sqrt{2\pi}} e^{-(-1)^2/2} = \frac{1}{\sqrt{2\pi}} e^{-1/2} > 0$

$f'(1) = \frac{-1}{\sqrt{2\pi}} e^{-1^2/2} = \frac{-1}{\sqrt{2\pi}} e^{-1/2} < 0$

$f'(x)$ is positive on $(-\infty, 0)$, so $f(x)$ is increasing. $f'(x)$ is negative on $(0, \infty)$, so $f(x)$ is decreasing.

65. $R(t) = \frac{20t}{t^2 + 100}$,

$R'(t) = \frac{(t^2+100)(20) - 20t(2t)}{(t^2+100)^2}$

$= \frac{20t^2 - 2000 - 40t^2}{(t^2+100)^2} = \frac{-20(t^2-100)}{(t^2+100)^2}$

$R'(t) = 0$ when $t = 10$. This is the only critical number. $R'(9) \approx .0116 > 0$ and $R'(11) \approx -.0086 < 0$. Hence a local maximum occurs at $t = 10$. The film length that received the highest rating is 10 minutes.

67. $K(x) = \frac{4x}{3x^2 + 27},\ x \ge 0$

$K'(x) = \frac{4(3x^2+27) - 4x(6x)}{(3x^2+27)^2} = \frac{-12x^2 + 108}{(3x^2+27)^2}$

$\frac{-12x^2 + 108}{(3x^2+27)^2} = 0 \Rightarrow -12x^2 + 108 = 0 \Rightarrow$

$x^2 = 9 \Rightarrow x = -3, 3$

Since $x \ge 0$, $x = 3$ is the critical point. Test $K'(x)$ at $x = 1, x = 4$.

$K'(1) = \frac{-12(1)^2 + 108}{(3(1)^2 + 27)^2} = \frac{96}{900} > 0$

$K'(4) = \frac{-12(4)^2 + 108}{(3(4)^2 + 27)^2} = \frac{-84}{5625} < 0$

a. $K'(x)$ is positive on [0, 3), so the drug concentration, $K(x)$, is increasing.

b. $K'(x)$ is negative on $(3, \infty)$, so $K(x)$ is decreasing.

69. $R(x) = .0245x^4 - .8x^3 + 8x^2 - 22x + 45$

Use a graphing utility to locate the value of x that give the local minima.

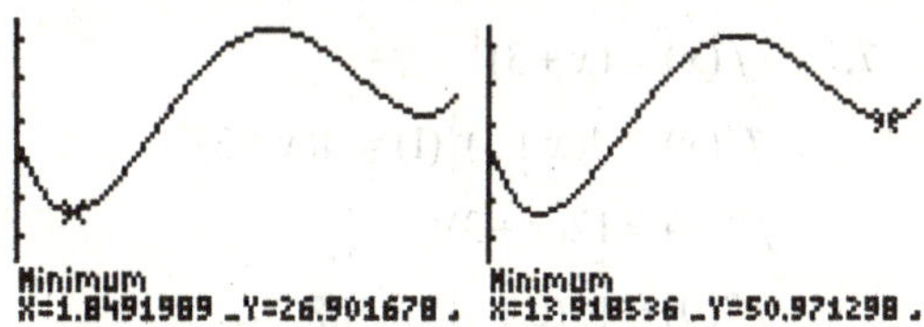

[0, 15] by [0, 100]

Therefore, the local minima occur at about (1.8, 26.9) and (13.9, 51.0).

Copyright © 2019 Pearson Education, Inc.

Section 12.2 The Second Derivative

1. $f(x)=x^3-6x^2+1$

$f'(x)=3x^2-12x$

$f''(x)$ is the derivative of $f'(x)$.

$f''(x)=6x-12$

$f''(0)=6(0)-12=-12$

$f''(2)=6(2)-12=0$

$f''(-3)=6(-3)-12=-30$

3. $f(x)=2e^x;\ f'(x)=2e^x;\ f''(x)=2e^x$

$f''(0)=2e^0=2(1)=2$

$f''(2)=2e^2$

$f''(-3)=2e^{-3}$ or $\dfrac{2}{e^3}$

5. $f(x)=6e^{2x}$

$f'(x)=6e^{2x}(2)=12e^{2x}$

$f''(x)=12e^{2x}(2)=24e^{2x}$

$f''(0)=24e^0=24$

$f''(2)=24e^4$

$f''(-3)=24e^{-6}=\dfrac{24}{e^6}$

7. $f(x)=(x+3)^4$

$f'(x)=4(x+3)^3(1)=4(x+3)^3$

$f''(x)=12(x+3)^2$

$f''(0)=12(0+3)^2=108$

$f''(2)=12(2+3)^2=300$

$f''(-3)=12(-3+3)^2=0$

9. $f(x)=\sqrt{4x+1}=(4x+1)^{1/2}$

$f'(x)=\dfrac{1}{2}(4x+1)^{-1/2}(4)=2(4x+1)^{-1/2}$

$f''(x)=\left(-\dfrac{1}{2}\right)2(4x+1)^{-3/2}(4)$

$=-4(4x+1)^{-3/2}$

or $-\dfrac{4}{(x+4)^{3/2}}$

$f''(0)=-\dfrac{4}{(4(0)+1)^{3/2}}=-\dfrac{4}{(1)^{3/2}}=-4$

$f''(2)=-\dfrac{4}{(4(2)+1)^{3/2}}=-\dfrac{4}{(9)^{3/2}}=-\dfrac{4}{27}$

$f''(-3)=-\dfrac{4}{(4(-3)+1)^{3/2}}=-\dfrac{4}{(-11)^{3/2}}=\text{und}$

11. $f(x)=-x^{2/3}$

$f'(x)=-\dfrac{2}{3}x^{-1/3}$

$f''(x)=\dfrac{2}{9}x^{-4/3}$ or $\dfrac{2}{9x^{4/3}}$

$f''(0)=\dfrac{2}{9(0)^{4/3}}$ not defined.

$f''(2)=\dfrac{2}{9(2)^{4/3}}=\dfrac{2}{9\cdot 2\cdot 2^{1/3}}=\dfrac{1}{9\cdot 2^{1/3}}$

$f''(-3)=\dfrac{2}{9(-3)^{4/3}}=\dfrac{2}{9(3)^{4/3}}$

13. $f(x)=\ln(2x+1)$

$f'(x)=\dfrac{2}{2x+1}=2(2x+1)^{-1}$

$f''(x)=-2(2x+1)^{-2}(2)=-\dfrac{4}{(2x+1)^2}$

$f''(0)=-\dfrac{4}{(2\cdot 0+1)^2}=-4$

$f''(2)=-\dfrac{4}{(2\cdot 2+1)^2}=-\dfrac{4}{25}$

$f''(-3)=$ Not defined, the value is not in the domain.

Copyright © 2019 Pearson Education, Inc.

15. $f(x) = xe^{-x}$

$f'(x) = x \cdot e^{-x}(-1) + e^{-x}(1) = e^{-x}(1-x)$

$f''(x) = e^{-x}(-1) + (1-x)e^{-x}(-1)$

$= -e^{-x} - e^{-x} + xe^{-x}$

$= e^{-x}(x-2)$

$f''(0) = e^{-0}(0-2) = -2$

$f''(2) = e^{-2}(2-2) = 0$

$f''(-3) = e^{3}(-3-2) = -5e^{3}$

17. $f(x) = \dfrac{x^2}{1+x}$

$f'(x) = \dfrac{(1+x)(2x) - x^2(1)}{(1+x)^2} = \dfrac{2x+x^2}{(1+x)^2}$

$f''(x) = \dfrac{(1+x)^2(2+2x) - \left(2x+x^2\right)(2)(1+x)}{(1+x)^4}$

$= \dfrac{(1+x)(2+2x) - \left(2x+x^2\right)(2)}{(1+x)^3}$

$= \dfrac{2}{(1+x)^3}$

$f''(0) = \dfrac{2}{(1+0)^3} = 2$

$f''(2) = \dfrac{2}{(1+2)^3} = \dfrac{2}{27}$

$f''(-3) = \dfrac{2}{(1-3)^3} = -\dfrac{1}{4}$

19. Answers vary; As the price $P(t)$ decreases, the slope of the graph is negative, so $P'(t)$ is negative. Since the price is decreasing faster and faster, the graph resembles the right-hand side of the graph in Figure 12.23(a). Here the slope is decreasing, so the second derivative $P''(t)$ is negative.

21. $s(t) = 6t^2 + 5t$

$v(t) = s'(t) = 12t + 5$

$a(t) = v'(t) = s''(t) = 12$

$v(0) = 12(0) + 5 = 5$ cm/sec

$v(4) = 12(4) + 5 = 53$ cm/sec

$a(0) = 12$ cm/sec^2

$a(4) = 12$ cm/sec^2

23. $s(t) = 3t^3 - 4t^2 + 8t - 9$

$v(t) = s'(t) = 9t^2 - 8t + 8$

$a(t) = v'(t) = s''(t) = 18t - 8$

$v(0) = 9(0)^2 - 8(0) + 8 = 8$ cm/sec

$v(4) = 9(4)^2 - 8(4) + 8 = 120$ cm/sec

$a(0) = 18(0) - 8 = -8$ cm/sec^2

$a(4) = 18(4) - 8 = 64$ cm/sec^2

25. $f(x) = x^3 + 3x - 5$

$f'(x) = 3x^2 + 3$

$f''(x) = 6x$

$f''(x) > 0$ if $x > 0$

$f''(x) < 0$ if $x < 0$

so $f''(x) > 0$ on $(0, \infty)$ and $f(x)$ is concave upward. $f''(x) < 0$ on $(-\infty, 0)$ and $f(x)$ is concave downward. $f''(x) = 0$ when $x = 0$ and so there is a point of inflection at $(0, -5)$.

27. $f(x) = x^3 + 4x^2 - 6x + 3$

$f'(x) = 3x^2 + 8x - 6$

$f''(x) = 6x + 8$

f is concave upward when

$f''(x) = 6x + 8 > 0 \Rightarrow x > -\dfrac{4}{3}$

f is concave downward when

$f''(x) = 6x + 8 < 0 \Rightarrow x < -\dfrac{4}{3}$

point of inflection at $x = -\dfrac{4}{3}$

$f\left(-\dfrac{4}{3}\right) = \left(-\dfrac{4}{3}\right)^3 + 4\left(-\dfrac{4}{3}\right)^2 - 6\left(-\dfrac{4}{3}\right) + 3$

$= \dfrac{425}{27}$

so f is concave upward on $\left(-\dfrac{4}{3}, \infty\right)$, concave downward on $\left(-\infty, -\dfrac{4}{3}\right)$, and has a point of inflection at $\left(-\dfrac{4}{3}, \dfrac{425}{27}\right)$.

Copyright © 2019 Pearson Education, Inc.

29. $f(x)=\frac{2}{x-4};\ f'(x)=\frac{-2}{(x-4)^2};$

$f''(x)=\frac{4}{(x-4)^3}$

$\frac{4}{(x-4)^3}>0$ when $(x-4)^3>0\Rightarrow x>4.$

$\frac{4}{(x-4)^3}<0$ when $(x-4)^3<0\Rightarrow x<4.$

$f(4)=\frac{2}{4-4}=\frac{2}{0}$ which does not exist.

So f is concave upward on $(4, \infty)$, concave downward on $(-\infty, 4)$, and has no points of inflection.

31. $f(x)=x^4+8x^3-30x^2+24x-3$

$f'(x)=4x^3+24x^2-60x+24$

$f''(x)=12x^2+48x-60$

$f''(x)=12x^2+48x-60=0\Rightarrow x=-5,1$

$f''(x)>0$ when $12x^2+48x-60>0\Rightarrow$

$12(x^2+4-5)>0\Rightarrow 12(x+5)(x-1)>0\Rightarrow$

$x+5>0$ and $x-1>0$, or $x+5<0$ and $x-1<0\Rightarrow x>-5$ and $x>1$ (which is $x>1$), or $x<-5$ and $x<1$ (which is $x<-5$)

$f''(x)<0$ when $12x^2+48x-60<0\Rightarrow$

$12(x^2+4-5)<0\Rightarrow 12(x+5)(x-1)<0\Rightarrow$

$x+5>0$ and $x-1<0$, or $x+5<0$ and $x-1>0\Rightarrow x>-5$ and $x<1$, or $x<-5$ and $x>1\Rightarrow -5<x<1.$

$f(-5)=(-5)^4+8(-5)^3-30(-5)^2+24(-5)-3$
$=-1248$

$f(1)=(1)^4+8(1)^3-30(1)^2+24(1)-3=0$

So f is concave upward on $(-\infty, -5)$ and $(1, \infty)$, concave downward on $(-5, 1)$, and has points of inflection at $(-5, -1248)$ and $(1, 0)$.

33. a. $R(x)=-x^3+30x^2+1000,\ 0\le x\le 15$

$R'(x)=-3x^2+60x$

$R''(x)=-6x+60$

b. $R''(x)=0$ when $-6x+60=0\Rightarrow x=10.$

$R(10)=-(10)^3+30(10)^2+1000$
$=3,000$

So the point of diminishing returns is $(10, 3000)$.

35. a. $R(x)=10,000-x^3+42x^2+800x,\ 0\le x\le 20$

$R'(x)=-3x^2+84x+800$

$R''(x)=-6x+84$

b. $R''(x)=0$ when $-6x+84=0\Rightarrow x=14.$

$R(14)=10,000-(14)^3+42(14)^2+800(14)$
$=26,688$

So the point of diminishing returns is $(14, 26,688)$.

37. $f(x)=-2x^3-3x^2+72x+1$

$f'(x)=-6x^2-6x+72$

$f''(x)=-12x-6$

$f'(x)=0\Rightarrow -6x^2-6x+72=0\Rightarrow$

$-6(x^2+x-12)=0\Rightarrow x=-4,3$

Use the second derivative test.

$f''(x)=-12x-6$

$f''(-4)=-12(-4)-6=42>0$

$f''(3)=-12(3)-6=-42<0$

$f(-4)=-2(-4)^3-3(-4)^2+72(-4)+1$
$=-207$

$f(3)=-2(3)^3-3(3)^2+72(3)+1$
$=136$

So $x=-4$ gives a local minimum with a function value of -207, and $x=3$ gives a local maximum with a function value of 136.

39. $f(x)=(x-3)^5$

$f'(x)=5(x-3)^4$

Solve $f'(x)=0$.

$5(x-3)^4=0\Rightarrow (x-3)^4=0\Rightarrow x=3$

Use the second derivative test.

$f''(x)=20(x-3)^3$

$f''(3)=20(3-3)^3=0$

For the critical number 3, the test gives no information about extrema. Use the first derivative test.

$f'(1)=5(1-3)^4=80>0$

$f'(5)=5(5-3)^4=80>0$

There are no local maxima or minima.

Copyright © 2019 Pearson Education, Inc.

41. $f(x) = x^4 - 8x^2$

$f'(x) = 4x^3 - 16x$

Solve for $f'(x) = 0$.

$4x^3 - 16x = 0 \Rightarrow 4x\left(x^2 - 4\right) = 0 \Rightarrow x = 0, \pm 2$

$x = 0$ or $x = 2$ or $x = -2$

Use the second derivative test.

$f''(x) = 12x^2 - 16$

$f''(0) = -16 < 0$

$f''(2) = 48 - 16 = 32 > 0$

$f''(-2) = 48 - 16 = 32 > 0$

Thus, $x = 0$ gives a local maximum, and $x = 2$ and $x = -2$ give local minima.

43. $f(x) = x + \dfrac{4}{x} = x + 4x^{-1}$

$f'(x) = 1 - 4x^{-2}$

Solve $f'(x) = 0$.

$1 - 4x^{-2} = 0 \Rightarrow 1 - \dfrac{4}{x^2} = 0 \Rightarrow 1 = \dfrac{4}{x^2} \Rightarrow$

$x^2 = 4 \Rightarrow x = \pm 2$

Use the second derivative test.

$f''(x) = 8x^{-3} = \dfrac{8}{x^3}$

$f''(2) = \dfrac{8}{(2)^3} = \dfrac{8}{8} = 1 > 0$

$f''(-2) = \dfrac{8}{(-2)^3} = \dfrac{8}{-8} = -1 < 0$

Thus, $x = 2$ gives a local minimum and $x = -2$ gives a local maximum.

45. $f(x) = \dfrac{x^2 + 9}{2x}$

$f'(x) = \dfrac{(2x)(2x) - (2)\left(x^2 + 9\right)}{(2x)^2} = \dfrac{4x^2 - 2x^2 - 18}{4x^2}$

$= \dfrac{2x^2 - 18}{4x^2} = \dfrac{x^2 - 9}{2x^2}$

Solve $f'(x) = 0$.

$\dfrac{x^2 - 9}{2x^2} = 0 \Rightarrow \dfrac{(x+3)(x-3)}{2x^2} = 0 \Rightarrow$

$(x+3)(x-3) = 0 \Rightarrow x = -3, 3$

Use the second derivative test.

$f''(x) = \dfrac{(4x)\left(4x^2\right) - (8x)\left(2x^2 - 18\right)}{\left(4x^2\right)^2}$

$= \dfrac{144x}{16x^4} = \dfrac{9}{x^3}$

$f''(3) = \dfrac{9}{(3)^3} = \dfrac{9}{27} = \dfrac{1}{3} > 0$

$f''(-3) = \dfrac{9}{(-3)^3} = -\dfrac{9}{27} = -\dfrac{1}{3} < 0$

Thus $x = 3$ gives a local minimum and $x = -3$ gives a local maximum.

47. $f(x) = \dfrac{2 - x}{2 + x}$

$f'(x) = \dfrac{(-1)(2 + x) - (1)(2 - x)}{(2 + x)^2}$

$= \dfrac{-2 - x - 2 + x}{(2 + x)^2} = \dfrac{-4}{(2 + x)^2}$

$f'(x)$ does not equal 0 for any value. $f'(-2)$ does not exist, but -2 is not a critical number because $f(-2)$ is undefined. Thus, there are no critical numbers and, consequently, there can be no local maxima or minima.

49. **a.** $f'(x) > 0$ and $f''(x) > 0$

These conditions indicate a point where the function is increasing and the graph is concave up. Point *E* satisfies these conditions.

b. $f'(x) < 0$ and $f''(x) > 0$

These conditions indicate a point where the function is decreasing and the graph concave up. Point *C* satisfies these conditions.

c. $f'(x) = 0$ and $f''(x) < 0$

These conditions indicate a critical point of the function where the graph is concave down. Points *A* and *F* satisfy these conditions.

d. $f'(x) = 0$ and $f''(x) > 0$

These conditions indicate a critical point of the function where the graph is concave up. Point *D* satisfies these conditions.

Copyright © 2019 Pearson Education, Inc.

e. $f'(x) < 0$ and $f''(x) = 0$
These conditions indicate a point where the function is decreasing and the same point is an inflection point. Point *B* satisfies these conditions.

51. $f(x) = 28e^{.02x}, \ 1 \le x \le 12$
$f'(x) = .56e^{.02x}$

a. $f'(x)$ is always positive, so the function is increasing over the defined interval, [1, 12].

b. $f'(5) = .56e^{.02(5)} \approx .62$
The stock is increasing at about \$.62 per week at the beginning of the 5th week.

c. $f'(10) = .56e^{.02(10)} \approx .68$
The stock is increasing at about \$.68 per week at the beginning of the 10th week.

53. $s(t) = -16t^2$
$v(t) = s'(t) = -32t$

a. $s'(3) = -32(3) = -96$ ft/sec

b. $s'(5) = -32(5) = -160$ ft/sec

c. $s'(8) = -32(8) = -256$ ft/sec

d. $a(t) = s''(t) = -32$ ft/sec^2

55. $s(t) = 256t - 16t^2$
$v(t) = s'(t) = 256 - 32t$
$a(t) = v'(t) = -32$
Maximum height occurs when $v = 0$.
So, $0 = 256 - 32t \Rightarrow 32t = 256 \Rightarrow t = 8$.
At $t = 8$ sec, maximum height is
$s(8) = 256(8) - 16(8)^2 = 1024$.
Maximum height is 1024 ft at 8 sec. The object hits the ground when $s = 0$. So,
$0 = 256t - 16t^2 \Rightarrow 0 = 16t(16 - t) \Rightarrow t = 0, 16$
Therefore, the object hits the ground at 16 sec.

57. $V(x) = -.1x^3 + 2.4x^2 + 1.2x + 15, \ 0 \le x \le 15$
The point of diminishing returns occurs at a point of inflection.
$V'(x) = -.3x^2 + 4.8x + 1.2$
$V''(x) = -.6x + 4.8$
$V''(x) = -.6x + 4.8 = 0 \Rightarrow x = 8$
Verify that this is a point of inflection:
$V''(9) = -.6(9) + 4.8 = -.6 < 0$
$V''(7) = -.6(7) + 4.8 = .6 > 0$
The point of diminishing returns occurs when $x = 8$ and is about \$127,000 or (8, 127).

59. a. $f(x) = -.002x^3 + .12x^2 - 2x + 14,$
$14 \le x \le 26$
$f'(x) = -.006x^2 + .24x - 2$
$f''(x) = -.012x + .24$

b. The point of diminishing returns occurs at a point of inflection.
$0 = -.012x + .24$
$x = 20$

c. Check points on both sides of the value found in part b:
$f''(21) = -.012(21) + .24 = -.012 < 0$
$f''(19) = -.012(19) + .24 = .012 > 0$
The interval on which the function is concave up will be when the second derivative is positive, or (14, 20).

d. The interval on which the function is concave down will be when the second derivative is negative, or (20, 26).

e. The stock was increasing at an increasing rate until week 20 and increasing at a decreasing rate thereafter.

61. a. $f(x) = \frac{1}{6}x^3 - \frac{7}{2}x^2 + \frac{45}{2}x - 12, \ 3 \le x \le 15$
$f'(x) = \frac{1}{2}x^2 - 7x + \frac{45}{2}$

b. The critical numbers occur when the first derivative is equal to 0.
$0 = x^2 - 14x + 45$
$0 = (x - 9)(x - 5)$
$x = 5, 9$

Copyright © 2019 Pearson Education, Inc.

c. $f''(x) = x - 7$

d. $f''(5) = 5 - 7 = -2 < 0$
$f''(9) = 9 - 7 = 2 > 0$
Therefore, a local maximum occurs in 2005 and a local minimum occurs in 2009.

63. a. $f(x) = 7877\ln(x) + 24{,}215, \quad 2 \le x \le 15$

$f'(x) = \dfrac{7877}{x}$

b. The function is always increasing over the given interval.

c. $f'(x) = 7877x^{-1}$

$f''(x) = -7877x^{-2} = -\dfrac{7877}{x^2}$

d. Concave down; it is increasing at a decreasing rate.

65. $R(x) = (200x)e^{-.01x}$

$R'(x) = (200x)e^{-.01x}(-.01) + 200e^{-.01x}$

$0 = (200x)e^{-.01x}(-.01) + 200e^{-.01x}$

$2e^{-.01x}(100 - x) = 0$

Since $e^{-.01x} \ne 0$, we have $100 - x = 0$ and so $x = 100$ is only critical number.

a. Test $R'(x)$ at $x = 90$ and $x = 110$.

$R'(90) = 2e^{-.01(90)}(100 - 90)$
$= 8.131 > 0$

$R'(110) = 2e^{-.01(110)}(100 - 110)$
$= -6.6574 < 0$

A local maximum occurs at $x = 100$ and so 100 phones should be sold to maximize revenue.

b. The maximum revenue is
$R(100) = (200 \cdot 100)e^{-.01(100)} \approx \$7357.59$

c. Average price per phone $= \dfrac{\$7357.59}{100}$
$= \$73.58$

Section 12.3 Optimization Applications

1. On $[0, 4]$ f has an absolute maximum at $x = 4$ and an absolute minimum at $x = 1$.

3. On $[-1, 2]$ f has an absolute maximum at $x = -1$ and an absolute minimum at $x = 1$.

5. On $[-8, 0]$ f has an absolute maximum at $x = -4$, and an absolute minimum at $x = -7$ and $x = -2$.

7. $f(x) = x^3 - x^2 - x + 10$ on $[0, 2]$

$f'(x) = 3x^2 - 2x - 1 = 0 \Rightarrow$
$(3x+1)(x-1) = 0 \Rightarrow x = -\frac{1}{3}, 1$

x	$f(x)$
0	10
1	9
2	12

So on $[0, 2]$ f has an absolute maximum at $x = 2$, and an absolute minimum at $x = 1$.

9. $f(x) = x^4 - 18x^2 + 1$ on $[-4, 4]$

$f'(x) = 4x^3 - 36x = 0 \Rightarrow 4x\left(x^2 - 9\right) = 0 \Rightarrow$
$4x(x+3)(x-3) = 0 \Rightarrow x = 0, 3, -3$

x	$f(x)$
−4	−31
−3	−80
0	1
3	−80
4	−31

So on $[-4, 4]$ f has an absolute maximum at $x = 0$, and an absolute minimum at $x = -3$ and $x = 3$.

Copyright © 2019 Pearson Education, Inc.

11. $f(x)=\frac{8+x}{8-x}$ on [4, 6]

$f'(x)=\frac{(8-x)(1)-(8+x)(-1)}{(8-x)^2}=\frac{16}{(8-x)^2}$

$f'(x)$ is never 0, but is undefined when $x = 8$. Since 8 is not [4, 6], check only 4 and 6.

x	$f(x)$
4	3
6	7

So on [4, 6], f has an absolute maximum at $x = 6$ and an absolute minimum at $x = 4$.

13. $f(x)=\frac{x}{x^2+4}$ on [0, 4]

$f'(x)=\frac{(x^2+4)(1)-x(2x)}{(x^2+4)^2}=\frac{-x^2+4}{(x^2+4)^2}=0$

when $x = -2$ or 2.
-2 is not in [0, 4], so check $x = 0, 2, 4$.

x	$f(x)$
0	0
2	$\frac{1}{4}$
4	$\frac{1}{5}$

So on [0, 4] f has an absolute maximum at $x = 2$ and an absolute minimum at $x = 0$.

15. $f(x)=(x^2+18)^{2/3}$ on [–3, 2]

$f'(x)=\frac{2}{3}(x^2+18)^{-1/3}(2x)=\frac{4x}{3(x^2+18)^{1/3}}$

$f'(x)=0$ when $x = 0$.

x	$f(x)$
–3	9
0	$3\sqrt[3]{12}\approx 6.87$
2	$22^{2/3}\approx 7.85$

So on [–3, 2] f has an absolute maximum at $x = -3$ and an absolute minimum at $x = 0$.

17. $f(x)=\frac{1}{\sqrt{x^2+1}}$ on [–1, 1]

$f(x)=(x^2+1)^{-1/2}$

$f'(x)=-\frac{1}{2}(x^2+1)^{-3/2}(2x)=\frac{-x}{(x^2+1)^{3/2}}=0$

when $x = 0$.

x	$f(x)$
–1	$\frac{1}{\sqrt{2}}$
0	1
1	$\frac{1}{\sqrt{2}}$

So on [–1, 1] f has an absolute maximum at $x = 0$, and an absolute minimum at $x = -1$ and $x = 1$.

19. $g(x)=2x^3-3x^2-12x+1$ on (0, 4)

$g'(x)=6x^2-6x-12=0$. Solve $g'(x)=0$.
$6(x-2)(x+1)=0 \Rightarrow x=2, -1$. $x = 2$ is the only critical number that lies in the given interval.
$g''(x)=12x-6$
$g''(2)=12>0$
Hence, f has a local minimum at $x = 2$. By the Critical Point Theorem, the absolute minimum of f on the interval (0, 4) occurs at $x = 2$.

21. $g(x)=3x-4x^{3/4}$ on $(0, \infty)$

$g'(x)=3-3x^{-1/4}$

Solve $g'(x)=0$. $g'(x)=3-\frac{3}{x^{1/4}}=0$

$\frac{3}{x^{1/4}}=3$

$x^{1/4}=1$

$x=1$

Since $x = 1$ is the only critical number that lies in the given interval.

$g''(x)=\frac{3}{4x^{5/4}}$

$g''(1)=\frac{3}{4(1)}>0$

Hence, g has a local minimumat $x = 1$. By the Critical Point Theorem, the absolute minimum of g on the interval $(0, \infty)$ occurs at $x = 1$ and there is no maximum.

Copyright © 2019 Pearson Education, Inc.

23. $g(x) = 2xe^{-x/3}$ on $(-\infty, \infty)$

$g'(x) = 2x \cdot e^{-x/3}\left(-\frac{1}{3}\right) + e^{-x/3}(2)$
$= e^{-x/3}\left(2 - \frac{2}{3}x\right)$

Solve $g'(x) = 0$. $g'(x) = e^{-x/3}\left(2 - \frac{2}{3}x\right) = 0$ Since $e^{-x/3} \neq 0$, $2 - \frac{2}{3}x = 0 \Rightarrow x = 3$ is the only critical number that lies in the given interval.

$g''(x) = e^{-x/3}\left(\frac{2}{9}x - \frac{4}{3}\right)$
$g''(3) = -.245 < 0$

Hence, g has a local maximum at $x = 3$. By the Critical Point Theorem, the absolute maximum of g on the interval $(-\infty, \infty)$ occurs at $x = 3$.

25. $g(x) = \dfrac{x}{e^x} = xe^{-x}$ on $(0, 3)$

$g'(x) = x \cdot e^{-x} \cdot (-1) + e^{-x}$
$= e^{-x}(-x+1)$

Solve $g'(x) = 0$. $g'(x) = e^{-x}(-x+1) = 0$ Since $e^{-x} \neq 0$, $-x + 1 = 0 \Rightarrow x = 1$ is the only critical number that lies in the given interval.

$g''(x) = (x-2)e^{-x}$
$g''(1) \approx -.3679 < 0$

Hence, g has a local maximum at $x = 1$. By the Critical Point Theorem, the absolute maximum of g on the interval $(0, 3)$ occurs at $x = 1$.

27. $\overline{C}(x) = .01x^2 - 1.5x + 175$ on $(0, 100)$

$\overline{C}'(x) = .02x - 1.5 = 0$. Solve $\overline{C}'(x) = 0$.

$.02x - 1.5 = 0 \Rightarrow x = 75$. $x = 75$ is the only critical number that lies in the given interval.

$\overline{C}''(x) = .02$
$\overline{C}''(75) = .02 > 0$

Hence, the average cost has a local minimum at $x = 75$. By the Critical Point Theorem, the absolute minimum of the average cost is $\overline{C}(75) = \$118.75$ per wheel when 75 wheels are produced.

29. $P(x) = 17x^2 - 306x + 2324$ on $[6, 15]$

$P'(x) = 34x - 306$. Solve $f'(x) = 0$.

$34x - 306 = 0 \Rightarrow x = 9$. $x = 9$ is the only critical number that lies in the given interval. Now substitute the critical number and the endpoints in the original equation to determine the minimum and maximum.

$P(6) = 17(6)^2 - 306(6) + 2324$
$= 1100$

$P(9) = 17(9)^2 - 306(9) + 2324$
$= 947$

$P(15) = 77(15)^2 - 306(15) + 2324$
$= 1559$

Therefore, the absolute maximum of \$1559 billion occurred when $x = 15$ and the minimum of \$947 billion occurred when $x = 9$.

31. **a.** The revenue function is $R(x) = 400x$.

b. The profit function is

$P(x) = R(x) - C(x)$
$= 400x - \left(525{,}000 - 30x + .012x^2\right)$
$= -.012x^2 + 430x - 525{,}000$

c. $P'(x) = -.024x + 430 = 0 \Rightarrow x \approx 17{,}917$

$P(17{,}917)$
$= -.012(17{,}917)^2 + 430(17{,}917) - 525{,}000$
$\approx \$3{,}327{,}083$

17,917 grills should be made to generate the maximum profit of \$3,327,083.

33. $E(x) = 2x^3 - 63x^2 + 624x + 816$ on $[3, 15]$

$E'(x) = 6x^2 - 126x + 624$. Solve $f'(x) = 0$.

$6x^2 - 126x + 624 = 0 \Rightarrow x = 8$ or $x = 13$. Both critical numbers lie in the given interval. Now substitute the critical numbers and the endpoints in the original equation to determine the minimum and maximum.

$E(3) = 2(3)^3 - 63(3)^2 + 624(3) + 816$
$= 2175$

$E(8) = 2(8)^3 - 63(3)^2 + 624(8) + 816$
$= 2800$

$E(13) = 2(13)^3 - 63(13)^2 + 624(13) + 816$
$= 2675$

$E(15) = 2(15)^3 - 63(15)^2 + 624(15) + 816$
$= 2751$

The minimum average consumer spending during the years 2003 – 2015 was \$2175 in 2003, and the maximum \$2800 in 2008.

Copyright © 2019 Pearson Education, Inc.

35. $N(t) = 20\left(\frac{t}{12} - \ln\left(\frac{t}{12}\right)\right) + 30, \ (1 \le t \le 15)$

$N'(t) = 20\left(\frac{1}{12} - \frac{1}{t}\right) = \frac{5}{3} - \frac{20}{t} = 0$ when $t = 12$

t	$N(t)$
1	81.365
12	50
15	50.537

a. $N(t)$ has an absolute minimum on [1, 15] when $t = 12$, so the number of bacteria is at a minimum after 12 days.

b. $N(12) = 50$, so the minimum number of bacteria is 50 per ml.

c. $N(t)$ has an absolute maximum on [1, 15] when $t = 1$, so the number of bacteria is at a maximum after 1 day.

d. $N(1) = 81.365$, so the maximum number of bacteria is 81.365 per ml.

37. $p(t) = 10te^{-t/8}, \ (0 \le t \le 40)$

a. $p'(t) = 10t\left(-\frac{1}{8}\right)e^{-t/8} + 10e^{-t/8}$

$= e^{-t/8}\left(-\frac{5}{4}t + 10\right)$

$0 = e^{-t/8}\left(-\frac{5}{4}t + 10\right) \Rightarrow 0 = -\frac{5}{4}t + 10 \Rightarrow$

$8 = t$

Test $p(t)$ at $x = 0$, $x = 8$, and $x = 40$.
$p(0) = 0$; $p(8) \approx 29.43$; $p(40) \approx 2.70$
The maximum number of people are infected after 8 days.

b. $p(8) = 10(8)e^{-8/8} = 29.43$
The maximum percent of people infected is 29.43%.

39. Let x = the length of a side of the square that is cut from each corner, in feet. Then the width of the box is $3 - 2x$, the length is $8 - 2x$, and the height is x. The height must not be more than $\frac{3}{2}$ ft because the width of the cardboard is 3 ft.

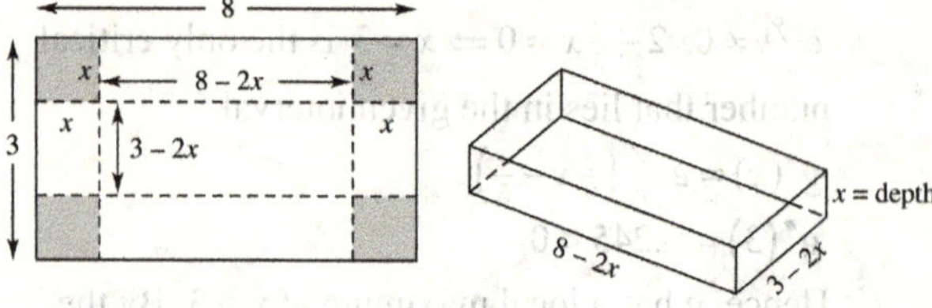

$V(x) = (8 - 2x)(3 - 2x)x = 24x - 22x^2 + 4x^3$

The domain is $\left(0, \frac{3}{2}\right)$.

$V'(x) = 24 - 44x + 12x^2$

Solve $V'(x) = 0$.

$24 - 44x + 12x^2 = 0 \Rightarrow 4\left(6 - 11x + 3x^2\right) = 0 \Rightarrow$

$(3 - x)(2 - 3x) = 0 \Rightarrow x = 3, \frac{2}{3}$

Because 3 is not in the domain, the only critical number of interest is $x = \frac{2}{3}$. Evaluate $V(x)$ at $x = \frac{2}{3}$ and at the endpoints of the domain.

x	$V(x)$
0	0
$\frac{2}{3}$	$\frac{200}{27} = 7\frac{11}{27}$ ← Maximum
$\frac{3}{2}$	0

The square should be $\frac{2}{3}$ ft by $\frac{2}{3}$ ft or 8 inches by 8 inches.

Copyright © 2019 Pearson Education, Inc.

41. Let x = a side of the base; h = height of the box. An equation for the volume of the box is

$V = 32 = x^2h$.

Thus, $h = \dfrac{32}{x^2}$.

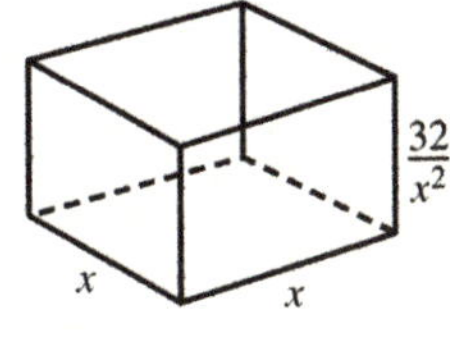

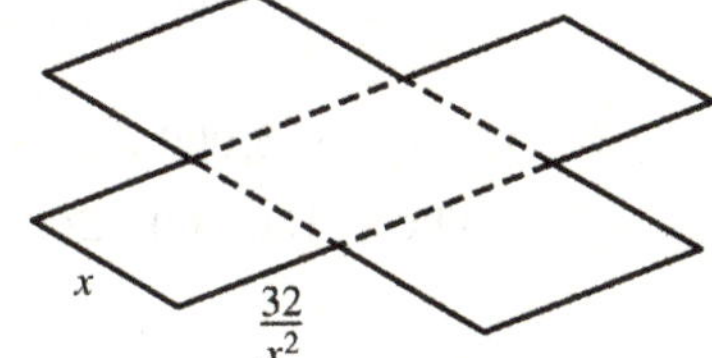

The box is open at the top, so the area of the surface material $S(x)$ in square inches is

$$S(x) = x^2 + 4xh = x^2 + 4x\left(\frac{32}{x^2}\right) = x^2 + \frac{128}{x}.$$

$$S'(x) = 2x - \frac{128}{x^2} = \frac{2x^3 - 128}{x^2}$$

Solve $S'(x) = 0$.

$$2x^3 - 128 = 0 \Rightarrow 2\left(x^3 - 64\right) = 0 \Rightarrow x = 4$$

$S''(x) = 2 + \dfrac{256}{x^3} > 0$ since $x > 0$. Thus, $x = 4$ minimizes the surface area. If $x = 4$,

$$h = \frac{32}{x^2} = \frac{32}{16} = 2.$$

The dimensions that will minimize surface area (amount of material) are 4 inches by 4 inches by 2 inches.

43. Let r = the radius of the top and bottom of the box; h = the height of the box.
The 130-cm ribbon must cover 4 radii on the top, 4 radii on the bottom, and 4 heights, and also includes 10 cm for the bow.

$$4r + 4r + 4h + 10 = 130 \Rightarrow 4h = 130 - 8r - 10 \Rightarrow$$

$$h = \frac{120 - 8r}{4} = 30 - 2r$$

The volume of the cylindrical box is

$V = \pi r^2 h$.

$V(r) = \pi r^2(30 - 2r) = 30\pi r^2 - 2\pi r^3$

$V'(r) = 60\pi r - 6\pi r^2$

When $V'(r) = 0, 60\pi r = 6\pi r^2 \Rightarrow 10 = r$.

$V''(r) = 60\pi - 12\pi r$

$V''(10) = 60\pi - 120\pi = -60\pi < 0$,

so $r = 10$ gives maximum volume.
When $r = 10$, $h = 30 - 2r = 30 - 20 = 10$.
The radius of the box with the largest possible volume is 10 cm. Its height is also 10 cm.

45. **a.** The length of the field in meters is $1200 - 2x$.

b. $A(x) = x(1200 - 2x) = 1200x - 2x^2$

c. $A'(x) = 1200 - 4x$

$1200 - 4x = 0 \Rightarrow 4(300 - x) = 0 \Rightarrow x = 300$.

$A''(x) = -4 < 0$ for all x.

The maximum area occurs at $x = 300$ m.

d. $A(300) = 1200(300) - 2(300)^2 = 180,000$

The maximum area is 180,000 sq m.

47. Let x = the length of the sides facing north and south, which costs \$3/ft.

Because $A = LW$, or $W = \dfrac{A}{L}$, $\dfrac{20,000}{x}$ is the length of each of the other sides, which costs \$6/ft. So $3x$ = the cost of each side, one facing north and one south. The cost for the other two sides is

$$6\left(\frac{20,000}{x}\right)$$

Let C be the total cost of the fencing.

$$C(x) = 2(3x) + 2(6)\left(\frac{20,000}{x}\right) = 6x + \frac{240,000}{x}$$

$$= \frac{6x^2 + 240,000}{x}$$

$$C'(x) = \frac{12x(x) - (6x^2 + 240,000)(1)}{x^2}$$

$$= \frac{6x^2 - 240,000}{x^2}$$

$6x^2 - 240,000 = 0 \Rightarrow x^2 = 40,000 \Rightarrow$

$x = 200$ or -200

Copyright © 2019 Pearson Education, Inc.

Discard the negative solution.

$$C''(x) = \frac{12x(x^2) - (6x^2 - 240,000)(2x)}{x^4}$$
$$= \frac{480,000x}{x^4} = \frac{480,000}{x^3} > 0$$

(since $x > 0$)

So C is minimized at $x = 200$.

$$C(200) = 6(200) + \frac{240,000}{200} = 2400$$

Therefore, the least expensive fence will be $2400.

49. Let x = width of printed material and y = length of printed material. Then, area of the printed material is $xy = 36 \Rightarrow y = \frac{36}{x}$.

Also, $x + 2$ = width of page and $y + 3$ = length of page.

$$A = (x+2)(y+3) = xy + 2y + 3x + 6$$
$$= 36 + 2\left(\frac{36}{x}\right) + 3x + 6 = 42 + 72x^{-1} + 3x$$

$$A' = \frac{-72}{x^2} + 3$$

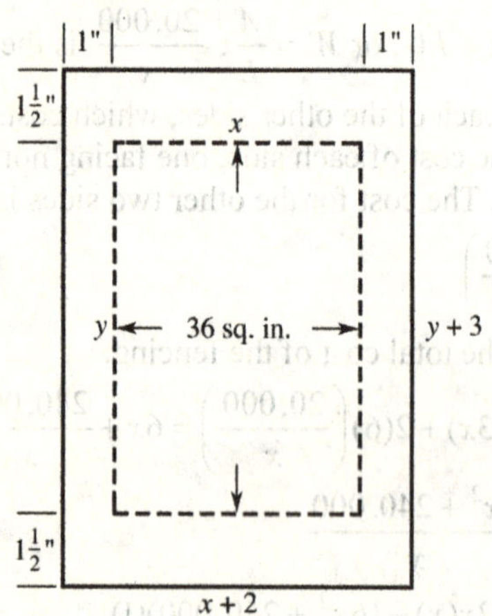

$$\frac{-72}{x^2} + 3 = 0 \Rightarrow x^2 = 24 \Rightarrow x = 2\sqrt{6}.$$

$A'' = \frac{144}{x^3} > 0$ when $x = 2\sqrt{6}$, which implies that A is minimized when $x = 2\sqrt{6}$.

$$y = \frac{36}{x} = \frac{36}{2\sqrt{6}} = \frac{18}{\sqrt{6}} = \frac{18\sqrt{6}}{6} = 3\sqrt{6}$$

Width of page $= x + 2 = 2\sqrt{6} + 2 \approx 6.90$ in.

Length of page $= y + 3 = 3\sqrt{6} + 3 \approx 10.35$ in.

51. a. Price in cents for x thousand bars is

$$P(x) = 100 - \frac{x}{10}.$$

Revenue (in cents) is

$$1000x \cdot p(x) = 1000x\left(100 - \frac{x}{10}\right) \Rightarrow$$
$$R(x) = 100,000x - 100x^2$$

b. $R'(x) = 100,000 - 200x$

When $100,000 - 200x = 0$,

$$x = 500.$$

$R''(x) = -200 < 0$ for all x.

The maximum revenue is attained when $x = 500$.

c. $R(500) = 100,000(500) - 100(500)^2$
$= 25,000,000$

The maximum revenue is 25,000,000 cents, or $250,000.

53.
$$f(x) = 55.44(54.3x - 143.9)e^{-.15x}$$
$$= 3010.392xe^{-.15x} - 7977.816e^{-.15x}$$
$$f'(x) = -451.5588xe^{-.15x} + 3010.392e^{-.15x} + 1196.6724e^{-.15x}$$
$$= -451.5588xe^{-.15x} + 4207.0644e^{-.15x}$$
$$-451.5588xe^{-.15x} + 4207.0644e^{-.15x} = 0 \Rightarrow$$
$$4207.0644e^{-.15x} = 451.5588xe^{-.15x} \Rightarrow$$
$$x = \frac{4207.0644}{451.5588} \approx 9.3$$

Verify that this is a maximum by testing $f'(0)$ and $f'(10)$. $f'(0) \approx 4207$; $f'(10) \approx -69$

According to the model, revenue reaches its maximum in 2009.

Copyright © 2019 Pearson Education, Inc.

55. We assume that the pigeon will fly straight to P, the location that minimizes the amount of energy used. Let x = the distance from P to A.

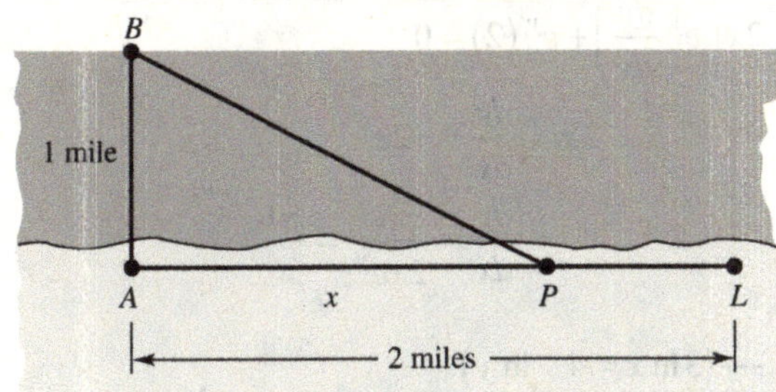

Energy used over land: 1 unit/mi

Energy used over water: $\frac{4}{3}$ units/mi

Distance over land: $2-x$ mi

Distance over water: $\sqrt{1+x^2}$ mi

Find the location of P to minimize energy used.

$$E(x)=1(2-x)+\frac{4}{3}\sqrt{1+x^2}$$

$$E'(x)=-1+\frac{4}{3}\left(\frac{1}{2}\right)\left(1+x^2\right)^{-1/2}(2x)$$

If $E'(x)=0$, then

$$\frac{4}{3}x\left(1+x^2\right)^{-1/2}=1\Rightarrow\frac{4x}{3\left(1+x^2\right)^{1/2}}=1\Rightarrow$$

$$\frac{4}{3}x=\left(1+x^2\right)^{1/2}\Rightarrow\frac{16}{9}x^2=1+x^2\Rightarrow x^2=\frac{9}{7}\Rightarrow$$

$$x=\frac{3}{\sqrt{7}}\text{ or }\frac{3\sqrt{7}}{7}.$$

$$E''(x)$$

$$=\frac{3\left(1+x^2\right)^{1/2}(4)-4x(3)\frac{1}{2}\left(1+x^2\right)^{-1/2}(2x)}{9\left(1+x^2\right)}$$

$$=\frac{12\left(1+x^2\right)^{1/2}-12x^2\left(1+x^2\right)^{-1/2}}{9\left(1+x^2\right)}$$

$$E''\left(\frac{3\sqrt{7}}{7}\right)>0\text{ so }x=\frac{3\sqrt{7}}{7}\text{ minimizes }E(x).$$

Point P is $\frac{3\sqrt{7}}{7}\approx 1.134$ mi from point A.

57. x = number of batches per year
M = 12,300 sinks annually
k = \$2.25, cost to store one unit for one year
a = \$350, cost to set up, or fixed cost
From Example 9, the annual number of batches that gives the minimu total production cost is

$$x=\sqrt{\frac{kM}{2a}}=\sqrt{\frac{(2.25)(12,300)}{2(350)}}$$

$$=\sqrt{39.54}\approx 6$$

Each year, 6 batches should be produced.

59. x = number of orders per year
M = 15,900 cases of beer annually
k = \$3.25, cost to store one case for one year
a = \$5.50, ordering cost

$$x=\sqrt{\frac{kM}{2a}}=\sqrt{\frac{(3.25)(15,900)}{2(5.50)}}$$

$$=\sqrt{4697.73}\approx 69$$

Each year, 69 batches should be produced.

61. x = number of booklets per year
k = \$.75, cost to store for one year
M = 50,000, annual demand
a = 80, cost to place an order (fixed cost)

$$x=\sqrt{\frac{kM}{2a}}=\sqrt{\frac{.75(50,000)}{2(80)}}\approx 15\text{ orders}$$

Since the annual demand is 50,000, and the number of orders is 15, the optimum number of copies per order is $\frac{50,000}{15}\approx 3333$ booklets per order.

Section 12.4 Implicit Differentiation

1. $\frac{d}{dx}(x+y=100)$

$$1+\frac{dy}{dx}=0$$

$$\frac{dy}{dx}=-1$$

Copyright © 2019 Pearson Education, Inc.

3. $\frac{d}{dx}\left(x^4 + y^4 = 13\right)$

$$4x^3 + 4y^3\frac{dy}{dx} = 0$$
$$4y^3\frac{dy}{dx} = -4x^3$$
$$\frac{dy}{dx} = \frac{-4x^3}{4y^3} = \frac{-x^3}{y^3}$$

5. $\frac{d}{dx}\left(10x^3 - 2y^5 = -6\right)$

$$30x^2 - 10y^4\frac{dy}{dx} = 0$$
$$-10y^4\frac{dy}{dx} = -30x^2$$
$$\frac{dy}{dx} = \frac{-30x^2}{-10y^4} = \frac{3x^2}{y^4}$$

7. $\frac{d}{dx}\left(x^2y^5 = 19\right)$

$$x^2\left(5y^4\frac{dy}{dx}\right) + y^5(2x) = 0$$
$$5x^2y^4\frac{dy}{dx} = -2xy^5$$
$$\frac{dy}{dx} = \frac{-2xy^5}{5x^2y^4} = \frac{-2y}{5x}$$

9. $\frac{d}{dx}\left(5x^{1/2} = 2y^{1/2} + 1\right)$

$$\frac{5}{2}x^{-1/2} = y^{-1/2}\frac{dy}{dx}$$
$$\frac{5}{2\sqrt{x}} = \frac{1}{\sqrt{y}}\frac{dy}{dx}$$
$$\frac{dy}{dx} = \frac{5\sqrt{y}}{2\sqrt{x}}$$

11. $\frac{d}{dx}\left(5x^2 = 1 - e^y\right)$

$$10x = -e^y\frac{dy}{dx}$$
$$\frac{dy}{dx} = -\frac{10x}{e^y}$$

13. $\frac{d}{dx}\left(2xe^y = 7\right)$

$$2x\left(e^y\frac{dy}{dx}\right) + e^y(2) = 0$$
$$2xe^y\frac{dy}{dx} = -2e^y$$
$$\frac{dy}{dx} = \frac{-2e^y}{2xe^y} = \frac{-1}{x}$$

15. $\frac{d}{dx}(3\ln x = 4 - \ln y)$

$$\frac{3}{x} = -\frac{1}{y}\frac{dy}{dx}$$
$$\frac{dy}{dx} = -\frac{3y}{x}$$

17. $\frac{d}{dx}(8y\ln x = 15)$

$$8y\left(\frac{1}{x}\right) + \ln x\left(8\frac{dy}{dx}\right) = 0$$
$$8\ln x\frac{dy}{dx} = -\frac{8y}{x}$$
$$\frac{dy}{dx} = \frac{-8y}{8x\ln x} = \frac{-y}{x\ln x}$$

19. $\frac{d}{dx}\left(2x^2 - 3y^3 = 5x - 3y\right)$

$$4x - 9y^2\frac{dy}{dx} = 5 - 3\frac{dy}{dx}$$
$$\frac{dy}{dx}\left(3 - 9y^2\right) = 5 - 4x$$
$$\frac{dy}{dx} = \frac{5 - 4x}{3 - 9y^2}$$

Now substitute the point into the equation.

$$\frac{dy}{dx} = \frac{5 - 4(0)}{3 - 9(1)^2} = -\frac{5}{6}$$

21. $\frac{d}{dx}\left(x^{3/2} = y^{2/3}\right)$

$$\frac{3}{2}x^{1/2} = \frac{2}{3}y^{-1/3}\frac{dy}{dx}$$
$$\frac{3\sqrt{x}}{2} = \frac{2}{3\sqrt[3]{y}}\frac{dy}{dx}$$
$$\frac{dy}{dx} = \frac{9\sqrt{x}\sqrt[3]{y}}{4}$$

Now substitute the point into the equation.

$$\frac{dy}{dx} = \frac{9\sqrt{1}\sqrt[3]{1}}{4} = \frac{9}{4}$$

Copyright © 2019 Pearson Education, Inc.

23. $\frac{d}{dx}\left(7x^2y^5 = 1 - x^2 - y\right)$

$$7x^2\left(5y^4\frac{dy}{dx}\right) + y^5(14x) = -2x - \frac{dy}{dx}$$
$$\frac{dy}{dx}\left(1 + 35x^2y^4\right) = -2x - 14xy^5$$
$$\frac{dy}{dx} = \frac{-2x - 14xy^5}{1 + 35x^2y^4}$$

Now substitute the point into the equation.

$$\frac{dy}{dx} = \frac{-2(0) - 14(0)(1)^5}{1 + 35(0)^2(1)^4} = 0$$

25. $\frac{d}{dx}\left(2xe^{5y} = 2\right)$

$$2x\left(5e^{5y}\frac{dy}{dx}\right) + e^{5y}(2) = 0$$
$$10xe^{5y}\frac{dy}{dx} = -2e^{5y}$$
$$\frac{dy}{dx} = \frac{-1}{5x}$$

Now substitute the point into the equation.

$$\frac{dy}{dx} = \frac{-1}{5(1)} = -\frac{1}{5}$$

27. $\frac{d}{dx}(x \ln y = 1)$

$$x\left(\frac{1}{y}\frac{dy}{dx}\right) + \ln y(1) = 0$$
$$\frac{x}{y}\frac{dy}{dx} = -\ln y$$
$$\frac{dy}{dx} = \frac{-y \ln y}{x}$$

Now substitute the point into the equation.

$$\frac{dy}{dx} = \frac{-e\ln(e)}{1} = -e$$

29. $\frac{d}{dx}\left(x^2 + y^2 = 25\right)$

$$2x + 2y\frac{dy}{dx} = 0$$
$$2y\frac{dy}{dx} = -2x$$
$$\frac{dy}{dx} = \frac{-x}{y}$$

Now substitute the point into the equation to find the slope.

$$\frac{dy}{dx} = \frac{-0}{5} = 0$$

Now use the point-slope formula to find the equation of the tangent line.

$$y - y_1 = m(x - x_1)$$
$$y - 5 = 0(x - 0)$$
$$y = 5$$

31. $\frac{d}{dx}\left(y^3 = 8x^4\right)$

$$3y^2\frac{dy}{dx} = 32x^3$$
$$\frac{dy}{dx} = \frac{32x^3}{3y^2}$$

Now substitute the point into the equation to find the slope.

$$\frac{dy}{dx} = \frac{32(1)^3}{3(2)^2} = \frac{8}{3}$$

Now use the point-slope formula to find the equation of the tangent line.

$$y - y_1 = m(x - x_1)$$
$$y - 2 = \frac{8}{3}(x - 1)$$
$$y - 2 = \frac{8}{3}x - \frac{8}{3}$$
$$y = \frac{8}{3}x - \frac{2}{3}$$

33. $\frac{d}{dx}\left(5x - xy + y^2 = 5\right)$

$$5 - x\frac{dy}{dx} + y(-1) + 2y\frac{dy}{dx} = 0$$
$$\frac{dy}{dx}(2y - x) = y - 5$$
$$\frac{dy}{dx} = \frac{y - 5}{2y - x}$$

Now substitute the point into the equation to find the slope.

$$\frac{dy}{dx} = \frac{1 - 5}{2(1) - 1} = -4$$

Now use the point-slope formula to find the equation of the tangent line.

$$y - y_1 = m(x - x_1)$$
$$y - 1 = -4(x - 1)$$
$$y - 1 = -4x + 4$$
$$y = -4x + 5$$

Copyright © 2019 Pearson Education, Inc.

35. **a.** $xe^y = 1$

$$e^y = \frac{1}{x}$$

$$y = \ln\left(\frac{1}{x}\right)$$

$$\frac{dy}{dx} = x\left(-x^{-2}\right)$$

$$\frac{dy}{dx} = -\frac{x}{x^2} = -\frac{1}{x}$$

b. $\frac{d}{dx}\left(xe^y = 1\right)$

$$xe^y\frac{dy}{dx} + e^y(1) = 0$$

$$xe^y\frac{dy}{dx} = -e^y$$

$$\frac{dy}{dx} = \frac{-e^y}{xe^y} = -\frac{1}{x}$$

c. They are the same.

37. $\frac{d}{dp}\left(p = 200 - \frac{1}{2}q\right)$

$$1 = -\frac{1}{2}\frac{dq}{dp}$$

$$\frac{dq}{dp} = -2$$

This means that if the price is increased by \$1, the quantity demanded decreases by approximately 2 tee times.

39. $\frac{d}{dp}\left(p = 5000 - \frac{1}{20}q^2\right)$

$$1 = -\frac{q}{10}\frac{dq}{dp}$$

$$\frac{dq}{dp} = \frac{-10}{q}$$

$$\frac{dq}{dp} = \frac{-10}{5} = -2$$

This means that if the price is increased by \$1, the quantity demanded decreases by approximately 2 suites.

41. $\frac{d}{dp}\left(p^2 + 100q^2 = 100\right)$

$$2p + 200q\frac{dq}{dp} = 0$$

$$200q\frac{dq}{dp} = -2p$$

$$\frac{dq}{dp} = \frac{-2p}{200q}$$

$$\frac{dq}{dp} = \frac{-2(6)}{200(.8)} \approx -.075 \text{ thousand}$$

This means that if the price is increased by \$1, the quantity demanded decreases by approximately 75 drinks.

43. **a.** Positive; if revenue grows, we would expect more employees would be hired.

b. $\frac{d}{dR}\left(2000R^2 = n^3\right)$

$$4000R = 3n^2\frac{dn}{dR}$$

$$\frac{dn}{dR} = \frac{4000R}{3n^2} \text{ thousand empolyees per billion dollars}$$

c. $\frac{dn}{dR} = \frac{4000(1287)}{3(320)^2}$

$$\frac{dn}{dR} = \frac{512,000}{307,200}$$

$$\frac{dn}{dR} \approx 1.667$$

This means that if revenue grows by \$1 billion, we expect a corresponding increase of approximately 1,667 employees (1.667 thousand employees).

45. **a.** $\frac{d}{dn}\left(3n^3 - 7000R^2 = 0\right)$

$$9n^2 - 14,000R\frac{dR}{dn} = 0$$

$$\frac{dR}{dn} = \frac{9n^2}{14,000R} \text{ billion dollars per thousand employees}$$

Copyright © 2019 Pearson Education, Inc.

b. $\frac{dR}{dn} = \frac{9(210)^2}{14{,}000(63)}$

$$\frac{dR}{dn} = \frac{396{,}900}{882{,}000}$$

$$\frac{dR}{dn} = .45$$

This means that if the number of employees grows by 1 thousand, we expect a corresponding increase of approximately \$450 million (\$.45 billion) in revenue.

47. $\frac{d}{dt}\left(Ae^{.2t} = 99\right)$

$$A\left(.2e^{.2t}\right) + e^{.2t}\frac{dA}{dt} = 0$$

$$e^{.2t}\frac{dA}{dt} = -.2Ae^{.2t}$$

$$\frac{dA}{dt} = -.2A$$

$$\frac{dA}{dt} = -.2(9) = -1.8$$

This means that if one more hour passes, approximately 1.8 mg of drug will be eliminated from the body.

Section 12.5 Related Rates

1. $\frac{d}{dt}\left(y^3 = 2x^4 + 6\right)$

$$3y^2\frac{dy}{dt} = 8x^3\frac{dx}{dt}$$

$$\frac{dy}{dt} = \frac{8x^3}{3y^2}\frac{dx}{dt}$$

Now substitute the known values into the equation to find the missing value.

$$\frac{dy}{dt} = \frac{8(1)^3}{3(2)^2}(3) = 2$$

3. $\frac{d}{dt}\left(2x^5 + x = y^2 - 1\right)$

$$10x^4\frac{dx}{dt} + \frac{dx}{dt} = 2y\frac{dy}{dt}$$

$$\frac{dx}{dt}\left(10x^4 + 1\right) = 2y\frac{dy}{dt}$$

$$\frac{dx}{dt} = \frac{2y}{10x^4 + 1}\frac{dy}{dt}$$

Now substitute the known values into the equation to find the missing value.

$$\frac{dx}{dt} = \frac{2(1)}{10(0)^4 + 1}(5) = 10$$

5. $\frac{d}{dt}(xy = x + y)$

$$x\frac{dy}{dt} + y\frac{dx}{dt} = \frac{dx}{dt} + \frac{dy}{dt}$$

$$\frac{dx}{dt}(y - 1) = \frac{dy}{dt}(1 - x)$$

$$\frac{dx}{dt} = \frac{1 - x}{y - 1}\frac{dy}{dt}$$

Now substitute the known values into the equation to find the missing value.

$$\frac{dx}{dt} = \frac{1 - 2}{2 - 1}(-1) = 1$$

7. $\frac{d}{dt}\left(x^3 + y^3 = 9\right)$

$$3x^2\frac{dx}{dt} + 3y^2\frac{dy}{dt} = 0$$

$$3y^2\frac{dy}{dt} = -3x^2\frac{dx}{dt}$$

$$\frac{dy}{dt} = \frac{-x^2}{y^2}\frac{dx}{dt}$$

Find the value for y.

$$(1)^3 + y^3 = 9$$

$$y^3 = 8$$

$$y = 2$$

Now substitute the known values into the equation to find the missing value.

$$\frac{dy}{dt} = \frac{-(1)^2}{(2)^2}(-4) = 1$$

9. $\frac{d}{dt}\left(x^3 = 9y^5 - 1\right)$

$$3x^2\frac{dx}{dt} = 45y^4\frac{dy}{dt}$$

$$\frac{dx}{dt} = \frac{15y^4}{x^2}\frac{dy}{dt}$$

Copyright © 2019 Pearson Education, Inc.

Find the value for y.

$$x^3 = 9(1)^5 - 1$$
$$x^3 = 8$$
$$x = 2$$

Now substitute the known values into the equation to find the missing value.

$$\frac{dx}{dt} = \frac{15(1)^4}{(2)^2}(1) = \frac{15}{4}$$

11. $\frac{d}{dt}(10n + 31R = 3500)$

$$10\frac{dn}{dt} + 31\frac{dR}{dt} = 0$$
$$\frac{dn}{dt} = \frac{31}{10}\frac{dR}{dt}$$

Now substitute $\frac{dR}{dt} = 848$ into the equation to find the missing value.

$$\frac{dn}{dt} = \frac{31}{10}(848) = 2,628.8$$

The rate of change of the number of employees with respect to time is approximately 2,629 employees.

13. $\frac{d}{dt}\left(2000R^2 = n^3\right)$

$$4000R\frac{dR}{dt} = 3n^2\frac{dn}{dt}$$
$$\frac{dn}{dt} = \frac{4000R}{3n^2}\frac{dR}{dt}$$

Now substitute $R = 54, n = 180,$ and $\frac{dR}{dt} = 7$ into the equation to find the missing value.

$$\frac{dn}{dt} = \frac{4000(54)}{3(180)^2}(7) \approx 15.566 \text{ thousand}$$

The rate of change of the number of employees with respect to time is approximately 15,566 employees per year.

15. a. If the turnover differential dropped, the winning percentage would be expected to decrease.

b. $\frac{d}{dt}(187W - 5T = 95)$

$$187\frac{dW}{dt} - 5\frac{dT}{dt} = 0$$
$$187\frac{dW}{dt} = 5\frac{dT}{dt}$$
$$\frac{dW}{dt} = \frac{5}{187}\frac{dT}{dt}$$

Now substitute $\frac{dT}{dt} = -2$ into the equation to find the missing value.

$$\frac{dW}{dt} = \frac{5}{187}(-2) \approx -.05$$

The rate of change of winning percentage with respect to time is approximately –.05.

17. $\frac{d}{dt}\left(P = -48x^2 + 576x - 528\right)$

$$\frac{dP}{dt} = -96x\frac{dx}{dt} + 576\frac{dx}{dt}$$

Now substitute $x = 4$ and $\frac{dx}{dt} = -1$ into the equation to find the missing value.

$$\frac{dP}{dt} = -96(4)(-1) + 576(-1) = -192$$

The rate of change of the profit is decreasing with respect to time is approximately \$192 per day.

19. $\frac{d}{dt}\left(25p^2 + 49x^2 = 1225\right)$

$$50p\frac{dp}{dt} + 98x\frac{dx}{dt} = 0$$
$$50p\frac{dp}{dt} = -98x\frac{dx}{dt}$$
$$\frac{dp}{dt} = -\frac{49x}{25p}\frac{dx}{dt}$$

Now substitute $x = 3,\ p = 5.60,$ and $\frac{dx}{dt} = .4$ into the equation to find the missing value.

$$\frac{dp}{dt} = -\frac{49(3)}{25(5.60)}(.4) \approx -.42$$

The rate of change of the price is decreasing with respect to time is approximately \$.42 per month.

Copyright © 2019 Pearson Education, Inc.

21. $\frac{d}{dt}(102\ln(R)+10C=586)$

$$\frac{102}{R}\frac{dR}{dt}+10\frac{dC}{dt}=0$$
$$\frac{dC}{dt}=-\frac{102}{10R}\frac{dR}{dt}$$

Now substitute $R=7$ and $\frac{dR}{dt}=1$ into the equation to find the missing value.

$$\frac{dC}{dt}=-\frac{102}{10(7)}(1)\approx-1.46$$

The rate of change of the percentage of electricity generated by coal with respect to time is approximately −1.46% per year.

23. $\frac{d}{dt}(A=\pi r^2)$

$$\frac{dA}{dt}=2\pi r\frac{dr}{dt}$$

Now substitute $r=3$ and $\frac{dr}{dt}=2$ into the equation to find the missing value.

$$\frac{dA}{dt}=2\pi(3)(2)\approx 37.7$$

The rate of change at which the puddle is increasing with respect to time is approximately 37.7 inches squared per minute.

25. $\frac{d}{dt}\left(V=\frac{4}{3}\pi r^3\right)$

$$\frac{dV}{dt}=4\pi r^2\frac{dr}{dt}$$

Now substitute $r=30$ and $\frac{dr}{dt}=.5$ into the equation to find the missing value.

$$\frac{dV}{dt}=4\pi(30)^2(.5)\approx 5655$$

The rate of change at which the volume of the tumor is increasing with respect to time is approximately 5655 mm^3 per month.

27. $\frac{d}{dt}\left(z^2=x^2+36\right)$

$$2z\frac{dz}{dt}=2x\frac{dx}{dt}$$
$$\frac{dz}{dt}=\frac{x}{z}\frac{dx}{dt}$$

Now substitute $x=8$, $z=10$, and $\frac{dx}{dt}=3$ into the equation to find the missing value.

$$\frac{dz}{dt}=\frac{8}{10}(3)=2.4$$

The rate of change of the distance between the quarterback and receiver is increasing with respect to time is approximately 2.4 yards per second.

29. $\frac{d}{dt}\left(W^2=9A^3\right)$

$$2W\frac{dW}{dt}=27A^2\frac{dA}{dt}$$
$$\frac{dW}{dt}=\frac{27A^2}{2W}\frac{dA}{dt}$$

Now substitute $W=24$, $A=4$, and $\frac{dA}{dt}=2$ into the equation to find the missing value.

$$\frac{dW}{dt}=\frac{27(4)^2}{2(24)}(2)=18$$

The rate of change of the weight is increasing with respect to time is approximately 18 pounds.

31. $\frac{d}{dt}\left(V=\frac{1}{3}\pi r^2h\right)$

$$\frac{dV}{dt}=\frac{1}{3}\pi r^2\frac{dh}{dt}+\frac{2}{3}\pi hr\frac{dr}{dt}$$

Find the value for h.

$h=2r$
$h=2(1)$
$h=2$

Now substitute $r=1$, $h=2$, $\frac{dr}{dt}=2$, and $\frac{dh}{dt}=2\cdot\frac{dr}{dt}=4$ into the equation to find the missing value.

$$\frac{dV}{dt}=\frac{1}{3}\pi(1)^2(4)+\frac{2}{3}\pi(2)(1)(2)\approx 12.6$$

The rate of change of the volume is increasing with respect to time is approximately 12.6 centimeters cubed per minute.

Copyright © 2019 Pearson Education, Inc.

Section 12.6 Curve Sketching

1. $f(x) = -x^2 - 10x - 25$

The y-intercept is $f(0) = -25$, but $(0, -25)$ is not a convenient point to plot. To find any x-intercepts, let $y = 0$.

$0 = -x^2 - 10x - 25 \Rightarrow 0 = -1\left(x^2 + 10x + 25\right) \Rightarrow$
$0 = -1(x+5)^2 \Rightarrow 0 = (x+5)^2 \Rightarrow 0 = x + 5 \Rightarrow$
$x = -5$

The only x-intercept is -5. This is a polynomial function, so there are no asymptotes.

Find $f'(x)$ and set it equal to zero to find critical numbers.

$f'(x) = -2x - 10 = -2(x+5) = 0$

Critical number: $x = -5$

$f''(x) = -2 < 0$ for all x

$f(-5) = 0$

The graph is concave downward on $(-\infty, \infty)$. Hence there is a local maximum of 0 at $x = -5$. There are no points of inflection. The graph is a parabola opening downward. The local maximum (which is also the absolute maximum) occurs at the vertex $(-5, 0)$. Plot some additional points and sketch the graph.

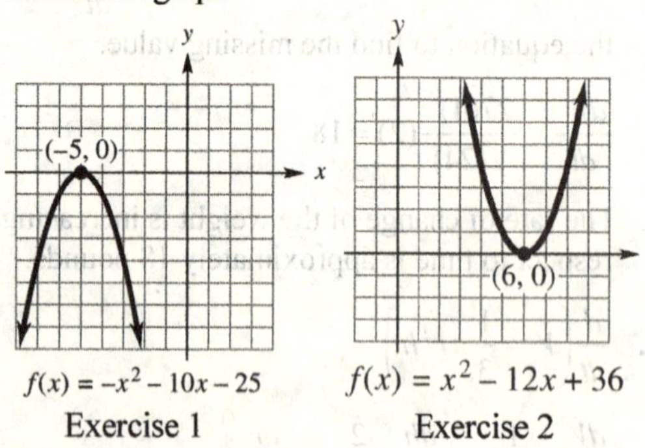

Exercise 1 Exercise 2

3. $f(x) = 3x^3 - 3x^2 + 1$

The y-intercept is $f(0) = 1$. It is not convenient to solve $f(x) = 0$ to find any x-intercepts.

$f'(x) = 9x^2 - 6x = 3x(3x - 2) = 0$

Critical numbers: $x = 0$ or $x = \frac{2}{3}$

$f''(x) = 18x - 6$

$f''(0) = -6 < 0$

$f''\left(\frac{2}{3}\right) = 6 > 0$

There is a local maximum of 1 at 0 and a local minimum of $\frac{5}{9}$ at $\frac{2}{3}$.

$f''(x) = 18x - 6 = 0 \Rightarrow x = \frac{1}{3}$

$f''(0) = -6 < 0$

$f''(1) = 12 > 0$

There is a point of inflection at $\left(\frac{1}{3}, \frac{7}{9}\right)$.

Plot the critical points, point of inflection, and a few additional points to sketch the graph.

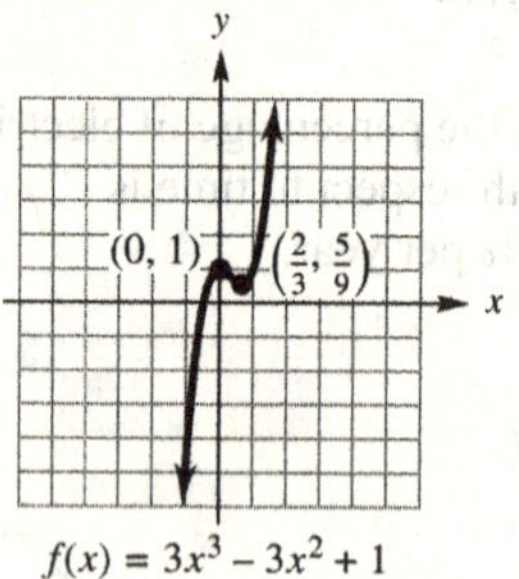

5. $f(x) = -2x^3 - 9x^2 + 108x - 10$

The y-intercept is $f(0) = -10$. It is not convenient to find the x-intercepts.

$f'(x) = -6x^2 - 18x + 108 = 0 \Rightarrow$
$-6\left(x^2 + 3x - 18\right) = 0 \Rightarrow (x+6)(x-3) = 0 \Rightarrow$
$x = -6, 3$ (critical numbers)

$f''(x) = -12x - 18$

$f''(-6) = 54 > 0$

$f''(3) = -54 < 0$

There is a local maximum of 179 at 3 and a local minimum of -550 at -6.

$f''(x) = -12x - 18 = 0 \Rightarrow x = -\frac{3}{2}$

$f''(-2) = 6 > 0$

$f''(-1) = -6 < 0$

There is a point of inflection at $\left(-\frac{3}{2}, -\frac{371}{2}\right)$.

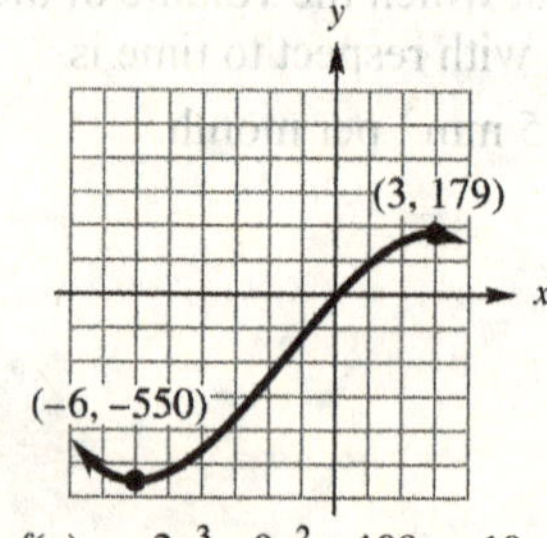

Copyright © 2019 Pearson Education, Inc.

7. $f(x) = 2x^3 + \frac{7}{2}x^2 - 5x + 3$

The y-intercept is $f(0) = 3$. It is not convenient to find any possible x-intercepts.

$f'(x) = 6x^2 + 7x - 5 = 0 \Rightarrow$

$(2x-1)(3x+5) = 0 \Rightarrow x = \frac{1}{2}, -\frac{5}{3}$ (critical numbers)

$f''(x) = 12x + 7$

$f''\left(\frac{1}{2}\right) = 13 > 0$

$f''\left(-\frac{5}{3}\right) = -13 < 0$

There is a local maximum of $\frac{637}{54}$ at $-\frac{5}{3}$ and a local minimum of $\frac{13}{8}$ at $\frac{1}{2}$.

$f''(x) = 12x + 7 = 0 \Rightarrow x = -\frac{7}{12}$

$f''(-1) = -5 < 0$

$f''(0) = 7 > 0$

There is a point of inflection at $\left(-\frac{7}{12}, \frac{2899}{432}\right)$.

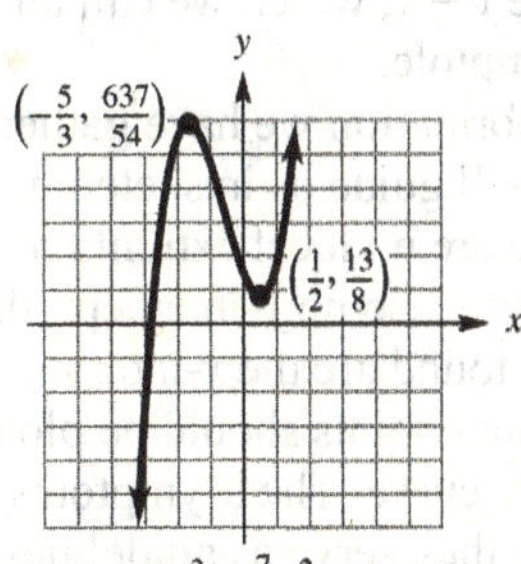

$f(x) = 2x^3 + \frac{7}{2}x^2 - 5x + 3$

9. $f(x) = (x+3)^4$

The y-intercept is $f(0) = 3^4 = 81$, but (0, 81) is not a convenient point to plot. Because the right-hand side is in factored form, it is easy to find any x-intercepts.

$0 = (x+3)^4 \Rightarrow x = -3$

The only x-intercept is –3.

$f'(x) = 4(x+3)^3(1) = 4(x+3)^3 = 0 \Rightarrow$
$x = -3$ (critical number)

$f''(x) = 12(x+3)^2$

$f''(-3) = 12(-3+3)^2 = 0$

The second derivative test fails, so use the first derivative test.

$f'(-2) = 4(-2+3)^3 = 4(1)^3 = 4 > 0$

$f'(-4) = 4(-4+3)^3 = 4(-1)^3 = -4 < 0$

$f(x)$ is increasing on $(-3, \infty)$ and decreasing on $(-\infty, -3)$. Thus, there is a local minimum at $x = -3$, which is also an absolute minimum.

$f''(x) = 12(x+3)^2 > 0$ for all $x \neq -3$, so the graph is always concave upward and there is no point of inflection.

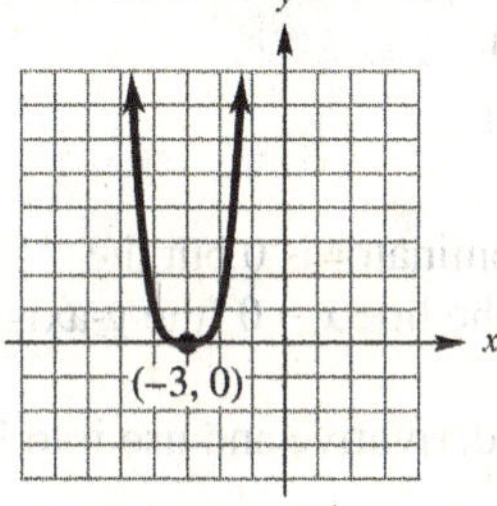

$f(x) = (x+3)^4$

11. $f(x) = x^4 - 18x^2 + 5$

The y-intercept is $f(0) = 5$. It is not convenient to find the x-intercepts.

$f'(x) = 4x^3 - 36x = 4x\left(x^2 - 9\right) = 0 \Rightarrow$
$x = 0, -3, 3$ (critical numbers)

$f''(x) = 12x^2 - 36$

$f''(0) = -36 < 0$

$f''(3) = 72 > 0$

$f''(-3) = 72 > 0$

There is a local maximum of 5 at $x = 0$ and a local minimum of –76 at $x = 3$ and $x = -3$.

$f''(x) = 12x^2 - 36 = 0 \Rightarrow x^2 = 3 \Rightarrow x = \pm\sqrt{3}$

$f''(-2) = 12 > 0$

$f''(-1) = -24 < 0$

There is a point of inflection at $\left(-\sqrt{3}, -40\right)$.

$f''(1) = -24 < 0$

$f''(2) = 12 > 0$

There is another point of inflection at $\left(\sqrt{3}, -40\right)$.

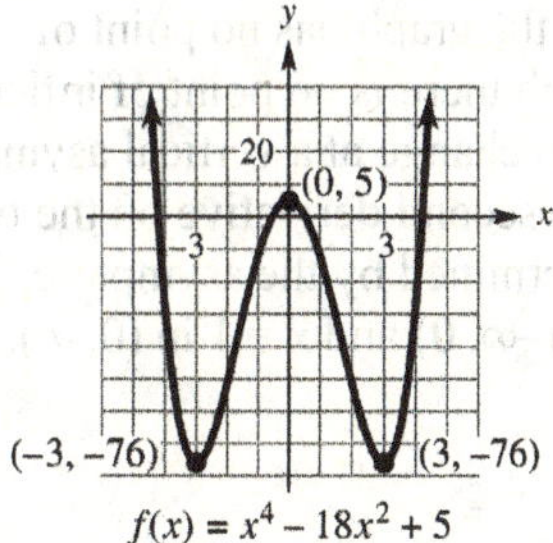

$f(x) = x^4 - 18x^2 + 5$

Copyright © 2019 Pearson Education, Inc.

13. $f(x) = x - \frac{1}{x}$

Step 1: There is no y-intercept since $f(0)$ is undefined. Find any x-intercepts by solving the equation $f(x) = 0$.

$$f(x) = x - \frac{1}{x} = 0 \Rightarrow x\left(\frac{x^2-1}{x}\right) = x \cdot 0 \Rightarrow$$

$$x^2 - 1 = 0 \Rightarrow (x+1)(x-1) = 0 \Rightarrow x = \pm 1$$

The x-intercepts are -1 and 1.

Step 2: f is a rational function. To find any vertical asymptotes, rewrite the function rule so that we have a single fraction.

$$f(x) = x - \frac{1}{x} = \frac{x^2-1}{x}$$

When $x = 0$, the denominator is 0 but the numerator is not, so the line $x = 0$ (the y-axis) is a vertical asymptote.

Step 3: Find the first derivative and use it to look for critical numbers.

$$f(x) = x - \frac{1}{x} = x - x^{-1}$$

$$f'(x) = 1 - (-1)x^{-2} = 1 + x^{-2} = 1 + \frac{1}{x^2}$$

The equation $f'(x) = 1 + \frac{1}{x^2} = 0$ has no real solution. Although $f'(0)$ does not exist, 0 is not a critical number because $f(0)$ also does not exist. Thus, there are no critical numbers and consequently no local extrema. The vertical asymptote at $x = 0$ (the y-axis) separates the graph into two branches. Note that $f'(x) = 1 + \frac{1}{x^2}$ is always positive, so the graph is increasing on both $(-\infty, 0)$ and $(0, \infty)$.

Step 4 We now consider the second derivative.

$$f''(x) = -2x^{-3} = -\frac{2}{x^3}$$

The equation $f''(x) = -\frac{2}{x^3} = 0$ has no solution.

Although $f''(x)$ does not exist at $x = 0$, there cannot be a point of inflection because $f(0)$ does not exist. (Recall that $x = 0$ is a vertical asymptote.) Thus, the graph has no point of inflection. Although there is no point of inflection, concavity may also change at a vertical asymptote, so we examine the second derivative on the two open intervals determined by the asymptote. We will test $x = -1$ in $(-\infty, 0)$ and $x = 1$ in $(0, \infty)$.

$$f''(-1) = -\frac{2}{(-1)^3} = 2 > 0$$

$$f''(1) = -\frac{2}{1^3} = -2 < 0$$

Thus, the graph is concave upward on $(-\infty, 0)$ and concave downward on $(0, \infty)$.

Step 5: To find any horizontal asymptote, consider the following limit.

$$\lim_{x\to\infty} x - \frac{1}{x} = \lim_{x\to\infty} \frac{x^2-1}{x} = \lim_{x\to\infty} \frac{\frac{x^2}{x^2} - \frac{1}{x^2}}{\frac{x}{x^2}}$$

$$= \lim_{x\to\infty} \frac{1 - \frac{1}{x^2}}{\frac{1}{x}} = \frac{1-0}{0} = \frac{1}{0}$$

Because division by 0 is undefined, this limit does not exist. Similarly, $\lim_{x\to\infty} f(x)$ does not exist.

Therefore, the graph has no horizontal asymptote.

Notice that as $x \to \infty$, $\frac{1}{x} \to 0$, so that

$f(x) = x - \frac{1}{x} \approx x$. Likewise, as $x \to -\infty$,

$f(x) = x - \frac{1}{x} \approx x$. This means that as x gets larger and larger in absolute value, the graph gets closer and closer to the line $y = x$, which we call an oblique or slant asymptote.

Step 6: All of the information we have gathered about this function will guide us in sketching the graph. Because there are no local extrema or points of inflection and also no y-intercept, the only points we have found are the x-intercepts -1 and 1. A few additional points should be plotted on each branch of the curve. The asymptotes are very helpful because they serve as guidelines for the graph.

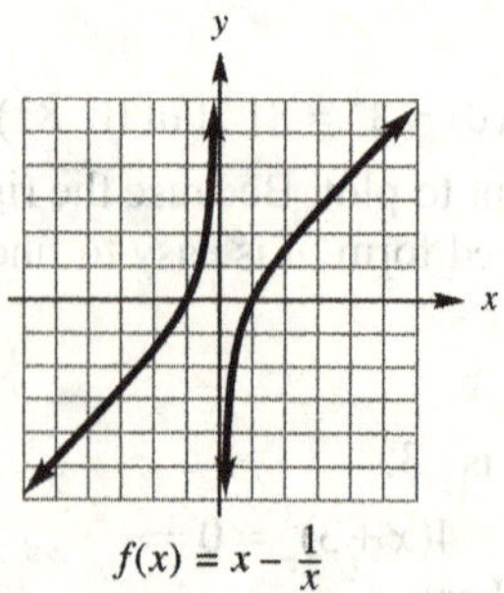

Copyright © 2019 Pearson Education, Inc.

15. $f(x)=\dfrac{x^2+25}{x}$

$f(0)$ is undefined, so there is no y-intercept.

$f(x)=0$ has no real solution because x^2+25 is always positive. Thus, there are no x-intercepts. When $x=0$, the denominator is 0 but the numerator is not, so the line $x=0$ (the y-axis) is a vertical asymptote. We now consider the first derivative and look for critical numbers.

$$f'(x)=\frac{x(2x)-\left(x^2+25\right)}{x^2}=\frac{x^2-25}{x^2}$$

$$f'(x)=\frac{(x+5)(x-5)}{x^2}=0\Rightarrow$$

$(x+5)(x-5)=0\Rightarrow x=\pm5$

Although $f'(0)$ does not exist, 0 is not a critical number because $f(0)$ also does not exist. Use the second derivative test to determine if there are local maxima or minima at the critical numbers −5 and 5.

$$f''(x)=\frac{x^2(2x)-2x\left(x^2-25\right)}{x^4}=\frac{50}{x^3}$$

$$f''(-5)=-\frac{2}{5}<0$$

$$f''(5)=\frac{2}{5}>0$$

Thus, there is a local maximum of −10 at $x=-5$ and a local minimum of 10 at $x=5$. f is increasing on $(-\infty,-5)$ and $(5,\infty)$ and decreasing on $(-5, 0)$ and $(0, 5)$.

$f''(x)=\dfrac{50}{x^3}\neq0$ for any x, so there are no points of inflection. This graph changes concavity at the vertical asymptote $x=0$. Earlier we found that $f''(-5)<0$ and $f''(5)>0$. This tells us that the graph is concave downward on $(-\infty, 0)$ and concave upward on $(0,\infty)$. Following the method shown in Example 1 and the solution for exercise 13, we determine that there is no horizontal asymptote. However, if we rewrite the function rule as $f(x)=x+\dfrac{25}{x}$,

we see that as x gets very large in absolute value, $\dfrac{25}{x}\to0$, and the graph approaches the line $y=x$, which is an oblique or slant asymptote.

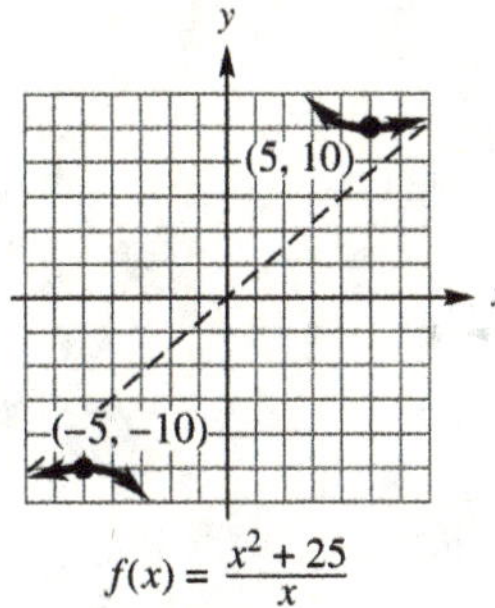

$f(x)=\dfrac{x^2+25}{x}$

17. $f(x)=\dfrac{x-1}{x+1}$

$f(0)=-1$, so the y-intercept is −1. $f(x)=0$ when $x-1=0$ or $x=1$, so there is one x-intercept, 1. When $x=-1$, the denominator is 0 but the numerator is not, so there is one vertical asymptote, the line $x=-1$.

$$f'(x)=\frac{(x+1)-(x-1)}{(x+1)^2}=\frac{2}{(x+1)^2}$$

$f'(x)$ is never zero.

$f'(x)$ fails to exist for $x=-1$, but −1 is not a critical number because $f(-1)$ does not exist. Thus, there are no critical points and there can be no local extrema. Because $f'(x)$ is always positive where it exists, the function is increasing on $(-\infty, -1)$ and $(-1,\infty)$.

$$f''(x)=\frac{(x+1)^2(0)-2(2)(x+1)}{(x+1)^4}$$

$$=\frac{-4(x+1)}{(x+1)^4}=\frac{-4}{(x+1)^3}$$

$f''(x)$ is never equal to 0, so there are no points of inflection. The concavity of the graph may change at the vertical asymptote, so we test numbers in the intervals $(-\infty,-1)$ and $(-1,\infty)$.

$$f''(-2)=\frac{-4}{(-2+1)^3}=4>0$$

$$f''(0)=\frac{-4}{(0+1)^3}=-4<0$$

Thus, f is concave upward on $(-\infty,-1)$ and concave downward on $(-1,\infty)$. Because this is a rational function, we now look for a horizontal asymptote.

$$\lim_{x\to\infty}\frac{x-1}{x+1}=\lim_{x\to\infty}\frac{\frac{x}{x}-\frac{1}{x}}{\frac{x}{x}+\frac{1}{x}}=\lim_{x\to\infty}\frac{1-\frac{1}{x}}{1+\frac{1}{x}}=\frac{1-0}{1+0}=1$$

Likewise, $\lim\limits_{x\to-\infty}f(x)=1$. Thus, there is a horizontal asymptote at $y=1$. The vertical and horizontal asymptotes act as guidelines for the graph.

Copyright © 2019 Pearson Education, Inc.

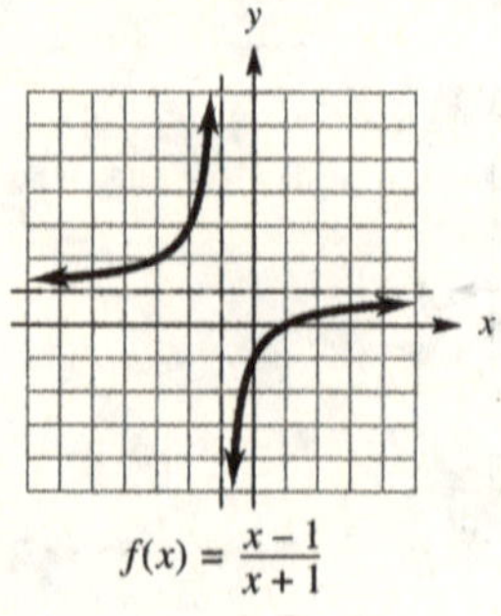

19. $y = x - \ln|x|$

$y' = 1 - \frac{1}{x} = \frac{x-1}{x}$

$y'' = \frac{1}{x^2}$

$y' = 0$ when $x = 1$. There is a vertical asymptote at $x = 0$. The critical numbers are 0 and 1. Test y' at $x = -1, \frac{1}{2}$, and $\frac{3}{2}$.

$y'(-1) = 2,\ y'\left(\frac{1}{2}\right) = -1,\ y'\left(\frac{3}{2}\right) = \frac{1}{3}.$

So y is increasing on $(-\infty, 0)$ and $(1, \infty)$, and decreasing on (0, 1). Since $y'' \neq 0$, there are no points of inflection. Also, y'' is always positive. Hence, the graph is always concave upward.

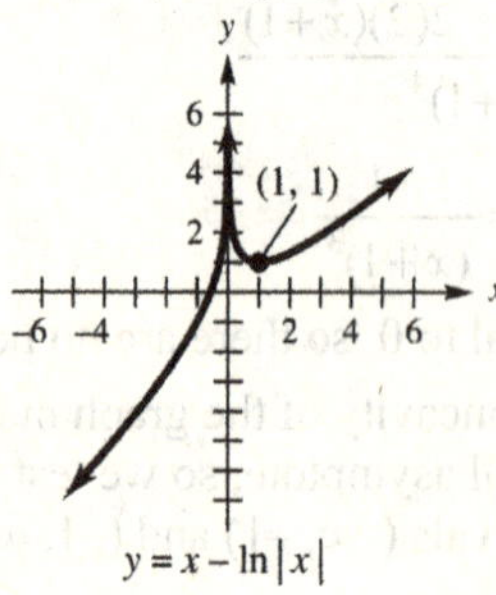

21. $y = xe^{-x}$

$y' = (1-x)e^{-x}$

$y'' = (x-2)e^{-x}$

Set $y' = 0$ and solve for x to get the critical numbers. Since $e^{-x} \neq 0$, the only critical number is $x = 1$. Test y' at $x = 0$ and $x = 2$.

$y'(0) = 1;\ y'(2) = -e^{-2}$

y is increasing on $(-\infty, 1)$ and decreasing on $(1, \infty)$. Set $y'' = 0$ to find points of inflection.

$y'' = 0 \Rightarrow (x-2)e^{-x} = 0$. An inflection point occurs at $x = 2$.

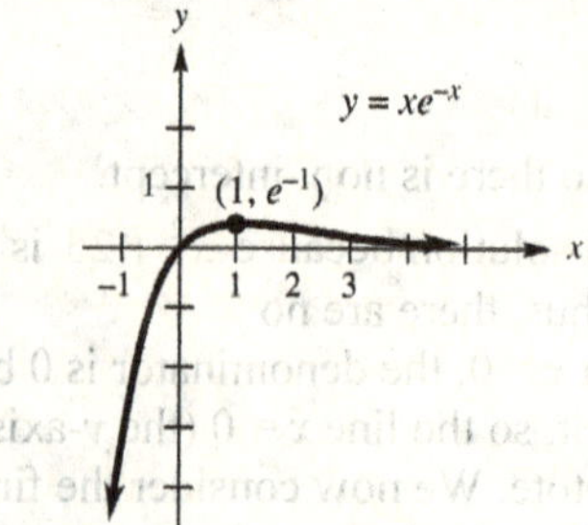

For Exercises 23–28, many correct answers are possible, including the following:

23.

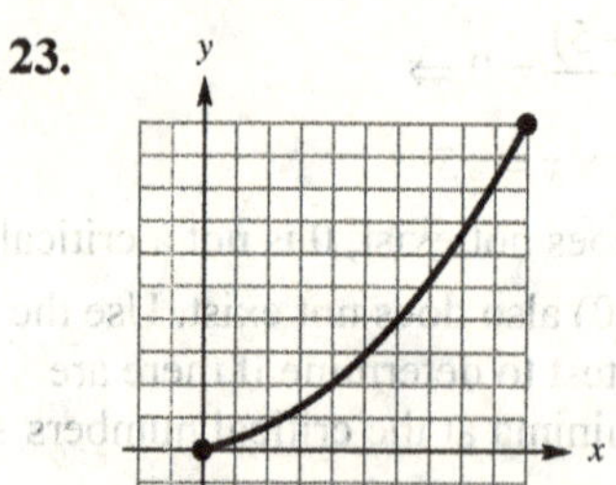

25.

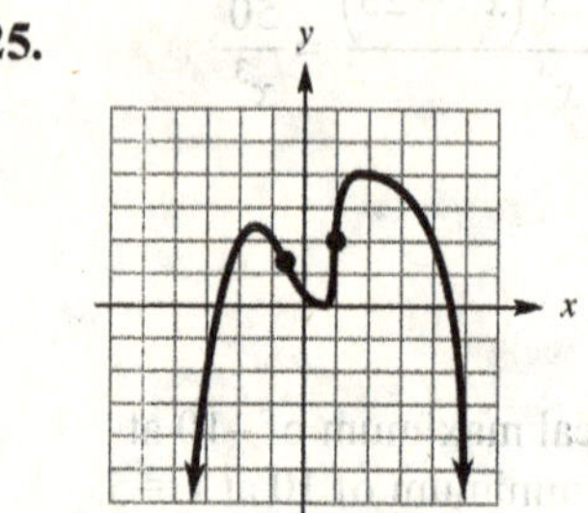

27.

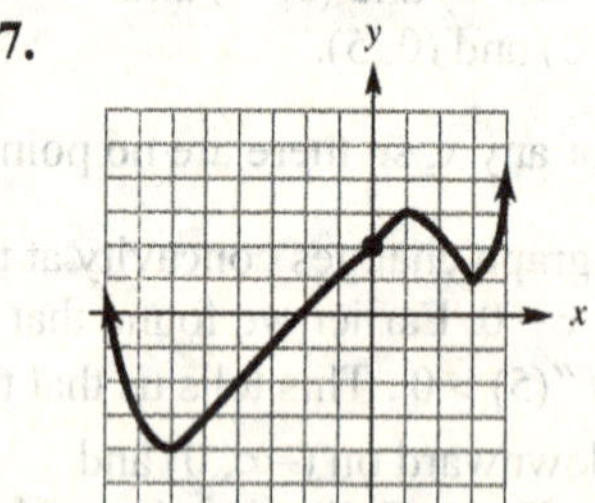

29. Smart cards; rate of growth of sales begins to decline at the point of inflection.

Copyright © 2019 Pearson Education, Inc.

31. a. $P(x) = -2.3x^2 + 55.2x - 252.6$
$P'(x) = -4.6x + 55.2$
$0 = -4.6x + 55.2$
$x = 12$

The local maximium profit occurred in 2012.

b. $P(12) = -2.3(12)^2 + 55.2(12) - 252.6$
$= 78.6$
Therefore the maximum profit was \$78.6 billion.

c. $P''(x) = -4.6$ therefore the second derivative is a constant so there is no point of inflection.

33. Plot $P(t) = t^3 - 18t^2 + 81t$ over the interval [0, 15], then find the maxima and minima.

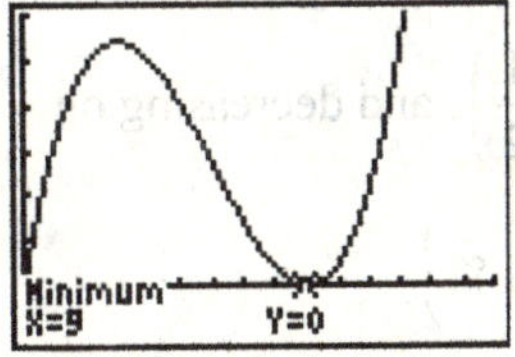

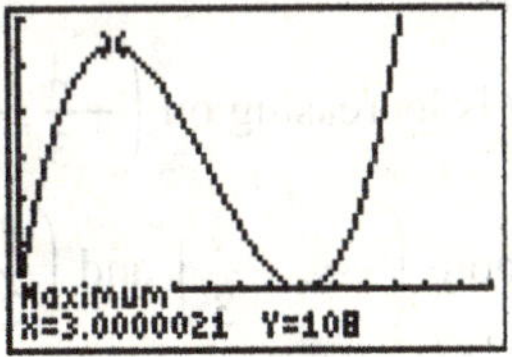

[0, 15] by [–20, 120]

Local minimum: (9, 0)
Local maximum: (3, 108)
Now plot $P''(x) = 6t - 36$ over [0, 15] and determine the zero.

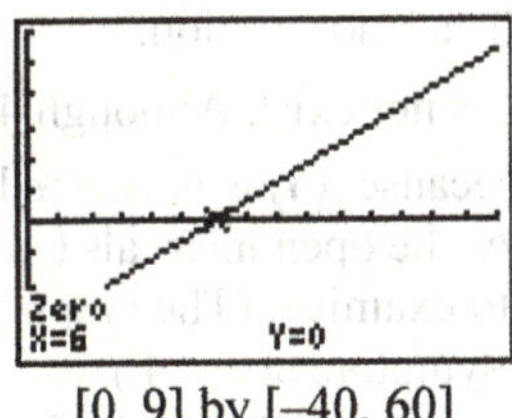

[0, 9] by [–40, 60]

Thus, there is an inflection point at $x = 6$.

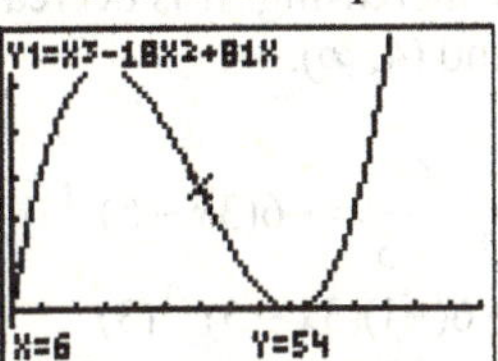

The inflection point is (6, 54).

35. Plot $S(x) = -.072x^3 + 2.45x^2 - 26.5x + 106$ over the interval [7, 15], then find the maxima and minima.

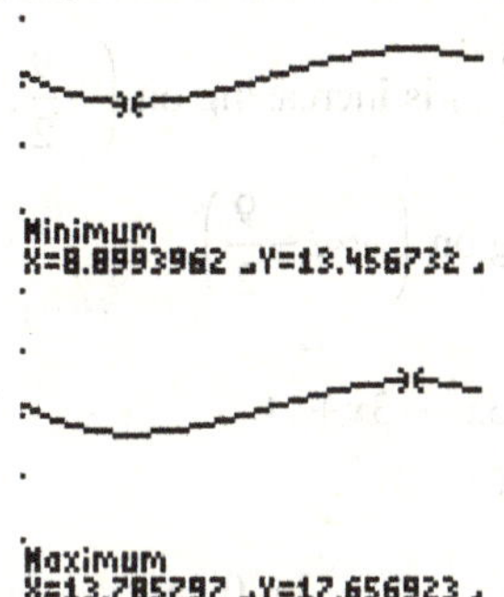

[7, 15] by [0, 25]

Local minimum: (8.9, 13.5)
Local maximum: (13.8, 17.7)

Chapter 12 Review Exercises

1. Answers vary; If $f'(x) > 0$ for each x in an interval, then f is increasing on the interval; if $f'(x) < 0$ for each x in an interval, then f is decreasing on the interval.

2. Answers vary; If either $f'(c) = 0$ or $f'(c)$ does not exist, there may be a local extremum at c. The first derivative test says that $f(c)$ is a local maximum if $f'(a) > 0$ and $f'(b) < 0$ and $a < c < b$. $f(c)$ is a local minimum if $f'(a) < 0$ and $f'(b) > 0$. If $f''(c)$ exists, the second derivative test says that if $f''(c) < 0$, then $f(c)$ is a local maximum. If $f''(c) > 0$, then $f(c)$ is a local minimum.

3. Answers vary; A local extremum is a maximum or minimum value of the function on an open interval. An absolute extremum is the largest or smallest possible value of the function. A local extremum can be an absolute extremum, but not necessarily.

4. Answers vary; The first derivative can be used to determine where a graph is increasing or decreasing. The second derivative can be used to determine where a graph is concave upward or concave downward.

Copyright © 2019 Pearson Education, Inc.

5. $f(x) = x^2 + 9x - 9;\ f'(x) = 2x + 9$

$f'(x) = 0 \Rightarrow 2x + 9 = 0 \Rightarrow x = -\frac{9}{2}$

$f''(x) = 2 > 0$ for all x. A local minimum occurs at $x = -\frac{9}{2}$. f is increasing on $\left(-\frac{9}{2}, \infty\right)$ and decreasing on $\left(-\infty, -\frac{9}{2}\right)$.

6. $f(x) = -3x^2 - 3x + 11$

$f'(x) = -6x - 3$

$f'(x) = 0 \Rightarrow -6x - 3 = 0 \Rightarrow x = -\frac{1}{2}$

$f''(x) = -6 < 0$ for x. A local maximum occurs at $x = -\frac{1}{2}$. f is increasing on $\left(-\infty, -\frac{1}{2}\right)$ and decreasing on $\left(-\frac{1}{2}, \infty\right)$.

7. $g(x) = 2x^3 - x^2 - 4x + 7$

$g'(x) = 6x^2 - 2x - 4$

$g'(x) = 0 \Rightarrow 6x^2 - 2x - 4 = 0 \Rightarrow$

$2(3x + 2)(x - 1) = 0 \Rightarrow x = -\frac{2}{3}, 1$

Use the second derivative test.

$g''(x) = 12x - 2$

$g''\left(-\frac{2}{3}\right) = -10 < 0;\ g''(1) = 10 > 0$

A local maximum occurs at $x = -\frac{2}{3}$. A local minimum occurs at $x = 1$. f is increasing on $\left(-\infty, -\frac{2}{3}\right)$ and $(1, \infty)$, and decreasing on $\left(-\frac{2}{3}, 1\right)$.

8. $g(x) = -4x^3 - 5x^2 + 8x + 1$

$g'(x) = -12x^2 - 10x + 8$

$g'(x) = 0 \Rightarrow -12x^2 - 10x + 8 = 0 \Rightarrow$

$-2(3x + 4)(2x - 1) = 0 \Rightarrow x = -\frac{4}{3}, \frac{1}{2}$

Use the second derivative test.

$g''(x) = -24x - 10$

$g''\left(-\frac{4}{3}\right) = 22 > 0;\ g''\left(\frac{1}{2}\right) = -22 < 0$

A local maximum occurs at $x = \frac{1}{2}$.

A local minimum occurs at $x = -\frac{4}{3}$.

f is increasing on $\left(-\frac{4}{3}, \frac{1}{2}\right)$, and decreasing on both $\left(-\infty, -\frac{4}{3}\right)$ and $\left(\frac{1}{2}, \infty\right)$.

9. $f(x) = \frac{4}{x - 4} = 4(x - 4)^{-1}$

$f'(x) = -4(x - 4)^{-2} = \frac{-4}{(x - 4)^2}$

$f'(x) = 0$ has no solution.

$f'(4)$ does not exist. Although 4 is not a critical number because $f(4)$ is undefined, this number determines the open intervals $(-\infty, 4)$ and $(4, \infty)$ we need to examine. (The graph will have a vertical asymptote at $x = 4$.)

$f'(2) = -1 < 0;\ f'(5) = -4 < 0$

f is never increasing. It is decreasing on $(-\infty, 4)$ and $(4, \infty)$.

10. $f(x) = \frac{-6}{3x - 5} = -6(3x - 5)^{-1}$

$f'(x) = -6(-1)(3x - 5)^{-2}(3)$

$= 18(3x - 5)^{-2} = \frac{18}{(3x - 5)^2}$

$f'(x) = 0$ has no solution.

$f'(x)$ does not exist if $x = \frac{5}{3}$, but this is not a critical number because $f\left(\frac{5}{3}\right)$ is undefined.

There are no critical numbers and there are no local extrema.

Test -1 in $\left(-\infty, \frac{5}{3}\right)$ and 2 in $\left(\frac{5}{3}, \infty\right)$.

Copyright © 2019 Pearson Education, Inc.

$f'(-1) = \frac{9}{32} > 0;\ f'(2) = 18 > 0$

f is never decreasing. It is increasing on $\left(-\infty, \frac{5}{3}\right)$ and $\left(\frac{5}{3}, \infty\right)$.

11. $f(x) = 2x^3 - 3x^2 - 36x + 10$
$f'(x) = 6x^2 - 6x - 36 = 0 \Rightarrow$
$6\left(x^2 - x - 6\right) = 0 \Rightarrow (x-3)(x+2) = 0 \Rightarrow$
$x = 3, -2$
Critical numbers: 3, –2
$f''(x) = 12x - 6$
$f''(3) = 30 > 0$, so there is a local minimum at $x = 3$.
$f''(-2) = -30 < 0$, so there is a local maximum at $x = -2$.
$f(-2) = 54;\ f(3) = -71$
f has a local maximum of 54 at $x = -2$ and a local minimum of –71 at $x = 3$.

12. $f(x) = 2x^3 - 3x^2 - 12x + 2$
$f'(x) = 6x^2 - 6x - 12 = 0 \Rightarrow x^2 - x - 2 = 0 \Rightarrow$
$(x-2)(x+1) = 0 \Rightarrow x = 2, -1$
–1 and 2 are the critical numbers.
$f''(x) = 12x - 6$
$f''(-1) = -18 < 0$
$f''(2) = 18 > 0$
$f(-1) = 9;\ f(2) = -18$
f has a local maximum of 9 at $x = -1$ and a local minimum of –18 at $x = 2$.

13. $f(x) = x^4 - \frac{8}{3}x^3 - 6x^2 + 2$
$f'(x) = 4x^3 - 8x^2 - 12x = 4x\left(x^2 - 2x - 3\right)$
$= 4x(x-3)(x+1)$
$f'(x) = 0$ when $x = -1$, 0, or 3, so these are the critical numbers.
$f(-1) = -\frac{1}{3};\ f(0) = 2;\ f(3) = -43$
$f''(x) = 12x^2 - 16x - 12$
$f''(3) = 48 > 0$
$f''(0) = -12 < 0$
$f''(-1) = 16 > 0$

A local minimum of $-\frac{1}{3}$ occurs at $x = -1$.
A local maximum of 2 occurs at $x = 0$.
A local minimum of –43 occurs at $x = 3$.

14. $f(x) = x \cdot e^x$
Find the derivative and set it equal to 0.
$f'(x) = xe^x + e^x = 0 \Rightarrow e^x(x+1) = 0 \Rightarrow x = -1$
The only critical number is –1.
$f''(x) = xe^x + e^x + e^x = xe^x + 2e^x$
$f''(-1) = -1e^{-1} + 2e^{-1} = -\frac{1}{e} + \frac{2}{e} = \frac{1}{e} > 0$
$f(-1) = -1e^{-1} = -\frac{1}{e} \approx -.368$

f has a local minimum of $-\frac{1}{e}$ at $x = -1$.

15. $f(x) = 3x \cdot e^{-x}$
$f'(x) = 3x\left(-e^{-x}\right) + 3e^{-x} = (3 - 3x)e^{-x}$
$f''(x) = (3 - 3x)\left(-e^{-x}\right) + (-3)e^{-x} = (3x - 6)e^{-x}$
$f'(x) = 0 \Rightarrow (3 - 3x)e^{-x} = 0 \Rightarrow 3 - 3x = 0 \Rightarrow$
$x = 1$
$x = 1$ is the critical number. $f(1) = 3e^{-1} = \frac{3}{e}$
$f''(1) = -3e^{-1} < 0$, so $x = 1$ is a local maximum.
A local maximum of $\frac{3}{e}$ occurs at $x = 1$.

16. $f(x) = \frac{e^x}{x-1}$
Find the first derivative and set it equal to zero.
$f'(x) = \frac{(x-1)e^x - e^x}{(x-1)^2} = 0 \Rightarrow$
$\frac{e^x(x-1-1)}{(x-1)^2} = 0 \Rightarrow \frac{e^x(x-2)}{(x-1)^2} = 0 \Rightarrow$
$x - 2 = 0 \Rightarrow x = 2$
$$f''(x) = \frac{(x-1)^2\left[e^x + (x-2)e^x\right]}{(x-1)^4} - \frac{e^x(x-2)(2)(x-1)}{(x-1)^4}$$
$f''(2) = \frac{1^2\left(e^2\right) - 0}{1^4} = e^2 > 0$
$f(2) = \frac{e^2}{1} = e^2$
A local minimum of e^2 occurs at $x = 2$.

Copyright © 2019 Pearson Education, Inc.

17. $f(x)=2x^5-5x^3+3x-1$

$f'(x)=10x^4-15x^2+3$

$f''(x)=40x^3-30x$

$f''(1)=40(1)^3-30(1)=10$

$f''(-2)=40(-2)^3-30(-2)=-260$

18. $f(x)=\frac{3-2x}{x+3}$

$f'(x)=\frac{(x+3)(-2)-(3-2x)(1)}{(x+3)^2}=\frac{-9}{(x+3)^2}$

$=-9(x+3)^{-2}$

$f''(x)=-9(-2)(x+3)^{-3}=\frac{18}{(x+3)^3}$

$f''(1)=\frac{18}{(1+3)^3}=\frac{18}{64}=\frac{9}{32}$

$f''(-2)=\frac{18}{(-2+3)^3}=\frac{18}{1}=18$

19. $f(x)=-5e^{2x}$

$f'(x)=-5e^{2x}(2)=-10e^{2x}$

$f''(x)=-10e^{2x}(2)=-20e^{2x}$

$f''(1)=-20e^{2(1)}=-20e^2$

$f''(-2)=-20e^{2(-2)}=-20e^{-4}=-\frac{20}{e^4}$

20. $f(x)=\ln|5x+2|$

$f'(x)=\frac{1}{5x+2}(5)=5(5x+2)^{-1}$

$f''(x)=5(-1)(5x+2)^{-2}(5)=-\frac{25}{(5x+2)^2}$

$f''(1)=-\frac{25}{[5(1)+2]^2}=-\frac{25}{49}$

$f''(-2)=-\frac{25}{[5(-2)+2]^2}=-\frac{25}{64}$

21. $f(x)=-2x^3-\frac{1}{2}x^2-x-3$

$f'(x)=-6x^2-x-1$

Solve $-6x^2-x-1=0$ for x:

$x=\frac{-(-1)\pm\sqrt{(-1)^2-4(-6)(-1)}}{2(-6)}=\frac{1\pm\sqrt{-23}}{-12},$

which has no real-number solution. Therefore there are no local extrema.

$f''(x)=-12x-1$

$f''(x)=0\Rightarrow -12x-1=0\Rightarrow x=-\frac{1}{12}$

$f(x)$ is concave upward on $\left(-\infty,-\frac{1}{12}\right)$ and concave downward on $\left(-\frac{1}{12},\infty\right)$.

$f\left(-\frac{1}{12}\right)=-2\left(-\frac{1}{12}\right)^3-\frac{1}{2}\left(-\frac{1}{12}\right)^2-\left(-\frac{1}{12}\right)-3$

≈ -2.92

$\left(-\frac{1}{12},-2.92\right)$ is an inflection point.

$f(0)=-2(0)^3-\frac{1}{2}(0)^2-0-3=-3,$ so the y-intercept is -3. It is not convenient to find the x-intercept. The function is a polynomial, so there are no asymptotes. Use all of the above information and plot some additional points to sketch the graph.

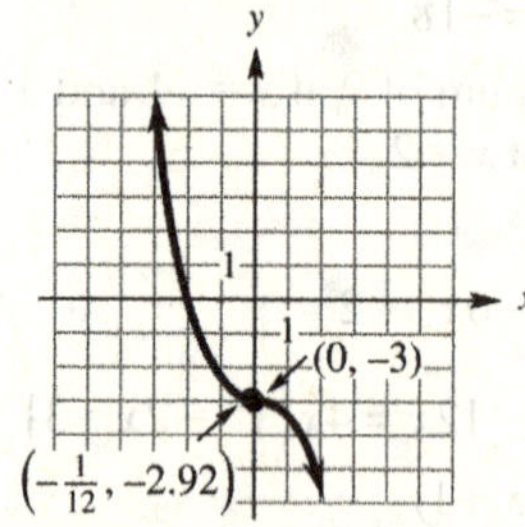

Copyright © 2019 Pearson Education, Inc.

22. $f(x)=-\frac{4}{3}x^3+x^2+30x-7$

$f'(x)=-4x^2+2x+30$

When ,

$f'(x)=0 \Rightarrow -2\left(2x^2-x-15\right)=0 \Rightarrow$

$(2x+5)(x-3)=0 \Rightarrow x=-\frac{5}{2}, 3$

$f\left(-\frac{5}{2}\right)=-\frac{4}{3}\left(-\frac{5}{2}\right)^3+\left(-\frac{5}{2}\right)^2+30\left(-\frac{5}{2}\right)-7$
≈ -54.9

$f(3)=-\frac{4}{3}(3)^3+3^2+30(3)-7=56$

$\left(-\frac{5}{2}, -54.9\right)$ and (3, 56) are critical points.

$f''(x)=-8x+2$

$f''\left(-\frac{5}{2}\right)=22>0$

There is a local minimum at $x=-\frac{5}{2}$.

$f''(3)=-22<0$

There is a local maximum at $x = 3$. $f(x)$ is increasing on $\left(-\frac{5}{2}, 3\right)$ and decreasing on $\left(-\infty, -\frac{5}{2}\right)$ and $(3, \infty)$.

$f''(x)=0 \Rightarrow -8x+2=0 \Rightarrow x=\frac{1}{4}$,

$f(x)$ is concave upward on $\left(-\infty, \frac{1}{4}\right)$ and concave downward on $\left(\frac{1}{4}, \infty\right)$.

$f\left(\frac{1}{4}\right)=-\frac{4}{3}\left(\frac{1}{4}\right)^3+\left(\frac{1}{4}\right)^2+30\left(\frac{1}{4}\right)-7 \approx .54$

$\left(\frac{1}{4}, .54\right)$ is an inflection point. $f(0) = -7$, so the y-intercept is 0. It is not convenient to find the x-intercept.

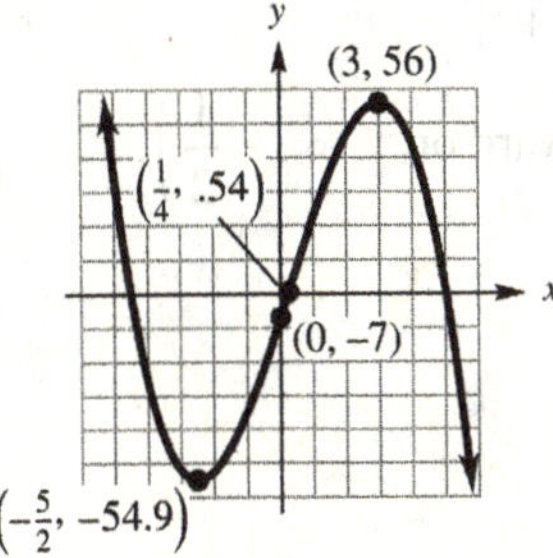

23. $f(x)=x^4-\frac{4}{3}x^3-4x^2+1$

$f'(x)=4x^3-4x^2-8x$

$f'(x)=0 \Rightarrow 4x\left(x^2-x-2\right)=0 \Rightarrow$

$x(x+1)(x-2)=0 \Rightarrow x=0, -1, 2$

-1, 0, and 2 are critical numbers.

$f(-1)=(-1)^4-\frac{4}{3}(-1)^3-4(-1)^2+1=-\frac{2}{3}$

$f(0)=1$

$f(2)=2^4-\frac{4}{3}(2)^3-4(2)^2+1=-\frac{29}{3}$

$\left(-1, -\frac{2}{3}\right)$, (0, 1), and $\left(2, -\frac{29}{3}\right)$ are critical points.

$f''(x)=12x^2-8x-8$

$f''(-1)=12>0$

There is a local minimum at $x=-1$.

$f''(0)=-8<0$

There is a local maximum at $x = 0$.

$f''(2)=24>0$

There is a local minimum at $x = 2$.

$f(x)$ is increasing on $(-1, 0)$ and $(2, \infty)$ and decreasing on $(-\infty, -1)$ and $(0, 2)$.

$f''(x)=0 \Rightarrow 4\left(3x^2-2x-2\right)=0 \Rightarrow$

$x=\frac{2\pm\sqrt{(-2)^2+24}}{6}=\frac{2\pm2\sqrt{7}}{6}=\frac{1\pm\sqrt{7}}{3}$

$f(x)$ is concave upward on $\left(-\infty, \frac{1-\sqrt{7}}{3}\right)$ and $\left(\frac{1+\sqrt{7}}{3}, \infty\right)$ and concave downward on $\left(\frac{1-\sqrt{7}}{3}, \frac{1+\sqrt{7}}{3}\right)$.

$f\left(\frac{1-\sqrt{7}}{3}\right)\approx .11$

$f\left(\frac{1+\sqrt{7}}{3}\right)\approx -5.12$

$\left(\frac{1-\sqrt{7}}{3}, .11\right)$ and $\left(\frac{1+\sqrt{7}}{3}, -5.12\right)$ are inflection points.

$f(0) = 1$, so the y-intercept is 1. The x-intercepts are not convenient to find but can be estimated from the graph. This is a polynomial function, so there are no asymptotes.

Copyright © 2019 Pearson Education, Inc.

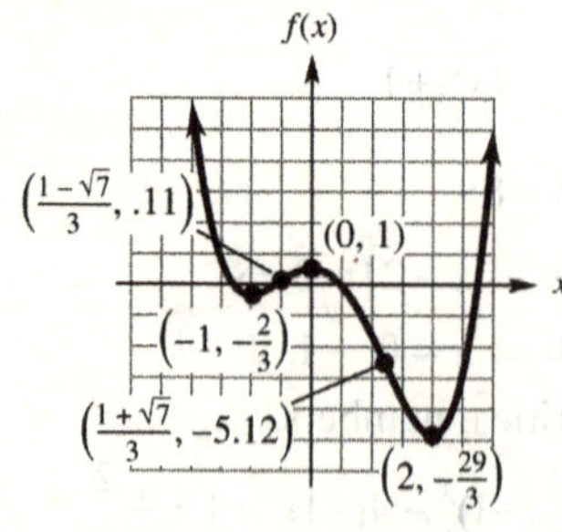

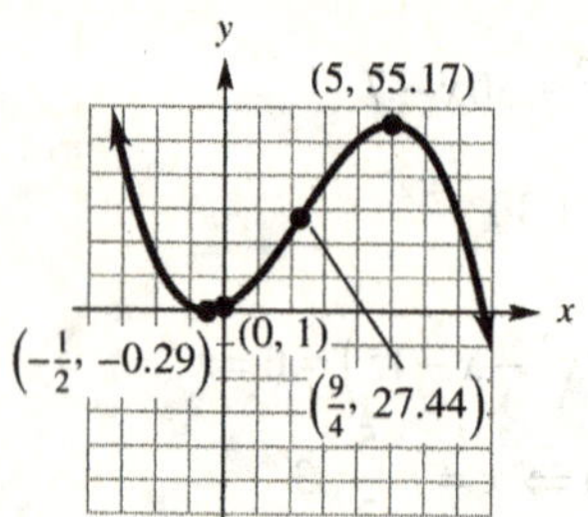

24. $f(x) = -\frac{2}{3}x^3 + \frac{9}{2}x^2 + 5x + 1$

$f'(x) = -2x^2 + 9x + 5$

$f'(x) = 0 \Rightarrow 2x^2 - 9x - 5 = 0 \Rightarrow$

$(2x+1)(x-5) = 0 \Rightarrow x = -\frac{1}{2}, 5$

$-\frac{1}{2}$ and 5 are critical numbers.

$f\left(-\frac{1}{2}\right) \approx -.29; f(5) \approx 55.17$

$\left(-\frac{1}{2}, -.29\right)$ and (5, 55.17) are critical points.

$f''(x) = -4x + 9$

$f''\left(-\frac{1}{2}\right) = 11 > 0$

There is a local minimum at $x = -\frac{1}{2}$.

$f''(5) = -11 < 0$

There is a local maximum at $x = 5$.

$f(x)$ is increasing on $\left(-\frac{1}{2}, 5\right)$ and decreasing on $\left(-\infty, -\frac{1}{2}\right)$ and $(5, \infty)$.

$f''(x) = 0 \Rightarrow -4x + 9 = 0 \Rightarrow x = \frac{9}{4}$,

$f(x)$ is concave upward on $\left(-\infty, \frac{9}{4}\right)$ and concave downward on $\left(\frac{9}{4}, \infty\right)$.

$f\left(\frac{9}{4}\right) \approx 27.44$

$\left(\frac{9}{4}, 27.44\right)$ is an inflection point. $f(0) = 1$, so the y-intercept is 1. It is not convenient to find the x-intercepts.

25. $f(x) = \frac{x-1}{2x+1}$

$f'(x) = \frac{(2x+1)-(x-1)(2)}{(2x+1)^2} = \frac{3}{(2x+1)^2}$

$= 3(2x+1)^{-2}$

$f'(x)$ is never zero. $f'\left(-\frac{1}{2}\right)$ does not exist, but since $f\left(-\frac{1}{2}\right)$ is undefined, $-\frac{1}{2}$ is not a critical number. (There is a vertical asymptote at $x = -\frac{1}{2}$.) This function has no critical numbers, so there can be no local extrema.

To determine where the function is increasing and decreasing, test points in the intervals $\left(-\infty, -\frac{1}{2}\right)$ and $\left(-\frac{1}{2}, \infty\right)$.

$f'(-1) = \frac{3}{[2(-1)+1]^2} = 2 > 0$

$f'(0) = \frac{3}{[2(0)+1]^2} = 2 > 0$

$f(x)$ is increasing on $\left(-\infty, -\frac{1}{2}\right)$ and on $\left(-\frac{1}{2}, \infty\right)$.

$f''(x) = 3(-2)(2x+1)^{-3}(2) = -\frac{12}{(2x+1)^3}$

Test $x = -1$ and $x = 0$ in $f''(x)$.

$f''(-1) = -\frac{12}{[2(-1)+1]^3} = 12 > 0$

$f(x)$ is concave upward on $\left(-\infty, -\frac{1}{2}\right)$.

Copyright © 2019 Pearson Education, Inc.

$$f''(0) = -\frac{12}{[2(0)+1]^3} = -12 < 0$$

$f(x)$ is concave downward on $\left(-\frac{1}{2}, \infty\right)$.

$f''(x)$ is never zero, so there are no inflection points.

When $f(x) = 0$,

$$f(x) = 0 \Rightarrow \frac{x-1}{2x+1} = 0 \Rightarrow x - 1 = 0 \Rightarrow x = 1$$

The only x-intercept is 1.

$f(0) = \frac{0-1}{2(0)+1} = -1$, so the y-intercept is -1.

This rational function has a vertical asymptote where the denominator is 0, that is, at $x = -\frac{1}{2}$. To find the horizontal asymptote, find the limit at infinity.

$$\lim_{x\to\infty} f(x) = \lim_{x\to\infty} \frac{x-1}{2x+1} = \lim_{x\to\infty} \frac{\frac{x}{x} - \frac{1}{x}}{\frac{2x}{x} + \frac{1}{x}}$$

$$= \lim_{x\to\infty} \frac{1-\frac{1}{x}}{2+\frac{1}{x}} = \frac{1-0}{2+0} = \frac{1}{2}$$

$y = \frac{1}{2}$ is a horizontal asymptote.

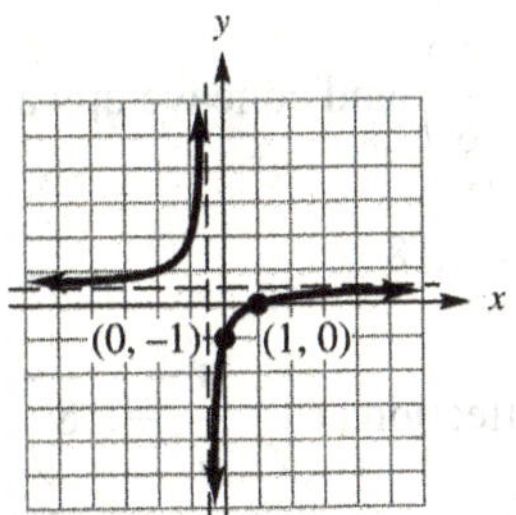

26. $f(x) = \frac{2x-5}{x+3}$

$$f'(x) = \frac{(x+3)(2) - (2x-5)}{(x+3)^2} = \frac{11}{(x+3)^2}$$

$$= 11(x+3)^{-2}$$

$f'(x)$ is never zero. -3 is not a critical number because $f(x)$ does not exist at $x = -3$. There are no local extrema. Test $x = -4$ and $x = -2$ in $f'(x)$.

$$f'(-4) = \frac{11}{(-4+3)^2} = 11 > 0$$

$$f(-2) = \frac{11}{(-2+3)^2} = 11 > 0$$

$f(x)$ is increasing on $(-\infty, -3)$ and on $(-3, \infty)$.

$$f''(x) = 11(-2)(x+3)^{-3} = -22(x+3)^{-3}$$

Test $x = -4$ and $x = -2$ in $f''(x)$.

$$f''(-4) = -22(-4+3)^{-3} = 22 > 0$$

$f(x)$ is concave upward on $(-\infty, -3)$.

$$f''(-2) = -22(-2+3)^{-3} = -22 < 0$$

$f(x)$ is concave downward on $(-3, \infty)$.

$f''(x)$ is never zero, so there are no inflection points.

If $f(x) = 0$,

$$f(x) = 0 \Rightarrow \frac{2x-5}{x+3} = 0 \Rightarrow 2x - 5 = 0 \Rightarrow x = \frac{5}{2}$$

The x-intercept is $\frac{5}{2}$.

$$f(0) = \frac{2 \cdot 0 - 5}{0+3} = -\frac{5}{3}$$

The y-intercept is $-\frac{5}{3}$.

This rational function has a vertical asymptote at $x = -3$.

$$\lim_{x\to\infty} f(x) = \lim_{x\to\infty} \frac{2x-5}{x+3} = \lim_{x\to\infty} \frac{\frac{2x}{x} - \frac{5}{x}}{\frac{x}{x} + \frac{3}{x}}$$

$$= \lim_{x\to\infty} \frac{2-\frac{5}{x}}{1+\frac{3}{x}} = \frac{2-0}{1+0} = 2$$

$y = 2$ is a horizontal asymptote.

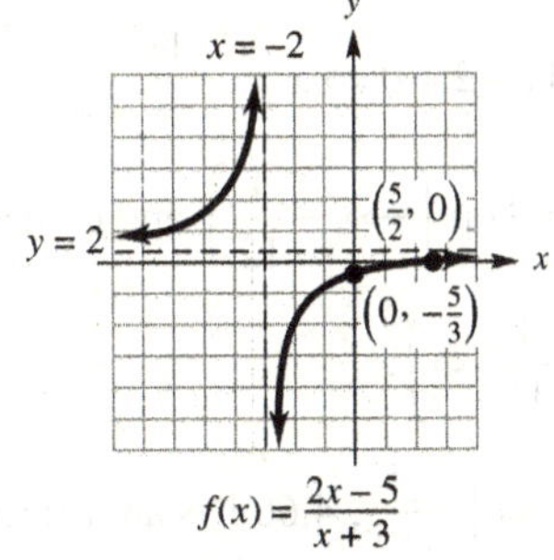

Copyright © 2019 Pearson Education, Inc.

27. $f(x) = -4x^3 - x^2 + 4x + 5$

$f'(x) = -12x^2 - 2x + 4$

$f'(x) = 0 \Rightarrow -2\left(6x^2 + x - 2\right) = 0 \Rightarrow$

$(3x + 2)(2x - 1) = 0$

$-\frac{2}{3}$ and $\frac{1}{2}$ are critical numbers.

$f\left(-\frac{2}{3}\right) \approx 3.07;\ f\left(\frac{1}{2}\right) = 6.25$

$\left(-\frac{2}{3}, 3.07\right)$ and $\left(\frac{1}{2}, 6.25\right)$ are critical points.

$f''(x) = -24x - 2$

$f''\left(-\frac{2}{3}\right) = 14 > 0$

There is a local minimum at $x = -\frac{2}{3}$.

$f''\left(\frac{1}{2}\right) = -14 < 0$

There is a local maximum at $x = \frac{1}{2}$.

f is increasing on $\left(-\frac{2}{3}, \frac{1}{2}\right)$ and decreasing on $\left(-\infty, -\frac{2}{3}\right)$ and $\left(\frac{1}{2}, \infty\right)$.

$f''(x) = 0 \Rightarrow -24x - 2 = 0 \Rightarrow x = -\frac{1}{12}$,

f is concave upward on $\left(-\infty, -\frac{1}{12}\right)$ and concave downward on $\left(-\frac{1}{12}, \infty\right)$.

$f\left(-\frac{1}{12}\right) \approx 4.66$, so $\left(-\frac{1}{12}, 4.66\right)$ is an inflection point. $f(0) = 5$, so the y-intercept is 5. It is not convenient to find the x-intercept.

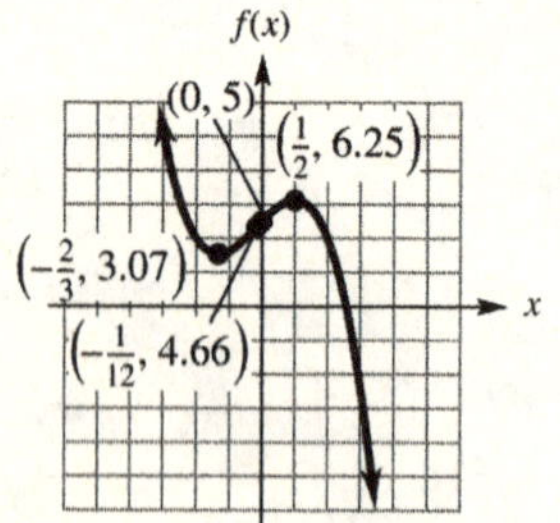

28. $f(x) = x^3 + \frac{5}{2}x^2 - 2x - 3$

$f'(x) = 3x^2 + 5x - 2 = (3x - 1)(x + 2) = 0 \Rightarrow$

$x = \frac{1}{3}, -2$

$\frac{1}{3}$ and -2 are critical numbers.

$f(-2) = 3;\ f\left(\frac{1}{3}\right) \approx -3.35$

$(-2, 3)$ and $\left(\frac{1}{3}, -3.35\right)$ are critical points.

$f''(x) = 6x + 5 = 0 \Rightarrow x = -\frac{5}{6}$

$f''\left(\frac{1}{3}\right) = 7 > 0$, so $f(x)$ has a local minimum at $x = \frac{1}{3}$.

$f''(-2) = -7 < 0$, so $f(x)$ has a local maximum at $x = -2$.

f is increasing on $(-\infty, -2)$ and $\left(\frac{1}{3}, \infty\right)$ and decreasing on $\left(-2, \frac{1}{3}\right)$. Thus, f is concave downward on $\left(-\infty, -\frac{5}{6}\right)$ and concave upward on $\left(-\frac{5}{6}, \infty\right)$. $f\left(-\frac{5}{6}\right) \approx -.18$

There is a point of inflection at $\left(-\frac{5}{6}, -.18\right)$. The y-intercept is -3. It is not convenient to find the x-intercepts.

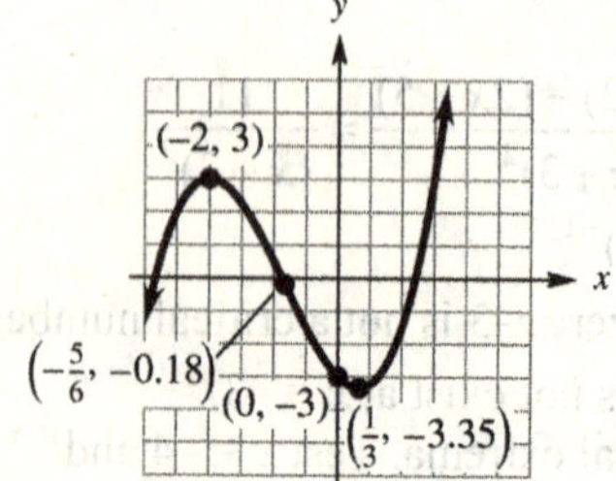

Copyright © 2019 Pearson Education, Inc.

29. $f(x) = x^4 + 2x^2$

$f'(x) = 4x^3 + 4x = 0 \Rightarrow 4x\left(x^2 + 1\right) = 0$

$x^2 + 1 = 0$ has no real solution, so $x = 0$ is the only critical number. $f(0) = 0$, so $(0, 0)$ is a critical number.

$f''(x) = 12x^2 + 4$

$f''(0) = 4 > 0$

There is a local minimum at $x = 0$. $f(x)$ is increasing on $(0, \infty)$ and decreasing on $(-\infty, 0)$.

$f''(x) = 12x^2 + 4 > 0$ for all x, so $f(x)$ is concave upward on $(-\infty, \infty)$.

$f''(x)$ is never zero, so there are no inflection points. $f(0) = 0$, so the y-intercept is 0. To find any x-intercepts, let $f(x) = 0$.

$f(x) = x^4 + 2x^2 = 0 \Rightarrow x^2\left(x^2 + 2\right) = 0$

The only real solution is $x = 0$, so the only x-intercept is 0.

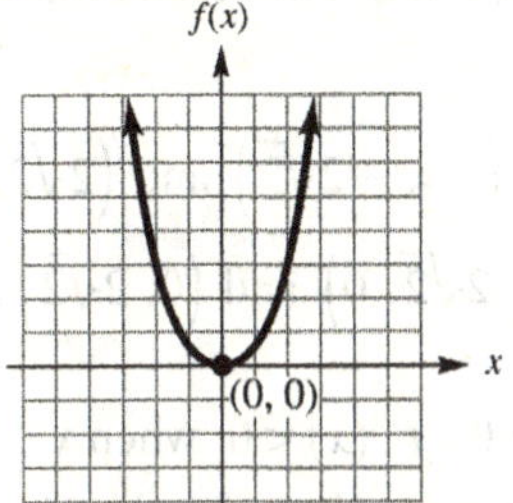

30. $f(x) = 6x^3 - x^4$

$f'(x) = 18x^2 - 4x^3 = 0 \Rightarrow 2x^2(9 - 2x) = 0 \Rightarrow$

$x = 0, \dfrac{9}{2}$

0 and $\dfrac{9}{2}$ are critical numbers.

$f(0) = 0;\ f\left(\dfrac{9}{2}\right) \approx 136.7$

$(0, 0)$ and $\left(\dfrac{9}{2}, 136.7\right)$ are critical points.

$f''(x) = 36x - 12x^2$

$f''(0) = 0$

There is neither a local minimum nor a local maximum at $x = 0$.

$f''\left(\dfrac{9}{2}\right) = -81 < 0$

There is a local maximum at $x = \dfrac{9}{2}$.

f is increasing on $\left(-\infty, \dfrac{9}{2}\right)$ and decreasing on $\left(\dfrac{9}{2}, \infty\right)$.

$f''(x) = 0 \Rightarrow 36x - 12x^2 = 0 \Rightarrow$
$12x(3 - x) = 0 \Rightarrow x = 0, 3$

Test $f''(x)$ at $x = -1$, $x = 1$, and $x = 4$.

$f''(-1) = -48 < 0$

f is concave downward on $(-\infty, 0)$.

$f''(1) = 24 > 0$

f is concave upward on $(0, 3)$.

$f''(4) = -48 < 0$

f is concave downward on $(3, \infty)$.

$f(3) = 81$

There are inflection points at $(0, 0)$ and $(3, 81)$. $f(0) = 0$, so the y-intercept is 0.

To find any x-intercepts, let $f(x) = 0$.

$f(x) = 6x^3 - x^4 = 0 \Rightarrow x^3(6 - x) = 0 \Rightarrow$
$x = 0, 6$

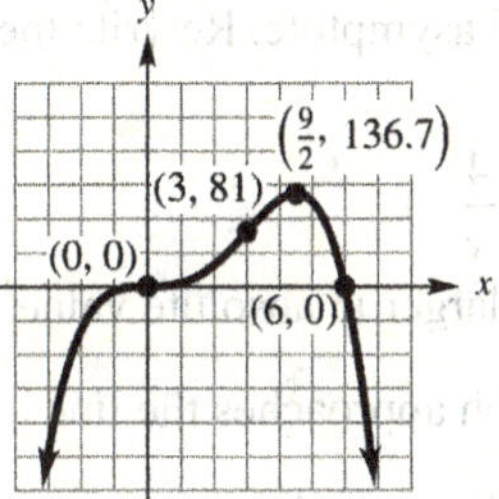

31. $f(x) = \dfrac{x^2 + 4}{x} = x + \dfrac{4}{x}$

$f'(x) = 1 - \dfrac{4}{x^2} = 0 \Rightarrow \dfrac{4}{x^2} = 1 \Rightarrow 4 = x^2 \Rightarrow$

$x = \pm 2$

-2 and 2 are critical numbers. Notice that $f(x)$ does not exist at $x = 0$. The line $x = 0$ is a vertical asymptote.

$f(-2) = -4;\ f(2) = 4$

$(-2, -4)$ and $(2, 4)$ are critical points.

$f''(x) = \dfrac{8}{x^3}$

$f''(-2) = -1 < 0$

There is a local maximum at $x = -2$.

$f''(2) = 1 > 0$

Copyright © 2019 Pearson Education, Inc.

There is a local minimum at $x = 2$. $f(x)$ is increasing on $(-\infty, -2)$ and $(2, \infty)$ and decreasing on $(-2, 0)$ and $(0, 2)$. $f''(x)$ is never zero, so there are no points of inflection. $f(x)$ is concave upward on $(0, \infty)$ and concave downward on $(-\infty, 0)$. Because $f(0)$ does not exist, there is no y-intercept. To find any x-intercepts, let $f(x) = 0$.

$$f(x) = \frac{x^2+4}{x} = 0 \Rightarrow x^2 + 4 = 0$$

This equation has no real solution, so there are no x-intercepts. When $x = 0$, the denominator of $\frac{x^2+4}{x}$ is equal to 0 but the numerator is not, so $x = 0$ (the y-axis) is a vertical asymptote.

We now determine whether this rational function has a horizontal asymptote.

$$\lim_{x\to\infty} \frac{x^2+4}{x} = \lim_{x\to\infty} \frac{\frac{x^2}{x^2}+\frac{4}{x^2}}{\frac{x}{x^2}} = \lim_{x\to\infty} \frac{1+\frac{4}{x^2}}{\frac{1}{x}}$$
$$= \frac{1+0}{0} = \frac{1}{0}$$

This expression is undefined, which indicates that there is no horizontal asymptote. Rewrite the function rule as

$$f(x) = \frac{x^2+4}{x} = x + \frac{4}{x}.$$

As x gets larger and larger in absolute value, $\frac{4}{x} \to 0$, and the graph approaches the line $y = x$, which is an oblique or slant asymptote.

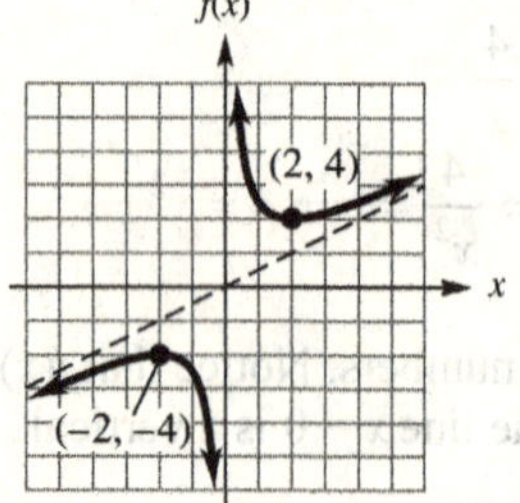

32. $f(x) = x + \frac{8}{x} = x + 8x^{-1}$

$f'(x) = 1 - 8x^{-2} = 1 - \frac{8}{x^2} = 0 \Rightarrow x^2 = 8 \Rightarrow$

$x = \pm 2\sqrt{2}$ Critical numbers

Notice that $f(x)$ does not exist at $x = 0$. The line $x = 0$ is a vertical asymptote.

$f(2\sqrt{2}) = 2\sqrt{2} + \frac{8}{2\sqrt{2}} = 2\sqrt{2} + 2\sqrt{2} = 4\sqrt{2}$

$f(-2\sqrt{2}) = -4\sqrt{2}$

Thus, $(2\sqrt{2}, 4\sqrt{2})$ and $(-2\sqrt{2}, -4\sqrt{2})$ are critical points.

$f''(x) = \frac{16}{x^3}$

$f''(2\sqrt{2}) > 0$, so $f(x)$ has a local minimum at $x = 2\sqrt{2}$.

$f''(-2\sqrt{2}) < 0$, so $f(x)$ has a local maximum at $x = -2\sqrt{2}$.

$f(x)$ is increasing on $(-\infty, -2\sqrt{2})$ and $(2\sqrt{2}, \infty)$ and decreasing on $(-2\sqrt{2}, 0)$ and $(0, 2\sqrt{2})$.

$f''(x) > 0$ when $x > 0$. $f''(x) < 0$ when $x < 0$. Thus, $f(x)$ is concave upward on $(0, \infty)$ and concave downward on $(-\infty, 0)$. There is no point of inflection at $x = 0$, even though $f(x)$ changes concavity because $f(0)$ is undefined. $f(0)$ is undefined, so there is no y-intercept.

To any x-intercepts, let $f(x) = 0$.

$f(x) = x + \frac{8}{x} = 0 \Rightarrow x\left(x + \frac{8}{x}\right) = x(0) \Rightarrow$

$x^2 + 8 = 0$

This equation has no real solution, so there are no x-intercepts. To look for vertical and horizontal asymptotes, we rewrite the function rule so that the right-hand side is a single fraction.

$f(x) = x + \frac{8}{x} = \frac{x^2+8}{x}$

Copyright © 2019 Pearson Education, Inc.

When $x = 0$, the denominator is 0 but the numerator is not, so $x = 0$ (the y-axis) is a vertical asymptote. Using the method shown in the solution for Exercise 31, we find that

$\lim_{x\to\infty} \frac{x^2+8}{x} = \frac{1}{0}$, which is undefined. This indicates that there is no horizontal asymptote. To find a possible slant asymptote, use the function rule in its original form. As $x \to \infty$ or $x \to -\infty$, $\frac{8}{x} \to 0$, so the graph will get closer and closer to the line $y = x$, which is a slant asymptote.

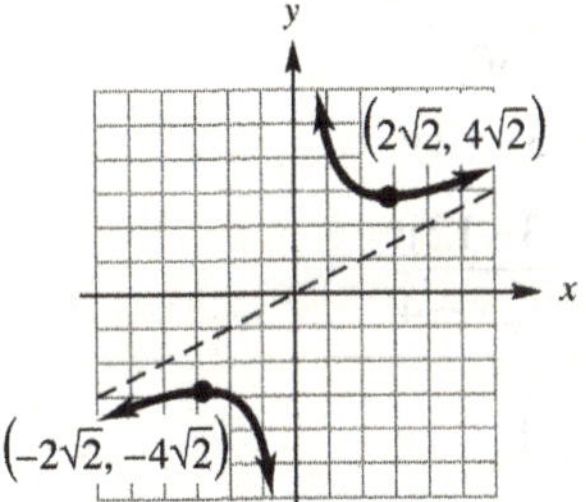

33. $f(x) = -x^2 + 6x + 1$; $[2, 4]$

$f'(x) = -2x + 6 = 0$ when $x = 3$

The absolute extrema must occur at the critical number 3 or at the endpoints.

$f(2) = 9$ Absolute minimum

$f(3) = 10$ Absolute maximum

$f(4) = 9$ Absolute minimum

The absolute maximum of 10 occurs at $x = 3$. The absolute minimum of 9 occurs at $x = 2$ and $x = 4$.

34. $f(x) = 4x^2 - 4x - 7$; $[-1, 3]$

$f'(x) = 8x - 4 = 0 \Rightarrow x = \frac{1}{2}$

x	-1	$\frac{1}{2}$	3
$f(x)$	1	-8	17

The absolute maximum of 17 occurs at $x = 3$ and the absolute minimum of -8 occurs at $x = \frac{1}{2}$.

35. $f(x) = x^3 + 2x^2 - 15x + 3$; $[-.5, 3.3]$

$f'(x) = 3x^2 + 4x - 15 = 0 \Rightarrow$

$(3x-5)(x+3) = 0 \Rightarrow x = \frac{5}{3}, -3$

Find the values of the function at the critical numbers and at the endpoints.

$f(-.5) = 10.875$

$f(3.3) = 11.217$ Absolute maximum

$f\left(\frac{5}{3}\right) = -\frac{319}{27}$ Absolute minimum

$f(2) = -11$

The absolute maximum of 11.217 occurs at $x = 3.3$.

The absolute minimum of $-\frac{319}{27}$ occurs at $x = \frac{5}{3}$.

36. $f(x) = -2x^3 - x^2 + 4x - 1$; $[-3, 1]$

$f'(x) = -6x^2 - 2x + 4 = -2(3x^2 + x - 2)$
$= -2(3x-2)(x+1)$

$f'(x) = 0$ when $x = -1$ or $x = \frac{2}{3}$,

so the critical numbers are -1 and $\frac{2}{3}$. Find the values of the function at the critical numbers and at the endpoint.

x	-3	-1	$\frac{2}{3}$	1
$f(x)$	32	-4	$\frac{17}{27}$	0

The absolute maximum of 32 occurs at $x = -3$ and the absolute minimum of -4 occurs at $x = -1$.

37. $f(x) = x^3 + 6x^2 - 15x + 1$; $(0, 9)$

$f'(x) = 3x^2 + 12x - 15 = 0 \Rightarrow$

$3(x+5)(x-1) = 0 \Rightarrow x = -5, 1$

Find the value of the second derivative at the critical point $x = 1$ to determine if the point is a max or min. The critical point $x = -5$ is not in the given interval, so it does not need to be tested.

$f''(x) = 6x + 12$

$f''(1) = 6(1) + 12 = 18 > 0$

The second derivative is greater than zero, therefore there is an absolute minimum at $x = 1$.

Copyright © 2019 Pearson Education, Inc.

38. $f(x) = x^3 + 6x^2 - 15x + 1;\ (-9, 0)$

$f'(x) = 3x^2 + 12x - 15 = 0 \Rightarrow$

$3(x+5)(x-1) = 0 \Rightarrow x = -5, 1$

Find the value of the second derivative at the critical point $x = -5$ to determine if the point is a max or min. The critical point $x = 1$ is not in the given interval, so it does not need to be tested.

$f''(x) = 6x + 12$

$f''(-5) = 6(-5) + 12 = -18 < 0$

The second derivative is less than zero, therefore there is an absolute maximum at $x = -5$.

39. $f(x) = xe^{-x/3};\ (-\infty, \infty)$

$f'(x) = e^{-x/3}\left(1 - \frac{x}{3}\right) = 0 \Rightarrow$

$1 - \frac{x}{3} = 0 \Rightarrow x = 3$

Find the value of the second derivative at the critical point $x = 3$ to determine if the point is a max or min.

$f''(x) = -\frac{1}{3}e^{-x/3}\left(2 + \frac{x}{3}\right)$

$f''(3) = -\frac{1}{3}e^{-3/3}\left(2 + \frac{3}{3}\right) = -\frac{1}{3e} < 0$

The second derivative is less than zero, therefore there is an absolute maximum at $x = 3$.

40. $f(x) = -5xe^{-2x};\ (-\infty, \infty)$

$f'(x) = e^{-2x}(10x - 5) = 0 \Rightarrow$

$10x - 5 = 0 \Rightarrow x = \frac{1}{2}$

Find the value of the second derivative at the critical point $x = \frac{1}{2}$ to determine if the point is a max or min.

$f''(x) = e^{-2x}(20 - 20x)$

$f''\left(\frac{1}{2}\right) = e^{-2.5}(20 - 20(.5)) = \frac{10}{e} > 0$

The second derivative is greater than zero, therefore there is an absolute minimum at $x = \frac{1}{2}$.

41. $\frac{d}{dx}(10xy = 2x + 5y)$

$10x\frac{dy}{dx} + y(10) = 2 + 5\frac{dy}{dx}$

$\frac{dy}{dx}(10x - 5) = 2 - 10y$

$\frac{dy}{dx} = \frac{2 - 10y}{10x - 5}$

42. $\frac{d}{dx}(12xy = 4y + 3x)$

$12x\frac{dy}{dx} + y(12) = 4\frac{dy}{dx} + 3$

$\frac{dy}{dx}(12x - 4) = 3 - 12y$

$\frac{dy}{dx} = \frac{3 - 12y}{12x - 4}$

43. $\frac{d}{dx}(ye^x - 10x = y)$

$ye^x + e^x\frac{dy}{dx} - 10 = \frac{dy}{dx}$

$\frac{dy}{dx}(e^x - 1) = 10 - ye^x$

$\frac{dy}{dx} = \frac{10 - ye^x}{e^x - 1}$

44. $\frac{d}{dx}(5ye^x + x^2 = -10y)$

$5ye^x + 5e^x\frac{dy}{dx} + 2x = -10\frac{dy}{dx}$

$\frac{dy}{dx}(5e^x + 10) = -2x - 5ye^x$

$\frac{dy}{dx} = \frac{-2x - 5ye^x}{5e^x + 10}$

45. $\frac{d}{dx}(5 - y^3 = 6 - 2x^6)$

$-3y^2\frac{dy}{dx} = -12x^5$

$\frac{dy}{dx} = \frac{-12x^5}{-3y^2} = \frac{4x^5}{y^2}$

Copyright © 2019 Pearson Education, Inc.

Now substitute the point into the equation to find the slope.

$$\frac{dy}{dx} = \frac{4(1)^5}{(1)^2} = 4$$

Now use the point-slope formula to find the equation of the tangent line.

$$y - y_1 = m(x - x_1)$$
$$y - 1 = 4(x - 1)$$
$$y - 1 = 4x - 4$$
$$y = 4x - 3$$

46. $\frac{d}{dx}\left(5 - y^2 = 6 - 2x^4\right)$

$$-2y\frac{dy}{dx} = -8x^3$$
$$\frac{dy}{dx} = \frac{-8x^3}{-2y} = \frac{4x^3}{y}$$

Now substitute the point into the equation to find the slope.

$$\frac{dy}{dx} = \frac{4(1)^3}{1} = 4$$

Now use the point-slope formula to find the equation of the tangent line.

$$y - y_1 = m(x - x_1)$$
$$y - 1 = 4(x - 1)$$
$$y - 1 = 4x - 4$$
$$y = 4x - 3$$

47. **a.** 40,000

b. 2008 – 2013

c. Absolute minimum attendance was 33,000 and the absolute maximum attendance was 41,000.

48. **a.** February; the Academy Awards are awarded in February

b. Decreasing

49. **a.** $f'(x) = .12432e^{.024x}$ therefore the first derivative will always be positive or increasing through 2030.

b. $f''(x) = .003e^{.024x}$ therefore the second derivative will always be positive or concave up through 2030. The graph is increasing and concave up, which means the population is increasing at a faster and faster rate.

50. **a.** $f'(x) = \frac{42}{x}$ therefore the first derivative will always be positive or increasing over the given period.

b. $f''(x) = -\frac{42}{x^2}$ therefore the second derivative will always be negative or concave down over the given period. The graph is increasing and concave down, which means the winning height is increasing at a slower and slower rate.

51. **a.** $I(x) = -68x^2 + 7480x - 118,000$

$I'(x) = -136x + 7480 = 0$ when $x = 55$

Determine the value of the first derivative to the left and right of the croitical value.

$I'(30) = -136(30) + 7480 = 3400$

$I'(60) = -136(60) + 7480 = -680$

a. The average adjusted gross income is increasing over the interval (25, 55).

b. The average adjusted gross income is decreasing over the interval (55, 75).

c. The absolute maximum will be

$$I(55) = -68(55)^2 + 7480(55) - 118,000 = \$87,700$$

occurred at $x = 55$.

52. $I(x) = -.02x^2 + 1.32x + 29.86$

$I'(x) = -.04x + 1.32 = 0$

when $x = 33$

The absolute extrema must occur at the critical number 33 or at the endpoints.

$I(0) = 29.86$

$I(33) = 51.64$

$I(60) = 37.06$

The absolute maximum stock price of 51.64 occurred 33 trading days after the IPO.

53. **a.** $P(0) = -.02(0)^2 + .32(0) + 127 \approx 127$

$P(20) = -.02(20)^2 + .32(20) + 127 \approx 125$

The population in Japan in the year 2000 was approximately 127 million and the population will be approximately 125 million in 2020.

Copyright © 2019 Pearson Education, Inc.

b. $P(x) = -.02x^2 + .32x + 127$

$P'(x) = -.04x + .32 = 0$ when $x = 8$

The absolute extrema must occur at the critical number or the endpoints.

$P(0) \approx 127$

$P(8) \approx 128$

$P(20) \approx 125$

During this period, Japan had an absolute minimum population of 125 million in 2020 and an absolute maximum of 128 million in 2008.

54. a. $W(59) = .011(59)^2 - 1.606(59) + 58.7 = 2$

$W(117) = .011(117)^2 - 1.606(117) + 58.7 = 21$

There were 2 female senators in 1959 and 21 in 2017.

b. $W(x) = .011x^2 - 1.606x + 58.7$

$W'(x) = .022x - 1.606 = 0$ when $x = 73$

The absolute extrema must occur at the critical number 73 or at the endpoints.

$f(59) = 2$

$f(73) = 0$ Absolute minimum

$f(117) = 21$ Absolute maximum

The absolute maximum of 21 female senators occured at $x = 117$ (or in 2017). The absolute minimum of 0 occurred at $x = 73$ (or in 1973).

55. a. $P(t) = -.2t^3 + 1.5t^2 - 2.4t + 48$

$P'(t) = -.6t^2 + 3t - 2.4 = 0$

when t = 1 and t = 4

The critical values occur when $t = 1$ and $t = 4$.

b and c. The local extrema must occur at the critical numbers 1 or 4.

$P(1) = 46.9$

$P(4) = 49.6$

The local minimum occurred when $t = 1$ (2009) and the maximum occurred when $t = 4$ (2012).

56.a. $P(t) = .1t^3 - 1.8t^2 + 8.1t + 93$

$P'(t) = .3t^2 - 3.6t + 8.1 = 0$

when t = 3 and t = 9

The critical values occur when $t = 3$ and $t = 9$.

b and c. The local extrema must occur at the critical numbers 3 or 9.

$P(3) = 103.8$

$P(9) = 93$

The local minimum occurred when $t = 9$ (September) and the maximum occurred when $t = 3$ (March).

57. Plot the graph.

$f(x) = -.02x^4 + .55x^3 - 4.8x^2 + 16x - 7$

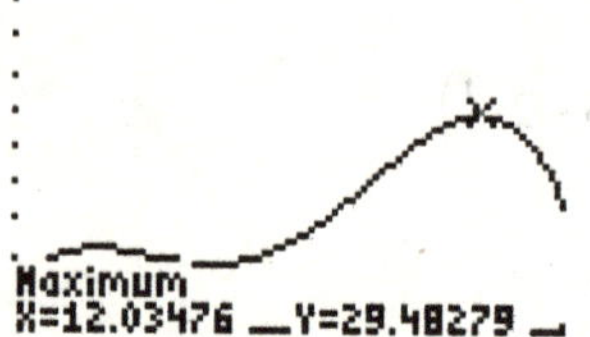

[1, 14] by [0, 50]

Absolute maxima: (12, 29)

The absolute maximum of 29% occurred in 2012.

58. Plot the graph.

$f(x) = -.31x^4 + 13.2x^3 - 206x^2 + 1394x - 3332$

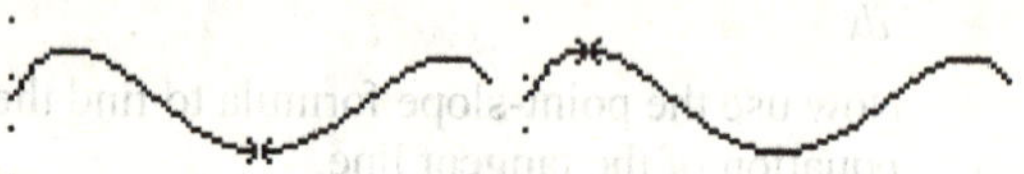

[7, 14] by [75 135]

Absolute minimum: (10.7, 106)

Absolute maxima: (7.9, 125)

The absolute minimum of 106 occurred in late 2010 and the absolute maximum of 125 occurred late in 2007.

59. Let x = width of fence and let y = length of fence. An equation describing the amount of fencing is

$500 = 2x + y \Rightarrow y = 500 - 2x$

$A = xy \Rightarrow A(x) = x(500 - 2x) = 500x - 2x^2$

$A'(x) = 500 - 4x = 0 \Rightarrow x = 125$

Then $y = 500 - 2(125) = 250$.

$A''(x) = -4 < 0$, so $A(125)$ is a maximum.

Dimensions for maximum area are 125 m by 250 m.

60. $500 = 2x + 2y \Rightarrow y = 250 - x$

$A = xy \Rightarrow A(x) = x(250 - x) = 250x - x^2$

$A'(x) = 250 - 2x = 0 \Rightarrow x = 125$

Then $y = 250 - 125 = 125$.

$A''(x) = -2 < 0$, so $A(125)$ is a maximum.

Dimensions for maximum area are 125 m by 125 m.

Copyright © 2019 Pearson Education, Inc.

61. Let x = length and width of a side of base and let h = height.
Volume = 27 cubic meters with a square base and top. Find height, length, and width for minimum surface area.

$$\text{Volume} = x^2h = 27 \Rightarrow h = \frac{27}{x^2}$$

$$\text{Surface area} = A(x) = 2x^2 + 4xh$$

$$A(x) = 2x^2 + 4x\left(\frac{27}{x^2}\right) = 2x^2 + 108x^{-1}$$

$$A'(x) = 4x - 108x^{-2}$$

$$A' = 0 \Rightarrow 4x - \frac{108}{x^2} = 0 \Rightarrow \frac{4x^3 - 108}{x^2} = 0 \Rightarrow$$

$$4x^3 = 108 \Rightarrow x^3 = 27 \Rightarrow x = 3$$

$$A''(x) = 4 + 2(108)x^{-3} = 4 + \frac{216}{x^3}$$

$$A''(3) = 12 > 0$$

The minimum occurs at $x = 3$ where $h = \frac{27}{3^2} = 3$.

The dimensions are 3 m by 3 m by 3 m.

62. Volume of open cylinder = 27π cubic inches. Find radius of bottom to minimize cost of material.
Volume of cylinder $= \pi r^2 h$
Surface area of cylinder open at one end $= 2\pi rh + \pi r^2$,

$$V = \pi r^2 h = 27\pi \Rightarrow h = \frac{27\pi}{\pi r^2}$$

$$A = 2\pi r\left(\frac{27\pi}{\pi r^2}\right) + \pi r^2 = 54\pi r^{-1} + \pi r^2$$

$$A' = -54\pi r^{-2} + 2\pi r = 0 \Rightarrow 2\pi r = \frac{54\pi}{r^2} \Rightarrow$$

$$r^3 = 27 \Rightarrow r = 3$$

$A'' = 108\pi r^{-3} + 2\pi > 0$, so for minimum cost, the radius should be 3 inches.

63. x = number of batches per year
M = 145,000 phones produced annually
k = \$3.25, cost to store one phone for one year
a = \$230, cost to produce each batch

$$x = \sqrt{\frac{kM}{2a}} = \sqrt{\frac{(3.25)(145{,}000)}{2(230)}} \approx 32$$

32 batches should be produced annually.

64. $\dfrac{145{,}000}{32} \approx 4531$
4531 phones should be produced for each batch.

65. x = number of batches per year
M = 250,000 pounds sold per year
k = 1.25, cost to store 1 pound for 1 year
a = 350, cost to produce each batch

$$x = \sqrt{\frac{kM}{2a}} = \sqrt{\frac{(1.25)(250{,}000)}{2(350)}} \approx 21$$

The company should produce 21 batches annually.

66. Each batch should contain $\dfrac{250{,}000}{21} \approx 11{,}905$ pounds.

67. $\dfrac{d}{dt}\left(2x^5 + x = y^2 - 1\right)$

$$10x^4\frac{dx}{dt} + \frac{dx}{dt} = 2y\frac{dy}{dt}$$

$$\frac{dx}{dt}\left(10x^4 + 1\right) = 2y\frac{dy}{dt}$$

$$\frac{dx}{dt} = \frac{2y}{10x^4 + 1}\frac{dy}{dt}$$

Now substitute the known values into the equation to find the missing value.

$$\frac{dx}{dt} = \frac{2(1)}{10(0)^4 + 1}(5) = 10$$

68. $\dfrac{d}{dt}\left(2x^5 + x = y^2 - 1\right)$

$$10x^4\frac{dx}{dt} + \frac{dx}{dt} = 2y\frac{dy}{dt}$$

$$\frac{dx}{dt}\left(10x^4 + 1\right) = 2y\frac{dy}{dt}$$

$$\frac{dy}{dt} = \frac{10x^4 + 1}{2y}\frac{dx}{dt}$$

Now substitute the known values into the equation to find the missing value.

$$\frac{dy}{dt} = \frac{10(0)^4 + 1}{2(1)}(10) = 5$$

69. a. $\dfrac{d}{dt}\left(F = \dfrac{9}{5}C + 32\right)$

$$\frac{dF}{dt} = \frac{9}{5}\frac{dC}{dt}$$

$$\frac{dC}{dt} = \frac{5}{9}\frac{dF}{dt}$$

Now substitute the known values into the equation to find the missing value.

Copyright © 2019 Pearson Education, Inc.

$$\frac{dC}{dt}=\frac{5}{9}(-4)\approx-2.2$$

The temperature was decreasing by approximately 2.2 Celcius degrees per hour.

b. $\frac{dF}{dt}=\frac{9}{5}(3)=5.4$

The temperature was increasing by 5.4 degrees Fahrenheit per hour.

70. $\frac{d}{dt}(4A+5H=880)$

$$4\frac{dA}{dt}+5\frac{dH}{dt}=0$$
$$5\frac{dH}{dt}=-4\frac{dA}{dt}$$
$$\frac{dH}{dt}=-\frac{4}{5}\frac{dA}{dt}$$

Now substitute the known values into the equation to find the missing value.

$$\frac{dH}{dt}=-\frac{4}{5}(1)=-.8$$

The heart rate should decrease by approximately .8 beats per minute.

71. $\frac{d}{dt}\left(P=-\frac{x^2}{360}+6x-300\right)$

$$\frac{dP}{dt}=-\frac{x}{180}\frac{dx}{dt}+6\frac{dx}{dt}$$

Now substitute the known values,

$x=450$ and $\frac{dx}{dt}=90,$ into the equation to find the missing value.

$$\frac{dP}{dt}=-\frac{450}{180}(90)+6(90)=315$$

The profit is increasing by \$315 per day.

72. $\frac{d}{dt}\left(3x+160p^2=1440\right)$

$$3\frac{dx}{dt}+320p\frac{dp}{dt}=0$$
$$3\frac{dx}{dt}=-320p\frac{dp}{dt}$$
$$\frac{dx}{dt}=-\frac{320p}{3}\frac{dp}{dt}$$

Now substitute the known values,

$p=2$ and $\frac{dp}{dt}=.5,$ into the equation to find the missing value.

$$\frac{dx}{dt}=-\frac{320(2)}{3}(.5)\approx-107$$

The demand is decreasing by 107 bottles.

73. $\frac{d}{dt}\left(x^2+y^2=z^2\right)$

$$2x\frac{dx}{dt}+2y\frac{dy}{dt}=2z\frac{dz}{dt}$$
$$x\frac{dx}{dt}+y\frac{dy}{dt}=z\frac{dz}{dt}$$

Now substitute the known values,

$x=24, y=10, z=26, \frac{dx}{dt}=221,$ and $\frac{dy}{dt}=0,$ into the equation to find the missing value.

$$24(221)+10(0)=26\frac{dz}{dt}$$
$$\frac{dz}{dt}=204$$

The distance between the spectator and competitor is increasing at 204 meters per minute.

74. $\frac{d}{dt}\left(x^2+y^2=z^2\right)$

$$2x\frac{dx}{dt}+2y\frac{dy}{dt}=2z\frac{dz}{dt}$$
$$x\frac{dx}{dt}+y\frac{dy}{dt}=z\frac{dz}{dt}$$

Now substitute the known values,

$x=45, y=90, z=45\sqrt{5}, \frac{dx}{dt}=-9,$ and $\frac{dy}{dt}=0,$ into the equation to find the missing value.

$$45(-9)+90(0)=45\sqrt{5}\frac{dz}{dt}$$
$$\frac{dz}{dt}\approx-4.0$$

The distance between the second base and the runner is decreasing y approximately 4.0 feet per second.

Copyright © 2019 Pearson Education, Inc.

Case 12: A Total Cost Model for a Training Program

1. $Z(m)=\dfrac{C_1}{m}+DtC_2+DC_3\left(\dfrac{m-1}{2}\right)$

$=C_1m^{-1}+DtC_2+\dfrac{DC_3m}{2}-\dfrac{DC_3}{2}$

$Z'(m)=-C_1m^{-2}+\dfrac{DC_3}{2}=-\dfrac{C_1}{m^2}+\dfrac{DC_3}{2}$

2. $Z'(m)=0\Rightarrow-\dfrac{C_1}{m^2}+\dfrac{DC_3}{2}=0\Rightarrow$

$-\dfrac{C_1}{m^2}=-\dfrac{DC_3}{2}\Rightarrow$

$2C_1=DC_3m^2\Rightarrow\sqrt{\dfrac{2C_1}{DC_3}}=m$ (m is positive)

3. Let $D=3$, $t=12$, $C_1=15{,}000$, $C_3=900$.

$m=\sqrt{\dfrac{2(15{,}000)}{(3)(900)}}=\sqrt{\dfrac{100}{9}}=\dfrac{10}{3}$

4. $m^+=4$ Whole number larger than $\dfrac{10}{3}$

$m^-=3$ Whole number smaller than $\dfrac{10}{3}$

5. Let $C_1=15{,}000$, $m^+=4$, $m^-=3$, $D=3$, $C_2=100$, $C_3=900$, $t=12$.

$Z(m)=\dfrac{C_1}{m}+DtC_2+DC_3\left(\dfrac{m-1}{2}\right)$

$Z\left(m^+\right)=Z(4)$

$=\dfrac{15{,}000}{4}+(3)(12)(100)+(3)(900)\left(\dfrac{4-1}{2}\right)$

$=3750+3600+4050=\$11{,}400$

$Z\left(m^-\right)=Z(3)$

$=\dfrac{15{,}000}{3}+(3)(12)(100)+(3)(900)\left(\dfrac{3-1}{2}\right)$

$=5000+3600+2700=\$11{,}300$

6. Since $Z\left(m^-\right)$ is smaller than $Z\left(m^+\right)$, the optimum value of Z is \$11,300. This occurs when $t=3$ months. Now $N=mD=(3)(3)=9$ so the number of trainees in a batch is 9

Copyright © 2019 Pearson Education, Inc.

Chapter 13 Integral Calculus

Section 13.1 Antiderivatives

1. If $F(x)$ and $G(x)$ are both antiderivatives of $f(x)$, then they differ only by a constant.

3. Answers vary.

5. $\int 12x\,dx = 12\int x\,dx = 12\left(\frac{x^2}{2}\right) + C = 6x^2 + C$

7. $\int 8p^3\,dp = 8\int p^3\,dp = 8\left(\frac{p^4}{4}\right) + C = 2p^4 + C$

9. $\int 105\,dx = 105\int 1\,dx = 105\int x^0\,dx$

$= 105\left(\frac{x^1}{1}\right) + C = 105x + C$

11. $\int (5z-1)\,dz = 5\int z\,dz - \int 1\,dz = 5\left(\frac{z^2}{2}\right) - \int z^0\,dz$

$= \frac{5z^2}{2} - \frac{z^1}{1} + C = \frac{5z^2}{2} - z + C$

13. $\int \left(z^2 - 4z + 6\right)dz = \int z^2\,dz - 4\int z\,dz + 6\int 1\,dz$

$= \frac{z^3}{3} - 4\left(\frac{z^2}{2}\right) + 6\int z^0\,dz$

$= \frac{z^3}{3} - 2z^2 + 6\left(\frac{z^1}{1}\right) + C$

$= \frac{z^3}{3} - 2z^2 + 6z + C$

15. $\int \left(x^3 - 14x^2 + 22x + 8\right)dx$

$= \int x^3\,dx - 14\int x^2\,dx + 22\int x\,dx + 8\int 1\,dx$

$= \frac{x^4}{4} - 14\left(\frac{x^3}{3}\right) + 22\left(\frac{x^2}{2}\right) + 8\int x^0\,dx$

$= \frac{x^4}{4} - \frac{14x^3}{3} + 11x^2 + 8\left(\frac{x^1}{1}\right) + C$

$= \frac{x^4}{4} - \frac{14x^3}{3} + 11x^2 + 8x + C$

17. $\int 6\sqrt{y}\,dy = 6\int y^{1/2}\,dy = 6\left(\frac{y^{3/2}}{\frac{3}{2}}\right) + C$

$= 6\left(\frac{2}{3}y^{3/2}\right) + C = 4y^{3/2} + C$

19. $\int \left(6t\sqrt{t} + 3\sqrt[4]{t}\right)dt = 6\int t \cdot t^{1/2}\,dt + 3\int t^{1/4}\,dt$

$= 6\int t^{3/2}\,dt + 3\int t^{1/4}\,dt$

$= 6\left(\frac{t^{5/2}}{\frac{5}{2}}\right) + 3\left(\frac{t^{5/4}}{\frac{5}{4}}\right) + C$

$= 6\left(\frac{2}{5}t^{5/2}\right) + 3\left(\frac{4}{5}t^{5/4}\right) + C$

$= \frac{12t^{5/2}}{5} + \frac{12t^{5/4}}{5} + C$

21. $\int \left(56t^{1/2} + 18t^{7/2}\right)dt = 56\int t^{1/2}\,dt + 18\int t^{7/2}\,dt$

$= 56\left(\frac{t^{3/2}}{\frac{3}{2}}\right) + 18\left(\frac{t^{9/2}}{\frac{9}{2}}\right) + C$

$= 56\left(\frac{2}{3}t^{3/2}\right) + 18\left(\frac{2}{9}t^{9/2}\right) + C$

$= \frac{112t^{3/2}}{3} + 4t^{9/2} + C$

23. $\int \frac{24}{x^3}\,dx = 24\int \frac{1}{x^3}\,dx = 24\int x^{-3}\,dx = 24\left(\frac{x^{-2}}{-2}\right) + C$

$= -12x^{-2} + C = -12\left(\frac{1}{x^2}\right) + C = -\frac{12}{x^2} + C$

25. $\int \left(\frac{1}{y^3} - \frac{2}{\sqrt{y}}\right)dy = \int \left(y^{-3} - 2y^{-1/2}\right)dy$

$= \int y^{-3}\,dy - 2\int y^{-1/2}\,dy$

$= \frac{y^{-2}}{-2} - 2\left(\frac{y^{1/2}}{\frac{1}{2}}\right) + C$

$= \frac{1}{-2y^2} - 2\left(2y^{1/2}\right) + C$

$= -\frac{1}{2y^2} - 4y^{1/2} + C$

Copyright © 2019 Pearson Education, Inc.

27. $\int (6x^{-3}+2x^{-1})\,dx = 6\int x^{-3}dx + 2\int x^{-1}dx$
$$= 6\left(\frac{x^{-2}}{-2}\right) + 2\ln|x| + C$$
$$= -3x^{-2} + 2\ln|x| + C$$

29. $\int 4e^{3u}\,du = 4\int e^{3u}\,du = 4\left(\frac{1}{3}e^{3u}\right) + C$
$$= \frac{4e^{3u}}{3} + C$$

31. $\int 3e^{-.8x}\,dx = \frac{3\left(e^{-.8x}\right)}{-.8} + C = \frac{-15e^{-.8x}}{4} + C$

33. $\int \left(\frac{6}{x} + 4e^{-.5x}\right)dx = 6\ln|x| + \frac{4e^{-.5x}}{-.5} + C$
$$= 6\ln|x| - 8e^{-.5x} + C$$

35. $\int \frac{1+2t^4}{t}\,dt = \int \left(\frac{1}{t} + \frac{2t^4}{t}\right)dt = \int \left(\frac{1}{t} + 2t^3\right)dt$
$$= \ln|t| + \frac{2t^4}{4} + C = \ln|t| + \frac{t^4}{2} + C$$

37. $\int \left(e^{2u} + \frac{4}{u}\right)du = \int e^{2u}\,du + \int \frac{4}{u}\,du$
$$= \frac{e^{2u}}{2} + 4\ln|u| + C$$

39. $\int (5x+1)^2\,dx = \int (25x^2 + 10x + 1)\,dx$
$$= 25\int x^2 dx + 10\int x\,dx + \int 1\,dx$$
$$= 25\left(\frac{x^3}{3}\right) + 10\left(\frac{x^2}{2}\right) + x + C$$
$$= \frac{25x^3}{3} + 5x^2 + x + C$$

41. $\int \frac{\sqrt{x}+1}{\sqrt[3]{x}}\,dx = \int \left(\frac{\sqrt{x}}{\sqrt[3]{x}} + \frac{1}{\sqrt[3]{x}}\right)dx$
$$= \int \left(x^{(1/2-1/3)} + x^{-1/3}\right)dx$$
$$= \int \left(x^{1/6} + x^{-1/3}\right)dx$$
$$= \frac{x^{7/6}}{\frac{7}{6}} + \frac{x^{2/3}}{\frac{2}{3}} + C$$
$$= \frac{6x^{7/6}}{7} + \frac{3x^{2/3}}{2} + C$$

43. $f'(x) = x^{2/3}$
$$f(x) = \int x^{2/3}dx = \frac{x^{5/3}}{\frac{5}{3}} + C = \frac{3}{5}x^{5/3} + C$$
$$= \frac{3x^{5/3}}{5} + C$$

If the point $\left(1, \frac{3}{5}\right)$ is on the curve, $f(1) = \frac{3}{5}$.

$$\frac{3}{5}(1)^{5/3} + C = \frac{3}{5} \Rightarrow \frac{3}{5} + C = \frac{3}{5} \Rightarrow C = 0$$

Therefore, the equation of the curve is

$$f(x) = \frac{3x^{5/3}}{5}.$$

45. a. $f'(x) = -.1x + .59$ and
$F(0) = 67.1$
$$F(x) = \int(-.1x + .59)\,dx$$
$$= -.1\int x\,dx + .59\int dx$$
$$= -.1\left(\frac{x^2}{2}\right) + .59x + C$$
$$= -.05x^2 + .59x + C$$
$$F(0) = 67.1 = -.05(0)^2 + .59(0) + C \Rightarrow$$
$$C = 67.1$$
$$F(x) = -.05x^2 + .59x + 67.1$$

b. $x = 2016 - 2000 = 16$
$$F(16) = -.05(16)^2 + .59(16) + 67.1$$
$$\approx 63.7$$
Home ownership was about 63.7% in 2016.

47. a. $P'(x) = p(x)$ and $P(0) = 29$
$$p(x) = \int \left(3e^{.12x}\right)dx$$
$$= 3\int e^{.12x}dx$$
$$= 3\left(\frac{e^{.12x}}{.12}\right) + C$$
$$= 25e^{.12x} + C$$
$$p(x) = 25e^{.12x} + C$$
$$29 = 25e^{.12(0)} + C$$
$$4 = C$$
$$p(x) = 25e^{.12x} + 4$$

b. $x = 12$
$$p(12) = 25e^{.12(12)} + 4$$
$$\approx 110$$
The stock price 12 month later was about \$110.

Copyright © 2019 Pearson Education, Inc.

49. $P'(x) = 4 - 6x + 3x^2$ and $P(0) = -40$

$$P(x) = \int (4 - 6x + 3x^2)\,dx$$
$$= 4\int 1\,dx - 6\int x\,dx + 3\int x^2 dx$$
$$= 4x - 6\left(\frac{x^2}{2}\right) + 3\left(\frac{x^3}{3}\right) + C$$
$$P(x) = 4x - 3x^2 + x^3 + C$$
$$-40 = 4(0) - 3(0)^2 + (0)^3 + C \Rightarrow -40 = C$$

Therefore,

$P(x) = x^3 - 3x^2 + 4x - 40$.

51. $C'(x) = x^{2/3} + 5,\ C(8) = 58$

$$C(x) = \int (x^{2/3} + 5)\,dx = \frac{3x^{5/3}}{5} + 5x + C$$
$$58 = \frac{3(8)^{5/3}}{5} + 5(8) + C \Rightarrow$$
$$58 = \frac{3(32)}{5} + 40 + C \Rightarrow$$
$$58 = \frac{96}{5} + 40 + C \Rightarrow \frac{-6}{5} = -1.2 = C$$

So, $C(x) = \dfrac{3x^{5/3}}{5} + 5x - 1.2$.

53. $C'(x) = x^2 - 2x + 3;\ C(3) = 25$

$$C(x) = \int (x^2 - 2x + 3)\,dx = \frac{x^3}{3} - \frac{2x^2}{2} + 3x + k$$
$$C(x) = \frac{x^3}{3} - x^2 + 3x + k$$
$$C(3) = \frac{3^3}{3} - 3^2 + 3(3) + k$$

Since $C(3) = 25$,

$25 = 9 - 9 + 9 + k \Rightarrow k = 16$

$$C(x) = \frac{x^3}{3} - x^2 + 3x + 16$$

55. $C'(x) = -\dfrac{40}{e^{.05x}} + 100,\ C(5) = 1400$

$$C(x) = \int\left(-\frac{40}{e^{.05x}} + 100\right)dx$$
$$= \frac{800}{e^{.05x}} + 100x + k$$
$$C(5) = \frac{800}{e^{.05(5)}} + 100(5) + k$$

Since $C(5) = 1400$

$$1400 = \frac{800}{e^{.25}} + 500 + k \Rightarrow k \approx 276.96$$
$$C(x) \approx \frac{800}{e^{.05x}} + 100x + 276.96$$

57. a. $F'(t) = f(t),\ F(17) = 326$

$$F(t) = \int \frac{66}{t}\,dt$$
$$= 66\int \frac{1}{t}\,dt$$
$$= 66\ln(t) + C$$
$$F(17) = 326 = 66\ln(17) + C \Rightarrow$$
$$C = 139$$
$$F(t) = 66\ln(t) + 139$$

b. $t = 47$

$F(47) = 66\ln(47) + 139 \approx 393$ million

The function estimates the projected population of the United States in 2047 will be about 393 million.

Section 13.2 Integration by Substitution

1. Integration by substitution is related to the chain rule for derivatives, but in reverse. Difficult integrals in which u replaces an expression in the integrand and the derivative of u also appear to suggest using integration by substitution.

3. $\int 3(12x - 1)^2\,dx$

Let $u = 12x - 1$. Then $du = 12\,dx$.

$$\int 3(12x-1)^2 dx = 3\int (12x-1)^2 dx$$
$$= 3\cdot\frac{1}{12}\cdot 12\int (12x-1)^2 dx$$
$$= \frac{3}{12}\int (12x-1)^2 12\,dx$$
$$= \frac{1}{4}\int u^2 du = \frac{1}{4}\left(\frac{u^3}{3}\right) + C$$
$$= \frac{u^3}{12} + C = \frac{(12x-1)^3}{12} + C$$

Copyright © 2019 Pearson Education, Inc.

5. $\int \frac{5}{(3t-6)^2}\,dt$

Let $u = 3t - 6$. Then $du = 3\,dt$.

$$\int \frac{5}{(3t-6)^2}\,dt = 5\int (3t-6)^{-2}\,dt$$
$$= 5\left(\frac{1}{3}\right)\int (3t-6)^{-2}(3)\,dt$$
$$= \frac{5}{3}\int u^{-2}\,du = \frac{5}{3}\left(\frac{u^{-1}}{-1}\right) + C$$
$$= -\frac{5}{3u} + C = -\frac{5}{3(3t-6)} + C$$

7. $\int \sqrt{5-2x}\,dx$

Let $u = 5-2x$. Then $du = -2dx$.

$$\int \sqrt{5-2x}\,dx$$
$$= \int (5-2x)^{1/2}\,dx$$
$$= -\frac{1}{2}\cdot -2\int (5-2x)^{1/2}\,dx$$
$$= -\frac{1}{2}\int (5-2x)^{1/2}(-2)dx$$
$$= -\frac{1}{2}\int u^{1/2}\,du = -\frac{1}{2}\left(\frac{u^{3/2}}{\frac{3}{2}}\right) + C = -\frac{1}{3}u^{3/2} + C$$
$$= -\frac{1}{3}(5-2x)^{3/2} + C$$

9. $\int r^2\sqrt{r^3+3}\,dr$

Let $u = r^3 + 3$. Then $du = 3r^2\,dr$.

$$\int r^2\sqrt{r^3+3}\,dr = \int \left(r^3+3\right)^{1/2} r^2\,dr$$
$$= \frac{1}{3}\cdot 3\int \left(r^3+3\right)^{1/2} r^2\,dr$$
$$= \frac{1}{3}\int \left(r^3+3\right)^{1/2} 3r^2\,dr$$
$$= \frac{1}{3}\int u^{1/2}\,du = \frac{1}{3}\left(\frac{u^{3/2}}{\frac{3}{2}}\right) + C$$
$$= \frac{1}{3}\left(\frac{2}{3}u^{3/2}\right) + C = \frac{2}{9}\left(r^3+3\right)^{3/2} + C$$
$$= \frac{2\left(r^3+3\right)^{3/2}}{9} + C$$

11. $\int \left(-4e^{5k}\right)dk$

Let $u = 5k$. Then $du = 5\,dk$.

$$\int \left(-4e^{5k}\right)dk = -4\int e^{5k}\,dk$$
$$= -4\cdot\frac{1}{5}\cdot 5\int e^{5k}\,dk = -\frac{4}{5}\int e^{5k}\,5\,dk$$
$$= -\frac{4}{5}\int e^u\,du = -\frac{4}{5}e^u + C$$
$$= -\frac{4}{5}e^{5k} + C$$

13. $\int 4w^3 e^{2w^4}\,dw$

Let $u = 2w^4$. Then $du = 8w^3\,dw$.

$$\int 4w^3 e^{2w^4}\,dw = 4\int e^{2w^4}\cdot w^3\,dw$$
$$= 4\cdot\frac{1}{8}\cdot 8\int e^{2w^4}\cdot w^3\,dw$$
$$= \frac{4}{8}\int e^{2w^4}\cdot 8w^3\,dw = \frac{1}{2}\int e^u\,du$$
$$= \frac{1}{2}e^u + C = \frac{e^{2w^4}}{2} + C$$

15. $\int (2-t)e^{4t-t^2}\,dt$

Let $u = 4t - t^2$.
Then $du = (4-2t)\,dt = 2(2-t)\,dt$.

$$\int (2-t)e^{4t-t^2}\,dt = \frac{1}{2}\cdot 2\int e^{4t-t^2}\cdot(2-t)dt$$
$$= \frac{1}{2}\int e^{4t-t^2}(4-2t)dt$$
$$= \frac{1}{2}\int e^u\,du = \frac{1}{2}e^u + C$$
$$= \frac{e^{4t-t^2}}{2} + C$$

17. $\int \frac{e^{\sqrt{y}}}{\sqrt{y}}\,dy = \int e^{y^{1/2}}\cdot y^{-1/2}\,dy$

Let $u = y^{1/2}$. Then $du = \frac{1}{2}y^{-1/2}\,dy$.

$$\int \frac{e^{\sqrt{y}}}{\sqrt{y}}\,dy = 2\cdot\frac{1}{2}\int e^{y^{1/2}}\cdot y^{-1/2}\,dy$$
$$= 2\int e^{y^{1/2}}\cdot\frac{1}{2}y^{-1/2}\,dy = 2\int e^u\,du$$
$$= 2e^u + C = 2e^{\sqrt{y}} + C$$

Copyright © 2019 Pearson Education, Inc.

19. $\int \frac{-5}{12+6x}dx$

Let $u = 12 + 6x$. Then $du = 6\,dx$.

$$\int \frac{-5}{12+6x}dx = -5\int \frac{1}{12+6x}dx$$
$$= -5\cdot\frac{1}{6}\cdot 6\int \frac{1}{12+6x}dx$$
$$= -\frac{5}{6}\int \frac{1}{12+6x}(6)\,dx = -\frac{5}{6}\int \frac{1}{u}du$$
$$= -\frac{5}{6}\ln|u| + C = -\frac{5\ln|12+6x|}{6} + C$$

21. $\int \frac{e^{2t}}{e^{2t}+1}dt$

Let $u = e^{2t}+1$. Then $du = e^{2t}(2)dt$.

$$\int \frac{e^{2t}}{e^{2t}+1}dt = \frac{1}{2}\cdot 2\int \frac{1}{e^{2t}+1}\cdot e^{2t}dt$$
$$= \frac{1}{2}\int \frac{1}{e^{2t}+1}\cdot 2e^{2t}dt = \frac{1}{2}\int \frac{1}{u}du$$
$$= \frac{1}{2}\ln|u| + C = \frac{1}{2}\ln\left|e^{2t}+1\right| + C$$

23. $\int \frac{x+2}{\left(2x^2+8x\right)^3}dx = \int \left(2x^2+8x\right)^{-3}(x+2)dx$

Let $u = 2x^2+8x$. Then $du = (4x+8)\,dx$.

$$\int \frac{x+2}{\left(2x^2+8x\right)^3}dx = \frac{1}{4}\cdot 4\int \left(2x^2+8x\right)^{-3}(x+2)dx$$
$$= \frac{1}{4}\int \left(2x^2+8x\right)^{-3}(4x+8)dx$$
$$= \frac{1}{4}\int u^{-3}du = \frac{1}{4}\cdot\frac{u^{-2}}{-2} + C$$
$$= -\frac{1}{8u^2} + C$$
$$= -\frac{1}{8\left(2x^2+8x\right)^2} + C$$

25. $\int \frac{1}{\sqrt{2x+9}}dx$

Let $w = 2x + 9$. Then $dw = 2\,dx$.

$$\int \frac{1}{\sqrt{2x+9}}dx = 1\left(\frac{1}{2}\right)(2)\int (2x+9)^{-1/2}dx$$
$$= \frac{1}{2}\int (2x+9)^{-1/2}\cdot 2\,du$$
$$= \frac{1}{2}\int w^{-1/2}dw = \frac{1}{2}\left(\frac{w^{1/2}}{\frac{1}{2}}\right) + C$$
$$= w^{1/2} + C = (2x+9)^{1/2} + C$$
$$= \sqrt{2x+9} + C$$

27. $\int \frac{x^2+1}{\left(x^3+3x\right)^{\frac{2}{3}}}dx = \frac{1}{3}\int \frac{3\left(x^2+1\right)dx}{\left(x^3+3x\right)^{\frac{2}{3}}}$

Let $u = x^3+3x$. Then

$du = \left(3x^2+3\right)dx = 3\left(x^2+1\right)dx$

$$\int \frac{x^2+1}{\left(x^3+3x\right)^{2/3}}dx = \frac{1}{3}\int \frac{du}{u^{2/3}} = \frac{1}{3}\int u^{-2/3}du$$
$$= \frac{1}{3}\left(\frac{u^{1/3}}{\frac{1}{3}}\right) + C = u^{1/3} + C$$
$$= \left(x^3+3x\right)^{1/3} + C$$

29. $\int \frac{6x+7}{3x^2+7x+8}dx = \int \frac{1}{3x^2+7x+8}(6x+7)dx$

Let $u = 3x^2+7x+8$. Then $du = (6x+7)dx$.

$$\int \frac{6x+7}{3x^2+7x+8}dx = \int \frac{1}{u}du = \ln|u| + C$$
$$= \ln\left|3x^2+7x+8\right| + C$$

31. $\int 2x\left(x^2+5\right)^3dx$

Let $u = x^2+5$. Then $du = 2x\,dx$.

$$\int 2x\left(x^2+5\right)^3dx = \int u^3du = \frac{u^4}{4} + C$$
$$= \frac{\left(x^2+5\right)^4}{4} + C$$

Copyright © 2019 Pearson Education, Inc.

33. $\int\left(\sqrt{x^2+12x}\right)(x+6)dx$

$= \int\left(x^2+12x\right)^{1/2}(x+6)dx$

Let $u = x^2+12x$. Then $du = 2x + 12dx$ or $(x+6)\,dx = \frac{du}{2}$.

$\int\left(\sqrt{x^2+12x}\right)(x+6)dx = \frac{1}{2}\int u^{1/2}du$

$= \frac{1}{2}\left(\frac{2}{3}\right)u^{3/2}+C$

$= \frac{\left(x^2+12x\right)^{3/2}}{3}+C$

35. $\int\frac{1}{x(\ln x)}dx = \int\frac{1}{\ln x}\cdot\frac{1}{x}dx$

Let $u = \ln x$. Then $du = \frac{1}{x}dx$.

$\int\frac{1}{x(\ln x)}dx = \int\frac{1}{u}du = \ln|u|+C = \ln|\ln x|+C$

37. $\int\frac{5u}{\sqrt{u-1}}du = \int 5u(u-1)^{-\frac{1}{2}}du$

Let $w = u - 1$. Then $dw = du$ and $u = w + 1$.

$\int\frac{5u}{\sqrt{u-1}}du = 5\int(w+1)w^{-1/2}dw$

$= 5\int w^{1/2}+w^{-1/2}dw$

$= \frac{5w^{3/2}}{\frac{3}{2}}+\frac{5w^{1/2}}{\frac{1}{2}}+C$

$= \frac{10}{3}w^{3/2}+10w^{1/2}+C$

$= \frac{10}{3}(u-1)^{3/2}+10(u-1)^{1/2}+C$

39. $\int t\sqrt{5t-1}dt = \frac{1}{5}\int 5t(5t-1)^{1/2}dt$

Let $u = 5t - 1$. Then $du = 5\,dt$ and $t = \frac{u+1}{5}$.

$\int t\sqrt{5t-1}dt = \frac{1}{5}\int\left(\frac{u+1}{5}\right)u^{1/2}du$

$= \frac{1}{25}\int\left(u^{3/2}+u^{1/2}\right)du$

$= \frac{1}{25}\left[\frac{u^{5/2}}{\frac{5}{2}}+\frac{u^{3/2}}{\frac{3}{2}}\right]+C$

$= \frac{1}{25}\left[\frac{2}{5}(5t-1)^{5/2}+\frac{2}{3}(5t-1)^{3/2}\right]+C$

$= \frac{2(5t-1)^{5/2}}{125}+\frac{2(5t-1)^{3/2}}{75}+C$

41. a. $C(x) = \int C'(x)dx = \int\frac{60x}{5x^2+1}dx$

Let $u = 5x^2+1$. Then $du = 10x\,dx$.

$C(x) = \int(5x^2+1)^{-1}6\cdot 10x\,dx = 6\int u^{-1}du$

$= 6\ln|u|+k$

Notice that for any x, $u \geq 0$.

$C(x) = 6\ln(5x^2+1)+k$

$C(0) = 6\ln(5(0)^2+1)+k \Rightarrow 10 = k$

$C(x) = 6\ln(5x^2+1)+10$

b. $C(5) = 6\ln(5(5)^2+1)+10 \approx 39.02$

Since this function is always increasing, only the cost at the 5th year is about $39.02 thousand, so, yes, they should add the new line.

43. a. $F'(x) = f(x),\ F(0) = 6192$

$F(x) = \int -185.85e^{-.03x}dx$

$= -185.85\left(-\frac{1}{.03}\right)e^{-.03x}+C$

$= 6195e^{-.03x}+C$

Since $F(0) = 6192$, we have

$6192 = 6195e^{-.03(0)}+C \Rightarrow C = -3$

$F(x) = 6195e^{-.03x}-3$

b. $x = 2020 - 2000 = 20$

$F(20) = 6195e^{-.03(20)} \approx 3397$

According to the model, there will be about 3397 stores 2020.

Copyright © 2019 Pearson Education, Inc.

45. a. $R'(x) = 50x\left(x^2+10{,}000\right)^{-3/4}$,

$R(0)=0$

$R(x) = \int 50x\left(x^2+10{,}000\right)^{-3/4}\,dx$

Let $u = x^2+10{,}000$. Then $du = 2x\,dx$.

$$R(x) = \int 50x\left(x^2+10{,}000\right)^{-3/4}\,dx$$
$$= 50\cdot\frac{1}{2}\int 2x\left(x^2+10{,}000\right)^{-3/4}\,dx$$
$$= 25\int u^{-3/4}\,du = 25\left(\frac{u^{1/4}}{1/4}\right)+C$$
$$= 100u^{1/4}+C$$
$$= 100\left(x^2+10{,}000\right)^{1/4}+C$$

Since $R(0)=0$, we have

$0 = 100\left(0^2+10{,}000\right)^{1/4}+C \Rightarrow$

$C = -1000.$

Thus, $R(x) = 100\left(x^2+10{,}000\right)^{1/4}-1000$

$= 100\sqrt[4]{x^2+10{,}000}-1000$

b. $R(1000) = 100\sqrt[4]{(1000)^2+10{,}000}-1000$

≈ 2170.15

The revenue from selling 1000 units is about \$2,170.15.

c. $100\sqrt[4]{x^2+10{,}000}-1000 = 2000 \Rightarrow$

$100\left(x^2+10{,}000\right)^{1/4} = 3000 \Rightarrow$

$\left(x^2+10{,}000\right)^{1/4} = 30 \Rightarrow$

$x^2+10{,}000 = 810{,}000 \Rightarrow$

$x^2 = 800{,}000 \Rightarrow x \approx 895$

About 895 units must be sold to produce revenue of at least \$2,000.

47. a. $F'(x) = f(x)$, $F(0) = 520$

$F(x) = \int \frac{53}{x+1}\,dx = 53\int (x+1)^{-1}\,dx$

Let $u = x+1$. Then $du = 1\,dx$.

$$53\int (x+1)^{-1}\,dx = 53\int u^{-1}\,du$$
$$= 53\ln u + C$$
$$= 53\ln|x+1| + C$$

Since $F(0) = 520$, we have

$520 = 53\ln|0+1| + C \Rightarrow 520 = C.$

Thus, $F(x) = 53\ln|x+1| + 520.$

b. $x = 2020-2000 = 20$

$F(20) = 53\ln|20+1| + 520 \approx 681$ thousand

In 2020, the projected number of mining jobs will be about 681,000.

Section 13.3 Integration by Parts

1. $\int\left(2e^{-35x}\right)dx$

Using integration by substitution, let $u = -35x$. Therefore, $du = -35dx$.

$$2\cdot-\frac{1}{35}\int\left(e^{-35x}\right)35dx$$
$$= -\frac{2}{35}\int e^u\,du$$
$$= -\frac{2}{35}e^u + c$$
$$= -\frac{2}{35}e^{-35x} + c$$

3. $\int(10xe^{5x})dx$

Using integration by parts, let $u = 10x$ and $dv = e^{5x}dx$. Therefore, $du = 10dx$ and $v = \frac{1}{5}e^{5x}$.

$$\int(10xe^{5x})dx$$
$$= (10x)\left(\frac{1}{5}e^{5x}\right) - \int\frac{1}{5}e^{5x}\cdot 10dx$$
$$= (10x)\left(\frac{1}{5}e^{5x}\right) - \int 2e^{5x}dx$$
$$= 2xe^{5x} - \frac{1}{5}(2e^{5x}+c)$$
$$= 2xe^{5x} - \frac{2}{5}e^{5x} + c$$

5. $\int(xe^{x/4})dx$

Using integration by parts, let $u = x$ and $dv = e^{x/4}dx$. Therefore, $du = dx$ and $v = 4e^{x/4}$.

$$\int(xe^{x/4})dx$$
$$= (x)\left(4e^{x/4}\right) - \int 4e^{x/4}dx$$
$$= 4xe^{x/4} - 4(4e^{x/4}+c)$$
$$= 4xe^{x/4} - 16e^{x/4} + c$$

Copyright © 2019 Pearson Education, Inc.

7. $\int\left(\frac{5x+1}{e^x}\right)dx$

Using integration by parts, let $u = 5x+1$ and $dv = e^{-x}dx$. Therefore, $du = 5dx$ and $v = -e^{-x}$.

$$\int\left(\frac{5x+1}{e^x}\right)dx$$
$$= (5x+1)\left(-e^{-x}\right) - \int -e^{-x}\cdot 5dx$$
$$= (5x+1)\left(-e^{-x}\right) - \int -5e^{-x}dx$$
$$= -5xe^{-x} - e^{-x} - 5e^{-x} + c$$
$$= -5xe^{-x} - 6e^{-x} + c$$

9. $\int(7xe^{1-3x})dx$

Using integration by parts, let $u = 7x$ and $dv = e^{1-3x}dx$. Therefore, $du = 7dx$ and $v = -\frac{1}{3}e^{1-3x}$.

$$\int(7xe^{1-3x})dx$$
$$= (7x)\left(-\frac{1}{3}e^{1-3x}\right) - \int -\frac{1}{3}e^{1-3x}\cdot 7dx$$
$$= (7x)\left(-\frac{1}{3}e^{1-3x}\right) + \int \frac{7}{3}e^{1-3x}dx$$
$$= -\frac{7}{3}xe^{1-3x} + \frac{7}{3}\left(-\frac{1}{3}e^{1-3x} + c\right)$$
$$= -\frac{7}{3}xe^{1-3x} - \frac{7}{9}e^{1-3x} + c$$

11. $\int((8x-3)e^{10x-17})dx$

Using integration by parts, let $u = 8x-3$ and $dv = e^{10x-17}dx$. Therefore, $du = 8dx$ and $v = \frac{1}{10}e^{10x-17}$.

$$\int((8x-3)e^{10x-17})dx$$
$$= (8x-3)\left(\frac{1}{10}e^{10x-17}\right) - \int \frac{1}{10}e^{10x-17}\cdot 8dx$$
$$= (8x-3)\left(\frac{1}{10}e^{10x-17}\right) - \int \frac{4}{5}e^{10x-17}dx$$
$$= \frac{4}{5}xe^{10x-17} - \frac{3}{10}e^{10x-17} - \frac{4}{5}\left(\frac{1}{10}e^{10x-17} + c\right)$$
$$= \frac{4}{5}xe^{10x-17} - \frac{19}{50}e^{10x-17} + c$$

13. $\int\left(4xe^{2x^2}\right)dx$

Using integration by substitution, let $u = 2x^2$. Therefore, $du = 4xdx$.

$$\int\left(e^{2x^2}\right)4xdx$$
$$= \int e^u du$$
$$= e^u + c$$
$$= e^{2x^2} + c$$

15. $\int(4x^2e^{2x})dx$

Using integration by parts, let $u = 4x^2$ and $dv = e^{2x}dx$. Therefore, $du = 8xdx$ and $v = \frac{1}{2}e^{2x}$.

$$\int(4x^2e^{2x})dx$$
$$= (4x^2)\left(\frac{1}{2}e^{2x}\right) - \int \frac{1}{2}e^{2x}\cdot 8xdx$$
$$= 2x^2e^{2x} - \int 4xe^{2x}dx$$

Using integration by parts a second time, let $u = 4x$ and $dv = e^{2x}dx$. Therefore, $du = 4dx$ and $v = \frac{1}{2}e^{2x}$.

$$2x^2e^{2x} - \int(4xe^{2x})dx$$
$$= 2x^2e^{2x} - \left(4x\left(\frac{1}{2}e^{2x}\right) - \int \frac{1}{2}e^{2x}\cdot 4dx\right)$$
$$= 2x^2e^{2x} - 2xe^{2x} + \int 2e^{2x}dx$$
$$= 2x^2e^{2x} - 2xe^{2x} + 2\left(\frac{1}{2}e^{2x} + c\right)$$
$$= 2x^2e^{2x} - 2xe^{2x} + e^{2x} + c$$

Copyright © 2019 Pearson Education, Inc.

17. $\int (x^2+1)e^x dx$

Using integration by parts, let $u = x^2+1$ and $dv = e^x dx$. Therefore, $du = 2xdx$ and $v = e^x$.

$$\int (x^2+1)e^x dx$$
$$= (x^2+1)\left(e^x\right) - \int e^x \cdot 2xdx$$
$$= x^2e^x + e^x - \int 2xe^x dx$$

Using integration by parts a second time, let $u = 2x$ and $dv = e^x dx$. Therefore, $du = 2dx$ and $v = e^x$.

$$x^2e^x + e^x - \int (2xe^x)dx$$
$$= x^2e^x + e^x - \left(2x\left(e^x\right) - \int e^x \cdot 2dx\right)$$
$$= x^2e^x + e^x - 2xe^x + \int 2e^x dx$$
$$= x^2e^x + e^x - 2xe^x + 2e^x + c$$
$$= x^2e^x - 2xe^x + 3e^x + c$$

19. $\int \left(xe^{x^2+1}\right)dx$

Using integration by substitution, let $u = x^2+1$. Therefore, $du = 2xdx$.

$$\frac{1}{2}\int \left(e^{x^2+1}\right)2xdx$$
$$= \frac{1}{2}\int e^u du$$
$$= \frac{1}{2}e^u + c$$
$$= \frac{1}{2}e^{x^2+1} + c$$

21. $\int (3\ln x - 7)dx$

Using integration by parts, let $u = 3\ln x - 7$ and $dv = dx$. Therefore, $du = \frac{3}{x}dx$ and $v = x$.

$$\int (3\ln x - 7)dx$$
$$= (3\ln x - 7)(x) - \int x \cdot \frac{3}{x}dx$$
$$= 3x\ln x - 7x - \int 3dx$$
$$= 3x\ln x - 7x - 3x + c$$
$$= 3x\ln x - 10x + c$$

23. $\int \ln(8x-5)dx$

Using the antiderivative of the Logarithmic function, let $u = 8x-5$. Therefore, $du = 8dx$.

$$\int \ln(8x-5)dx$$
$$= \frac{1}{8}((8x-5)\ln(8x-5) - (8x-5)) + c$$

25. $\int 26x\ln xdx$

Using integration by parts, let $u = \ln x$ and $dv = 26xdx$. Therefore, $du = \frac{1}{x}dx$ and $v = 13x^2$.

$$\int 26x\ln xdx$$
$$= \ln x(13x^2) - \int 13x^2 \cdot \frac{1}{x}dx$$
$$= 13x^2\ln x - \int 13xdx$$
$$= 13x^2\ln x - \frac{13}{2}x^2 + c$$

27. $\int x^3\ln xdx$

Using integration by parts, let $u = \ln x$ and $dv = x^3dx$. Therefore, $du = \frac{1}{x}dx$ and $v = \frac{1}{4}x^4$.

$$\int x^3\ln xdx$$
$$= \ln x\left(\frac{1}{4}x^4\right) - \int \frac{1}{4}x^4 \cdot \frac{1}{x}dx$$
$$= \frac{1}{4}x^4\ln x - \frac{1}{4}\int x^3dx$$
$$= \frac{1}{4}x^4\ln x - \frac{1}{4}\left(\frac{1}{4}x^4\right) + c$$
$$= \frac{1}{4}x^4\ln x - \frac{1}{16}x^4 + c$$

Copyright © 2019 Pearson Education, Inc.

29. $\int 9\sqrt{x}\ln x dx$

Using integration by parts, let $u = \ln x$ and $dv = 9x^{1/2}dx$. Therefore, $du = \frac{1}{x}dx$ and $v = 6x^{3/2}$.

$$\begin{aligned}&\int 9\sqrt{x}\ln x dx\\ &= \ln x\left(6x^{3/2}\right) - \int 6x^{3/2}\cdot\frac{1}{x}dx\\ &= 6x^{3/2}\ln x - 6\int x^{1/2}dx\\ &= 6x^{3/2}\ln x - 6\left(\frac{2}{3}x^{3/2}\right) + c\\ &= 6x^{3/2}\ln x - 4x^{3/2} + c\end{aligned}$$

31. $\int (9-2x)\ln x dx$

Using integration by parts, let $u = \ln x$ and $dv = (9-2x)dx$. Therefore, $du = \frac{1}{x}dx$ and $v = 9x - x^2$.

$$\begin{aligned}&\int (9-2x)\ln x dx\\ &= \ln x\left(9x - x^2\right) - \int (9-x^2)\cdot\frac{1}{x}dx\\ &= 9x\ln x - x^2\ln x - \int (9-x)dx\\ &= 9x\ln x - x^2\ln x - \left(9x - \tfrac{1}{2}x^2\right) + c\\ &= \left(9x - x^2\right)\ln x - \left(9x - \tfrac{1}{2}x^2\right) + c\end{aligned}$$

33. $\int (18x^2 + 12x + 4)\ln x dx$

Using integration by parts, let $u = \ln x$ and $dv = (18x^2 + 12x + 4)dx$. Therefore, $du = \frac{1}{x}dx$ and $v = 6x^3 + 6x^2 + 4x$.

$$\begin{aligned}&\int (18x^2 + 12x + 4)\ln x dx\\ &= \ln x\left(6x^3 + 6x^2 + 4x\right) - \int (6x^3 + 6x^2 + 4x)\cdot\frac{1}{x}dx\\ &= \ln x\left(6x^3 + 6x^2 + 4x\right) - \int (6x^2 + 6x + 4)dx\\ &= \ln x\left(6x^3 + 6x^2 + 4x\right) - \left(2x^3 + 3x^2 + 4x\right) + c\end{aligned}$$

35. $\int 15(x+1)\sqrt{x+2}dx$

Using integration by parts, let $u = 15x + 15$ and $dv = (x+2)^{1/2}dx$. Therefore, $du = 15dx$ and $v = \frac{2}{3}(x+2)^{3/2}$.

$$\begin{aligned}&\int 15(x+1)\sqrt{x+2}dx\\ &= 15(x+1)\left(\tfrac{2}{3}(x+2)^{3/2}\right) - \int \left(\tfrac{2}{3}(x+2)^{3/2}\right)\cdot 15dx\\ &= 10(x+1)(x+2)^{3/2} - 10\int (x+2)^{3/2}dx\\ &= 10(x+1)(x+2)^{3/2} - 10\left(\tfrac{2}{5}(x+2)^{5/2}\right) + c\\ &= 10(x+1)(x+2)^{3/2} - 4(x+2)^{5/2} + c\end{aligned}$$

37. a. The car started out driving 80 ft/sec, sped up to 95 ft/sec after 4 seconds, and slowed back down to about 80 ft/sec after 30 seconds.

b. $\int 80 + 10te^{-t/4}dt$

Using integration by parts for the second portion of the integral, let $u = 10t$ and $dv = e^{-t/4}dt$. Therefore, $du = 10dt$ and $v = -4e^{-t/4}$.

$$\begin{aligned}s(t) &= \int 80 + 10te^{-t/4}dt\\ s(t) &= \int 80dt + \int 10te^{-t/4}dt\\ &= 80t + \left(10t(-4e^{-t/4}) - \int -4e^{-t/4}(10)dt\right)\\ &= 80t - 40te^{-t/4} + 40\int e^{-t/4}dt\\ &= 80t - 40te^{-t/4} + 40\left(-4e^{-t/4}\right) + c\\ &= 80t - 40te^{-t/4} - 160e^{-t/4} + c\end{aligned}$$

Now find the constant using the fact that at $t = 0$, $s(t) = 0$;

$$\begin{aligned}0 &= 80(0) - 40(0)e^0 - 160e^0 + c\\ c &= 160\end{aligned}$$

Therefore, the equation will be $s(t) = 80t - 40te^{-t/4} - 160e^{-t/4} + 160$.

Copyright © 2019 Pearson Education, Inc.

39. **a.** $\int 18\ln(3t+1)dt$

Using the antiderivative of the Logarithmic function, let $u = 3t+1$. Therefore, $du = 3dx$.

$$\int 18\ln(3t+1)dt$$
$$= \frac{18}{3}((3t+1)\ln(3t+1)-(3t+1))+c$$
$$= 6((3t+1)\ln(3t+1)-(3t+1))+c$$
$$= (18t+6)\ln(3t+1)-18t-6+c$$

Now find the constant using the fact that at $t = 0$, $s(t) = 0$;

$$0 = (18(0)+6)\ln(3(0)+1)-18(0)-6+c$$
$$c = 6$$

Therefore, the equation will be
$$s(t) = (18t+6)\ln(3t+1)-18t-6+6$$
$$s(t) = (18t+6)\ln(3t+1)-18t$$

b. Let $t = 30$, then
$$s(30) = (18(30)+6)\ln(3(30)+1)-18(30)$$
$$s(30) \approx 1923$$

The car is about 1923 feet from the starting position after 30 seconds.

c. Let $t = 30$, then

$$v(30) = 18\ln((3(30)+1)$$
$$v(30) \approx 81 \text{ ft/sec}$$

$$\frac{81\text{ feet}}{1\text{ second}}\cdot\frac{3600\text{ seconds}}{1\text{ hour}}\cdot\frac{1\text{ mile}}{5280\text{ feet}} \approx 55\text{mph}$$

The car is travelling about 81 ft/sec or about 55 mph.

41. $\int t^2e^{-2t}dt$

Using integration by parts, let $u = t^2$ and $dv = e^{-2t}dt$. Therefore, $du = 2tdt$ and $v = -\frac{1}{2}e^{-2t}$.

$$\int t^2e^{-2t}dt$$
$$= t^2\left(-\frac{1}{2}e^{-2t}\right)-\int -\frac{1}{2}e^{-2t}\cdot 2tdt$$
$$= -\frac{1}{2}t^2e^{-2t}+\int te^{-2t}dt$$

Using integration by parts a second time, let $u = t$ and $dv = e^{-2t}dt$. Therefore, $du = dt$ and $v = -\frac{1}{2}e^{-2t}$.

$$-\tfrac{1}{2}t^2e^{-2t}+\int (te^{-2t})dt$$
$$= -\tfrac{1}{2}t^2e^{-2t}+\left(t\left(-\tfrac{1}{2}e^{-2t}\right)-\int -\tfrac{1}{2}e^{-2t}dt\right)$$
$$= -\tfrac{1}{2}t^2e^{-2t}-\tfrac{1}{2}te^{-2t}+\int \tfrac{1}{2}e^{-2t}dt$$
$$= -\tfrac{1}{2}t^2e^{-2t}-\tfrac{1}{2}te^{-2t}+\tfrac{1}{2}\cdot-\tfrac{1}{2}e^{-2t}+c$$
$$= -\tfrac{1}{2}t^2e^{-2t}-\tfrac{1}{2}te^{-2t}-\tfrac{1}{4}e^{-2t}+c$$

Now find the constant using the fact that at $t = 0$, $s(t) = 0$;

$$0 = -\tfrac{1}{2}(0)^2e^0-\tfrac{1}{2}(0)e^0-\tfrac{1}{4}e^0+c$$
$$c = \tfrac{1}{4}$$

Therefore, the equation will be
$$s(t) = -\tfrac{1}{2}t^2e^{-2t}-\tfrac{1}{2}te^{-2t}-\tfrac{1}{4}e^{-2t}+\tfrac{1}{4}$$

43. **a.** $\int 2xe^{-x/7}dx$

Using integration by parts, let $u = 2x$ and $dv = e^{-x/7}dx$. Therefore, $du = 2dx$ and $v = -7e^{-x/7}$.

$$F(x) = \int 2xe^{-x/7}dx$$
$$= \left(2x(-7e^{-x/7})-\int -7e^{-x/7}(2)dx\right)$$
$$= -14xe^{-x/7}+14\int e^{-x/7}dx$$
$$= -14xe^{-x/7}+14\left(-7e^{-x/7}\right)+c$$
$$= -14xe^{-x/7}-98e^{-x/7}+c$$

Now find the constant using the fact that at $x = 0$, $F(x) = 12$;

$$12 = -14(0)e^0-98e^0+c$$
$$c = 110$$

Therefore, the equation will be $F(x) = -14xe^{-x/7}-98e^{-x/7}+110$.

b. Let $x = 28$ (4 weeks times 7days), then
$$F(28) = -14(28)e^{-28/7}-98e^{-28/7}+110$$
$$F(28) \approx 101$$

There would be about 101 people infected after 4 weeks.

Copyright © 2019 Pearson Education, Inc.

45. $\int\left(\frac{5x+1}{e^x}\right)dx$

Using integration by parts, let $u = 5x+1$ and $dv = e^{-x}dx$. Therefore, $du = 5dx$ and $v = -e^{-x}$.

$$\int\left(\frac{5x+1}{e^x}\right)dx$$
$$= (5x+1)\left(-e^{-x}\right) - \int -e^{-x}\cdot 5dx$$
$$= (5x+1)\left(-e^{-x}\right) - \int -5e^{-x}dx$$
$$= -5xe^{-x} - e^{-x} - 5e^{-x} + c$$
$$= -5xe^{-x} - 6e^{-x} + c$$

Now find the constant using the fact that at $x = 0$, $C(x) = 10$;

$$10 = -5(0)e^0 - 6e^0 + c$$
$$c = 16$$

Therefore, the equation will be $C(x) = -5xe^{-x} - 6e^{-x} + 16.$

47.a. $f(t) = (75t + 2700)(15e^{-.02t})$
$= (1125t + 40,500)e^{-.02t}$

b. $\int((1125t+40,500)e^{-.02t})dt$

Using integration by parts, let $u = 1125t + 40,500$ and $dv = e^{-.02t}dt$.
Therefore, $du = 1125dt$ and $v = -50e^{-.02t}$.

$$\int((1125t+40,500)e^{-.02t})dt$$
$$= (1125t+40,500)\left(-50e^{-.02t}\right) - \int -50e^{-.02t}\cdot 1125dt$$
$$= (1125t+40,500)\left(-50e^{-.02t}\right) + \int 56,250e^{-.02t}dt$$
$$= -56,250te^{-.02t} - 2,025,000e^{-.02t} + 56,250(-50e^{-.02t} + c)$$
$$= -56,250te^{-.02t} - 2,025,000e^{-.02t} - 2,812,500e^{-.02t} + c$$
$$= -56,250te^{-.02t} - 4,837,500e^{-.02t} + c$$

Now find the constant using the fact that at $t = 0$, $F(x) = 0$;

$$0 = -56,250(0)e^0 - 4,837,500e^0 + c$$
$$c = 4,837,500$$

Therefore, the equation will be
$F(t) = -56,250te^{-.02t} - 4,837,500e^{-.02t} + 4,837,500.$

49. $\int(43\ln(x)+124)dx$

Using the antiderivative of the Logarithmic function for the first part of the integral, let $u = x$. Therefore, $du = dx$.

$$\int 42\ln(x)+124dx$$
$$= \int 42\ln(x)dx + \int 124dx$$
$$= 42\left(x\ln(x) - x\right) + 124x + c$$
$$= 42x\ln(x) - 42x + 124x + c$$
$$= 42x\ln(x) + 82x + c$$

Now find the constant using the fact that at $x = 1$, $F(x) = 0$;

$$0 = 42(1)\ln(1) + 82(1) + c$$
$$c = -82$$

Therefore, the equation will be
$F(x) = 42x\ln(x) + 82x - 82$

Section 13.4 Area and the Definite Integral

1. The total usage of electricity is the total area between the graph of the rate function and the x-axis from $x = 0$ (midnight) to $x = 24$ (the next midnight). Approximate this area by using 12 rectangles each with base of length 2 and heights determined by the graph.

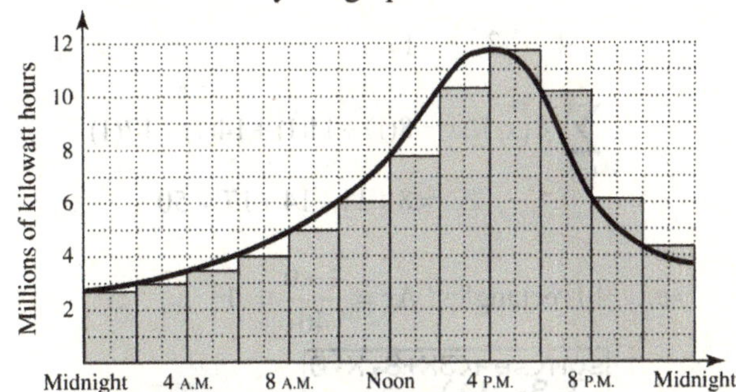

Estimate the area to be the sum
$2\cdot 3 + 2\cdot 3 + 2\cdot 3.5 + 2\cdot 4 + 2\cdot 5 + 2\cdot 6 + 2\cdot 8 + 2\cdot 10 + 2\cdot 11.5 + 2\cdot 10 + 2\cdot 6 + 2\cdot 4.5 = 151$
The total usage of electricity is about 149 kWh. (Your answer may vary depending on how you interpreted the height of each rectangle from the graph.)

Copyright © 2019 Pearson Education, Inc.

3. Approximate enrollment by finding the sum of the areas of 12 rectangles, each with base of length 1 and heights determined by the graph.
$1 \cdot 3.16 + 1 \cdot 3.09 + 1 \cdot 2.99 + 1 \cdot 2.98 + 1 \cdot 3.03 + 1 \cdot 3.06 + 1 \cdot 3.10 + 1 \cdot 3.15 + 1 \cdot 3.19 + 1 \cdot 3.22 + 1 \cdot 3.26 + 1 \cdot 3.31 = 37.54$
Enrollment from 2010–2022 will be about 37.54 million students.

5. Answers vary; The indefinite integral $\int f(x)dx$ denotes a set of functions, whereas the definite integral represents a number.

7. a. $f(x) = 3x + 8$; $[0, 4]$

two rectangles: $\Delta x = \frac{4-0}{2} = 2$

i	x_i	$f(x_i)$
1	0	8
2	2	14

$$\sum_{i=1}^{2} f(x_i)\Delta x = 8(2) + 14(2) = 16 + 28 = 44$$

b. four rectangles: $\Delta x = \frac{4-0}{4} = 1$

i	x_i	$f(x_i)$
1	0	8
2	1	11
3	2	14
4	3	17

$$\sum_{i=1}^{4} f(x_i)\Delta x = 8(1) + 11(1) + 14(1) + 17(1)$$
$$= 8 + 11 + 14 + 17 = 50$$

c. 40 rectangles: $\Delta x = \frac{4-0}{40} = .1$

```
sum(seq(3X+8,X,0
,3.9,.1))
             554
Ans*.1
            55.4
```

$$\sum_{i=1}^{40} f(x_i)\Delta x = 55.4$$

9. a. $f(x) = 4 - x^2$; $[-2, 2]$

two rectangles: $\Delta x = \frac{2-(-2)}{2} = 2$

i	x_i	$f(x_i)$
1	−2	0
2	0	4

$$\sum_{x=1}^{2} f(x_i)\Delta x = 0(2) + 4(2) = 8$$

b. four rectangles: $\Delta x = \frac{2-(-2)}{4} = 1$

i	x_i	$f(x_i)$
1	−2	0
2	−1	3
3	0	4
4	1	3

$$\sum_{i=1}^{4} f(x_i)\Delta x = 0(1) + 3(1) + 4(1) + 3(1) = 10$$

c. 40 rectangles: $\Delta x = \frac{2-(-2)}{40} = .1$

```
sum(seq(4-X²,X,-
2,1.9,.1))
           106.6
Ans*.1
           10.66
```

$$\sum_{i=1}^{40} f(x_i)\Delta x = 10.66$$

11. a. $f(x) = e^{2x} - .5$; $[0, 2]$

two rectangles: $\Delta x = \frac{2-0}{2} = 1$

i	x_i	$f(x_i)$
1	0	.5
2	1	6.89

$$\sum_{i=1}^{2} f(x_i)\Delta x = .5(1) + 6.89(1) = 7.39$$

Copyright © 2019 Pearson Education, Inc.

b. four rectangles: $\Delta x = \frac{2-0}{4} = 0.5$

i	x_i	$f(x_i)$
1	0	.5
2	.5	2.218
3	1	6.889
4	1.5	19.589

$$\sum_{i=1}^{4} f(x_i)\Delta x$$
$$= .25 + 1.109 + 3.4445 + 9.7945 = 14.60$$

c. 40 rectangles: $\Delta x = \frac{2-0}{40} = .05$

```
sum(seq(e^(2X)-.
5,X,0,1.95,.05))
         489.6290021
Ans*.05
         24.48145011
```

$$\sum_{i=1}^{40} f(x_i)\Delta x = 24.48$$

13. a. $f(x) = \frac{60}{x+5}$ from $x = 1$ to $x = 9$

For $n = 2$ rectangles: $x = \frac{2-(-2)}{2} = 2$.

i	x_i	$f(x_i)$
1	−2	$\frac{60}{3} = 20$
2	0	$\frac{60}{5} = 12$

$$A = \sum_{i=1}^{2} f(x_i)\Delta x = 20(2) + 12(2) = 64$$

b. For $n = 4$ rectangles: $\Delta x = \frac{2-(-2)}{4} = 1$.

i	x_i	$f(x_i)$
1	−2	$\frac{60}{3} = 20$
2	−1	$\frac{60}{4} = 15$
3	0	$\frac{60}{5} = 12$
4	1	$\frac{60}{6} = 10$

$$A = \sum_{i=1}^{4} f(x_i)\Delta x$$
$$= 20(1) + 15(1) + 12(1) + 10(1) = 57$$

c. 40 rectangles: $\Delta x = \frac{2-(-2)}{40} = .1$

```
sum(seq(60/(X+5)
,X,-2,1.9,.1))
         514.1383475
Ans*.1
         51.41383475
```

$$\sum_{i=1}^{40} f(x_i)\Delta x \approx 51.41$$

15. $\int_0^5 (5-x)dx$

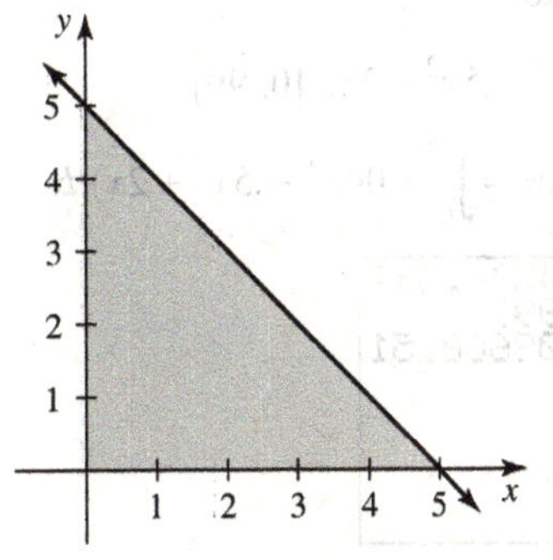

Copyright © 2019 Pearson Education, Inc.

$\int_0^5 (5-x)\,dx$ is the area of a triangle with base = 5 and height = 5.

Area $= \frac{1}{2}$ (base)(height) $= \frac{1}{2}(5)(5) = 12.5$

Exercises 17–22 were worked using a TI-84 Plus graphing calculator. Similar results can be obtained using other calculators.

17. $\int_{-5}^{0} \left(x^3 + 6x^2 - 10x + 2\right) dx$

```
fnInt(X³+6X²-10X
+2,X,-5,0)
           228.75
```

19. $\int_2^7 5\ln\left(2x^2+1\right) dx$

```
fnInt(5ln(2X²+1)
,X,2,7)
      90.54808787
```

21. $\int_0^3 4x^2e^{-3x}\,dx$

```
fnInt(4X²e^(-3X)
,X,0,3)
        .29444972
```

Exercises 23–26 were worked using a TI-84 Plus graphing calculator. Similar results can be obtained using other calculators.

23. $R'(x) = .04x^3 - .5x^2 + 2x,\ [0, 99]$

Total revenue $= \int_0^{99} (.04x^3 - .5x^2 + 2x)dx$

```
fnInt(.04X³-.5X²
+2X,X,0,99)
        808680.51
```

The total revenue over the period is $808,680.51.

25. $R'(x) = \frac{175}{1+3e^{-.7x}} - 20,\ [0, 310]$

Total Revenue $= \int_0^{310} \left(\frac{175}{1+3e^{-.7x}} - 20\right) dx$

```
fnInt(175/(1+3e^
(-.7X))-20,X,0,3
10)
      47703.42641
```

The total revenue over the period is about $47,703.43.

27. **a.** At $t = 1$, the red car was travelling at the same speed as the blue car, but it had travelled a longer distance.

b. At $t = 2$, the red car was travelling slower, but it had travelled the same distance.

c. At $t = 3$, the red car was travelling slower, and had travelled a shorter distance.

29. Read values of the function on the graph for every 5 sec from $t = 0$ to $t = 20$.

These are speeds in miles per hour, so multiply these values by $\frac{5280}{3600}$ to get the speeds in feet per second. Then to estimate the distance traveled, find the sum of areas of rectangles with heights the sides of the rectangles and with widths of 5 sec. The total distance traveled is approximately

$$\frac{5280}{3600}[0(5) + 50(5) + 80(5) + 90(5)]$$
$$= \frac{5280}{3600}(1100) \approx 1613.$$

The total distance traveled is approximately 1613 ft.

Copyright © 2019 Pearson Education, Inc.

Section 13.5 The Fundamental Theorem of Calculus

1. $\int_{-1}^{3}\left(6x^2-7x+8\right)dx$
$$=\left(2x^3-\frac{7x^2}{2}+8x\right)\Big|_{-1}^{3}$$
$$=\left[2(3)^3-\frac{7(3)^2}{2}+8(3)\right]-\left[2(-1)^3-\frac{7(-1)^2}{2}+8(-1)\right]$$
$$=\left(54-\frac{63}{2}+24\right)-\left(-2-\frac{7}{2}-8\right)$$
$$=\frac{93}{2}-\left(-\frac{27}{2}\right)=\frac{120}{2}=60$$

3. $\int_{0}^{2}3\sqrt{4u+1}du$
Let $w = 4u + 1$. Then $dw = 4\ du$. If $u = 2$, $w = 9$. If $u = 0$, $w = 1$.
$$\int_0^2 3\sqrt{4u+1}du=\frac{3}{4}\int_0^2(4u+1)^{1/2}4\,du$$
$$=\frac{3}{4}\int_1^9 w^{1/2}dw=\frac{3}{4}\left(\frac{w^{3/2}}{\frac{3}{2}}\right)\Big|_1^9$$
$$=\frac{3}{4}\cdot\frac{2}{3}w^{3/2}\Big|_1^9$$
$$=\frac{1}{2}(9)^{3/2}-\frac{1}{2}(1)^{3/2}$$
$$=\frac{27}{2}-\frac{1}{2}=13$$

5. $\int_0^1 3\left(t^{1/2}+4t\right)dt=3\left(\frac{t^{3/2}}{3/2}+2t^2\right)\Big|_0^1$
$$=3\left(\frac{1^{3/2}}{3/2}+2(1)^2\right)-3\left(\frac{0^{3/2}}{3/2}+2(0)^2\right)$$
$$=3\left(\frac{2}{3}+2\right)-0=8$$

7. $\int_1^4\left(10y\sqrt{y}+3\sqrt{y}\right)dy=\int_1^4\left(10y^{3/2}+3y^{1/2}\right)dy$
$$=\left(10\cdot\frac{y^{5/2}}{\frac{5}{2}}+3\cdot\frac{y^{3/2}}{\frac{3}{2}}\right)\Big|_1^4$$
$$=\left(4y^{5/2}+2y^{3/2}\right)\Big|_1^4$$
$$=\left[4(4)^{5/2}+2(4)^{3/2}\right]-\left[4(1)^{5/2}+2(1)^{3/2}\right]$$
$$=(128+16)-(4+2)$$
$$=138$$

9. $\int_4^7\frac{11}{(x-3)^2}dx=\int_4^7 11(x-3)^{-2}dx$
Let $u = x - 3$. Then $du = dx$. If $x = 7$, $u = 4$. If $x = 4$, $u = 1$.
$$\int_4^7\frac{11}{(x-3)^2}dx=\int_1^4 11u^{-2}du=11\cdot\frac{u^{-1}}{-1}\Big|_1^4$$
$$=-\frac{11}{u}\Big|_1^4=-\frac{11}{4}-\left(-\frac{11}{1}\right)$$
$$=-\frac{11}{4}+\frac{44}{4}=\frac{33}{4}=8\frac{1}{4}$$

11. $\int_1^5\left(6n^{-1}+n^{-4}\right)dn=\int_1^5 6n^{-1}dn+\int_1^5 n^{-4}dn$
$$=6\int_1^5 n^{-1}dn+\int_1^5 n^{-4}dn$$
$$=6\left(\ln n\right)\Big|_1^5+\frac{n^{-3}}{-3}\Big|_1^5$$
$$=6\left(\ln 5-\ln 1\right)-\frac{1}{3}\left(5^{-3}-1^{-3}\right)$$
$$=6\ln 5-\frac{1}{3}\left(\frac{1}{125}-1\right)\approx 9.9873$$

13. $\int_0^5\left(6e^{2x}\right)dx=\frac{6e^{2x}}{2}\Big|_0^5$
$$=3e^{2x}\Big|_0^5$$
$$=3e^{10}-3e^0$$
$$=3e^{10}-3$$

Copyright © 2019 Pearson Education, Inc.

15. $\int_2^3 \left(.1e^{-.1A} + \frac{3}{A}\right) dA$

$= 0.1\int_2^3 e^{-.1A} dA + 3\int_2^3 \frac{1}{A} dA$

$= 0.1 \cdot \left.\frac{e^{-.1A}}{-.1}\right|_2^3 + 3\ln|A|\Big|_2^3 = -e^{-.1A}\Big|_2^3 + 3\ln|A|\Big|_2^3$

$= -e^{-.3} - \left(-e^{-.2}\right) + 3\ln 3 - 3\ln 2$

$= e^{-.2} - e^{-.3} + 3\ln 3 - 3\ln 2 \approx 1.294$

17. $\int_{-1}^0 y\left(2y^2 - 3\right)^5 dy$

Let $u = 2y^2 - 3$. Then $du = 4y\, dy$. If $y = 0$, then $u = -3$. If $y = -1$, then $u = -1$.

$\int_{-1}^0 y\left(2y^2 - 3\right)^5 dy = \frac{1}{4}\int_{-1}^0 \left(2y^2 - 3\right)^5 4y\, dy$

$= \frac{1}{4}\int_{-1}^{-3} u^5 du = \frac{1}{4} \cdot \left.\frac{u^6}{6}\right|_{-1}^{-3}$

$= \frac{1}{24}(-3)^6 - \frac{1}{24}(-1)^6$

$= \frac{729}{24} - \frac{1}{24} = \frac{728}{24} = \frac{91}{3}$

19. $\int_1^{64} \frac{\sqrt{z} - 2}{\sqrt[3]{z}} dz$

$= \int_1^{64} \left(\frac{z^{1/2}}{z^{1/3}} - 2z^{-1/3}\right) dz$

$= \int_1^{64} z^{1/6} dz - 2\int_1^{64} z^{-1/3} dz$

$= \left.\frac{z^{7/6}}{\frac{7}{6}}\right|_1^{64} - 2 \cdot \left.\frac{z^{2/3}}{\frac{2}{3}}\right|_1^{64} = \left.\frac{6z^{7/6}}{7}\right|_1^{64} - 3z^{2/3}\Big|_1^{64}$

$= \frac{6(64)^{7/6}}{7} - \frac{6(1)^{7/6}}{7} - 3\left(64^{2/3} - 1^{2/3}\right)$

$= \frac{6(128)}{7} - \frac{6}{7} - 3(16 - 1) = \frac{447}{7} \approx 63.857$

21. $\int_1^3 \frac{\ln x}{4x} dx = \frac{1}{4}\int_1^3 \ln x \cdot \frac{1}{x} dx$

Let $u = \ln x$. Then $du = \frac{1}{x} dx$. If $x = 3$, then $u = \ln 3$. If $x = 1$, $u = \ln 1 = 0$.

$\int_1^3 \frac{\ln x}{4x} dx = \frac{1}{4}\int_0^{\ln 3} u\, du = \left(\frac{1}{4}\right) \cdot \left.\frac{u^2}{2}\right|_0^{\ln 3}$

$= \frac{(\ln 3)^2}{8} - \frac{0^2}{8} = \frac{(\ln 3)^2}{8} \approx .1509$

23. $\int_0^8 x^{1/3}\sqrt{x^{4/3} + 9}\, dx$

Let $u = x^{4/3} + 9$. Then $du = \frac{4}{3}x^{1/3} dx$. If $x = 8$, then $u = 25$. If $x = 0$, then $u = 9$.

$\int_0^8 x^{1/3}\sqrt{x^{4/3} + 9}\, dx$

$= \frac{3}{4}\int_0^8 \left(x^{4/3} + 9\right)^{1/2} \cdot \frac{4}{3}x^{1/3} dx$

$= \frac{3}{4}\int_9^{25} u^{1/2} du = \frac{3}{4} \cdot \left.\frac{u^{3/2}}{\frac{3}{2}}\right|_9^{25}$

$= \frac{3}{4} \cdot \frac{2}{3} u^{3/2}\Big|_9^{25} = \frac{1}{2}(25)^{3/2} - \frac{1}{2}(9)^{3/2} = 49$

25. $\int_0^1 \frac{4e^t}{\left(3 + e^t\right)^2} dt$

Let $u = 3 + e^t$. Then $du = e^t dt$. If $t = 1$, then $u = 3 + e$. If $t = 0$, $u = 4$.

$\int_0^1 \frac{4e^t}{\left(3 + e^t\right)^2} dt = 4\int_0^1 \left(3 + e^t\right)^{-2} e^t dt$

$= 4\int_4^{3+e} u^{-2} du = \left.\frac{4u^{-1}}{-1}\right|_4^{3+e}$

$= -\left.\frac{4}{u}\right|_4^{3+e} = -\frac{4}{3 + e} - \left(-\frac{4}{4}\right)$

$\approx .3005$

Copyright © 2019 Pearson Education, Inc.

27. $\int_1^{49} \frac{\left(1+\sqrt{x}\right)^{4/3}}{\sqrt{x}} dx = \int_1^{49} \left(1+x^{1/2}\right)^{4/3} x^{-1/2} dx$

Let $u = 1 + x^{1/2}$. Then $du = \frac{1}{2} x^{-1/2} dx$.

If $x = 49$, $u = 8$. If $x = 1$, $u = 2$.

$$\begin{aligned}\int_1^{49} \frac{\left(1+\sqrt{x}\right)^{4/3}}{\sqrt{x}} dx &= \int_1^{49} \left(1+x^{1/2}\right)^{4/3} x^{-1/2} dx \\ &= 2\int_1^{49} \left(1+x^{1/2}\right)^{4/3} \frac{1}{2} x^{-1/2} dx \\ &= 2\int_2^8 u^{4/3} du = 2 \cdot \frac{u^{7/3}}{\frac{7}{3}}\Bigg|_2^8 \\ &= 2 \cdot \frac{3}{7} u^{7/3}\Big|_2^8 \\ &= \frac{6}{7}(8)^{7/3} - \frac{6}{7}(2)^{7/3} \\ &\approx 105.3946\end{aligned}$$

29. $\int_2^3 \frac{4x^3+6}{x^4+6x+9} dx$

Let $u = x^4 + 6x + 9$. Then $du = \left(4x^3+6\right) dx$.

If $x = 3$, then $u = 3^4 + 6(3) + 9 = 108$. If $x = 2$, then $u = 2^4 + 6(2) + 9 = 37$.

$$\begin{aligned}\int_2^3 \frac{4x^3+6}{x^4+6x+9} dx &= \int_{37}^{108} \frac{1}{u} du = \ln |u| \Big|_{37}^{108} \\ &= \ln 108 - \ln 37 \approx 1.0712\end{aligned}$$

31. A negative definite integral for the first year and a half would indicate a loss for that period. The overall profit for the two-year period is represented by $\int_0^2 \left(6x^2 - 7x - 3\right) dx$.

33. $$\begin{aligned}A &= \int_0^2 \left(2 - .5x^2\right) dx \\ &= \left(2x - \frac{.5x^3}{3}\right)\Bigg|_0^2 \\ &= \left(4 - \frac{4}{3}\right) - 0 \\ &= \frac{8}{3}\end{aligned}$$

35. $$\begin{aligned}A &= \int_{-1}^0 \left(x^2 - 2x\right) dx + \left|\int_0^2 \left(x^2 - 2x\right) dx\right| \\ &= \left(\frac{x^3}{3} - x^2\right)\Bigg|_{-1}^0 + \left|\left(\frac{x^3}{3} - x^2\right)\Bigg|_0^2\right| \\ &= 0 - \left(-\frac{1}{3} - 1\right) + \left|\left(\frac{8}{3} - 4\right) - 0\right| \\ &= \frac{4}{3} + \left|-\frac{4}{3}\right| = \frac{8}{3} \approx 2.67\end{aligned}$$

37. $f(x) = 9 - x^2$; $[0, 4]$

The graph of f is above the x-axis on $[0, 3)$ and below the x-axis on $(3, 4]$.

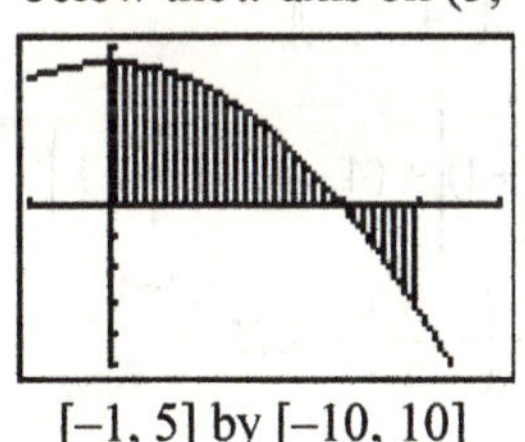

$[-1, 5]$ by $[-10, 10]$

$$\begin{aligned}A &= \int_0^3 \left(9 - x^2\right) dx + \left|\int_3^4 \left(9 - x^2\right) dx\right| \\ &= \left(9x - \frac{x^3}{3}\right)\Bigg|_0^3 + \left|\left(9x - \frac{x^3}{3}\right)\Bigg|_3^4\right| \\ &= (27 - 9) - 0 + \left|(36 - \frac{64}{3}) - (27 - 9)\right| \\ &= 18 + \left|\frac{44}{3} - 18\right| = \frac{64}{3}\end{aligned}$$

Copyright © 2019 Pearson Education, Inc.

39. $f(x)=x^3-1;[-1,2]$

The graph of f is below the x-axis on $[-1, 1)$ and above the x-axis on $(1, 2]$.

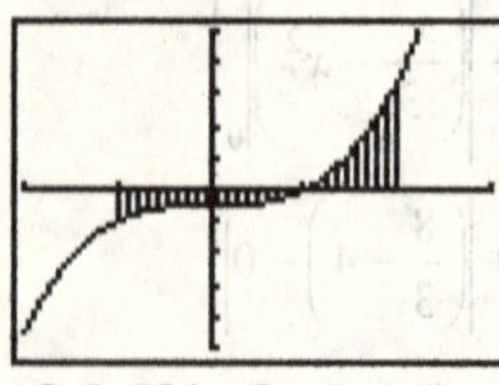

$[-2, 3]$ by $[-10, 10]$

$$A=\left|\int_{-1}^{1}\left(x^3-1\right)dx\right|+\int_{1}^{2}\left(x^3-1\right)dx$$

$$=\left|\left(\frac{x^4}{4}-x\right)\Bigg|_{-1}^{1}\right|+\left(\frac{x^4}{4}-x\right)\Bigg|_{1}^{2}$$

$$=\left|\left(\frac{1}{4}-1\right)-\left(\frac{1}{4}+1\right)\right|+(4-2)-\left(\frac{1}{4}-1\right)$$

$$=\left|-\frac{3}{4}-\frac{5}{4}\right|+2-\left(-\frac{3}{4}\right)=4.75$$

41. $f(x)=e^{2x}-1;[-2,1]$

Solve $f(x)=0$ to determine where the graph crosses the x-axis.

$e^{2x}-1=0\Rightarrow e^{2x}=1\Rightarrow 2x\ln e=\ln 1\Rightarrow$
$2x=0\Rightarrow x=0$

The graph crosses the x-axis at 0 in the given interval $[-2, 1]$. The total area is

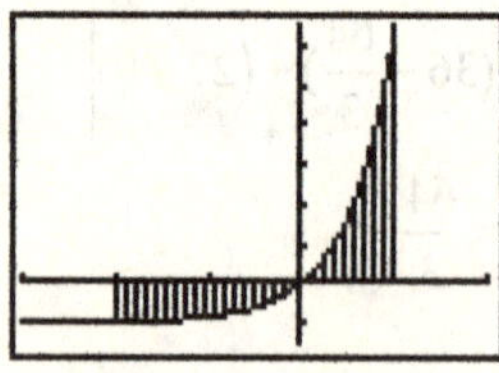

$[-3, 2]$ by $[-1.5, 6.5]$

$$\left|\int_{-2}^{0}\left(e^{2x}-1\right)dx\right|+\int_{0}^{1}\left(e^{2x}-1\right)dx$$

$$=\left|\left(\frac{e^{2x}}{2}-x\right)\Bigg|_{-2}^{0}\right|+\left(\frac{e^{2x}}{2}-x\right)\Bigg|_{0}^{1}$$

$$=\left|\left(\frac{1}{2}-0\right)-\left(\frac{e^{-4}}{2}+2\right)\right|+\left(\frac{e^{2}}{2}-1\right)-\left(\frac{1}{2}-0\right)$$

$$=\left|-\frac{3}{2}-\frac{e^{-4}}{2}\right|+\frac{e^2}{2}-1-\frac{1}{2}$$

$$=\left|-\frac{3}{2}-\frac{e^{-4}}{2}\right|+\frac{e^2}{2}-\frac{3}{2}\approx 3.7037$$

43. $f(x)=x^2e^{-x^3/2}dx$ on $[0, 3]$

The graph of f is above the x-axis on $[0, 3]$.

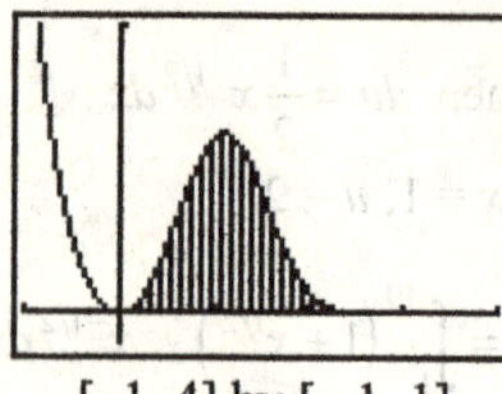

$[-1, 4]$ by $[-.1, 1]$

Let $u=-\frac{1}{2}x^3$. Then $du=-\frac{3}{2}x^2dx$. If $x=3$, then $u=-\frac{27}{2}$. If $x=0$, then $u=0$.

$$\int_0^3 x^2e^{-x^3/2}dx=-\frac{2}{3}\int_0^3 x^2e^{-x^3/2}\left(-\frac{3}{2}\right)dx$$

$$=-\frac{2}{3}\int_0^{-27/2}e^u\,du=-\frac{2}{3}e^u\Big|_0^{-27/2}$$

$$=-\frac{2}{3}\left(e^{-27/2}-1\right)\approx .6667$$

45. $f(x)=\frac{1}{x};[1,e]$

$\frac{1}{x}=0$ has no solution, so the graph does not cross the x-axis in the given interval $[1, e]$.

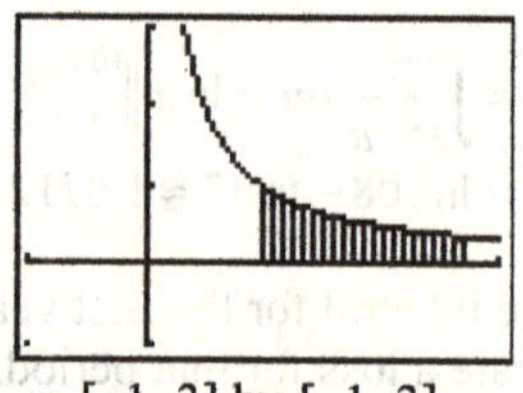

$[-1, 3]$ by $[-1, 3]$

$$\int_1^e \frac{1}{x}dx=\ln x\big|_1^e=\ln e-\ln 1=1$$

47. $f(x)=\frac{12(\ln x)^3}{x}$ on $[1, 4]$

The graph of f is above the x-axis on $[0, 3]$.

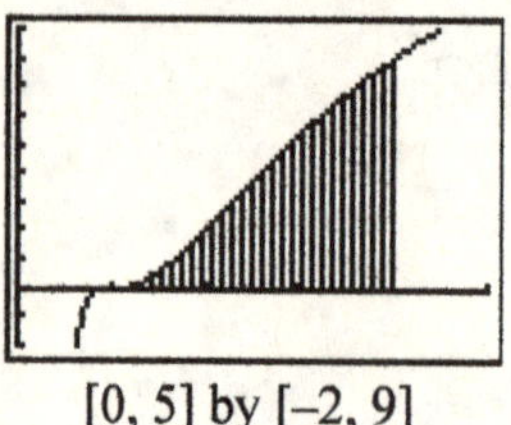

$[0, 5]$ by $[-2, 9]$

Copyright © 2019 Pearson Education, Inc.

$$\int_1^4 \frac{12(\ln x)^3}{x}\,dx$$

Let $u = \ln x$. Then $du = \frac{1}{x}dx$. If $x = 4$, then $u = \ln 4$. If $x = 1$, then $u = 0$.

$$\int_1^4 \frac{12(\ln x)^3}{x}\,dx = 12\int_0^{\ln 4} u^3\,du = 12\frac{u^4}{4}\Big|_0^{\ln 4}$$
$$= 3\left((\ln 4)^4 - 0\right) \approx 11.08$$

49. $f(x) = -3\sqrt{x-5}$ on $[5,9]$.

This function is below the x-axis for the entire interval so the integral will be negative.

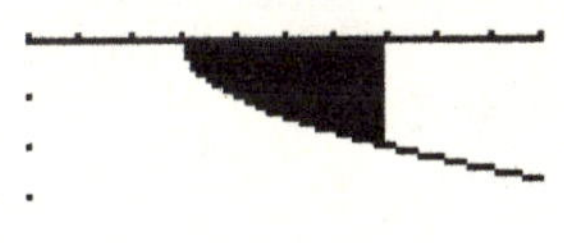

[2, 12] by [–15, 5]

Let $u = x - 5$. Then $du = dx$.

If $x = 5$, then $u = 0$, and if $x = 9$, then $u = 4$.

$$-\int_5^9 -3(x-5)^{1/2}\,dx = 3\int_0^4 u^{1/2}\,du = 2u^{3/2}\Big|_0^4$$
$$= 2\left[(4)^{3/2} - (0)^{3/2}\right] = 16$$

51. $E(x) = 4x + 2$ is the rate of expenditure per day.

a. The total expenditure is hundreds of dollars in 10 days is

$$\int_0^{10}(4x+2)\,dx = \left(\frac{4x^2}{2} + 2x\right)\Big|_0^{10}$$
$$= 2(100) + 20 - 0 = 220$$

Therefore, since $220(100) = 22{,}000$, the total expenditure is \$22,000.

b. From the tenth to the twenty-fifth day:

$$\int_{10}^{25}(4x+2)\,dx = \left(\frac{4x^2}{2} + 2x\right)\Big|_{10}^{25}$$
$$= [2(625) + 50] - [2(100) + 20]$$
$$= 1080$$

That is, \$108,000 is spent.

53. $f'(x) = 114e^{.086x}$

The total sales between the start of 2000 and the start of 2014 is given by

$$\int_0^{15}(114e^{.086x})\,dx = \left(\frac{114}{.086}e^{.086x}\right)\Big|_0^{15}$$
$$= \frac{114}{.086}(e^{.086(15)} - e^0)$$
$$\approx 3490$$

The total sales is about \$3490 billion.

55. a. $\int_{20}^{30}\left(-.81x^2 + 38x + 3904\right)dx$

$$= \left(\frac{-.81x^3}{3} + \frac{38x^2}{2} + 3904x\right)\Big|_{20}^{30}$$
$$= 126{,}930 - 83{,}520 = 43{,}410 \text{ thousand}$$

There are about 43,410,000 people between 20 and 30.

b. $\int_0^{90}\left(-.81x^2 + 38x + 3904\right)dx$

$$= \left(\frac{-.81x^3}{3} + \frac{38x^2}{2} + 3904x\right)\Big|_0^{90}$$
$$= 308{,}430 - 0 = 308{,}430 \text{ thousand}$$

There are about 308,430,000 people between 0 and 90.

c. $\int_{10}^{30}\left(-.81x^2 + 38x + 3904\right)dx$

$$= \left(\frac{-.81x^3}{3} + \frac{38x^2}{2} + 3904x\right)\Big|_0^{90}$$
$$= 126{,}930 - 40{,}670 = 86{,}260 \text{ thousand}$$

There are about 86,260,000 people that are considered millennials.

Copyright © 2019 Pearson Education, Inc.

57. $f(t)=\dfrac{6000(.3+.28t)}{\left(1+.3t+.14t^2\right)^2}$

$$\int_0^3 \frac{6000(.3+.28t)}{\left(1+.3t+.14t^2\right)^2}$$

Let $u=1+.3t+.14t^2$ and $du=(.3+.28t)\,dt$.

$$\int \frac{6000(.3+.28t)}{\left(1+.3t+.14t^2\right)^2}=6000\int\frac{du}{u^2}$$

$$=6000\left(\frac{u^{-1}}{-1}\right)+C$$

$$=-\frac{6000}{u}+C=-\frac{6000}{1+.3t+.14t^2}+C$$

$$\int_0^3 \frac{6000(.3+.28t)}{(1+.3t+.14t^2)^2}dt$$

$$=-\frac{6000}{1+.3t+.14t^2}\Bigg|_0^3$$

$$=-\frac{6000}{1+.3(3)+.14(3)^2}-\left(-\frac{6000}{1+.3(0)+.14(0)^2}\right)$$

$$\approx -1898.734-(-6000)\approx 4101.27$$

Total depreciation at the end of 3 years is about \$4101.27.

59. $L'(t)=\dfrac{70\ln(t+1)}{t+1}$

Consider $\displaystyle\int\frac{70\ln(t+1)}{t+1}dt.$

Let $u=\ln(t+1)$. Then $du=\dfrac{1}{t+1}dt$.

$$\int\frac{70\ln(t+1)}{(t+1)}=70\int\ln(t+1)\frac{1}{t+1}dt=70\int u\,du=70\left(\frac{u^2}{2}\right)+C=35u^2+C=35[\ln(t+1)]^2+C$$

a. The total number of barrels leaked on the first day is

$$\int_0^{24}\frac{70\ln(t+1)}{t+1}=35[\ln(t+1)]^2\Big|_0^{24}=35(\ln 25)^2-35(\ln 1)^2=35(\ln 25)^2\approx 363 \text{ barrels.}$$

b. The total number of barrels leaked on the second day is

$$\int_{24}^{48}\frac{70\ln(t+1)}{t+1}dt=35[\ln(t+1)]^2\Big|_{24}^{48}=35(\ln 49)^2-35(\ln 25)^2\approx 167 \text{ barrels.}$$

Copyright © 2019 Pearson Education, Inc.

61. $P'(x) = xe^{-.04x}$

$P(x) = \int (xe^{-.04x})dx$

Using integration by parts, let $u = x$ and $dv = e^{-.04x}dx$. Therefore, $du = dx$ and $v = -25e^{-.04x}$.

$$\begin{aligned} P(x) &= \int (xe^{-.04x})dx \\ &= (x)(-25e^{.04x}) - \int -25e^{-.04x}dx \\ &= -25xe^{-.04x} + (-625e^{-.04x} + c) \\ &= -25xe^{-.04x} - 625e^{-.04x} + c \end{aligned}$$

Given that $P(0) = 0$, the function will be $P(x) = -25xe^{.04x} - 625e^{-.04x} + 625$.

$$\begin{aligned} P(100) &= -25(100)e^{.04(100)} - 625e^{-.04(100)} + 625 \\ &\approx 567.76 \end{aligned}$$

The total profit from the sale of 100 units is $576.76.

63. $\int_0^5 \frac{1}{3}e^{-x/3}dx = -e^{-x/3}\Big|_0^5 = -e^{-5/3} - \left(-e^{-0/3}\right)$

$= -e^{-1} + 1 \approx -.189 + 1 = .811$

The probability that the first customer arrives in the first 5 minutes is about .811.

Section 13.6 Applications of Integrals

1. $M(x) = 60\left(1 + x^2\right)$

The total maintenance charge on a two-year lease is given by

$$\int_0^2 60\left(1 + x^2\right)dx$$

$$= 60\int_0^2 \left(1 + x^2\right)dx = 60\left(x + \frac{x^3}{3}\right)\Bigg|_0^2$$

$$= 60\left(2 + \frac{2^3}{3} - 0\right) = 60\left(2 + \frac{8}{3}\right) = \$280$$

Monthly addition for maintenance:

$$\frac{280}{24} = \$11.67$$

3. Rate of savings: $S(t) = 1000(t + 2)$

During the first year, total savings were

$$\int_0^1 1000(t+2)dt = 1000\int_0^1 (t+2)dt$$

$$= 1000\left(\frac{t^2}{2} + 2t\right)\Bigg|_0^1$$

$$= 1000\left(\frac{1}{2} + 2 - 0\right) = \$2500$$

During the first 6 years total savings were

$$\int_0^6 1000(t+2)dt = 100\int_0^6 (t+2)dt$$

$$= 1000\left(\frac{t^2}{2} + 2t\right)\Bigg|_0^6$$

$$= 1000\left(\frac{36}{2} + 12 - 0\right)$$

$$= \$30,000$$

5. $f'(t) = -1.38t^2 + 34t - 123$

$$\begin{aligned} f &= \int_{10}^{17} \left(-1.38t^2 + 34t - 123\right)dt \\ &= \left(-.46t^3 + 17t^2 - 123t\right)\Big|_{10}^{17} \\ &= \left[.46\left(17^3\right) + 17(17^2) - 123(17)\right] \\ &\quad - \left[.46\left(10^3\right) + 17(10^2) - 123(10)\right] \\ &\approx 552 \end{aligned}$$

Total revenue generated between the start of 2010 and the start of 2017 was about $552 billion.

7. $f(t) = 340e^{.22t}$

$$\begin{aligned} F(t) &= \int_{10}^{16} \left(340e^{.22t}\right)dt \\ &= \left(\tfrac{340}{.22}e^{.22t}\right)\Big|_{10}^{16} \\ &= \left[\tfrac{340}{.22}e^{.22(16)} - \tfrac{340}{.22}e^{.22(10)}\right] \\ &\approx 38,265 \end{aligned}$$

The total revenue generated between the start of 2010 and the start of 2016 was about $38,265 million ($38.265 billion).

Copyright © 2019 Pearson Education, Inc.

9. **a.** To find the point of intersection, set the two equations equal to each other and solve the result

$$f(x) = g(x)$$
$$x^2 = 8 - 2x$$
$$x^2 + 2x - 8 = 0$$
$$(x+4)(x-2) = 0$$
$$x = -4 \text{ and } x = 2$$

Therefore, the point of intersection will be $x = 2$ only because the graph is defined on the interval [0, 3].

b. To find the integral from [0, 2], we will use the function *f(x)* subtracted from *g(x)*.

$$A(x) = \int_0^2 \left(8 - 2x - x^2\right) dx$$
$$= \left(8x - x^2 - \tfrac{1}{3}x^3\right)\Big|_0^2$$
$$= \left[8(2) - (2)^2 - \tfrac{1}{3}(2)^3 - 0\right]$$
$$= \frac{28}{3}$$

c. To find the integral from [2, 3], we will use the function *g(x)* subtracted from *f(x)*.

$$B(x) = \int_2^3 \left(x^2 + 2x - 8\right) dx$$
$$= \left(\tfrac{1}{3}x^3 + x^2 - 8x\right)\Big|_2^3$$
$$= \left[\tfrac{1}{3}(3)^3 + (3)^2 - 8(3)\right] - \left[\tfrac{1}{3}(2)^3 + (2)^2 - 8(2)\right]$$
$$= \frac{10}{3}$$

d. To find the integral from [0, 3], we will use the solutions to parts (b) and (c) and add the answers together.

$$C(x) = A(x) + B(x)$$
$$= \frac{28}{3} + \frac{10}{3} = \frac{38}{3}$$

11. **a.** To find the point of intersection, set the two equations equal to each other and solve the result

$$f(x) = g(x)$$
$$\frac{x}{2} = \sqrt{x}$$
$$x = 2\sqrt{x}$$
$$x^2 = 4x$$
$$x(x-4) = 0$$
$$x = 0 \text{ and } x = 4$$

Therefore, the point of intersection will be $x = 4$ only because the graph is defined on the interval [0, 9].

b. To find the integral from [0, 4], we will use the function *f(x)* subtracted from *g(x)*.

$$A(x) = \int_0^4 \left(x^{1/2} - \tfrac{1}{2}x\right) dx$$
$$= \left(\tfrac{2}{3}x^{3/2} - \tfrac{1}{4}x^2\right)\Big|_0^4$$
$$= \left[\tfrac{2}{3}(4)^{3/2} - \tfrac{1}{4}(4)^2 - 0\right]$$
$$= \frac{4}{3}$$

c. To find the integral from [4, 9], we will use the function *g(x)* subtracted from *f(x)*.

$$B(x) = \int_4^9 \left(\tfrac{1}{2}x - x^{1/2}\right) dx$$
$$= \left(\tfrac{1}{4}x^2 - \tfrac{2}{3}x^{3/2}\right)\Big|_4^9$$
$$= \left[\tfrac{1}{4}(9)^2 - \tfrac{2}{3}(9)^{3/2}\right] - \left[\tfrac{1}{4}(4)^2 - \tfrac{2}{3}(4)^{3/2}\right]$$
$$= \frac{43}{12}$$

d. To find the integral from [0, 9], we will use the solutions to parts (b) and (c) and add the answers together.

$$C(x) = A(x) + B(x)$$
$$= \frac{4}{3} + \frac{43}{12} = \frac{59}{12}$$

13. $y = 3x$ and $y = x^2 - 4$ from $x = -1$ to $x = 3$

On the interval, $3x \geq x^2 - 4$

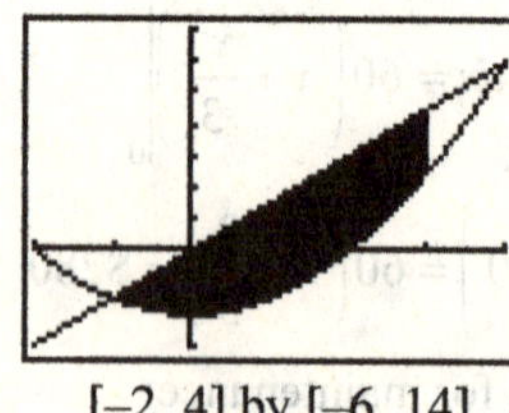

[−2, 4] by [−6, 14]

Copyright © 2019 Pearson Education, Inc.

$$\int_{-1}^{3}\left(3x-(x^2-4)\right)dx$$
$$=\int_{-1}^{3}\left(-x^2+3x+4\right)dx$$
$$=-\frac{x^3}{3}+\frac{3x^2}{2}+4x\Big|_{-1}^{3}$$
$$=-\frac{(3)^3}{3}+\frac{3(3)^2}{2}+4(3)-\left(-\frac{(-1)^3}{3}+\frac{3(-1)^2}{2}+4(-1)\right)$$
$$=\frac{33}{2}-\left(-\frac{13}{6}\right)=\frac{56}{3}$$

15. $y=x^2$ and $y=x^3$ from $x=0$ to $x=1/2$

On the interval, $x^2\ge x^3$

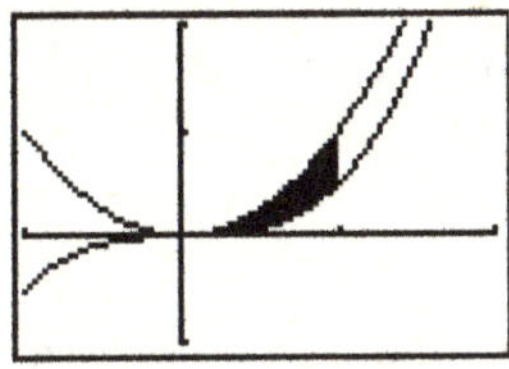

[–.5, 1] by [–.25, .5]

$$\int_0^{1/2}(x^2-x^3)dx=\left(\frac{x^3}{3}-\frac{x^4}{4}\right)\Big|_0^{1/2}$$
$$=\left(\frac{1}{24}-\frac{1}{64}\right)-0=\frac{5}{192}$$

17. $y=e^x$ and $y=x$ from $x=0$ to $x=1$

On the interval, $e^x\ge x$

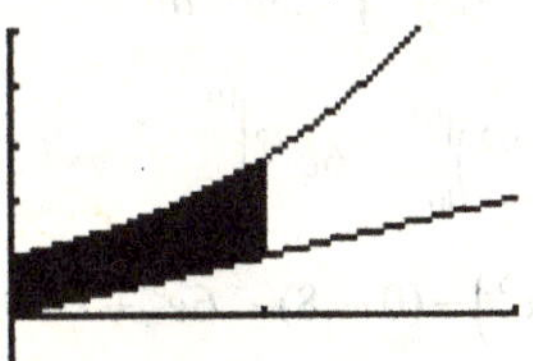

[0, 2] by [–1, 5]

$$\int_0^1(e^x-x)dx=\left(e^x-\frac{x^2}{2}\right)\Big|_0^1$$
$$=\left(e^1-\frac{(1)^2}{2}\right)-\left(e^0-\frac{(0)^2}{2}\right)$$
$$\approx 1.22$$

Exercises 19–22 were done using a TI-84 Plus graphing calculator. Similar results can be obtained using other graphing calculators.

19. $y=\ln x$ and $y=2xe^x$, $[1, 4]$

$$A=\int_1^4\left(2xe^x-\ln x\right)dx$$

```
fnInt(2Xe^(X)-ln
(X),X,1,4)
        325.0437228
```

The area is approximately 325.04.

21. $y=\sqrt{9-x^2}$ and $y=\sqrt{x+1}$, $[-1, 2]$

$$A=\int_{-1}^{2}\left(\sqrt{9-x^2}-\sqrt{x+1}\right)dx$$

```
fnInt(√(9-X²)-√(
X+1),X,-1,2)
        4.999220084
```

The area is approximately 4.999.

Copyright © 2019 Pearson Education, Inc.

23. $S'(x) = 150 - x^2$

$C'(x) = x^2 + \frac{11}{4}x$

a. Find the value where the first derivatives are equal to each other.

$$150 - x^2 = x^2 + \frac{11}{4}x$$
$$0 = 2x^2 + \frac{11}{4}x - 150$$
$$0 = 8x^2 + 11x - 600$$

Solving the above equation, we find $x = 8$ or $x \approx -9.375$.
Thus, the last year it will be profitable is during the 8th yr.

b. The total net savings during the first year is given by

$$\int_0^1 (150 - x^2)\,dx - \int_0^1 \left(x^2 + \frac{11}{4}x\right)dx = \left(150x - \frac{x^3}{3}\right)\Bigg|_0^1 - \left(\frac{x^3}{3} + \frac{11}{4}\cdot\frac{x^2}{2}\right)\Bigg|_0^1$$
$$= \left(150 - \frac{1}{3}\right) - 0 - \left[\left(\frac{1}{3} + \frac{11}{8}\right) - 0\right] \approx 147.96$$

The total is about \$147.96.

c. The total net savings over the entire period of use is given by

$$\int_0^8 (150 - x^2)\,dx - \int_0^8 \left(x^2 + \frac{11}{4}x\right)dx = \left(150x - \frac{x^3}{3}\right)\Bigg|_0^8 - \left(\frac{x^3}{3} + \frac{11}{4}\cdot\frac{x^2}{2}\right)\Bigg|_0^8$$
$$= \left(1200 - \frac{512}{3}\right) - 0 - \left[\left(\frac{512}{3} + \frac{11}{4}\cdot\frac{64}{2}\right) - 0\right] \approx 770.6667$$

The total is about \$771.

25. a. Find the value where the equations are equal to each other.

$$104 - .4e^{\frac{t}{2}} = .3e^{\frac{t}{2}}$$
$$104 = .7e^{\frac{t}{2}}$$
$$\frac{104}{.7} = e^{\frac{t}{2}}$$
$$\ln\left(\frac{104}{.7}\right) = \frac{t}{2}$$
$$10 = t$$

After 10 years, the new process will no longer be profitable.

b. The total net savings is given by

$$\int_0^{10} \left(104 - .4e^{t/2}\right)dt - \int_0^{10} .3e^{t/2}\,dt$$
$$= \left(104t - .8e^{t/2}\right)\Big|_0^{10} - .6e^{t/2}\Big|_0^{10}$$
$$= \left(1040 - .8e^5\right) - (0 - .8) - .6e^5 + .6$$
$$\approx 833.622$$

In thousands of dollars, the total savings will be approximately \$833,622.

Copyright © 2019 Pearson Education, Inc.

27. a. The rate of consumption will equal the rate of production when $\frac{20}{1.2t+1.6} = t+.8.$

$$\frac{20}{1.2t+1.6} = t+.8 \Rightarrow$$
$$20 = (1.2t+1.6)(t+.8) \Rightarrow$$
$$20 = 1.2t^2 + 2.56t + 1.28 \Rightarrow$$
$$0 = 1.2t^2 + 2.56t - 18.72$$

Applying the quadratic formula we have, $t = 3.02$ or $t = -5.16$. Reject $t = -5.16$. In about 3.02 years the rates will be equal.

b. The total excess production before consumption and production are equal is given by

$$\int_0^{3.02} \frac{20}{1.2t+1.6}\,dt - \int_0^{3.02} (t+.8)dt$$
$$= \frac{20}{1.2}\ln|1.2t+1.6|\Big|_0^{3.02} - \left(\frac{t^2}{2}+.8t\right)\Big|_0^{3.02}$$
$$= \left[\frac{20}{1.2}\ln 5.224 - \frac{20}{1.2}\ln 1.6\right] - \left[\frac{(3.02)^2}{2} + .8(3.02) - 0\right]$$
$$\approx 12.74$$

The total is about 12.74 trillion gallons.

29. a. The total pollution in the lake after 20 hours is given by $\int_0^{20} 8\left(1-e^{-.25t}\right)dt - \int_0^{20} .2t\,dt$

$$= \int_0^{20}\left[8\left(1-e^{-.25t}\right) - .2t\right]dt.$$

```
fnInt(8(1-e^(-.2
5X))-.2X,X,0,20)
          88.2156143
```

The total pollution in the lake after 20 hours was about 88.22 gallons.

b. The rate at which pollution enters the lake will be equal to the rate at which pollution is being removed when $f(t) = g(t)$.

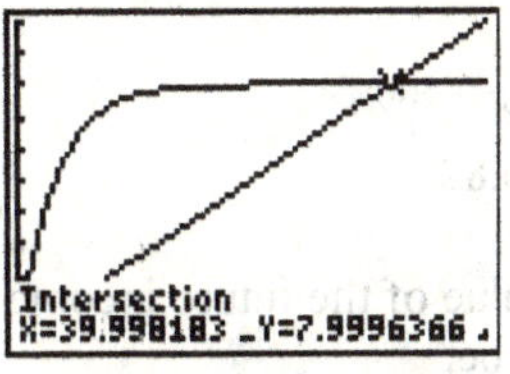

[0, 50] by [0, 10]

The rate at which pollution enters the lake will be equal to the rate at which pollution is being removed at about 40 hours.

c. The total pollution in the lake after 40 hours is given by $\int_0^{40} 8\left(1-e^{-.25t}\right)dt - \int_0^{40} .2t\,dt$

$$= \int_0^{40}\left[8\left(1-e^{-.25t}\right) - .2t\right]dt.$$

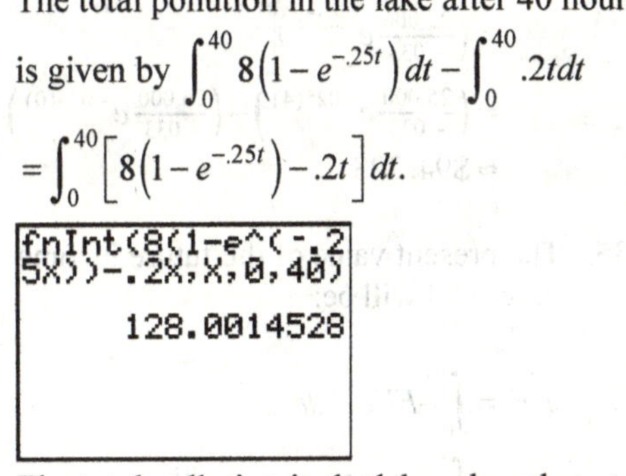

The total pollution in the lake when the rate at which pollution enters the lake will be equal to the rate at which pollution is being removed (40 hours) is about 128 gallons.

d. All the pollution will be removed from the lake when $\int_0^t 8\left(1-e^{-.25t}\right)dt - \int_0^t .2t\,dt = 0.$

$$\int_0^x 8\left(1-e^{-.25t}\right)dt - \int_0^x .2t\,dt = 0$$
$$\int_0^x 8\left(1-e^{-.25t}\right)dt = \int_0^x .2t\,dt$$
$$\left(32e^{-.25t} + 8t\right)\Big|_0^x = .1t^2\Big|_0^x$$
$$\left(32e^{-.25x} + 8x - 32\right) = .1x^2$$

Using a graphing calculator, we have

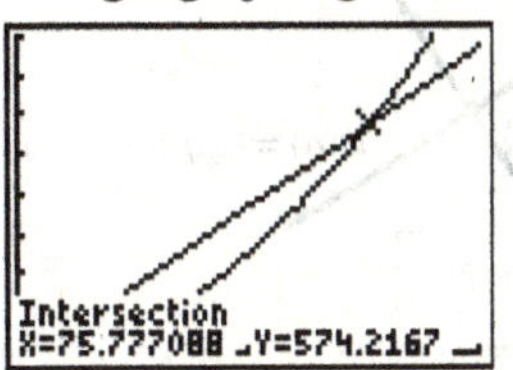

All pollution will be removed from the lake in about 75.78 hours.

Copyright © 2019 Pearson Education, Inc.

31. The present value of the single future payment described will be:

$$PV = FVe^{-rt}$$
$$= 40{,}000e^{-.015(8)}$$
$$\approx \$35{,}476.82$$

33. The present value of the future income stream described will be:

$$PV = \int_0^4 FVe^{-rt}\,dt$$
$$= \int_0^4 25{,}000e^{-.033t}\,dt$$
$$= \left(\frac{25{,}000}{-.033}e^{-.033t}\right)\Bigg|_0^4$$
$$= \left(\frac{25{,}000}{-.033}e^{-.025(4)}\right) - \left(\frac{25{,}000}{-.033}e^{-.025(0)}\right)$$
$$\approx \$94{,}223$$

35. The present value of the future income stream described will be:

$$PV = \int_0^{10} FVe^{-rt}\,dt$$
$$= \int_0^{10} (5000+500t)e^{-.02t}\,dt$$
$$= \left(\begin{array}{l}(5000+500t)(-50e^{-.02t})\\ \quad -1{,}250{,}000e^{-.02t}\end{array}\right)\Bigg|_0^{10}$$
$$= \left(\begin{array}{l}(5000+500(10))(-50e^{-.02(10)})\\ \quad -1{,}250{,}000e^{-.02(10)}\end{array}\right) - \left(\begin{array}{l}(5000+500(0))(-50e^{-.02(0)})\\ \quad -1{,}250{,}000e^{-.02(0)}\end{array}\right)$$
$$\approx \$67{,}221.20$$

37. a. We graph the linear functions $S(q) = \frac{7}{5}q$ and $D(q) = -\frac{3}{5}q + 10$

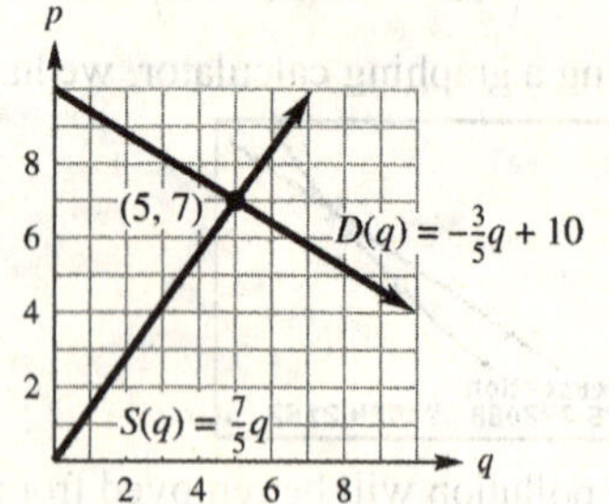

b. $S(q) = D(q) \Rightarrow \frac{7}{5}q = -\frac{3}{5}q + 10 \Rightarrow$
$7q = -3q + 50 \Rightarrow q = 5$
$S(5) = 7$
The equilibrium point is (5, 7).

c. The consumers' surplus is given by

$$\int_0^5 \left[\left(-\frac{3}{5}q+10\right) - 7\right]dq = \int_0^5 \left(-\frac{3}{5}q+3\right)dq$$
$$= \left(-\frac{3}{5}\cdot\frac{q^2}{2} + 3q\right)\Bigg|_0^5$$
$$= \left(-\frac{15}{2}+15\right) - 0$$
$$= 7.5$$

The consumers' surplus is \$7.50.

d. The producers' surplus is given by

$$\int_0^5 \left(7 - \frac{7}{5}q\right)dq = \left(7q - \frac{7}{5}\cdot\frac{q^2}{2}\right)\Bigg|_0^5$$
$$= \left(35 - \frac{35}{2}\right) - 0 = 17.5$$

The producers' surplus is \$17.50.

39. If $S(q) = 100 + 3q^{3/2} + q^{5/2}$,
$S(9) = 100 + 81 + 243 = 424$.
Therefore, the producers' surplus is given by

$$\int_0^9 \left[424 - \left(100 + 3q^{3/2} + q^{5/2}\right)\right]dq$$
$$= \int_0^9 \left(324 - 3q^{3/2} - q^{5/2}\right)dq$$
$$= \left(324q - 3\cdot\frac{q^{5/2}}{\frac{5}{2}} - \frac{q^{7/2}}{\frac{7}{2}}\right)\Bigg|_0^9$$
$$= \left(324q - \frac{6}{5}q^{5/2} - \frac{2}{7}q^{7/2}\right)\Bigg|_0^9$$
$$\approx (2916 - 291.6 - 624.86) - 0 = 1999.54$$

The producers' surplus is \$1999.54.

Copyright © 2019 Pearson Education, Inc.

41. If $D(q)=\dfrac{15,500}{(3.2q+7)^3}$, then

$$D(5)=\frac{15,500}{(16+7)^3}=1.2739.$$

Therefore, consumers' surplus is given by

$$\int_0^5\left[\frac{15,500}{(3.2q+7)^3}-1.2739\right]dq.$$

Let $u=3.2q+7$. Then $du=3.2\,dq$. If $q=5$, $u=23$. If $q=0$, $u=7$.

$$\int_0^5\left[\frac{15,500}{(3.2q+7)^3}-1.2739\right]dq$$
$$=\frac{1}{3.2}\int_0^5\left[15,500(3.2q+7)^{-3}-1.2739\right]3.2\,dq$$
$$=\frac{1}{3.2}\int_7^{23}\left(15,500u^{-3}-1.2739\right)du$$
$$=\frac{1}{3.2}\left(15,500\cdot\frac{u^{-2}}{-2}-1.2739u\right)\Bigg|_7^{23}$$
$$=\frac{1}{3.2}\left(-\frac{7750}{u^2}-1.2739u\right)\Bigg|_7^{23}$$
$$=\frac{1}{3.2}\left(-\frac{7750}{529}-29.2997\right)$$
$$-\frac{1}{3.2}\left(-\frac{7750}{49}-8.9173\right)=38.50$$

The consumers' surplus is \$38.50.

43. a. Graph the exponential functions $S(q)=e^{q/2}-1$ and $D(q)=400-e^{q/2}$.

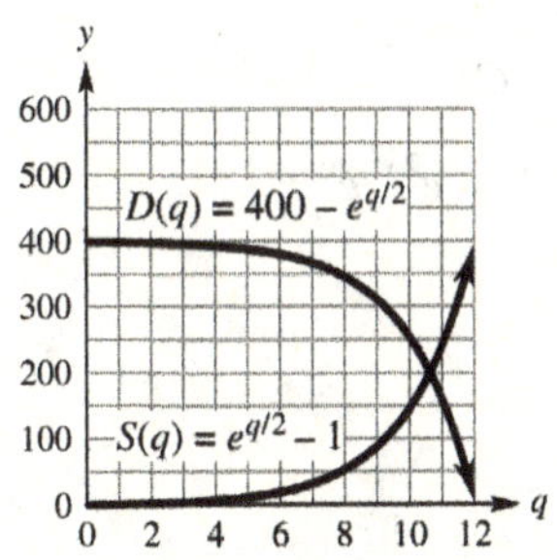

b. $S(q)=D(q)$

$$e^{q/2}-1=400-e^{q/2}\Rightarrow 2e^{q/2}=401\Rightarrow$$
$$e^{q/2}=200.5\Rightarrow\frac{q}{2}=\ln 200.5\Rightarrow$$
$$q=2\ln 200.5\approx 10.60$$
$$S(2\ln 200.5)=e^{(2\ln 200.5)/2}-1=e^{\ln 200.5}-1$$
$$=200.5-1=199.5$$

The equilibrium point is (10.60, 199.50).

c. The consumers' surplus is given by

$$\int_0^{10.60}\left[\left(400-e^{q/2}\right)-199.50\right]dq$$
$$=\int_0^{10.60}\left(200.50-e^{q/2}\right)dq$$
$$=\left(200.50q-2e^{q/2}\right)\Big|_0^{10.60}$$
$$=\left[200.50(10.60)-2e^{10.60/2}\right]-(0-2)$$
$$\approx 1726.63.$$

The consumers' surplus is \$1726.63.

d. The producers' surplus is given by

$$\int_0^{10.60}\left(200.50-e^{q/2}\right)dq$$
$$=\left(200.50q-2e^{q/2}\right)\Big|_0^{10.60}$$
$$=\left[200.50(10.60)-2e^{10.60/2}\right]-(0-2)$$
$$\approx 1726.63.$$

The producers' surplus is \$1726.63.

Section 13.7 Differential Equations

1. $\dfrac{dy}{dx}=-3x^2+7x$

$$dy=\left(-3x^2+7x\right)dx$$
$$\int dy=\int\left(-3x^2+7x\right)dx$$
$$y=-x^3+\frac{7}{2}x^2+C$$

3. $\dfrac{dy}{dx}=6e^{-2x}$

$$dy=6e^{-2x}dx$$
$$\int dy=\int 6e^{-2x}dx$$
$$y=-3e^{-2x}+C$$

Copyright © 2019 Pearson Education, Inc.

5. $3x^3 - 2\frac{dy}{dx} = 0 \Rightarrow -2\frac{dy}{dx} = -3x^3 \Rightarrow$
$\frac{dy}{dx} = \frac{3}{2}x^3 \Rightarrow y = \int \frac{3}{2}x^3\,dx \Rightarrow y = \frac{3}{2}\int x^3\,dx \Rightarrow$
$y = \frac{3}{2}\cdot\frac{x^4}{4} + C \Rightarrow y = \frac{3x^4}{8} + C$

7. $5y^4\frac{dy}{dx} = 6x \Rightarrow 5y^4\,dy = 6x\,dx \Rightarrow$
$y^5 = 3x^2 + C_1 \Rightarrow y = \sqrt[5]{3x^2 + C}$

9. $\frac{dy}{dx} = \frac{1}{21y^2 + 8} \Rightarrow$
$(21y^2 + 8)\,dy = dx \Rightarrow$
$7y^3 + 8y = x + C$

11. $\frac{dy}{dx} = \frac{10x}{e^y} \Rightarrow$
$e^y\,dy = 10x\,dx \Rightarrow$
$e^y = 5x^2 + C$
$y = \ln(5x^2 + C)$

13. $\frac{dy}{dx} = \frac{1}{3y^2} \Rightarrow dx = 3y^2\,dy \Rightarrow$
$\int dx = 3\int y^2\,dy \Rightarrow x + C = \frac{3y^3}{3} \Rightarrow$
$y = \sqrt[3]{x + C}$

15. $\frac{dy}{dx} = -3y \Rightarrow$
$\frac{dy}{y} = -3\,dx \Rightarrow$
$\ln|y| = -3x + C_1 \Rightarrow$
$y = e^{-3x+C_1} = e^{3x}e^{C_1}$
$y = Me^{-3x}$

17. $\frac{dy}{dx} = y - 7$
$\frac{dy}{y-7} = dx$
$\ln|y - 7| = x + C$
$|y - 7| = e^{x+C} = e^x e^C$
$y - 7 = Me^x \Rightarrow y = 7 + Me^x$

19. $\frac{dy}{dx} = 1 - y$
$\frac{1}{1-y}\,dy = dx$
$\int \frac{1}{1-y}\,dy = \int dx$
$-\ln|1 - y| = x + C$
$\ln|1 - y| = -x + C$
$|1 - y| = e^{-x+C}$
$|1 - y| = e^{-x}e^C$
$1 - y = \pm e^C e^{-x}$
$1 - y = Me^{-x} \Rightarrow y = Me^{-x} + 1$

21. $\frac{dy}{dx} = 4xy$
$\frac{dy}{y} = 4x\,dx$
$\int \frac{dy}{y} = 4\int x\,dx$
$\ln|y| = 4\frac{x^2}{2} + C$
$\ln|y| = 2x^2 + C$
$|y| = e^{2x^2+C} = e^{2x^2}e^c$
$y = Me^{2x^2}$

Copyright © 2019 Pearson Education, Inc.

23. $\frac{dy}{dx}+2x=3x^2$; $y=4$ when $x=0$.

$\frac{dy}{dx}=3x^2-2x \Rightarrow y=\int(3x^2-2x)\,dx \Rightarrow$

$y=y=x^3-x^2+C$

The general solution is

$y=x^3-x^2+C$.

When $x=0$, $y=4$,

$4=0^3-0^2+C \Rightarrow C=4$

Thus, the particular solution is

$y=x^3-x^2+4$.

25. $\frac{dy}{dx}=3.7y$; $y=91$ when $x=0$

$\frac{dy}{y}=3.7dx \Rightarrow$

$\ln|y|=3.7x+C \Rightarrow$

$y=e^{3.7x+C} \Rightarrow$

$y=e^{3.7x}e^C=Me^{3.7x}$

$y=91$ when $x=0$, so

$91=Me^{3.7(0)} \Rightarrow M=91$

Thus, the particular solution is $y=91e^{3.7x}$

27. $\frac{dy}{dx}=\frac{x^2}{y}$; $y=4$ when $x=0$.

$y\,dy=x^2dx \Rightarrow \int y\,dy=\int x^2dx$

$\Rightarrow \frac{y^2}{2}=\frac{x^3}{3}+C \Rightarrow y^2=\frac{2x^3}{3}+C$

Let $x=0$ and $y=4$.

$16=\frac{2(0)}{3}+C \Rightarrow C=16$

The particular solution is $y^2=\frac{2x^3}{3}+16$.

29. $(5x+3)y=\frac{dy}{dx}$; $y=1$ when $x=0$.

$(5x+3)dx=\frac{dy}{y} \Rightarrow \int(5x+3)dx=\int\frac{dy}{y} \Rightarrow$

$\frac{5x^2}{2}+3x=\ln|y|+C$

If $x=0$ and $y=1$, $0=\ln|1|+C \Rightarrow 0=C$.

We now have $\frac{5x^2}{2}+3x=\ln|y|$.

Rewrite this equation in exponential form. The particular solution is $y=e^{2.5x^2+3x}$.

31. $\frac{dy}{dx}=\frac{7x+1}{y-3}$; $y=4$ when $x=0$.

$(y-3)dy=(7x+1)dx$

$\int(y-3)dy=\int(7x+1)dx$

$\frac{y^2}{2}-3y=\frac{7x^2}{2}+x+C$

If $x=0$ and $y=4$, $\frac{16}{2}-12=C \Rightarrow -4=C$.

The particular solution is $\frac{y^2}{2}-3y=\frac{7x^2}{2}+x-4$

or $y^2-6y=7x^2+2x-8$.

33. Answers vary.

35. $\frac{dy}{dx}=\frac{3x^2-4x}{25} \Rightarrow$

$y=\int\frac{3x^2-4x}{25}dx=\frac{1}{25}\int(3x^2-4x)dx \Rightarrow$

$y=\frac{1}{25}(x^3-2x^2)+C$

Since production is 2300 units when investment is \$5000, we have

$23=\frac{1}{25}(5^3-2(5)^2)+C \Rightarrow C=20.$

Thus, $y=\frac{1}{25}(x^3-2x^2)+20.$

a. $y=\frac{1}{25}(8^3-2(8)^2)+20=35.36$

The production level is 3536 units if the investment is \$8000.

b. $y=\frac{1}{25}(11^3-2(11)^2)-20=63.56$

The production level is 6356 units if the investment is \$11,000.

37. $\frac{dy}{dt}=.23y$; $t=0$ and $y=.3$.

a. $\frac{dy}{dt}=.23y \Rightarrow \frac{dy}{y}=.23\,dt \Rightarrow$

$\int\frac{dy}{y}=\int.23\,dt \Rightarrow$

$\ln|y|=.23t+C$

$y=Ce^{.23t}$

When $t=0$, $y=.3$ so $.3=Ce^0 \Rightarrow C=.3$,

and $y=.3e^{.23t}$.

Copyright © 2019 Pearson Education, Inc.

b. If $t = 15$, then
$y = .3e^{.23(15)} \Rightarrow y = .3e^{3.45}$
$y \approx 9.5$
The percentage of the population in low-income countries that were internet users in 2015 was about 9.5%.

c. If $y = 10$, then
$10 = .3e^{.23t} \Rightarrow \frac{10}{.3} = e^{.23t} \Rightarrow$
$\ln \frac{10}{.3} = .23t$
$t = \frac{\ln \frac{10}{.3}}{.23} \approx 15.2$
The percentage of the population in low-income countries reached 10% in about 15.2 years (or sometime in 2016).

39. $\frac{dy}{dt} = 8t$; $t = 0$ and $y = 50$.

a. $\frac{dy}{dt} = 8t \Rightarrow dy = 8t\,dt \Rightarrow$
$\int dy = \int 8t\,dt \Rightarrow$
$y = 4t^2 + C$
When $t = 0$, $y = 50$ so
$50 = 4(0)^2 + C \Rightarrow C = 50$, and
$y = 4t^2 + 50$.

b. If $y = 450$, then
$450 = 4t^2 + 50 \Rightarrow 400 = 4t^2 \Rightarrow$
$100 = t^2$
$t = 10$
The number of bacteria will reach 450 (thousand) in 10 days.

41. a. $\frac{dy}{dt} = -.15y$

b. $\frac{dy}{dt} = -.15y \Rightarrow \frac{dy}{y} = -.15dt \Rightarrow$
$\int \frac{dy}{y} = \int -.15dt \Rightarrow \ln y = -.15t + C \Rightarrow$
$y = e^{-.15t+C} = e^{-.15t}e^{C} \Rightarrow$
$y = Me^{-.15t}$

c. At $t = 0$, $M = 1$ because when $t = 0$, $e^{-.15t} = 1$. So we have
$.25 = e^{-.15t} \Rightarrow \ln(.25) = -.15t \Rightarrow$
$t = \frac{\ln(.25)}{-.15} \approx 9.2$
Sales will decrease to 25% of their original level in about 9.2 years.

43. At $t = 0$, $M = 1$. So we have
$2 = e^{.02t} \Rightarrow \ln(2) = .02t \Rightarrow$
$t = \frac{\ln(2)}{.02} \approx 34.7$
It will take about 34.7 years for a $1 item to cost twice as much if inflation grows continuously at a rate of 2%.

45. a. $\frac{dy}{dt} = ky \Rightarrow \int \frac{dy}{y} = \int k\,dt \Rightarrow$
$\ln|y| = kt + C \Rightarrow e^{\ln|y|} = e^{kt+C} \Rightarrow$
$y = \pm(e^{kt})(e^{C}) = Me^{kt}$
If $y = 1$ when $t = 0$ and $y = 5$ when $t = 2$, we have the following system of equations.
$1 = Me^{k(0)}$ (1)
$5 = Me^{2k}$ (2)
From equation (1),
$1 = M(1) \Rightarrow M = 1$.
Substitute 1 for M in equation (2) to find k.
$5 = (1)e^{2k} \Rightarrow e^{2k} = 5 \Rightarrow 2k \ln e = \ln 5 \Rightarrow$
$k = \frac{\ln 5}{2} \approx .8$

b. If $k = .8$ and $M = 1$, then $y = e^{.8t}$.
When $t = 3$, $y = e^{.8(3)} \approx 11$.
In 3 days, 11 people heard the rumor.

c. When $t = 5$, $y = e^{.8(5)} \approx 55$.
In 5 days, 55 people heard the rumor.

d. When $t = 10$, $y = e^{.8(10)} \approx 2981$.
In 10 days, 2981 people heard the rumor.

Copyright © 2019 Pearson Education, Inc.

47. $\frac{dA}{dt} = Ai \Rightarrow \frac{dA}{A} = i\,dt \Rightarrow \int \frac{dA}{A} = i\int dt \Rightarrow$
$\ln A = it + C \Rightarrow A = e^{it+C} = e^{it}e^{c} = Me^{it}$
Since the initial investment is \$5000, we have $A = 5000e^{it}$.
$20{,}000 = 5000e^{i(24)} \Rightarrow 4 = e^{24i} \Rightarrow \ln 4 = 24i \Rightarrow$
$i = \frac{\ln 4}{24} = \frac{\ln(2^2)}{24} = \frac{2\ln 2}{24} = \frac{\ln 2}{12}$
The answer is (d).

49. a. $\frac{dy}{dx} = 30 - y$; $y = 1$ thousand when $x = 0$

$$\frac{dy}{30-y} = dx$$
$$\int \frac{dy}{30-y} = \int dx$$
$$-\ln|30-y| = x + C$$
$$\ln|30-y| = -x + C$$
$$|30-y| = e^{-x+C}$$
$$30 - y = Me^{-x}$$
$$y = 30 - Me^{-x}$$
$$1 = 30 - M$$
$$M = 29$$
$$y = 30 - 29e^{-x}$$

b. If $x = 10$, $y = 30 - 29e^{-10} \approx 29.999$.
About 29.999 thousand or 29,999 bacteria are present.

Chapter 13 Review Exercises

1. $\int (x^2 - 8x - 7)\,dx = \frac{x^3}{3} - 8\cdot\frac{x^2}{2} - 7x + C = \frac{x^3}{3} - 4x^2 - 7x + C$

2. $\int (6 - x^2)\,dx = 6x - \frac{x^3}{3} + C$

3. $\int 7\sqrt{x}\,dx = 7\int x^{1/2}\,dx = 7\cdot\frac{x^{3/2}}{\frac{3}{2}} + C = 7\cdot\frac{2}{3}x^{3/2} + C = \frac{14x^{3/2}}{3} + C$

4. $\int \frac{\sqrt{x}}{8}\,dx = \frac{1}{8}\int x^{1/2} + C = \frac{1}{8}\cdot\frac{x^{3/2}}{\frac{3}{2}} + C = \frac{1}{8}\cdot\frac{2}{3}x^{3/2} + C = \frac{x^{3/2}}{12} + C$

5. $\int \left(x^{1/2} + 3x^{-2/3}\right)dx = \frac{x^{3/2}}{\frac{3}{2}} + 3\cdot\frac{x^{1/3}}{\frac{1}{3}} + C = \frac{2}{3}x^{3/2} + 3\cdot\frac{3}{1}x^{1/3} + C = \frac{2x^{3/2}}{3} + 9x^{1/3} + C$

6. $\int \left(8x^{4/3} + x^{-1/2}\right)dx = 8\cdot\frac{x^{7/3}}{\frac{7}{3}} + \frac{x^{1/2}}{\frac{1}{2}} + C = 8\cdot\frac{3}{7}x^{7/3} + 2x^{1/2} + C = \frac{24x^{7/3}}{7} + 2x^{1/2} + C$

7. $\int \frac{-6}{x^3}\,dx = -6\int x^{-3}\,dx = -6\cdot\frac{x^{-2}}{-2} + C = 3x^{-2} + C$

8. $\int \frac{9}{x^4}\,dx = 9\int x^{-4}\,dx = 9\cdot\frac{x^{-3}}{-3} + C = -\frac{3}{x^3} + C$

9. $\int -3e^{2x}\,dx = -3\int e^{2x}\,dx = -3\cdot\frac{1}{2}e^{2x} + C = -\frac{3e^{2x}}{2} + C$

10. $\int 8e^{-x}\,dx = 8\int e^{-x}\,dx = 8\cdot\frac{1}{-1}e^{-x} + C = -8e^{-x} + C$

11. $\int \frac{10}{x-3}\,dx = 10\int \frac{1}{x-3}\,dx = 10\ln|x-3| + C$

12. $\int \frac{-14}{x+2}\,dx = -14\int \frac{1}{x+2}\,dx = -14\ln|x+2| + C$

13. $\int \frac{1}{3-x}\,dx$
Let $u = 3 - x$. Then $du = -dx$.
$\int \frac{1}{3-x}\,dx = -\int \frac{1}{u}\,du = -\ln|u| + C = -\ln|3-x| + C$

Copyright © 2019 Pearson Education, Inc.

14. $\int \frac{5}{2-x}dx$

Let $u = 2 - x$. Then $du = -dx$.

$$\int \frac{5}{2-x}dx = -5\int \frac{1}{u}du$$
$$= -5\ln|u| + C$$
$$= -5\ln|2-x| + C$$

15. $\int \frac{3x}{x^2-1}dx$

Let $u = x^2 - 1$. Then $du = 2x\,dx$.

$$\int \frac{3x}{x^2-1}dx = 3\cdot\frac{1}{2}\int\frac{1}{x^2-1}2x\,dx$$
$$= \frac{3}{2}\int\frac{1}{u}du = \frac{3}{2}\ln|u| + C$$
$$= \frac{3\ln|x^2-1|}{2} + C$$

16. $\int \frac{-6x}{4-x^2}dx$

Let $u = 4 - x^2$. Then $du = -2x\,dx$.

$$\int \frac{-6x}{4-x^2}dx = 6\cdot\frac{1}{2}\int\frac{1}{4-x^2}(-2x)dx = 3\int\frac{1}{u}du$$
$$= 3\ln|u| + C = 3\ln|4-x^2| + C$$

17. $\int \frac{12x^2dx}{\left(x^3+5\right)^4} = 12\cdot\frac{1}{3}\int\frac{3x^2dx}{\left(x^3+5\right)^4}$

Let $u = x^3 + 5$. Then $du = 3x^2dx$.

$$\int \frac{12x^2dx}{\left(x^3+5\right)^4} = 12\cdot\frac{1}{3}\int\frac{du}{u^4} = 4\int u^{-4}du$$
$$= 4\left(\frac{u^{-3}}{-3}\right) + C = -\frac{4}{3}\left(x^3+5\right)^{-3} + C$$

18. $\int\left(x^2-5x\right)^4(2x-5)dx$

Let $u = x^2 - 5x$. Then $du = (2x - 5)dx$.

$$\int\left(x^2-5x\right)^4(2x-5)dx = \int u^4du = \frac{u^5}{5} + C$$
$$= \frac{\left(x^2-5x\right)^5}{5} + C$$

19. $\int \frac{4x-5}{2x^2-5x}dx$

Let $u = 2x^2 - 5x$. Then $du = (4x - 5)dx$.

$$\int \frac{4x-5}{2x^2-5x}dx = \int\frac{du}{u} = \ln|u| + C$$
$$= \ln|2x^2-5x| + C$$

20. $\int \frac{8(2x+9)}{x^2+9x+1}dx$

Let $u = x^2 + 9x + 1$. Then $du = (2x + 9)\,dx$.

$$\int \frac{8(2x+9)}{x^2+9x+1}dx = 8\int\frac{du}{u} = 8\ln|u| + C$$
$$= 8\ln|x^2+9x+1| + C$$

21. $\int 5xe^{3x^2}dx$

Let $u = 3x^2$. Then $du = 6x\,dx$.

$$\int 5xe^{3x^2}dx = \frac{1}{6}\cdot 5\int e^{3x^2}\cdot 6x\,dx = \frac{5}{6}\int e^u du$$
$$= \frac{5}{6}e^u + C = \frac{5}{6}e^{3x^2} + C = \frac{5e^{3x^2}}{6} + C$$

22. $\int 4xe^{-3x^2+7}dx = 4\int xe^{-3x^2+7}dx$

Let $u = -3x^2 + 7$. Then $du = -6xdx$.

$$\int 4xe^{-3x^2+7}dx = 4\left(-\frac{1}{6}\right)\int e^u du$$
$$= -\frac{2}{3}e^u + C = -\frac{2}{3}e^{-3x^2+7} + C$$

23. $\int \frac{x^2}{e^{3x^3}}dx = \int x^2e^{-3x^3}dx = -\frac{1}{9}\int -9x^2e^{-3x^3}dx$

Let $u = -3x^3$. Then $du = -9x^2dx$.

$$\int \frac{x^2}{e^{3x^3}}dx = -\frac{1}{9}\int -9x^2e^{-3x^3}dx = -\frac{1}{9}\int e^u du$$
$$= -\frac{1}{9}e^u + C = -\frac{e^{-3x^3}}{9} + C$$

24. $\int \frac{-2x}{e^{x^2}}dx$

Let $u = -x^2$. Then $du = -2x\,dx$.

$$\int \frac{-2x}{e^{x^2}}dx = \int e^{-x^2}\cdot -2x\,dx = \int e^u du$$
$$= e^u + C$$
$$= e^{-x^2} + C$$

Copyright © 2019 Pearson Education, Inc.

25. $\int \frac{3(\ln x)^5}{x}\,dx = 3\int \frac{(\ln x)^5}{x}\,dx$

Let $u = \ln x$. Then $du = \frac{1}{x}\,dx$.

$$\int \frac{3(\ln x)^5}{x}\,dx = 3\int u^5\,du = 3\left(\frac{u^6}{6}\right)+C$$
$$= \frac{u^6}{2}+C = \frac{(\ln x)^6}{2}+C$$

26. $\int \frac{10(\ln(5x+3))^2}{5x+3}\,dx = 10\int \frac{(\ln(5x+3))^2}{5x+3}\,dx$

Let $u = \ln(5x+3)$. Then $du = \left(\frac{1}{5x+3}\cdot 5\right)dx$.

$$\int \frac{10(\ln(5x+3))^2}{5x+3}\,dx = 10\cdot\frac{1}{5}\int u^2\,du = 2\frac{u^3}{3}+C$$
$$= \frac{2}{3}\left[\ln(5x+3)\right]^3 + C$$

27. $\int \sqrt{2x+9}\,dx$

Let $u = 2x+9$. Then $du = 2\,dx$.

$$\int \sqrt{2x+9}\,dx = \frac{1}{2}\int (2x+9)^{1/2}\cdot 2\,dx = \frac{1}{2}\int u^{1/2}\,du$$
$$= \frac{1}{2}\frac{u^{3/2}}{\frac{3}{2}}+C$$
$$= \frac{1}{3}(2x+9)^{3/2}+C$$

28. $\int \sqrt{1+5x}\,dx$

Let $u = 1+5x$. Then $du = 5\,dx$.

$$\int \sqrt{1+5x}\,dx = \frac{1}{5}\int (1+5x)^{1/2}\cdot 5\,dx = \frac{1}{5}\int u^{1/2}\,du$$
$$= \frac{1}{5}\frac{u^{3/2}}{\frac{3}{2}}+C$$
$$= \frac{2}{15}(1+5x)^{3/2}+C$$

29. $\int (7xe^{4-x})dx$

Using integration by parts, let $u = 7x$ and $dv = e^{4-x}\,dx$. Therefore, $du = 7dx$ and $v = -e^{4-x}$.

$$\int (7xe^{4-x})dx$$
$$= (7x)\left(-e^{4-x}\right) - \int -e^{4-x}\cdot 7dx$$
$$= (7x)\left(-e^{4-x}\right) + \int 7e^{4-x}dx$$
$$= -7xe^{4-x} + 7(-e^{4-x}+c)$$
$$= -7xe^{4-x} - 7e^{4-x} + c$$

30. $\int (xe^{3x-5})dx$

Using integration by parts, let $u = x$ and $dv = e^{3x-5}dx$. Therefore, $du = dx$ and $v = \frac{1}{3}e^{3x-5}$.

$$\int (xe^{3x-5})dx$$
$$= (x)\left(\frac{1}{3}e^{3x-5}\right) - \int \frac{1}{3}e^{3x-5}dx$$
$$= (x)\left(\frac{1}{3}e^{3x-5}\right) - \int \frac{1}{3}e^{3x-5}dx$$
$$= \frac{1}{3}xe^{3x-5} - \frac{1}{3}\left(\frac{1}{3}e^{3x-5}+c\right)$$
$$= \frac{1}{3}xe^{3x-5} - \frac{1}{9}e^{3x-5} + c$$

31. $\int (28x^2e^{2x})dx$

Using integration by parts, let $u = 28x^2$ and $dv = e^{2x}dx$. Therefore, $du = 56xdx$ and $v = \frac{1}{2}e^{2x}$.

$$\int (28x^2e^{2x})dx$$
$$= (28x^2)\left(\frac{1}{2}e^{2x}\right) - \int \frac{1}{2}e^{2x}\cdot 56xdx$$
$$= 14x^2e^{2x} - \int 28xe^{2x}dx$$

Copyright © 2019 Pearson Education, Inc.

Using integration by parts a second time, let $u = 28x$ and $dv = e^{2x}dx$. Therefore, $du = 28dx$ and $v = \frac{1}{2}e^{2x}$.

$$14x^2e^{2x} - \int(28xe^{2x})dx$$
$$= 14x^2e^{2x} - \left(28x\left(\frac{1}{2}e^{2x}\right) - \int\frac{1}{2}e^{2x} \cdot 28dx\right)$$
$$= 14x^2e^{2x} - 14xe^{2x} + \int 14e^{2x}dx$$
$$= 14x^2e^{2x} - 14xe^{2x} + 14(\frac{1}{2}e^{2x} + c)$$
$$= 14x^2e^{2x} - 14xe^{2x} + 7e^{2x} + c$$

32. $\int(9x^2e^{-x})dx$

Using integration by parts, let $u = 9x^2$ and $dv = e^{-x}dx$. Therefore, $du = 18xdx$ and $v = -e^{-x}$.

$$\int(9x^2e^{-x})dx$$
$$= (9x^2)\left(-e^{-x}\right) - \int -e^{-x} \cdot 18xdx$$
$$= -9x^2e^{2x} + \int 18xe^{-x}dx$$

Using integration by parts a second time, let $u = 18x$ and $dv = e^{-x}dx$. Therefore, $du = 18dx$ and $v = -e^{-x}$.

$$-9x^2e^{-x} + \int(18xe^{-x})dx$$
$$= -9x^2e^{-x} + \left(18x\left(-e^{-x}\right) - \int -e^{-x} \cdot 18dx\right)$$
$$= -9x^2e^{-x} - 18xe^{-x} + \int 18e^{-x}dx$$
$$= -9x^2e^{-x} - 18xe^{-x} + 18(-e^{-x} + c)$$
$$= -9x^2e^{-x} - 18xe^{-x} - 18e^{-x} + c$$

33. $\int 6\ln(6x+1)dx$

Using the antiderivative of the Logarithmic function, let $u = 6x+1$. Therefore, $du = 6dx$.

$$\int 6\ln(6x+1)dx$$
$$= \frac{6}{6}((6x+1)\ln(6x+1) - (6x+1)) + c$$
$$= (6x+1)\ln(6x+1) - (6x+1) + c$$

34. $\int \ln(8-x)dx$

Using the antiderivative of the Logarithmic function, let $u = 8-x$. Therefore, $du = -dx$.

$$\int \ln(8-x)dx$$
$$= -\frac{1}{1}((8-x)\ln(8-x) - (8-x)) + c$$
$$= -(8-x)\ln(8-x) + (8-x) + c$$

35.

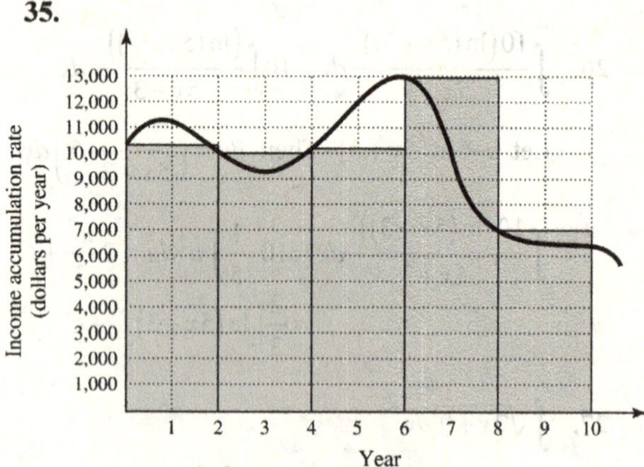

Left Endpoint	Height of rectangle (function values)
0	10,500
2	10,000
4	10,200
6	13,000
8	7000

The total area is approximately
10,500(2) + 10,000(2) + 10,200(2) + 13,000(2) + 7000(2) = 101,400.
The total income is approximately \$101,400.

36. $f(x) = 2x + 3$ from $x = 0$ to $x = 4$

$$\Delta x = \frac{4-0}{4} = 1$$

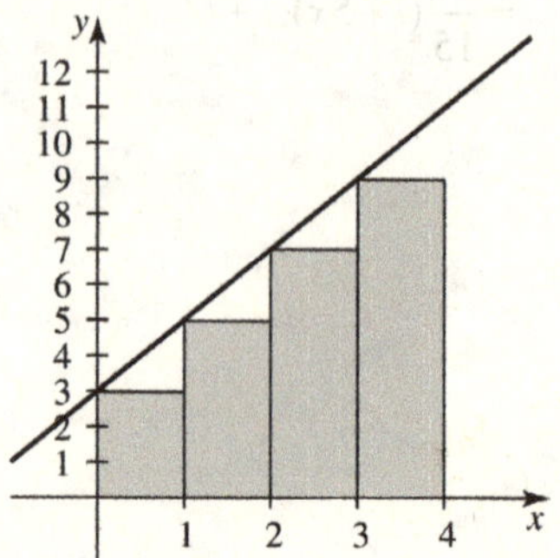

Copyright © 2019 Pearson Education, Inc.

i	x_i	$f(x_i)$
1	0	3
2	1	5
3	2	7
4	3	9

$$\sum_{i=1}^{4} f(x_i)\Delta x = 3(1)+5(1)+7(1)+9(1) = 24$$

37. a. $f(x) = 16x^2 - x^4 + 2$ from $x = -2$ to $x = 3$

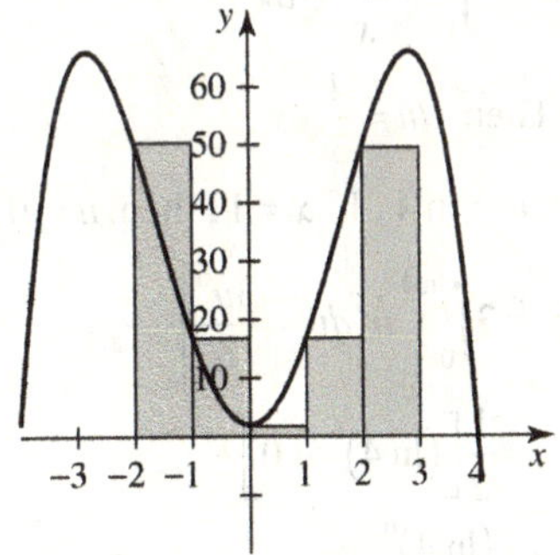

$$\text{Area} = f(-2)+f(-1)+f(0)+f(1)+f(2)$$
$$= 50+17+2+17+50 = 136$$

b.

```
fnInt(16X²-X^4+2
,X,-2,3)
          141.6666667
```

The area is approximately 141.667.

38. a. $g(x) = -x^4 + 12x^2 + x + 5$ from $x = -3$ to $x = 3$

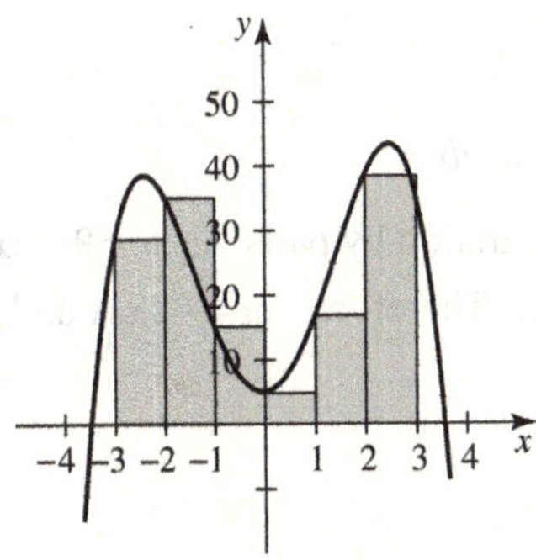

$$\text{Area} = f(-3)+f(-2)+f(-1)$$
$$+ f(0)+f(1)+f(2)$$
$$= 29+35+15+5+17+39 = 140$$

b.

```
fnInt(-X^4+12X²-
X+5,X,-3,3)
            148.8
```

The area is approximately 148.8.

39. $\int_0^4 (x^3 - x^2)\,dx = \left(\frac{x^4}{4} - \frac{x^3}{3}\right)\Bigg|_0^4$

$$= \left(\frac{256}{4} - \frac{64}{3}\right) - 0 = \frac{128}{3}$$

40. $\int_0^1 (5 - 10x + 15x^2)\,dx = (5x - 5x^2 + 5x^3)\Big|_0^1$

$$= (5-5+5) - 0 = 5$$

41. $\int_2^5 (6x^{-2} + x^{-3})\,dx = \left(\frac{6x^{-1}}{-1} + \frac{x^{-2}}{-2}\right)\Bigg|_2^5$

$$= \left(-\frac{6}{x} - \frac{1}{2x^2}\right)\Bigg|_2^5$$
$$= \left(-\frac{6}{5} - \frac{1}{50}\right) - \left(-3 - \frac{1}{8}\right)$$
$$= \frac{381}{200}$$

42. $\int_2^3 (5x^{-2} + 7x^{-4})\,dx$

$$= \left(\frac{5x^{-1}}{-1} + \frac{7x^{-3}}{-3}\right)\Bigg|_2^3 = \left(\frac{-5}{x} - \frac{7}{3x^3}\right)\Bigg|_2^3$$
$$= \left(\frac{-5}{3} - \frac{7}{81}\right) - \left(\frac{-5}{2} - \frac{7}{24}\right) = \frac{673}{648} \approx 1.039$$

43. $\int_1^5 15x^{-1}dx = 15\int_1^5 \frac{dx}{x} = 15\ln|x|\Big|_1^5$

$$= 15\ln 5 - 15\ln 1 = 15\ln 5 \approx 24.1416$$

44. $\int_1^6 \frac{8x^{-1}}{3}\,dx = \frac{8}{3}\int_1^6 \frac{1}{x}\,dx = \frac{8}{3}(\ln x)\Big|_1^6$

$$= \frac{8}{3}(\ln 6 - \ln 1) = \frac{8}{3}\ln 6 \approx 4.778$$

45. $\int_0^4 2e^{-5x}\,dx = \frac{2e^{-5x}}{-5}\Bigg|_0^4 = -\frac{2}{5}e^{-20} - \frac{2}{5}e^0$

$$= -\frac{2}{5}e^{-20} - \frac{2}{5} \approx .40$$

Copyright © 2019 Pearson Education, Inc.

46. $\int_1^2 \frac{7}{2}e^{4x}\,dx = \frac{7}{2}\cdot\frac{e^{4x}}{4}\Big|_1^2 = \frac{7e^{4x}}{8}\Big|_1^2$

$= \frac{7\left(e^8 - e^4\right)}{8} \approx 2560.5649$

47. $\int_{\sqrt{5}}^5 2x\sqrt{x^2-3}\,dx$

Let $u = x^2 - 3$. Then $du = 2x\,dx$.
If $x = 5$, $u = 22$. If $x = \sqrt{5}$, $u = 2$.

$\int_{\sqrt{5}}^5 2x\sqrt{x^2-3}\,dx = \int_{\sqrt{5}}^5 \sqrt{x^2-3}\cdot 2x\,dx$

$= \int_2^{22} u^{1/2}\,du = \frac{u^{3/2}}{\frac{3}{2}}\Big|_2^{22}$

$= \frac{2}{3}\left(22^{3/2} - 2^{3/2}\right) \approx 66.907$

48. $\int_0^1 x\sqrt{5x^2+4}\,dx$

Let $u = 5x^2 + 4$. Then $du = 10x\,dx$.
If $x = 1$, $u = 9$. If $x = 0$, $u = 4$.

$\int_0^1 x\sqrt{5x^2+4}\,dx = \frac{1}{10}\int_0^1 \left(5x^2+4\right)^{1/2}\cdot 10x\,dx$

$= \frac{1}{10}\int_4^9 u^{1/2}\,du = \frac{1}{10}\cdot\frac{u^{3/2}}{\frac{3}{2}}\Big|_4^9$

$= \frac{2}{30}\left(9^{3/2} - 4^{3/2}\right) = \frac{19}{15}$

49. $\int_1^6 \frac{8x+12}{x^2+3x+9}\,dx = 4\int_1^6 \frac{2x+3}{x^2+3x+9}\,dx$

Let $u = x^2 + 3x + 9$. Then $du = (2x+3)\,dx$.
If $x = 6$, then $u = 63$. If $x = 1$, then $u = 13$.

$4\int_1^6 \frac{2x+3}{x^2+3x+9}\,dx = 4\int_{13}^{63}\frac{1}{u}\,du = 4\ln|u|\Big|_{13}^{63}$

$= 4(\ln 63 - \ln 13)$

$= 4\ln\frac{63}{13} \approx 6.313$

50. $\int_0^3 \frac{6x}{x^2+1}\,dx$

Let $u = x^2 + 1$. Then $du = 2x\,dx$.
If $x = 0$, then $u = 1$. If $x = 3$, then $u = 10$.

$\int_0^3 \frac{6x}{x^2+1}\,dx = 3\int_1^{10}\frac{1}{u}\,du = 3\ln|u|\Big|_1^{10}$

$= 3(\ln 10 - \ln 1)$

$= 3\ln 10$

51. $\int_1^e \frac{9(\ln x)^2}{x}\,dx = 9\int_1^e \frac{(\ln x)^2}{x}\,dx$

Let $u = \ln x$. Then $du = \frac{1}{x}\,dx$.
If $x = 1$, then $u = \ln 1 = 0$. If $x = e$, then $u = 1$.

$\int_1^e \frac{9(\ln x)^2}{x}\,dx = 9\int_0^1 u^2\,du = \frac{9u^3}{3}\Big|_0^1$

$= 3\left[1^3 - 0^3\right]$

$= 3$

52. $\int_1^4 \frac{3(\ln x)^5}{x}\,dx = 3\int_1^4 \frac{(\ln x)^5}{x}\,dx$

Let $u = \ln x$. Then $du = \frac{1}{x}\,dx$.
If $x = 4$, then $u = \ln 4$. If $x = 1$, then $u = 0$.

$\int_1^4 \frac{3(\ln x)^5}{x}\,dx = 3\int_0^{\ln 4} u^5\,du = \frac{3u^6}{6}\Big|_0^{\ln 4}$

$= \frac{1}{2}\left[(\ln 4)^6 - 0\right]$

$= \frac{(\ln 4)^6}{2} \approx 3.549$

53. $\int_0^1 (3x+7)e^x\,dx$

Using integration by parts, let $u = 3x + 7$ and $dv = e^x\,dx$. Therefore, $du = 3x dx$ and $v = e^x$.

$\int_0^1 (3x+7)e^x\,dx$

$= (3x+7)e^x\Big|_0^1 - \int_0^1 e^x\cdot 3dx$

$= (3x+7)e^x - 3e^x\Big|_0^1$

$= \left[(3(1)+7)e^1 - 3e^1\right] - \left[(3(0)+7)e^0 - 3e^0\right]$

$= 7e - 4$

54. $\int_0^5 (9-2x)e^x\,dx$

Using integration by parts, let $u = 9 - 2x$ and $dv = e^x\,dx$. Therefore, $du = -2x dx$ and $v = e^x$.

Copyright © 2019 Pearson Education, Inc.

$$\int_0^5 (9-2x)e^x\,dx$$
$$=(9-2x)e^x\Big|_0^5-\int_0^5 e^x\cdot -2dx$$
$$=(9-2x)e^x+2e^x\Big|_0^5$$
$$=\left[(9-2(5))e^5+2e^5\right]-\left[(9-2(0))e^0+2e^0\right]$$
$$=e^5-11$$

55. $\int_0^3 4x^2e^x\,dx$

Using integration by parts, let $u=4x^2$ and $dv=e^x\,dx$. Therefore, $du=8x dx$ and $v=e^x$.

$$\int_0^3 4x^2e^x\,dx$$
$$=4x^2e^x\Big|_0^3-\int_0^3 e^x(8xdx)$$
$$=4x^2e^x\Big|_0^3-\int_0^3 8xe^x\,dx$$

Using integration by parts a second time, let $u=8x$ and $dv=e^x\,dx$. Therefore, $du=8dx$ and $v=e^x$.

$$=4x^2e^x\Big|_0^3-\int_0^3 8xe^x\,dx$$
$$=(4x^2e^x-8xe^x)\Big|_0^3+\int_0^3 8e^x\,dx$$
$$=(4x^2e^x-8xe^x+8e^x)\Big|_0^3$$
$$=\left[4(9)e^3-8(3)e^3+8e^3\right]$$
$$-\left[4(0)e^0-8(0)e^0+8e^0\right]$$
$$=20e^3-8$$

56. $\int_0^1 5x^2e^x\,dx$

Using integration by parts, let $u=5x^2$ and $dv=e^x\,dx$. Therefore, $du=10x dx$ and $v=e^x$.

$$\int_0^1 5x^2e^x\,dx$$
$$=5x^2e^x\Big|_0^1-\int_0^1 e^x(10xdx)$$
$$=5x^2e^x\Big|_0^1-\int_0^1 10xe^x\,dx$$

Using integration by parts a second time, let $u=10x$ and $dv=e^x\,dx$. Therefore, $du=10dx$ and $v=e^x$.

$$=5x^2e^x\Big|_0^1-\int_0^1 10xe^x\,dx$$
$$=(5x^2e^x-10xe^x)\Big|_0^1+\int_0^1 10e^x\,dx$$
$$=(5x^2e^x-10xe^x+10e^x)\Big|_0^1$$
$$=\left[5(1)e^1-10(1)e^1+10e^1\right]$$
$$-\left[5(0)e^0-10(0)e^0+10e^0\right]$$
$$=5e-10$$

57. $\int_{1/8}^{e/8} \ln(8x)dx$

Using the antiderivative of the Logarithmic function, let $u=8x$. Therefore, $du=8dx$.

$$\int_{1/8}^{e/8} \ln(8x)dx$$
$$=\frac{1}{8}(8x\ln(8x)-8x)\Big|_{1/8}^{e/8}$$
$$=(x\ln(8x)-x)\Big|_{1/8}^{e/8}$$
$$=\left[\frac{e}{8}\ln(e)-\frac{e}{8}\right]-\left[\frac{1}{8}\ln(1)-\frac{1}{8}\right]$$
$$=\frac{1}{8}$$

58. $\int_1^e 8\ln(x)dx$

Using the antiderivative of the Logarithmic function, let $u=x$. Therefore, $du=dx$.

$$\int_1^e 8\ln(x)dx$$
$$=8(x\ln(x)-x)\Big|_1^e$$
$$=(8x\ln(x)-8x)\Big|_1^e$$
$$=[8e\ln(e)-8e]-[8(1)\ln(1)-8(1)]$$
$$=8$$

59. $f(x)=e^x$; $[0, 2]$

$$A=\int_0^2 e^x\,dx=e^x\Big|_0^2=e^2-e^0=e^2-1\approx 6.3891$$

60. $f(x)=1+e^{-x}$; $[0, 4]$

$$A=\int_0^4\left(1+e^{-x}\right)dx=\left(x+\frac{1}{-1}e^{-x}\right)\Big|_0^4$$
$$=\left(4-e^{-4}\right)-(0-1)=5-e^{-4}\approx 4.982$$

Copyright © 2019 Pearson Education, Inc.

61. $C'(x) = 10 - 5x$; fixed cost is \$4. The fixed cost tells us that $C(0) = 4$.

$C(x) = \int (10 - 5x)dx$

$C(x) = 10x - \frac{5x^2}{2} + k$

$4 = 10(0) - \frac{5}{2}(0)^2 + k \Rightarrow k = 4$

Thus, the cost function is

$C(x) = 10x - \frac{5x^2}{2} + 4$.

62. $C'(x) = 2x + 3x^2$; 2 units cost \$12. Since, 2 units cost \$12, we have $C(2) = 12$.

$C(x) = \int \left(2x + 3x^2\right) dx$

$C(x) = x^2 + x^3 + k \Rightarrow 12 = 2^2 + 2^3 + k \Rightarrow$

$12 = 4 + 8 + k \Rightarrow k = 0$

Thus, the cost function is $C(x) = x^2 + x^3$.

63. $C'(x) = 3\sqrt{2x-1}$; 13 units cost \$270. Since 13 units cost \$270, we have $C(13) = 270$.

$C(x) = \int 3\sqrt{2x-1}dx$

Let $u = 2x - 1$. Then $du = 2\,dx$.

$C(x) = 3 \cdot \frac{1}{2} \int \sqrt{2x-1} \cdot 2\,dx = \frac{3}{2} \int u^{1/2} du \Rightarrow$

$C(x) = \frac{3}{2} \cdot \frac{u^{3/2}}{\frac{3}{2}} + k \Rightarrow C(x) = u^{3/2} + k \Rightarrow$

$C(x) = (2x-1)^{3/2} + k \Rightarrow 270 = 25^{3/2} + k \Rightarrow$
$k = 145$

Thus, the cost function is

$C(x) = (2x-1)^{3/2} + 145$.

64. $C'(x) = \frac{6}{x+1}$; fixed cost is \$18. Since the fixed cost is \$18, $C(0) = 18$.

$C(x) = \int \frac{6}{x+1} dx = 6 \ln|x+1| + k \Rightarrow$

$18 = 6 \ln 1 + k \Rightarrow k = 18$

Thus, the cost function is $C(x) = 6 \ln|x+1| + 18$.

65. $f(t) = 1288t - 10{,}576$

Since we are seeking the total revenue from the beginning of 2012 to the start of 2017, we use the interval [12, 17).

$\int_{12}^{17} (1288t - 10{,}576)\,dt$

$= \left[644t^2 - 10{,}576t\right]_{12}^{17}$

$= \left[644(17)^2 - 10{,}576(17)\right]$

$\quad - \left[644(12)^2 - 10{,}576(12)\right]$

$= 6324 - (-34{,}176) = 40{,}500$ million

The total revenue from the beginning of 2012 to the start of 2017 was about \$40,500 million or \$40.5 billion.

66. $f(t) = 17.2t + 238.8$

Since we are seeking the total projected spending from the beginning of 2016 to the start of 2026, we use the interval [16, 26).

$\int_{16}^{26} (17.2t + 238.8)\,dt$

$= \left[8.6t^2 + 238.8t\right]_{16}^{26}$

$= \left[8.6(26)^2 + 238.8(26)\right]$

$\quad - \left[8.6(16)^2 + 238.8(16)\right]$

$= 12{,}022.4 - 6{,}022.4 = 6{,}000$ billion

The total projected spending from the beginning of 2016 to the start of 2026 was about \$6,000 billion or \$6 trillion.

67. $f(t) = 32t^2 - 715t + 5240$

$\int_{10}^{17} \left(32t^2 - 715t + 5240\right) dx$

$= \left[\left(\frac{32t^3}{3}\right) - \left(\frac{715t^2}{2}\right) + 5240t\right]_{10}^{17}$

$\approx 38{,}167.83 - 27{,}316.67$
$\approx 10{,}851$ million

The total profits generated from the start of 2007 to the start of 2017 was \$10,851 million (\$10.851 billion).

Copyright © 2019 Pearson Education, Inc.

68. $f(t) = 2.27t^3 - 82t^2 + 700t + 36,200$

$$\int_0^{20} \left(2.27t^3 - 82t^2 + 700t + 36,200\right) dt$$

$$= \left[\left(\frac{2.27t^4}{4}\right) - \left(\frac{82t^3}{3}\right) + \left(\frac{700t^2}{2}\right) + 36,200t\right]\Bigg|_0^{20}$$

$$\approx 736,133 - 0 \approx 736,133 \text{ trillion}$$

The total projected petroleum consumption from 2015 to 2035 is projected to be 736,133 trillion BTU.

69. $S' = \sqrt{x} + 3$

$$S(x) = \int_0^9 \left(x^{1/2} + 3\right) dx = \left(\frac{x^{3/2}}{\frac{3}{2}} + 3x\right)\Bigg|_0^9$$

$$= \frac{2}{3}(9)^{3/2} + 27 = 45$$

Total sales are 45(1000), or 45,000 units.

70. $\int_0^{10} \left(150 - \sqrt{3.2t + 4}\right) dt$

$$= \int_0^{10} 150\, dt - \int_0^{10} \sqrt{3.2t + 4}\, dt$$

Let $u = 3.2t + 4$ and $du = 3.2\, dt$ or $\frac{1}{3.2} du = dt.$

$$\int_0^{10} 150\, dt - \int_0^{10} \sqrt{3.2t + 4}\, dt$$

$$= 150t\Big|_0^{10} - \int_0^{10} \sqrt{u}\, \frac{1}{3.2} du$$

$$= 1500 - \frac{1}{3.2} \frac{u^{1.5}}{1.5}$$

$$\approx 1500 - .2083(3.2t + 4)^{1.5}\Big|_0^{10}$$

$$\approx 1500 - \left(.2083(3.2 \cdot 10 + 4)^{1.5} - .2083(3.2 \cdot 0 + 4)^{1.5}\right)$$

$$\approx 1456.66$$

In the first 10 months there were about 1457 additional spiders.

71. $S'(x) = 225 - x^2$

$C'(x) = x^2 + 25x + 150$

$S'(x) = C'(x) \Rightarrow 225 - x^2 = x^2 + 25x + 150 \Rightarrow$
$2x^2 + 25x - 75 = 0 \Rightarrow (2x - 5)(x + 15) = 0 \Rightarrow$
$x = \frac{5}{2}, -15$

Discard the negative solution. The company should use the machinery for 2.5 years.

$$\int_0^{2.5} \left[\left(225 - x^2\right) - \left(x^2 + 25x + 150\right)\right] dx$$

$$= \int_0^{2.5} \left(-2x^2 - 25x + 75\right) dx$$

$$= \left(\frac{-2x^3}{3} - \frac{25x^2}{2} + 75x\right)\Bigg|_0^{2.5}$$

$$= \frac{-2(2.5)^3}{3} - \frac{25(2.5)^2}{2} + 75(2.5)$$

$$\approx 98.95833$$

The net savings are \$98,958.33.

72. $S'(x) = 60 - 4x$

$C'(x) = x^2 + x + 10$

$S'(x) = C'(x) \Rightarrow 60 - 4x = x^2 + x + 10 \Rightarrow$
$x^2 + 5x - 50 = 0 \Rightarrow (x - 5)(x + 10) = 0 \Rightarrow$
$x = 5, -10$

Discard the negative solution. The company should use the machinery for 5 years.

$$\int_0^5 \left[(60 - 4x) - \left(x^2 + x + 10\right)\right] dx$$

$$= \int_0^5 \left(-x^2 - 5x + 50\right) dx$$

$$= \left(\frac{-x^3}{3} - \frac{5x^2}{2} + 50x\right)\Bigg|_0^5$$

$$= \frac{-(5)^3}{3} - \frac{5(5)^2}{2} + 50(5)$$

$$\approx 145.8333$$

The net savings are \$145,833.

Copyright © 2019 Pearson Education, Inc.

73. $S(q) = q^2 + 5q + 100$

$D(q) = 350 - q^2$

$$S(q) = D(q)$$

$$q^2 + 5q + 100 = 350 - q^2$$

$$2q^2 + 5q - 250 = 0$$

$$(2q+25)(q-10) = 0$$

$$q = \frac{-25}{2} \text{ or } q = 10$$

Discard the negative solution.

$S(10) = 10^2 + 5(10) + 100 = 250$

The equilibrium point is (10, 250).

The producers' surplus is given by

$$\int_0^{10}\left[250 - \left(q^2 + 5q + 100\right)\right]dq$$

$$= \int_0^{10}\left(150 - q^2 - 5q\right)dq$$

$$= \left(150q - \frac{q^3}{3} - \frac{5q^2}{2}\right)\Bigg|_0^{10} \approx 916.67$$

The producers' surplus is \$916.67.

74. The consumers' surplus is given by

$$\int_0^{10}\left[\left(350 - q^2\right) - 250\right]dq$$

$$= \int_0^{10}\left(100 - q^2\right)dq = \left(100q - \frac{q^3}{3}\right)\Bigg|_0^{10}$$

$$= \left[100(10) - \frac{10^3}{3}\right] - 0 \approx 666.67$$

The consumers' surplus is \$666.67.

75. $I'(t) = \frac{100t}{t^2 + 2}$

The total number of infected people is given by

$$\int_0^4 \frac{100t}{t^2+2}\,dt.$$

Let $u = t^2 + 2$. Then $du = 2t\,dt$. If $t = 4$, $u = 18$. If $t = 0$, $u = 2$.

$$\int_0^4 \frac{100t}{t^2+2}\,dt = 50\int_0^4 \frac{2t\,dt}{t^2+2} = 50\int_2^{18}\frac{du}{u}$$

$$= 50\ln|u|\Big|_2^{18} = 50\ln 18 - 50\ln 2$$

$$= 50\ln\left(\frac{18}{2}\right) \approx 109.86$$

Approximately 110 people will be infected.

76. a.
$$\int_1^{10}\left(\frac{4}{3t} + \frac{2}{t^2}\right)dt = \int_1^{10}\left(\frac{4}{3}t^{-1} + 2t^{-2}\right)dt$$

$$= \left(\frac{4}{3}\ln t - 2t^{-1}\right)\Bigg|_1^{10}$$

$$= \left[\frac{4}{3}\ln(10) - \frac{2}{10}\right] - \left[\frac{4}{3}\ln(1) - \frac{2}{1}\right]$$

$$\approx 4.87$$

The total amount of the drug released from 1 to 10 hours is about 4.87 mg.

b.
$$\int_{10}^{20}\left(\frac{4}{3t} + \frac{2}{t^2}\right)dt = \int_{10}^{20}\left(\frac{4}{3}t^{-1} + 2t^{-2}\right)dt$$

$$= \left(\frac{4}{3}\ln t - 2t^{-1}\right)\Bigg|_{10}^{20}$$

$$= \left[\frac{4}{3}\ln(20) - \frac{2}{20}\right] - \left[\frac{4}{3}\ln(10) - \frac{2}{10}\right]$$

$$\approx 1.02$$

The total amount of the drug released from 10 to 20 hours is about 1.02 mg.

77. a.
$$\int_0^2\left(\frac{1}{2}e^{-x/2}\right)dx = -e^{-x/2}\Big|_0^2 = -e^{-1} - \left(-e^0\right)$$

$$\approx .6321$$

The probability that a customer in line waits less than 2 minutes is about .6321.

b.
$$1 - \int_0^3\left(\frac{1}{2}e^{-x/2}\right)dx = 1 - \left[-e^{-x/2}\Big|_0^3\right]$$

$$= 1 - \left[-e^{-3/2} - \left(-e^0\right)\right]$$

$$\approx .2231$$

The probability that a customer in line waits more than 3 minutes is about .2231.

78. a.
$$\int_1^2 .44e^{-.44t}\,dt = -e^{-.44t}\Big|_1^2 = -e^{-.88} - \left(-e^{-.44}\right)$$

$$\approx .2293$$

The probability that the person will take between 1 and 2 seconds between blinks is about .2293.

Copyright © 2019 Pearson Education, Inc.

b. $1-\int_0^4 .44e^{-.44t}\,dt = 1-\left[-e^{-.44t}\Big|_0^4\right]$
$= 1-\left[-e^{-1.76}-\left(-e^{0}\right)\right]$
$\approx .1720$

The probability that the person will take more than 4 seconds between blinks is about .1720.

79. $\frac{dy}{dx} = -.07y \Rightarrow$

$\frac{dy}{y} = -.07\,dx \Rightarrow$

$\ln|y| = -.07x + C_1 \Rightarrow$

$y = e^{-.07x+C_1} = e^{-.07x}e^{C_1}$

$y = Me^{-.07x}$

80. $\frac{dy}{dx} = \frac{y}{13} \Rightarrow$

$\frac{dy}{y} = \frac{1}{13}dx \Rightarrow$

$\ln|y| = \frac{x}{13} + C_1 \Rightarrow$

$y = e^{x/13+C_1} = e^{x/13}e^{C_1}$

$y = Me^{x/13}$

81. $\frac{dy}{dx} = \frac{1-20x^4}{3y^2} \Rightarrow$

$3y^2\,dy = (1-20x^4)\,dx \Rightarrow$

$y^3 = x - 4x^5 + C \Rightarrow$

$y = \sqrt[3]{x-4x^5+C}$

82. $\frac{dy}{dx} = \frac{14e^{2x}}{y^2+5} \Rightarrow$

$(y^2+5)dy = 14e^{2x}\,dx \Rightarrow$

$\frac{1}{3}y^3 + 5y = 7e^{2x} + C$

83. $\frac{dy}{dx} = 5\left(e^{-x}-1\right)$; $y = 17$ when $x = 0$

$dy = 5\left(e^{-x}-1\right)dx \Rightarrow \int dy = \int 5\left(e^{-x}-1\right)dx \Rightarrow$

$y = 5\left[\frac{e^{-x}}{-1} - x\right] + C$

The general solution is $y = -5e^{-x} - 5x + C$.

Let $x = 0$ and $y = 17$.

$17 = -5e^0 - 5(0) + C \Rightarrow 17 = -5 + C \Rightarrow C = 22$

The particular solution is $y = -5e^{-x} - 5x + 22$.

84. $\frac{dy}{dx} = \frac{x}{x^2-3} + 7$; $y = 52$ when $x = 2$

$y = \int\left(\frac{x}{x^2-3}\right) + 7\,dx = \frac{1}{2}\int\left(\frac{2x}{x^2-3}\right) + 7\,dx$

The general solution is $y = \frac{1}{2}\ln\left|x^2-3\right| + 7x + C$.

Let $x = 2$ and $y = 52$.

$52 = \frac{1}{2}\ln|4-3| + 14 + C \Rightarrow 38 = \frac{1}{2}\ln 1 + C \Rightarrow$

$C = 38$

The particular solution is

$y = \frac{1}{2}\ln\left|x^2-3\right| + 7x + 38$.

85. $(5-2x)y = \frac{dy}{dx}$; $y = 2$ when $x = 0$

$(5-2x)dx = \frac{dy}{y} \Rightarrow \int(5-2x)dx = \int\frac{dy}{y}$

The general solution is $5x - x^2 = \ln|y| + C$.

Let $x = 0$ and $y = 2$.

$0 = \ln 2 + C \Rightarrow C = -\ln 2 \Rightarrow$

$5x - x^2 = \ln|y| - \ln 2 = \ln\left|\frac{y}{2}\right| \Rightarrow$

$e^{5x-x^2} = \left|\frac{y}{2}\right| \Rightarrow 2e^{5x-x^2} = |y|$

The particular solution is $y = 2e^{5x-x^2}$.

Copyright © 2019 Pearson Education, Inc.

86. $\sqrt{x}\dfrac{dy}{dx} = xy$; $y = 4$ when $x = 1$

Write $\sqrt{x}$ as $x^{1/2}$ and separate the variables.

$y = \dfrac{x^{1/2}}{x}\dfrac{dy}{dx} \Rightarrow \dfrac{1}{y}dy = x^{1/2}dx \Rightarrow$

$\int \dfrac{1}{y}dy = \int x^{1/2}dx \Rightarrow \ln|y| = \dfrac{2}{3}x^{3/2} + C$

Let $x = 1$ and $y = 4$.

$\ln 4 = \dfrac{2}{3}(1)^{3/2} + C \Rightarrow C = \ln 4 - \dfrac{2}{3} \approx .7196$

$\ln|y| \approx \dfrac{2}{3}x^{3/2} + .7196 \Rightarrow |y| \approx e^{.7196}e^{\frac{2x^{3/2}}{3}} \Rightarrow$

$y \approx \pm 2.054e^{\frac{2x^{3/2}}{3}}$

Since $y = 4$ when $x = 1$, we must use the positive solution: $y = 2.054e^{\frac{2x^{3/2}}{3}}$

87. a. $\dfrac{dy}{dt} = .02y \Rightarrow$

$\dfrac{dy}{y} = .02\,dt \Rightarrow$

$\ln|y| = .02t + C_1 \Rightarrow$

$y = e^{.02t + C_1} = e^{.02t}e^{C_1}$

$y = Me^{.02t}$

The initial price of the guitar was \$13,000, therefore the equation will be

$y = 13{,}000e^{.02t}$.

b. Let $t = 3$,

$y = 13{,}000e^{.02(3)}$

$\approx 13{,}804$

The value of the guitar in 2020 will be about \$13,804.

c. Let $y = 15{,}000$,

$15{,}000 = 13{,}000e^{.02t}$

$\dfrac{15}{13} = e^{.02t}$

$\ln\left(\dfrac{15}{13}\right) = .02t$

$t \approx 7.2$

The value of the guitar will be \$15,000 in about 7.2 years.

88. a. $\dfrac{dy}{dt} = .058y \Rightarrow$

$\dfrac{dy}{y} = .058\,dt \Rightarrow$

$\ln|y| = .058t + C_1 \Rightarrow$

$y = e^{.058t + C_1} = e^{.058t}e^{C_1}$

$y = Me^{.058t}$

The initial number of Wi-fi sessions was 22 thousand, therefore the equation will be

$y = 22e^{.058t}$.

b. Let $t = 7$,

$y = 22e^{.058(7)}$

≈ 33

There were about 33 thousand (33,000) Wi-fi sessions when $t = 7$.

c. Let $y = 300$,

$300 = 22e^{.058t}$

$\dfrac{300}{22} = e^{.058t}$

$\ln\left(\dfrac{150}{11}\right) = .058t$

$t \approx 45$

The number of Wi-fi sessions will be about 300,000 in about 45 months or 3.75 years.

89. $\dfrac{dy}{dx} = .05(70 - y)$; $x = 0$ and $y = 27$.

a.

$\dfrac{dy}{dx} = .05(70 - y)$

$\dfrac{1}{70 - y}dy = .05dx$

$\int \dfrac{1}{70 - y}dy = .05\int dx$

$-\ln|70 - y| = .05x + C$

$\ln|70 - y| = -.05x + C$

$|70 - y| = e^{-.05x + C}$

$|70 - y| = e^{-.05x}e^{C}$

$70 - y = \pm e^{C}e^{-.05x}$

$70 - y = Ce^{-.05x} \Rightarrow y = Ce^{-.05x} + 70$

When $x = 0$, $y = 27$ so

$27 = Ce^{0} + 70 \Rightarrow C = -43$, and

$y = 70 - 43e^{-.05x}$.

Copyright © 2019 Pearson Education, Inc.

b. If $x = 50$, then

$y = 70 - 43e^{-.05(50)} \Rightarrow y = 70 - 43e^{-2.5}$

$y \approx 66\%$.

The projected percent of 3- to 5- year-old children that will be enrolled in preprimary programs in 2015 will be approximately 66%.

90. $\frac{dT}{dt} = k(T - T_F) \Rightarrow \frac{dT}{T - T_F} = k\,dt \Rightarrow$

$\int \frac{dT}{T - T_F} = \int k\,dt \Rightarrow \ln|T - T_F| = kt + C \Rightarrow$

$|T - T_F| = e^{kt+C} \Rightarrow |T - T_F| = e^{kt} \cdot e^C \Rightarrow$

$T - T_F = \pm e^C e^{kt} \Rightarrow T = T_F + Me^{kt}$

$T_F = 300°$, $T(0) = 40°$, and $T(1) = 150°$

$T = 300 + Me^{kt} \Rightarrow 40 = 300 + Me^0 \Rightarrow$

$M = -260$

$T = 300 - 260e^{kt} \Rightarrow 150 = 300 - 260e^{k \cdot 1} \Rightarrow$

$-150 = -260e^k \Rightarrow e^k = \frac{150}{260} \Rightarrow$

$k = \ln\left(\frac{150}{260}\right) \approx -.550 \Rightarrow T = 300 - 260e^{-.550t}$

a. After 2.5 hours, the temperature is

$T = 300 - 260e^{-.550(2.5)} \approx 234.26$

The temperature is approximately 234°.

b. If $T = 250$,

$250 = 300 - 260e^{-.550t} \Rightarrow$

$-50 = -260e^{-.550t} \Rightarrow e^{-.550t} = \frac{50}{260} \Rightarrow$

$-.550t = \ln\left(\frac{5}{26}\right) \Rightarrow t = \frac{\ln\left(\frac{5}{26}\right)}{-.550} \approx 2.998$

The temperature will be 250° in approximately 3 hours.

Case 13 Bounded Population Growth

1. Answers vary; Verify $\frac{1}{P(C-P)} = \frac{\frac{1}{C}}{P} + \frac{\frac{1}{C}}{C-P}$

$$\frac{\frac{1}{C}}{P} + \frac{\frac{1}{C}}{C-P} = \frac{\left(\frac{1}{C}\right)(C-P)}{P(C-P)} + \frac{\left(\frac{1}{C}\right)P}{P(C-P)}$$

$$= \frac{\left(\frac{1}{C}\right)(C-P) + \left(\frac{1}{C}\right)P}{P(C-P)}$$

$$= \frac{\left(\frac{1}{C}\right)C - \left(\frac{1}{C}\right)P + \left(\frac{1}{C}\right)P}{P(C-P)}$$

$$= \frac{1-0}{P(C-P)} = \frac{1}{P(C-P)}$$

2. Answers vary; $P(0) = 10$

$\frac{1}{200}\ln(P) - \frac{1}{200}\ln(200 - P) = .005t + K \Rightarrow$

$\ln P - \ln(200 - P) = t + 200K \Rightarrow$

$e^{\ln(P) - \ln(200-P)} = e^{t+200K} \Rightarrow$

$\frac{P}{200 - P} = e^{200K}e^t \Rightarrow$

$P = 200e^{200K}e^t - Pe^{200K}e^t \Rightarrow$

$P + P(e^{200K}e^t) = 200e^{200K}e^t \Rightarrow$

$P = \frac{200e^{200K}e^t}{1 + e^{200K}e^t} \Rightarrow P = \frac{200e^t}{e^{-200K} + e^t}$

Let $A = e^{-200K}$. Then $P(t) = \frac{200e^t}{A + e^t}$

Since $P(0) = \frac{200}{A+1}$ must equal 10, it follows that $A = 19$ and $P(t) = \frac{200e^t}{e^t + 19}$.

3. a. $P(0) = 2$, $C = 10$, $k = .007$

$\frac{dP}{dt} = .007P(10 - P)$

$\int \frac{1}{P(10-P)}\,dP = \int .007\,dt \Rightarrow$

$\frac{1}{10}\ln(P) - \frac{1}{10}\ln(10 - P) = .007t + K$

$\ln(P) - \ln(10 - P) = .07t + 10K$

Performing similar calculations as in Exercise 2,

$P = \frac{10e^{.07t}}{e^{-10K} + e^{.07t}}$

$P(0) = \frac{10}{e^{-10K} + 1} \Rightarrow 2 = \frac{10}{e^{-10K} + 1} \Rightarrow$

$e^{-10K} + 1 = 5 \Rightarrow e^{-10K} = 4 \Rightarrow$

$-10K = \ln 4 \Rightarrow K = \frac{\ln 4}{-10}$

Rewrite P with the value of K.

$P(t) = \frac{10e^{.07t}}{4 + e^{.07t}}$

b. Now find $P(t)$ when $t = 38$.

$P(38) = \frac{10e^{.07(38)}}{4 + e^{.07(38)}} \approx 7.8$

The seal population in 2018 will be approximately 7.8 million.

Copyright © 2019 Pearson Education, Inc.

c. Now find t when $P(t) = 8$.

$$8 = \frac{10e^{.07t}}{4+e^{.07t}} \Rightarrow 8(4+e^{.07t}) = 10e^{.07t} \Rightarrow$$
$$32+8e^{.07t} = 10e^{.07t} \Rightarrow 32 = 2e^{.07t} \Rightarrow$$
$$16 = e^{.07t} \Rightarrow \ln 16 = .07t \Rightarrow$$
$$t = \frac{\ln 16}{.07} \approx 40$$

The population will reach 8 million after about 40 years or in the year 2020.

4. a. $P(0) = 20$, $C = 50$, $k = .065$

$$\frac{dP}{dt} = .065P(50-P)$$
$$\int \frac{1}{P(50-P)}\,dP = \int .065\,dt \Rightarrow$$
$$\frac{1}{50}\ln(P) - \frac{1}{50}\ln(50-P) = .065t + K$$
$$\ln(P) - \ln(50-P) = 3.25t + 50K$$

Performing similar calculations as in Exercise 2,

$$P = \frac{50e^{3.25t}}{e^{-50K}+e^{3.25t}}$$
$$P(0) = \frac{50}{e^{-50K}+1} \Rightarrow 20 = \frac{50}{e^{-50K}+1} \Rightarrow$$
$$e^{-50K}+1 = 2.5 \Rightarrow e^{-50K} = 1.5 \Rightarrow$$
$$-50K = \ln 1.5 \Rightarrow K = \frac{\ln 1.5}{-50}$$

Rewrite P with the value of K.

$$P(t) = \frac{50e^{.065t}}{e^{.065t}+1.5}$$

b. Now find $P(t)$ when $t = 24$, $t = 40$, and $t = 44$.

$$P(24) = \frac{50e^{.065(24)}}{e^{.065(24)}+1.5} \approx 38$$
$$P(40) = \frac{50e^{.065(40)}}{e^{.065(40)}+1.5} \approx 45$$
$$P(44) = \frac{50e^{.065(44)}}{e^{.065(44)}+1.5} \approx 46$$

The percentage of Olympians that were or will be women in 2000, 2016, and 2020 will be 38%, 45%, and 46% respectively.

Copyright © 2019 Pearson Education, Inc.

Chapter 14 Multivariate Calculus

Section 14.1 Functions of Several Variables

1. $f(x, y) = 5x + 2y - 4$

$$f(2, -1) = 5(2) + 2(-1) - 4 = 10 - 2 - 4 = 4$$
$$f(-4, 1) = 5(-4) + 2(1) - 4 = -20 + 2 - 4 = -22$$
$$f(-2, -3) = 5(-2) + 2(-3) - 4 = -10 - 6 - 4 = -20$$
$$f(0, 8) = 5(0) + 2(8) - 4 = 0 + 16 - 4 = 12$$

3. $f(x, y) = 2x^2 - xy + y^2$

$$f(2, -1) = 2(2)^2 - 2(-1) + (-1)^2 = 8 + 2 + 1 = 11$$
$$f(-4, 1) = 2(-4)^2 - (-4)1 + 1^2 = 32 + 4 + 1 = 37$$
$$f(-2, -3) = 2(-2)^2 - (-2)(-3) + (-3)^2 = 8 - 6 + 9 = 11$$
$$f(0, 8) = 2(0)^2 - 0(8) + 8^2 = 0 - 0 + 64 = 64$$

5. $f(x, y) = \sqrt{y^2 + 2x^2}$

$$f(2, -1) = \sqrt{(-1)^2 + 2(2)^2} = \sqrt{1+8} = \sqrt{9} = 3$$
$$f(-4, 1) = \sqrt{(1)^2 + 2(-4)^2} = \sqrt{1+32} = \sqrt{33}$$
$$f(-2, -3) = \sqrt{(-3)^2 + 2(-2)^2} = \sqrt{9+8} = \sqrt{17}$$
$$f(0, 8) = \sqrt{(8)^2 + 2(0)^2} = \sqrt{64} = 8$$

7. $f(x, y) = e^x - 7y^3$
The domain is the set of all ordered pairs (x, y) such that x and y are real numbers.

9. $f(x, y) = \sqrt{x} - \sqrt{y}$
The domain is the set of all ordered pairs (x, y) such that x and y are nonnegative real numbers.

11. $h(u, v, w) = \left|\dfrac{u+v}{v-w}\right|$
The domain is the set of all ordered triples (u, v, w) such that u, v, and w are real numbers, and $v \neq w$.

13. $f(x, y) = \ln(x) + y - 5$
The domain is the set of all ordered pairs (x, y) such that $x > 0$ and y is any real number.

15. $P(l, w) = 2l + 2w$

17. Let x = length of decorative fencing and y = length of privacy fencing.
$C(x, y) =$ total cost.
Then $C(x, y) = 40x + 14y$.

19. Let x = units of labor, y = units of materials, and z = units of capital.
$C(x, y, z) =$ total cost.
Then $C(x, y, z) = 200x + 100y + 50z$.

21. The *xy*-, *xz*- and *yz*- traces of a graph are the curves that result when a surface that is the graph of a function of two variables, $z = f(x, y)$, is cut by the *xy*-, *xz*- and *yz*- planes, respectively.

23. $3x + 2y + z = 12$
Let $x = 0$ and $y = 0$. Then $z = 12$. The point (0, 0, 12) is on the graph. Let $x = 0$ and $z = 0$. Then $y = 6$. The point (0, 6, 0) is on the graph. Let $y = 0$ and $z = 0$. Then $x = 4$. The point (4, 0, 0) is on the graph. The graph is a plane. We sketch the portion in the first octant through these three points.

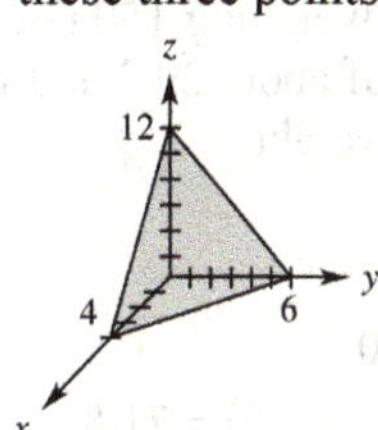

Copyright © 2019 Pearson Education, Inc.

25. $x+y=4$
We let $x = 0$ and find $y = 4$. The point (0, 4, 0) is on the graph. We let $y = 0$ and find $x = 4$. The point (4, 0, 0) is on the graph. Since there is no z-intercept, the plane is parallel to the z-axis. We sketch the portion in the first octant through these two points.

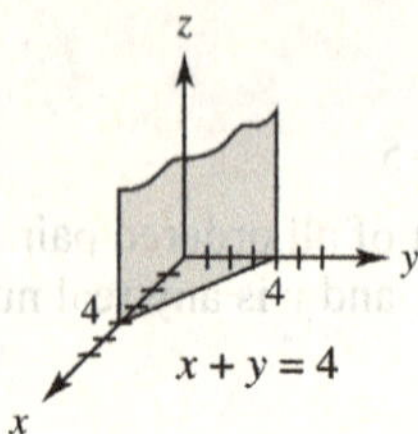

27. $z=4$
The z-intercept is (0, 0, 4). Since there is no x-intercept or y-intercept, the plane is parallel to the xy-plane. We sketch the portion in the first octant through the point (0, 0, 4).

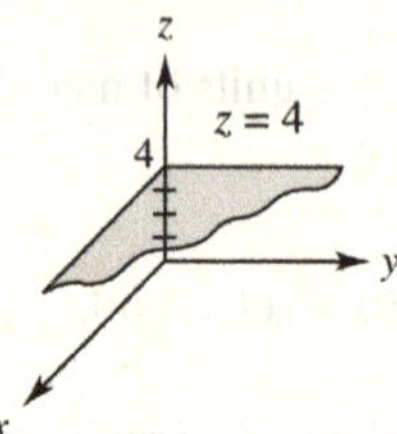

29. a. Let $h = 60$ inches and $w = 100$ pounds.
$$BMI = \frac{703(100)}{60^2} \approx 19.5$$
A 5 feet tall person weighing 100 pounds has a body mass index of about 19.5 and would not be considered overweight.

b. Let $h = 60$ inches and $w = 150$ pounds.
$$BMI = \frac{703(150)}{60^2} \approx 29.3$$
A 5 feet tall person weighing 150 pounds has a body mass index of about 29.3 and would be considered overweight.

31. a. $f(d,w) = 15 + .4(d+w)$
If $d = 72$ and $w = 70$,
$f(72,70) = 15 + .4(72+70) = 71.8$
A temperature humidity index of 71.8 would feel comfortable.

b. $f(d,w) = 15 + .4(d+w)$
If $d = 72$ and $w = 78$,
$f(72,78) = 15 + .4(72+78) = 75$
A temperature humidity index of 75 would feel uncomfortable.

33. a. Let $v = 1{,}000$, $m = 10$, and $r = .05$.
$$p = \frac{1{,}000}{(1+.05)^{10}} \approx 613.91$$
The price of the zero-coupon bond under the given conditions would be about $613.91.

b. Let $v = 1{,}000$, $m = 10$, and $r = .06$.
$$p = \frac{1{,}000}{(1+.06)^{10}} \approx 558.39$$
The price of the zero-coupon bond under the given conditions would be about $558.39.

35. $$M(n, i, t) = \frac{(1+.05)^{25}(1-.33)+.33}{\left[1+(1-.33).05\right]^{25}} \approx 1.14$$
Since $M > 1$, the IRA grows faster.

37. Let $l = 28$ and $o = 1$.
$$\begin{aligned} f(l,o) &= 349.1 - 5.17(28) + 36.6(1) - .45(28)(1) \\ &= 349.1 - 144.76 + 36.6 - 12.6 \\ &\approx 228 \end{aligned}$$
The number of deaths in Florida from malignant melanoma of the skin is about 228 people.

39. a. Let p = number of people in line and n = the number of items ahead of you.
$f(p,n) = 41p + 3n$

b. Let $p = 5$ and $n = 50$.
$f(5,50) = 41(5) + 3(50) = 355$
The wait in a line with 5 people and 50 items will be 355 seconds.

c. Let $p = 5$ and $n = 50$.
$f(2,80) = 41(2) + 3(80) = 322$
The wait in a line with 2 people and 80 items will be 322 seconds.

d. No; the express lane would be slower by 33 seconds.

Copyright © 2019 Pearson Education, Inc.

41. $z = x^2 + y^2$

Find the traces.

xy-trace: $0 = x^2 + y^2$

xz-trace: $z = x^2$

yz-trace: $z = y^2$

These traces match with graph (c), the correct graph.

43. $x^2 - y^2 = z$

xy-trace: $x^2 - y^2 = 0$

yz-trace: $y^2 = -z$

xz-trace: $x^2 = z$

These traces match with graph (e), the correct graph.

45. $\frac{x^2}{16} + \frac{y^2}{25} + \frac{z^2}{4} = 1$

xy-trace: $\frac{x^2}{16} + \frac{y^2}{25} = 1$

yz-trace: $\frac{y^2}{25} + \frac{z^2}{4} = 1$

xz-trace: $\frac{x^2}{16} + \frac{z^2}{4} = 1$

These traces are all ellipses, which match with graph (b), the correct graph.

47. $z = x^{.7}y^{.3}$ and $z = 500 = x^{.7}y^{.3} = 500$

$$y^{.3} = \frac{500}{x^{.7}} \Rightarrow y^{3/10} = \frac{500}{x^{7/10}} \Rightarrow$$

$$y = \left(\frac{500}{x^{7/10}}\right)^{10/3} = \frac{500^{10/3}}{x^{7/3}} \approx \frac{9.9 \times 10^8}{x^{7/3}} \approx \frac{10^9}{x^{7/3}}$$

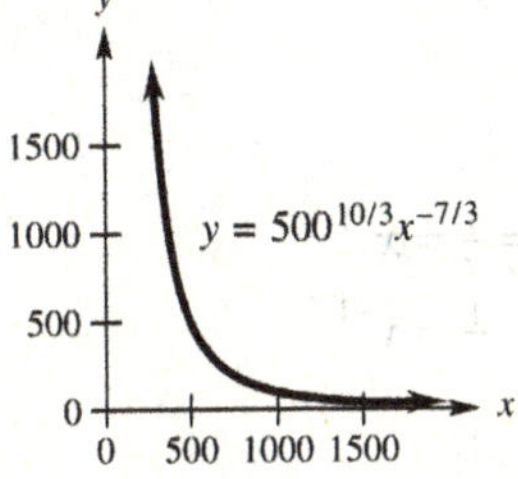

Section 14.2 Partial Derivatives

1. $z = f(x, y) = 8x^3 - 4x^2y + 9y^2$

a. $\frac{\partial z}{\partial x} = 24x^2 - 8xy + 0 = 24x^2 - 8xy$

b. $\frac{\partial z}{\partial y} = 0 - 4x^2 + 18y = -4x^2 + 18y$

c. $f_x(2, 3) = 24(2)^2 - 8(2)(3) = 48$

d. $f_y(1, -2) = -4(1)^2 + 18(-2) = -40$

3. $f(x, y) = 6y^3 - 4xy + 5$

$f_x = 0 - 4y = -4y$

$f_y = 18y^2 - 4x$

$f_x(2, -1) = -4(-1) = 4$

$f_y(-4, 3) = 18(3^2) - 4(-4) = 178$

5. $f(x, y) = e^{2x+y}$

$f_x = e^{2x+y}(2) = 2e^{2x+y}$

$f_y = e^{2x+y}(1) = e^{2x+y}$

$f_x(2, -1) = 2e^{2(2)+(-1)} = 2e^3$

$f_y(-4, 3) = e^{2(-4)+3} = e^{-5}$

7. $f(x, y) = \ln(x - 2y)$

$f_x = \frac{1}{x-2y}(1) = \frac{1}{x-2y}$

$f_y = \frac{1}{x-2y}(-2) = -\frac{2}{x-2y}$

$f_x(2, -1) = \frac{1}{2-2(-1)} = \frac{1}{4}$

$f_y(-4, 3) = -\frac{2}{-4-2(3)} = -\frac{2}{-10} = \frac{1}{5}$

9. $f(x, y) = \sqrt{8x^2y+4} = (8x^2y+4)^{1/2}$

$f_x = \frac{1}{2}(8x^2y+4)^{-1/2}(16xy) = \frac{8xy}{\sqrt{8x^2y+4}}$

$f_y = \frac{1}{2}(8x^2y+4)^{-1/2}(8x^2) = \frac{4x^2}{\sqrt{8x^2y+4}}$

Copyright © 2019 Pearson Education, Inc.

11. $f(x, y) = x^2e^{2xy}$
$f_x = x^2 \cdot 2ye^{2xy} + 2xe^{2xy} = 2x^2ye^{2xy} + 2xe^{2xy}$
$f_y = x^2 \cdot 2xe^{2xy} = 2x^3e^{2xy}$
$f_x(2, -1) = 2(2)^2(-1)e^{2(2)(-1)} + 2(2)e^{2(2)(-1)}$
$= -8e^{-4} + 4e^{-4} = -4e^{-4}$
$f_y(-4, 3) = 2(-4)^3e^{2(-4)(3)} = -128e^{-24}$

13. $f(x, y) = \dfrac{3x}{1+6y}$

Use the quotient rule to find $f_x(x, y)$ and $f_y(x, y)$.

$$f_x = \frac{(1+6y)(3)-(3x)(0)}{(1+6y)^2} = \frac{(1+6y)3}{(1+6y)^2} = \frac{3}{1+6y}$$

$$f_y = \frac{(1+6y)(0)-(3x)(6)}{(1+6y)^2} = \frac{-18x}{(1+6y)^2}$$

15. $f(x, y) = 4x^2y^2 - 16x^2 + 4y$
$f_x = 8xy^2 - 32x$
$f_y = 8x^2y + 4$
$f_{xx} = 8y^2 - 32$
$f_{xy} = 16xy$
$f_{yy} = 8x^2$
$f_{yx} = 16xy$

17. $h(x, y) = -3y^2 - 4x^2y^2 + 7xy^2$
$h_x = 0 - 8xy^2 + 7y^2 = -8xy^2 + 7y^2$
$h_y = -6y - 8x^2y + 14xy$
$h_{xx} = -8y^2 + 0 = -8y^2$
$h_{xy} = -16xy + 14y$
$h_{yy} = -6 - 8x^2 + 14x$
$h_{yx} = 0 - 16xy + 14y = -16xy + 14y$

19. $f(x, y) = x^3y^{-1}$
$f_x = 3x^2y^{-1}$
$f_y = -x^3y^{-2}$
$f_{xx} = 6xy^{-1}$
$f_{xy} = -3x^2y^{-2}$
$f_{yy} = 2x^3y^{-3}$
$f_{yx} = -3x^2y^{-2}$

21. $f(x, y) = x^{-1} - 5y^{-2}$
$f_x = -x^{-2}$
$f_y = 10y^{-3}$
$f_{xx} = 2x^{-3} = \dfrac{2}{x^3}$
$f_{xy} = 0$
$f_{yy} = -30y^{-4} = -\dfrac{30}{y^4}$
$f_{yx} = 0$

23. $f(x, y) = e^{4y-3x}$
$f_x = -3e^{4y-3x}$
$f_y = 4e^{4y-3x}$
$f_{xx} = (-3)(-3e^{4y-3x}) = 9e^{4y-3x}$
$f_{xy} = (4)(-3e^{4y-3x}) = -12e^{4y-3x}$
$f_{yy} = (4)(4e^{4y-3x}) = 16e^{4y-3x}$
$f_{yx} = (-3)(4e^{4y-3x}) = -12e^{4y-3x}$

25. $z = 4xe^y$
$z_x = 4e^y, \quad z_y = 4xe^y$
$z_{xx} = 0, \quad z_{yy} = 4xe^y$
$z_{xy} = 4e^y, \quad z_{yx} = 4e^y$

27. $f(x, y) = \ln(3xy)$
$f_x = \frac{1}{3xy}(3y) = \frac{1}{x} = x^{-1}$
$f_y = \frac{1}{3xy}(3x) = \frac{1}{y} = y^{-1}$
$f_{xx} = -x^{-2} = -\frac{1}{x^2}$
$f_{xy} = 0$
$f_{yy} = -y^{-2} = -\frac{1}{y^2}$
$f_{yx} = 0$

Copyright © 2019 Pearson Education, Inc.

29. $f(x, y) = 3x(y-1)^{-1}$

$f_x = 3(y-1)^{-1}$

$f_y = -3x(y-1)^{-2}$

$f_{xx} = 0$

$f_{xy} = -3(y-1)^{-2} = \frac{-3}{(y-1)^2}$

$f_{yy} = 6x(y-1)^{-3} = \frac{6x}{(y-1)^3}$

$f_{yx} = -3(y-1)^{-2} = \frac{-3}{(y-1)^2}$

31. $z = x\ln(xy) = x(\ln x + \ln y) = x\ln x + x\ln y$

$z_x = x\left(\frac{1}{x}\right) + 1(\ln x) + \ln y = 1 + \ln x + \ln y$

$z_{xx} = 0 + \frac{1}{x} + 0 = \frac{1}{x}$

$z_y = 0 + x\left(\frac{1}{y}\right) = \frac{x}{y}$

$z_{xy} = 0 + 0 + \frac{1}{y} = \frac{1}{y}$

$z_{yy} = -\frac{x}{y^2}$

33. $f(x, y) = x\ln(xy)$

From exercise 31, $f_{xy} = \frac{1}{y}$ and $f_{yy} = -\frac{x}{y^2}$.

Therefore, $f_{xy}(1, 2) = \frac{1}{2}$ and

$f_{yy}(3, 1) = -\frac{3}{1^2} = -3.$

35. $f(x, y, z) = x^2 + yz + z^4$

$f_x = 2x$

$f_y = z$

$f_z = y + 4z^3$

$f_{yz} = 1$

$f_y(3, 4, -2) = -2$

$f_{yz}(1, -1, 0) = 1$

37. $f(x, y, z) = 9ye^{3xz^4}$

$f_x = 9ye^{3xz^4}(3z^4) = 27yz^4e^{3xz^4}$

$f_y = 9e^{3xz^4}$

$f_z = 9ye^{3xz^4}(12xz^3) = 108xyz^3e^{3xz^4}$

$f_{yz} = 9e^{3xz^4}(12xz^3) = 108xz^3e^{3xz^4}$

39. A function with three independent variables has three partial derivatives. For a function $f(x, y, z)$, the 3 partial derivatives are f_x, f_y and f_z. In the second-order partial derivatives, each of the two subscripts may be x, y, or z, so there will be $3 \cdot 3 = 9$ possibilities. For a function $f(x, y, z)$, the 9 second-order partial derivatives are f_{xx}, f_{xy}, f_{xz}, f_{yx}, f_{yy}, f_{yz}, f_{zx}, f_{zy} and f_{zz}.

41. $C(x, y) = 500x + 1100y + 20{,}000$

a. $C_x(x, y) = 500$ The marginal cost of producing one additional aluminum frame when the production of carbon fiber models is constant is \$500.

b. $C_y(x, y) = 1100$ The marginal cost of producing one additional carbon fiber frame when the production of aluminum models is constant is \$1100.

43. $C(x, y) = 600x + 120y + 4000$

a. In $C(x, y)$, 4000 represents the fixed costs, that is, the cost when no items have been produced.

b. $\frac{\partial C}{\partial x} = \$600$, $\frac{\partial C}{\partial y} = \$120$

$\frac{\partial C}{\partial x}$ is the marginal cost of producing 1 unit of chargers when the production of covers remains constant. $\frac{\partial C}{\partial y}$ is the marginal cost of producing 1 unit of covers when the production of chargers remains constant.

45. $P(x, y) = 3x^2 + 4y^2 - 4xy$, where x is units of tranquilizer and y is units of antibiotic.

a. $P_x(x, y) = 6x - 4y$

$P_x(9, 5) = 6(9) - 4(5) = 34$

The approximate change in profit if x increases from 9 to 10 while y remains constant is \$34.

Copyright © 2019 Pearson Education, Inc.

b. $P_y(x,y)=8y-4x$
$P_y(9,5)=8(5)-4(9)=4$
The approximate change in revenue if y increases from 5 to 6 while x remains constant is \$4.

47. $f(p,i)=99p-.5pi-.0025p^2$

a. If $p=19,400$ and $i=8$, then
$$f(19400,8) = 99(19,400)-.5(19,400)(8) -.0025(19,400)^2 = 1,920,600-77,600-940,900 = 902,100$$
The weekly sales are \$902,100.

b. $f(p,i)=99p-.5pi-.0025p^2$
$f_p(p,i)=99-.5i-.005p$
$f_i(p,i)=-.5p$
$f_p(p,i)$ is the rate at which weekly sales are changing per unit change in price while the interest rate remains constant.
$f_i(p,i)$ is the rate at which weekly sales are changing per unit change in interest rate while price remains constant.

c. $f_i(19400,8)=-.5(19400)=-9700$
Changing the interest rate from 8% to 9% would decrease weekly sales by \$9700.

49. $A=P(1+i)^n$

a. Positive; $A_p=(1+i)^n$

b. Positive; $A_i=nP(1+i)^{n-1}$

51. $A=Pe^{rt}$

a. Positive; $A_p=e^{rt}$

b. Positive; $A_r=tPe^{rt}$

c. Positive; $A_t=rPe^{rt}$

53. $f(x,y)=3x^{1/3}y^{2/3}$, where x is labor and y is capital.

a. $f_x=x^{-2/3}y^{2/3}=\dfrac{y^{2/3}}{x^{2/3}}=\left(\dfrac{y}{x}\right)^{2/3}$
$$f_x(64,125)=\left(\frac{125}{64}\right)^{2/3}=\frac{25}{16}=1.5625$$
which is the approximate change in production (in thousands of units) for a 1-unit change in labor.
$$f_y=2x^{1/3}y^{-1/3}=2\left(\frac{x^{1/3}}{y^{1/3}}\right)=2\left(\frac{x}{y}\right)^{1/3}$$
$$f_y(64,125)=2\left(\frac{64}{125}\right)^{1/3}=\frac{8}{5}=1.6$$
which is the approximate change in production (in thousands of units) for a 1-unit change in capital.

b. An increase of approximately $\frac{25}{16}(1000)\approx 1563$ batteries would result.

c. An increase of approximately $\frac{8}{5}(1000)=1600$ batteries would be the effect of increasing capital to 126 while holding labor at 64. This is the better option.

55. $z=x^{.7}y^{.3}$, where x is labor, y is capital.
Marginal productivity of labor is
$$\frac{\partial z}{\partial x}=.7x^{-.3}y^{.3}=\frac{.7y^{.3}}{x^{.3}}.$$
Marginal productivity of capital is
$$\frac{\partial z}{\partial y}=.3x^{.7}y^{-.7}=\frac{.3x^{.7}}{y^{.7}}.$$

57. a. Let $w=220$ and $h=74$.
$$B=\frac{703(220)}{(74)^2}\approx 28.24$$

b. $$\frac{\partial B}{\partial w}=\frac{703}{h^2}$$
$$\frac{\partial B}{\partial h}=-2(703w)h^{-3}=\frac{-1406w}{h^3}$$

Copyright © 2019 Pearson Education, Inc.

59. $A(M,H) = .202M^{.425}H^{.725}$

a. $A_M(M,H) = .202(.425)M^{-.575}H^{.725}$
$= .08585M^{-.575}H^{.725}$
$A_M(72,1.8) = .08585(72)^{-.575}(1.8)^{.725}$
$= .0112\,\text{m}^2/\text{g}$

b. $A_H(M,H) = .202M^{.425}(.725)H^{-.275}$
$A_H(70,1.6) = .14645M^{.425}H^{-.275}(0.1)$
$= .14645(70)^{.425}(1.6)^{-.275}(.01)$
$= .0783\,\text{m}^2/\text{m}$

61. a. $p(33,\ 10) = 1 + \frac{33}{33}\left(1 - e^{-.1386(10)}\right) \approx 1.75$ atm

b. Find $p_t(l,\ t)$.

$$p(l,\ t) = 1 + \frac{l}{33}\left(1 - e^{-.1386t}\right)$$
$$p_l(l,\ t) = \frac{1}{33} - \frac{e^{-.1386t}}{33}$$
$$p_l(33,\ 10) = \frac{1}{33} - \frac{e^{-.1386(10)}}{33} \approx .0227$$
$$p_t(l,\ t) = -\frac{l}{33}(-.1386)e^{-.1386t} = \frac{.1386l}{33}e^{-.1386t}$$
$$p_t(33,\ 10) = \frac{.1386(33)}{33}e^{-.1386(10)} \approx .0347$$

c. Let $l = 66$ feet and solve for t.

$2.15 = 1 + \frac{66}{33}\left(1 - e^{-.1386t}\right) \Rightarrow$
$1.15 = 2\left(1 - e^{-.1386t}\right) \Rightarrow -.85 = -2e^{-.1386t} \Rightarrow$
$.425 = e^{-.1386t} \Rightarrow \ln .425 = -.1386t \Rightarrow$
$t \approx 6.17$ minutes

Section 14.3 Extrema of Functions of Several Variables

1. $f(x,\ y) = 6x^2 + 6y^2 + 6xy + 36x - 5$
$f_x = 12x + 0 + 6y + 36 - 0 = 12x + 6y + 36$
$f_y = 0 + 12y + 6x - 0 - 0 = 6x + 12y$

Set each of these two partial derivatives equal to zero, forming a system of linear equations.

$12x + 6y + 36 = 0 \quad (1)$
$6x + 12y \quad = 0 \quad (2)$

To solve this system by the elimination method, rewrite equation (1) in the form of $ax + by = c$ and multiply equation (2) by –2; then add the results.

$$\begin{aligned} 12x + 6y &= -36 \quad (1) \\ -12x - 24y &= 0 \quad (2) \\ \hline -18y &= -36 \Rightarrow y = 2 \end{aligned}$$

Substitute 2 for y in equation (2).
$6x + 12(2) = 0 \Rightarrow 6x = -24 \Rightarrow x = -4$
The critical point is (–4, 2).

3. $f(x,\ y) = 9xy - x^3 - y^3 - 6$
$f_x = 9y - 3x^2 - 0 - 0 = -3x^2 + 9y$
$f_y = 9x - 0 - 3y^2 - 0 = 9x - 3y^2$

Set each of these two partial derivatives equal to zero, forming a system of linear equations.

$-3x^2 + 9y = 0 \quad (1)$
$9x - 3y^2 = 0 \quad (2)$

To solve this system by the substitution method, solve equation (2) for x and substitute the result in equation (1).

$$x = \frac{1}{3}y^2$$
$$9y - 3\left(\frac{1}{3}y^2\right)^2 = 0$$
$$9y - \frac{1}{3}y^4 = 0$$
$$\frac{1}{3}y(27 - y^3) = 0$$
$$\frac{1}{3}y(3 - y)(9 + 3y + y^2) = 0$$
$$y = 0 \text{ and } y = 3$$

Substitute 0 and 3 for y in equation (2).
$9x - 3(0)^2 = 0 \Rightarrow 9x = 0 \Rightarrow x = 0$
$9x - 3(3)^2 = 0 \Rightarrow 9x = 27 \Rightarrow x = 3$
The critical points are (0, 0) and (3, 3).

Copyright © 2019 Pearson Education, Inc.

5. $f(x, y) = 2x^2 + 4xy + 6y^2 - 8x - 10$
$f_x = 4x + 4y + 0 - 8 - 0 = 4x + 4y - 8$
$f_y = 0 + 4x + 12y - 0 - 0 = 4x + 12y$
Set each of these two partial derivatives equal to zero, forming a system of linear equations.
$4x + 4y - 8 = 0$ (1)
$4x + 12y = 0$ (2)
To solve this system by the elimination method, rewrite equation (1) in the form of $ax + by = c$ and multiply equation (2) by –1; then add the results.
$4x + 4y = 8$ (1)
$-4x - 12y = 0$ (2)
$-8y = 8 \Rightarrow y = -1$
Substitute –1 for y in equation (2).
$4x + 12(-1) = 0 \Rightarrow 4x = 12 \Rightarrow x = 3$
The critical point is (3, –1).
$f_{xx} = 4,\ f_{yy} = 12,$ and $f_{xy} = 4$
$$M = f_{xx}(3, -1) \cdot f_{yy}(3, -1) - \left[f_{xy}(3, -1)\right]^2$$
$$= 4 \cdot 12 - 4^2 = 32$$
Since $M > 0$ and $f_{xx}(3, -1) > 0$, we have a local minimum at (3, –1).
$$f(3, -1) = 2(3)^2 + 4(3)(-1) + 6(-1)^2 - 8(3) - 10$$
$$= 18 - 12 + 6 - 24 - 10 = -22$$
There is a local minimum of –22 at (3, –1).

7. $f(x, y) = x^2 - xy + y^2 + 2x + 2y + 6$
$f_x = 2x - y + 0 + 2 + 0 + 0 = 2x - y + 2$
$f_y = 0 - x + 2y + 0 + 2 + 0 = -x + 2y + 2$

Set $f_x = 0$ and $f_y = 0$. Solve the resulting system of linear equations by the elimination method.
$$\begin{matrix} 2x - y + 2 = 0 \\ -x + 2y + 2 = 0 \end{matrix} \Rightarrow \begin{matrix} 2x - y + 2 = 0 \\ -2x + 4y + 4 = 0 \end{matrix} \Rightarrow$$

$3y + 6 = 0 \Rightarrow 3y = -6 \Rightarrow y = -2$
$-x + 2(-2) + 2 = 0 \Rightarrow x = -2$
The critical point is (–2, –2).
$f_{xx} = 2,\ f_{yy} = 2,$ and $f_{xy} = -1$
$$M = f_{xx}(-2, -2) \cdot f_{yy}(-2, -2) - \left[f_{xy}(-2, -2)\right]^2$$
$$= 2 \cdot 2 - (-1)^2 = 3$$
Since $M > 0$ and $f_{xx}(-2, -2) > 0$, we have a local minimum at (–2, –2).
$$f(-2, -2) = (-2)^2 - (-2)(-2) + (-2)^2 + 2(-2) + 2(-2) + 6$$
$$= 4 - 4 + 4 - 4 - 4 + 6 = 2$$
There is a local minimum of 2 at (–2, –2).

9. $f(x, y) = 3xy + 6y - 5x$
$f_x = 3y - 5;\ f_y = 3x + 6$
Solve the system of equations.
$$\begin{matrix} 3y - 5 = 0 \\ 3x + 6 = 0 \end{matrix} \Rightarrow y = \frac{5}{3},\ x = -2$$
The critical point is $\left(-2, \frac{5}{3}\right)$.
$f_{xx} = 0$, $f_{yy} = 0$, and $f_{xy} = 3$
$$M = f_{xx}\left(-2, \frac{5}{3}\right) \cdot f_{yy}\left(-2, \frac{5}{3}\right) - \left[f_{xy}\left(-2, \frac{5}{3}\right)\right]^2$$
$$= 0 \cdot 0 - 3^2 = -9$$
Since $M < 0$, we have a saddle point at $\left(-2, \frac{5}{3}\right)$

11. $f(x, y) = 4xy - 10x^2 - 4y^2 + 8x + 8y + 9$
$f_x = 4y - 20x + 8$
$f_y = 4x - 8y + 8$
Solve the system of equations.
$$\begin{matrix} 4y - 20x + 8 = 0 \\ 4x - 8y + 8 = 0 \end{matrix} \Rightarrow$$
$4y - 20x + 8 = 0$ (1)
$-4y + 2x + 4 = 0$ (2)
$-18x + 12 = 0 \Rightarrow x = \frac{2}{3}$
$$4y - 20\left(\frac{2}{3}\right) + 8 = 0 \Rightarrow y = \frac{4}{3}$$
The critical point is $\left(\frac{2}{3}, \frac{4}{3}\right)$.
$f_{xx} = -20,\ f_{yy} = -8,\ f_{xy} = 4$
For $\left(\frac{2}{3}, \frac{4}{3}\right)$, $M = (-20)(-8) - 16 = 144 > 0$.
Since $f_{xx} < 0$, a local maximum occurs at $\left(\frac{2}{3}, \frac{4}{3}\right)$.

Copyright © 2019 Pearson Education, Inc.

$$f\left(\frac{2}{3}, \frac{4}{3}\right) = 4\left(\frac{2}{3}\right)\left(\frac{4}{3}\right) - 10\left(\frac{2}{3}\right)^2 - 4\left(\frac{4}{3}\right)^2 + 8\left(\frac{2}{3}\right) + 8\left(\frac{4}{3}\right) + 9 = 17$$

There is a local maximum of 17 at $\left(\frac{2}{3}, \frac{4}{3}\right)$.

13. $f(x, y) = x^2 + xy + y^2 - 3x - 5$

$f_x = 2x + y - 3$

$f_y = x + 2y$

Solve the system of equations.

$$\begin{aligned} 2x + \ y - 3 &= 0 \\ x + 2y \quad &= 0 \end{aligned} \Rightarrow$$

$$\begin{aligned} 2x + \ y &= 3 \quad (1) \\ -2x - 4y &= 0 \quad (2) \\ \hline -3y &= 3 \Rightarrow y = -1 \end{aligned}$$

Substitute -1 for y in equation (1) to find x.

$2x + (-1) = 3 \Rightarrow 2x = 4 \Rightarrow x = 2$

Here $f_{xx} = 2,\ f_{yy} = 2$, and $f_{xy} = 1$.

$M = 2(2) - 1^2 = 3 > 0$ and $f_{xx} > 0$.

There is a local minimum at $(2, -1)$.

$f(2, -1) = (2)^2 + 2(-1) + (-1)^2 - 3(2) - 5 = -8$

There is a local minimum of -8 at $(2, -1)$.

15. $f(x, y) = x^2 - y^2 - 2x + 4y - 7$

$f_x = 2x - 2$

$f_y = -2y + 4$

If $2x - 2 = 0,\ x = 1$.

If $-2y + 4 = 0,\ y = 2$.

The critical point is $(1, 2)$.

$f_{xx} = 2,\ f_{yy} = -2,\ f_{xy} = 0$

For $(1, 2)$, $M = -4 - 0 < 0$.

There is a saddle point at $(1, 2)$.

17. $f(x, y) = 2x^3 + 2y^2 - 12xy + 15$

$f_x = 6x^2 + 0 - 12y + 0 = 6x^2 - 12y$

$f_y = 0 + 4y - 12x + 0 = 4y - 12x$

Setting $f_x = 0$ and $f_y = 0$, we obtain the following nonlinear system.

$6x^2 - 12y = 0 \quad (1)$

$4y - 12x = 0 \quad (2)$

We will solve the system by the substitution method. Simplify equation (1) by dividing both sides by 6. Solve equation (2) for y.

$x^2 - 2y = 0 \quad (3)$

$y = 3x \quad (4)$

Now substitute $3x$ for y in equation (3) and solve the resulting equation for x.

$x^2 - 2(3x) = 0 \Rightarrow x(x - 6) = 0 \Rightarrow x = 0, 6$

If $x = 0,\ y = 0$. If $x = 6,\ y = 18$.

There are two critical points: $(0, 0)$ and $(6, 18)$.

$f_{xx} = 12x,\ f_{yy} = 4,$ and $f_{xy} = -12$

At $(0, 0)$,

$$M = f_{xx}(0, 0) \cdot f_{yy}(0, 0) - \left[f_{xy}(0, 0)\right]^2 = 0(4) - (-12)^2 = -144$$

Since $M < 0$, there is a saddle point at $(0, 0)$.

At $(6, 18)$,

$$M = f_{xx}(6, 18) \cdot f_{yy}(6, 18) - \left[f_{xy}(6, 18)\right]^2 = 72(4) - (-12)^2 = 144$$

Since $M > 0$ and $f_{xx}(6, 18) > 0$, there is a local minimum at $(6, 18)$.

$f(6, 18) = 2(6)^3 + 2(18)^2 - 12(6)(18) + 15 = -201$

There is a local minimum of -201 at $(6, 18)$.

Copyright © 2019 Pearson Education, Inc.

19. $f(x, y) = x^2 + 4y^3 - 6xy - 1$
$f_x = 2x - 6y$
$f_y = 12y^2 - 6x$
As in Exercise 15, setting $f_x = 0$ and $f_y = 0$ yields a nonlinear system of equations which we can solve by the substitution method.

$$\begin{matrix} 2x - 6y = 0 \\ 12y^2 - 6x = 0 \end{matrix} \Rightarrow \begin{matrix} x = 3y \\ 2y^2 - x = 0 \end{matrix} \Rightarrow$$

$2y^2 - 3y = 0 \Rightarrow y(2y-3) = 0 \Rightarrow y = 0, \frac{3}{2}$

If $y = 0, x = 0$. If $y = \frac{3}{2}, x = \frac{9}{2}$.

There are two critical points, $(0, 0)$ and $\left(\frac{9}{2}, \frac{3}{2}\right)$.

$f_{xx} = 2$, $f_{yy} = 24y$, and $f_{xy} = -6$
At $(0, 0)$,

$$M = f_{xx}(0, 0) \cdot f_{yy}(0, 0) - \left[f_{xy}(0, 0)\right]^2$$
$$= 2(0) - (-6)^2 = -36$$

Since $M < 0$, there is a saddle point at $(0, 0)$.

At $\left(\frac{9}{2}, \frac{3}{2}\right)$,

$$M = f_{xx}\left(\frac{9}{2}, \frac{3}{2}\right) \cdot f_{yy}\left(\frac{9}{2}, \frac{3}{2}\right) - \left[f_{xy}\left(\frac{9}{2}, \frac{3}{2}\right)\right]^2$$
$$= 2\left(24 \cdot \frac{3}{2}\right) - (-6)^2 = 36$$

Since $M > 0$ and $f_{xx}\left(\frac{9}{2}, \frac{3}{2}\right) > 0$, there is a local minimum at $\left(\frac{9}{2}, \frac{3}{2}\right)$.

$$f\left(\frac{9}{2}, \frac{3}{2}\right) = \left(\frac{9}{2}\right)^2 + 4\left(\frac{3}{2}\right)^3 - 6\left(\frac{9}{2}\right)\left(\frac{3}{2}\right) - 1$$
$$= -\frac{31}{4}$$

There is a local minimum of -7.75 at $(4.5, 1.5)$.

21. $f(x, y) = e^{xy}$
$f_x = ye^{xy}$, $f_y = xe^{xy}$
Solve the system of equations.

$$\begin{matrix} ye^{xy} = 0 \\ xe^{xy} = 0 \end{matrix} \Rightarrow ye^{xy} = xe^{xy} \Rightarrow x = y = 0$$

The critical point is $(0, 0)$.
$f_{xx} = y^2e^{xy}$
$f_{yy} = x^2e^{xy}$
$f_{xy} = e^{xy} + xye^{xy}$
For $(0, 0)$,
$M = 0 \cdot 0 - \left(e^0\right)^2 = -1 < 0$.
There is a saddle point at $(0, 0)$.

23. $z = f(x, y) = -3xy + x^3 - y^3 + \frac{1}{8}$

$f_x = -3y + 3x^2$
$f_y = -3x - 3y^2$
Solve the system $f_x = 0$, $f_y = 0$.
$-3y + 3x^2 = 0$
$-3x - 3y^2 = 0$
Solve the first equation for y.
$-3y + 3x^2 = 0 \Rightarrow 3y = 3x^2 \Rightarrow y = x^2$
Substitute into the second equation and solve for x.

$-3x - 3\left(x^2\right)^2 = 0 \Rightarrow -3x - 3x^4 = 0 \Rightarrow$
$-3x\left(1 + x^3\right) = 0 \Rightarrow x = 0, -1$
Then, $y = 0$ or $y = 1$.
The critical points are $(0, 0)$ and $(-1, 1)$.
$f_{xx} = 6x$, $f_{yy} = -6y$, $f_{xy} = -3$
For $(0, 0)$,

$$M = f_{xx}(0, 0) \cdot f_{yy}(0, 0) - \left[f_{xy}(0, 0)\right]^2$$
$$= 0(0) - (-3)^2 = -9 < 0.$$

So $(0, 0)$ is a saddle point.

Copyright © 2019 Pearson Education, Inc.

For (–1, 1),

$M = f_{xx}(-1, 1) \cdot f_{yy}(-1, 1) - \left[f_{xy}(-1, 1) \right]^2$

$= (-6)(-6) - [-3]^2 = 36 - 9 = 27 > 0$

So (–1, 1) is a local maximum.

$f(-1, 1) = -3(-1)(1) + (-1)^3 - (1)^3 + \frac{1}{8} = \frac{9}{8}$

There is a relative maximum of $\frac{9}{8}$ or $1\frac{1}{8}$ at (–1, 1). This is graph (a).

25. $z = f(x, y) = y^4 - 2y^2 + x^2 - \frac{17}{16}$

$f_x = 2x$

$f_y = 4y^3 - 4y$

Solve the system $f_x = 0,\ f_y = 0$.

$2x = 0 \quad (1)$

$4y^3 - 4y = 0 \quad (2)$

Equation (1) gives $2x = 0 \Rightarrow x = 0$.
Equation (2) gives
$4y(y^2 - 1) = 0 \Rightarrow y = 0, 1, -1$.
The critical points are (0, 0), (0, 1), and (0, –1).

$f_{xx} = 2,\ f_{yy} = 12y^2 - 4,\ f_{xy} = 0$

For (0, 0), $M = (2)(-4) - 0^2 = -8 < 0$.
There is a saddle point at (0, 0).
For (0, 1), $M = (2)(8) - 0^2 = 16 > 0$.
$f_{xx} = 2 > 0$, so (0, 1) is a local minimum.
For (0, –1),

$M = (2)(8) - 0^2 = 16 > 0$

$f_{xx} = 2 > 0$

So (0, –1) is a local minimum.

$f(0, 1) = (1)^4 - 2(1)^2 + (0)^2 - \frac{17}{16} = -\frac{33}{16}$

$f(0, -1) = (-1)^4 - 2(-1)^2 + (0)^2 - \frac{17}{16} = -\frac{33}{16}$

A local minimum of $-\frac{33}{16}$ or $-2\frac{1}{16}$ occurs at (0, 1) and (0, –1). This is graph (b).

27. $f(x, y) = -x^4 + y^4 + 2x^2 - 2y^2 + \frac{1}{16}$

$f_x = -4x^3 + 4x$

$f_y = 4y^3 - 4y$

Solve $f_x = 0,\ f_y = 0$.

$-4x^3 + 4x = 0 \quad (1)$

$4y^3 - 4y = 0 \quad (2)$

$-4x^3 + 4x = 0 \Rightarrow -4x\left(x^2 - 1\right) = 0 \Rightarrow x = 0, \pm 1$

$4y^3 - 4y = 0 \Rightarrow 4y\left(y^2 - 1\right) = 0 \Rightarrow y = 0, \pm 1$

Because any of the 3 x-values can be paired with any of the 3 y-values, there are 9 critical points. The critical points are (0, 0), (0, –1), (0, 1), (–1, 0), (–1, –1), (–1, 1), (1, 0), (1, –1), and (1, 1).

$f_{xx} = -12x^2 + 4,\ f_{yy} = 12y^2 - 4,\ f_{xy} = 0$

For (0, 0),
$M = (4)(-4) - 0 = -16 < 0$: Saddle point
For (0, –1),
$M = (4)(8) - 0 = 32 > 0$ and $f_{xx} = 4$: Local minimum
For (0, 1),
$M = (4)(8) - 0 = 32 > 0$ and $f_{xx} = 4$: Local minimum
For (–1, 0),
$M = (-8)(-4) - 0 = 32 > 0$ and $f_{xx} = -8$:Local maximum
For (–1, –1), $M = (-8)(8) - 0 = -64 < 0$: Saddle point
For (–1, 1), $M = (-8)(8) - 0 = -64 < 0$: Saddle point
For (1, 0), $M = (-8)(-4) - 0 = 32 > 0$ and $f_{xx} = -8$: Local maximum
For (1, –1), $M = (-8)(8) - 0 = -64 < 0$: Saddle point
For (1, 1), $M = (-8)(8) - 0 = -64 < 0$: Saddle point

Copyright © 2019 Pearson Education, Inc.

The saddle points are at (0, 0), (–1, –1), (–1, 1), (1, –1), and (1, 1).
The local maxima are at (1, 0) and (–1, 0).
The local minima are at (0, 1) and (0, –1).

$f(1, 0) = -(1)^4 + (0)^4 + 2(1)^2 - 2(0)^2 + \frac{1}{16} = \frac{17}{16}$

$f(-1, 0) = -(-1)^4 + (0)^4 + 2(-1)^2 - 2(0)^2 + \frac{1}{16}$

$= \frac{17}{16}$

A local maximum of $\frac{17}{16}$ occurs at (1, 0) and (–1, 0).

$f(0, 1) = -(0)^4 + (1)^4 + 2(0)^2 - 2(1)^2 + \frac{1}{16} = -\frac{15}{16}$

$f(0, -1) = -(0)^4 + (-1)^4 + 2(0)^2 - 2(-1)^2 + \frac{1}{16}$

$= -\frac{15}{16}$

A local minimum of $-\frac{15}{16}$ occurs at (0, 1) and (0, –1). This is graph (e).

29. $L(x, y) = \frac{3}{2}x^2 + y^2 - 2x - 2y - 2xy + 68$, where x is the number of skilled hours and y is the number of semi-skilled hours.

$L_x = 3x - 2 - 2y;\ L_y = 2y - 2 - 2x$

$$\begin{array}{ll} 3x - 2 - 2y = 0 & (1) \\ \underline{-2x - 2 + 2y = 0} & (2) \\ x - 4 \quad = 0 & \Rightarrow x = 4 \end{array}$$

Substitute 4 for x in equation (2).

$-2x - 2 + 2y = 0 \Rightarrow -2(4) - 2 + 2y = 0 \Rightarrow y = 5$

$L_{xx} = 3,\ L_{yy} = 2,\ L_{xy} = -2$

$M = 3(2) - (-2)^2 = 2 > 0$ and $L_{xx} > 0$.

The local minimum at (4, 5) is

$L(4, 5) = \frac{3}{2}(4)^2 + (5)^2 - 2(4) - 2(5)$
$\quad - 2(4)(5) + 68 = 59$

So the minimum labor charge was \$59 when $x = 4$ and $y = 5$.

31. $P(x, y) = 600 - 2x^3 + 6xy - y^2$

$P_x = 0 - 6x^2 + 6y - 0 = -6x^2 + 6y$

$P_y = 0 - 0 + 6x - 2y = 6x - 2y$

Set $P_x = 0$ and $P_y = 0$. Solve the resulting nonlinear system by the substitution method.

$$\begin{array}{l} -6x^2 + 6y = 0 \\ 6x - 2y = 0 \end{array} \Rightarrow \begin{array}{l} x^2 - y = 0 \\ y = 3x \end{array} \Rightarrow$$

$x^2 - (3x) = 0 \Rightarrow x(x - 3) = 0 \Rightarrow x = 0, 3$

If $x = 0, y = 0$. If $x = 3, y = 9$.

The critical points are (0, 0) and (3, 9). We ignore (0, 0) since it makes no sense in the problem.

$P_{xx} = -12x,\ P_{yy} = -2$, and $P_{xy} = 6$

At (3, 9),

$M = P_{xx}(3, 9) \cdot P_{yy}(3, 9) - \left[P_{xy}(3, 9)\right]^2$
$= (-36)(-2) - (6)^2 = 36$

Since $M > 0$ and $P_{xx}(3, 9) < 0$, there is a local maximum at (3, 9).

$P(3, 9) = 800 - 2(3)^3 + 12(3)(9) - (9)^2$
$= 627$

The maximum profit of \$627 per day occurs when the cost of a unit of a gallon of gas is \$3 and the cost of a unit of labor is \$9.

33. $C(x, y) = 2x^2 + 3y^2 - 2xy + 2x - 126y + 3800$

$C_x = 4x - 2y + 2;\ C_y = 6y - 2x - 126$

$$\begin{array}{ll} 4x - 2y + 2 = 0 & (1) \\ 6y - 2x - 126 = 0 & (2) \end{array} \Rightarrow \begin{array}{l} 2x - y + 1 = 0 \\ -2x + 6y - 126 = 0 \end{array} \Rightarrow$$

$5y - 125 = 0 \Rightarrow y = 25$

Substitute 25 for y in equation (1).

$4x - 2(25) + 2 = 0 \Rightarrow x = 12$

Critical point (12, 25).

$C_{xx} = 4,\ C_{yy} = 6,\ C_{xy} = -2$

For (12, 25), $M = 4(6) - 4 = 20 > 0$.

Since $C_{xx} > 0$, 12 units of electrical tape and 25 units of packing tape should be produced to yield a minimum cost.

$C(12, 25) = 2(12)^2 + 3(25)^2 - 2(12 \cdot 25)$
$\quad + 2(12) - 126(25) + 3800$
$= 2237$

The minimum cost is 2237.

Copyright © 2019 Pearson Education, Inc.

35. The volume is $V = xyz = 27$, so $z = \frac{27}{xy}$.

The surface area is

$$S = 2xy + 2yz + 2xz$$
$$= 2xy + 2y\left(\frac{27}{xy}\right) + 2x\left(\frac{27}{xy}\right)$$
$$= 2xy + \frac{54}{x} + \frac{54}{y}$$

$$S_x = 2y - \frac{54}{x^2},\ S_y = 2x - \frac{54}{y^2}$$

Let $S_x = 0$ and $S_y = 0$ to obtain the following system.

$$2y - \frac{54}{x^2} = 0 \quad (1)$$

$$2x - \frac{54}{y^2} = 0 \quad (2)$$

Solve equation (1) for y.

$$2y = \frac{54}{x^2} \Rightarrow y = \frac{27}{x^2}$$

Substitute this expression for y into equation (2).

$$2x - \frac{54}{\left(\frac{27}{x^2}\right)^2} = 0 \Rightarrow x = \frac{27}{\frac{(27)^2}{x^4}} \Rightarrow 27x = x^4 \Rightarrow$$

$$0 = x^4 - 27x \Rightarrow 0 = x\left(x^3 - 27\right) \Rightarrow x = 0, 3$$

If $x = 0$, $y = \frac{27}{0}$, which is undefined.

(If $x = 0$, there would be no box, so this is not relevant.)

If $x = 3$, $y = \frac{27}{x^2} = \frac{27}{3^2} = \frac{27}{9} = 3$.

Thus, the critical point is (3, 3).

$$S_{xx} = \frac{108}{x^3},\ S_{yy} = \frac{108}{y^3},\ S_{xy} = 2$$

$$M = (4)(4) - 2^2 = 12 > 0;\ S_{xx} = \frac{108}{27} = 4 > 0$$

A minimum surface area will occur when

$x = 3$, $y = 3$, and $z = \frac{27}{3(3)} = 3$.

The dimensions of the box will be 3 m by 3 m by 3 m.

37. $V = LWH$ and $2W + 2H + L = 108$

$$L = 108 - 2W - 2H$$
$$V(H, W) = (108 - 2W - 2H)WH$$
$$= 108WH - 2W^2H - 2WH^2$$

$$V_H = 108W - 2W^2 - 4HW$$
$$V_W = 108H - 4WH - 2H^2$$

$$\left.\begin{array}{l}108W - 2W^2 - 4HW = 0\\ 108H - 4WH - 2H^2 = 0\end{array}\right\} \Rightarrow$$

$$\left.\begin{array}{l}2W(54 - W - 2H) = 0\\ 2H(54 - 2W - H) = 0\end{array}\right\} \Rightarrow$$

$2W = 0$ or $54 - W - 2H = 0$

and

$2H = 0$ or $54 - 2W - H = 0$

Clearly $W = 0$ or $H = 0$ makes no sense in the problem, so we have the following system.

$$W + 2H = 54$$
$$2W + H = 54$$

Solve this system by the elimination method.

$$\begin{array}{r}-2W - 4H = -108\\ 2W + H = 54\\ \hline -3H = -54 \Rightarrow H = 18\end{array}$$

If $H = 18$, $2W + 18 = 54$ and $W = 18$. The only critical point we need consider is (18, 18).

$$V_{HH} = -4W,\ V_{WW} = -4H$$
$$V_{HW} = 108 - 4W - 4H$$

$$M = V_{HH}(18, 18) \cdot V_{WW}(18, 18) - \left[V_{HW}(18, 18)\right]^2$$
$$= (-72)(-72) - [108 - 4(18) - 4(18)]^2 = 3888$$

Since $M > 0$ and $V_{HH}(18, 18) < 0$, we have a local maximum at (18, 18).

$L = 108 - 2(18) - 2(18) = 36$

The dimensions of the box with maximum volume are 18 inches by 18 inches by 36 inches.

Copyright © 2019 Pearson Education, Inc.

39. $V = \pi r^2 h$ and girth $= 2\pi r + h$
Since $2\pi r + h = 108$, $h = 108 - 2\pi r$

$$V = \pi r^2(108 - 2\pi r)$$
$$V = 108\pi r^2 - 2\pi^2 r^3$$
$$\frac{dV}{dr} = 216\pi r - 6\pi^2 r^2$$
$$216\pi r - 6\pi^2 r^2 = 0 \Rightarrow 6\pi r(36 - \pi r) = 0 \Rightarrow$$
$$r = 0, \frac{36}{\pi}$$

Disregard $r = 0$ since it makes no sense in the problem.

If $r < \frac{36}{\pi}$, then $\frac{dV}{dr} > 0$.

If $r > \frac{36}{\pi}$, then $\frac{dV}{dr} < 0$.

There is a local maximum when $r = \frac{36}{\pi}$.

$$h = 108 - 2\pi\left(\frac{36}{\pi}\right) = 108 - 72 = 36$$

The circular tube with maximum volume has a radius of $\frac{36}{\pi}$ in (approximately 11.5 in) and a length of 36 in.

41. $P(x, y) = R(x, y) - C(x, y)$

$$= (2xy + 2y + 12) - (2x^2 + y^2)$$
$$= 2xy + 2y + 12 - 2x^2 - y^2$$
$$P_x(x, y) = 2y + 0 + 0 - 4x - 0$$
$$= 2y - 4x$$
$$P_y(x, y) = 2x + 2 + 0 - 0 - 2y$$
$$= 2x + 2 - 2y$$

$$\begin{array}{ll} 2y - 4x = 0 & (1) \\ 2x + 2 - 2y = 0 & (2) \end{array} \Rightarrow \begin{array}{ll} y = 2x & (3) \\ x - y = -1 & (4) \end{array}$$

Substitute $2x$ for y in equation (4).
$x - 2x = -1 \Rightarrow x = 1$
If $x = 1$, $y = 2$.
The critical point is (1, 2).
$P_{xx} = -4$, $P_{yy} = -2$, and $P_{xy} = 2$

$$M = P_{xx}(1, 2) \cdot P_{yy}(1, 2) - \left[P_{xy}(1, 2)\right]^2$$
$$= (-4)(-2) - (2)^2 = 4$$

Since $M > 0$ and $P_{xx}(1, 2) < 0$, we have a local maximum at (1, 2).

$$P(1, 2) = 2(1)(2) + 2(2) + 12 - 2(1)^2 - (2)^2 = 14$$

A maximum profit of $1400 occurs when 1000 tons of grade A and 2000 tons of grade B ore are used.

43. $P(x, y) = 36xy - x^3 - 8y^3$

$$P_x(x, y) = 36y - 3x^2$$
$$P_y(x, y) = 36x - 24y^2$$

$$\begin{array}{l} 36y - 3x^2 = 0 \\ 36x - 24y^2 = 0 \end{array} \Rightarrow \begin{array}{ll} y = \frac{x^2}{12} & (1) \\ 3x - 2y^2 = 0 & (2) \end{array}$$

Substitute $\frac{x^2}{12}$ for y in equation (2).

$$3x - 2\left(\frac{x^2}{12}\right)^2 = 0 \Rightarrow 3x - 2\left(\frac{x^4}{144}\right) = 0 \Rightarrow$$
$$3x - \frac{x^4}{72} = 0 \Rightarrow x\left(3 - \frac{x^3}{72}\right) = 0 \Rightarrow x = 0, 6$$

Disregard $x = 0$ since it makes no sense in the problem. If $x = 6$, then $y = \frac{6^2}{12} = \frac{36}{12} = 3$.

Consider the critical point $(6, 3)$.

$$P_{xx}(x, y) = -6x \Rightarrow P_{xx}(6, 3) = -6(6) = -36$$
$$P_{yy}(x, y) = -48y \Rightarrow P_{yy}(6, 3) = -48(3) = -144$$
$$P_{xy}(x, y) = 36 \Rightarrow P_{xy}(6, 3) = 36$$
$$M = (-36)\cdot(-144) - [36]^2 = 3888$$

Since $M > 0$ and $P_{xx} < 0$, there is a local maximum at (6, 3).

$$P(6, 3) = 36(6)(3) - (6)^3 - 8(3)^3 = 216.$$

Profit is a maximum of $216,000 when 6 tons of steel and 3 tons of aluminum are used.

Copyright © 2019 Pearson Education, Inc.

Section 14.4 Lagrange Multipliers

1. Maximize $f(x,y)=2xy$ subject to $x+y=20$.

The Lagrange function is:

$$F(x,y,\lambda)=f(x,y)-\lambda\cdot g(x,y)$$
$$=2xy-\lambda x-\lambda y+20\lambda$$

Compute the partial derivatives:

$$F_x(x,y,\lambda)=2y-\lambda$$
$$F_y(x,y,\lambda)=2x-\lambda$$
$$F_\lambda(x,y,\lambda)=-x-y+20$$

Set each partial derivative to zero to obtain the system:

$$2y-\lambda=0$$
$$2x-\lambda=0$$
$$-x-y+20=0$$

Then solve the system:

$$\left.\begin{array}{l}2y-\lambda=0\\2x-\lambda=0\end{array}\right\}\Rightarrow\left.\begin{array}{l}\lambda=2y\\\lambda=2x\end{array}\right\}\Rightarrow\begin{array}{l}2x=2y\\x=y\end{array}$$
$$-x-y+20=0$$
$$-x-x+20=0\Rightarrow-2x=-20\Rightarrow x=10$$

Since $y=x$, $y=10$ and $\lambda=2x=2\cdot10=20$.

So, the maximum value of $f(x,y)=2xy$ subject to $x+y=20$ occurs when $x=10$ and when $y=10$. The maximum value is $f(10,10)=2\cdot10\cdot10=200$.

3. Maximize $f(x,y)=xy^2$ subject to $x+4y=15$.

The Lagrange function is:

$$F(x,y,\lambda)=f(x,y)-\lambda\cdot g(x,y)$$
$$=xy^2-\lambda x-4\lambda y+15\lambda$$

Compute the partial derivatives:

$$F_x(x,y,\lambda)=y^2-\lambda$$
$$F_y(x,y,\lambda)=2xy-4\lambda$$
$$F_\lambda(x,y,\lambda)=-x-4y+15$$

Set each partial derivative to zero to obtain the system:

$$y^2-\lambda=0$$
$$2xy-4\lambda=0$$
$$-x-4y+15=0$$

Then solve the system:

$$\left.\begin{array}{l}y^2-\lambda=0\\2xy-4\lambda=0\end{array}\right\}\Rightarrow\left.\begin{array}{l}y^2=\lambda\\2xy=4\lambda\end{array}\right\}\Rightarrow\left.\begin{array}{l}4y^2=4\lambda\\2xy=4\lambda\end{array}\right\}\Rightarrow$$
$$4y^2-2xy=0\Rightarrow4y^2=2xy\Rightarrow2y=x$$
$$-x-4y+15=0\Rightarrow-2y-4y+15=0\Rightarrow$$
$$-6y=-15\Rightarrow y=\frac{5}{2}$$

Since $2y=x$, $x=5$ and $\lambda=y^2=\left(\frac{5}{2}\right)^2=\frac{25}{4}$.

So, the maximum value of $f(x,y)=xy^2$ subject to $x+4y=15$ occurs when $x=5$ and $y=\frac{5}{2}$. The maximum value is

$$f\left(5,\frac{5}{2}\right)=5\left(\frac{5}{2}\right)^2=\frac{125}{4}.$$

5. Minimize $f(x,y)=x^2+2y^2-xy$ subject to $x+y=8$. The Lagrange function is:

$$F(x,y,\lambda)=f(x,y)-\lambda\cdot g(x,y)$$
$$=x^2+2y^2-xy-x\lambda-y\lambda+8\lambda$$

Compute the partial derivatives:

$$F_x(x,y,\lambda)=2x-y-\lambda$$
$$F_y(x,y,\lambda)=4y-x-\lambda$$
$$F_\lambda(x,y,\lambda)=-x-y+8$$

Set each partial derivative to zero to obtain the system:

$$2x-y-\lambda=0$$
$$4y-x-\lambda=0$$
$$-x-y+8=0$$

Then solve the system:

$$\left.\begin{array}{l}\lambda=2x-y\\\lambda=4y-x\end{array}\right\}\Rightarrow2x-y=4y-x\Rightarrow3x=5y\Rightarrow$$
$$y=\frac{3}{5}x$$
$$-x-\frac{3}{5}x+8=0\Rightarrow-\frac{8}{5}x+8=0\Rightarrow x=5$$

Since $y=\frac{3}{5}x$, $y=\frac{3}{5}(5)=3$. So, the minimum value of $f(x,y)=x^2+2y^2-xy$ subject to $x+y=8$ occurs when $x=5$ and when $y=3$. The minimum value is

$$f(5,3)=5^2+2(3)^2-(5)(3)=28.$$

Copyright © 2019 Pearson Education, Inc.

7. Maximize $f(x,y)=x^2-10y^2$ subject to $x-y=18$. The Lagrange function is:

$$F(x,y,\lambda)=f(x,y)-\lambda\cdot g(x,y)$$
$$=x^2-10y^2-x\lambda+y\lambda+18\lambda$$

Compute the partial derivatives:

$$F_x(x,y,\lambda)=2x-\lambda$$
$$F_y(x,y,\lambda)=-20y+\lambda$$
$$F_\lambda(x,y,\lambda)=-x+y+18$$

Set each partial derivative to zero to obtain the system:

$$2x-\lambda=0$$
$$-20y+\lambda=0$$
$$-x+y+18=0$$

Then solve the system:

$$\left.\begin{array}{l}\lambda=2x\\ \lambda=20y\end{array}\right\}\Rightarrow \begin{array}{l}2x=20y\\ x=10y\end{array}$$

$$-(10y)+y+18=0\Rightarrow -9y+18=0\Rightarrow y=2$$

Since $x=10y$, $x=10(2)=20$. So, the maximum value of $f(x,y)=x^2-10y^2$ subject to $x-y=18$ occurs when $x=20$ and when $y=2$. The maximum value is

$$f(20,2)=20^2-10(2)^2=360.$$

9. Maximize $f(x,y)=e^{-x^2-y^2}$ subject to $x+y=2$. The Lagrange function is:

$$F(x,y,\lambda)=f(x,y)-\lambda\cdot g(x,y)$$
$$=e^{-x^2-y^2}-\lambda x-\lambda y+2\lambda$$

Compute the partial derivatives:

$$F_x(x,y,\lambda)=-2xe^{-x^2-y^2}-\lambda$$
$$F_y(x,y,\lambda)=-2ye^{-x^2-y^2}-\lambda$$
$$F_\lambda(x,y,\lambda)=-x-y+2$$

Set each partial derivative to zero to obtain the system:

$$-2xe^{-x^2-y^2}-\lambda=0$$
$$-2ye^{-x^2-y^2}-\lambda=0$$
$$-x-y+2=0$$

Then solve the system:

$$\left.\begin{array}{l}-2xe^{-x^2-y^2}-\lambda=0\\ -2ye^{-x^2-y^2}-\lambda=0\end{array}\right\}\Rightarrow \left.\begin{array}{l}\lambda=-2xe^{-x^2-y^2}\\ \lambda=-2ye^{-x^2-y^2}\end{array}\right\}$$

$$\Rightarrow \begin{array}{l}-2x=-2y\\ x=y\end{array}$$

$$-x-y+2=0$$
$$-x-x+2=0\Rightarrow -2x=-2\Rightarrow x=1$$

Since $y=x$, $y=1$. So, the maximum value of $f(x,y)=e^{-x^2-y^2}$ subject to $x+y=2$ occurs when $x=1$ and when $y=1$. The maximum value is $f(1,1)=e^{-1-1}=e^{-2}$.

11. Maximize $f(x,y,z)=xyz^2$ subject to $x+y+z=6$. The Lagrange function is:

$$F(x,y,z,\lambda)=f(x,y,z)-\lambda\cdot g(x,y,z)$$
$$=xyz^2-x\lambda-y\lambda-z\lambda+6\lambda$$

Compute the partial derivatives:

$$F_x(x,y,z,\lambda)=yz^2-\lambda$$
$$F_y(x,y,z,\lambda)=xz^2-\lambda$$
$$F_z(x,y,z,\lambda)=2xyz-\lambda$$
$$F_\lambda(x,y,z,\lambda)=-x-y-z+6$$

Set each partial derivative to zero to obtain the system:

$$yz^2-\lambda=0$$
$$xz^2-\lambda=0$$
$$2xyz-\lambda=0$$
$$-x-y-z+6=0$$

Then solve the system:

$$\left.\begin{array}{l}\lambda=yz^2\\ \lambda=xz^2\\ \lambda=2xyz\end{array}\right\}\Rightarrow \begin{array}{l}yz^2=xz^2\Rightarrow y=x\\ xz^2=2xyz\Rightarrow z=2y\end{array}$$

$$-y-y-2y+6=0\Rightarrow -4y=-6\Rightarrow y=\frac{3}{2}$$

Since $x=y$, $x=\frac{3}{2}$ and since $z=2y$,

$$z=2\left(\frac{3}{2}\right)=3.$$

So, the maximum value of $f(x,y,z)=xyz^2$ subject to $x+y+z=6$ occurs when $x=\frac{3}{2}$, $y=\frac{3}{2}$, and when $z=3$. The maximum value is

$$f\left(\frac{3}{2},\frac{3}{2},3\right)=\left(\frac{3}{2}\right)\left(\frac{3}{2}\right)(3)^2=\frac{81}{4}.$$

Copyright © 2019 Pearson Education, Inc.

13. Maximize $f(x,y)=2xy^2$ subject to $x+y=21$.
The Lagrange function is:
$F(x,y,\lambda)=f(x,y)-\lambda\cdot g(x,y)$
$=2xy^2-\lambda x-\lambda y+21\lambda$
Compute the partial derivatives:
$F_x(x,y,\lambda)=2y^2-\lambda$
$F_y(x,y,\lambda)=4xy-\lambda$
$F_\lambda(x,y,\lambda)=-x-y+21$
Set each partial derivative to zero to obtain the system:
$2y^2-\lambda=0$
$4xy-\lambda=0$
$-x-y+21=0$
Then solve the system:
$\left.\begin{array}{l}\lambda=2y^2\\ \lambda=4xy\end{array}\right\}\Rightarrow 2y^2=4xy\Rightarrow y=2x$
$-x-(2x)+21=0\Rightarrow -3x=-21\Rightarrow x=7$
Since $y=2x$, $y=2(7)=14$. So, the maximum value of $f(x,y)=2xy^2$ subject to $x+y=21$ occurs when $x=7$ and $y=14$. The maximum value is $f(7,14)=2(7)(14)^2=2744$.

15. Maximize $f(x,y,z)=xyz$ subject to $x+y+z=102$. The Lagrange function is:
$F(x,y,z,\lambda)=f(x,y,z)-\lambda\cdot g(x,y,z)$
$=xyz-\lambda x-\lambda y-\lambda z+102\lambda$
Compute the partial derivatives:
$F_x(x,y,z,\lambda)=yz-\lambda$
$F_y(x,y,z,\lambda)=xz-\lambda$
$F_z(x,y,z,\lambda)=xy-\lambda$
$F_\lambda(x,y,z,\lambda)=-x-y-z+102$
Set each partial derivative to zero to obtain the system:
$yz-\lambda=0$
$xz-\lambda=0$
$xy-\lambda=0$
$-x-y-z+102=0$
Then solve the system:
$\left.\begin{array}{l}\lambda=yz\\ \lambda=xz\\ \lambda=xy\end{array}\right\}\Rightarrow \begin{array}{l}x=y\\ y=z\end{array}$
$x=y=z$
$-x-x-x+102=0\Rightarrow x=34$
Since $x=y=z$, $y=34$ and $z=34$.

So, the maximum value of $f(x,y)=xyz$ subject to $x+y+z=102$ occurs when $x=34$, $y=34$, and $z=34$. The maximum value is $f(34, 34, 34)=(34)(34)(34)=39{,}304$.

17. Answers vary.

19. Answers vary.

21. Maximize $P(x,y)=-x^2-y^2+4x+8y$ subject to $x+y=6$. The Lagrange function is:
$F(x,y,\lambda)=f(x,y)-\lambda\cdot g(x,y)$
$=-x^2-y^2+4x+8y-\lambda x-\lambda y+6\lambda$
Compute the partial derivatives:
$F_x(x,y,\lambda)=-2x+4-\lambda$
$F_y(x,y,\lambda)=-2y+8-\lambda$
$F_\lambda(x,y,\lambda)=-x-y+6$
Set each partial derivative to zero to obtain the system:
$-2x+4-\lambda=0$ (1)
$-2y+8-\lambda=0$ (2)
$-x-y+6=0$ (3)

Then solve the system:
$\left.\begin{array}{l}\lambda=-2x+4\\ \lambda=-2y+8\end{array}\right\}\Rightarrow -2x+4=-2y+8\Rightarrow$
$-2x=-2y+4\Rightarrow x=y-2$
Using equation (3), we have
$-x-y+6=0\Rightarrow -(y-2)-y+6=0\Rightarrow y=4$
Since $x=y-2$, $x=4-2=2$. Profit is maximized when 2 automobile radiators and 4 generator radiators are sold. Maximum profit is
$P(2,4)=-(2)^2-(4)^2+4(2)+8(4)=\$20$.

Copyright © 2019 Pearson Education, Inc.

23. Maximize $f(x,y)=3x^{1/3}y^{2/3}$ subject to $80x+150y=40{,}000$. The Lagrange function is:

$$F(x,y,\lambda)=f(x,y)-\lambda\cdot g(x,y)$$
$$=3x^{1/3}y^{2/3}-80\lambda x-150\lambda y+40{,}000\lambda$$

Compute the partial derivatives:

$$F_x(x,y,\lambda)=x^{-2/3}y^{2/3}-80\lambda$$
$$F_y(x,y,\lambda)=2x^{1/3}y^{-1/3}-150\lambda$$
$$F_\lambda(x,y,\lambda)=-80x-150y+40{,}000$$

Set each partial derivative to zero to obtain the system:

$$x^{-2/3}y^{2/3}-80\lambda=0$$
$$2x^{1/3}y^{-1/3}-150\lambda=0$$
$$-80x-150y+40{,}000=0$$

Then solve the system:

$$\left.\begin{aligned}\lambda&=\frac{x^{-2/3}y^{2/3}}{80}\\ \lambda&=\frac{x^{1/3}y^{-1/3}}{75}\end{aligned}\right\}\Rightarrow\frac{x^{-2/3}y^{2/3}}{80}=\frac{x^{1/3}y^{-1/3}}{75}\Rightarrow$$

$$\Rightarrow\frac{y^{2/3}}{80x^{2/3}}=\frac{x^{1/3}}{75y^{1/3}}\Rightarrow 75y=80x\Rightarrow y=\frac{16}{15}x$$

$$-80x-150\left(\frac{16}{15}x\right)+40{,}000=0\Rightarrow$$
$$-80x-160x+40{,}000=0\Rightarrow$$
$$-240x=-40{,}000\Rightarrow x\approx 166.67$$

Since $y=\frac{16}{15}x$, $y\approx 177.78$. Maximum production occurs when approximately 166.67 units of labor and approximately 177.78 units of capital are expended.

25. Let x = width and y = length. Maximize $A(x,y)=xy$ subject to $4x+6y+6x+4y=1200\Rightarrow 10x+10y=1200$.

The Lagrange function is:

$$F(x,y,\lambda)=f(x,y)-\lambda\cdot g(x,y)$$
$$=xy-10\lambda x-10\lambda y+1200\lambda$$

Compute the partial derivatives:

$$F_x(x,y,\lambda)=y-10\lambda$$
$$F_y(x,y,\lambda)=x-10\lambda$$
$$F_\lambda(x,y,\lambda)=-10x-10y+1200$$

Set each partial derivative to zero to obtain the system:

$$y-10\lambda=0$$
$$x-10\lambda=0$$
$$-10x-10y+1200=0$$

Then solve the system:

$$\left.\begin{aligned}10\lambda&=y\\10\lambda&=x\end{aligned}\right\}\Rightarrow x=y$$

$$-10x-10x+1200=0\Rightarrow -20x=-1200\Rightarrow$$
$$x=60$$

Since $y=x$, $y=60$. So, the maximum area is produced when $x=60$ feet and $y=60$ feet.

The area is $A(60,60)=60\cdot 60=3600\text{ ft}^2$.

27. Let x = width and let y = length. Maximize $A(x,y)=xy$ subject to $2x+2y=200$.

The Lagrange function is:

$$F(x,y,\lambda)=f(x,y)-\lambda\cdot g(x,y)$$
$$=xy-2\lambda x-2\lambda y+200\lambda$$

Compute the partial derivatives:

$$F_x(x,y,\lambda)=y-2\lambda$$
$$F_y(x,y,\lambda)=x-2\lambda$$
$$F_\lambda(x,y,\lambda)=-2x-2y+200$$

Set each partial derivative to zero to obtain the system:

$$y-2\lambda=0$$
$$x-2\lambda=0$$
$$-2x-2y+200=0$$

Then solve the system:

$$\left.\begin{aligned}2\lambda&=x\\2\lambda&=y\end{aligned}\right\}\Rightarrow x=y$$

$$-2x-2x+200=0\Rightarrow -4x=-200\Rightarrow x=50$$

Since $y=x$, $y=50$. The maximum area of the rectangular field occurs when the length and width are both 50 meters, resulting in an area of 2500 m^2.

Copyright © 2019 Pearson Education, Inc.

29. Let h = height of cylinder and r = radius. Minimize the surface area,
$S(h,r) = 2\pi r^2 + 2hr\pi$ subject to
$V(h,r) = h\pi r^2 = 250\pi$.
The Lagrange function is:

$$F(h,r,\lambda) = f(h,r) - \lambda \cdot g(h,r)$$
$$= 2\pi r^2 + 2\pi hr - \pi\lambda hr^2 + 250\pi\lambda$$

Compute the partial derivatives:

$$F_h(h,r,\lambda) = 2\pi r - \pi r^2\lambda$$
$$F_r(h,r,\lambda) = 4\pi r + 2\pi h - 2\pi hr\lambda$$
$$F_\lambda(h,r,\lambda) = -\pi hr^2 + 250\pi$$

Set each partial derivative to zero to obtain the system:

$$2\pi r - \pi r^2\lambda = 0$$
$$4\pi r + 2\pi h - 2\pi hr\lambda = 0$$
$$-\pi hr^2 + 250\pi = 0$$

Then solve the system:

$$\left.\begin{array}{l} \pi r^2\lambda = 2\pi r \\ \lambda = \dfrac{2}{r} \\ 2\pi hr\lambda = 4\pi r + 2\pi h \\ hr\lambda = 2r + h \\ \lambda = \dfrac{2r+h}{hr} \end{array}\right\} \Rightarrow \frac{2}{r} = \frac{2r+h}{hr} \Rightarrow$$

$$2h = 2r + h \Rightarrow h = 2r$$
$$-\pi(2r)r^2 + 250\pi = 0 \Rightarrow -2r^3 = -250 \Rightarrow$$
$$r^3 = 125 \Rightarrow r = 5$$

Since $h = 2r$, $h = 2(5) = 10$. The cylindrical can holding 250π in^3 has a minimum surface area if the radius is 5 inches and the height is 10 inches.

31. Let l = length, w = width, and h = height of the box. Maximize volume, $V(l,w,h) = lwh$ subject to the constraint, surface area,
$S(l,w,h) = 2lh + 2wh + lw = 500$.
The Lagrange function is:

$$F(l,w,h,\lambda) = f(l,w,h) - \lambda \cdot g(l,w,h)$$
$$= lwh - 2lh\lambda - 2wh\lambda - lw\lambda + 500\lambda$$

Compute the partial derivatives:

$$F_l(l,w,h,\lambda) = wh - 2h\lambda - w\lambda$$
$$F_w(l,w,h,\lambda) = lh - 2h\lambda - l\lambda$$
$$F_h(l,w,h,\lambda) = lw - 2l\lambda - 2w\lambda$$
$$F_\lambda(l,w,h,\lambda) = -2lh - 2wh - lw + 500$$

Set each partial derivative to zero to obtain the system:

$$wh - 2h\lambda - w\lambda = 0$$
$$lh - 2h\lambda - l\lambda = 0$$
$$lw - 2l\lambda - 2w\lambda = 0$$
$$-2lh - 2wh - lw + 500 = 0$$

Then solve the system:

$$\left.\begin{array}{l} \lambda = \dfrac{wh}{2h+w} \\ \lambda = \dfrac{lh}{2h+l} \\ \lambda = \dfrac{lw}{2l+2w} \end{array}\right\} \Rightarrow \left.\begin{array}{l} \dfrac{wh}{2h+w} = \dfrac{lh}{2h+l} \\ \dfrac{lh}{2h+l} = \dfrac{lw}{2l+2w} \end{array}\right\} \Rightarrow$$

$$\left.\begin{array}{l} 2wh + wl = 2hl + lw \Rightarrow w = l \\ 2lh + 2wh = 2wh + wl \Rightarrow 2h = w \end{array}\right\} \Rightarrow w = l = 2h$$

$$-2(2h)h - 2(2h)(h) - (2h)(2h) + 500 = 0 \Rightarrow$$
$$-4h^2 - 4h^2 - 4h^2 = -500 \Rightarrow -12h^2 = -500 \Rightarrow$$
$$h^2 = \frac{125}{3} \Rightarrow h \approx 6.45$$

Since $w = l = 2h$, $w = l \approx 2(6.45) \approx 12.91$. To maximize the volume, the dimensions of the box should be about 12.91 m × 12.91 m × 6.45 m.

Copyright © 2019 Pearson Education, Inc.

33. Let l = length, w = width, and h = height of the aquarium. Minimize the surface area, $S(l, w, h) = 2lh + 2wh + lw$, subject to the constraint, $V(l, w, h) = lwh = 32$. The Lagrange function is:

$$F(l, w, h, \lambda) = f(l, w, h) - \lambda \cdot g(l, w, h)$$
$$= 2lh + 2wh + lw - \lambda lwh + 32\lambda$$

Compute the partial derivatives:

$$F_l(l, w, h, \lambda) = 2h + w - \lambda wh$$
$$F_w(l, w, h, \lambda) = 2h + l - \lambda lh$$
$$F_h(l, w, h, \lambda) = 2l + 2w - \lambda lw$$
$$F_\lambda(l, w, h, \lambda) = -lwh + 32$$

Set each partial derivative to zero to obtain the system:

$$2h + w - \lambda wh = 0$$
$$2h + l - \lambda lh = 0$$
$$2l + 2w - \lambda lw = 0$$
$$-lwh + 32 = 0$$

Then solve the system:

$$\left.\begin{array}{l} \lambda = \dfrac{2h + 2w}{wh} \\ \lambda = \dfrac{2h + l}{lh} \\ \lambda = \dfrac{2l + 2w}{lw} \end{array}\right\} \Rightarrow \left.\begin{array}{l} \dfrac{2h + w}{wh} = \dfrac{2h + l}{lh} \\ \dfrac{2h + l}{lh} = \dfrac{2l + 2w}{lw} \end{array}\right\} \Rightarrow$$

$$\left.\begin{array}{l} 2hl + wl = 2hw + lw \Rightarrow w = l \\ 2hw + lw = 2lh + 2wh \Rightarrow w = 2h \end{array}\right\} \Rightarrow l = w = 2h$$

$$-(2h)(2h)h = -32 \Rightarrow 4h^3 = 32 \Rightarrow$$
$$h^3 = 8 \Rightarrow h = 2$$

Since $w = 2h$, $w = 2(2) = 4$, and $l = 4$. The minimum surface area of the aquarium occurs when the dimensions are 4 ft × 4 ft × 2 ft.

35. a. Area of ends $= 2yz$.

Cost of ends $= (2yz)(20) = 40yz$.

Cost of rest of the box = 0.

Total cost to construct the box $= 0 + 40yz = 40yz$.

Volume of the box $= xyz$.

Number of trips $= \left(\dfrac{400}{xyz}\right)(0.1) = \dfrac{40}{xyz}$.

Total cost to construct box and transport material $= C(x, y, z) = \dfrac{40}{xyz} + 40yz$.

Constraint: $xy + 2xz = 4$.

b. To use the Solver feature of Excel, set up the initial spreadsheet as follows. Enter the formula for $f(x, y, z)$ in cell D2, and enter the formula for the constraint in cell E2.

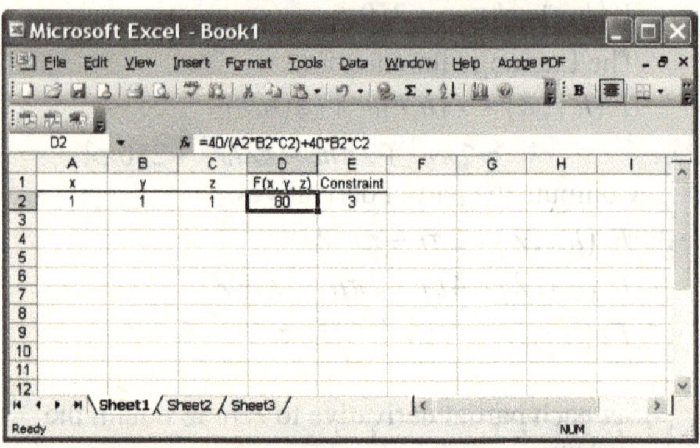

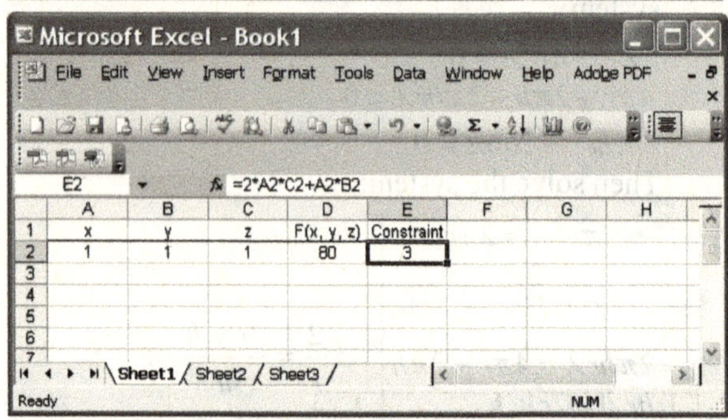

Set the Solver parameters as follows:

Solver Parameters

Set Target Cell: $D$2

Equal To: ○ Max ⦿ Min ○ Value of: 0

By Changing Cells:

$A$2:$C$2

Subject to the Constraints:

$E$2 = 4

Solve | Close | Guess | Options | Add | Change | Reset All | Delete | Help

Excel returns the following values.

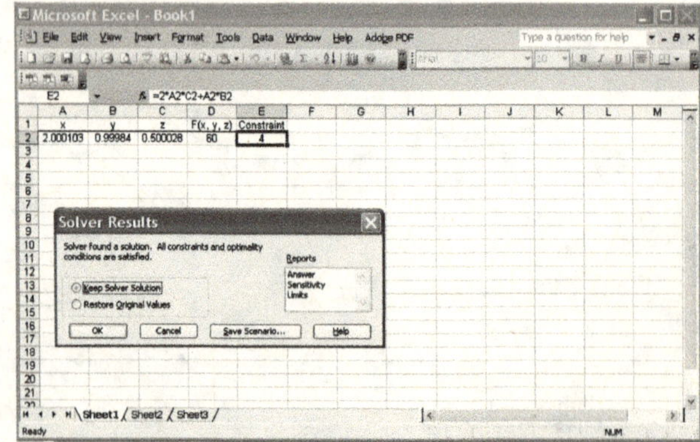

Copyright © 2019 Pearson Education, Inc.

To minimize the cost, the dimensions of the box should be 2 yd × 1 yd × .5 yd.

Chapter 14 Review Exercises

1. $f(x, y) = 6y^2 - 5xy + 2x$
$f(-3, 1) = 6(1)^2 - 5(-3)(1) + 2(-3) = 15$
$f(5, -2) = 6(-2)^2 - 5(5)(-2) + 2(5) = 84$

2. $f(x, y) = -3x + 2x^2y^2 + 5y$
$f(-3, 1) = -3(-3) + 2(-3)^2(1)^2 + 5(1) = 32$
$f(5, -2) = -3(5) + 2(5)^2(-2)^2 + 5(-2) = 175$

3. $f(x, y) = \dfrac{2x-4}{x+3y}$
$f(-3, 1) = \dfrac{2(-3)-4}{(-3)+3(1)} = \dfrac{-10}{0}$, undefined
$f(5, -2) = \dfrac{2(5)-4}{5+3(-2)} = \dfrac{6}{-1} = -6$

4. $f(x, y) = \dfrac{12y-6x}{x+3y}$
$f(-3, 1) = \dfrac{12(1)-6(-3)}{(-3)+3(1)} = \dfrac{30}{0}$, undefined
$f(5, -2) = \dfrac{12(-2)-6(5)}{5+3(-2)} = \dfrac{-54}{-1} = 54$

5. $f(x, y) = 7x\sqrt{y}$
$f(-3, 1) = 7(-3)\sqrt{1} = -21$
$f(5, -2) = 7(5)\sqrt{-2}$ = undefined

6. $f(x, y) = x\sqrt{x^2 + y^2}$
$f(-3, 1) = -3\sqrt{(-3)^2 + 1^2} = -3\sqrt{10}$
$f(5, -2) = 5\sqrt{5^2 + (-2)^2} = 5\sqrt{29}$

7. Answers vary. Sample answer:
The graph of $2x + y + 4z = 12$ is a plane passing through (6, 0, 0), (0, 12, 0), and (0, 0, 3).

8. Answers vary. Sample answer:
The graph of $y = 2$ is a plane parallel to the xz-plane passing through (0, 2, 0).

9. $x + 2y + 4z = 4$
Let $x = 0$ and $y = 0$. Then $z = 1$. The point (0, 0, 1) is on the graph. Let $x = 0$ and $z = 0$. Then $y = 2$. The point (0, 2, 0) is on the graph. Let $y = 0$ and $z = 0$. Then $x = 4$. The point (4, 0, 0) is on the graph. The graph is a plane. We sketch the portion in the first octant through these three points.

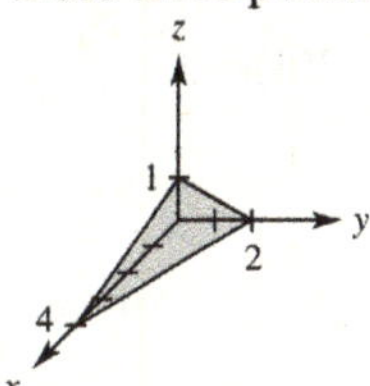

10. $3x + 2y = 6$
We let $x = 0$ and find $y = 3$. The point (0, 3, 0) is on the graph. We let $y = 0$ and find $x = 2$. The point (2, 0, 0) is on the graph. Since there is no z-intercept, the plane is parallel to the z-axis. We sketch the portion in the first octant through these two points.

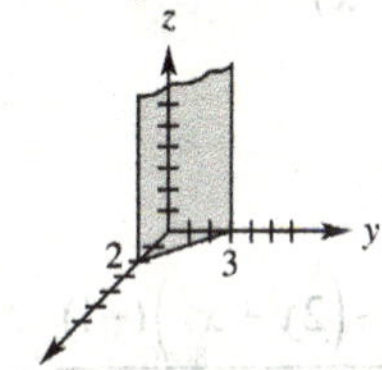

11. $4x + 5y = 20$
We let $x = 0$ and find $y = 4$. The point (0, 4, 0) is on the graph. We let $y = 0$ and find $x = 5$. The point (5, 0, 0) is on the graph. Since there is no z-intercept, the plane is parallel to the z-axis. We sketch the portion in the first octant through these two points.

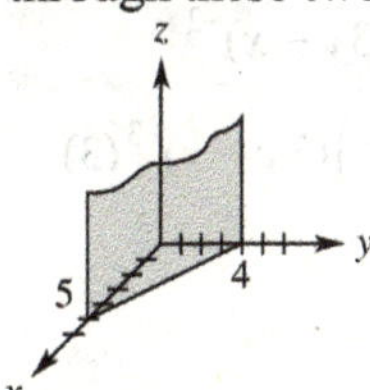

12. $x = 6$
The x-intercept is (6, 0, 0). Since there is no y-intercept or z-intercept, the plane is parallel to the yz-plane. We sketch the portion in the first octant through the point (6, 0, 0).

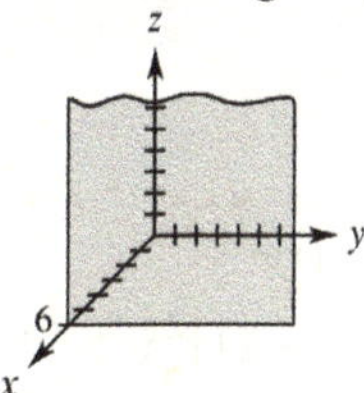

Copyright © 2019 Pearson Education, Inc.

13. $z = f(x, y) = -2x^2 + 5xy + y^2$

a. $\frac{\partial z}{\partial x} = -4x + 5y$

b. $\frac{\partial z}{\partial y} = 5x + 2y$

$\frac{\partial z}{\partial y}(-1, 4) = 5(-1) + 2(4) = 3$

c. $f_{xy} = 5$

$f_{xy}(2, -1) = 5$

14. $z = f(x, y) = \frac{2y + x^2}{3y - x}$

Use the quotient rule to find the partial derivatives.

a. $\frac{\partial z}{\partial y} = \frac{(3y - x)(2) - (2y + x^2)(3)}{(3y - x)^2}$

$= \frac{-2x - 3x^2}{(3y - x)^2}$

b. $\frac{\partial z}{\partial x} = \frac{(3y - x)(2x) - (2y + x^2)(-1)}{(3y - x)^2}$

$= \frac{6xy - x^2 + 2y}{(3y - x)^2}$

$\frac{\partial z}{\partial x}(0, 2) = \frac{6(0)(2) - (0)^2 + 2(2)}{[3(2) - 0]^2} = \frac{4}{36} = \frac{1}{9}$

c. $f_y = (-2x - 3x^2)(3y - x)^{-2}$

$f_{yy} = -2(-2x - 3x^2)(3y - x)^{-3}(3)$

$= \frac{6(2x + 3x^2)}{(3y - x)^3}$

$f_{yy}(-1, 0) = \frac{6[2(-1) + 3(-1)^2]}{[3(0) - (-1)]^3} = \frac{6(1)}{1} = 6$

15. Answers vary.

16. Answers vary.

17. $f(x, y) = 3y - 7x^2y^3$

$f_x = 0 - 14xy^3 = -14xy^3$

$f_y = 3 - 7 \cdot 3x^2y^2 = 3 - 21x^2y^2$

18. $f(x, y) = 4x^3y + 12xy^3$

$f_x = 4 \cdot 3x^2y + 12 \cdot 1 \cdot y^3 = 12x^2y + 12y^3$

$f_y = 4x^3 \cdot 1 + 12 \cdot 3xy^2 = 4x^3 + 36xy^2$

19. $f(x, y) = \sqrt{3x^2 + 2y^2} = (3x^2 + 2y^2)^{\frac{1}{2}}$

$f_x = \frac{1}{2}(3x^2 + 2y^2)^{-\frac{1}{2}} \cdot 6x = \frac{3x}{\sqrt{3x^2 + 2y^2}}$

$f_y = \frac{1}{2}(3x^2 + 2y^2)^{-\frac{1}{2}} \cdot 4y = \frac{2y}{\sqrt{3x^2 + 2y^2}}$

20. $f(x, y) = \sqrt{1 - 4xy^3} = (1 - 4xy^3)^{\frac{1}{2}}$

$f_x = \frac{1}{2}(1 - 4xy^3)^{-\frac{1}{2}} \cdot -4y^3 = \frac{-2y^3}{\sqrt{1 - 4xy^3}}$

$f_y = \frac{1}{2}(1 - 4xy^3)^{-\frac{1}{2}} \cdot -12xy^2 = \frac{-6xy^2}{\sqrt{1 - 4xy^3}}$

21. $f(x, y) = \frac{-x^2y}{5x + 8y^3}$

$f_x = \frac{(5x + 8y^3)(-2xy) - (-x^2y)(5)}{(5x + 8y^3)^2}$

$= \frac{-10x^2y - 16xy^4 + 5x^2y}{(5x + 8y^3)^2} = \frac{-5x^2y - 16xy^4}{(5x + 8y^3)^2}$

$f_y = \frac{(5x + 8y^3)(-x^2) - (-x^2y)(24y^2)}{(5x + 8y^3)^2}$

$= \frac{-5x^3 - 8x^2y^3 + 24x^2y^3}{(5x + 8y^3)^2}$

$= \frac{-5x^3 + 16x^2y^3}{(5x + 8y^3)^2}$

Copyright © 2019 Pearson Education, Inc.

22. $f(x, y) = \dfrac{3x^2 - 2y^2}{x^2 + 4y^2}$

$$f_x = \frac{(x^2 + 4y^2)(6x) - (3x^2 - 2y^2)(2x)}{(x^2 + 4y^2)^2}$$

$$= \frac{6x^3 + 24xy^2 - 6x^3 + 4xy^2}{(x^2 + 4y^2)^2} = \frac{28xy^2}{(x^2 + 4y^2)^2}$$

$$f_y = \frac{(x^2 + 4y^2)(-4y) - (3x^2 - 2y^2)(8y)}{(x^2 + 4y^2)^2}$$

$$= \frac{-4x^2y - 16y^3 - 24x^2y + 16y^3}{(x^2 + 4y^2)^2}$$

$$= -\frac{28x^2y}{(x^2 + 4y^2)^2}$$

23. $f(x, y) = x^3e^{3y}$

$f_x = 3x^2e^{3y}$

$f_y = x^3 \cdot 3e^{3y} = 3x^3e^{3y}$

24. $f(x, y) = (y+1)^2 e^{2x+y}$

$f_x = (y+1)^2 \cdot 2e^{2x+y} = 2(y+1)^2 e^{2x+y}$

$f_y = (y+1)^2 \cdot 1 \cdot e^{2x+y} + 2(y+1)^1 \cdot 1 \cdot e^{2x+y}$

$= (y+1)^2 e^{2x+y} + 2(y+1)e^{2x+y}$

$= (y+1)e^{2x+y}[(y+1) + 2]$

$= (y+1)(y+3)e^{2x+y}$

25. $f(x, y) = \ln|x^2 - 4y^3|$

$f_x = \dfrac{2x}{x^2 - 4y^3}$

$f_y = \dfrac{-12y^2}{x^2 - 4y^3}$

26. $f(x, y) = \ln|1 + x^3y^2|$

$f_x = \dfrac{0 + 3x^2y^2}{1 + x^3y^2} = \dfrac{3x^2y^2}{1 + x^3y^2}$

$f_y = \dfrac{0 + 2x^3y}{1 + x^3y^2} = \dfrac{2x^3y}{1 + x^3y^2}$

27. $f(x, y) = 4x^3y^2 - 8xy$

$f_x = 12x^2y^2 - 8y$

$f_{xx} = 24xy^2$

$f_{xy} = 24x^2y - 8$

28. $f(x, y) = -6xy^3 + 2x^2y$

$f_x = -6y^3 + 4xy$

$f_{xx} = 4y$

$f_{xy} = -18y^2 + 4x$

29. $f(x, y) = \dfrac{2x + y}{x - 2y}$

$$f_x = \frac{(x - 2y) \cdot 2 - (2x + y) \cdot 1}{(x - 2y)^2} = -\frac{5y}{(x - 2y)^2}$$

$$f_{xx} = \frac{(x - 2y)^2(0) - (-5y)(2)(x - 2y)}{(x - 2y)^4}$$

$$= \frac{10y}{(x - 2y)^3}$$

$$f_{xy} = \frac{(x - 2y)^2(-5) - (-5y)(-4)(x - 2y)}{(x - 2y)^4}$$

$$= \frac{(x - 2y)\left[(x - 2y)(-5) - (-5y)(-4)\right]}{(x - 2y)^4}$$

$$= \frac{-5x - 10y}{(x - 2y)^3}$$

30. $f(x, y) = \dfrac{3x + y}{x - 1}$

$$f_x = \frac{(x - 1) \cdot 3 - (3x + y) \cdot 1}{(x - 1)^2} = \frac{-3 - y}{(x - 1)^2}$$

$= (-3 - y)(x - 1)^{-2}$

$f_{xx} = -2(-3 - y)(x - 1)^{-3} = \dfrac{2(3 + y)}{(x - 1)^3}$

$f_{xy} = \dfrac{-1}{(x - 1)^2}$

31. $f(x, y) = x^2e^y$

$f_x = 2xe^y$

$f_{xx} = 2e^y, \quad f_{xy} = 2xe^y$

32. $f(x, y) = ye^{x^2}$

$f_x = 2xye^{x^2}$

$f_{xx} = 2xy \cdot 2xe^{x^2} + e^{x^2} \cdot 2y = 2ye^{x^2}(2x^2 + 1)$

$f_{xy} = 2xe^{x^2}$

Copyright © 2019 Pearson Education, Inc.

33. $f(x, y) = \ln(2 - x^2y)$

$$f_x = \frac{1}{2-x^2y}\cdot(-2xy) = \frac{2xy}{x^2y-2}$$

$$f_{xx} = \frac{(x^2y-2)2y - 2xy(2xy)}{(x^2y-2)^2}$$
$$= \frac{2y\left[(x^2y-2)-2x^2y\right]}{(x^2y-2)^2} = \frac{2y(-x^2y-2)}{(x^2y-2)^2}$$
$$= \frac{-2x^2y^2-4y}{(x^2y-2)^2}$$

$$f_{xy} = \frac{(x^2y-2)2x - (2xy)x^2}{(x^2y-2)^2}$$
$$= \frac{2x\left[(x^2y-2)-x^2y\right]}{(x^2y-2)^2} = \frac{2x(-2)}{(x^2y-2)^2}$$
$$= -\frac{4x}{(x^2y-2)^2}$$

34. $f(x, y) = \ln(1+3xy^2)$

$$f_x = \frac{1}{1+3xy^2}\cdot 3y^2 = \frac{3y^2}{1+3xy^2} = 3y^2\left(1+3xy^2\right)^{-1}$$

$$f_{xx} = 3y^2\cdot\left(-3y^2\right)\left(1+3xy^2\right)^{-2} = \frac{-9y^4}{\left(1+3xy^2\right)^2}$$

$$f_{xy} = \frac{\left(1+3xy^2\right)\cdot 6y - 3y^2(6xy)}{\left(1+3xy^2\right)^2} = \frac{6y}{\left(1+3xy^2\right)^2}$$

35. $C(x, y) = 4x^2 + 5y^2 - 4xy + 50$

a. $C(10, 8) = 4(10)^2 + 5(8)^2 - 4(10)(8) + 50$
$= 450$
The cost when 10 hours of labor and 8 gallons of paint are used is $450.

b. $C(12, 10) = 4(12)^2 + 5(10)^2 - 4(12)(10) + 50$
$= 646$
The cost when 12 hours of labor and 10 gallons of paint are used is $646.

c. $C(14, 14) = 4(14)^2 + 5(14)^2 - 4(14)(14) + 50$
$= 1030$
The cost when 14 hours of labor and 14 gallons of paint are used is $1030.

36. Total area = area bottom and top + area of ends + area of sides, so
$F(L, W, H) = 2LW + 2WH + 2LH$.

37. $C(x, y) = \frac{x}{4} + y + \frac{xy}{4} + 1$

a. $\frac{\partial c}{\partial x} = \frac{1}{4} + \frac{y}{4}$
$\frac{\partial c}{\partial x}(32, 1) = \frac{1}{4} + \frac{1}{4} = \frac{2}{4} = \frac{1}{2}$

b. $\frac{\partial c}{\partial y} = 1 + \frac{x}{4}$
$\frac{\partial c}{\partial y}(32, 1) = 1 + \frac{32}{4} = 1 + 8 = 9$

38. $z = x^{.7}y^{.3}$

a. $\frac{\partial z}{\partial x} = .7x^{.7-1}y^{.3} = \frac{.7y^{.3}}{x^{.3}}$

b. $\frac{\partial z}{\partial y} = .3x^{.7}y^{.3-1} = \frac{.3x^{.7}}{y^{.7}}$

39. $z = f(x, y) = x^2 + 2y^2 - 4y$
$f_x = 2x$
$f_y = 4y - 4$
Set each of these two partial derivatives equal to zero forming a system of linear equations.
$$\left.\begin{array}{r}2x = 0\\ 4y-4 = 0\end{array}\right\} \Rightarrow x = 0,\ y = 1$$
Thus, the critical point is $(0, 1)$.
$f_{xx} = 2,\ f_{yy} = 4,\ f_{xy} = 0$
$$M = f_{xx}(0,1)\cdot f_{yy}(0,1) - \left[f_{xy}(0,1)\right]^2$$
$$= 2\cdot 4 - 0 = 8$$
Since $M > 0$ and $f_{xx}(0,1) > 0$, we have a local minimum at $(0, 1)$.

$z = f(0, 1) = 0^2 + 2(1)^2 - 4(1) = -2$

There is a local minimum of –2 at $(0, 1)$.

Copyright © 2019 Pearson Education, Inc.

40. $z = f(x,y) = x^2 + y^2 + 9x - 8y + 1$
$f_x = 2x + 9$
$f_y = 2y - 8$
Set each of these two partial derivatives equal to zero forming a system of linear equations.
$$\left.\begin{array}{l}2x+9=0\\2y-8=0\end{array}\right\} \Rightarrow x = -\left(\frac{9}{2}\right),\ y = 4$$
Thus, the critical point is $\left(-\frac{9}{2}, 4\right)$.
$f_{xx} = 2,\ f_{yy} = 2,\ f_{xy} = 0$
$$M = f_{xx}\left(-\frac{9}{2},4\right)\cdot f_{yy}\left(-\frac{9}{2},4\right) - \left[f_{xy}\left(-\frac{9}{2},4\right)\right]^2$$
$$= 2\cdot 2 - 0 = 4$$
Since $M > 0$ and $f_{xx}\left(-\frac{9}{2},4\right) > 0$, we have a local minimum at $\left(-\frac{9}{2},4\right)$.
$$z = f\left(-\frac{9}{2},4\right)$$
$$= \left(-\frac{9}{2}\right)^2 + (4)^2 + 9\left(-\frac{9}{2}\right) - 8(4) + 1 = -\frac{141}{4}$$
There is a local minimum of $-\frac{141}{4}$ at $\left(-\frac{9}{2},4\right)$.

41. $f(x,y) = x^2 + 5xy - 10x + 3y^2 - 12y$
$f_x = 2x + 5y - 10$
$f_y = 5x + 6y - 12$
Set each of these two partial derivatives equal to zero forming a system of linear equations.
$$\left.\begin{array}{l}2x+5y-10=0\\5x+6y-12=0\end{array}\right\} \Rightarrow \left.\begin{array}{l}12x+30y-60=0\\-25x-30y+60=0\end{array}\right\} \Rightarrow$$
$13x = 0 \Rightarrow x = 0$
$2(0) + 5y - 10 = 0 \Rightarrow y = 2$
The critical point is $(0,2)$.
$f_{xx} = 2,\ f_{yy} = 6,\ f_{xy} = 5$
$$M = f_{xx}(0,2)\cdot f_{yy}(0,2) - \left[f_{xy}(0,2)\right]^2$$
$$= 2\cdot 6 - (5)^2 = 12 - 25 = -13$$
Since $M < 0$, we have a saddle point at $(0,2)$.

42. $z = f(x,y) = x^3 - 8y^2 + 6xy + 4$
$f_x = 3x^2 + 6y$
$f_y = -16y + 6x$
Set each of these two partial derivatives equal to zero forming a system of linear equations.
$$\left.\begin{array}{l}3x^2+6y=0\\-16y+6x=0\end{array}\right\} \Rightarrow -16y + 6x = 0 \Rightarrow x = \frac{8}{3}y$$
$$3\left(\frac{8}{3}y\right)^2 + 6y = 0 \Rightarrow \frac{64}{3}y^2 + 6y = 0 \Rightarrow$$
$$y\left(\frac{64}{3}y + 6\right) = 0 \Rightarrow y = 0 \text{ or } y = -\frac{9}{32}$$
If $y = 0$, then $x = 0$.
If $y = -\frac{9}{32}$, then $x = -\frac{3}{4}$.
There are two critical points: $(0,0)$ and $\left(-\frac{3}{4}, -\frac{9}{32}\right)$.
$f_{xx} = 6x,\ f_{yy} = -16,\ f_{xy} = 6$
At $(0,0)$
$$M = f_{xx}(0,0)\cdot f_{yy}(0,0) - \left[f_{xy}(0,0)\right]^2$$
$$= (6)(0)\cdot(-16) - (6)^2 = -36$$
Since $M < 0$ there is a saddle point at $(0,0)$.
At $\left(-\frac{3}{4}, -\frac{9}{32}\right)$
$$M = f_{xx}\left(-\frac{3}{4},-\frac{9}{32}\right)\cdot f_{yy}\left(-\frac{3}{4},-\frac{9}{32}\right) - \left[f_{xy}\left(-\frac{3}{4},-\frac{9}{32}\right)\right]^2$$
$$= (6)\left(-\frac{3}{4}\right)\cdot(-16) - (6)^2 = 72 - 36 = 36$$
Since $M > 0$ and $f_{xx}\left(-\frac{3}{4},-\frac{9}{32}\right) < 0$, there is a local maximum at $\left(-\frac{3}{4},-\frac{9}{32}\right)$.
$$f\left(-\frac{3}{4},-\frac{9}{32}\right)$$
$$= \left(-\frac{3}{4}\right)^3 - 8\left(-\frac{9}{32}\right)^2 + 6\left(-\frac{3}{4}\right)\left(-\frac{9}{32}\right) + 4$$
$$= \frac{539}{128} \approx 4.21$$
There is a local maximum of 4.21 at $\left(-\frac{3}{4},-\frac{9}{32}\right)$.

Copyright © 2019 Pearson Education, Inc.

43. $z = f(x, y) = x^3 + y^2 + 2xy - 4x - 3y - 2$
$f_x = 3x^2 + 2y - 4$
$f_y = 2y + 2x - 3$
Set each of these two partial derivatives equal to zero forming a system of nonlinear equations.
$$\left.\begin{array}{r} 3x^2 + 2y - 4 = 0 \\ 2y + 2x - 3 = 0 \end{array}\right\} \Rightarrow 2y = 3 - 2x$$
Substitute 3 – 2x for 2y in the first equation.
$3x^2 + 3 - 2x - 4 = 0 \Rightarrow 3x^2 - 2x - 1 = 0 \Rightarrow$
$(3x+1)(x-1) = 0 \Rightarrow x = -\frac{1}{3}, 1$
If $x = -\frac{1}{3},\ 2y = 3 - 2\left(-\frac{1}{3}\right) = \frac{11}{3} \Rightarrow y = \frac{11}{6}$.
If $x = 1,\ 2y = 3 - 2(1) = 1 \Rightarrow y = \frac{1}{2}$.
There are two critical points: $\left(-\frac{1}{3}, \frac{11}{6}\right)$ and $\left(1, \frac{1}{2}\right)$.
$f_{xx} = 6x,\ f_{yy} = 2,\ f_{xy} = 2$
At $\left(-\frac{1}{3}, \frac{11}{6}\right)$
$$M = f_{xx}\left(-\frac{1}{3}, \frac{11}{6}\right) \cdot f_{yy}\left(-\frac{1}{3}, \frac{11}{6}\right) - \left[f_{xy}\left(-\frac{1}{3}, \frac{11}{6}\right)\right]^2$$
$$= (-2)(2) - (2)^2 = -4 - 4 = -8$$
Since $M < 0$ there is a saddle point at $\left(-\frac{1}{3}, \frac{11}{6}\right)$.
At $\left(1, \frac{1}{2}\right)$
$$M = f_{xx}\left(1, \frac{1}{2}\right) \cdot f_{yy}\left(1, \frac{1}{2}\right) - \left[f_{xy}\left(1, \frac{1}{2}\right)\right]^2$$
$$= 6(1) \cdot 2 - (2)^2 = 12 - 4 = 8$$
$$z = f\left(1, \frac{1}{2}\right)$$
$$= (1)^3 + \left(\frac{1}{2}\right)^2 + 2(1)\left(\frac{1}{2}\right) - 4(1) - 3\left(\frac{1}{2}\right) - 2$$
$$= -\frac{21}{4} \approx -5.25$$
Since $M > 0$ and $f_{xx}\left(1, \frac{1}{2}\right) = 6 > 0$, there is a local minimum of –5.25 at $\left(1, \frac{1}{2}\right)$.

44. $z = f(x, y) = 7x^2 + y^2 - 3x + 6y - 5xy$
$f_x = 14x - 3 - 5y$
$f_y = 2y + 6 - 5x$
Set each of these two partial derivatives equal to zero forming a system of linear equations.
$$\left.\begin{array}{r} 14x - 3 - 5y = 0 \\ 2y + 6 - 5x = 0 \end{array}\right\} \Rightarrow \left.\begin{array}{r} 14x - 5y = 3 \\ -5x + 2y = -6 \end{array}\right\} \Rightarrow$$
$$\left.\begin{array}{r} 28x - 10y = 6 \\ -25x + 10y = -30 \end{array}\right\} \Rightarrow 3x = -24 \Rightarrow x = -8$$
$2y + 6 - 5(-8) = 0 \Rightarrow y = -23$
The critical point is $(-8, -23)$.
$f_{xx} = 14,\ f_{yy} = 2,\ f_{xy} = -5$
$$M = f_{xx}(-8, -23) \cdot f_{yy}(-8, -23) - \left[f_{xy}(-8, -23)\right]^2$$
$$= (14) \cdot (2) - (-5)^2 = 3$$
Since $M > 0$ and $f_{xx}(-8, -23) > 0$, we have a local minimum at $(-8, -23)$.
$$z = f(-8, -23)$$
$$= 7(-8)^2 + (-23)^2 - 3(-8) + 6(-23) - 5(-8)(-23)$$
$$= -57$$
There is a local minimum of –57 at $(-8, -23)$.

45. $f(x, y) = e^{x^2 + y^2}$
$f_x = 2xe^{x^2+y^2},\ f_y = 2ye^{x^2+y^2}$
Solve the system of equations.
$2xe^{x^2+y^2} = 0$
$2ye^{x^2+y^2} = 0$
$\Rightarrow 2ye^{x^2+y^2} = 2xe^{x^2+y^2} \Rightarrow x = y = 0$
The critical point is (0, 0).
$f_{xx} = 2e^{x^2+y^2}(2x^2 + 1)$
$f_{yy} = 2e^{x^2+y^2}(2y^2 + 1)$
$f_{xy} = 4xye^{x^2+y^2}$
For (0, 0),
$M = 2 \cdot 2 - (0)^2 = 4 > 0$. Since $M > 0$ and $f_{xx}(0,0) = 2e^0(2(0)^2 + 1) = 2 > 0$, there is a local minimum at (0, 0).

Copyright © 2019 Pearson Education, Inc.

46. $f(x, y) = e^{xy}$

$f_x = ye^{xy},\ f_y = xe^{xy}$

Solve the system of equations.

$$\left.\begin{array}{l} ye^{xy} = 0 \\ xe^{xy} = 0 \end{array}\right. \Rightarrow ye^{xy} = xe^{xy} \Rightarrow x = y = 0$$

The critical point is (0, 0).

$f_{xx} = y^2e^{xy}$

$f_{yy} = x^2e^{xy}$

$f_{xy} = e^{xy} + xye^{xy}$

For (0, 0),

$M = 0 \cdot 0 - (e^0)^2 = -1 < 0.$

There is a saddle point at (0, 0).

47. **a.** $c(x, y) = x^2 + 5y^2 + 4xy - 70x - 164y + 1800$

$c_x = 2x + 4y - 70$

$c_y = 10y + 4x - 164$

$$\left.\begin{array}{l} 2x + 4y - 70 = 0 \\ 4x + 10y - 164 = 0 \end{array}\right\} \Rightarrow \left.\begin{array}{l} -4x - 8y + 140 = 0 \\ 4x + 10y - 164 = 0 \end{array}\right\} \Rightarrow$$

$2y - 24 = 0 \Rightarrow y = 12$

$2x + 4(12) - 70 = 0 \Rightarrow x = 11$

$c_{xx} = 2,\ c_{yy} = 10,\ c_{xy} = 4$

At (11, 12), $M = 2 \cdot 10 - 4^2 = 4 > 0$ and $c_{xx} > 0$.

A local minimum occurs at (11, 12).

b. The minimum cost is

$c(11, 12)$

$= (11)^2 + 5(12)^2 + 4(11)(12) - 70(11) - 164(12) + 1800$

$= 121 + 720 + 528 - 770 - 1968 + 1800$

$= \$431$

48. $P(x, y)$

$= .01(-x^2 + 3xy + 160x - 5y^2 + 200y + 2600)$

a. $x + y = 280 \Rightarrow y = 280 - x$

$P(x, y) = .01[-x^2 + 3x(280 - x) + 160x - 5(280 - x)^2 + 200(280 - x) + 2600]$

$= .01[-x^2 + 840x - 3x^2 + 160x - 5(78,400 - 560x + x^2) + 56,000 - 200x + 2600]$

$= .01(-x^2 + 840x - 3x^2 + 160x - 392,000 + 2800x - 5x^2 + 56,000 - 200x + 2600)$

$P(x, y) = .01(-9x^2 + 3600x - 333,400)$

$P'(x, y) = .01(-18x + 3600)$

$.01(-18x + 3600) = 0 \Rightarrow x = 200$

If $x < 200$, $P' > 0$. If $x > 200$, $P' < 0$.

There is a maximum when $x = 200$.

If $x = 200$, $y = 280 - 200 = 80$.

$P(200, 80) = .01[-(200)^2 + 3(200)(80) + 160(200) - 5(80)^2 + 200(80) + 2600]$

$= 266$

\$200 spent on fertilizer and \$80 spent on seed will produce a maximum profit of \$266 per acre.

b. $P_x = .01(-2x + 3y + 160)$

$P_y = .01(3x - 10y + 200)$

$$\left.\begin{array}{l} -2x + 3y + 160 = 0 \\ 3x - 10y + 200 = 0 \end{array}\right\} \Rightarrow$$

$$\left.\begin{array}{l} -2x + 3y = -160 \\ 3x - 10y = -200 \end{array}\right\} \Rightarrow \left.\begin{array}{l} -6x + 9y = -480 \\ 6x - 20y = -400 \end{array}\right\} \Rightarrow$$

$-11y = -880 \Rightarrow y = 80$

$-2x + 3(80) = -160 \Rightarrow -2x = -400 \Rightarrow x = 200$

The critical point is (200, 80).

$P_{xx} = .01(-2) = -.02$

$P_{yy} = .01(-10) = -.1$

$P_{xy} = .01(3) = .03$

Copyright © 2019 Pearson Education, Inc.

At (200, 80),
$M = P_{xx}(200, 80) \cdot P_{yy}(200, 80) - [P_{xy}(200, 80)]^2$
$= (-.02)(-.1) - (.03)^2 = .0011$
Since $M > 0$ and $P_{xx}(200, 80) < 0$, there is a local maximum at (200, 80). This is the same as in part (a).

c. In (a), we saw that $P(200, 80) = 266$. \$200 spent on fertilizer and \$80 spent on seed will produce a maximum profit of \$266 per acre.

49. Minimize $f(x, y) = x^2 + y^2$ subject to $x = y + 2$. The Lagrange function is:
$F(x, y, \lambda) = f(x, y) - \lambda \cdot g(x, y)$
$= x^2 + y^2 - \lambda x + \lambda y + 2\lambda$
Compute the partial derivatives:
$F_x(x, y, \lambda) = 2x - \lambda$
$F_y(x, y, \lambda) = 2y + \lambda$
$F_\lambda(x, y, \lambda) = -x + y + 2$
Set each partial derivative to zero to obtain the system:
$2x - \lambda = 0$
$2y + \lambda = 0$
$-x + y + 2 = 0$
Then solve the system:
$\left.\begin{matrix} \lambda = 2x \\ \lambda = -2y \end{matrix}\right\} \Rightarrow 2x = -2y \Rightarrow x = -y$
$-(-y) + y + 2 = 0 \Rightarrow 2y = -2 \Rightarrow y = -1$
Since $x = -y$, $x = -(-1) = 1$. So, the minimum value of $f(x, y) = x^2 + y^2$ subject to $x = y + 2$ occurs when $x = 1$ and $y = -1$. The minimum value is $f(1, -1) = (1)^2 + (-1)^2 = 2$.

50. Minimize and maximize $f(x, y) = x^2 y$ subject to $x + y = 4$. The Lagrange function is:
$F(x, y, \lambda) = f(x, y) - \lambda \cdot g(x, y)$
$= x^2 y - \lambda x - \lambda y + 4\lambda$
Compute the partial derivatives:
$F_x(x, y, \lambda) = 2xy - \lambda$
$F_y(x, y, \lambda) = x^2 - \lambda$
$F_\lambda(x, y, \lambda) = -x - y + 4$
Set each partial derivative to zero to obtain the system:
$2xy - \lambda = 0$
$x^2 - \lambda = 0$
$-x - y + 4 = 0$
Then solve the system:
$\left.\begin{matrix} \lambda = 2xy \\ \lambda = x^2 \end{matrix}\right\} \Rightarrow 2xy = x^2 \Rightarrow 2y = x.$
If $x = 2y$, then $-(2y) - y + 4 = 0 \Rightarrow y = \frac{4}{3}$.
Then $x = \frac{8}{3}$.
So, the maximum value of $f(x, y) = x^2 y$ subject to $x + y = 4$ occurs when $x = \frac{8}{3}$ and when $y = \frac{4}{3}$. The maximum value is
$f\left(\frac{8}{3}, \frac{4}{3}\right) = \left(\frac{8}{3}\right)^2 \left(\frac{4}{3}\right) = \frac{256}{27}$.

51. Maximize $f(x, y) = x^2 y$ subject to $x + y = 80$. The Lagrange function is:
$F(x, y, \lambda) = f(x, y) - \lambda \cdot g(x, y)$
$= x^2 y - \lambda x - \lambda y + 80\lambda$
Compute the partial derivatives:
$F_x(x, y, \lambda) = 2xy - \lambda$
$F_y(x, y, \lambda) = x^2 - \lambda$
$F_\lambda(x, y, \lambda) = -x - y + 80$
Set each partial derivative to zero to obtain the system:
$2xy - \lambda = 0$
$x^2 - \lambda = 0$
$-x - y + 80 = 0$

Copyright © 2019 Pearson Education, Inc.

Then solve the system:

$$\left.\begin{aligned}\lambda &= 2xy\\ \lambda &= x^2\end{aligned}\right\} \Rightarrow 2xy = x^2 \Rightarrow 2y = x$$

$$-2y - y + 80 = 0 \Rightarrow -3y = -80 \Rightarrow y = \frac{80}{3}$$

Since $x = 2y$, $x = 2\left(\frac{80}{3}\right) = \frac{160}{3}$. So, the maximum value of $f(x, y) = x^2 y$ subject to $x + y = 80$ occurs when $x = \frac{160}{3}$ and when $y = \frac{80}{3}$. The maximum value is

$$f\left(\frac{160}{3}, \frac{80}{3}\right) = \frac{2{,}048{,}000}{27} \approx 75{,}852.$$

52. Maximize $f(x, y) = xy^2$ subject to $x + y = 50$.

The Lagrange function is:

$$\begin{aligned}F(x, y, \lambda) &= f(x, y) - \lambda \cdot g(x, y)\\ &= xy^2 - \lambda x - \lambda y + 50\lambda\end{aligned}$$

Compute the partial derivatives:

$$F_x(x, y, \lambda) = y^2 - \lambda$$
$$F_y(x, y, \lambda) = 2xy - \lambda$$
$$F_\lambda(x, y, \lambda) = -x - y + 50$$

Set each partial derivative to zero to obtain the system:

$$\begin{aligned}y^2 - \lambda &= 0\\ 2xy - \lambda &= 0\\ -x - y + 50 &= 0\end{aligned}$$

Then solve the system:

$$\left.\begin{aligned}\lambda &= 2xy\\ \lambda &= y^2\end{aligned}\right\} \Rightarrow 2xy = y^2 \Rightarrow y = 2x$$

$$-x - (2x) + 50 = 0 \Rightarrow -3x + 50 = 0 \Rightarrow x = \frac{50}{3}$$

Since $y = 2x$, $y = 2\left(\frac{50}{3}\right) = \frac{100}{3}$. So, the maximum value of $f(x, y) = xy^2$ subject to $x + y = 50$ occurs when $y = \frac{100}{3}$ and when $x = \frac{50}{3}$. The maximum value is

$$f\left(\frac{50}{3}, \frac{100}{3}\right) = \left(\frac{50}{3}\right)\left(\frac{100}{3}\right)^2 = \frac{500{,}000}{27}.$$

53. Answers vary

Copyright © 2019 Pearson Education, Inc.

54. Maximize $P(x,y)=.01\left(-x^2+3xy+160x-5y^2+200y+2600\right)$ subject to $x+y=280$

The Lagrange function is:

$$F(x,y,\lambda)=f(x,y)-\lambda\cdot g(x,y)=.01\left(-x^2+3xy+160x-5y^2+200y+2600\right)-\lambda x-\lambda y+280\lambda$$
$$=-.01x^2+.03xy+1.6x-.05y^2+2y+26-\lambda x-\lambda y+280\lambda$$

Compute the partial derivatives:
$$F_x(x,y,\lambda)=-.02x+.03y+1.6-\lambda$$
$$F_y(x,y,\lambda)=.03x-.1y+2-\lambda$$
$$F_\lambda(x,y,\lambda)=-x-y+280$$

Set each partial derivatives to zero to obtain the system:

$$-.02x+.03y+1.6-\lambda=0$$
$$.03x-.1y+2-\lambda=0$$
$$-x-y+280=0$$

Solving the system using the elimination method (or using technology), we obtain $x=200$ and $y=80$.

$P(200,80)=.01(-(200)^2+3(200)(80)+160(200)-5(80)^2+200(80)+2600)=266$

$P(x,y)$ has a maximum of $266 when $200 is spent on fertilizer and when $80 is spent on hybrid seed.

Case 14 Global Warming and the Method of Least Squares

1. $f_m(m,b)=-105{,}129+23{,}031{,}712m+11{,}752b$

$f_b(m,b)=-53.58+11{,}752m+6b$

$f_{mm}(m,b)=23{,}031{,}712$

$f_{bb}(m,b)=6$

$f_{mb}(m,b)=11{,}752$

$f_{mm}f_{bb}-[f_{mb}]^2=80{,}768$

At $m=.01,\ b=-10.85,$ we have $f_m=f_b=0,\ f_{mm}>0,$ and $f_{mm}f_{bb}-[f_{mb}]^2>0,$ so f has a minimum at $(.01,-10.85)$.

2.

$$m=\frac{3(1900\cdot 8.36+1960\cdot 8.47+2016\cdot 9.96)-(1900+1960+2016)(8.36+8.47+9.96)}{3(1900^2+1960^2+2016^2)-(1900+1960+2016)^2}=.01$$

$$b=\frac{8.36+8.47+9.96}{3}-.01\left(\frac{1900+1960+2016}{3}\right)=-10.85$$

3. Using summation notation,

$$m=\frac{4\sum xy-\sum x\sum y}{4\sum x^2-\left(\sum x\right)^2}=.01$$

$$b=\frac{\sum y}{4}-m\frac{\sum x}{4}=-17.55$$

Answers vary. The slope .01 of the least-squares line still suggests a gradual rise in the neighborhood of .01 degrees per year, but the pattern of data points suggests that the global land temperature fell during the first half of the century, and then rose during the second half at something like .015°C/year.

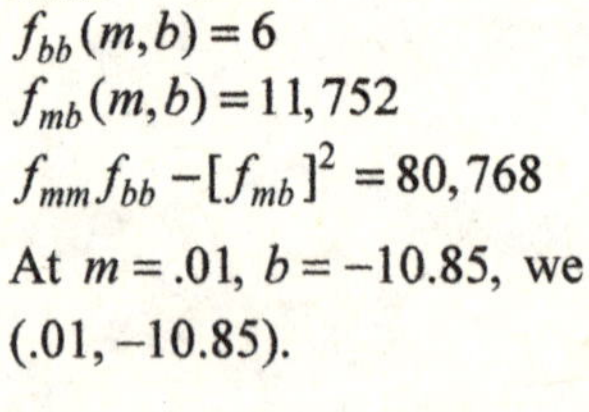

Copyright © 2019 Pearson Education, Inc.